Orbital Mechanics for Engineering Students

To my parents, Rondo and Geraldine, and my wife, Connie Dee

ORBITAL MECHANICS FOR ENGINEERING STUDENTS

HOWARD D. CURTIS
Embry-Riddle Aeronautical University
Daytona Beach, Florida

ELSEVIER
BUTTERWORTH
HEINEMANN

AMSTERDAM • BOSTON • HEIDELBERG • LONDON • NEW YORK • OXFORD
PARIS • SAN DIEGO • SAN FRANCISCO • SINGAPORE • SYDNEY • TOKYO

Elsevier Butterworth-Heinemann
Linacre House, Jordan Hill, Oxford OX2 8DP
30 Corporate Drive, Burlington, MA 01803

First published 2005
Reprinted 2005

Copyright © 2005, Howard D. Curtis. All rights reserved

The right of Howard D. Curtis to be identified as the author of this work has been
asserted in accordance with the Copyright, Designs and Patents Act 1988

No part of this publication may be reproduced in any material form (including
photocopying or storing in any medium by electronic means and whether
or not transiently or incidentally to some other use of this publication) without
the written permission of the copyright holder except in accordance with the
provisions of the Copyright, Designs and Patents Act 1988 or under the terms of
a licence issued by the Copyright Licensing Agency Ltd, 90 Tottenham Court Road,
London, England W1T 4LP. Applications for the copyright holder's written
permission to reproduce any part of this publication should be addressed
to the publisher

Permissions may be sought directly from Elsevier's Science & Technology Rights
Department in Oxford, UK: phone: (+44) 1865 843830, fax: (+44) 1865 853333,
e-mail: permissions@elsevier.co.uk. You may also complete your request on-line via
the Elsevier homepage (http://www.elsevier.com), by selecting 'Customer Support'
and then 'Obtaining Permissions'

British Library Cataloguing in Publication Data
A catalogue record for this book is available from the British Library

Library of Congress Cataloguing in Publication Data
A catalogue record for this book is available from the Library of Congress

ISBN 0 7506 6169 0

For information on all Elsevier Butterworth-Heinemann
publications visit our website at www.bh.com

Working together to grow
libraries in developing countries

www.elsevier.com | www.bookaid.org | www.sabre.org

ELSEVIER   BOOK AID International   Sabre Foundation

Typeset by Charon Tec Pvt. Ltd, Chennai, India
www.charontec.com
Printed and bound in Great Britain by Biddles Ltd., King's Lynn, Norfolk

Contents

Preface ... xi
Supplements to the text ... xv

Chapter 1

Dynamics of point masses ... 1

1.1 Introduction ... 1
1.2 Kinematics ... 2
1.3 Mass, force and Newton's law of gravitation ... 7
1.4 Newton's law of motion ... 10
1.5 Time derivatives of moving vectors ... 15
1.6 Relative motion ... 20
Problems ... 29

Chapter 2

The two-body problem ... 33

2.1 Introduction ... 33
2.2 Equations of motion in an inertial frame ... 34
2.3 Equations of relative motion ... 37
2.4 Angular momentum and the orbit formulas ... 42
2.5 The energy law ... 50
2.6 Circular orbits ($e = 0$) ... 51
2.7 Elliptical orbits ($0 < e < 1$) ... 55
2.8 Parabolic trajectories ($e = 1$) ... 65
2.9 Hyperbolic trajectories ($e > 1$) ... 69
2.10 Perifocal frame ... 76
2.11 The Lagrange coefficients ... 78
2.12 Restricted three-body problem ... 89
    2.12.1 Lagrange points ... 92
    2.12.2 Jacobi constant ... 96
Problems ... 101

Chapter 3

Orbital position as a function of time ... 107

3.1 Introduction ... 107
3.2 Time since periapsis ... 108

3.3	**CIRCULAR ORBITS**	108
3.4	**ELLIPTICAL ORBITS**	109
3.5	**PARABOLIC TRAJECTORIES**	124
3.6	**HYPERBOLIC TRAJECTORIES**	125
3.7	**UNIVERSAL VARIABLES**	134
	PROBLEMS	145

CHAPTER 4

ORBITS IN THREE DIMENSIONS — 149

4.1	**INTRODUCTION**	149
4.2	**GEOCENTRIC RIGHT ASCENSION–DECLINATION FRAME**	150
4.3	**STATE VECTOR AND THE GEOCENTRIC EQUATORIAL FRAME**	154
4.4	**ORBITAL ELEMENTS AND THE STATE VECTOR**	158
4.5	**COORDINATE TRANSFORMATION**	164
4.6	**TRANSFORMATION BETWEEN GEOCENTRIC EQUATORIAL AND PERIFOCAL FRAMES**	172
4.7	**EFFECTS OF THE EARTH'S OBLATENESS**	177
	PROBLEMS	187

CHAPTER 5

PRELIMINARY ORBIT DETERMINATION — 193

5.1	**INTRODUCTION**	193
5.2	**GIBBS' METHOD OF ORBIT DETERMINATION FROM THREE POSITION VECTORS**	194
5.3	**LAMBERT'S PROBLEM**	202
5.4	**SIDEREAL TIME**	213
5.5	**TOPOCENTRIC COORDINATE SYSTEM**	218
5.6	**TOPOCENTRIC EQUATORIAL COORDINATE SYSTEM**	221
5.7	**TOPOCENTRIC HORIZON COORDINATE SYSTEM**	223
5.8	**ORBIT DETERMINATION FROM ANGLE AND RANGE MEASUREMENTS**	228
5.9	**ANGLES-ONLY PRELIMINARY ORBIT DETERMINATION**	235
5.10	**GAUSS'S METHOD OF PRELIMINARY ORBIT DETERMINATION**	236
	PROBLEMS	250

CHAPTER 6

ORBITAL MANEUVERS — 255

6.1	**INTRODUCTION**	255
6.2	**IMPULSIVE MANEUVERS**	256
6.3	**HOHMANN TRANSFER**	257

6.4	BI-ELLIPTIC HOHMANN TRANSFER	264
6.5	PHASING MANEUVERS	268
6.6	NON-HOHMANN TRANSFERS WITH A COMMON APSE LINE	273
6.7	APSE LINE ROTATION	279
6.8	CHASE MANEUVERS	285
6.9	PLANE CHANGE MANEUVERS	290
	PROBLEMS	304

CHAPTER 7

RELATIVE MOTION AND RENDEZVOUS 315

7.1	INTRODUCTION	315
7.2	RELATIVE MOTION IN ORBIT	316
7.3	LINEARIZATION OF THE EQUATIONS OF RELATIVE MOTION IN ORBIT	322
7.4	CLOHESSY–WILTSHIRE EQUATIONS	324
7.5	TWO-IMPULSE RENDEZVOUS MANEUVERS	330
7.6	RELATIVE MOTION IN CLOSE-PROXIMITY CIRCULAR ORBITS	338
	PROBLEMS	340

CHAPTER 8

INTERPLANETARY TRAJECTORIES 347

8.1	INTRODUCTION	347
8.2	INTERPLANETARY HOHMANN TRANSFERS	348
8.3	RENDEZVOUS OPPORTUNITIES	349
8.4	SPHERE OF INFLUENCE	354
8.5	METHOD OF PATCHED CONICS	359
8.6	PLANETARY DEPARTURE	360
8.7	SENSITIVITY ANALYSIS	366
8.8	PLANETARY RENDEZVOUS	368
8.9	PLANETARY FLYBY	375
8.10	PLANETARY EPHEMERIS	387
8.11	NON-HOHMANN INTERPLANETARY TRAJECTORIES	391
	PROBLEMS	398

CHAPTER 9

RIGID-BODY DYNAMICS 399

9.1	INTRODUCTION	399
9.2	KINEMATICS	400
9.3	EQUATIONS OF TRANSLATIONAL MOTION	408
9.4	EQUATIONS OF ROTATIONAL MOTION	410

9.5	**MOMENTS OF INERTIA**		414
	9.5.1 Parallel axis theorem		428
9.6	**EULER'S EQUATIONS**		435
9.7	**KINETIC ENERGY**		441
9.8	**THE SPINNING TOP**		443
9.9	**EULER ANGLES**		448
9.10	**YAW, PITCH AND ROLL ANGLES**		459
	PROBLEMS		463

CHAPTER 10

SATELLITE ATTITUDE DYNAMICS — 475

10.1	**INTRODUCTION**	475
10.2	**TORQUE-FREE MOTION**	476
10.3	**STABILITY OF TORQUE-FREE MOTION**	486
10.4	**DUAL-SPIN SPACECRAFT**	491
10.5	**NUTATION DAMPER**	495
10.6	**CONING MANEUVER**	503
10.7	**ATTITUDE CONTROL THRUSTERS**	506
10.8	**YO-YO DESPIN MECHANISM**	509
10.9	**GYROSCOPIC ATTITUDE CONTROL**	516
10.10	**GRAVITY-GRADIENT STABILIZATION**	530
	PROBLEMS	543

CHAPTER 11

ROCKET VEHICLE DYNAMICS — 551

11.1	**INTRODUCTION**	551
11.2	**EQUATIONS OF MOTION**	552
11.3	**THE THRUST EQUATION**	555
11.4	**ROCKET PERFORMANCE**	557
11.5	**RESTRICTED STAGING IN FIELD-FREE SPACE**	560
11.6	**OPTIMAL STAGING**	570
	11.6.1 Lagrange multiplier	570
	PROBLEMS	578

REFERENCES AND FURTHER READING — 581

APPENDIX A

PHYSICAL DATA — 583

APPENDIX B

A ROAD MAP — 585

APPENDIX C

NUMERICAL INTEGRATION OF THE n-BODY EQUATIONS OF MOTION — 587

- C.1 FUNCTION FILE `accel_3body.m` — 590
- C.2 SCRIPT FILE `threebody.m` — 592

APPENDIX D

MATLAB ALGORITHMS — 595

- D.1 INTRODUCTION — 596
- D.2 ALGORITHM 3.1: SOLUTION OF KEPLER'S EQUATION BY NEWTON'S METHOD — 596
- D.3 ALGORITHM 3.2: SOLUTION OF KEPLER'S EQUATION FOR THE HYPERBOLA USING NEWTON'S METHOD — 598
- D.4 CALCULATION OF THE STUMPFF FUNCTIONS $S(z)$ AND $C(z)$ — 600
- D.5 ALGORITHM 3.3: SOLUTION OF THE UNIVERSAL KEPLER'S EQUATION USING NEWTON'S METHOD — 601
- D.6 CALCULATION OF THE LAGRANGE COEFFICIENTS f AND g AND THEIR TIME DERIVATIVES — 603
- D.7 ALGORITHM 3.4: CALCULATION OF THE STATE VECTOR $(\mathbf{r}, \mathbf{v})$ GIVEN THE INITIAL STATE VECTOR $(\mathbf{r}_0, \mathbf{v}_0)$ AND THE TIME LAPSE Δt — 604
- D.8 ALGORITHM 4.1: CALCULATION OF THE ORBITAL ELEMENTS FROM THE STATE VECTOR — 606
- D.9 ALGORITHM 4.2: CALCULATION OF THE STATE VECTOR FROM THE ORBITAL ELEMENTS — 610
- D.10 ALGORITHM 5.1: GIBBS' METHOD OF PRELIMINARY ORBIT DETERMINATION — 613
- D.11 ALGORITHM 5.2: SOLUTION OF LAMBERT'S PROBLEM — 616
- D.12 CALCULATION OF JULIAN DAY NUMBER AT 0 HR UT — 621
- D.13 ALGORITHM 5.3: CALCULATION OF LOCAL SIDEREAL TIME — 623
- D.14 ALGORITHM 5.4: CALCULATION OF THE STATE VECTOR FROM MEASUREMENTS OF RANGE, ANGULAR POSITION AND THEIR RATES — 626
- D.15 ALGORITHMS 5.5 AND 5.6: GAUSS'S METHOD OF PRELIMINARY ORBIT DETERMINATION WITH ITERATIVE IMPROVEMENT — 631
- D.16 CONVERTING THE NUMERICAL DESIGNATION OF A MONTH OR A PLANET INTO ITS NAME — 640
- D.17 ALGORITHM 8.1: CALCULATION OF THE STATE VECTOR OF A PLANET AT A GIVEN EPOCH — 641
- D.18 ALGORITHM 8.2: CALCULATION OF THE SPACECRAFT TRAJECTORY FROM PLANET 1 TO PLANET 2 — 648

APPENDIX E

GRAVITATIONAL POTENTIAL ENERGY OF A SPHERE — 657

INDEX — 661

PREFACE

This textbook evolved from a formal set of notes developed over nearly ten years of teaching an introductory course in orbital mechanics for aerospace engineering students. These undergraduate students had no prior formal experience in the subject, but had completed courses in physics, dynamics and mathematics through differential equations and applied linear algebra. That is the background I have presumed for readers of this book.

This is by no means a grand, descriptive survey of the entire subject of astronautics. It is a foundations text, a springboard to advanced study of the subject. I focus on the physical phenomena and analytical procedures required to understand and predict, to first order, the behavior of orbiting spacecraft. I have tried to make the book readable for undergraduates, and in so doing I do not shy away from rigor where it is needed for understanding. Spacecraft operations that take place in earth orbit are considered as are interplanetary missions. The important topic of spacecraft control systems is omitted. However, the material in this book and a course in control theory provide the basis for the study of spacecraft attitude control.

A brief perusal of the Contents shows that there are more than enough topics to cover in a single semester or term. Chapter 1 is a review of vector kinematics in three dimensions and of Newton's laws of motion and gravitation. It also focuses on the issue of relative motion, crucial to the topics of rendezvous and satellite attitude dynamics. Chapter 2 presents the vector-based solution of the classical two-body problem, coming up with a host of practical formulas for orbit and trajectory analysis. The restricted three-body problem is covered in order to introduce the notion of Lagrange points. Chapter 3 derives Kepler's equations, which relate position to time for the different kinds of orbits. The concept of 'universal variables' is introduced. Chapter 4 is devoted to describing orbits in three dimensions and accounting for the major effects of the earth's oblate, non-spherical shape. Chapter 5 is an introduction to preliminary orbit determination, including Gibbs' and Gauss's methods and the solution of Lambert's problem. Auxiliary topics include topocentric coordinate systems, Julian day numbering and sidereal time. Chapter 6 presents the common means of transferring from one orbit to another by impulsive delta-v maneuvers, including Hohmann transfers, phasing orbits and plane changes. Chapter 7 derives and employs the equations of relative motion required to understand and design two-impulse rendezvous maneuvers. Chapter 8 explores the basics of interplanetary mission analysis. Chapter 9 presents those elements of rigid-body dynamics required to characterize the attitude of an orbiting satellite. Chapter 10 describes the methods of controlling, changing and stabilizing the attitude of spacecraft by means of thrusters, gyros and other devices. Finally, Chapter 11 is a brief introduction to the characteristics and design of multi-stage launch vehicles.

Chapters 1 through 4 form the core of a first orbital mechanics course. The time devoted to Chapter 1 depends on the background of the student. It might be surveyed briefly and used thereafter simply as a reference. What follows Chapter 4 depends on the objectives of the course.

Chapters 5 through 8 carry on with the subject of orbital mechanics. Chapter 6 on orbital maneuvers should be included in any case. Coverage of Chapters 5, 7 and 8 is optional. However, if all of Chapter 8 on interplanetary missions is to form a part of the course, then the solution of Lambert's problem (Section 5.3) must be studied beforehand.

Chapters 9 and 10 must be covered if the course objectives include an introduction to satellite dynamics. In that case Chapters 5, 7 and 8 would probably not be studied in depth.

Chapter 11 is optional if the engineering curriculum requires a separate course in propulsion, including rocket dynamics.

To understand the material and to solve problems requires using a lot of undergraduate mathematics. Mathematics, of course, is the language of engineering. Students must not forget that Sir Isaac Newton had to invent calculus so he could solve orbital mechanics problems precisely. Newton (1642–1727) was an English physicist and mathematician, whose 1687 publication *Mathematical Principles of Natural Philosophy* ('the *Principia*') is one of the most influential scientific works of all time. It must be noted that the German mathematician Gottfried Wilhelm von Leibniz (1646–1716) is credited with inventing infinitesimal calculus independently of Newton in the 1670s.

In addition to honing their math skills, students are urged to take advantage of computers (which, incidentally, use the binary numeral system developed by Leibniz). There are many commercially available mathematics software packages for personal computers. Wherever possible they should be used to relieve the burden of repetitive and tedious calculations. Computer programming skills can and should be put to good use in the study of orbital mechanics. Elementary MATLAB® programs (M-files) appear at the end of this book to illustrate how some of the procedures developed in the text can be implemented in software. All of the scripts were developed using MATLAB version 5.0 and were successfully tested using version 6.5 (release 13). Information about MATLAB, which is a registered trademark of The MathWorks, Inc., may be obtained from:

The MathWorks, Inc.
3 Apple Hill Drive
Natick, MA, 01760-2098 USA
Tel: 508-647-7000
Fax: 508-647-7101
E-mail: info@mathworks.com
Web: www.mathworks.com

The text contains many detailed explanations and worked-out examples. Their purpose is not to overwhelm but to elucidate. It is always assumed that the material is being seen for the first time and, wherever possible, solution details are provided so as to leave little to the reader's imagination. There are some exceptions to this objective, deemed necessary to maintain the focus and control the size of the book. For example, in Chapter 6, the notion of specific impulse is laid on the table as a means of rating rocket motor performance and to show precisely how delta-v is related to propellant expenditure. In Chapter 10 Routh–Hurwitz stability criteria are used without proof to show quantitatively that a particular satellite configuration is, indeed, stable. Specific impulse is covered in more detail in Chapter 11, and the stability of linear systems is treated in depth in books on control theory. See, for example, Nise (2003) and Ogata (2001).

Supplementary material appears in the appendices at the end of the book. Appendix A lists physical data for use throughout the text. Appendix B is a 'road map' to guide the reader through Chapters 1, 2 and 3. Appendix C shows how to set up the n-body equations of motion and program them in MATLAB. Appendix D lists the MATLAB implementations of algorithms presented in several of the chapters. Appendix E shows that the gravitational field of a spherically symmetric body is the same as if the mass were concentrated at its center.

The field of astronautics is rich and vast. References cited throughout this text are listed at the end of the book. Also listed are other books on the subject that might be of interest to those seeking additional insights.

I wish to thank colleagues who provided helpful criticism and advice during the development of this book. Yechiel Crispin and Charles Eastlake were sources for ideas about what should appear in the summary chapter on rocket dynamics. Habib Eslami, Lakshmanan Narayanaswami, Mahmut Reyhanoglu and Axel Rohde all used the evolving manuscript as either a text or a reference in their space mechanics courses. Based on their classroom experiences, they gave me valuable feedback in the form of corrections, recommendations and much-needed encouragement. Tony Hagar voluntarily and thoroughly reviewed the entire manuscript and made a number of suggestions, nearly all of which were incorporated into the final version of the text.

I am indebted to those who reviewed the manuscript for the publisher for their many suggestions on how the book could be improved and what additional topics might be included.

Finally, let me acknowledge how especially grateful I am to the students who, throughout the evolution of the book, reported they found it to be a helpful and understandable introduction to space mechanics.

Howard D. Curtis
Embry-Riddle Aeronautical University
Daytona Beach, Florida

Supplements to the text

For the student:

- Copies of the MATLAB programs (M-files) that appear in Appendix D can be downloaded from the companion website accompanying this book. To access these please visit http://books.elsevier.com/companions and follow the instructions on screen.

For the instructor:

- A full *Instructor's Solutions Manual* is available for adopting tutors, which provides complete worked-out solutions to the problems set at the end of each chapter. To access these please visit http://books.elsevier.com/manuals and follow the instructions on screen.

CHAPTER 1

Dynamics of point masses

Chapter outline

1.1	Introduction	1
1.2	Kinematics	2
1.3	Mass, force and Newton's law of gravitation	7
1.4	Newton's law of motion	10
1.5	Time derivatives of moving vectors	15
1.6	Relative motion	20
Problems		29

1.1 Introduction

This chapter serves as a self-contained reference on the kinematics and dynamics of point masses as well as some basic vector operations. The notation and concepts summarized here will be used in the following chapters. Those familiar with the vector-based dynamics of particles can simply page through the chapter and then refer back to it later as necessary. Those who need a bit more in the way of review will find the chapter contains all of the material they need in order to follow the development of orbital mechanics topics in the upcoming chapters.

We begin with the problem of describing the curvilinear motion of particles in three dimensions. The concepts of force and mass are considered next, along with Newton's inverse-square law of gravitation. This is followed by a presentation of Newton's second law of motion ('force equals mass times acceleration') and the important concept of angular momentum.

As a prelude to describing motion relative to moving frames of reference, we develop formulas for calculating the time derivatives of moving vectors. These are applied to the computation of relative velocity and acceleration. Example problems illustrate the use of these results as does a detailed consideration of how the earth's rotation and curvature influence our measurements of velocity and acceleration. This brings in the curious concept of Coriolis force. Embedded in exercises at the end of the chapter is practice in verifying several fundamental vector identities that will be employed frequently throughout the book.

1.2 KINEMATICS

To track the motion of a particle P through Euclidean space we need a frame of reference, consisting of a clock and a cartesian coordinate system. The clock keeps track of time t and the xyz axes of the cartesian coordinate system are used to locate the spatial position of the particle. In non-relativistic mechanics, a single 'universal' clock serves for all possible cartesian coordinate systems. So when we refer to a frame of reference we need think only of the mutually orthogonal axes themselves.

The unit of time used throughout this book is the second (s). The unit of length is the meter (m), but the kilometer (km) will be the length unit of choice when large distances and velocities are involved. Conversion factors between kilometers, miles and nautical miles are listed in Table A.3.

Given a frame of reference, the position of the particle P at a time t is defined by the position vector $\mathbf{r}(t)$ extending from the origin O of the frame out to P itself, as illustrated in Figure 1.1. (Vectors will always be indicated by boldface type.) The

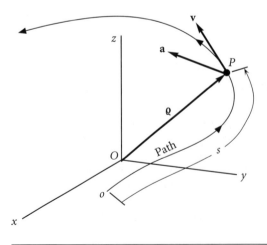

Figure 1.1 Position, velocity and acceleration vectors.

components of **r**(t) are just the x, y and z coordinates,

$$\mathbf{r}(t) = x(t)\hat{\mathbf{i}} + y(t)\hat{\mathbf{j}} + z(t)\hat{\mathbf{k}}$$

$\hat{\mathbf{i}}, \hat{\mathbf{j}}$ and $\hat{\mathbf{k}}$ are the unit vectors which point in the positive direction of the x, y and z axes, respectively. Any vector written with the overhead hat (e.g., $\hat{\mathbf{a}}$) is to be considered a vector of unit dimensionless magnitude.

The distance of P from the origin is the magnitude or length of **r**, denoted $\|\mathbf{r}\|$ or just r,

$$\|\mathbf{r}\| = r = \sqrt{x^2 + y^2 + z^2}$$

The magnitude of **r**, or any vector **A** for that matter, can also be computed by means of the dot product operation,

$$r = \sqrt{\mathbf{r} \cdot \mathbf{r}} \qquad \|\mathbf{A}\| = \sqrt{\mathbf{A} \cdot \mathbf{A}}$$

The velocity **v** and acceleration **a** of the particle are the first and second time derivatives of the position vector,

$$\mathbf{v}(t) = \frac{dx(t)}{dt}\hat{\mathbf{i}} + \frac{dy(t)}{dt}\hat{\mathbf{j}} + \frac{dz(t)}{dt}\hat{\mathbf{k}} = v_x(t)\hat{\mathbf{i}} + v_y(t)\hat{\mathbf{j}} + v_z(t)\hat{\mathbf{k}}$$

$$\mathbf{a}(t) = \frac{dv_x(t)}{dt}\hat{\mathbf{i}} + \frac{dv_y(t)}{dt}\hat{\mathbf{j}} + \frac{dv_z(t)}{dt}\hat{\mathbf{k}} = a_x(t)\hat{\mathbf{i}} + a_y(t)\hat{\mathbf{j}} + a_z(t)\hat{\mathbf{k}}$$

It is convenient to represent the time derivative by means of an overhead dot. In this shorthand notation, if ( ) is any quantity, then

$$\dot{(\,)} \equiv \frac{d(\,)}{dt} \qquad \ddot{(\,)} \equiv \frac{d^2(\,)}{dt^2} \qquad \dddot{(\,)} \equiv \frac{d^3(\,)}{dt^3}, \text{etc.}$$

Thus, for example,

$$\mathbf{v} = \dot{\mathbf{r}}$$
$$\mathbf{a} = \dot{\mathbf{v}} = \ddot{\mathbf{r}}$$
$$v_x = \dot{x} \qquad v_y = \dot{y} \qquad v_z = \dot{z}$$
$$a_x = \dot{v}_x = \ddot{x} \qquad a_y = \dot{v}_y = \ddot{y} \qquad a_z = \dot{v}_z = \ddot{z}$$

The locus of points that a particle occupies as it moves through space is called its path or trajectory. If the path is a straight line, then the motion is rectilinear. Otherwise, the path is curved, and the motion is called curvilinear. The velocity vector **v** is tangent to the path. If $\hat{\mathbf{u}}_t$ is the unit vector tangent to the trajectory, then

$$\mathbf{v} = v\hat{\mathbf{u}}_t$$

where v, the speed, is the magnitude of the velocity **v**. The distance ds that P travels along its path in the time interval dt is obtained from the speed by

$$ds = v\, dt$$

In other words,

$$v = \dot{s}$$

The distance s, measured along the path from some starting point, is what the odometers in our automobiles record. Of course, $\dot{s}$, our speed along the road, is indicated by the dial of the speedometer.

Note carefully that $v \neq \dot{r}$, i.e., the magnitude of the derivative of $\mathbf{r}$ does not equal the derivative of the magnitude of $\mathbf{r}$.

EXAMPLE 1.1

The position vector in meters is given as a function of time in seconds as

$$\mathbf{r} = (8t^2 + 7t + 6)\hat{\mathbf{i}} + (5t^3 + 4)\hat{\mathbf{j}} + (0.3t^4 + 2t^2 + 1)\hat{\mathbf{k}} \text{ (m)} \qquad (a)$$

At $t = 10$ seconds, calculate v (the magnitude of the derivative of $\mathbf{r}$) and $\dot{r}$ (the derivative of the magnitude of $\mathbf{r}$).

The velocity $\mathbf{v}$ is found by differentiating the given position vector with respect to time,

$$\mathbf{v} = \frac{d\mathbf{r}}{dt} = (16t + 7)\hat{\mathbf{i}} + 15t^2\hat{\mathbf{j}} + (1.2t^3 + 4t)\hat{\mathbf{k}}$$

The magnitude of this vector is the square root of the sum of the squares of its components,

$$\|\mathbf{v}\| = (1.44t^6 + 234.6t^4 + 272t^2 + 224t + 49)^{\frac{1}{2}}$$

Evaluating this at $t = 10$ s, we get

$$v = 1953.3 \text{ m/s}$$

Calculating the magnitude of $\mathbf{r}$ in (a), leads to

$$\|\mathbf{r}\| = (0.09t^8 + 26.2t^6 + 68.6t^4 + 152t^3 + 149t^2 + 84t + 53)^{\frac{1}{2}}$$

Differentiating this expression with respect to time,

$$\dot{r} = \frac{dr}{dt} = \frac{0.36t^7 + 78.6t^5 + 137.2t^3 + 228t^2 + 149t + 42}{(0.09t^8 + 26.2t^6 + 68.6t^4 + 152t^3 + 149t^2 + 84t + 53)^{\frac{1}{2}}}$$

Substituting $t = 10$ s, yields

$$\dot{r} = 1935.5 \text{ m/s}$$

If $\mathbf{v}$ is given, then we can find the components of the unit tangent $\hat{\mathbf{u}}_t$ in the cartesian coordinate frame of reference

$$\hat{\mathbf{u}}_t = \frac{\mathbf{v}}{\|\mathbf{v}\|} = \frac{v_x}{v}\hat{\mathbf{i}} + \frac{v_y}{v}\hat{\mathbf{j}} + \frac{v_z}{v}\hat{\mathbf{k}} \quad \left(v = \sqrt{v_x^2 + v_y^2 + v_z^2}\right)$$

The acceleration may be written,

$$\mathbf{a} = a_t \hat{\mathbf{u}}_t + a_n \hat{\mathbf{u}}_n$$

where a_t and a_n are the tangential and normal components of acceleration, given by

$$a_t = \dot{v}\,(= \ddot{s}) \qquad a_n = \frac{v^2}{\varrho} \tag{1.1}$$

ϱ is the radius of curvature, which is the distance from the particle P to the center of curvature of the path at that point. The unit principal normal $\hat{\mathbf{u}}_n$ is perpendicular to $\hat{\mathbf{u}}_t$ and points towards the center of curvature C, as shown in Figure 1.2. Therefore, the position of C relative to P, denoted $\mathbf{r}_{C/P}$, is

$$\mathbf{r}_{C/P} = \varrho \hat{\mathbf{u}}_n$$

The orthogonal unit vectors $\hat{\mathbf{u}}_t$ and $\hat{\mathbf{u}}_n$ form a plane called the osculating plane. The unit normal to the osculating plane is $\hat{\mathbf{u}}_b$, the binormal, and it is obtained from $\hat{\mathbf{u}}_t$ and $\hat{\mathbf{u}}_n$ by taking their cross product,

$$\hat{\mathbf{u}}_b = \hat{\mathbf{u}}_t \times \hat{\mathbf{u}}_n$$

The center of curvature lies in the osculating plane. When the particle P moves an incremental distance ds the radial from the center of curvature to the path sweeps out a small angle $d\phi$, measured in the osculating plane. The relationship between this angle and ds is

$$ds = \varrho\, d\phi$$

so that $\dot{s} = \varrho \dot{\phi}$, or

$$\dot{\phi} = \frac{v}{\varrho} \tag{1.2}$$

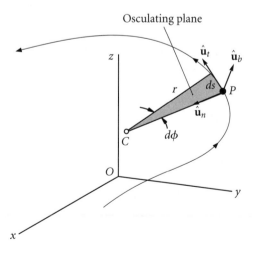

Figure 1.2 Orthogonal triad of unit vectors associated with the moving point P.

EXAMPLE 1.2

Relative to a cartesian coordinate system, the position, velocity and acceleration of a particle relative at a given instant are

$$\mathbf{r} = 250\hat{\mathbf{i}} + 630\hat{\mathbf{j}} + 430\hat{\mathbf{k}} \text{ (m)}$$

$$\mathbf{v} = 90\hat{\mathbf{i}} + 125\hat{\mathbf{j}} + 170\hat{\mathbf{k}} \text{ (m/s)}$$

$$\mathbf{a} = 16\hat{\mathbf{i}} + 125\hat{\mathbf{j}} + 30\hat{\mathbf{k}} \text{ (m/s}^2\text{)}$$

Find the coordinates of the center of curvature at that instant.

First, we calculate the speed v,

$$v = \|\mathbf{v}\| = \sqrt{90^2 + 125^2 + 170^2} = 229.4 \text{ m/s}$$

The unit tangent is, therefore,

$$\hat{\mathbf{u}}_t = \frac{\mathbf{v}}{v} = \frac{90\hat{\mathbf{i}} + 125\hat{\mathbf{j}} + 170\hat{\mathbf{k}}}{797.4} = 0.3923\hat{\mathbf{i}} + 0.5449\hat{\mathbf{j}} + 0.7411\hat{\mathbf{k}}$$

We project the acceleration vector onto the direction of the tangent to get its tangential component a_t,

$$a_t = \mathbf{a} \cdot \hat{\mathbf{u}}_t = (16\hat{\mathbf{i}} + 125\hat{\mathbf{j}} + 30\hat{\mathbf{k}}) \cdot (0.3923\hat{\mathbf{i}} + 0.5449\hat{\mathbf{j}} + 0.7411\hat{\mathbf{k}}) = 96.62 \text{ m/s}^2$$

The magnitude of **a** is

$$a = \sqrt{16^2 + 125^2 + 30^2} = 129.5 \text{ m/s}^2$$

Since $\mathbf{a} = a_t\hat{\mathbf{u}}_t + a_n\hat{\mathbf{u}}_n$ and $\hat{\mathbf{u}}_t$ and $\hat{\mathbf{u}}_n$ are perpendicular to each other, it follows that $a^2 = a_t^2 + a_n^2$, which means

$$a_n = \sqrt{a^2 - a_t^2} = \sqrt{129.5^2 - 96.62^2} = 86.29 \text{ m/s}^2$$

Hence,

$$\hat{\mathbf{u}}_n = \frac{1}{a_n}(\mathbf{a} - a_t\hat{\mathbf{u}}_t)$$

$$= \frac{1}{86.29}[(16\hat{\mathbf{i}} + 125\hat{\mathbf{j}} + 30\hat{\mathbf{k}}) - 96.62(0.3923\hat{\mathbf{i}} + 0.5449\hat{\mathbf{j}} + 0.7411\hat{\mathbf{k}})]$$

$$= -0.2539\hat{\mathbf{i}} + 0.8385\hat{\mathbf{j}} - 0.4821\hat{\mathbf{k}}$$

The equation $a_n = v^2/\varrho$ can now be solved for ϱ to yield

$$\varrho = \frac{v^2}{a_n} = \frac{229.4^2}{86.29} = 609.9 \text{ m}$$

Let $\mathbf{r}_C$ be the position vector of the center of curvature C. Then

$$\begin{aligned}\mathbf{r}_C &= \mathbf{r} + \mathbf{r}_{C/P} \\ &= \mathbf{r} + \varrho\hat{\mathbf{u}}_n = 250\hat{\mathbf{i}} + 630\hat{\mathbf{j}} + 430\hat{\mathbf{k}} + 609.9(-0.2539\hat{\mathbf{i}} + 0.8385\hat{\mathbf{j}} - 0.4821\hat{\mathbf{k}}) \\ &= 95.16\hat{\mathbf{i}} + 1141\hat{\mathbf{j}} + 136.0\hat{\mathbf{k}}\ (\text{m})\end{aligned}$$

That is, the coordinates of C are

$$x = 95.16\ \text{m} \qquad y = 1141\ \text{m} \qquad z = 136.0\ \text{m}$$

1.3 Mass, force and Newton's law of gravitation

Mass, like length and time, is a primitive physical concept: it cannot be defined in terms of any other physical concept. Mass is simply the quantity of matter. More practically, mass is a measure of the inertia of a body. Inertia is an object's resistance to changing its state of motion. The larger its inertia (the greater its mass), the more difficult it is to set a body into motion or bring it to rest. The unit of mass is the kilogram (kg).

Force is the action of one physical body on another, either through direct contact or through a distance. Gravity is an example of force acting through a distance, as are magnetism and the force between charged particles. The gravitational force between two masses m_1 and m_2 having a distance r between their centers is

$$F_g = G\frac{m_1 m_2}{r^2} \tag{1.3}$$

This is Newton's law of gravity, in which G, the universal gravitational constant, has the value $6.6742 \times 10^{11}\ \text{m}^3/\text{kg} \cdot \text{s}^2$. Due to the inverse-square dependence on distance, the force of gravity rapidly diminishes with the amount of separation between the two masses. In any case, the force of gravity is minuscule unless at least one of the masses is extremely big.

The force of a large mass (such as the earth) on a mass many orders of magnitude smaller (such as a person) is called weight, W. If the mass of the large object is M and that of the relatively tiny one is m, then the weight of the small body is

$$W = G\frac{Mm}{r^2} = m\left(\frac{GM}{r^2}\right)$$

or

$$W = mg \tag{1.4}$$

where

$$g = \frac{GM}{r^2} \tag{1.5}$$

g has units of acceleration (m/s²) and is called the acceleration of gravity. If planetary gravity is the only force acting on a body, then the body is said to be in free fall. The force of gravity draws a freely falling object towards the center of attraction (e.g., center of the earth) with an acceleration g. Under ordinary conditions, we sense our own weight by feeling contact forces acting on us in opposition to the force of gravity. In free fall there are, by definition, no contact forces, so there can be no sense of weight. Even though the weight is not zero, a person in free fall experiences weightlessness, or the absence of gravity.

Let us evaluate Equation 1.5 at the surface of the earth, whose radius according to Table A.1 is 6378 km. Letting g_0 represent the standard sea-level value of g, we get

$$g_0 = \frac{GM}{R_E^2} \qquad (1.6)$$

In SI units,

$$g_0 = 9.807 \text{ m/s} \qquad (1.7)$$

Substituting Equation 1.6 into Equation 1.5 and letting z represent the distance above the earth's surface, so that $r = R_E + z$, we obtain

$$g = g_0 \frac{R_E^2}{(R_E + z)^2} = \frac{g_0}{(1 + z/R_E)^2} \qquad (1.8)$$

Commercial airliners cruise at altitudes on the order of 10 kilometers (six miles). At that height, Equation 1.8 reveals that g (and hence weight) is only three-tenths of a percent less than its sea-level value. Thus, under ordinary conditions, we ignore the variation of g with altitude. A plot of Equation 1.8 out to a height of 1000 km (the upper limit of low-earth orbit operations) is shown in Figure 1.3. The variation of g over that range is significant. Even so, at space station altitude (300 km), weight is only about 10 percent less that it is on the earth's surface. The astronauts experience weightlessness, but they clearly are not weightless.

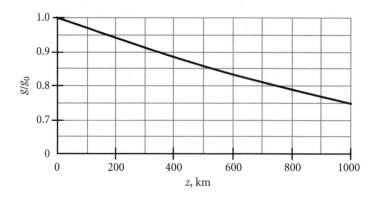

Figure 1.3 Variation of the acceleration of gravity with altitude.

EXAMPLE 1.3

Show that in the absence of an atmosphere, the shape of a low altitude ballistic trajectory is a parabola. Assume the acceleration of gravity g is constant and neglect the earth's curvature.

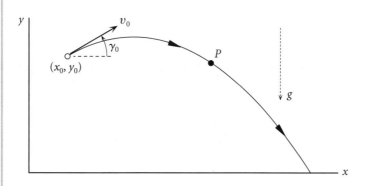

Figure 1.4 Flight of a low altitude projectile in free fall (no atmosphere).

Figure 1.4 shows a projectile launched at $t = 0$ with a speed v_0 at a flight path angle γ_0 from the point with coordinates (x_0, y_0). Since the projectile is in free fall after launch, its only acceleration is that of gravity in the negative y-direction:

$$\ddot{x} = 0$$
$$\ddot{y} = -g$$

Integrating with respect to time and applying the initial conditions leads to

$$x = x_0 + (v_0 \cos \gamma_0)t \tag{a}$$

$$y = y_0 + (v_0 \sin \gamma_0)t - \frac{1}{2}gt^2 \tag{b}$$

Solving (a) for t and substituting the result into (b) yields

$$y = y_0 + (x - x_0)\tan \gamma_0 - \frac{1}{2}\frac{g}{v_0 \cos \gamma_0}(x - x_0)^2 \tag{c}$$

This is the equation of a second-degree curve, a parabola, as sketched in Figure 1.4.

EXAMPLE 1.4

An airplane flies a parabolic trajectory like that in Figure 1.4 so that the passengers will experience free fall (weightlessness). What is the required variation of the flight path angle γ with speed v? Ignore the curvature of the earth.

Figure 1.5 reveals that for a 'flat' earth, $d\gamma = -d\phi$, i.e.,

$$\dot{\gamma} = -\dot{\phi}$$

(Example 1.4 continued)

It follows from Equation 1.2 that

$$\varrho\dot{\gamma} = -v \tag{1.9}$$

The normal acceleration a_n is just the component of the gravitational acceleration g in the direction of the unit principal normal to the curve (from P towards C). From Figure 1.5, then,

$$a_n = g\cos\gamma \tag{a}$$

Substituting Equation 1.1 into (a) and solving for the radius of curvature yields

$$\varrho = \frac{v^2}{g\cos\gamma} \tag{b}$$

Combining Equations 1.9 and (b), we find the time rate of change of the flight path angle,

$$\dot{\gamma} = -\frac{g\cos\gamma}{v}$$

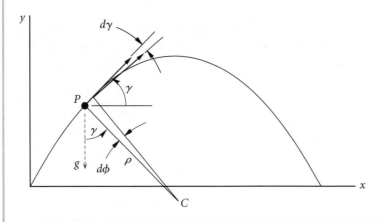

Figure 1.5 Relationship between $d\gamma$ and $d\phi$ for a 'flat' earth.

1.4 Newton's law of motion

Force is not a primitive concept like mass because it is intimately connected with the concepts of motion and inertia. In fact, the only way to alter the motion of a body is to exert a force on it. The degree to which the motion is altered is a measure of the force. This is quantified by Newton's second law of motion. If the resultant or net force on a body of mass m is $\mathbf{F}_{net}$, then

$$\mathbf{F}_{net} = m\mathbf{a} \tag{1.10}$$

1.4 Newton's law of motion

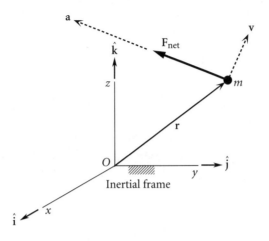

Figure 1.6 The absolute acceleration of a particle is in the direction of the net force.

In this equation, **a** is the absolute acceleration of the center of mass. The absolute acceleration is measured in a frame of reference which itself has neither translational nor rotational acceleration relative to the fixed stars. Such a reference is called an absolute or inertial frame of reference.

Force, then, is related to the primitive concepts of mass, length and time by Newton's second law. The unit of force, appropriately, is the Newton, which is the force required to impart an acceleration of 1 m/s^2 to a mass of 1 kg. A mass of one kilogram therefore weighs 9.81 Newtons at the earth's surface. The kilogram is not a unit of force.

Confusion can arise when mass is expressed in units of force, as frequently occurs in US engineering practice. In common parlance either the pound or the ton (2000 pounds) is more likely to be used to express the mass. The pound of mass is officially defined precisely in terms of the kilogram as shown in Table A.3. Since one pound of mass weighs one pound of force where the standard sea-level acceleration of gravity ($g_0 = 9.80665 \text{ m/s}^2$) exists, we can use Newton's second law to relate the pound of force to the Newton:

$$1 \text{ lb (force)} = 0.4536 \text{ kg} \times 9.807 \text{ m/s}^2$$

$$= 4.448 \text{ N}$$

The slug is the quantity of matter accelerated at one foot per second^2 by a force of one pound. We can again use Newton's second law to relate the slug to the kilogram. Noting the relationship between feet and meters in Table A.3, we find

$$1 \text{ slug} = \frac{1 \text{ lb}}{1 \text{ ft/s}^2} = \frac{4.448 \text{ N}}{0.3048 \text{ m/s}^2} = 14.59 \frac{\text{kg} \cdot \text{m/s}^2}{\text{m/s}^2}$$

$$= 14.59 \text{ kg}$$

EXAMPLE 1.5

On a NASA mission the space shuttle Atlantis orbiter was reported to weigh 239 255 lb just prior to lift-off. On orbit 18 at an altitude of about 350 km, the orbiter's weight was reported to be 236 900 lb. (a) What was the mass, in kilograms, of Atlantis on the launch pad and in orbit? (b) If no mass were lost between launch and orbit 18, what would have been the weight of Atlantis in pounds?

(a) The given data illustrates the common use of weight in pounds as a measure of mass. The 'weights' given are actually the mass in pounds of mass. Therefore, prior to launch

$$m_{\text{launch pad}} = 239\,255 \text{ lb (mass)} \times \frac{0.4536 \text{ kg}}{1 \text{ lb (mass)}} = \underline{108\,500 \text{ kg}}$$

In orbit,

$$m_{\text{orbit 18}} = 236\,900 \text{ lb (mass)} \times \frac{0.4536 \text{ kg}}{1 \text{ lb (mass)}} = \underline{107\,500 \text{ kg}}$$

The decrease in mass is the propellant expended by the orbital maneuvering and reaction control rockets on the orbiter.

(b) Since the space shuttle launch pad at Kennedy Space Center is essentially at sea level, the launch-pad weight of Atlantis in lb (force) is numerically equal to its mass in lb (mass). With no change in mass, the force of gravity at 350 km would be, according to Equation 1.8,

$$W = 239\,255 \text{ lb (force)} \times \left(\frac{1}{1 + \frac{350}{6378}}\right)^2 = \underline{215\,000 \text{ lb (force)}}$$

The integral of a force $\mathbf{F}$ over a time interval is called the impulse I of the force,

$$\mathbf{I} = \int_{t_1}^{t_2} \mathbf{F}\, dt \qquad (1.11)$$

From Equation 1.10 it is apparent that if the mass is constant, then

$$\mathbf{I}_{\text{net}} = \int_{t_1}^{t_2} m\frac{d\mathbf{v}}{dt}\, dt = m\mathbf{v}_2 - m\mathbf{v}_1 \qquad (1.12)$$

That is, the net impulse on a body yields a change $m\Delta\mathbf{v}$ in its linear momentum, so that

$$\Delta\mathbf{v} = \frac{\mathbf{I}_{\text{net}}}{m} \qquad (1.13)$$

If $\mathbf{F}_{\text{net}}$ is constant, then $\mathbf{I}_{\text{net}} = \mathbf{F}_{\text{net}}\Delta t$, in which case Equation 1.13 becomes

$$\Delta\mathbf{v} = \frac{\mathbf{F}_{\text{net}}}{m}\Delta t \quad (\text{if } \mathbf{F}_{\text{net}} \text{ is constant}) \qquad (1.14)$$

Let us conclude this section by introducing the concept of angular momentum. The moment of the net force about O in Figure 1.6 is

$$\mathbf{M}_{O_{\text{net}}} = \mathbf{r} \times \mathbf{F}_{\text{net}}$$

Substituting Equation 1.10 yields

$$\mathbf{M}_{O_{\text{net}}} = \mathbf{r} \times m\mathbf{a} = \mathbf{r} \times m\frac{d\mathbf{v}}{dt} \quad (1.15)$$

But, keeping in mind that the mass is constant,

$$\mathbf{r} \times m\frac{d\mathbf{v}}{dt} = \frac{d}{dt}(\mathbf{r} \times m\mathbf{v}) - \left(\frac{d\mathbf{r}}{dt} \times m\mathbf{v}\right) = \frac{d}{dt}(\mathbf{r} \times m\mathbf{v}) - (\mathbf{v} \times m\mathbf{v})$$

Since $\mathbf{v} \times m\mathbf{v} = m(\mathbf{v} \times \mathbf{v}) = \mathbf{0}$, it follows that Equation 1.15 can be written

$$\mathbf{M}_{O_{\text{net}}} = \frac{d\mathbf{H}_O}{dt} \quad (1.16)$$

where $\mathbf{H}_O$ is the angular momentum about O,

$$\mathbf{H}_O = \mathbf{r} \times m\mathbf{v} \quad (1.17)$$

Thus, just as the net force on a particle changes its linear momentum $m\mathbf{v}$, the moment of that force about a fixed point changes the moment of its linear momentum about that point. Integrating Equation 1.16 with respect to time yields

$$\int_{t_1}^{t_2} \mathbf{M}_{O_{\text{net}}} dt = \mathbf{H}_{O_2} - \mathbf{H}_{O_1} \quad (1.18)$$

The integral on the left is the net angular impulse. This angular impulse–momentum equation is the rotational analog of the linear impulse–momentum relation given above in Equation 1.12.

EXAMPLE 1.6

A particle of mass m is attached to point O by an inextensible string of length l. Initially the string is slack when m is moving to the left with a speed v_o in the position shown. Calculate the speed of m just after the string becomes taut. Also, compute the

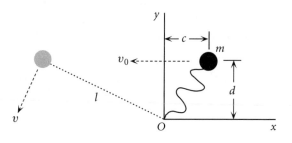

Figure 1.7 Particle attached to O by an inextensible string.

(Example 1.6 continued) average force in the string over the small time interval Δt required to change the direction of the particle's motion.

Initially, the position and velocity of the particle are

$$\mathbf{r}_1 = c\hat{\mathbf{i}} + d\hat{\mathbf{j}} \qquad \mathbf{v}_1 = -v_0\hat{\mathbf{i}}$$

The angular momentum is

$$\mathbf{H}_1 = \mathbf{r}_1 \times m\mathbf{v}_1 = \begin{vmatrix} \hat{\mathbf{i}} & \hat{\mathbf{j}} & \hat{\mathbf{k}} \\ c & d & 0 \\ -mv_0 & 0 & 0 \end{vmatrix} = mv_0 d\hat{\mathbf{k}} \qquad (a)$$

Just after the string becomes taut

$$\mathbf{r}_2 = -\sqrt{l^2 - d^2}\,\hat{\mathbf{i}} + d\hat{\mathbf{j}} \qquad \mathbf{v}_2 = v_x\hat{\mathbf{i}} + v_y\hat{\mathbf{j}} \qquad (b)$$

and the angular momentum is

$$\mathbf{H}_2 = \mathbf{r}_2 \times m\mathbf{v}_2 = \begin{vmatrix} \hat{\mathbf{i}} & \hat{\mathbf{j}} & \hat{\mathbf{k}} \\ -\sqrt{l^2 - d^2} & d & 0 \\ v_x & v_y & 0 \end{vmatrix} = \left(-mv_x d - mv_y\sqrt{l^2 - d^2}\right)\hat{\mathbf{k}} \qquad (c)$$

Initially the force exerted on m by the slack string is zero. When the string becomes taut, the force exerted on m passes through O. Therefore, the moment of the net force on m about O remains zero. According to Equation 1.18,

$$\mathbf{H}_2 = \mathbf{H}_1$$

Substituting (a) and (c) yields

$$v_x d + \sqrt{l^2 - d^2}\, v_y = -v_0 d \qquad (d)$$

The string is inextensible, so the component of the velocity of m along the string must be zero:

$$\mathbf{v}_2 \cdot \mathbf{r}_2 = 0$$

Substituting $\mathbf{v}_2$ and $\mathbf{r}_2$ from (b) and solving for v_y we get

$$v_y = v_x\sqrt{\frac{l^2}{d^2} - 1} \qquad (e)$$

Solving (d) and (e) for v_x and v_y leads to

$$v_x = -\frac{d^2}{l^2} v_0 \qquad v_y = -\sqrt{1 - \frac{d^2}{l^2}}\,\frac{d}{l} v_0 \qquad (f)$$

Thus, the speed, $v = \sqrt{v_x^2 + v_y^2}$, after the string becomes taut is

$$\underline{v = \frac{d}{l} v_0}$$

From Equation 1.12, the impulse on m during the time it takes the string to become taut is

$$\mathbf{I} = m(\mathbf{v}_2 - \mathbf{v}_1) = m\left[\left(-\frac{d^2}{l^2}v_0\hat{\mathbf{i}} - \sqrt{1-\frac{d^2}{l^2}}\frac{d}{l}v_0\hat{\mathbf{j}}\right) - (-v_0\hat{\mathbf{i}})\right]$$

$$= \left(1 - \frac{d^2}{l^2}\right)mv_0\hat{\mathbf{i}} - \sqrt{1-\frac{d^2}{l^2}}\frac{d}{l}mv_0\hat{\mathbf{j}}$$

The magnitude of this impulse, which is directed along the string, is

$$I = \sqrt{1 - \frac{d^2}{l^2}}\,mv_0$$

Hence, the average force in the string during the small time interval Δt required to change the direction of the velocity vector turns out to be

$$F_{\text{avg}} = \frac{I}{\Delta t} = \sqrt{1 - \frac{d^2}{l^2}}\frac{mv_0}{\Delta t}$$

1.5 Time derivatives of moving vectors

Figure 1.8(a) shows a vector **A** inscribed in a rigid body B that is in motion relative to an inertial frame of reference (a rigid, cartesian coordinate system which is fixed relative to the fixed stars). The magnitude of **A** is fixed. The body B is shown at two times, separated by the differential time interval dt. At time $t + dt$ the orientation of

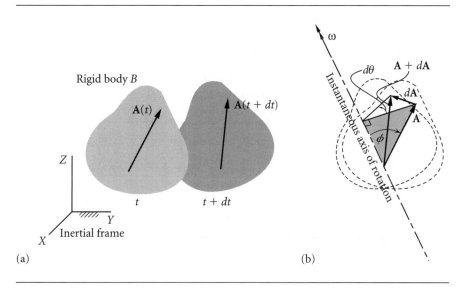

Figure 1.8 Displacement of a rigid body.

vector **A** differs slightly from that at time t, but its magnitude is the same. According to one of the many theorems of the prolific eighteenth century Swiss mathematician Leonhard Euler (1707–1783), there is a unique axis of rotation about which B and, therefore, **A** rotates during the differential time interval. If we shift the two vectors $\mathbf{A}(t)$ and $\mathbf{A}(t+dt)$ to the same point on the axis of rotation, so that they are tail-to-tail as shown in Figure 1.8(b), we can assess the difference $d\mathbf{A}$ between them caused by the infinitesimal rotation. Remember that shifting a vector to a parallel line does not change the vector. The rotation of the body B is measured in the plane perpendicular to the instantaneous axis of rotation. The amount of rotation is the angle $d\theta$ through which a line element normal to the rotation axis turns in the time interval dt. In Figure 1.8(b) that line element is the component of **A** normal to the axis of rotation. We can express the difference $d\mathbf{A}$ between $\mathbf{A}(t)$ and $\mathbf{A}(t+dt)$ as

$$d\mathbf{A} = \overbrace{[(\|\mathbf{A}\| \cdot \sin\phi)d\theta]}^{\text{magnitude of } d\mathbf{A}} \hat{\mathbf{n}} \qquad (1.19)$$

where $\hat{\mathbf{n}}$ is the unit normal to the plane defined by **A** and the axis of rotation, and it points in the direction of the rotation. The angle ϕ is the inclination of **A** to the rotation axis. By definition,

$$d\theta = \|\boldsymbol{\omega}\|dt \qquad (1.20)$$

where $\boldsymbol{\omega}$ is the angular velocity vector, which points along the instantaneous axis of rotation and its direction is given by the right-hand rule. That is, wrapping the right hand around the axis of rotation, with the fingers pointing in the direction of $d\theta$, results in the thumb's defining the direction of $\boldsymbol{\omega}$. This is evident in Figure 1.8(b). It should be pointed out that the time derivative of $\boldsymbol{\omega}$ is the angular acceleration, usually given the symbol $\boldsymbol{\alpha}$. Thus,

$$\boldsymbol{\alpha} = \frac{d\boldsymbol{\omega}}{dt} \qquad (1.21)$$

Substituting Equation 1.20 into Equation 1.19, we get

$$d\mathbf{A} = \|\mathbf{A}\| \cdot \sin\phi \|\boldsymbol{\omega}\|dt \cdot \hat{\mathbf{n}} = (\|\boldsymbol{\omega}\| \cdot \|\mathbf{A}\| \cdot \sin\phi)\,\hat{\mathbf{n}}\,dt \qquad (1.22)$$

By definition of the cross product, $\boldsymbol{\omega} \times \mathbf{A}$ is the product of the magnitude of $\boldsymbol{\omega}$, the magnitude of **A**, the sine of the angle between $\boldsymbol{\omega}$ and **A** and the unit vector normal to the plane of $\boldsymbol{\omega}$ and **A**, in the rotation direction. That is,

$$\boldsymbol{\omega} \times \mathbf{A} = \|\boldsymbol{\omega}\| \cdot \|\mathbf{A}\| \cdot \sin\phi \cdot \hat{\mathbf{n}} \qquad (1.23)$$

Substituting Equation 1.23 into Equation 1.22 yields

$$d\mathbf{A} = \boldsymbol{\omega} \times \mathbf{A}\,dt$$

Dividing through by dt, we finally obtain

$$\frac{d\mathbf{A}}{dt} = \boldsymbol{\omega} \times \mathbf{A} \qquad (1.24)$$

Equation 1.24 is a formula we can use to compute the time derivative of any vector of constant magnitude.

1.5 Time derivatives of moving vectors

EXAMPLE 1.7

Calculate the second time derivative of a vector **A** of constant magnitude, expressing the result in terms of **ω** and its derivatives and **A**.

Differentiating Equation 1.24 with respect to time, we get

$$\frac{d^2\mathbf{A}}{dt^2} = \frac{d}{dt}\frac{d\mathbf{A}}{dt} = \frac{d}{dt}(\boldsymbol{\omega} \times \mathbf{A}) = \frac{d\boldsymbol{\omega}}{dt} \times \mathbf{A} + \boldsymbol{\omega} \times \frac{d\mathbf{A}}{dt}$$

Using Equations 1.21 and 1.24, this can be written

$$\frac{d^2\mathbf{A}}{dt^2} = \boldsymbol{\alpha} \times \mathbf{A} + \boldsymbol{\omega} \times (\boldsymbol{\omega} \times \mathbf{A}) \tag{1.25}$$

EXAMPLE 1.8

Calculate the third derivative of a vector **A** of constant magnitude, expressing the result in terms of **ω** and its derivatives and **A**.

$$\frac{d^3\mathbf{A}}{dt^3} = \frac{d}{dt}\frac{d^2\mathbf{A}}{dt^2} = \frac{d}{dt}[\boldsymbol{\alpha} \times \mathbf{A} + \boldsymbol{\omega} \times (\boldsymbol{\omega} \times \mathbf{A})]$$

$$= \frac{d}{dt}(\boldsymbol{\alpha} \times \mathbf{A}) + \frac{d}{dt}[\boldsymbol{\omega} \times (\boldsymbol{\omega} \times \mathbf{A})]$$

$$= \left(\frac{d\boldsymbol{\alpha}}{dt} \times \mathbf{A} + \boldsymbol{\alpha} \times \frac{d\mathbf{A}}{dt}\right) + \left[\frac{d\boldsymbol{\omega}}{dt} \times (\boldsymbol{\omega} \times \mathbf{A}) + \boldsymbol{\omega} \times \frac{d}{dt}(\boldsymbol{\omega} \times \mathbf{A})\right]$$

$$= \left[\frac{d\boldsymbol{\alpha}}{dt} \times \mathbf{A} + \boldsymbol{\alpha} \times (\boldsymbol{\omega} \times \mathbf{A})\right] + \left[\boldsymbol{\alpha} \times (\boldsymbol{\omega} \times \mathbf{A}) + \boldsymbol{\omega} \times \left(\frac{d\boldsymbol{\omega}}{dt} \times \mathbf{A} + \boldsymbol{\omega} \times \frac{d\mathbf{A}}{dt}\right)\right]$$

$$= \left[\frac{d\boldsymbol{\alpha}}{dt} \times \mathbf{A} + \boldsymbol{\alpha} \times (\boldsymbol{\omega} \times \mathbf{A})\right] + \{\boldsymbol{\alpha} \times (\boldsymbol{\omega} \times \mathbf{A}) + \boldsymbol{\omega} \times [\boldsymbol{\alpha} \times \mathbf{A} + \boldsymbol{\omega} \times (\boldsymbol{\omega} \times \mathbf{A})]\}$$

$$= \frac{d\boldsymbol{\alpha}}{dt} \times \mathbf{A} + \boldsymbol{\alpha} \times (\boldsymbol{\omega} \times \mathbf{A}) + \boldsymbol{\alpha} \times (\boldsymbol{\omega} \times \mathbf{A}) + \boldsymbol{\omega} \times (\boldsymbol{\alpha} \times \mathbf{A}) + \boldsymbol{\omega} \times [\boldsymbol{\omega} \times (\boldsymbol{\omega} \times \mathbf{A})]$$

$$= \frac{d\boldsymbol{\alpha}}{dt} \times \mathbf{A} + 2\boldsymbol{\alpha} \times (\boldsymbol{\omega} \times \mathbf{A}) + \boldsymbol{\omega} \times (\boldsymbol{\alpha} \times \mathbf{A}) + \boldsymbol{\omega} \times [\boldsymbol{\omega} \times (\boldsymbol{\omega} \times \mathbf{A})]$$

$$\frac{d^3\mathbf{A}}{dt^3} = \frac{d\boldsymbol{\alpha}}{dt} \times \mathbf{A} + 2\boldsymbol{\alpha} \times (\boldsymbol{\omega} \times \mathbf{A}) + \boldsymbol{\omega} \times [\boldsymbol{\alpha} \times \mathbf{A} + \boldsymbol{\omega} \times (\boldsymbol{\omega} \times \mathbf{A})]$$

Let *XYZ* be a rigid inertial frame of reference and *xyz* a rigid moving frame of reference, as shown in Figure 1.9. The moving frame can be moving (translating and rotating) freely of its own accord, or it can be imagined to be attached to a physical object, such as a car, an airplane or a spacecraft. Kinematic quantities measured relative to the fixed inertial frame will be called absolute (e.g., absolute acceleration), and those measured relative to the moving system will be called relative (e.g., relative acceleration). The unit vectors along the inertial *XYZ* system are $\hat{\mathbf{I}}, \hat{\mathbf{J}}$ and $\hat{\mathbf{K}}$, whereas those of the moving *xyz* system are $\hat{\mathbf{i}}, \hat{\mathbf{j}}$ and $\hat{\mathbf{k}}$. The motion of the moving frame is arbitrary, and its absolute angular velocity is **Ω**. If, however, the moving frame is rigidly attached to an object, so that it not only translates but rotates with it, then the

18 Chapter 1 *Dynamics of point masses*

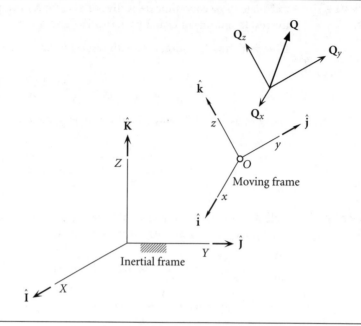

Figure 1.9 Fixed (inertial) and moving rigid frames of reference.

frame is called a body frame and the axes are referred to as body axes. A body frame clearly has the same angular velocity as the body to which it is bound.

Let **Q** be any time-dependent vector. Resolved into components along the inertial frame of reference, it is expressed analytically as

$$\mathbf{Q} = Q_X \hat{\mathbf{I}} + Q_Y \hat{\mathbf{J}} + Q_Z \hat{\mathbf{K}}$$

where Q_X, Q_Y and Q_Z are functions of time. Since $\hat{\mathbf{I}}$, $\hat{\mathbf{J}}$ and $\hat{\mathbf{K}}$ are fixed, the time derivative of **Q** is simply given by

$$\frac{d\mathbf{Q}}{dt} = \frac{dQ_X}{dt}\hat{\mathbf{I}} + \frac{dQ_Y}{dt}\hat{\mathbf{J}} + \frac{dQ_Z}{dt}\hat{\mathbf{K}}$$

dQ_X/dt, dQ_Y/dt and dQ_Z/dt are the components of the absolute time derivative of **Q**.

Q may also be resolved into components along the moving *xyz* frame, so that, at any instant,

$$\mathbf{Q} = Q_x \hat{\mathbf{i}} + Q_y \hat{\mathbf{j}} + Q_z \hat{\mathbf{k}} \tag{1.26}$$

Using this expression to calculate the time derivative of **Q** yields

$$\frac{d\mathbf{Q}}{dt} = \frac{dQ_x}{dt}\hat{\mathbf{i}} + \frac{dQ_y}{dt}\hat{\mathbf{j}} + \frac{dQ_z}{dt}\hat{\mathbf{k}} + Q_x \frac{d\hat{\mathbf{i}}}{dt} + Q_y \frac{d\hat{\mathbf{j}}}{dt} + Q_z \frac{d\hat{\mathbf{k}}}{dt} \tag{1.27}$$

The unit vectors $\hat{\mathbf{i}}$, $\hat{\mathbf{j}}$ and $\hat{\mathbf{k}}$ are not fixed in space, but are continuously changing direction; therefore, their time derivatives are not zero. They obviously have a constant magnitude (unity) and, being attached to the *xyz* frame, they all have the angular velocity $\mathbf{\Omega}$. It follows from Equation 1.24 that

$$\frac{d\hat{\mathbf{i}}}{dt} = \mathbf{\Omega} \times \hat{\mathbf{i}} \qquad \frac{d\hat{\mathbf{j}}}{dt} = \mathbf{\Omega} \times \hat{\mathbf{j}} \qquad \frac{d\hat{\mathbf{k}}}{dt} = \mathbf{\Omega} \times \hat{\mathbf{k}}$$

Substituting these on the right-hand side of Equation 1.27 yields

$$\begin{aligned}
\frac{d\mathbf{Q}}{dt} &= \frac{dQ_x}{dt}\hat{\mathbf{i}} + \frac{dQ_y}{dt}\hat{\mathbf{j}} + \frac{dQ_z}{dt}\hat{\mathbf{k}} + Q_x(\mathbf{\Omega} \times \hat{\mathbf{i}}) + Q_y(\mathbf{\Omega} \times \hat{\mathbf{j}}) + Q_z(\mathbf{\Omega} \times \hat{\mathbf{k}}) \\
&= \frac{dQ_x}{dt}\hat{\mathbf{i}} + \frac{dQ_y}{dt}\hat{\mathbf{j}} + \frac{dQ_z}{dt}\hat{\mathbf{k}} + (\mathbf{\Omega} \times Q_x\hat{\mathbf{i}}) + (\mathbf{\Omega} \times Q_y\hat{\mathbf{j}}) + (\mathbf{\Omega} \times Q_z\hat{\mathbf{k}}) \\
&= \frac{dQ_x}{dt}\hat{\mathbf{i}} + \frac{dQ_y}{dt}\hat{\mathbf{j}} + \frac{dQ_z}{dt}\hat{\mathbf{k}} + \mathbf{\Omega} \times (Q_x\hat{\mathbf{i}} + Q_y\hat{\mathbf{j}} + Q_z\hat{\mathbf{k}})
\end{aligned}$$

In view of Equation 1.26, this can be written

$$\frac{d\mathbf{Q}}{dt} = \left.\frac{d\mathbf{Q}}{dt}\right)_{\text{rel}} + \mathbf{\Omega} \times \mathbf{Q} \tag{1.28}$$

where

$$\left.\frac{d\mathbf{Q}}{dt}\right)_{\text{rel}} = \frac{dQ_x}{dt}\hat{\mathbf{i}} + \frac{dQ_y}{dt}\hat{\mathbf{j}} + \frac{dQ_z}{dt}\hat{\mathbf{k}} \tag{1.29}$$

$d\mathbf{Q}/dt)_{\text{rel}}$ is the time derivative of $\mathbf{Q}$ relative to the moving frame. Equation 1.28 shows how the absolute time derivative is obtained from the relative time derivative. Clearly, $d\mathbf{Q}/dt = d\mathbf{Q}/dt)_{\text{rel}}$ only when the moving frame is in pure translation ($\mathbf{\Omega} = \mathbf{0}$).

Equation 1.28 can be used recursively to compute higher order time derivatives. Thus, differentiating Equation 1.28 with respect to t, we get

$$\frac{d^2\mathbf{Q}}{dt^2} = \left.\frac{d}{dt}\frac{d\mathbf{Q}}{dt}\right)_{\text{rel}} + \frac{d\mathbf{\Omega}}{dt} \times \mathbf{Q} + \mathbf{\Omega} \times \frac{d\mathbf{Q}}{dt}$$

Using Equation 1.28 in the last term yields

$$\frac{d^2\mathbf{Q}}{dt^2} = \left.\frac{d}{dt}\frac{d\mathbf{Q}}{dt}\right)_{\text{rel}} + \frac{d\mathbf{\Omega}}{dt} \times \mathbf{Q} + \mathbf{\Omega} \times \left[\left.\frac{d\mathbf{Q}}{dt}\right)_{\text{rel}} + \mathbf{\Omega} \times \mathbf{Q}\right] \tag{1.30}$$

Equation 1.28 also implies that

$$\left.\frac{d}{dt}\frac{d\mathbf{Q}}{dt}\right)_{\text{rel}} = \left.\frac{d^2\mathbf{Q}}{dt^2}\right)_{\text{rel}} + \mathbf{\Omega} \times \left.\frac{d\mathbf{Q}}{dt}\right)_{\text{rel}} \tag{1.31}$$

where

$$\left.\frac{d^2\mathbf{Q}}{dt^2}\right)_{\text{rel}} = \frac{d^2Q_x}{dt^2}\hat{\mathbf{i}} + \frac{d^2Q_y}{dt^2}\hat{\mathbf{j}} + \frac{d^2Q_z}{dt^2}\hat{\mathbf{k}}$$

Substituting Equation 1.31 into Equation 1.30 yields

$$\frac{d^2\mathbf{Q}}{dt^2} = \left[\left.\frac{d^2\mathbf{Q}}{dt^2}\right)_{\text{rel}} + \mathbf{\Omega} \times \left.\frac{d\mathbf{Q}}{dt}\right)_{\text{rel}}\right] + \frac{d\mathbf{\Omega}}{dt} \times \mathbf{Q} + \mathbf{\Omega} \times \left[\left.\frac{d\mathbf{Q}}{dt}\right)_{\text{rel}} + \mathbf{\Omega} \times \mathbf{Q}\right] \tag{1.32}$$

Collecting terms, this becomes

$$\frac{d^2\mathbf{Q}}{dt^2} = \left.\frac{d^2\mathbf{Q}}{dt^2}\right)_{\text{rel}} + \dot{\mathbf{\Omega}} \times \mathbf{Q} + \mathbf{\Omega} \times (\mathbf{\Omega} \times \mathbf{Q}) + 2\mathbf{\Omega} \times \left.\frac{d\mathbf{Q}}{dt}\right)_{\text{rel}}$$

where $\dot{\mathbf{\Omega}} \equiv d\mathbf{\Omega}/dt$ is the absolute angular acceleration of the xyz frame. Formulas for higher order time derivatives are found in a similar fashion.

1.6 Relative motion

Let P be a particle in arbitrary motion. The absolute position vector of P is $\mathbf{r}$ and the position of P relative to the moving frame is $\mathbf{r}_{\text{rel}}$. If $\mathbf{r}_O$ is the absolute position of the origin of the moving frame, then it is clear from Figure 1.10 that

$$\mathbf{r} = \mathbf{r}_O + \mathbf{r}_{\text{rel}} \tag{1.33}$$

Since $\mathbf{r}_{\text{rel}}$ is measured in the moving frame,

$$\mathbf{r}_{\text{rel}} = x\hat{\mathbf{i}} + y\hat{\mathbf{j}} + z\hat{\mathbf{k}} \tag{1.34}$$

where x, y and z are the coordinates of P relative to the moving reference.

The absolute velocity $\mathbf{v}$ of P is $d\mathbf{r}/dt$, so that from Equation 1.33 we have

$$\mathbf{v} = \mathbf{v}_O + \frac{d\mathbf{r}_{\text{rel}}}{dt} \tag{1.35}$$

where $\mathbf{v}_O = d\mathbf{r}_O/dt$ is the (absolute) velocity of the origin of the xyz frame. From Equation 1.28, we can write

$$\frac{d\mathbf{r}_{\text{rel}}}{dt} = \mathbf{v}_{\text{rel}} + \mathbf{\Omega} \times \mathbf{r}_{\text{rel}} \tag{1.36}$$

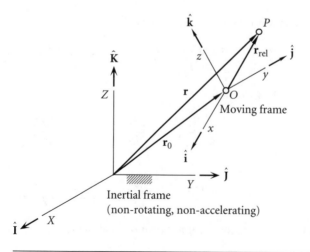

Figure 1.10 Absolute and relative position vectors.

1.6 Relative motion

where $\mathbf{v}_{rel}$ is the velocity of P relative to the xyz frame:

$$\mathbf{v}_{rel} = \left.\frac{d\mathbf{r}_{rel}}{dt}\right)_{rel} = \frac{dx}{dt}\hat{\mathbf{i}} + \frac{dy}{dt}\hat{\mathbf{j}} + \frac{dz}{dt}\hat{\mathbf{k}} \qquad (1.37)$$

Substituting Equation 1.36 into Equation 1.35 yields

$$\mathbf{v} = \mathbf{v}_O + \mathbf{\Omega} \times \mathbf{r}_{rel} + \mathbf{v}_{rel} \qquad (1.38)$$

The absolute acceleration $\mathbf{a}$ of P is $d\mathbf{v}/dt$, so that from Equation 1.35 we have

$$\mathbf{a} = \mathbf{a}_O + \frac{d^2\mathbf{r}_{rel}}{dt^2} \qquad (1.39)$$

where $\mathbf{a}_O = d\mathbf{v}_O/dt$ is the absolute acceleration of the origin of the xyz frame. We evaluate the second term on the right using Equation 1.32:

$$\frac{d^2\mathbf{r}_{rel}}{dt^2} = \left.\frac{d^2\mathbf{r}_{rel}}{dt^2}\right)_{rel} + \dot{\mathbf{\Omega}} \times \mathbf{r}_{rel} + \mathbf{\Omega} \times (\mathbf{\Omega} \times \mathbf{r}_{rel}) + 2\mathbf{\Omega} \times \left.\frac{d\mathbf{r}_{rel}}{dt}\right)_{rel} \qquad (1.40)$$

Since $\mathbf{v}_{rel} = d\mathbf{r}_{rel}/dt)_{rel}$ and $\mathbf{a}_{rel} = d^2\mathbf{r}_{rel}/dt^2)_{rel}$, this can be written

$$\frac{d^2\mathbf{r}_{rel}}{dt^2} = \mathbf{a}_{rel} + \dot{\mathbf{\Omega}} \times \mathbf{r}_{rel} + \mathbf{\Omega} \times (\mathbf{\Omega} \times \mathbf{r}_{rel}) + 2\mathbf{\Omega} \times \mathbf{v}_{rel} \qquad (1.41)$$

Upon substituting this result into Equation 1.39, we find

$$\mathbf{a} = \mathbf{a}_O + \dot{\mathbf{\Omega}} \times \mathbf{r}_{rel} + \mathbf{\Omega} \times (\mathbf{\Omega} \times \mathbf{r}_{rel}) + 2\mathbf{\Omega} \times \mathbf{v}_{rel} + \mathbf{a}_{rel} \qquad (1.42)$$

The cross product $2\mathbf{\Omega} \times \mathbf{v}_{rel}$ is called the Coriolis acceleration after Gustave Gaspard de Coriolis (1792–1843), the French mathematician who introduced this term (Coriolis, 1835). For obvious reasons, Equation 1.42 is sometimes referred to as the five-term acceleration formula.

EXAMPLE 1.9

At a given instant, the absolute position, velocity and acceleration of the origin O of a moving frame are

$$\mathbf{r}_O = 100\hat{\mathbf{I}} + 200\hat{\mathbf{J}} + 300\hat{\mathbf{K}} \text{ (m)}$$
$$\mathbf{v}_O = -50\hat{\mathbf{I}} + 30\hat{\mathbf{J}} - 10\hat{\mathbf{K}} \text{ (m/s)} \qquad \text{(given)} \qquad \text{(a)}$$
$$\mathbf{a}_O = -15\hat{\mathbf{I}} + 40\hat{\mathbf{J}} + 25\hat{\mathbf{K}} \text{ (m/s}^2\text{)}$$

The angular velocity and acceleration of the moving frame are

$$\mathbf{\Omega} = 1.0\hat{\mathbf{I}} - 0.4\hat{\mathbf{J}} + 0.6\hat{\mathbf{K}} \text{ (rad/s)}$$
$$\dot{\mathbf{\Omega}} = -1.0\hat{\mathbf{I}} + 0.3\hat{\mathbf{J}} - 0.4\hat{\mathbf{K}} \text{ (rad/s}^2\text{)} \qquad \text{(given)} \qquad \text{(b)}$$

The unit vectors of the moving frame are

$$\hat{\mathbf{i}} = 0.5571\hat{\mathbf{I}} + 0.7428\hat{\mathbf{J}} + 0.3714\hat{\mathbf{K}}$$
$$\hat{\mathbf{j}} = -0.06331\hat{\mathbf{I}} + 0.4839\hat{\mathbf{J}} - 0.8728\hat{\mathbf{K}} \qquad \text{(given)} \qquad \text{(c)}$$
$$\hat{\mathbf{k}} = -0.8280\hat{\mathbf{I}} + 0.4627\hat{\mathbf{J}} + 0.3166\hat{\mathbf{K}}$$

(Example 1.9 continued)

The absolute position, velocity and acceleration of P are

$$\mathbf{r} = 300\hat{\mathbf{I}} - 100\hat{\mathbf{J}} + 150\hat{\mathbf{K}} \text{ (m)}$$
$$\mathbf{v} = 70\hat{\mathbf{I}} + 25\hat{\mathbf{J}} - 20\hat{\mathbf{K}} \text{ (m/s)} \quad \text{(given)} \quad (d)$$
$$\mathbf{a} = 7.5\hat{\mathbf{I}} - 8.5\hat{\mathbf{J}} + 6.0\hat{\mathbf{K}} \text{ (m/s}^2\text{)}$$

Find the velocity $\mathbf{v}_{rel}$ and acceleration $\mathbf{a}_{rel}$ of P relative to the moving frame.

First use Equations (c) to solve for $\hat{\mathbf{I}}, \hat{\mathbf{J}}$ and $\hat{\mathbf{K}}$ in terms of $\hat{\mathbf{i}}, \hat{\mathbf{j}}$ and $\hat{\mathbf{k}}$ (three equations in three unknowns):

$$\hat{\mathbf{I}} = 0.5571\hat{\mathbf{i}} - 0.06331\hat{\mathbf{j}} - 0.8280\hat{\mathbf{k}}$$
$$\hat{\mathbf{J}} = 0.7428\hat{\mathbf{i}} + 0.4839\hat{\mathbf{j}} + 0.4627\hat{\mathbf{k}} \quad (e)$$
$$\hat{\mathbf{K}} = 0.3714\hat{\mathbf{i}} - 0.8728\hat{\mathbf{j}} + 0.3166\hat{\mathbf{k}}$$

The relative position vector is

$$\mathbf{r}_{rel} = \mathbf{r} - \mathbf{r}_O = (300\hat{\mathbf{I}} - 100\hat{\mathbf{J}} + 150\hat{\mathbf{K}}) - (100\hat{\mathbf{I}} + 200\hat{\mathbf{J}} + 300\hat{\mathbf{K}})$$
$$= 200\hat{\mathbf{I}} - 300\hat{\mathbf{J}} - 150\hat{\mathbf{K}} \text{ (m)} \quad (f)$$

From Equation 1.38, the relative velocity vector is

$$\mathbf{v}_{rel} = \mathbf{v} - \mathbf{v}_O - \mathbf{\Omega} \times \mathbf{r}_{rel}$$

$$= (70\hat{\mathbf{I}} + 25\hat{\mathbf{J}} - 20\hat{\mathbf{K}}) - (-50\hat{\mathbf{I}} + 30\hat{\mathbf{J}} - 10\hat{\mathbf{K}}) - \begin{vmatrix} \hat{\mathbf{I}} & \hat{\mathbf{J}} & \hat{\mathbf{K}} \\ 1.0 & -0.4 & 0.6 \\ 200 & -300 & -150 \end{vmatrix}$$

$$= (70\hat{\mathbf{I}} + 25\hat{\mathbf{J}} - 20\hat{\mathbf{K}}) - (-50\hat{\mathbf{I}} + 30\hat{\mathbf{J}} - 10\hat{\mathbf{K}}) - (240\hat{\mathbf{I}} + 270\hat{\mathbf{J}} - 220\hat{\mathbf{K}})$$

or

$$\underline{\mathbf{v}_{rel} = -120\hat{\mathbf{I}} - 275\hat{\mathbf{J}} + 210\hat{\mathbf{K}} \text{ (m/s)}} \quad (g)$$

To obtain the components of the relative velocity along the axes of the moving frame, substitute Equations (e) into Equation (g).

$$\mathbf{v}_{rel} = -120(0.5571\mathbf{i} - 0.06331\mathbf{j} - 0.8280\mathbf{k})$$
$$-275(0.7428\mathbf{i} + 0.4839\mathbf{j} + 0.4627\mathbf{k}) + 210(0.3714\mathbf{i} - 0.8728\mathbf{j} + 0.3166\mathbf{k})$$

so that

$$\underline{\mathbf{v}_{rel} = -193.1\hat{\mathbf{i}} - 308.8\hat{\mathbf{j}} + 38.60\hat{\mathbf{k}} \text{ (m/s)}} \quad (h)$$

Alternatively,

$$\underline{\mathbf{v}_{rel} = 366.2\hat{\mathbf{u}}_v \text{ (m/s)}}, \quad \text{where } \hat{\mathbf{u}}_v = -0.5272\hat{\mathbf{i}} - 0.8432\hat{\mathbf{j}} + 0.1005\hat{\mathbf{k}} \quad (i)$$

To find the relative acceleration, we use the five-term acceleration formula, Equation 1.42:

$$\mathbf{a}_{rel} = \mathbf{a} - \mathbf{a}_O - \dot{\boldsymbol{\Omega}} \times \mathbf{r}_{rel} - \boldsymbol{\Omega} \times (\boldsymbol{\Omega} \times \mathbf{r}_{rel}) - 2(\boldsymbol{\Omega} \times \mathbf{v}_{rel})$$

$$= \mathbf{a} - \mathbf{a}_O - \begin{vmatrix} \hat{\mathbf{I}} & \hat{\mathbf{J}} & \hat{\mathbf{K}} \\ -1.0 & 0.3 & -0.4 \\ 200 & -300 & -150 \end{vmatrix} - \boldsymbol{\Omega}$$

$$\times \begin{vmatrix} \hat{\mathbf{I}} & \hat{\mathbf{J}} & \hat{\mathbf{K}} \\ 1.0 & -0.4 & 0.6 \\ 200 & -300 & -150 \end{vmatrix} - 2 \begin{vmatrix} \hat{\mathbf{I}} & \hat{\mathbf{J}} & \hat{\mathbf{K}} \\ 1.0 & -0.4 & 0.6 \\ -120 & -275 & 210 \end{vmatrix}$$

$$= \mathbf{a} - \mathbf{a}_O - (-165\hat{\mathbf{I}} - 230\hat{\mathbf{J}} + 240\hat{\mathbf{K}}) - \begin{vmatrix} \hat{\mathbf{I}} & \hat{\mathbf{J}} & \hat{\mathbf{K}} \\ 1.0 & -0.4 & 0.6 \\ 240 & 270 & -220 \end{vmatrix}$$

$$- (162\hat{\mathbf{I}} - 564\hat{\mathbf{J}} - 646\hat{\mathbf{K}})$$

$$= (7.5\hat{\mathbf{I}} - 8.5\hat{\mathbf{J}} + 6\hat{\mathbf{K}}) - (-15\hat{\mathbf{I}} + 40\hat{\mathbf{J}} + 25\hat{\mathbf{K}})$$

$$- (-165\hat{\mathbf{I}} - 230\hat{\mathbf{J}} + 240\hat{\mathbf{K}}) - (-74\hat{\mathbf{I}} + 364\hat{\mathbf{J}} + 366\hat{\mathbf{K}})$$

$$- (162\hat{\mathbf{I}} - 564\hat{\mathbf{J}} - 646\hat{\mathbf{K}})$$

$$\mathbf{a}_{rel} = 99.5\hat{\mathbf{I}} + 381.5\hat{\mathbf{J}} + 21.0\hat{\mathbf{K}} \; (\text{m/s}^2) \tag{j}$$

The components of the relative acceleration along the axes of the moving frame are found by substituting Equations (e) into Equation (j):

$$\mathbf{a}_{rel} = 99.5(0.5571\hat{\mathbf{i}} - 0.06331\hat{\mathbf{j}} - 0.8280\hat{\mathbf{k}})$$

$$+ 381.5(0.7428\hat{\mathbf{i}} + 0.4839\hat{\mathbf{j}} + 0.4627\hat{\mathbf{k}}) + 21.0(0.3714\hat{\mathbf{i}} - 0.8728\hat{\mathbf{j}} + 0.3166\hat{\mathbf{k}})$$

$$\underline{\mathbf{a}_{rel} = 346.6\hat{\mathbf{i}} + 160.0\hat{\mathbf{j}} + 100.8\hat{\mathbf{k}} \; (\text{m/s}^2)} \tag{k}$$

or

$$\underline{\mathbf{a}_{rel} = 394.8\hat{\mathbf{u}}_a \; (\text{m/s}^2)}, \quad \text{where } \hat{\mathbf{u}}_a = 0.8778\hat{\mathbf{i}} + 0.4052\hat{\mathbf{j}} + 0.2553\hat{\mathbf{k}} \tag{l}$$

Figure 1.11 shows the non-rotating inertial frame of reference XYZ with its origin at the center C of the earth, which we shall assume to be a sphere. That assumption will be relaxed in Chapter 5. Embedded in the earth and rotating with it is the orthogonal $x'y'z'$ frame, also centered at C, with the z' axis parallel to Z, the earth's axis of rotation. The x' axis intersects the equator at the prime meridian (zero degrees longitude), which passes through Greenwich in London, England. The angle between X and x' is θ_g, and the rate of increase of θ_g is just the angular velocity Ω of the earth. P is a particle (e.g., an airplane, spacecraft, etc.), which is moving in an arbitrary fashion above the surface of the earth. $\mathbf{r}_{rel}$ is the position vector of P relative to C in the rotating $x'y'z'$ system. At a given instant, P is directly over point O, which lies on

24 Chapter 1 *Dynamics of point masses*

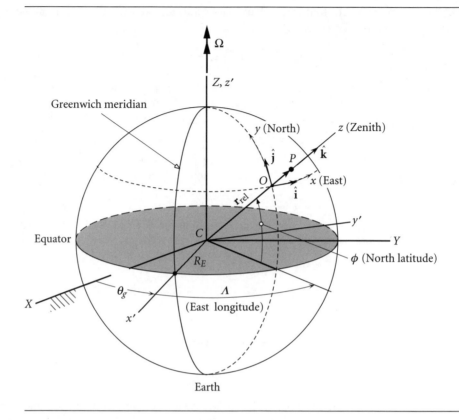

Figure 1.11 Earth-centered inertial frame (XYZ); earth-centered non-inertial $x'y'z'$ frame embedded in and rotating with the earth; and a non-inertial, topocentric-horizon frame xyz attached to a point O on the earth's surface.

the earth's surface at longitude Λ and latitude ϕ. Point O coincides instantaneously with the origin of what is known as a topocentric-horizon coordinate system xyz. For our purposes x and y are measured positive eastward and northward along the local latitude and meridian, respectively, through O. The tangent plane to the earth's surface at O is the local horizon. The z axis is the local vertical (straight up) and it is directed radially outward from the center of the earth. The unit vectors of the xyz frame are $\hat{\mathbf{i}}\hat{\mathbf{j}}\hat{\mathbf{k}}$, as indicated in Figure 1.11. Keep in mind that O remains directly below P, so that as P moves, so do the xyz axes. Thus, the $\hat{\mathbf{i}}\hat{\mathbf{j}}\hat{\mathbf{k}}$ triad, which are the unit vectors of a spherical coordinate system, vary in direction as P changes location, thereby accounting for the curvature of the earth.

Let us find the absolute velocity and acceleration of P. It is convenient first to obtain the velocity and acceleration of P relative to the non-rotating earth, and then use Equations 1.38 and 1.42 to calculate their inertial values.

The relative position vector can be written

$$\mathbf{r}_{\text{rel}} = (R_E + z)\hat{\mathbf{k}} \qquad (1.43)$$

where R_E is the radius of the earth and z is the height of P above the earth (i.e., its altitude). The time derivative of $\mathbf{r}_{\text{rel}}$ is the velocity $\mathbf{v}_{\text{rel}}$ relative to the non-rotating earth,

$$\mathbf{v}_{\text{rel}} = \frac{d\mathbf{r}_{\text{rel}}}{dt} = \dot{z}\hat{\mathbf{k}} + (R_E + z)\frac{d\hat{\mathbf{k}}}{dt} \quad (1.44)$$

To calculate $d\hat{\mathbf{k}}/dt$, we must use Equation 1.24. The angular velocity $\boldsymbol{\omega}$ of the xyz frame relative to the non-rotating earth is found in terms of the rates of change of latitude ϕ and longitude Λ,

$$\boldsymbol{\omega} = -\dot{\phi}\hat{\mathbf{i}} + \dot{\Lambda}\cos\phi\hat{\mathbf{j}} + \dot{\Lambda}\sin\phi\hat{\mathbf{k}} \quad (1.45)$$

Thus,

$$\frac{d\hat{\mathbf{k}}}{dt} = \boldsymbol{\omega} \times \hat{\mathbf{k}} = \dot{\Lambda}\cos\phi\hat{\mathbf{i}} + \dot{\phi}\hat{\mathbf{j}} \quad (1.46)$$

Let us also record the following for future use:

$$\frac{d\hat{\mathbf{j}}}{dt} = \boldsymbol{\omega} \times \hat{\mathbf{j}} = -\dot{\Lambda}\sin\phi\hat{\mathbf{j}} - \dot{\phi}\hat{\mathbf{k}} \quad (1.47)$$

$$\frac{d\hat{\mathbf{i}}}{dt} = \boldsymbol{\omega} \times \hat{\mathbf{i}} = \dot{\Lambda}\sin\phi\hat{\mathbf{j}} - \dot{\Lambda}\cos\phi\hat{\mathbf{k}} \quad (1.48)$$

Substituting Equation 1.46 into Equation 1.44 yields

$$\mathbf{v}_{\text{rel}} = \dot{x}\hat{\mathbf{i}} + \dot{y}\hat{\mathbf{j}} + \dot{z}\hat{\mathbf{k}} \quad (1.49\text{a})$$

where

$$\dot{x} = (R_E + z)\dot{\Lambda}\cos\phi \qquad \dot{y} = (R_E + z)\dot{\phi} \quad (1.49\text{b})$$

It is convenient to use these results to express the rates of change of latitude and longitude in terms of the components of relative velocity over the earth's surface,

$$\dot{\phi} = \frac{\dot{y}}{R_E + z} \qquad \dot{\Lambda} = \frac{\dot{x}}{(R_E + z)\cos\phi} \quad (1.50)$$

The time derivatives of these two expressions are

$$\ddot{\phi} = \frac{(R_E + z)\ddot{y} - \dot{y}\dot{z}}{(R_E + z)^2} \qquad \ddot{\Lambda} = \frac{(R_E + z)\ddot{x}\cos\phi - (\dot{z}\cos\phi - \dot{y}\sin\phi)\dot{x}}{(R_E + z)^2 \cos^2\phi} \quad (1.51)$$

The acceleration of P relative to the non-rotating earth is found by taking the time derivative of $\mathbf{v}_{\text{rel}}$. From Equation 1.49 we thereby obtain

$$\mathbf{a}_{\text{rel}} = \ddot{x}\hat{\mathbf{i}} + \ddot{y}\hat{\mathbf{j}} + \ddot{z}\hat{\mathbf{k}} + \dot{x}\frac{d\hat{\mathbf{i}}}{dt} + \dot{y}\frac{d\hat{\mathbf{j}}}{dt} + \dot{z}\frac{d\hat{\mathbf{k}}}{dt}$$

$$= [\dot{z}\dot{\Lambda}\cos\phi + (R_E + z)\ddot{\Lambda}\cos\phi - (R_E + z)\dot{\phi}\dot{\Lambda}\sin\phi]\hat{\mathbf{i}}$$

$$+ [\dot{z}\dot{\phi} + (R_E + z)\ddot{\phi}]\hat{\mathbf{j}} + \ddot{z}\hat{\mathbf{k}} + (R_E + z)\dot{\Lambda}\cos\phi(\boldsymbol{\omega} \times \hat{\mathbf{i}})$$

$$+ (R_E + z)\dot{\phi}(\boldsymbol{\omega} \times \hat{\mathbf{j}}) + \dot{z}(\boldsymbol{\omega} \times \hat{\mathbf{k}})$$

Substituting Equations 1.46 through 1.48 together with 1.50 and 1.51 into this expression yields, upon simplification,

$$\mathbf{a}_{rel} = \left[\ddot{x} + \frac{\dot{x}(\dot{z} - \dot{y}\tan\phi)}{R_E + z}\right]\hat{\mathbf{i}} + \left(\ddot{y} + \frac{\dot{y}\dot{z} + \dot{x}^2\tan\phi}{R_E + z}\right)\hat{\mathbf{j}} + \left(\ddot{z} - \frac{\dot{x}^2 + \dot{y}^2}{R_E + z}\right)\hat{\mathbf{k}} \quad (1.52)$$

Observe that the curvature of the earth's surface is neglected by letting $R_E + z$ become infinitely large, in which case

$$\mathbf{a}_{rel})_{\text{neglecting earth's curvature}} = \ddot{x}\hat{\mathbf{i}} + \ddot{y}\hat{\mathbf{j}} + \ddot{z}\hat{\mathbf{k}}$$

That is, for a 'flat earth', the components of the relative acceleration vector are just the derivatives of the components of the relative velocity vector.

For the absolute velocity we have, according to Equation 1.38,

$$\mathbf{v} = \mathbf{v}_C + \boldsymbol{\Omega} \times \mathbf{r}_{rel} + \mathbf{v}_{rel} \quad (1.53)$$

From Figure 1.11 it can be seen that $\hat{\mathbf{K}} = \cos\phi\hat{\mathbf{j}} + \sin\phi\hat{\mathbf{k}}$, which means the angular velocity of the earth is

$$\boldsymbol{\Omega} = \Omega\hat{\mathbf{K}} = \Omega\cos\phi\hat{\mathbf{j}} + \Omega\sin\phi\hat{\mathbf{k}} \quad (1.54)$$

Substituting this, together with Equations 1.43 and 1.49a and the fact that $\mathbf{v}_C = \mathbf{0}$, into Equation 1.53 yields

$$\mathbf{v} = [\dot{x} + \Omega(R_E + z)\cos\phi]\hat{\mathbf{i}} + \dot{y}\hat{\mathbf{j}} + \dot{z}\hat{\mathbf{k}} \quad (1.55)$$

From Equation 1.42 the absolute acceleration of P is

$$\mathbf{a} = \mathbf{a}_C + \dot{\boldsymbol{\Omega}} \times \mathbf{r}_{rel} + \boldsymbol{\Omega} \times (\boldsymbol{\Omega} \times \mathbf{r}_{rel}) + 2\boldsymbol{\Omega} \times \mathbf{v}_{rel} + \mathbf{a}_{rel}$$

Since $\mathbf{a}_C = \dot{\boldsymbol{\Omega}} = \mathbf{0}$, we find, upon substituting Equations 1.43, 1.49a, 1.52 and 1.54, that

$$\begin{aligned}\mathbf{a} = &\left[\ddot{x} + \frac{\dot{x}(\dot{z} - \dot{y}\tan\phi)}{R_E + z} + 2\Omega(\dot{z}\cos\phi - \dot{y}\sin\phi)\right]\hat{\mathbf{i}} \\ &+ \left\{\ddot{y} + \frac{\dot{y}\dot{z} + \dot{x}^2\tan\phi}{R_E + z} + \Omega\sin\phi[\Omega(R_E + z)\cos\phi + 2\dot{x}]\right\}\hat{\mathbf{j}} \\ &+ \left\{\ddot{z} - \frac{\dot{x}^2 + \dot{y}^2}{R_E + z} - \Omega\cos\phi[\Omega(R_E + z)\cos\phi + 2\dot{x}]\right\}\hat{\mathbf{k}}\end{aligned} \quad (1.56)$$

Some special cases of Equations 1.55 and 1.56 follow.

Straight and level, unaccelerated flight: $\dot{z} = \ddot{z} = \ddot{x} = \ddot{y} = 0$

$$\mathbf{v} = [\dot{x} + \Omega(R_E + z)\cos\phi]\hat{\mathbf{i}} + \dot{y}\hat{\mathbf{j}} \tag{1.57a}$$

$$\mathbf{a} = -\left[\frac{\dot{x}\dot{y}\tan\phi}{R_E + z} + 2\Omega\dot{y}\sin\phi\right]\hat{\mathbf{i}}$$

$$+ \left\{\frac{\dot{x}^2 \tan\phi}{R_E + z} + \Omega\sin\phi[\Omega(R_E + z)\cos\phi + 2\dot{x}]\right\}\hat{\mathbf{j}}$$

$$- \left\{\frac{\dot{x}^2 + \dot{y}^2}{R_E + z} + \Omega\cos\phi[\Omega(R_E + z)\cos\phi + 2\dot{x}]\right\}\hat{\mathbf{k}} \tag{1.57b}$$

Flight due north (y) at constant speed and altitude: $\dot{z} = \ddot{z} = \dot{x} = \ddot{x} = \ddot{y} = 0$

$$\mathbf{v} = \Omega(R_E + z)\cos\phi\hat{\mathbf{i}} + \dot{y}\hat{\mathbf{j}} \tag{1.58a}$$

$$\mathbf{a} = -2\Omega\dot{y}\sin\phi\hat{\mathbf{i}} + \Omega^2(R_E + z)\sin\phi\cos\phi\hat{\mathbf{j}}$$

$$- \left[\frac{\dot{y}^2}{R_E + z} + \Omega^2(R_E + z)\cos^2\phi\right]\hat{\mathbf{k}} \tag{1.58b}$$

Flight due east (x) at constant speed and altitude: $\dot{z} = \ddot{z} = \ddot{x} = \dot{y} = \ddot{y} = 0$

$$\mathbf{v} = [\dot{x} + \Omega(R_E + z)\cos\phi]\hat{\mathbf{i}} \tag{1.59a}$$

$$\mathbf{a} = \left\{\frac{\dot{x}^2 \tan\phi}{R_E + z} + \Omega\sin\phi[\Omega(R_E + z)\cos\phi + 2\dot{x}]\right\}\hat{\mathbf{j}}$$

$$- \left\{\frac{\dot{x}^2}{R_E + z} + \Omega\cos\phi[\Omega(R_E + z)\cos\phi + 2\dot{x}]\right\}\hat{\mathbf{k}} \tag{1.59b}$$

Flight straight up (z): $\dot{x} = \ddot{x} = \dot{y} = \ddot{y} = 0$

$$\mathbf{v} = \Omega(R_E + z)\cos\phi\hat{\mathbf{i}} + \dot{z}\hat{\mathbf{k}} \tag{1.60a}$$

$$\mathbf{a} = 2\Omega(\dot{z}\cos\phi)\hat{\mathbf{i}} + \Omega^2(R_E + z)\sin\phi\cos\phi\hat{\mathbf{j}}$$

$$+ [\ddot{z} - \Omega^2(R_E + z)\cos^2\phi]\hat{\mathbf{k}} \tag{1.60b}$$

Stationary: $\dot{x} = \ddot{x} = \dot{y} = \ddot{y} = \dot{z} = \ddot{z} = 0$

$$\mathbf{v} = \Omega(R_E + z)\cos\phi\hat{\mathbf{i}} \tag{1.61a}$$

$$\mathbf{a} = \Omega^2(R_E + z)\sin\phi\cos\phi\hat{\mathbf{j}} - \Omega^2(R_E + z)\cos^2\phi\hat{\mathbf{k}} \tag{1.61b}$$

EXAMPLE 1.10

An airplane of mass 70 000 kg is traveling due north at latitude 30° north, at an altitude of 10 km (32 800 ft) with a speed of 300 m/s (671 mph). Calculate (a) the components of the absolute velocity and acceleration along the axes of the topocentric-horizon reference frame, and (b) the net force on the airplane.

(Example 1.10 continued)

(a) First, using the sidereal rotation period of the earth in Table A.1, we note that the earth's angular velocity is

$$\Omega = \frac{2\pi \text{ rad}}{\text{sidereal day}} = \frac{2\pi \text{ rad}}{23.93 \text{ hr}} = \frac{2\pi \text{ rad}}{86\,160 \text{ s}} = 7.292 \times 10^{-5} \text{ rad/s} \quad (a)$$

From Equation 1.58a, the absolute velocity is

$$\mathbf{v} = \Omega(R_E + z)\cos\phi \hat{\mathbf{i}} + \dot{y}\hat{\mathbf{j}} = \left[(7.292 \times 10^{-5}) \cdot (6378 + 10) \cdot 10^3 \cos 30°\right]\hat{\mathbf{i}} + 300\hat{\mathbf{j}}$$

or

$$\mathbf{v} = 403.4\hat{\mathbf{i}} + 300\hat{\mathbf{j}} \text{ (m/s)}$$

The 403.4 m/s (901 mph) component of velocity to the east (x direction) is due entirely to the earth's rotation.

From Equation 1.58b$_2$, the absolute acceleration is

$$\mathbf{a} = -2\Omega\dot{y}\sin\phi\hat{\mathbf{i}} + \Omega^2(R_E+z)\sin\phi\cos\phi\hat{\mathbf{j}} - \left[\frac{\dot{y}^2}{R_E+z} + \Omega^2(R_E+z)\cos^2\phi\right]\hat{\mathbf{k}}$$

$$= -2(7.292 \times 10^{-5}) \cdot 300 \cdot \sin 30°\hat{\mathbf{i}}$$

$$+ (7.292 \times 10^{-5})^2 \cdot (6378 + 10) \cdot 10^3 \cdot \sin 30° \cdot \cos 30°\hat{\mathbf{j}}$$

$$- \left[\frac{300^2}{(6378+10) \cdot 10^3} + (7.292 \times 10^{-5})^2 \cdot (6378+10) \cdot 10^3 \cdot \cos^2 30°\right]\hat{\mathbf{k}}$$

or

$$\mathbf{a} = -0.02187\hat{\mathbf{i}} + 0.01471\hat{\mathbf{j}} - 0.03956\hat{\mathbf{k}} \text{ (m/s}^2\text{)} \quad (a)$$

The westward acceleration of 0.02187 m/s^2 is the Coriolis acceleration.

(b) Since the acceleration in part (a) is the absolute acceleration, we can use it in Newton's law to calculate the net force on the airplane,

$$\mathbf{F}_{net} = m\mathbf{a} = 70\,000(-0.02187\hat{\mathbf{i}} + 0.01471\hat{\mathbf{j}} - 0.03956\hat{\mathbf{k}})$$

$$= -1531\hat{\mathbf{i}} + 1029\hat{\mathbf{j}} - 2769\hat{\mathbf{k}} \text{ (N)}$$

Figure 1.12 shows the components of this relatively small force. The forward and downward forces are in the directions of the airplane's centripetal acceleration, caused by the earth's rotation and, in the case of the downward force, by the earth's curvature as well. The westward force is in the direction of the Coriolis acceleration, which is due to the combined effects of the earth's rotation and the motion of the airplane. These net external forces must exist if the airplane is to fly the prescribed path.

In the vertical direction, the net force is that of the upward lift L of the wings plus the downward weight W of the aircraft, so that

$$F_{net})_z = L - W = -2769 \quad \Rightarrow \quad L = W - 2769 \text{ (N)}$$

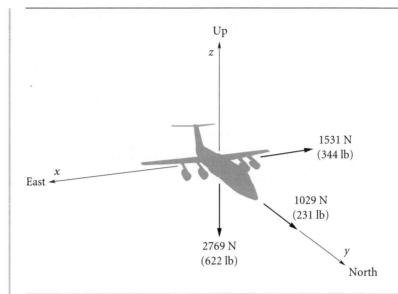

Figure 1.12 Components of the net force on the airplane.

Thus, the effect of the earth's rotation and curvature is to apparently produce an outward *centrifugal force*, reducing the weight of the airplane a bit, in this case by about 0.4 percent. The fictitious centrifugal force also increases the apparent drag in the flight direction by 1029 N. That is, in the flight direction

$$F_{net})_y = T - D = -2769\,\text{N}$$

where T is the thrust and D is the drag. Hence

$$T = D + 1029\,(\text{N})$$

The 1531 N force to the left, produced by crabbing the airplane very slightly in that direction, is required to balance the fictitious Coriolis force which would otherwise cause the airplane to deviate to the right of its flight path.

Problems

1.1 Given the three vectors $\mathbf{A} = A_x\hat{\mathbf{i}} + A_y\hat{\mathbf{j}} + A_z\hat{\mathbf{k}}$ $\mathbf{B} = B_x\hat{\mathbf{i}} + B_y\hat{\mathbf{j}} + B_z\hat{\mathbf{k}}$ $\mathbf{C} = C_x\hat{\mathbf{i}} + C_y\hat{\mathbf{j}} + C_z\hat{\mathbf{k}}$ show, analytically, that
(a) $\mathbf{A} \cdot \mathbf{A} = A^2$
(b) $\mathbf{A} \cdot (\mathbf{B} \times \mathbf{C}) = (\mathbf{A} \times \mathbf{B}) \cdot \mathbf{C}$ (interchangeability of the 'dot' and 'cross')
(c) $\mathbf{A} \times (\mathbf{B} \times \mathbf{C}) = \mathbf{B}(\mathbf{A} \cdot \mathbf{C}) - \mathbf{C}(\mathbf{A} \cdot \mathbf{B})$ (the *bac – cab* rule)

(Simply compute the expressions on each side of the = signs and demonstrate conclusively that they are the same. Do *not* substitute numbers to 'prove' your point. Use the fact that the cartesian coordinate unit vectors $\hat{\mathbf{i}}, \hat{\mathbf{j}}$ and $\hat{\mathbf{k}}$ form a right-handed orthogonal triad, so that

$$\hat{\mathbf{i}} \cdot \hat{\mathbf{j}} = \hat{\mathbf{i}} \cdot \hat{\mathbf{k}} = \hat{\mathbf{j}} \cdot \hat{\mathbf{k}} = 0 \qquad \hat{\mathbf{i}} \cdot \hat{\mathbf{i}} = \hat{\mathbf{j}} \cdot \hat{\mathbf{j}} = \hat{\mathbf{k}} \cdot \hat{\mathbf{k}} = 1$$

$$\hat{\mathbf{i}} \times \hat{\mathbf{j}} = \hat{\mathbf{k}} \qquad \hat{\mathbf{j}} \times \hat{\mathbf{k}} = \hat{\mathbf{i}} \qquad \hat{\mathbf{k}} \times \hat{\mathbf{i}} = \hat{\mathbf{j}} \qquad (\hat{\mathbf{i}} \times \hat{\mathbf{k}} = -\hat{\mathbf{j}} \quad \hat{\mathbf{j}} \times \hat{\mathbf{i}} = -\hat{\mathbf{k}} \quad \hat{\mathbf{k}} \times \hat{\mathbf{j}} = -\hat{\mathbf{i}})$$

Also,

$$\hat{\mathbf{i}} \times \hat{\mathbf{i}} = \hat{\mathbf{j}} \times \hat{\mathbf{j}} = \hat{\mathbf{k}} \times \hat{\mathbf{k}} = 0$$

1.2 Use just the vector identities in parts (a) and (b) of Exercise 1.1 to show that

$$(\mathbf{A} \times \mathbf{B}) \cdot (\mathbf{C} \times \mathbf{D}) = (\mathbf{A} \cdot \mathbf{C})(\mathbf{B} \cdot \mathbf{D}) - (\mathbf{A} \cdot \mathbf{D})(\mathbf{B} \cdot \mathbf{C})$$

1.3 The absolute position, velocity and acceleration of O are

$$\mathbf{r}_O = 300\hat{\mathbf{I}} + 200\hat{\mathbf{J}} + 100\hat{\mathbf{K}} \text{ (m)}$$
$$\mathbf{v}_O = -10\hat{\mathbf{I}} + 30\hat{\mathbf{J}} - 50\hat{\mathbf{K}} \text{ (m/s)}$$
$$\mathbf{a}_O = 25\hat{\mathbf{I}} + 40\hat{\mathbf{J}} - 15\hat{\mathbf{K}} \text{ (m/s}^2\text{)}$$

The angular velocity and acceleration of the moving frame are

$$\boldsymbol{\Omega} = 0.6\hat{\mathbf{I}} - 0.4\hat{\mathbf{J}} + 1.0\hat{\mathbf{K}} \text{ (rad/s)}$$
$$\dot{\boldsymbol{\Omega}} = -0.4\hat{\mathbf{I}} + 0.3\hat{\mathbf{J}} - 1.0\hat{\mathbf{K}} \text{ (rad/s}^2\text{)}$$

The unit vectors of the moving frame are

$$\hat{\mathbf{i}} = 0.57735\hat{\mathbf{I}} + 0.57735\hat{\mathbf{J}} + 0.57735\hat{\mathbf{K}}$$
$$\hat{\mathbf{j}} = -0.74296\hat{\mathbf{I}} + 0.66475\hat{\mathbf{J}} + 0.078206\hat{\mathbf{K}}$$
$$\hat{\mathbf{k}} = -0.33864\hat{\mathbf{I}} - 0.47410\hat{\mathbf{J}} + 0.81274\hat{\mathbf{K}}$$

The absolute position of P is

$$\mathbf{r} = 150\hat{\mathbf{I}} - 200\hat{\mathbf{J}} + 300\hat{\mathbf{K}} \text{ (m)}$$

The velocity and acceleration of P relative to the moving frame are

$$\mathbf{v}_{\text{rel}} = -20\hat{\mathbf{i}} + 25\hat{\mathbf{j}} + 70\hat{\mathbf{k}} \text{ (m/s)} \qquad \mathbf{a}_{\text{rel}} = 7.5\hat{\mathbf{i}} - 8.5\hat{\mathbf{j}} + 6.0\hat{\mathbf{k}} \text{ (m/s}^2\text{)}$$

Calculate the absolute velocity $\mathbf{v}_P$ and acceleration $\mathbf{a}_P$ of P.
{Ans.: $\mathbf{v}_P = 478.7\hat{\mathbf{u}}_v$ (m/s), $\hat{\mathbf{u}}_v = 0.5352\hat{\mathbf{I}} - 0.5601\hat{\mathbf{J}} - 0.6324\hat{\mathbf{K}}$;
$\mathbf{a}_P = 616.3\hat{\mathbf{u}}_a$ (m/s^2), $\hat{\mathbf{u}}_a = 0.1655\hat{\mathbf{I}} + 0.9759\hat{\mathbf{J}} + 0.1424\hat{\mathbf{K}}$}

1.4 $\mathbf{F}$ is a force vector of fixed magnitude embedded on a rigid body in plane motion (in the xy plane). At a given instant, $\boldsymbol{\omega} = 3\hat{\mathbf{k}}$ rad/s, $\dot{\boldsymbol{\omega}} = -2\hat{\mathbf{k}}$ rad/s^2, $\ddot{\boldsymbol{\omega}} = 0$ and $\mathbf{F} = 10\hat{\mathbf{i}}$ N. At that instant, calculate $\ddot{\mathbf{F}}$.
{Ans.: $\ddot{\mathbf{F}} = 180\hat{\mathbf{i}} - 270\hat{\mathbf{j}}$ N/s^3}

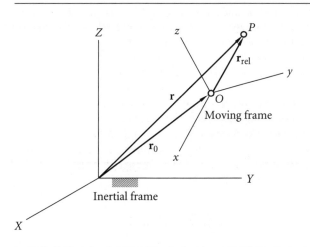

Figure P.1.3

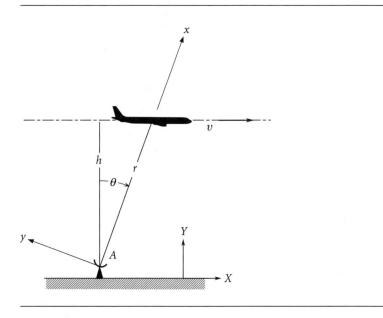

Figure P.1.5

1.5 An airplane in level flight at an altitude h and a uniform speed v passes directly over a radar tracking station A. Calculate the angular velocity $\dot{\theta}$ and angular acceleration of the radar antenna $\ddot{\theta}$ as well as the rate $\dot{r}$ at which the airplane is moving away from the antenna. Use the equations of this chapter (rather than polar coordinates, which you can use to check your work). Attach the inertial frame of reference to the ground and assume a non-rotating earth. Attach the moving frame to the antenna, with the x axis pointing always from the antenna towards the airplane.
{Ans.: (a) $\dot{\theta} = v\cos^2\theta/h$; (b) $\ddot{\theta} = -2v^2\cos^3\theta\sin\theta/h^2$; (c) $v_{\text{rel}} = v\sin\theta$}

1.6 At 30° north latitude, a 1000 kg (2205 lb) car travels due north at a constant speed of 100 km/hr (62 mph) on a level road at sea level. Taking into account the earth's rotation, calculate the lateral (sideways) force of the road on the car, and the normal force of the road on the car.
{Ans.: $F_{lateral} = 2.026$ N, to the left (west); $N = 9784$ N}

1.7 At 29° north latitude, what is the deviation d from the vertical of a plumb bob at the end of a 30 m string, due to the earth's rotation?
{Ans.: 44.1 mm to the south}

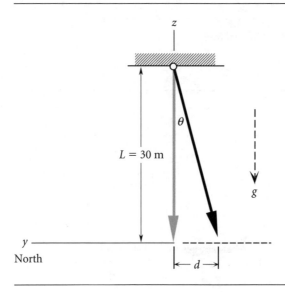

Figure P.1.7

CHAPTER 2

THE TWO-BODY PROBLEM

CHAPTER OUTLINE

2.1	Introduction	33
2.2	Equations of motion in an inertial frame	34
2.3	Equations of relative motion	37
2.4	Angular momentum and the orbit formulas	42
2.5	The energy law	50
2.6	Circular orbits ($e=0$)	51
2.7	Elliptical Orbits ($0 < e < 1$)	55
2.8	Parabolic trajectories ($e=1$)	65
2.9	Hyperbolic trajectories ($e>1$)	69
2.10	Perifocal frame	76
2.11	The Lagrange coefficients	78
2.12	Restricted three-body problem	89
	2.12.1 Lagrange points	92
	2.12.2 Jacobi constant	96
Problems		101

2.1 INTRODUCTION

This chapter presents the vector-based approach to the classical problem of determining the motion of two bodies due solely to their own mutual gravitational attraction. We show that the path of one of the masses relative to the other is a conic section (circle, ellipse, parabola or hyperbola) whose shape is determined by the eccentricity. Several fundamental properties of the different types of orbits are developed with the aid of the laws of conservation of angular momentum and energy. These properties include the period of elliptical orbits, the escape velocity associated with parabolic paths and the characteristic energy of hyperbolic trajectories. Following the presentation of the four types of orbits, the perifocal frame is introduced. This frame of reference is used to describe orbits in three dimensions, which is the subject of Chapter 4.

In this chapter the perifocal frame provides the backdrop for developing the Lagrange f and g coefficients. By means of the Lagrange f and g coefficients, the position and velocity on a trajectory can be found in terms of the position and velocity at an initial time. These functions are needed in the orbit determination algorithms of Lambert and Gauss presented in Chapter 5.

The chapter concludes with a discussion of the restricted three-body problem in order to provide a basis for understanding of the concepts of Lagrange points as well as the Jacobi constant. This material is optional.

In studying this chapter it would be well from time to time to review the road map provided in Appendix B.

2.2 Equations of motion in an inertial frame

Figure 2.1 shows two point masses acted upon only by the mutual force of gravity between them. The positions of their centers of mass are shown relative to an inertial frame of reference XYZ. The origin O of the frame may move with constant velocity (relative to the fixed stars), but the axes do not rotate. Each of the two bodies is acted upon by the gravitational attraction of the other. $\mathbf{F}_{12}$ is the force exerted on m_1 by m_2, and $\mathbf{F}_{21}$ is the force exerted on m_2 by m_1.

The position vector $\mathbf{R}_G$ of the center of mass G of the system in Figure 2.1(a) is, defined by the formula

$$\mathbf{R}_G = \frac{m_1 \mathbf{R}_1 + m_2 \mathbf{R}_2}{m_1 + m_2} \tag{2.1}$$

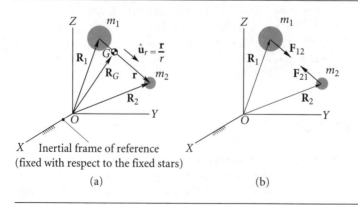

Figure 2.1 (a) Two masses located in an inertial frame. (b) Free-body diagrams.

Therefore, the absolute velocity and the absolute acceleration of G are

$$\mathbf{v}_G = \dot{\mathbf{R}}_G = \frac{m_1 \dot{\mathbf{R}}_1 + m_2 \dot{\mathbf{R}}_2}{m_1 + m_2} \qquad (2.2)$$

$$\mathbf{a}_G = \ddot{\mathbf{R}}_G = \frac{m_1 \ddot{\mathbf{R}}_1 + m_2 \ddot{\mathbf{R}}_2}{m_1 + m_2} \qquad (2.3)$$

The adjective 'absolute' means that the quantities are measured relative to an inertial frame of reference.

Let $\mathbf{r}$ be the position vector of m_2 relative to m_1. Then

$$\mathbf{r} = \mathbf{R}_2 - \mathbf{R}_1 \qquad (2.4)$$

Furthermore, let $\hat{\mathbf{u}}_r$ be the unit vector pointing from m_1 towards m_2, so that

$$\hat{\mathbf{u}}_r = \frac{\mathbf{r}}{r} \qquad (2.5)$$

where $r = \|\mathbf{r}\|$, the magnitude of $\mathbf{r}$. The body m_1 is acted upon only by the force of gravitational attraction towards m_2. The force of gravitational attraction, F_g, which acts along the line joining the centers of mass of m_1 and m_2, is given by Equation 1.3. The force exerted on m_2 by m_1 is

$$\mathbf{F}_{21} = \frac{Gm_1 m_2}{r^2}(-\hat{\mathbf{u}}_r) = -\frac{Gm_1 m_2}{r^2}\hat{\mathbf{u}}_r \qquad (2.6)$$

where $-\hat{\mathbf{u}}_r$ accounts for the fact that the force vector $\mathbf{F}_{21}$ is directed from m_2 towards m_1. (Do not confuse the symbol G, used in this context to represent the universal gravitational constant, with its use elsewhere in the book to denote the center of mass.) Newton's second law of motion as applied to body m_2 is $\mathbf{F}_{21} = m_2 \ddot{\mathbf{R}}_2$, where $\ddot{\mathbf{R}}_2$ is the absolute acceleration of m_2. Thus

$$-\frac{Gm_1 m_2}{r^2}\hat{\mathbf{u}}_r = m_2 \ddot{\mathbf{R}}_2 \qquad (2.7)$$

By Newton's third law (the action–reaction principle), $\mathbf{F}_{12} = -\mathbf{F}_{21}$, so that for m_1 we have

$$\frac{Gm_1 m_2}{r^2}\hat{\mathbf{u}}_r = m_1 \ddot{\mathbf{R}}_1 \qquad (2.8)$$

Equations 2.7 and 2.8 are the equations of motion of the two bodies in inertial space. By adding each side of these equations together, we find $m_1 \ddot{\mathbf{R}}_1 + m_2 \ddot{\mathbf{R}}_2 = 0$. According to Equation 2.3, that means the acceleration of the center of mass G of the system of two bodies m_1 and m_2 is zero. G moves with a constant velocity $\mathbf{v}_G$ in a straight line, so that its position vector relative to XYZ given by

$$\mathbf{R}_G = \mathbf{R}_{G_0} + \mathbf{v}_G t \qquad (2.9)$$

where $\mathbf{R}_{G_0}$ is the position of G at time $t = 0$. The center of mass of a two-body system may therefore serve as the origin of an inertial frame.

EXAMPLE 2.1

Use the equations of motion to show why orbiting astronauts experience weightlessness.

We sense weight by feeling the contact forces that develop wherever our body is supported. Consider an astronaut of mass m_A strapped into the space shuttle of mass m_S, in orbit about the earth. The distance between the center of the earth and the spacecraft is r, and the mass of the earth is M_E. Since the only external force on the space shuttle is that of gravity, $\mathbf{F}_S)_g$, the equation of motion of the shuttle is

$$\mathbf{F}_S)_g = m_S \mathbf{a}_S \tag{a}$$

According to Equation 2.6,

$$\mathbf{F}_S)_g = -\frac{GM_E m_S}{r^2}\hat{\mathbf{u}}_r \tag{b}$$

where $\hat{\mathbf{u}}_r$ is the unit vector pointing outward from the earth to the orbiting space shuttle. Thus, (a) and (b) imply

$$\mathbf{a}_S = -\frac{GM_E}{r^2}\hat{\mathbf{u}}_r \tag{c}$$

The equation of motion of the astronaut is

$$\mathbf{F}_A)_g + \mathbf{C}_A = m_A \mathbf{a}_A \tag{d}$$

where $\mathbf{F}_A)_g$ is the force of gravity on (i.e., the weight of) the astronaut, $\mathbf{C}_A$ is the net contact force on the astronaut from restraints (e.g., seat, seat belt), and $\mathbf{a}_A$ is the astronaut's acceleration. According to Equation 2.6,

$$\mathbf{F}_A)_g = -\frac{GM_E m_A}{r^2}\hat{\mathbf{u}}_r \tag{e}$$

Since the astronaut is moving with the shuttle we have, noting (c),

$$\mathbf{a}_A = \mathbf{a}_S = -\frac{GM_E}{r^2}\hat{\mathbf{u}}_r \tag{f}$$

Substituting (e) and (f) into (d) yields

$$-\frac{GM_E m_A}{r^2}\hat{\mathbf{u}}_r + \mathbf{C}_A = m_A\left(-\frac{GM_E}{r^2}\hat{\mathbf{u}}_r\right)$$

from which it is clear that $\mathbf{C}_A = \mathbf{0}$. The net contact force on the astronaut is zero. With no reaction to the force of gravity exerted on the body, there is no sensation of weight.

The potential energy V of the gravitational force in Equation 2.6 is given by

$$V = -\frac{Gm_1 m_2}{r} \tag{2.10}$$

A force can be obtained from its potential energy function by means of the gradient operator,

$$\mathbf{F} = -\nabla V \tag{2.11}$$

where, in cartesian coordinates,

$$\nabla = \frac{\partial}{\partial x}\hat{\mathbf{i}} + \frac{\partial}{\partial y}\hat{\mathbf{j}} + \frac{\partial}{\partial z}\hat{\mathbf{k}} \qquad (2.12)$$

In Appendix E it is shown that the gravitational potential, and hence the gravitational force, outside of a sphere with a spherically symmetric mass distribution M is the same as that of a point mass M located at the center of the sphere. Therefore, the two-body problem applies not just to point masses but also to spherical bodies (as long, of course, as they do not come into contact!).

2.3 Equations of relative motion

Let us now multiply Equation 2.7 by m_1 and Equation 2.8 by m_2 to obtain

$$-\frac{Gm_1^2 m_2}{r^2}\hat{\mathbf{u}}_r = m_1 m_2 \ddot{\mathbf{R}}_2$$

$$\frac{Gm_1 m_2^2}{r^2}\hat{\mathbf{u}}_r = m_1 m_2 \ddot{\mathbf{R}}_1$$

Subtracting the second of these two equations from the first yields

$$m_1 m_2 (\ddot{\mathbf{R}}_2 - \ddot{\mathbf{R}}_1) = -\frac{Gm_1 m_2}{r^2}(m_1 + m_2)\hat{\mathbf{u}}_r$$

Canceling the common factor $m_1 m_2$ and using Equation 2.4 yields

$$\ddot{\mathbf{r}} = -\frac{G(m_1 + m_2)}{r^2}\hat{\mathbf{u}}_r \qquad (2.13)$$

Let the gravitational μ parameter be defined as

$$\mu = G(m_1 + m_2) \qquad (2.14)$$

The units of μ are $km^3 s^{-2}$. Using Equation 2.14 together with Equation 2.5, we can write Equation 2.13 as

$$\ddot{\mathbf{r}} = -\frac{\mu}{r^3}\mathbf{r} \qquad (2.15)$$

This is the second order differential equation that governs the motion of m_2 relative to m_1. It has two vector constants of integration, each having three scalar components. Therefore, Equation 2.15 has six constants of integration. Note that interchanging the roles of m_1 and m_2 in all of the above amounts to simply multiplying Equation 2.15 through by -1, which, of course, changes nothing. Thus, the motion of m_2 as seen from m_1 is precisely the same as the motion of m_1 as seen from m_2.

The relative position vector $\mathbf{r}$ in Equation 2.15 was defined in the inertial frame (Equation 2.4). It is convenient, however, to measure the components of $\mathbf{r}$ in a frame of reference attached to and moving with m_1. In a co-moving reference frame, such as the xyz system illustrated in Figure 2.2, $\mathbf{r}$ has the expression

$$\mathbf{r} = x\hat{\mathbf{i}} + y\hat{\mathbf{j}} + z\hat{\mathbf{k}}$$

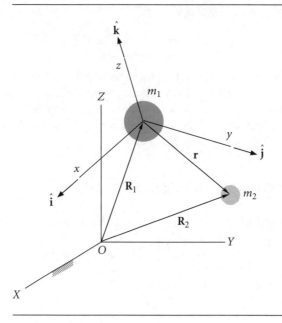

Figure 2.2 Moving reference frame *xyz* attached to the center of mass of m_1.

The relative velocity $\dot{\mathbf{r}}_{\text{rel}}$ and acceleration $\ddot{\mathbf{r}}_{\text{rel}}$ in the co-moving frame are found by simply taking the derivatives of the coefficients of the unit vectors, which themselves are fixed in the *xyz* system. Thus

$$\dot{\mathbf{r}}_{\text{rel}} = \dot{x}\hat{\mathbf{i}} + \dot{y}\hat{\mathbf{j}} + \dot{z}\hat{\mathbf{k}} \quad \ddot{\mathbf{r}}_{\text{rel}} = \ddot{x}\hat{\mathbf{i}} + \ddot{y}\hat{\mathbf{j}} + \ddot{z}\hat{\mathbf{k}}$$

From Equation 1.40 we know that the relationship between absolute acceleration $\ddot{\mathbf{r}}$ and relative acceleration $\ddot{\mathbf{r}}_{\text{rel}}$ is

$$\ddot{\mathbf{r}} = \ddot{\mathbf{r}}_{\text{rel}} + \dot{\boldsymbol{\Omega}} \times \mathbf{r} + \boldsymbol{\Omega} \times (\boldsymbol{\Omega} \times \mathbf{r}) + 2\boldsymbol{\Omega} \times \dot{\mathbf{r}}_{\text{rel}}$$

where $\boldsymbol{\Omega}$ and $\dot{\boldsymbol{\Omega}}$ are the angular velocity and angular acceleration of the moving frame of reference. Thus $\ddot{\mathbf{r}} = \ddot{\mathbf{r}}_{\text{rel}}$ only if $\boldsymbol{\Omega} = \dot{\boldsymbol{\Omega}} = 0$. That is to say, the relative acceleration may be used on the left of Equation 2.15 as long as the co-moving frame in which it is measured is not rotating.

As an example of two-body motion, consider two identical, isolated bodies m_1 and m_2 positioned in an inertial frame of reference, as shown in Figure 2.3. At time $t = 0$, m_1 is at rest at the origin of the frame, whereas m_2, to the right of m_1, has a velocity $\mathbf{v}_o$ directed upward to the right, making a 45° angle with the *X* axis. The subsequent motion of the two bodies, which is due solely to their mutual gravitational attraction, is determined relative to the inertial frame by means of Equations 2.7 and 2.8. Figure 2.3 is a computer-generated solution of those equations. The motion is rather complex. Nevertheless, at any time t, m_1 and m_2 lie in the *XY* plane, equidistant and in opposite directions from their center of mass *G*, whose straight-line path is also shown in Figure 2.3. The very same motion appears rather less complex when viewed from m_1, as the computer simulation reveals in Figure 2.4(a). Figure 2.4(a)

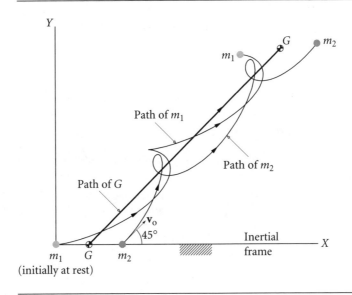

Figure 2.3 The motion of two identical bodies acted on only by their mutual gravitational attraction, as viewed from the inertial frame of reference.

represents the solution to Equation 2.15, and we see that, relative to m_1, m_2 follows what appears to be an elliptical path. (So does the center of mass.) Figure 2.4(b) reveals that both m_1 and m_2 follow elliptical paths around the center of mass.

Since the center of mass has zero acceleration, we can use it as an inertial reference frame. Let $\mathbf{r}_1$ and $\mathbf{r}_2$ be the position vectors of m_1 and m_2, respectively, relative to the center of mass G in Figure 2.1. The equation of motion of m_2 relative to the center of mass is

$$-G\frac{m_1 m_2}{r^2}\hat{\mathbf{u}}_r = m_2 \ddot{\mathbf{r}}_2 \tag{2.16}$$

where, as before, $\mathbf{r}$ is the position vector of m_2 relative to m_1. In terms of $\mathbf{r}_1$ and $\mathbf{r}_2$,

$$\mathbf{r} = \mathbf{r}_2 - \mathbf{r}_1$$

Since the position vector of the center of mass relative to itself is zero, it follows from Equation 2.1 that

$$m_1 \mathbf{r}_1 + m_2 \mathbf{r}_2 = 0$$

Therefore,

$$\mathbf{r}_1 = -\frac{m_2}{m_1}\mathbf{r}_2$$

so that

$$\mathbf{r} = \frac{m_1 + m_2}{m_1}\mathbf{r}_2$$

Substituting this back into Equation 2.16 and using the fact that $\hat{\mathbf{u}}_r = \mathbf{r}_2/r_2$, we get

$$-G\frac{m_1^3 m_2}{(m_1 + m_2)^2 r_2^3}\mathbf{r}_2 = m_2 \ddot{\mathbf{r}}_2$$

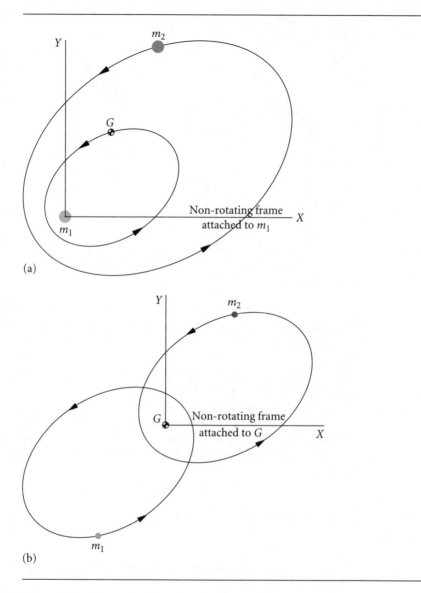

Figure 2.4 The motion in Figure 2.3, (a) as viewed relative to m_1 (or m_2); (b) as viewed from the center of mass.

which, upon simplification, becomes

$$-\left(\frac{m_1}{m_1+m_2}\right)\frac{\mu}{r_2^3}\mathbf{r}_2 = \ddot{\mathbf{r}}_2 \qquad (2.17)$$

where μ is given by Equation 2.14. If we let

$$\mu' = \left(\frac{m_1}{m_1+m_2}\right)^3 \mu$$

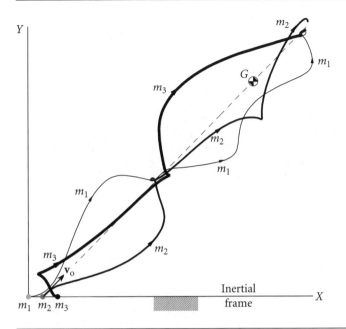

Figure 2.5 The motion of three identical masses as seen from the inertial frame in which m_1 and m_3 are initially at rest, while m_2 has an initial velocity $\mathbf{v}_0$ directed upwards and to the right, as shown.

then Equation 2.17 reduces to

$$\ddot{\mathbf{r}}_2 = -\frac{\mu'}{r_2^3}\mathbf{r}_2$$

which is identical in form to Equation 2.15.

In a similar fashion, the equation of motion of m_1 relative to the center of mass is found to be

$$\ddot{\mathbf{r}}_1 = -\frac{\mu''}{r_1^3}\mathbf{r}_1$$

in which

$$\mu'' = \left(\frac{m_2}{m_1 + m_2}\right)^3 \mu$$

Since the equations of motion of either particle relative to the center of mass have the same form as the equations of motion relative to either one of the bodies, m_1 or m_2, it follows that the relative motion as viewed from these different perspectives must be similar, as illustrated in Figure 2.4.

One may wonder what the motion looks like if there are more than two bodies moving under the influence only of their mutual gravitational attraction. The *n*-body problem with $n > 2$ has no closed form solution, which is complex and chaotic in nature. We can use a computer simulation (see Appendix C.1) to get an idea of the motion for some special cases. Figure 2.5 shows the motion of three equal masses,

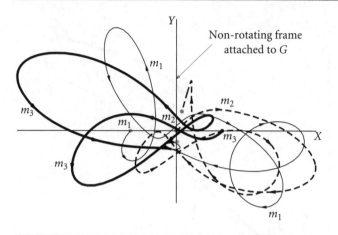

Figure 2.6 The same motion as Figure 2.5, as viewed from the inertial frame attached to the center of mass G.

equally spaced initially along the X axis of an inertial frame. The center mass has an initial velocity, while the other two are at rest. As time progresses, we see no periodic behavior as was evident in the two-body motion in Figure 2.3. The chaos is more obvious if the motion is viewed from the center of mass of the three-body system, as shown in Figure 2.6. The computer simulation from which these figures were taken shows that the masses eventually collide.

2.4 ANGULAR MOMENTUM AND THE ORBIT FORMULAS

The angular momentum of body m_2 relative to m_1 is the moment of m_2's relative linear momentum $m_2\dot{\mathbf{r}}$ (cf. Equation 1.17),

$$\mathbf{H}_{2/1} = \mathbf{r} \times m_2\dot{\mathbf{r}}$$

where $\dot{\mathbf{r}} = \mathbf{v}$ is the velocity of m_2 relative to m_1. Let us divide this equation through by m_2 and let $\mathbf{h} = \mathbf{H}_{2/1}/m_2$, so that

$$\mathbf{h} = \mathbf{r} \times \dot{\mathbf{r}} \tag{2.18}$$

$\mathbf{h}$ is the relative angular momentum of m_2 per unit mass, that is, the specific relative angular momentum. The units of $\mathbf{h}$ are km² s⁻¹.

Taking the time derivative of $\mathbf{h}$ yields

$$\frac{d\mathbf{h}}{dt} = \dot{\mathbf{r}} \times \dot{\mathbf{r}} + \mathbf{r} \times \ddot{\mathbf{r}}$$

But $\dot{\mathbf{r}} \times \dot{\mathbf{r}} = 0$. Furthermore, $\ddot{\mathbf{r}} = -(\mu/r^3)\mathbf{r}$, according to Equation 2.15, so that

$$\mathbf{r} \times \ddot{\mathbf{r}} = \mathbf{r} \times \left(-\frac{\mu}{r^3}\mathbf{r}\right) = -\frac{\mu}{r^3}(\mathbf{r} \times \mathbf{r}) = 0$$

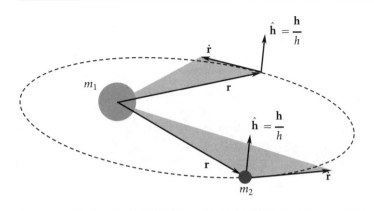

Figure 2.7 The path of m_2 around m_1 lies in a plane whose normal is defined by **h**.

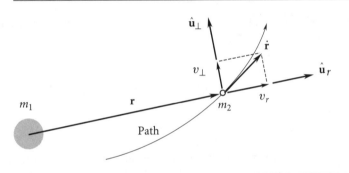

Figure 2.8 Components of the velocity of m_2, viewed above the plane of the orbit.

Therefore,

$$\frac{d\mathbf{h}}{dt} = 0 \quad (\text{or } \mathbf{r} \times \dot{\mathbf{r}} = \text{constant}) \tag{2.19}$$

At any given time, the position vector **r** and the velocity vector $\dot{\mathbf{r}}$ lie in the same plane, as illustrated in Figure 2.7. Their cross product $\mathbf{r} \times \dot{\mathbf{r}}$ is perpendicular to that plane. Since $\mathbf{r} \times \dot{\mathbf{r}} = \mathbf{h}$, the unit vector normal to the plane is

$$\hat{\mathbf{h}} = \frac{\mathbf{h}}{h} \tag{2.20}$$

But, according to Equation 2.19, this unit vector is constant. Thus, the path of m_2 around m_1 lies in a single plane.

Since the orbit of m_2 around m_1 forms a plane, it is convenient to orient oneself above that plane and look down upon the path, as shown in Figure 2.8. Let us resolve the relative velocity vector $\dot{\mathbf{r}}$ into components $\mathbf{v}_r = v_r \hat{\mathbf{u}}_r$ and $\mathbf{v}_\perp = v_\perp \hat{\mathbf{u}}_\perp$ along the outward radial from m_1 and perpendicular to it, respectively, where $\hat{\mathbf{u}}_r$ and $\hat{\mathbf{u}}_\perp$ are the radial and perpendicular (azimuthal) unit vectors. Then we can write Equation 2.18

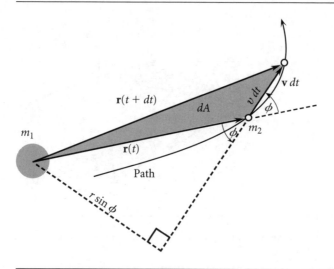

Figure 2.9 Differential area dA swept out by the relative position vector $\mathbf{r}$ during time interval dt.

as

$$\mathbf{h} = r\hat{\mathbf{u}}_r \times (v_r\hat{\mathbf{u}}_r + v_\perp\hat{\mathbf{u}}_\perp) = rv_\perp \hat{\mathbf{h}}$$

That is,

$$h = rv_\perp \tag{2.21}$$

Clearly, the angular momentum depends only on the azimuth component of the relative velocity.

During the differential time interval dt the position vector $\mathbf{r}$ sweeps out an area dA, as shown in Figure 2.9. From the figure it is clear that the triangular area dA is given by

$$dA = \frac{1}{2} \times \text{base} \times \text{altitude} = \frac{1}{2} \times v\,dt \times r\sin\phi = \frac{1}{2}r(v\sin\phi)dt = \frac{1}{2}rv_\perp dt$$

Therefore, using Equation 2.21 we have

$$\frac{dA}{dt} = \frac{h}{2} \tag{2.22}$$

dA/dt is called the areal velocity, and according to Equation 2.22 it is constant. Named after the German astronomer Johannes Kepler (1571–1630), this result is known as Kepler's second law: equal areas are swept out in equal times.

Before proceeding with an effort to integrate Equation 2.15, recall the vector identity known as the *bac − cab* rule:

$$\mathbf{A} \times (\mathbf{B} \times \mathbf{C}) = \mathbf{B}(\mathbf{A} \cdot \mathbf{C}) - \mathbf{C}(\mathbf{A} \cdot \mathbf{B}) \tag{2.23}$$

Recall as well that

$$\mathbf{r} \cdot \mathbf{r} = r^2 \tag{2.24}$$

so that
$$\frac{d}{dt}(\mathbf{r} \cdot \mathbf{r}) = 2r\frac{dr}{dt}$$

But
$$\frac{d}{dt}(\mathbf{r} \cdot \mathbf{r}) = \mathbf{r} \cdot \frac{d\mathbf{r}}{dt} + \frac{d\mathbf{r}}{dt} \cdot \mathbf{r} = 2\mathbf{r} \cdot \frac{d\mathbf{r}}{dt}$$

Thus, we obtain the important identity
$$\mathbf{r} \cdot \dot{\mathbf{r}} = r\dot{r} \qquad (2.25a)$$

Since $\dot{\mathbf{r}} = \mathbf{v}$ and $r = \|\mathbf{r}\|$, this can be written alternatively as
$$\mathbf{r} \cdot \mathbf{v} = \|\mathbf{r}\|\frac{d\|\mathbf{r}\|}{dt} \qquad (2.25b)$$

Now let us take the cross product of both sides of Equation 2.15 [$\ddot{\mathbf{r}} = -(\mu/r^3)\mathbf{r}$] with the specific angular momentum $\mathbf{h}$:
$$\ddot{\mathbf{r}} \times \mathbf{h} = -\frac{\mu}{r^3}\mathbf{r} \times \mathbf{h} \qquad (2.26)$$

Since $\frac{d}{dt}(\dot{\mathbf{r}} \times \mathbf{h}) = \ddot{\mathbf{r}} \times \mathbf{h} + \dot{\mathbf{r}} \times \dot{\mathbf{h}}$, the left-hand side can be written
$$\ddot{\mathbf{r}} \times \mathbf{h} = \frac{d}{dt}(\dot{\mathbf{r}} \times \mathbf{h}) - \dot{\mathbf{r}} \times \dot{\mathbf{h}}$$

But according to Equation 2.19, the angular momentum is constant ($\dot{\mathbf{h}} = 0$), so this reduces to
$$\ddot{\mathbf{r}} \times \mathbf{h} = \frac{d}{dt}(\dot{\mathbf{r}} \times \mathbf{h}) \qquad (2.27)$$

The right-hand side of Equation 2.26 can be transformed by the following sequence of substitutions:
$$\frac{1}{r^3}\mathbf{r} \times \mathbf{h} = \frac{1}{r^3}[\mathbf{r} \times (\mathbf{r} \times \dot{\mathbf{r}})] \qquad \text{(Equation 2.18 }[\mathbf{h} = \mathbf{r} \times \dot{\mathbf{r}}])$$
$$= \frac{1}{r^3}[\mathbf{r}(\mathbf{r} \cdot \dot{\mathbf{r}}) - \dot{\mathbf{r}}(\mathbf{r} \cdot \mathbf{r})] \qquad \text{(Equation 2.23 }[bac - cab \text{ rule}])$$
$$= \frac{1}{r^3}[\mathbf{r}(r\dot{r}) - \dot{\mathbf{r}}r^2] \qquad \text{(Equations 2.24 and 2.25)}$$
$$= \frac{\mathbf{r}\dot{r} - \dot{\mathbf{r}}r}{r^2}$$

But
$$\frac{d}{dt}\left(\frac{\mathbf{r}}{r}\right) = \frac{\dot{\mathbf{r}}r - \mathbf{r}\dot{r}}{r^2} = -\frac{\mathbf{r}\dot{r} - \dot{\mathbf{r}}r}{r^2}$$

Therefore
$$\frac{1}{r^3}\mathbf{r} \times \mathbf{h} = -\frac{d}{dt}\left(\frac{\mathbf{r}}{r}\right) \qquad (2.28)$$

Substituting Equations 2.27 and 2.28 into Equation 2.26, we get

$$\frac{d}{dt}(\dot{\mathbf{r}} \times \mathbf{h}) = \frac{d}{dt}\left(\mu \frac{\mathbf{r}}{r}\right)$$

or

$$\frac{d}{dt}\left(\dot{\mathbf{r}} \times \mathbf{h} - \mu \frac{\mathbf{r}}{r}\right) = 0$$

That is,

$$\dot{\mathbf{r}} \times \mathbf{h} - \mu \frac{\mathbf{r}}{r} = \mathbf{C} \qquad (2.29)$$

where the vector $\mathbf{C}$ is an arbitrary constant of integration having the dimensions of μ. Equation 2.29 is the first integral of the equation of motion, $\ddot{\mathbf{r}} = -(\mu/r^3)\mathbf{r}$. Taking the dot product of both sides of Equation 2.29 with the vector $\mathbf{h}$ yields

$$(\dot{\mathbf{r}} \times \mathbf{h}) \cdot \mathbf{h} - \mu \frac{\mathbf{r} \cdot \mathbf{h}}{r} = \mathbf{C} \cdot \mathbf{h}$$

Since $\dot{\mathbf{r}} \times \mathbf{h}$ is perpendicular to both $\dot{\mathbf{r}}$ and $\mathbf{h}$, it follows that $(\dot{\mathbf{r}} \times \mathbf{h}) \cdot \mathbf{h} = 0$. Likewise, since $\mathbf{h} = \mathbf{r} \times \dot{\mathbf{r}}$ is perpendicular to both $\mathbf{r}$ and $\dot{\mathbf{r}}$, it is true that $\mathbf{r} \cdot \mathbf{h} = 0$. Therefore, we have $\mathbf{C} \cdot \mathbf{h} = 0$, i.e., $\mathbf{C}$ is perpendicular to $\mathbf{h}$, which is normal to the orbital plane. That of course means $\mathbf{C}$ must lie in the orbital plane.

Let us rearrange Equation 2.29 and write it as

$$\frac{\mathbf{r}}{r} + \mathbf{e} = \frac{\dot{\mathbf{r}} \times \mathbf{h}}{\mu} \qquad (2.30)$$

where $\mathbf{e} = \mathbf{C}/\mu$. The dimensionless vector $\mathbf{e}$ is called the eccentricity vector. The line defined by the vector $\mathbf{e}$ is commonly called the apse line. In order to obtain a scalar equation, let us take the dot product of both sides of Equation 2.30 with $\mathbf{r}$:

$$\frac{\mathbf{r} \cdot \mathbf{r}}{r} + \mathbf{r} \cdot \mathbf{e} = \frac{\mathbf{r} \cdot (\dot{\mathbf{r}} \times \mathbf{h})}{\mu} \qquad (2.31)$$

In order to simplify the right-hand side, we can employ the useful vector identity, known as the interchange of the dot and the cross,

$$\mathbf{A} \cdot (\mathbf{B} \times \mathbf{C}) = (\mathbf{A} \times \mathbf{B}) \cdot \mathbf{C} \qquad (2.32)$$

to obtain

$$\mathbf{r} \cdot (\dot{\mathbf{r}} \times \mathbf{h}) = (\mathbf{r} \times \dot{\mathbf{r}}) \cdot \mathbf{h} = \mathbf{h} \cdot \mathbf{h} = h^2 \qquad (2.33)$$

Substituting this expression into the right-hand side of Equation 2.31, and substituting $\mathbf{r} \cdot \mathbf{r} = r^2$ on the left yields

$$r + \mathbf{r} \cdot \mathbf{e} = \frac{h^2}{\mu} \qquad (2.34)$$

Observe that by following the steps leading from Equation 2.30 to 2.34 we have lost track of the variable time. This occurred at Equation 2.33, because h is constant. Finally, from the definition of the dot product we have

$$\mathbf{r} \cdot \mathbf{e} = re \cos \theta$$

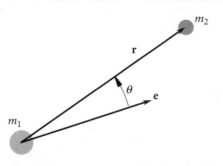

Figure 2.10 The true anomaly θ is the angle between the eccentricity vector **e** and the position vector **r**.

in which e is the eccentricity (the magnitude of the eccentricity vector **e**) and θ is the true anomaly. θ is the angle between the fixed vector **e** and the variable position vector **r**, as illustrated in Figure 2.10. (Other symbols used to represent true anomaly include ν, f, υ and ϕ.) In terms of the eccentricity and the true anomaly, we may therefore write Equation 2.34 as

$$r + re\cos\theta = \frac{h^2}{\mu}$$

or

$$r = \frac{h^2}{\mu}\frac{1}{1+e\cos\theta} \qquad (2.35)$$

This is the orbit equation, and it defines the path of the body m_2 around m_1, relative to m_1. Remember that μ, h, and e are constants. Observe as well that there is no significance to negative values of eccentricity; i.e., $e \geq 0$. Since the orbit equation describes conic sections, including ellipses, it is a mathematical statement of Kepler's first law, namely, that the planets follow elliptical paths around the sun. Two-body orbits are often referred to as Keplerian orbits.

In Section 2.3 it was pointed out that integration of the equation of relative motion, Equation 2.15, leads to six constants of integration. In this section it would seem that we have arrived at those constants, namely the three components of the angular momentum **h** and the three components of the eccentricity vector **e**. However, we showed that **h** is perpendicular to **e**. This places a condition, namely $\mathbf{h} \cdot \mathbf{e} = 0$, on the components of **h** and **e**, so that we really have just five independent constants of integration. The sixth constant of the motion will arise when we work time back into the picture in the next chapter.

The angular velocity of the position vector **r** is $\dot\theta$, the rate of change of the true anomaly. The component of velocity normal to the position vector is found in terms of the angular velocity by the formula

$$v_\perp = r\dot\theta \qquad (2.36)$$

Substituting this into Equation 2.21 ($h = rv_\perp$) yields the specific angular momentum in terms of the angular velocity,

$$h = r^2\dot\theta \qquad (2.37)$$

It is convenient to have formulas for computing the radial and azimuth components of velocity shown in Figure 2.11. From $h = rv_\perp$ we of course obtain

$$v_\perp = \frac{h}{r}$$

Substituting r from Equation 2.35 readily yields

$$v_\perp = \frac{\mu}{h}(1 + e\cos\theta) \qquad (2.38)$$

Since $v_r = \dot{r}$, we take the derivative of Equation 2.35 to get

$$\dot{r} = \frac{dr}{dt} = \frac{h^2}{\mu}\left[-\frac{e(-\dot\theta\sin\theta)}{(1+e\cos\theta)^2}\right] = \frac{h^2}{\mu}\frac{e\sin\theta}{(1+e\cos\theta)^2}\frac{h}{r^2}$$

where we made use of the fact that $\dot\theta = h/r^2$, from Equation 2.37. Substituting Equation 2.35 once again and simplifying finally yields

$$v_r = \frac{\mu}{h}e\sin\theta \qquad (2.39)$$

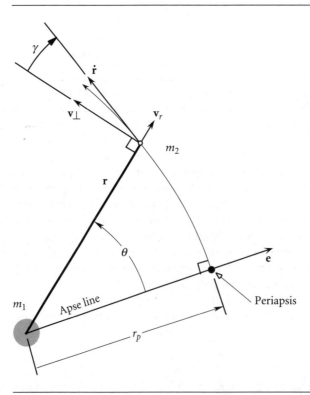

Figure 2.11 Position and velocity of m_2 in polar coordinates centered at m_1, with the eccentricity vector being the reference for true anomaly (polar angle) θ. γ is the flight path angle.

We see from Equation 2.35 that m_2 comes closest to m_1 (r is smallest) when $\theta = 0$ (unless $e = 0$, in which case the distance between m_1 and m_2 is constant). The point of closest approach lies on the apse line and is called periapsis. The distance r_p to periapsis, as shown in Figure 2.11, is obtained by setting the true anomaly equal to zero,

$$r_p = \frac{h^2}{\mu} \frac{1}{1+e} \tag{2.40}$$

Clearly, $v_r = 0$ at periapsis.

The flight path angle γ is also illustrated in Figure 2.11. It is the angle that the velocity vector $\mathbf{v} = \dot{\mathbf{r}}$ makes with the normal to the position vector. The normal to the position vector points in the direction of $\mathbf{v}_\perp$, and it is called the local horizon. From Figure 2.11 it is clear that

$$\tan \gamma = \frac{v_r}{v_\perp} \tag{2.41}$$

Substituting Equations 2.38 and 2.39 leads at once to the expression

$$\tan \gamma = \frac{e \sin \theta}{1 + e \cos \theta} \tag{2.42}$$

Since $\cos(-\theta) = \cos \theta$, the trajectory described by the orbit equation is symmetric about the apse line, as illustrated in Figure 2.12, which also shows a chord, the straight line connecting any two points on the orbit. The latus rectum is the chord through the center of attraction perpendicular to the apse line. By symmetry, the center of attraction divides the latus rectum into two equal parts, each of length p, known historically as the semi-latus rectum. In modern parlance, p is called the parameter of the orbit. From Equation 2.35 it is apparent that

$$p = \frac{h^2}{\mu} \tag{2.43}$$

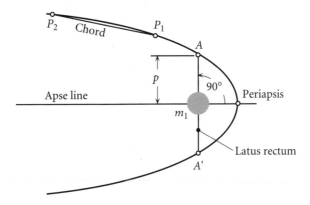

Figure 2.12 Illustration of latus rectum, semi-latus rectum p, and the chord between any two points on an orbit.

Since the path of m_2 around m_1 lies in a plane, for the time being we will for simplicity continue to view the trajectory from above the plane. Unless there is reason to do otherwise, we will assume that the eccentricity vector points to the right and that m_2 moves counterclockwise around m_1, which means that the true anomaly is measured positive counterclockwise, consistent with the usual polar coordinate sign convention.

2.5 THE ENERGY LAW

By taking the cross product of Equation 2.15, $\ddot{\mathbf{r}} = -(\mu/r^3)\mathbf{r}$ (Newton's second law of motion), with the specific relative angular momentum per unit mass $\mathbf{h}$, we were led to the vector Equation 2.29, and from that we obtained the orbit formula, Equation 2.35. Now let us see what results from taking the *dot* product of Equation 2.15 with the relative linear momentum per unit mass. The relative linear momentum per unit mass is just the relative velocity,

$$\frac{m_2 \dot{\mathbf{r}}}{m_2} = \dot{\mathbf{r}}$$

Thus, carrying out the dot product in Equation 2.15 yields

$$\ddot{\mathbf{r}} \cdot \dot{\mathbf{r}} = -\mu \frac{\mathbf{r} \cdot \dot{\mathbf{r}}}{r^3} \tag{2.44}$$

For the left-hand side we observe that

$$\ddot{\mathbf{r}} \cdot \dot{\mathbf{r}} = \frac{1}{2} \frac{d}{dt}(\dot{\mathbf{r}} \cdot \dot{\mathbf{r}}) = \frac{1}{2} \frac{d}{dt}(\mathbf{v} \cdot \mathbf{v}) = \frac{1}{2} \frac{d}{dt}(v^2) = \frac{d}{dt}\left(\frac{v^2}{2}\right) \tag{2.45}$$

For the right-hand side of Equation 2.44 we have, recalling that $\mathbf{r} \cdot \mathbf{r} = r^2$ and

$$d(1/r)/dt = (-1/r^2)(dr/dt),$$

$$\mu \frac{\mathbf{r} \cdot \dot{\mathbf{r}}}{r^3} = \mu \frac{r\dot{r}}{r^3} = \mu \frac{\dot{r}}{r^2} = -\frac{d}{dt}\left(\frac{\mu}{r}\right) \tag{2.46}$$

Substituting Equations 2.45 and 2.46 into Equation 2.44 yields

$$\frac{d}{dt}\left(\frac{v^2}{2} - \frac{\mu}{r}\right) = 0$$

or

$$\frac{v^2}{2} - \frac{\mu}{r} = \varepsilon \quad \text{(constant)} \tag{2.47}$$

where ε is a constant. $v^2/2$ is the relative kinetic energy per unit mass. $(-\mu/r)$ is the potential energy per unit mass of the body m_2 in the gravitational field of m_1. The total mechanical energy per unit mass ε is the sum of the kinetic and potential energies per unit mass. Equation 2.47 is a statement of conservation of energy, namely, that the specific mechanical energy is the same at all points of the trajectory. Equation 2.47 is also known as the *vis-viva* ('living force') equation. Since ε is constant, let us evaluate it at periapsis ($\theta = 0$),

$$\varepsilon = \varepsilon_p = \frac{v_p^2}{2} - \frac{\mu}{r_p} \qquad (2.48)$$

where r_p and v_p are the position and speed at periapsis. Since $v_r = 0$ at periapsis, we have $v_p = v_\perp = h/r_p$. Thus,

$$\varepsilon = \frac{1}{2}\frac{h^2}{r_p^2} - \frac{\mu}{r_p} \qquad (2.49)$$

Substituting Equation 2.40 into 2.49 yields a formula for the orbital specific energy in terms of the orbital constants h and e,

$$\varepsilon = -\frac{1}{2}\frac{\mu^2}{h^2}(1 - e^2) \qquad (2.50)$$

Clearly, the orbital energy is not an independent orbital parameter.

Note that the mechanical energy $\mathcal{E}$ of a satellite of mass m_1 is obtained from the specific energy ε by the formula

$$\mathcal{E} = m_1 \varepsilon \qquad (2.51)$$

2.6 Circular orbits ($e = 0$)

Setting $e = 0$ in the orbital equation $r = (h^2/\mu)/(1 + e\cos\theta)$ yields

$$r = \frac{h^2}{\mu} \qquad (2.52)$$

That is, $r = $ constant, which means the orbit of m_2 around m_1 is a circle. Since $\dot{r} = 0$, it follows that $v = v_\perp$ so that the angular momentum formula $h = rv_\perp$ becomes simply $h = rv$ for a circular orbit. Substituting this expression for h into Equation 2.52 and solving for v yields the velocity of a circular orbit,

$$v_{\text{circular}} = \sqrt{\frac{\mu}{r}} \qquad (2.53)$$

The time T required for one orbit is known as the period. Because the speed is constant, the period of a circular orbit is easy to compute:

$$T = \frac{\text{circumference}}{\text{speed}} = \frac{2\pi r}{\sqrt{\dfrac{\mu}{r}}}$$

so that

$$T_{\text{circular}} = \frac{2\pi}{\sqrt{\mu}} r^{\frac{3}{2}} \qquad (2.54)$$

The specific energy of a circular orbit is found by setting $e=0$ in Equation 2.50,

$$\varepsilon = -\frac{1}{2}\frac{\mu^2}{h^2}$$

Employing Equation 2.52 yields

$$\varepsilon_{circular} = -\frac{\mu}{2r} \qquad (2.55)$$

Obviously, the energy of a circular orbit is negative. As the radius goes up, the energy becomes less negative, i.e., it increases. In other words, the higher the orbit, the greater its energy.

To launch a satellite from the surface of the earth into a circular orbit requires increasing its specific mechanical energy ε. This energy comes from the rocket motors of the launch vehicle. Since the mechanical energy of a satellite of mass m is $\mathcal{E} = m\varepsilon$, a propulsion system that can place a large mass in a low earth orbit can place a smaller mass in a higher earth orbit.

The space shuttle orbiters are the largest man-made satellites so far placed in orbit with a single launch vehicle. For example, on NASA mission STS-82 in February 1997, the orbiter Discovery rendezvoused with the Hubble space telescope to repair and refurbish it. The altitude of the nearly circular orbit was 580 km (360 miles). Discovery's orbital mass early in the mission was 106 000 kg (117 tons). That was only 6 percent of the total mass of the shuttle prior to launch (comprising the orbiter's dry mass, plus that of its payload and fuel, plus the two solid rocket boosters, plus the external fuel tank filled with liquid hydrogen and oxygen). This mass of about 2 million kilograms (2200 tons) was lifted off the launch pad by a total thrust in the vicinity of 35 000 kN (7.8 million pounds). Eighty-five percent of the thrust was furnished by the solid rocket boosters (SRBs), which were depleted and jettisoned about two minutes into the flight. The remaining thrust came from the three liquid rockets (space shuttle main engines, or SSMEs) on the orbiter. These were fueled by the external tank which was jettisoned just after the SSMEs were shut down at MECO (main engine cut off), about eight and a half minutes after lift-off.

Manned orbital spacecraft and a host of unmanned remote sensing, imaging and navigation satellites occupy nominally circular, low-earth orbits. A low-earth orbit (LEO) is one whose altitude lies between about 150 km (100 miles) and about 1000 km (600 miles). An LEO is well above the nominal outer limits of the drag-producing atmosphere (about 80 km or 50 miles), and well below the hazardous Van Allen radiation belts, the innermost of which begins at about 2400 km (1500 miles).

Nearly all of our applications of the orbital equations will be to the analysis of man-made spacecraft, all of which have a mass that is insignificant compared to the sun and planets. For example, since the earth is nearly 20 orders of magnitude more massive than the largest conceivable artificial satellite, the center of mass of the two-body system lies at the center of the earth and μ in Equation 3.14 becomes

$$\mu = G(m_{earth} + \cancel{m_{satellite}}) = Gm_{earth}$$

The value of the earth's gravitational parameter to be used throughout this book is found in Table A.2,

$$\mu_{earth} = 398\,600 \text{ km}^3/\text{s}^2 \qquad (2.56)$$

2.6 Circular orbits ($e = 0$)

EXAMPLE 2.2

Plot the speed v and period T of a satellite in circular LEO as a function of altitude z.

Equations 2.53 and 2.54 give the speed and period, respectively, of the satellite:

$$v = \sqrt{\frac{\mu}{r}} = \sqrt{\frac{\mu}{R_E + z}} = \sqrt{\frac{398\,600}{6378 + z}} \qquad T = \frac{2\pi}{\sqrt{\mu}} r^{\frac{3}{2}} = \frac{2\pi}{\sqrt{398\,600}} (6378 + z)^{\frac{3}{2}}$$

These relations are graphed in Figure 2.13.

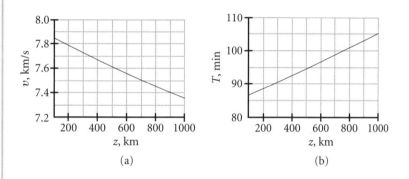

Figure 2.13 Circular orbital speed (a) and period (b) as a function of altitude.

If a satellite remains always above the same point on the earth's equator, then it is in a circular, geostationary equatorial orbit or *GEO*. For GEO, the radial from the center of the earth to the satellite must have the same angular velocity as the earth itself, namely, 2π radians per sidereal day. The sidereal day is the time it takes the earth to complete one rotation relative to inertial space (the fixed stars). The ordinary 24-hour day, or synodic day, is the time it takes the sun to apparently rotate once around the earth, from high noon one day to high noon the next. The synodic and sidereal days would be identical if the earth stood still in space. However, while the earth makes one absolute rotation around its axis, it advances $2\pi/365.26$ radians along its solar orbit. Therefore, its inertial angular velocity ω_E is $[(2\pi + 2\pi/365.26)\text{radians}]/(24 \text{ hours})$; i.e.,

$$\omega_E = 72.9217 \times 10^{-6} \text{ rad/s} \qquad (2.57)$$

Communications satellites and global weather satellites are placed in geostationary orbit because of the large portion of the earth's surface visible from that altitude and the fact that ground stations do not have to track the satellite, which appears motionless in the sky.

EXAMPLE 2.3

Calculate the altitude z_{GEO} and speed v_{GEO} of a geostationary earth satellite.

The speed of the satellite in its circular GEO of radius r_{GEO} is

$$v_{\text{GEO}} = \sqrt{\frac{\mu}{r_{\text{GEO}}}} \qquad (a)$$

(Example 2.3 continued)

On the other hand, the speed v_{GEO} along its circular path is related to the absolute angular velocity ω_E of the earth by the kinematics formula

$$v_{GEO} = \omega_E r_{GEO}$$

Equating these two expressions and solving for r_{GEO} yields

$$r_{GEO} = \sqrt[3]{\frac{\mu}{\omega_E^2}}$$

Substituting Equation 2.56, we get

$$r_{GEO} = \sqrt[3]{\frac{398\,600}{(72.9217 \times 10^{-6})^2}} = 42\,164\,\text{km} \tag{2.58}$$

Therefore, the distance of the satellite above the earth's surface is

$$z_{GEO} = r_{GEO} - R_E = 42\,164 - 6378 = \underline{35\,786\,\text{km}} \quad (22\,241\,\text{mi})$$

Substituting Equation 2.58 into (a) yields the speed,

$$v_{GEO} = \sqrt{\frac{398\,600}{42\,164}} = \underline{3.075\,\text{km/s}} \tag{2.59}$$

EXAMPLE 2.4

Calculate the maximum latitude and the percentage of the earth's surface visible from GEO.

To find the maximum viewable latitude ϕ, use Figure 2.14, from which it is apparent that

$$\phi = \cos^{-1}\frac{R_E}{r} \tag{a}$$

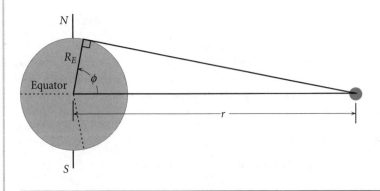

Figure 2.14 Satellite in GEO.

where $R_E = 6378$ km and, according to Equation 2.57, $r = 42\,164$ km. Therefore

$$\phi = \cos^{-1}\frac{6378}{42\,164} = \underline{81.30°} \quad \text{Maximum visible north or south latitude.} \quad \text{(b)}$$

The surface area S visible from GEO is the shaded region illustrated in Figure 2.15. It can be shown that the area S is given by

$$S = 2\pi R_E^2 (1 - \cos\phi)$$

Therefore, the percentage of the hemisphere visible from GEO is

$$\frac{S}{2\pi R_E^2} \times 100 = (1 - \cos 81.30°) \times 100 = 84.9\%$$

which of course means that $\underline{42.4}$ percent of the total surface of the earth can be seen from GEO.

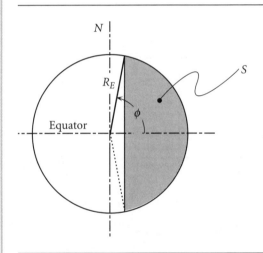

Figure 2.15 Surface area S visible from GEO.

Figure 2.16 is a photograph taken from geosynchronous equatorial orbit by one of the National Oceanic and Atmospheric Administation's Geostationary Operational Environmental Satellites (GOES).

2.7 ELLIPTICAL ORBITS ($0 < e < 1$)

If $0 < e < 1$, then the denominator of Equation 2.35 varies with the true anomaly θ, but it remains positive, never becoming zero. Therefore, the relative position vector remains bounded, having its smallest magnitude at periapsis r_p, given by Equation 2.40. The maximum value of r is reached when the denominator of $r = (h^2/\mu)/(1 + e\cos\theta)$ obtains its minimum value, which occurs at $\theta = 180°$. That

Figure 2.16 The view from GEO. *NASA-Goddard Space Flight Center, data from NOAA GOES.*

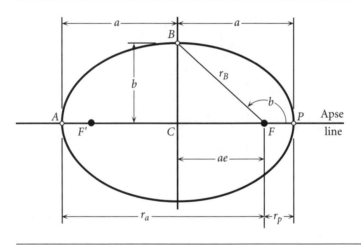

Figure 2.17 Elliptical orbit. m_1 is at the focus F. F' is the unoccupied empty focus.

point is called the apoapsis, and its radial coordinate, denoted r_a, is

$$r_a = \frac{h^2}{\mu} \frac{1}{1-e} \qquad (2.60)$$

The curve defined by Equation 2.35 in this case is an ellipse.

Let $2a$ be the distance measured along the apse line from periapsis P to apoapsis A, as illustrated in Figure 2.17. Then

$$2a = r_p + r_a$$

2.7 Elliptical orbits ($0 < e < 1$)

Substituting Equations 2.40 and 2.61 into this expression we get

$$a = \frac{h^2}{\mu} \frac{1}{1 - e^2} \tag{2.61}$$

a is the semimajor axis of the ellipse. Solving Equation 2.61 for h^2/μ and putting the result into Equation 2.35 yields an alternative form of the orbit equation,

$$r = a \frac{1 - e^2}{1 + e \cos \theta} \tag{2.62}$$

In Figure 2.17, let F denote the location of the body m_1, which is the origin of the r, θ polar coordinate system. The center C of the ellipse is the point lying midway between the apoapsis and periapsis. The distance CF from C to F is

$$CF = a - FP = a - r_p$$

But from Equation 2.62,

$$r_p = a(1 - e) \tag{2.63}$$

Therefore, $CF = ae$, as indicated in Figure 2.17.

Let B be the point on the orbit which lies directly above C, on the perpendicular bisector of AP. The distance b from C to B is the semiminor axis. If the true anomaly of point B is β, then according to Equation 2.62, the radial coordinate of B is

$$r_B = a \frac{1 - e^2}{1 + e \cos \beta} \tag{2.64}$$

The projection of r_B onto the apse line is ae; i.e.,

$$ae = r_B \cos(180 - \beta) = -r_B \cos \beta = -\left(a \frac{1 - e^2}{1 + e \cos \beta}\right) \cos \beta$$

Solving this expression for e, we obtain

$$e = -\cos \beta \tag{2.65}$$

Substituting this result into Equation 2.64 reveals the interesting fact that

$$r_B = a$$

According to the Pythagorean theorem,

$$b^2 = r_B^2 - (ae)^2 = a^2 - a^2 e^2$$

which means the semiminor axis is found in terms of the semimajor axis and the eccentricity of the ellipse as

$$b = a\sqrt{1 - e^2} \tag{2.66}$$

Let an xy cartesian coordinate system be centered at C, as shown in Figure 2.18. In terms of r and θ, we see from the figure that the x coordinate of a point on

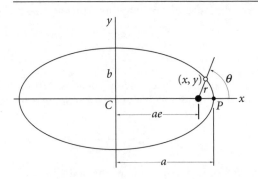

Figure 2.18 Cartesian coordinate description of the orbit.

the orbit is

$$x = ae + r\cos\theta = ae + \left(a\frac{1-e^2}{1+e\cos\theta}\right)\cos\theta = a\frac{e+\cos\theta}{1+e\cos\theta}$$

From this we have

$$\frac{x}{a} = \frac{e+\cos\theta}{1+e\cos\theta} \tag{2.67}$$

For the y coordinate we have, making use of Equation 2.66,

$$y = r\sin\theta = \left(a\frac{1-e^2}{1+e\cos\theta}\right)\sin\theta = b\frac{\sqrt{1-e^2}}{1+e\cos\theta}\sin\theta$$

Therefore,

$$\frac{y}{b} = \frac{\sqrt{1-e^2}}{1+e\cos\theta}\sin\theta \tag{2.68}$$

Using Equations 2.67 and 2.68, we find

$$\frac{x^2}{a^2} + \frac{y^2}{b^2} = \frac{1}{(1+e\cos\theta)^2}\left[(e+\cos\theta)^2 + (1-e^2)\sin^2\theta\right]$$

$$= \frac{1}{(1+e\cos\theta)^2}\left[e^2 + 2e\cos\theta + \cos^2\theta + \sin^2\theta - e^2\sin^2\theta\right]$$

$$= \frac{1}{(1+e\cos\theta)^2}\left[e^2 + 2e\cos\theta + 1 - e^2\sin^2\theta\right]$$

$$= \frac{1}{(1+e\cos\theta)^2}\left[e^2(1-\sin^2\theta) + 2e\cos\theta + 1\right]$$

$$= \frac{1}{(1+e\cos\theta)^2}\left[e^2\cos^2\theta + 2e\cos\theta + 1\right]$$

$$= \frac{1}{(1+e\cos\theta)^2}(1+e\cos\theta)^2$$

That is,

$$\frac{x^2}{a^2} + \frac{y^2}{b^2} = 1 \tag{2.69}$$

This is the familiar cartesian coordinate formula for an ellipse centered at the origin, with x intercepts at $\pm a$ and y intercepts at $\pm b$. If $a = b$, Equation 2.69 describes a circle, which is really an ellipse whose eccentricity is zero.

The specific energy of an elliptical orbit is negative, and it is found by substituting the specific angular momentum and eccentricity into Equation 2.50,

$$\varepsilon = -\frac{1}{2}\frac{\mu^2}{h^2}(1-e^2)$$

However, according to Equation 2.61, $h^2 = \mu a(1-e^2)$, so that

$$\varepsilon = -\frac{\mu}{2a} \tag{2.70}$$

This shows that the specific energy is independent of the eccentricity and depends only on the semimajor axis of the ellipse. For an elliptical orbit, the conservation of energy (Equation 2.47) may therefore be written

$$\frac{v^2}{2} - \frac{\mu}{r} = -\frac{\mu}{2a} \tag{2.71}$$

The area of an ellipse is found in terms of its semimajor and semiminor axes by the formula $A = \pi ab$ (which reduces to the formula for the area of a circle if $a = b$). To find the period T of the elliptical orbit, we employ Kepler's second law, $dA/dt = h/2$, to obtain

$$\Delta A = \frac{h}{2}\Delta t$$

For one complete revolution, $\Delta A = \pi ab$ and $\Delta t = T$. Thus, $\pi ab = (h/2)T$, or

$$T = \frac{2\pi ab}{h}$$

Substituting Equations 2.61 and 2.66, we get

$$T = \frac{2\pi}{h}a^2\sqrt{1-e^2} = \frac{2\pi}{h}\left(\frac{h^2}{\mu}\frac{1}{1-e^2}\right)^2\sqrt{1-e^2}$$

so that the formula for the period of an elliptical orbit, in terms of the orbital parameters h and e, becomes

$$T = \frac{2\pi}{\mu^2}\left(\frac{h}{\sqrt{1-e^2}}\right)^3 \tag{2.72}$$

We can once again appeal to Equation 2.61 to substitute $h = \sqrt{\mu a(1-e^2)}$ into this equation, thereby obtaining an alternative expression for the period,

$$T = \frac{2\pi}{\sqrt{\mu}}a^{\frac{3}{2}} \tag{2.73}$$

This expression, which is identical to that of a circular orbit of radius a (Equation 2.54), reveals that, like the energy, the period of an elliptical orbit is independent

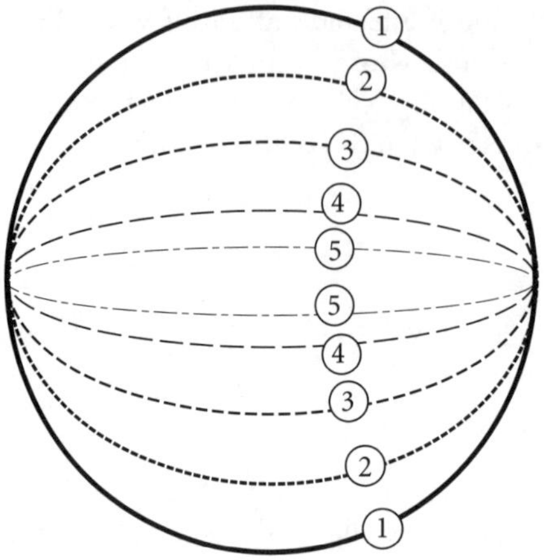

Figure 2.19 Since all five ellipses have the same major axis, their periods and energies are identical.

of the eccentricity (see Figure 2.19). Equation 2.73 embodies Kepler's third law: the period of a planet is proportional to the three-halves power of its semimajor axis.

Finally, observe that dividing Equation 2.40 by Equation 2.60 yields

$$\frac{r_p}{r_a} = \frac{1-e}{1+e}$$

Solving this for e results in a useful formula for calculating the eccentricity of an elliptical orbit, namely,

$$e = \frac{r_a - r_p}{r_a + r_p} \tag{2.74}$$

From Figure 2.17 it is apparent that $r_a - r_p = \overline{F'F}$, the distance between the foci. As previously noted, $r_a + r_p = 2a$. Thus, Equation 2.74 has the geometrical interpretation,

$$eccentricity = \frac{\text{distance between the foci}}{\text{length of the major axis}}$$

What is the average distance of m_2 from m_1 in the course of one complete orbit? To answer this question, we divide the range of the true anomaly (2π) into n equal segments $\Delta\theta$, so that

$$n = \frac{2\pi}{\Delta\theta}$$

We then use $r = (h^2/\mu)/(1 + e\cos\theta)$ to evaluate $r(\theta)$ at the n equally spaced values of true anomaly, starting at periapsis:

$$\theta_1 = 0, \quad \theta_2 = \Delta\theta, \quad \theta_3 = 2\Delta\theta, \ldots, \theta_n = (n-1)\Delta\theta$$

2.7 Elliptical orbits ($0 < e < 1$)

The average of this set of n values of r is given by

$$\bar{r}_\theta = \frac{1}{n}\sum_{i=1}^{n} r(\theta_i) = \frac{\Delta\theta}{2\pi}\sum_{i=1}^{n} r(\theta_i) = \frac{1}{2\pi}\sum_{i=1}^{n} r(\theta_i)\Delta\theta \quad (2.75)$$

Now let n become very large, so that $\Delta\theta$ becomes very small. In the limit as $n \to \infty$, Equation 2.75 becomes

$$\bar{r}_\theta = \frac{1}{2\pi}\int_0^{2\pi} r(\theta)d\theta \quad (2.76)$$

Substituting Equation 2.62 into the integrand yields

$$\bar{r}_\theta = \frac{1}{2\pi}a(1-e^2)\int_0^{2\pi} \frac{d\theta}{1+e\cos\theta}$$

The integral in this expression can be found in integral tables (e.g., Beyer, 1991), from which we obtain

$$\bar{r}_\theta = \frac{1}{2\pi}a(1-e^2)\left(\frac{2\pi}{\sqrt{1-e^2}}\right) = a\sqrt{1-e^2} \quad (2.77)$$

Comparing this result with Equation 2.66, we see that the true-anomaly-averaged orbital radius equals the length of the semiminor axis b of the ellipse. Thus, the semimajor axis, which is the average of the maximum and minimum distances from the focus, is not the mean distance. Since, from Equation 2.62, $r_p = a(1-e)$ and $r_a = a(1+e)$, Equation 2.77 also implies that

$$\bar{r}_\theta = \sqrt{r_p r_a} \quad (2.78)$$

The mean distance is the one-half power of the product of the maximum and minimum distances from the focus and not one-half their sum.

EXAMPLE 2.5

An earth satellite is in an orbit with perigee altitude $z_p = 400$ km and an eccentricity $e = 0.6$. Find (a) the perigee velocity, v_p; (b) the apogee radius, r_a; (c) the semimajor axis, a; (d) the true-anomaly-averaged radius $\bar{r}_\theta$; (e) the apogee velocity; (f) the period of the orbit; (g) the true anomaly when $r = \bar{r}_\theta$; (h) the satellite speed when $r = \bar{r}_\theta$; (i) the flight path angle γ when $r = \bar{r}_\theta$; (j) the maximum flight path angle γ_{max} and the true anomaly at which it occurs.

The strategy is always to go after the primary orbital parameters, eccentricity and angular momentum, first. In this problem we are given the eccentricity, so we will first seek h. Recall from Equation 2.56 that $\mu = 398\,600$ km^3/s^2 and also that $R_E = 6378$ km.

(a) The perigee radius is

$$r_p = R_E + z_p = 6378 + 400 = 6778 \text{ km}$$

Evaluating the orbit formula, Equation 2.35, at $\theta = 0$ (perigee), we get

$$r_p = \frac{h^2}{\mu}\frac{1}{1+e}$$

(Example 2.5 continued)

We use this to evaluate the angular momentum

$$6778 = \frac{h^2}{398\,600} \frac{1}{1+0.6}$$

$$h = 65\,750 \text{ km}^2/\text{s}$$

Now we can find the perigee velocity using the angular momentum formula, Equation 2.21:

$$v_p = v_\perp)_{\text{perigee}} = \frac{h}{r_p} = \frac{65\,750}{6778} = \underline{9.700 \text{ km/s}}$$

(b) The apogee radius is found by evaluating the orbit equation at $\theta = 180°$ (apogee):

$$r_a = \frac{h^2}{\mu} \frac{1}{1-e} = \frac{65\,750^2}{398\,600} \frac{1}{1-0.6} = \underline{27\,110 \text{ km}}$$

(c) The semimajor axis is the average of the perigee and apogee radii:

$$a = \frac{r_p + r_a}{2} = \frac{6778 + 27\,110}{2} = \underline{16\,940 \text{ km}}$$

(d) The azimuth-averaged radius is given by Equation 2.78:

$$\bar{r}_\theta = \sqrt{r_p r_a} = \sqrt{6778 \cdot 27\,110} = \underline{13\,560 \text{ km}}$$

(e) The apogee velocity, like that at perigee, is obtained from the angular momentum formula,

$$v_a = v_\perp)_{\text{apogee}} = \frac{h}{r_a} = \frac{65\,750}{27\,110} = \underline{2.425 \text{ km/s}}$$

(f) To find the orbit period, use Equation 2.73

$$T = \frac{2\pi}{\mu^2} \left(\frac{h}{\sqrt{1-e^2}}\right)^3 = \frac{2\pi}{398\,600^2} \left(\frac{65\,750}{\sqrt{1-0.6^2}}\right)^3 = 21\,950 \text{ s} = \underline{6.098 \text{ hr}}$$

(g) To find the true anomaly when $r = \bar{r}_\theta$, we again use the orbit formula

$$\bar{r}_\theta = \frac{h^2}{\mu} \frac{1}{1+e\cos\theta}$$

$$13\,560 = \frac{65\,750^2}{398\,600} \frac{1}{1+0.6\cos\theta}$$

$$\cos\theta = -0.3333$$

This means

$\theta = \underline{109.5°}$, where the satellite passes through $\bar{r}_\theta$ on its way *from* perigee

and

$\theta = \underline{250.5°}$, where the satellite passes through $\bar{r}_\theta$ on its way *towards* perigee

(h) To find the speed of the satellite when $r = \bar{r}_\theta$, we first calculate the radial and transverse components of velocity:

$$v_\perp = \frac{h}{\bar{r}_\theta} = \frac{65\,750}{13\,560} = 4.850 \text{ km/s}$$

For the radial velocity component, use Equation 2.38,

$$v_r = \frac{\mu}{h} e \sin\theta = \frac{398\,600}{65\,750} \cdot 0.6 \cdot \sin(109.5°) = 3.430 \text{ km/s}$$

or

$$v_r = \frac{\mu}{h} e \sin\theta = \frac{398\,600}{65\,750} \cdot 0.6 \cdot \sin(250.5°) = -3.430 \text{ km/s}$$

The magnitude of the velocity can now be found as

$$v = \sqrt{v_r^2 + v_\perp^2} = \sqrt{3.430^2 + 4.850^2} = \underline{5.940 \text{ km/s}}$$

We could have obtained the speed v more directly by using conservation of energy (Equation 2.71), since the semimajor axis is available from part (c) above. However, we would still need to compute v_r and $v_\perp$ in order to solve the next part of this problem.

(i) Use Equation 2.39 to calculate the flight path angle at $r = \bar{r}_\theta$,

$$\gamma = \tan^{-1} \frac{v_r}{v_\perp} = \tan^{-1} \frac{3.430}{4.850} = \underline{35.26°} \text{ at } \theta = 109.5°$$

γ is positive, meaning the velocity vector is above the local horizon, indicating the spacecraft is flying away from the attracting force. At $\theta = 250.5°$, where the spacecraft is flying towards perigee, $\gamma = -35.26°$. Since the satellite is approaching the attracting body, the velocity vector lies below the local horizon, as indicated by the minus sign.

(j) Equation 2.42 gives the flight path angle in terms of the true anomaly,

$$\gamma = \tan^{-1} \frac{e \sin\theta}{1 + e \cos\theta} \qquad (a)$$

To find where γ is a maximum, we must take the derivative of this expression with respect to θ and set the result equal to zero. Using the rules of calculus,

$$\frac{d\gamma}{d\theta} = \frac{1}{1 + \left(\frac{e \sin\theta}{1 + e \cos\theta}\right)^2} \frac{d}{d\theta}\left(\frac{e \sin\theta}{1 + e \cos\theta}\right) = \frac{e(e + \cos\theta)}{(1 + e \cos\theta)^2 + e^2 \sin^2\theta}$$

For $e < 1$, the denominator is positive for all values of θ. Therefore, $d\gamma/d\theta = 0$ only if the numerator vanishes, that is, if $\cos\theta = -e$. Recall from Equation 2.65 that this true anomaly locates the end-point of the minor axis of the ellipse. The maximum positive flight path angle therefore occurs at the true anomaly,

$$\theta = \cos^{-1}(-0.6) = \underline{126.9°}$$

(Example 2.5 continued)

Substituting this into (a), we find the value of the flight path angle to be

$$\gamma_{max} = \tan^{-1} \frac{0.6 \sin 126.9°}{1 + 0.6 \cos 126.9°} = \underline{36.87°}$$

After attaining this greatest magnitude, the flight path angle starts to decrease steadily towards its value at apogee (zero).

EXAMPLE 2.6

At two points on a geocentric orbit the altitude and true anomaly are $z_1 = 1545$ km, $\theta_1 = 126°$ and $z_2 = 852$ km, $\theta_2 = 58°$, respectively. Find (a) the eccentricity; (b) the altitude of perigee; (c) the semimajor axis; and (d) the period.

(a) The radii of the two points are

$$r_1 = R_E + z_1 = 6378 + 1545 = 7923 \text{ km}$$

$$r_2 = R_E + z_2 = 6378 + 852 = 7230 \text{ km}$$

Applying the orbit formula, Equation 2.35, to both of these points yields two equations for the two primary orbital parameters, angular momentum h and eccentricity e:

$$r_1 = \frac{h^2}{\mu} \frac{1}{1 + e \cos \theta_1}$$

$$7923 = \frac{h^2}{398\,600} \frac{1}{1 + e \cos 126°}$$

$$h^2 = 3.158 \times 10^9 - 1.856 \times 10^9 e \qquad \text{(a)}$$

$$r_2 = \frac{h^2}{\mu} \frac{1}{1 + e \cos \theta_2}$$

$$7230 = \frac{h^2}{398\,600} \frac{1}{1 + e \cos 58°}$$

$$h^2 = 2.882 \times 10^9 + 1.527 \times 10^9 e \qquad \text{(b)}$$

Equating (a) and (b), the two expressions for h^2, yields a single equation for the eccentricity e,

$$3.158 \times 10^9 - 1.856 \times 10^9 e = 2.882 \times 10^9 + 1.527 \times 10^9 e \Rightarrow 3.384 \times 10^9 e$$

$$= 276.2 \times 10^6$$

Therefore,

$$\underline{e = 0.08164} \text{ (an ellipse)} \qquad \text{(c)}$$

(b) By substituting the eccentricity back into (a) [or (b)] we find the angular momentum,

$$h^2 = 3.158 \times 10^9 - 1.856 \times 10^9 \cdot 0.08164 \Rightarrow h = 54\,830 \text{ km}^2/\text{s} \qquad \text{(d)}$$

Now we can use the orbit equation to obtain the perigee radius

$$r_p = \frac{h^2}{\mu}\frac{1}{1+e\cos(0)} = \frac{54\,830^2}{398\,600}\frac{1}{1+0.08164} = \underline{6974\,\text{km}}$$

and perigee altitude

$$z_p = r_p - R_E = 6974 - 6378 = \underline{595.5\,\text{km}}$$

(c) The semimajor axis can be found after we calculate the apogee radius by means of the orbit equation, just as we did for perigee radius:

$$r_a = \frac{h^2}{\mu}\frac{1}{1+e\cos(180°)} = \frac{54\,830^2}{398\,600}\frac{1}{1-0.08164} = 8213\,\text{km}$$

Hence

$$a = \frac{r_p + r_a}{2} = \frac{8213 + 6974}{2} = \underline{7593\,\text{km}}$$

(d) Since the semimajor axis is available, it is convenient to use Equation 2.74 to find the period:

$$T = \frac{2\pi}{\sqrt{\mu}}a^{\frac{3}{2}} = \frac{2\pi}{\sqrt{398\,600}}7593^{\frac{3}{2}} = 6585\,\text{s} = \underline{1.829\,\text{hr}}$$

2.8 PARABOLIC TRAJECTORIES ($e=1$)

If the eccentricity equals 1, then the orbit equation (Equation 2.35) becomes

$$r = \frac{h^2}{\mu}\frac{1}{1+\cos\theta} \qquad (2.79)$$

As the true anomaly θ approaches 180°, the denominator approaches zero, so that r tends towards infinity. According to Equation 2.50, the energy of a trajectory for which $e=1$ is zero, so that for a parabolic trajectory the conservation of energy, Equation 2.47, is

$$\frac{v^2}{2} - \frac{\mu}{r} = 0$$

In other words, the speed anywhere on a parabolic path is

$$v = \sqrt{\frac{2\mu}{r}} \qquad (2.80)$$

If the body m_2 is launched on a parabolic trajectory, it will coast to infinity, arriving there with zero velocity relative to m_1. It will not return. Parabolic paths are therefore called escape trajectories. At a given distance r from m_1, the escape velocity is given by Equation 2.80,

$$v_{esc} = \sqrt{\frac{2\mu}{r}} \qquad (2.81)$$

Let v_o be the speed of a satellite in a circular orbit of radius r. Then from Equations 2.53 and 2.81 we have

$$v_{esc} = \sqrt{2}v_o \qquad (2.82)$$

That is, to escape from a circular orbit requires a velocity boost of 41.4 percent. However, remember our assumption is that m_1 and m_2 are the only objects in the universe. A spacecraft launched from earth with velocity v_{esc} (relative to the earth) will not coast to infinity (i.e., leave the solar system) because it will eventually succumb to the gravitational influence of the sun and, in fact, end up in the same orbit as earth. This will be discussed in more detail in Chapter 8.

For the parabola, Equation 2.42 for the flight path angle takes the form

$$\tan \gamma = \frac{\sin \theta}{1 + \cos \theta}$$

Using the trigonometric identities

$$\sin \theta = 2 \sin \frac{\theta}{2} \cos \frac{\theta}{2}$$

$$\cos \theta = \cos^2 \frac{\theta}{2} - \sin^2 \frac{\theta}{2} = 2 \cos^2 \frac{\theta}{2} - 1$$

we can write

$$\tan \gamma = \frac{2 \sin \frac{\theta}{2} \cos \frac{\theta}{2}}{2 \cos^2 \frac{\theta}{2}} = \frac{\sin \frac{\theta}{2}}{\cos \frac{\theta}{2}} = \tan \frac{\theta}{2}$$

It follows that

$$\gamma = \frac{\theta}{2} \qquad (2.83)$$

That is, on parabolic trajectories the flight path angle is one-half the true anomaly.

Recall that the parameter p of an orbit is given by Equation 2.43. Let us substitute that expression into Equation 2.79 and then plot $r = 2a/(1 + \cos \theta)$ in a cartesian coordinate system centered at the focus, as illustrated in Figure 2.21. From the figure it is clear that

$$x = r \cos \theta = p \frac{\cos \theta}{1 + \cos \theta} \qquad (2.84a)$$

$$y = r \sin \theta = p \frac{\sin \theta}{1 + \cos \theta} \qquad (2.84b)$$

2.8 Parabolic trajectories ($e=1$)

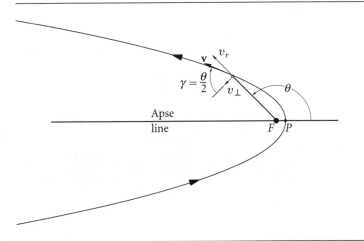

Figure 2.20 Parabolic trajectory around the focus F.

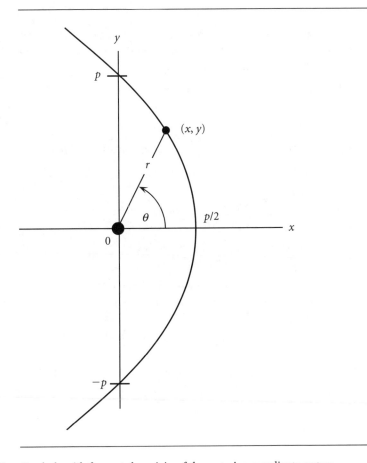

Figure 2.21 Parabola with focus at the origin of the cartesian coordinate system.

68 Chapter 2 *The two-body problem*

Therefore

$$\frac{x}{p/2} + \left(\frac{y}{p}\right)^2 = 2\frac{\cos\theta}{1+\cos\theta} + \frac{\sin^2\theta}{(1+\cos\theta)^2}$$

Working to simplify the right-hand side, we get

$$\frac{x}{p/2} + \left(\frac{y}{p}\right)^2 = \frac{2\cos\theta(1+\cos\theta) + \sin^2\theta}{(1+\cos\theta)^2} = \frac{2\cos\theta + 2\cos^2\theta + (1-\cos^2\theta)}{(1+\cos\theta)^2}$$

$$= \frac{1 + 2\cos\theta + \cos^2\theta}{(1+\cos\theta)^2} = \frac{(1+\cos\theta)^2}{(1+\cos\theta)^2} = 1$$

It follows that

$$x = \frac{p}{2} - \frac{y^2}{2p} \tag{2.85}$$

This is the equation of a parabola in a cartesian coordinate system whose origin serves as the focus.

EXAMPLE 2.7

The perigee of a satellite in a parabolic geocentric trajectory is 7000 km. Find the distance d between points P_1 and P_2 on the orbit which are 8000 km and 16 000 km, respectively, from the center of the earth.

First, let us calculate the angular momentum of the satellite by evaluating the orbit equation at perigee,

$$r_p = \frac{h^2}{\mu}\frac{1}{1+\cos(0)} = \frac{h^2}{2\mu}$$

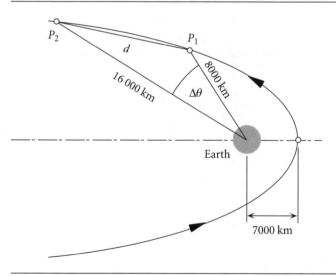

Figure 2.22 Parabolic geocentric trajectory.

from which

$$h = \sqrt{2\mu r_p} = \sqrt{2 \cdot 398\,600 \cdot 7000} = 74\,700 \text{ km}^2/\text{s} \qquad (a)$$

To find the length of the chord $\overline{P_1 P_2}$, we must use the law of cosines from trigonometry,

$$d^2 = 8000^2 + 16\,000^2 - 2 \cdot 8000 \cdot 16\,000 \cos \Delta\theta \qquad (b)$$

The true anomalies of points P_1 and P_2 are found using the orbit equation:

$$8000 = \frac{74\,700^2}{398\,600} \frac{1}{1 + \cos\theta_1} \quad \Rightarrow \quad \cos\theta_1 = 0.75 \quad \Rightarrow \quad \theta_1 = 41.41°$$

$$16\,000 = \frac{74\,700^2}{398\,600} \frac{1}{1 + \cos\theta_2} \quad \Rightarrow \quad \cos\theta_2 = -0.125 \quad \Rightarrow \quad \theta_2 = 97.18°$$

Therefore, $\Delta\theta = 97.18° - 41.41° = 55.78°$, so that (b) yields

$$d = 13\,270 \text{ km} \qquad (c)$$

2.9 Hyperbolic trajectories ($e > 1$)

If $e > 1$, the orbit formula,

$$r = \frac{h^2}{\mu} \frac{1}{1 + e\cos\theta} \qquad (2.86)$$

describes the geometry of the hyperbola shown in Figure 2.23. The system consists of two symmetric curves. One of them is occupied by the orbiting body, the other one is its empty, mathematical image. Clearly, the denominator of Equation 2.86 goes to zero when $\cos\theta = -1/e$. We denote this value of true anomaly

$$\theta_\infty = \cos^{-1}(-1/e) \qquad (2.87)$$

since the radial distance approaches infinity as the true anomaly approaches θ_∞. θ_∞ is known as the true anomaly of the asymptote. Observe that θ_∞ lies between 90° and 180°. From trigonometry it follows that

$$\sin\theta_\infty = \frac{\sqrt{e^2 - 1}}{e} \qquad (2.88)$$

For $-\theta_\infty < \theta < \theta_\infty$, the physical trajectory is the occupied hyperbola I shown on the left in Figure 2.23. For $\theta_\infty < \theta < (360° - \theta_\infty)$, hyperbola II – the vacant orbit around the empty focus F' – is traced out. (The vacant orbit is physically impossible, because it would require a *repulsive* gravitational force.) Periapsis P lies on the apse line on the physical hyperbola I, whereas apoapsis A lies on the apse line on the vacant orbit. The point halfway between periapsis and apoapsis is the center C of the hyperbola. The asymptotes of the hyperbola are the straight lines towards which the curves tend

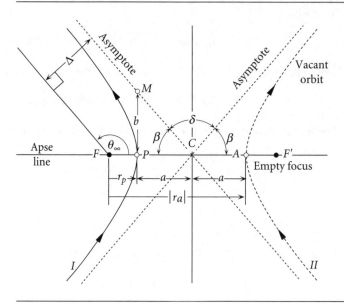

Figure 2.23 Hyperbolic trajectory.

as they approach infinity. The asymptotes intersect at C, making an acute angle β with the apse line, where $\beta = 180° - \theta_\infty$. Therefore, $\cos\beta = -\cos\theta_\infty$, which means

$$\beta = \cos^{-1}(1/e) \qquad (2.89)$$

The angle δ between the asymptotes is called the turn angle. This is the angle through which the velocity vector of the orbiting body is rotated as it rounds the attracting body at F and heads back towards infinity. From the figure we see that $\delta = 180° - 2\beta$, so that

$$\sin\frac{\delta}{2} = \sin\left(\frac{180° - 2\beta}{2}\right) = \sin(90° - \beta) = \cos\beta \stackrel{\text{Eq. 2.89}}{=} \frac{1}{e}$$

or

$$\delta = 2\sin^{-1}(1/e) \qquad (2.90)$$

The distance r_p from the focus F to the periapsis is given by Equation 2.40,

$$r_a = \frac{h^2}{\mu}\frac{1}{1+e} \qquad (2.91)$$

Just as for an ellipse, the radial coordinate r_a of apoapsis is found by setting $\theta = 180°$ in Equation 2.35,

$$r_a = \frac{h^2}{\mu}\frac{1}{1-e} \qquad (2.92)$$

Observe that r_a is negative, since $e > 1$ for the hyperbola. That means the apoapse lies to the right of the focus F. From Figure 2.23 we see that the distance $2a$ from periapse

P to apoapse A is

$$2a = |r_a| - r_p = -r_a - r_p$$

Substituting Equations 2.91 and 2.92 yields

$$2a = -\frac{h^2}{\mu}\left(\frac{1}{1-e} + \frac{1}{1+e}\right)$$

From this it follows that a, the semimajor axis of the hyperbola, is given by an expression which is nearly identical to that for an ellipse (Equation 2.62),

$$a = \frac{h^2}{\mu}\frac{1}{e^2 - 1} \tag{2.93}$$

Therefore, Equation 2.86 may be written for the hyperbola

$$r = a\frac{e^2 - 1}{1 + e\cos\theta} \tag{2.94}$$

This formula is analogous to Equation 2.63 for the elliptical orbit. Furthermore, from Equation 2.94 it follows that

$$r_p = a(e - 1) \tag{2.95a}$$

$$r_a = -a(e + 1) \tag{2.95b}$$

The distance b from periapsis to an asymptote, measured perpendicular to the apse line, is the semiminor axis of the hyperbola. From Figure 2.23, we see that the length b of the semiminor axis $\overline{PM}$ is

$$b = a\tan\beta = a\frac{\sin\beta}{\cos\beta} = a\frac{\sin(180 - \theta_\infty)}{\cos(180 - \theta_\infty)} = a\frac{\sin\theta_\infty}{-\cos\theta_\infty} = a\frac{\frac{\sqrt{e^2-1}}{e}}{-\left(-\frac{1}{e}\right)}$$

so that for the hyperbola,

$$b = a\sqrt{e^2 - 1} \tag{2.96}$$

This relation is analogous to Equation 2.67 for the semiminor axis of an ellipse.

The distance Δ between the asymptote and a parallel line through the focus is called the aiming radius, which is illustrated in Figure 2.23. From that figure we see that

$$\Delta = (r_p + a)\sin\beta$$
$$\quad = ae\sin\beta \qquad \text{(Equation 2.95a)}$$
$$\quad = ae\frac{\sqrt{e^2 - 1}}{e} \qquad \text{(Equation 2.89)}$$
$$\quad = ae\sin\theta_\infty \qquad \text{(Equation 2.88)}$$
$$\quad = ae\sqrt{1 - \cos^2\theta_\infty} \qquad \text{(trig identity)}$$
$$\quad = ae\sqrt{1 - \frac{1}{e^2}} \qquad \text{(Equation 2.87)}$$

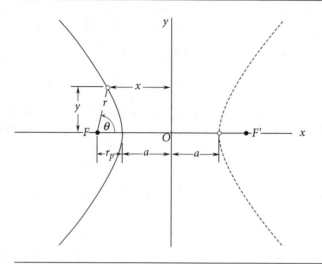

Figure 2.24 Plot of Equation 2.93 in a cartesian coordinate system with origin O midway between the two foci.

or

$$\Delta = a\sqrt{e^2 - 1} \qquad (2.97)$$

Comparing this result with Equation 2.96, it is clear that the aiming radius equals the length of the semiminor axis of the hyperbola.

As with the ellipse and the parabola, we can express the polar form of the equation of the hyperbola in a cartesian coordinate system whose origin is in this case midway between the two foci, as illustrated in Figure 2.24. From the figure it is apparent that

$$x = -a - r_p + r \cos \theta \qquad (2.98a)$$

$$y = r \sin \theta \qquad (2.98b)$$

Using Equations 2.94 and 2.95a in 2.98a, we obtain

$$x = -a - a(e-1) + a \frac{e^2 - 1}{1 + e \cos \theta} \cos \theta = -a \frac{e + \cos \theta}{1 + e \cos \theta}$$

Substituting Equations 2.94 and 2.96 into 2.98b yields

$$y = \frac{b}{\sqrt{e^2 - 1}} \frac{e^2 - 1}{1 + e \cos \theta} \sin \theta = b \frac{\sqrt{e^2 - 1} \sin \theta}{1 + e \cos \theta}$$

It follows that

$$\frac{x^2}{a^2} - \frac{y^2}{b^2} = \left(\frac{e + \cos \theta}{1 + e \cos \theta} \right)^2 - \left(\frac{\sqrt{e^2 - 1} \sin \theta}{1 + e \cos \theta} \right)^2$$

$$= \frac{e^2 + 2e \cos \theta + \cos^2 \theta - (e^2 - 1)(1 - \cos^2 \theta)}{(1 + e \cos \theta)^2}$$

$$= \frac{1 + 2e \cos \theta + e^2 \cos^2 \theta}{(1 + e \cos \theta)^2} = \frac{(1 + e \cos \theta)^2}{(1 + e \cos \theta)^2}$$

That is,

$$\frac{x^2}{a^2} - \frac{y^2}{b^2} = 1 \qquad (2.99)$$

This is the familiar equation of a hyperbola which is symmetric about the x and y axes, with intercepts on the x axis.

The specific energy of the hyperbolic trajectory is given by Equation 2.50. Substituting Equation 2.93 into that expression yields

$$\varepsilon = \frac{\mu}{2a} \qquad (2.100)$$

The specific energy of a hyperbolic orbit is clearly positive and independent of the eccentricity. The conservation of energy for a hyperbolic trajectory is

$$\frac{v^2}{2} - \frac{\mu}{r} = \frac{\mu}{2a} \qquad (2.101)$$

Let v_∞ denote the speed at which a body on a hyperbolic path arrives at infinity. According to Equation 2.101

$$v_\infty = \sqrt{\frac{\mu}{a}} \qquad (2.102)$$

v_∞ is called the hyperbolic excess speed. In terms of v_∞ we may write Equation 2.101 as

$$\frac{v^2}{2} - \frac{\mu}{r} = \frac{v_\infty^2}{2}$$

Substituting the expression for escape speed, $v_{esc} = \sqrt{2\mu/r}$ (Equation 2.81), we obtain for a hyperbolic trajectory

$$v^2 = v_{esc}^2 + v_\infty^2 \qquad (2.103)$$

This equation clearly shows that the hyperbolic excess speed v_∞ represents the excess kinetic energy over that which is required to simply escape from the center of attraction. The square of v_∞ is denoted C_3, and is known as the characteristic energy,

$$C_3 = v_\infty^2 \qquad (2.104)$$

C_3 is a measure of the energy required for an interplanetary mission and C_3 is also a measure of the maximum energy a launch vehicle can impart to a spacecraft of a given mass. Obviously, to match a launch vehicle with a mission, $C_3)_{\text{launchvehicle}} > C_3)_{\text{mission}}$.

Note that the hyperbolic excess speed can also be obtained from Equations 2.39 and 2.88,

$$v_\infty = \frac{\mu}{h} e \sin\theta_\infty = \frac{\mu}{h}\sqrt{e^2 - 1} \qquad (2.105)$$

Finally, for purposes of comparison, Figure 2.25 shows a range of trajectories, from a circle through hyperbolas, all having a common focus and periapsis. The parabola is the demarcation between the closed, negative energy orbits (ellipses) and open, positive energy orbits (hyperbolas).

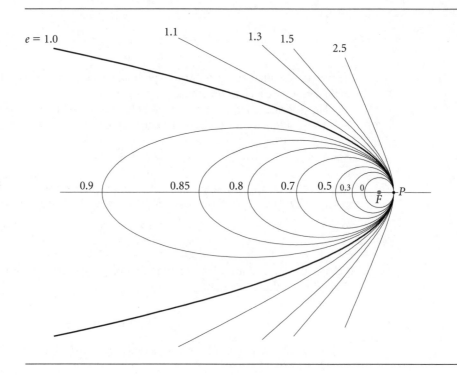

Figure 2.25 Orbits of various eccentricities, having a common focus F and periapsis P.

EXAMPLE 2.8

At a given point of a spacecraft's geocentric trajectory, the radius is 14 600 km, the speed is 8.6 km/s, and the flight path angle is 50°. Show that the path is a hyperbola and calculate the following: (a) C_3, (b) angular momentum, (c) true anomaly, (d) eccentricity, (e) radius of perigee, (f) turn angle, (g) semimajor axis, and (h) aiming radius.

To determine the type of the trajectory, calculate the escape speed at the given radius:

$$v_{esc} = \sqrt{\frac{2\mu}{r}} = \sqrt{\frac{2 \cdot 398\,600}{14\,600}} = 7.389 \text{ km/s}$$

Since the escape speed is less than the spacecraft's speed of 8.6 km/s, the path is a hyperbola.

(a) The hyperbolic excess velocity v_∞ is found from Equation 2.103,

$$v_\infty^2 = v^2 - v_{esc}^2 = 8.6^2 - 7.389^2 = 19.36 \text{ km}^2/\text{s}^2$$

From Equation 2.104 it follows that

$$C_3 = 19.36 \text{ km}^2/\text{s}^2$$

(b) Knowing the speed and the flight path angle, we can obtain both v_r and $v_\perp$:

$$v_r = v \sin\gamma = 8.6 \sin 50° = 6.588 \text{ km/s} \qquad \text{(a)}$$

2.9 Hyperbolic trajectories ($e > 1$) 75

$$v_\perp = v \cos \gamma = 8.6 \cdot \cos 50° = 5.528 \text{ km/s} \quad \text{(b)}$$

Then Equation 2.21 provides us with the angular momentum,

$$h = r v_\perp = 14\,600 \cdot 5.528 = \underline{80\,710 \text{ km}^2/\text{s}} \quad \text{(c)}$$

(c) Evaluating the orbit equation at the given location on the trajectory, we get

$$14\,600 = \frac{80\,710^2}{398\,600} \frac{1}{1 + e \cos \theta}$$

from which

$$e \cos \theta = 0.1193 \quad \text{(d)}$$

The radial component of velocity is given by Equation 2.39, $v_r = \mu e \sin \theta / h$, so that with (a) and (c), we obtain

$$6.588 = \frac{398\,600}{80\,170} e \sin \theta$$

or

$$e \sin \theta = 1.334 \quad \text{(e)}$$

Computing the ratio of (e) to (d) yields

$$\tan \theta = \frac{1.334}{0.1193} = 11.18 \Rightarrow \underline{\theta = 84.89°}$$

(d) We substitute the true anomaly back into either (d) or (e) to find the eccentricity,

$$\underline{e = 1.339}$$

(e) The radius of perigee can now be found from the orbit equation,

$$r_p = \frac{h^2}{\mu} \frac{1}{1 + e \cos(0)} = \frac{80\,710^2}{398\,600} \frac{1}{1 + 1.339} = \underline{6986 \text{ km}}$$

(f) The formula for turn angle is Equation 2.90, from which

$$\delta = 2 \sin^{-1}\left(\frac{1}{e}\right) = 2 \sin^{-1}\left(\frac{1}{1.339}\right) = \underline{96.60°}$$

(g) The semimajor axis of the hyperbola is found in Equation 2.93,

$$a = \frac{h^2}{\mu} \frac{1}{e^2 - 1} = \frac{80\,710^2}{398\,600} \frac{1}{1.339^2 - 1} = \underline{20\,590 \text{ km}}$$

(h) According to Equations 2.96 and 2.97, the aiming radius is

$$\Delta = a\sqrt{e^2 - 1} = 20\,590\sqrt{1.339^2 - 1} = \underline{18\,340 \text{ km}}$$

2.10 PERIFOCAL FRAME

The perifocal frame is the 'natural frame' for an orbit. It is centered at the focus of the orbit. Its $\bar{x}\bar{y}$ plane is the plane of the orbit, and its x axis is directed from the focus through periapse, as illustrated in Figure 2.26. The unit vector along the $\bar{x}$ axis (the apse line) is denoted $\hat{\mathbf{p}}$. The $\bar{y}$ axis, with unit vector $\hat{\mathbf{q}}$, lies at 90° true anomaly to the $\bar{x}$ axis. The $\bar{z}$ axis is normal to the plane of the orbit in the direction of the angular momentum vector $\mathbf{h}$. The $\bar{z}$ unit vector is $\hat{\mathbf{w}}$,

$$\hat{\mathbf{w}} = \frac{\mathbf{h}}{h} \qquad (2.106)$$

In the perifocal frame, the position vector $\mathbf{r}$ is written (see Figure 2.27)

$$\mathbf{r} = \bar{x}\hat{\mathbf{p}} + \bar{y}\hat{\mathbf{q}} \qquad (2.107)$$

where

$$\bar{x} = r\cos\theta \qquad \bar{y} = r\sin\theta \qquad (2.108)$$

and r, the magnitude of $\mathbf{r}$, is given by the orbit equation, $r = (h^2/\mu)[1/(1+e\cos\theta)]$. Thus, we may write Equation 2.107 as

$$\mathbf{r} = \frac{h^2}{\mu}\frac{1}{1+e\cos\theta}(\cos\theta\hat{\mathbf{p}} + \sin\theta\hat{\mathbf{q}}) \qquad (2.109)$$

The velocity is found by taking the time derivative of $\mathbf{r}$,

$$\mathbf{v} = \dot{\mathbf{r}} = \dot{\bar{x}}\hat{\mathbf{p}} + \dot{\bar{y}}\hat{\mathbf{q}} \qquad (2.110)$$

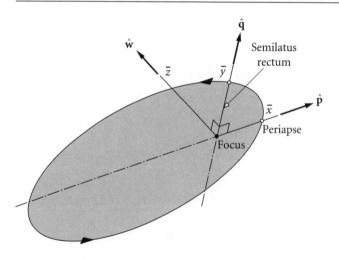

Figure 2.26 Perifocal frame $\hat{\mathbf{p}}\hat{\mathbf{q}}\hat{\mathbf{w}}$.

From Equations 2.108 we obtain

$$\dot{\bar{x}} = \dot{r}\cos\theta - r\dot{\theta}\sin\theta$$
$$\dot{\bar{y}} = \dot{r}\sin\theta + r\dot{\theta}\cos\theta \quad (2.111)$$

$\dot{r}$ is the radial component of velocity, v_r. Therefore, according to Equation 2.39,

$$\dot{r} = \frac{\mu}{h}e\sin\theta \quad (2.112)$$

From Equations 2.36 and 2.38 we have

$$r\dot{\theta} = v_\perp = \frac{\mu}{h}(1 + e\cos\theta) \quad (2.113)$$

Substituting Equations 2.112 and 2.113 into 2.111 and simplifying the results yields

$$\dot{\bar{x}} = -\frac{\mu}{h}\sin\theta$$
$$\dot{\bar{y}} = \frac{\mu}{h}(e + \cos\theta) \quad (2.114)$$

Hence, Equation 2.110 becomes

$$\mathbf{v} = \frac{\mu}{h}[-\sin\theta\hat{\mathbf{p}} + (e + \cos\theta)\hat{\mathbf{q}}] \quad (2.115)$$

Formulating the kinematics of orbital motion in the perifocal frame, as we have done here, is a prelude to the study of orbits in three dimensions (Chapter 4). We also need Equations 2.107 and 2.110 in the next section.

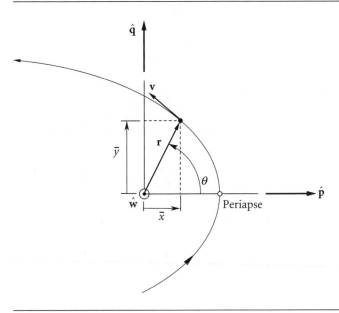

Figure 2.27 Position and velocity relative to the perifocal frame.

2.11 THE LAGRANGE COEFFICIENTS

In this section we will establish what may seem intuitively obvious: if the position and velocity of an orbiting body are known at a given instant, then the position and velocity at any later time are found in terms of the initial values. Let us start with Equations 2.107 and 2.110,

$$\mathbf{r} = \bar{x}\hat{\mathbf{p}} + \bar{y}\hat{\mathbf{q}} \tag{2.116}$$

$$\mathbf{v} = \dot{\mathbf{r}} = \dot{\bar{x}}\hat{\mathbf{p}} + \dot{\bar{y}}\hat{\mathbf{q}} \tag{2.117}$$

Attach a subscript 'zero' to quantities evaluated at time $t = t_0$. Then the expressions for $\mathbf{r}$ and $\mathbf{v}$ evaluated at $t = t_0$ are

$$\mathbf{r}_0 = \bar{x}_0\hat{\mathbf{p}} + \bar{y}_0\hat{\mathbf{q}} \tag{2.118}$$

$$\mathbf{v}_0 = \dot{\bar{x}}_0\hat{\mathbf{p}} + \dot{\bar{y}}_0\hat{\mathbf{q}} \tag{2.119}$$

The angular momentum $\mathbf{h}$ is constant, so let us calculate it using the initial conditions. Substituting Equations 2.118 and 2.119 into Equation 2.18 yields

$$\mathbf{h} = \mathbf{r}_0 \times \mathbf{v}_0 = \begin{vmatrix} \hat{\mathbf{p}} & \hat{\mathbf{q}} & \hat{\mathbf{w}} \\ \bar{x}_0 & \bar{y}_0 & 0 \\ \dot{\bar{x}}_0 & \dot{\bar{y}}_0 & 0 \end{vmatrix} = \hat{\mathbf{w}}(\bar{x}_0\dot{\bar{y}}_0 - \bar{y}_0\dot{\bar{x}}_0) \tag{2.120}$$

Recall that $\hat{\mathbf{w}}$ is the unit vector in the direction of $\mathbf{h}$ (Equation 2.106). Therefore, the coefficient of $\hat{\mathbf{w}}$ on the right of Equation 2.120 must be the magnitude of the angular momentum. That is,

$$h = \bar{x}_0\dot{\bar{y}}_0 - \bar{y}_0\dot{\bar{x}}_0 \tag{2.121}$$

Now let us solve the two vector equations (2.118) and (2.119) for the unit vectors $\hat{\mathbf{p}}$ and $\hat{\mathbf{q}}$ in terms of $\mathbf{r}_0$ and $\mathbf{v}_0$. From (2.118) we get

$$\hat{\mathbf{q}} = \frac{1}{\bar{y}_0}\mathbf{r}_0 - \frac{\bar{x}_0}{\bar{y}_0}\hat{\mathbf{p}} \tag{2.122}$$

Substituting this into Equation (2.119), combining terms and using Equation 2.121 yields

$$\mathbf{v}_0 = \dot{\bar{x}}_0\hat{\mathbf{p}} + \dot{\bar{y}}_0\left(\frac{1}{\bar{y}_0}\mathbf{r}_0 - \frac{\bar{x}_0}{\bar{y}_0}\hat{\mathbf{p}}\right) = \frac{\bar{y}_0\dot{\bar{x}}_0 - \bar{x}_0\dot{\bar{y}}_0}{\bar{y}_0}\hat{\mathbf{p}} + \frac{\dot{\bar{y}}_0}{\bar{y}_0}\mathbf{r}_0 = -\frac{h}{\bar{y}_0}\hat{\mathbf{p}} + \frac{\dot{\bar{y}}_0}{\bar{y}_0}\mathbf{r}_0$$

Solve this for $\hat{\mathbf{p}}$ to obtain

$$\hat{\mathbf{p}} = \frac{\dot{\bar{y}}_0}{h}\mathbf{r}_0 - \frac{\bar{y}_0}{h}\mathbf{v}_0 \tag{2.123}$$

Putting this result back into Equation 2.122 gives

$$\hat{\mathbf{q}} = \frac{1}{\bar{y}_0}\mathbf{r}_0 - \frac{\bar{x}_0}{\bar{y}_0}\left(\frac{\dot{\bar{y}}_0}{h}\mathbf{r}_0 - \frac{\bar{y}_0}{h}\mathbf{v}_0\right) = \frac{h - \bar{x}_0\dot{\bar{y}}_0}{\bar{y}_0}\mathbf{r}_0 + \frac{\bar{x}_0}{h}\mathbf{v}_0$$

Upon replacing h by the right-hand side of Equation 2.121 we get

$$\hat{\mathbf{q}} = -\frac{\dot{\bar{x}}_0}{h}\mathbf{r}_0 + \frac{\bar{x}_0}{h}\mathbf{v}_0 \qquad (2.124)$$

Equations 2.123 and 2.124 give $\hat{\mathbf{p}}$ and $\hat{\mathbf{q}}$ in terms of the initial position and velocity. Substituting those two expressions back into Equations 2.116 and 2.117 yields, respectively

$$\mathbf{r} = \bar{x}\left(\frac{\dot{\bar{y}}_0}{h}\mathbf{r}_0 - \frac{\bar{y}_0}{h}\mathbf{v}_0\right) + \bar{y}\left(-\frac{\dot{\bar{x}}_0}{h}\mathbf{r}_0 + \frac{\bar{x}_0}{h}\mathbf{v}_0\right) = \frac{\bar{x}\dot{\bar{y}}_0 - \bar{y}\dot{\bar{x}}_0}{h}\mathbf{r}_0 + \frac{-\bar{x}\bar{y}_0 + \bar{y}\bar{x}_0}{h}\mathbf{v}_0$$

$$\mathbf{v} = \dot{\bar{x}}\left(\frac{\dot{\bar{y}}_0}{h}\mathbf{r}_0 - \frac{\bar{y}_0}{h}\mathbf{v}_0\right) + \dot{\bar{y}}\left(-\frac{\dot{\bar{x}}_0}{h}\mathbf{r}_0 + \frac{\bar{x}_0}{h}\mathbf{v}_0\right) = \frac{\dot{\bar{x}}\dot{\bar{y}}_0 - \dot{\bar{y}}\dot{\bar{x}}_0}{h}\mathbf{r}_0 + \frac{-\dot{\bar{x}}\bar{y}_0 + \dot{\bar{y}}\bar{x}_0}{h}\mathbf{v}_0$$

Therefore,

$$\mathbf{r} = f\mathbf{r}_0 + g\mathbf{v}_0 \qquad (2.125)$$

$$\mathbf{v} = \dot{f}\mathbf{r}_0 + \dot{g}\mathbf{v}_0 \qquad (2.126)$$

where f and g are given by

$$f = \frac{\bar{x}\dot{\bar{y}}_0 - \bar{y}\dot{\bar{x}}_0}{h} \qquad (2.127a)$$

$$g = \frac{-\bar{x}\bar{y}_0 + \bar{y}\bar{x}_0}{h} \qquad (2.127b)$$

together with their time derivatives

$$\dot{f} = \frac{\dot{\bar{x}}\dot{\bar{y}}_0 - \dot{\bar{y}}\dot{\bar{x}}_0}{h} \qquad (2.128a)$$

$$\dot{g} = \frac{-\dot{\bar{x}}\bar{y}_0 + \dot{\bar{y}}\bar{x}_0}{h} \qquad (2.128b)$$

The f and g functions are referred to as the Lagrange coefficients after Joseph-Louis Lagrange (1736–1813), a French mathematical physicist whose numerous contributions include calculations of planetary motion.

From Equations 2.125 and 2.126 we see that the position and velocity vectors $\mathbf{r}$ and $\mathbf{v}$ are indeed linear combinations of the initial position and velocity vectors. The Lagrange coefficients and their time derivatives in these expressions are themselves functions of time and the initial conditions.

Before proceeding, let us show that the conservation of angular momentum $\mathbf{h}$ imposes a condition on f and g and their time derivatives $\dot{f}$ and $\dot{g}$. Calculate $\mathbf{h}$ using Equations 2.125 and 2.126,

$$\mathbf{h} = \mathbf{r} \times \mathbf{v} = (f\mathbf{r}_0 + g\mathbf{v}_0) \times (\dot{f}\mathbf{r}_0 + \dot{g}\mathbf{v}_0)$$

Expanding the right-hand side yields

$$\mathbf{h} = (f\mathbf{r}_0 \times \dot{f}\mathbf{r}_0) + (f\mathbf{r}_0 \times \dot{g}\mathbf{v}_0) + (g\mathbf{v}_0 \times \dot{f}\mathbf{r}_0) + (g\mathbf{v}_0 \times \dot{g}\mathbf{v}_0)$$

Factoring out the scalars $f, g, \dot{f}$ and $\dot{g}$, we get

$$\mathbf{h} = f\dot{f}(\mathbf{r}_0 \times \mathbf{r}_0) + f\dot{g}(\mathbf{r}_0 \times \mathbf{v}_0) + \dot{f}g(\mathbf{v}_0 \times \mathbf{r}_0) + g\dot{g}(\mathbf{v}_0 \times \mathbf{v}_0)$$

But

$$\mathbf{r}_0 \times \mathbf{r}_0 = \mathbf{v}_0 \times \mathbf{v}_0 = 0, \quad \text{so}$$

$$\mathbf{h} = f\dot{g}(\mathbf{r}_0 \times \mathbf{v}_0) + \dot{f}g(\mathbf{v}_0 \times \mathbf{r}_0)$$

Since

$$\mathbf{v}_0 \times \mathbf{r}_0 = -(\mathbf{r}_0 \times \mathbf{v}_0)$$

this reduces to

$$\mathbf{h} = (f\dot{g} - \dot{f}g)(\mathbf{r}_0 \times \mathbf{v}_0)$$

or

$$\mathbf{h} = (f\dot{g} - \dot{f}g)\mathbf{h}_0$$

where $\mathbf{h}_0 = \mathbf{r}_0 \times \mathbf{v}_0$, which is the angular momentum at $t = t_0$. But the angular momentum is constant (recall Equation 2.19), which means $\mathbf{h} = \mathbf{h}_0$, so that

$$\mathbf{h} = (f\dot{g} - \dot{f}g)\mathbf{h}$$

Since $\mathbf{h}$ cannot be zero (unless the body is traveling in a straight line towards the center of attraction), it follows that

$$f\dot{g} - \dot{f}g = 1 \quad \text{(conservation of angular momentum)} \quad (2.129)$$

Thus, if any three of the functions $f, g, \dot{f}$ and $\dot{g}$ are known, the fourth may be found from Equation 2.129.

Let us use Equations 2.127 and 2.128 to evaluate the Lagrange coefficients and their time derivatives in terms of the true anomaly. First of all, note that evaluating Equations 2.108 at time $t = t_0$ yields

$$\begin{aligned} \bar{x}_0 &= r_0 \cos\theta_0 \\ \bar{y}_0 &= r_0 \sin\theta_0 \end{aligned} \quad (2.130)$$

Likewise, from Equations 2.114 we get

$$\begin{aligned} \dot{\bar{x}}_0 &= -\frac{\mu}{h}\sin\theta_0 \\ \dot{\bar{y}}_0 &= \frac{\mu}{h}(e + \cos\theta_0) \end{aligned} \quad (2.131)$$

To evaluate the function f, we substitute Equations 2.108 and 2.131 into Equation 2.127a,

$$\begin{aligned} f &= \frac{\bar{x}\dot{\bar{y}}_0 - \bar{y}\dot{\bar{x}}_0}{h} \\ &= \frac{1}{h}\left\{[r\cos\theta]\left[\frac{\mu}{h}(e + \cos\theta_0)\right] - [r\sin\theta]\left[-\frac{\mu}{h}\sin\theta_0\right]\right\} \\ &= \frac{\mu r}{h^2}[e\cos\theta + (\cos\theta\cos\theta_0 + \sin\theta\sin\theta_0)] \end{aligned} \quad (2.132)$$

2.11 The Lagrange coefficients

If we invoke the trig identity

$$\cos(\theta - \theta_0) = \cos\theta\cos\theta_0 + \sin\theta\sin\theta_0 \tag{2.133}$$

and let $\Delta\theta$ represent the difference between the current and initial true anomalies,

$$\Delta\theta = \theta - \theta_0 \tag{2.134}$$

then Equation 2.132 reduces to

$$f = \frac{\mu r}{h^2}(e\cos\theta + \cos\Delta\theta) \tag{2.135}$$

Finally, from Equation 2.35, we have

$$e\cos\theta = \frac{h^2}{\mu r} - 1 \tag{2.136}$$

Substituting this into Equation 2.135 leads to

$$f = 1 - \frac{\mu r}{h^2}(1 - \cos\Delta\theta) \tag{2.137}$$

We obtain r from the orbit formula, Equation 2.35, in which the true anomaly θ appears, whereas the difference in the true anomalies occurs on the right-hand side of Equation 2.137. However, we can express the orbit equation in terms of the difference in true anomalies as follows. From Equation 2.134 we have $\theta = \theta_0 + \Delta\theta$, which means we can write the orbit equation as

$$r = \frac{h^2}{\mu}\frac{1}{1 + e\cos(\theta_0 + \Delta\theta)} \tag{2.138}$$

By replacing θ_0 by $-\Delta\theta$ in Equation 2.133, Equation 2.138 becomes

$$r = \frac{h^2}{\mu}\frac{1}{1 + e\cos\theta_0\cos\Delta\theta - e\sin\theta_0\sin\Delta\theta} \tag{2.139}$$

To remove θ_0 from this expression, observe first of all that Equation 2.136 implies that, at $t = t_0$,

$$e\cos\theta_0 = \frac{h^2}{\mu r_0} - 1 \tag{2.140}$$

Furthermore, from Equation 2.39 for the radial velocity we obtain

$$e\sin\theta_0 = \frac{hv_{r0}}{\mu} \tag{2.141}$$

Substituting Equations 2.140 and 2.141 into 2.139 yields

$$r = \frac{h^2}{\mu}\frac{1}{1 + \left(\frac{h^2}{\mu r_0} - 1\right)\cos\Delta\theta - \frac{hv_{r0}}{\mu}\sin\Delta\theta} \tag{2.142}$$

Using this form of the orbit equation, we can find r in terms of the initial conditions and the change in the true anomaly. Thus f in Equation 2.137 depends only on $\Delta\theta$.

The Lagrange coefficient g is found by substituting Equations 2.108 and 2.130 into Equation 2.127b,

$$\begin{aligned} g &= \frac{-\bar{x}\bar{y}_0 + \bar{y}\bar{x}_0}{h} \\ &= \frac{1}{h}[(-r\cos\theta)(r_0\sin\theta_0) + (r\sin\theta)(r_0\cos\theta_0)] \\ &= \frac{rr_0}{h}(\sin\theta\cos\theta_0 - \cos\theta\sin\theta_0) \end{aligned} \quad (2.143)$$

Making use of the trig identity

$$\sin(\theta - \theta_0) = \sin\theta\cos\theta_0 - \cos\theta\sin\theta_0$$

together with Equation 2.134, we find

$$g = \frac{rr_0}{h}\sin(\Delta\theta) \quad (2.144)$$

To obtain $\dot{g}$, substitute Equations 2.114 and 2.130 into Equation 2.128b,

$$\begin{aligned} \dot{g} &= \frac{-\dot{\bar{x}}\bar{y}_0 + \dot{\bar{y}}\bar{x}_0}{h} \\ &= \frac{1}{h}\left\{-\left[-\frac{\mu}{h}\sin\theta\right][r_0\sin\theta_0] + \left[\frac{\mu}{h}(e+\cos\theta)\right](r_0\cos\theta_0)\right\} \\ &= \frac{\mu r_0}{h^2}[e\cos\theta_0 + (\cos\theta\cos\theta_0 + \sin\theta\sin\theta_0)] \end{aligned}$$

With the aid of Equations 2.133 and 2.140, this reduces to

$$\dot{g} = 1 - \frac{\mu r_0}{h^2}(1 - \cos\Delta\theta) \quad (2.145)$$

$\dot{f}$ can be found using Equation 2.129. Thus

$$\dot{f} = \frac{1}{g}(f\dot{g} - 1) \quad (2.146)$$

Substituting Equations 2.137, 2.143 and 2.145 results in

$$\begin{aligned} \dot{f} &= \frac{1}{\frac{rr_0}{h}\sin\Delta\theta}\left\{\left[1 - \frac{\mu r}{h^2}(1 - \cos\Delta\theta)\right]\left[1 - \frac{\mu r_0}{h^2}(1 - \cos\Delta\theta)\right] - 1\right\} \\ &= \frac{1}{\frac{rr_0}{h}\sin\Delta\theta}\frac{h^2\mu rr_0}{h^4}\left[(1 - \cos\Delta\theta)^2\frac{\mu}{h^2} - (1 - \cos\Delta\theta)\left(\frac{1}{r_0} + \frac{1}{r}\right)\right] \end{aligned}$$

or

$$\dot{f} = \frac{\mu}{h}\frac{1 - \cos\Delta\theta}{\sin\Delta\theta}\left[\frac{\mu}{h^2}(1 - \cos\Delta\theta) - \frac{1}{r_0} - \frac{1}{r}\right] \quad (2.147)$$

2.11 The Lagrange coefficients

To summarize, the Lagrange coefficients in terms of the change in true anomaly are

$$f = 1 - \frac{\mu r}{h^2}(1 - \cos \Delta \theta) \tag{2.148a}$$

$$g = \frac{r r_0}{h} \sin \Delta \theta \tag{2.148b}$$

$$\dot{f} = \frac{\mu}{h} \frac{1 - \cos \Delta \theta}{\sin \Delta \theta} \left[\frac{\mu}{h^2}(1 - \cos \Delta \theta) - \frac{1}{r_0} - \frac{1}{r} \right] \tag{2.148c}$$

$$\dot{g} = 1 - \frac{\mu r_0}{h^2}(1 - \cos \Delta \theta) \tag{2.148d}$$

where r is given by Equation 2.142.

Observe that using the Lagrange coefficients to determine the position and velocity from the initial conditions does not require knowing the type of orbit we are dealing with (ellipse, parabola, hyperbola), since the eccentricity does not appear in Equations 2.142 and 2.148. However, the initial position and velocity give us that information. From $\mathbf{r}_0$ and $\mathbf{v}_0$ we obtain the angular momentum $h = |\mathbf{r}_0 \times \mathbf{v}_0|$. The initial radius r_0 is just the magnitude of the vector $\mathbf{r}_0$. The initial radial velocity v_{r0} is the projection of $\mathbf{v}_0$ onto the direction of $\mathbf{r}_0$,

$$v_{r0} = \mathbf{v}_0 \cdot \frac{\mathbf{r}_0}{r_0}$$

From Equations 2.35 and 2.39 we have

$$r_0 = \frac{h^2}{\mu} \frac{1}{1 + e \cos \theta_0} \qquad v_{r0} = \frac{\mu}{h} e \sin \theta_0 \tag{2.149}$$

These two equations can be solved for the eccentricity e and the true anomaly of the initial point θ_0.

EXAMPLE 2.9

An earth satellite moves in the xy plane of an inertial frame with origin at the earth's center. Relative to that frame, the position and velocity of the satellite at time t_0 are

$$\mathbf{r}_0 = 8182.4 \hat{\mathbf{i}} - 6865.9 \hat{\mathbf{j}} \text{ (km)}$$
$$\mathbf{v}_0 = 0.47572 \hat{\mathbf{i}} + 8.8116 \hat{\mathbf{j}} \text{ (km/s)} \tag{a}$$

Compute the position and velocity vectors after the satellite has traveled through a true anomaly of 120°.

First, use $\mathbf{r}_0$ and $\mathbf{v}_0$ to calculate the angular momentum of the satellite:

$$\mathbf{h} = \mathbf{r}_0 \times \mathbf{v}_0 = \begin{vmatrix} \hat{\mathbf{i}} & \hat{\mathbf{j}} & \hat{\mathbf{k}} \\ 8182.4 & -6865.9 & 0 \\ 0.47572 & 8.8116 & 0 \end{vmatrix} = 75\,366 \hat{\mathbf{k}} \text{ (km}^2/\text{s)}$$

so that

$$h = 75\,366 \text{ km}^2/\text{s} \tag{b}$$

(*Example 2.9 continued*)

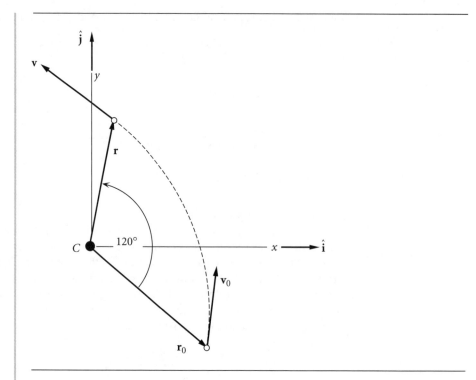

Figure 2.28 The initial and final position and velocity vectors.

The magnitude of the position vector $\mathbf{r}_0$ is

$$r_0 = \sqrt{\mathbf{r}_0 \cdot \mathbf{r}_0} = 10\,861 \text{ km} \qquad \text{(c)}$$

The initial radial velocity v_{r0} is found by projecting the velocity $\mathbf{v}_0$ onto the unit vector in the radial direction $\mathbf{r}_0$,

$$v_{r0} = \mathbf{v}_0 \cdot \frac{\mathbf{r}_0}{r_0} = \frac{(0.47572\hat{\mathbf{i}} + 8.8116\hat{\mathbf{j}}) \cdot (8182.4\hat{\mathbf{i}} - 6865.9\hat{\mathbf{j}})}{10\,681} = -5.2996 \text{ km/s} \qquad \text{(d)}$$

The final distance r is obtained from Equation 2.142,

$$r = \frac{h^2}{\mu} \frac{1}{1 + \left(\dfrac{h^2}{\mu r_0} - 1\right)\cos\Delta\theta - \dfrac{h v_{r0}}{\mu}\sin\Delta\theta}$$

$$= \frac{75\,366^2}{398\,600} \frac{1}{1 + \left(\dfrac{75\,366^2}{398\,600 \cdot 10\,681} - 1\right)\cos 120° - \dfrac{75\,366 \cdot (-5.2995)}{398\,600}\sin 120°}$$

so that

$$r = 8378.8 \text{ km} \qquad \text{(e)}$$

2.11 The Lagrange coefficients

Now we can evaluate the Lagrange coefficients in Equations 2.148:

$$f = 1 - \frac{\mu r}{h^2}(1 - \cos \Delta\theta)$$

$$= 1 - \frac{398\,600 \cdot 8378.9}{75\,366^2}(1 - \cos 120°) = 0.11802 \text{ (dimensionless)} \qquad (f)$$

$$g = \frac{r r_0}{h} \sin(\Delta\theta)$$

$$= \frac{8378.9 \cdot 10\,681}{75\,366} \sin(120°) = 1028.4 \text{ s} \qquad (g)$$

$$\dot{f} = \frac{\mu}{h} \frac{1 - \cos \Delta\theta}{\sin \Delta\theta} \left[\frac{\mu}{h^2}(1 - \cos \Delta\theta) - \frac{1}{r_0} - \frac{1}{r}\right]$$

$$= \frac{398\,600}{75\,366} \frac{1 - \cos 120°}{\sin 120°} \left[\frac{398\,600}{75\,366^2}(1 - \cos 120°) - \frac{1}{10\,681} - \frac{1}{8378.9}\right] \qquad (h)$$

$$= -9.8665 \times 10^{-4} \text{ (dimensionless)}$$

$$\dot{g} = 1 - \frac{\mu r_0}{h^2}(1 - \cos \Delta\theta)$$

$$= 1 - \frac{398\,600 \cdot 10\,681}{75\,366^2}(1 - \cos 120°) = -0.12432 \text{ (dimensionless)} \qquad (i)$$

At this point we have all that is required to find the final position and velocity vectors. From Equation 2.125 we have

$$\mathbf{r} = f\mathbf{r}_0 + g\mathbf{v}_0$$

Substituting Equations (a), (f) and (g), we get

$$\mathbf{r} = 0.11802(8182.4\hat{\mathbf{i}} - 6865.9\hat{\mathbf{j}}) + 1028.4(0.47572\hat{\mathbf{i}} + 8.8116\hat{\mathbf{j}})$$

$$= \underline{1454.9\hat{\mathbf{i}} + 8251.6\hat{\mathbf{j}}} \text{ (km)}$$

Likewise, according to Equation 2.126,

$$\mathbf{v} = \dot{f}\mathbf{r}_0 + \dot{g}\mathbf{v}_0$$

Substituting Equations (a), (h) and (i) yields

$$\mathbf{v} = (-9.8665 \times 10^{-4})(8182.4\hat{\mathbf{i}} - 6865.9\hat{\mathbf{j}}) + (-0.12435)(0.47572\hat{\mathbf{i}} + 8.8116\hat{\mathbf{j}})$$

or

$$\mathbf{v} = \underline{-8.1323\hat{\mathbf{i}} + 5.6785\hat{\mathbf{j}}} \text{ (km/s)}$$

In order to use the Lagrange coefficients to find the position and velocity as a function of time, we need to come up with a relation between $\Delta\theta$ and time. We will deal with that complex problem in the next chapter. Meanwhile, for times t which are close to the initial time t_0, we can obtain polynomial expressions for f and g in which the variable $\Delta\theta$ is replaced by the time interval $\Delta t = t - t_0$.

To do so, we expand the position vector $\mathbf{r}(t)$, considered to be a function of time, in a Taylor series about $t = t_0$. By definition, the Taylor series is given by

$$\mathbf{r}(t) = \sum_{n=0}^{\infty} \frac{1}{n!} \mathbf{r}^{(n)}(t_0)(t - t_0)^n \qquad (2.150)$$

where $\mathbf{r}^{(n)}(t_0)$ is the nth time derivative of $\mathbf{r}(t)$, evaluated at t_0,

$$\mathbf{r}^{(n)}(t_0) = \left(\frac{d^n \mathbf{r}}{dt^n}\right)_{t=t_0} \qquad (2.151)$$

Let us truncate this infinite series at five terms. Then, to that degree of approximation,

$$\mathbf{r}(t) = \mathbf{r}(t_0) + \left(\frac{d\mathbf{r}}{dt}\right)_{t=t_0} \Delta t + \frac{1}{2}\left(\frac{d^2\mathbf{r}}{dt^2}\right)_{t=t_0} \Delta t^2 + \frac{1}{6}\left(\frac{d^3\mathbf{r}}{dt^3}\right)_{t=t_0} \Delta t^3$$

$$+ \frac{1}{24}\left(\frac{d^4\mathbf{r}}{dt^4}\right)_{t=t_0} \Delta t^4 \qquad (2.152)$$

where $\Delta t = t - t_0$. To evaluate the four derivatives, we note first that $(d\mathbf{r}/dt)_{t=t_0}$ is just the velocity $\mathbf{v}_0$ at $t = t_0$,

$$\left(\frac{d\mathbf{r}}{dt}\right)_{t=t_0} = \mathbf{v}_0 \qquad (2.153)$$

$(d^2\mathbf{r}/dt^2)_{t=t_0}$ is evaluated using Equation 2.15,

$$\ddot{\mathbf{r}} = -\frac{\mu}{r^3}\mathbf{r} \qquad (2.154)$$

Thus,

$$\left(\frac{d^2\mathbf{r}}{dt^2}\right)_{t=t_0} = -\frac{\mu}{r_0^3}\mathbf{r}_0 \qquad (2.155)$$

$(d^3\mathbf{r}/dt^3)_{t=t_0}$ is evaluated by differentiating Equation 2.154,

$$\frac{d^3\mathbf{r}}{dt^3} = -\mu \frac{d}{dt}\left(\frac{\mathbf{r}}{r^3}\right) = -\mu\left(\frac{r^3\mathbf{v} - 3rr^2\dot{r}}{r^6}\right) = -\mu\frac{\mathbf{v}}{r^3} + 3\mu\frac{\dot{r}\mathbf{r}}{r^4} \qquad (2.156)$$

From Equation 2.25a we have

$$\dot{r} = \frac{\mathbf{r} \cdot \mathbf{v}}{r} \qquad (2.157)$$

Hence, Equation 2.156, evaluated at $t = t_0$, is

$$\left(\frac{d^3\mathbf{r}}{dt^3}\right)_{t=t_0} = -\mu\frac{\mathbf{v}_0}{r_0^3} + 3\mu\frac{\mathbf{r}_0 \cdot \mathbf{v}_0}{r_0^5}\mathbf{r}_0 \qquad (2.158)$$

Finally, $(d^4\mathbf{r}/dt^4)_{t=t_0}$ is found by first differentiating Equation 2.156,

$$\frac{d^4\mathbf{r}}{dt^4} = \frac{d}{dt}\left(-\mu\frac{\dot{\mathbf{r}}}{r^3} + 3\mu\frac{\dot{r}\mathbf{r}}{r^4}\right) = -\mu\left(\frac{r^3\ddot{\mathbf{r}} - 3r^2\dot{r}\dot{\mathbf{r}}}{r^6}\right) + 3\mu\left[\frac{r^4(\ddot{r}\mathbf{r} + \dot{r}\dot{\mathbf{r}}) - 4r^3\dot{r}^2\mathbf{r}}{r^8}\right] \quad (2.159)$$

$\ddot{r}$ is found in terms of $\mathbf{r}$ and $\mathbf{v}$ by differentiating Equation 2.157 and making use of Equation 2.154. This leads to the expression

$$\ddot{r} = \frac{d}{dt}\left(\frac{\mathbf{r}\cdot\dot{\mathbf{r}}}{r}\right) = \frac{v^2}{r} - \frac{\mu}{r^2} - \frac{(\mathbf{r}\cdot\mathbf{v})^2}{r^3} \quad (2.160)$$

Substituting Equations 2.154, 2.157 and 2.160 into Equation 2.159, combining terms and evaluating the result at $t = t_0$ yields

$$\left(\frac{d^4\mathbf{r}}{dt^4}\right)_{t=t_0} = \left[-2\frac{\mu^2}{r_0^6} + 3\mu\frac{v_0^2}{r_0^5} - 15\mu\frac{(\mathbf{r}_0\cdot\mathbf{v}_0)^2}{r_0^7}\right]\mathbf{r}_0 + 6\mu\frac{(\mathbf{r}_0\cdot\mathbf{v}_0)}{r_0^5}\mathbf{v}_0 \quad (2.161)$$

After substituting Equations 2.153, 2.155, 2.158 and 2.161 into Equation 2.152 and rearranging terms, we obtain

$$\mathbf{r}(t) = \left\{1 - \frac{\mu}{2r_0^3}\Delta t^2 + \frac{\mu}{2}\frac{\mathbf{r}_0\cdot\mathbf{v}_0}{r_0^5}\Delta t^3 + \frac{\mu}{24}\left[-2\frac{\mu}{r_0^6} + 3\frac{v_0^2}{r_0^5} - 15\frac{(\mathbf{r}_0\cdot\mathbf{v}_0)^2}{r_0^7}\right]\Delta t^4\right\}\mathbf{r}_0$$

$$+ \left[\Delta t - \frac{1}{6}\frac{\mu}{r_0^3}\Delta t^3 + \frac{\mu}{4}\frac{(\mathbf{r}_0\cdot\mathbf{v}_0)}{r_0^5}\Delta t^4\right]\mathbf{v}_0 \quad (2.162)$$

Comparing this expression with Equation 2.125, we see that, to the fourth order in Δt,

$$f = 1 - \frac{\mu}{2r_0^3}\Delta t^2 + \frac{\mu}{2}\frac{\mathbf{r}_0\cdot\mathbf{v}_0}{r_0^5}\Delta t^3 + \frac{\mu}{24}\left[-2\frac{\mu}{r_0^6} + 3\frac{v_0^2}{r_0^5} - 15\frac{(\mathbf{r}_0\cdot\mathbf{v}_0)^2}{r_0^7}\right]\Delta t^4$$

$$g = \Delta t - \frac{1}{6}\frac{\mu}{r_0^3}\Delta t^3 + \frac{\mu}{4}\frac{(\mathbf{r}_0\cdot\mathbf{v}_0)}{r_0^5}\Delta t^4 \quad (2.163)$$

For small values of elapsed time Δt these f and g series may be used to calculate the position of an orbiting body from the initial conditions.

EXAMPLE 2.10

The orbit of an earth satellite has an eccentricity $e = 0.2$ and a perigee radius of 7000 km. Starting at perigee, plot the radial distance as a function of time using the f and g series and compare the curve with the exact solution.

Since the satellite starts at perigee, $t_0 = 0$ and we have, using the perifocal frame,

$$\mathbf{r}_0 = 7000\hat{\mathbf{p}} \text{ (km)} \quad (a)$$

The orbit equation evaluated at perigee is Equation 2.40, which in the present case becomes

$$7000 = \frac{h^2}{398\,600}\frac{1}{1 + 0.2}$$

(Example 2.10 continued)

Solving for the angular momentum, we get $h = 57\,864\text{ km}^2/\text{s}$. Then, using the angular momentum formula, Equation 2.21, we find that the speed at perigee is $v_0 = 8.2663$ km/s, so that

$$\mathbf{v}_0 = 8.2663\hat{\mathbf{q}} \text{ (km/s)} \tag{b}$$

Clearly, $\mathbf{r}_0 \cdot \mathbf{v}_0 = 0$. Hence, with $\mu = 398\,600\text{ km}^3/\text{s}^2$, the Lagrange series in Equation 2.163 become

$$f = 1 - 5.8105(10^{-7})t^2 + 9.0032(10^{-14})t^4$$

$$g = t - 1.9368(10^{-7})t^3$$

where the units of t are seconds. Substituting f and g into Equation 2.125 yields

$$\mathbf{r} = [1 - 5.8105(10^{-7})t^2 + 9.0032(10^{-14})t^4](7000\hat{\mathbf{p}}) + [t - 1.9368(10^{-7})t^3](8.2663\hat{\mathbf{q}})$$

From this we obtain

$$r = \|\mathbf{r}\|$$
$$= \sqrt{49(10^6) + 11.389t^2 - 1.103(10^{-6})t^4 - 2.5633(10^{-12})t^6 + 3.9718(10^{-19})t^8} \tag{c}$$

For the exact solution of r versus time we must appeal to the methods presented in the next chapter. The exact solution and the series solution [Equation (c)] are plotted in Figure 2.29. As can be seen, the series solution begins to seriously diverge from the exact solution after about ten minutes.

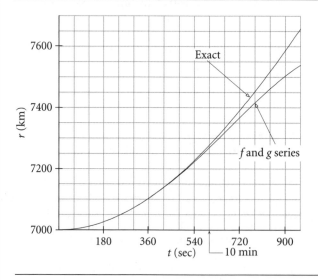

Figure 2.29 Exact and series solutions for the radial position of the satellite.

If we include terms of fifth and higher order in the f and g series, Equations 2.163, then the approximate solution in the above example will agree with the exact solution

2.12 Restricted three-body problem

for a longer time interval than that indicated in Figure 2.29. However, there is a time interval beyond which the series solution will diverge from the exact one no matter how many terms we include. This time interval is called the radius of convergence. According to Bond and Allman (1996), for the elliptical orbit of Example 2.10, the radius of convergence is 1700 seconds (not quite half an hour), which is one-fifth of the period of that orbit. This further illustrates the fact that the series form of the Lagrange coefficients is applicable only over small time intervals. For arbitrary time intervals the closed form of these functions, presented in Chapter 3, must be employed.

Consider two bodies m_1 and m_2 moving under the action of just their mutual gravitation, and let their orbit around each other be a circle of radius r_{12}. Consider a non-inertial, co-moving frame of reference xyz whose origin lies at the center of mass G of the two-body system, with the x axis directed towards m_2, as shown in Figure 2.30. The y axis lies in the orbital plane, to which the z axis is perpendicular. In this frame of reference, m_1 and m_2 appear to be at rest.

The constant, inertial angular velocity $\boldsymbol{\Omega}$ is given by

$$\boldsymbol{\Omega} = \Omega \hat{\mathbf{k}} \tag{2.164}$$

where

$$\Omega = \frac{2\pi}{T}$$

and T is the period of the orbit (Equation 2.54),

$$T = 2\pi \frac{r_{12}^{\frac{3}{2}}}{\sqrt{\mu}}$$

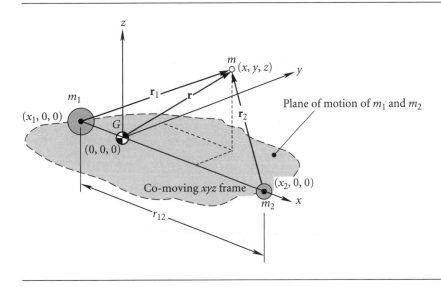

Figure 2.30 Primary bodies m_1 and m_2 in circular orbit around each other, plus a secondary mass m.

Thus

$$\Omega = \sqrt{\frac{\mu}{r_{12}^3}} \qquad (2.165)$$

Recall that if M is the total mass of the system,

$$M = m_1 + m_2 \qquad (2.166)$$

then

$$\mu = GM \qquad (2.167)$$

m_1 and m_2 lie in the orbital plane, so their y and z coordinates are zero. To determine their locations on the x axis, we use the definition of the center of mass (Equation 2.1) to write

$$m_1 x_1 + m_2 x_2 = 0$$

Since m_2 is at a distance r_{12} from m_1 in the positive x direction, it is also true that

$$x_2 = x_1 + r_{12}$$

From these two equations we obtain

$$x_1 = -\pi_2 r_{12} \qquad (2.168a)$$
$$x_2 = \pi_1 r_{12} \qquad (2.168b)$$

where the dimensionless mass ratios π_1 and π_2 are given by

$$\pi_1 = \frac{m_1}{m_1 + m_2}$$
$$\pi_2 = \frac{m_2}{m_1 + m_2} \qquad (2.169)$$

We now introduce a third body of mass m, which is vanishingly small compared to the primary masses m_1 and m_2 – like the mass of a spacecraft compared to that of a planet or moon of the solar system. This is called the restricted three-body problem, because the mass m is assumed to be so small that it has no effect on the motion of the primary bodies. We are interested in the motion of m due to the gravitational fields of m_1 and m_2. Unlike the two-body problem, there is no general, closed form solution for this motion. However, we can set up the equations of motion and draw some general conclusions from them.

In the co-moving coordinate system, the position vector of the secondary mass m relative to m_1 is given by

$$\mathbf{r}_1 = (x - x_1)\hat{\mathbf{i}} + y\hat{\mathbf{j}} + z\hat{\mathbf{k}} = (x + \pi_2 r_{12})\hat{\mathbf{i}} + y\hat{\mathbf{j}} + z\hat{\mathbf{k}} \qquad (2.170)$$

Relative to m_2 the position of m is

$$\mathbf{r}_2 = (x - \pi_1 r_{12})\hat{\mathbf{i}} + y\hat{\mathbf{j}} + z\hat{\mathbf{k}} \qquad (2.171)$$

Finally, the position vector of the secondary body relative to the center of mass is

$$\mathbf{r} = x\hat{\mathbf{i}} + y\hat{\mathbf{j}} + z\hat{\mathbf{k}} \tag{2.172}$$

The inertial velocity of m is found by taking the time derivative of Equation 2.172.

However, relative to inertial space, the xyz coordinate system is rotating with the angular velocity $\mathbf{\Omega}$, so that the time derivatives of the unit vectors $\hat{\mathbf{i}}$ and $\hat{\mathbf{j}}$ are not zero. To account for the rotating frame, we use Equation 1.38 to obtain

$$\dot{\mathbf{r}} = \mathbf{v}_G + \mathbf{\Omega} \times \mathbf{r} + \mathbf{v}_{rel} \tag{2.173}$$

$\mathbf{v}_G$ is the inertial velocity of the center of mass (the origin of the xyz frame), and $\mathbf{v}_{rel}$ is the velocity of m as measured in the moving xyz frame, namely,

$$\mathbf{v}_{rel} = \dot{x}\hat{\mathbf{i}} + \dot{y}\hat{\mathbf{j}} + \dot{z}\hat{\mathbf{k}} \tag{2.174}$$

The absolute acceleration of m is found using the 'five-term' relative acceleration formula, Equation 1.42,

$$\ddot{\mathbf{r}} = \mathbf{a}_G + \dot{\mathbf{\Omega}} \times \mathbf{r} + \mathbf{\Omega} \times (\mathbf{\Omega} \times \mathbf{r}) + 2\mathbf{\Omega} \times \mathbf{v}_{rel} + \mathbf{a}_{rel} \tag{2.175}$$

Recall from Section 2.2 that the velocity $\mathbf{v}_G$ of the center of mass is constant, so that $\mathbf{a}_G = 0$. Furthermore, $\dot{\mathbf{\Omega}} = 0$ since the angular velocity of the circular orbit is constant. Therefore, Equation 2.175 reduces to

$$\ddot{\mathbf{r}} = \mathbf{\Omega} \times (\mathbf{\Omega} \times \mathbf{r}) + 2\mathbf{\Omega} \times \mathbf{v}_{rel} + \mathbf{a}_{rel} \tag{2.176}$$

where

$$\mathbf{a}_{rel} = \ddot{x}\hat{\mathbf{i}} + \ddot{y}\hat{\mathbf{j}} + \ddot{z}\hat{\mathbf{k}} \tag{2.177}$$

Substituting Equations 2.164, 2.172, 2.174 and 2.177 into Equation 2.176 yields

$$\ddot{\mathbf{r}} = (\Omega\hat{\mathbf{k}}) \times \left[(\Omega\hat{\mathbf{k}}) \times (x\hat{\mathbf{i}} + y\hat{\mathbf{j}} + z\hat{\mathbf{k}})\right] + 2(\Omega\hat{\mathbf{k}}) \times (\dot{x}\hat{\mathbf{i}} + \dot{y}\hat{\mathbf{j}} + \dot{z}\hat{\mathbf{k}}) + \ddot{x}\hat{\mathbf{i}} + \ddot{y}\hat{\mathbf{j}} + \ddot{z}\hat{\mathbf{k}}$$

$$= \left[-\Omega^2(x\hat{\mathbf{i}} + y\hat{\mathbf{j}})\right] + (2\Omega\dot{x}\hat{\mathbf{j}} - 2\Omega\dot{y}\hat{\mathbf{i}}) + \ddot{x}\hat{\mathbf{i}} + \ddot{y}\hat{\mathbf{j}} + \ddot{z}\hat{\mathbf{k}}$$

Collecting terms, we find

$$\ddot{\mathbf{r}} = (\ddot{x} - 2\Omega\dot{y} - \Omega^2 x)\hat{\mathbf{i}} + (\ddot{y} + 2\Omega\dot{x} - \Omega^2 y)\hat{\mathbf{j}} + \ddot{z}\hat{\mathbf{k}} \tag{2.178}$$

Now that we have an expression for the inertial acceleration in terms of quantities measured in the rotating frame, let us observe that Newton's second law for the secondary body is

$$m\ddot{\mathbf{r}} = \mathbf{F}_1 + \mathbf{F}_2 \tag{2.179}$$

$\mathbf{F}_1$ and $\mathbf{F}_2$ are the gravitational forces exerted on m by m_1 and m_2, respectively. Recalling Equation 2.6, we have

$$\mathbf{F}_1 = -\frac{Gm_1 m}{r_1^2}\mathbf{u}_{r_1} = -\frac{\mu_1 m}{r_1^3}\mathbf{r}_1$$

$$\mathbf{F}_2 = -\frac{Gm_2 m}{r_2^2}\mathbf{u}_{r_2} = -\frac{\mu_2 m}{r_2^3}\mathbf{r}_2 \tag{2.180}$$

where
$$\mu_1 = Gm_1 \quad \mu_2 = Gm_2 \tag{2.181}$$

Substituting Equations 2.180 into 2.179 and canceling out m yields

$$\ddot{\mathbf{r}} = -\frac{\mu_1}{r_1^3}\mathbf{r}_1 - \frac{\mu_2}{r_2^3}\mathbf{r}_2 \tag{2.182}$$

Finally, we substitute Equation 2.178 on the left and Equations 2.170 and 2.171 on the right to obtain

$$(\ddot{x} - 2\Omega\dot{y} - \Omega^2 x)\hat{\mathbf{i}} + (\ddot{y} + 2\Omega\dot{x} - \Omega^2 y)\hat{\mathbf{j}} + \ddot{z}\hat{\mathbf{k}} = -\frac{\mu_1}{r_1^3}\left[(x + \pi_2 r_{12})\hat{\mathbf{i}} + y\hat{\mathbf{j}} + z\hat{\mathbf{k}}\right]$$
$$- \frac{\mu_2}{r_2^3}\left[(x - \pi_1 r_{12})\hat{\mathbf{i}} + y\hat{\mathbf{j}} + z\hat{\mathbf{k}}\right]$$

Equating the coefficients of $\hat{\mathbf{i}}, \hat{\mathbf{j}}$ and $\hat{\mathbf{k}}$ on each side of this equation yields the three scalar equations of motion for the restricted three-body problem:

$$\ddot{x} - 2\Omega\dot{y} - \Omega^2 x = -\frac{\mu_1}{r_1^3}(x + \pi_2 r_{12}) - \frac{\mu_2}{r_2^3}(x - \pi_1 r_{12}) \tag{2.183a}$$

$$\ddot{y} + 2\Omega\dot{x} - \Omega^2 y = -\frac{\mu_1}{r_1^3}y - \frac{\mu_2}{r_2^3}y \tag{2.183b}$$

$$\ddot{z} = -\frac{\mu_1}{r_1^3}z - \frac{\mu_2}{r_2^3}z \tag{2.183c}$$

2.12.1 LAGRANGE POINTS

Although Equations 2.183 have no closed form analytical solution, we can use them to determine the location of the equilibrium points. These are the locations in space where the secondary mass m would have zero velocity and zero acceleration, i.e., where m would appear permanently at rest relative to m_1 and m_2 (and therefore appear to an inertial observer to move in circular orbits around m_1 and m_2). Once placed at an equilibrium point (also called libration point or Lagrange point), a body will presumably stay there. The equilibrium points are therefore defined by the conditions

$$\dot{x} = \dot{y} = \dot{z} = 0 \quad \text{and} \quad \ddot{x} = \ddot{y} = \ddot{z} = 0$$

Substituting these conditions into Equations 2.183 yields

$$-\Omega^2 x = -\frac{\mu_1}{r_1^3}(x + \pi_2 r_{12}) - \frac{\mu_2}{r_2^3}(x - \pi_1 r_{12}) \tag{2.184a}$$

$$-\Omega^2 y = -\frac{\mu_1}{r_1^3}y - \frac{\mu_2}{r_2^3}y \tag{2.184b}$$

$$0 = -\frac{\mu_1}{r_1^3}z - \frac{\mu_2}{r_2^3}z \tag{2.184c}$$

2.12 Restricted three-body problem

From Equation 2.184c we have

$$\left(\frac{\mu_1}{r_1^3} + \frac{\mu_2}{r_2^3}\right)z = 0 \tag{2.185}$$

Since $\mu_1/r_1^3 > 0$ and $\mu_2/r_2^3 > 0$, it must therefore be true that $z=0$. That is, the equilibrium points lie in the orbital plane.

From Equations 2.169 it is clear that

$$\pi_1 = 1 - \pi_2 \tag{2.186}$$

Using this, along with Equation 2.165, and assuming $y \neq 0$, we can write Equations 2.184a and 2.184b as

$$(1 - \pi_2)(x + \pi_2 r_{12})\frac{1}{r_1^3} + \pi_2(x + \pi_2 r_{12} - r_{12})\frac{1}{r_2^3} = \frac{x}{r_{12}^3}$$

$$(1 - \pi_2)\frac{1}{r_1^3} + \pi_2\frac{1}{r_2^3} = \frac{1}{r_{12}^3} \tag{2.187}$$

where we made use of the fact that

$$\pi_1 = \mu_1/\mu \quad \pi_2 = \mu_2/\mu \tag{2.188}$$

Treating Equations 2.187 as two linear equations in $1/r_1^3$ and $1/r_2^3$, we solve them simultaneously to find that

$$\frac{1}{r_1^3} = \frac{1}{r_2^3} = \frac{1}{r_{12}^3}$$

or

$$r_1 = r_2 = r_{12} \tag{2.189}$$

Using this result, together with $z=0$ and Equation 2.186, we obtain from Equations 2.170 and 2.171, respectively,

$$r_{12}^2 = (x + \pi_2 r_{12})^2 + y^2 \tag{2.190}$$

$$r_{12}^2 = (x + \pi_2 r_{12} - r_{12})^2 + y^2 \tag{2.191}$$

Equating the right-hand sides of these two equations leads at once to the conclusion that

$$x = \frac{r_{12}}{2} - \pi_2 r_{12} \tag{2.192}$$

Substituting this result into Equation 2.190 or 2.191 and solving for y yields

$$y = \pm\frac{\sqrt{3}}{2}r_{12}$$

We have thus found two of the equilibrium points, the Lagrange points L_4 and L_5. As Equation 2.189 shows, these points are the same distance r_{12} from the primary bodies m_1 and m_2 that the primary bodies are from each other, and in the co-moving coordinate system their coordinates are

$$L_4, L_5 : x = \frac{r_{12}}{2} - \pi_2 r_{12}, \quad y = \pm \frac{\sqrt{3}}{2} r_{12}, \quad z = 0 \qquad (2.193)$$

Therefore, the two primary bodies and these two Lagrange points lie at the vertices of equilateral triangles, as illustrated in Figure 2.32.

The remaining equilibrium points are found by setting $y = 0$ as well as $z = 0$, which satisfy both Equations 2.184b and 2.184c. For these values, Equations 2.170 and 2.171 become

$$\mathbf{r}_1 = (x + \pi_2 r_{12})\hat{\mathbf{i}}$$

$$\mathbf{r}_2 = (x - \pi_1 r_{12})\hat{\mathbf{i}} = (x + \pi_2 r_{12} - r_{12})\hat{\mathbf{i}}$$

Therefore

$$r_1 = |x + \pi_2 r_{12}|$$

$$r_2 = |x + \pi_2 r_{12} - r_{12}|$$

Substituting these together with Equations 2.165, 2.186 and 2.188 into Equation 2.184a yields

$$\frac{1 - \pi_2}{|x + \pi_2 r_{12}|^3}(x + \pi_2 r_{12}) + \frac{\pi_2}{|x + \pi_2 r_{12} - r_{12}|^3}(x + \pi_2 r_{12} - r_{12}) - \frac{1}{r_{12}^3}x = 0 \quad (2.194)$$

Further simplification is obtained by non-dimensionalizing x,

$$\xi = \frac{x}{r_{12}}$$

In terms of ξ, Equation 2.194 becomes $f(\xi) = 0$, where

$$f(\xi) = \frac{1 - \pi_2}{|\xi + \pi_2|^3}(\xi + \pi_2) + \frac{\pi_2}{|\xi + \pi_2 - 1|^3}(\xi + \pi_2 - 1) - \xi \qquad (2.195)$$

The roots of $f(\xi) = 0$ yields the other equilibrium points besides L_4 and L_5. To find them first requires specifying a value for the mass ratio π_2, and then using a numerical technique to obtain the roots for that particular value. For example, let the two primary bodies m_1 and m_2 be the earth and the moon, respectively. Then

$$\begin{aligned} m_1 &= 5.974 \times 10^{24} \text{ kg} \\ m_2 &= 7.348 \times 10^{22} \text{ kg} \\ r_{12} &= 3.844 \times 10^5 \text{ km} \end{aligned} \qquad (2.196)$$

(from Table A.1) using this data, we find

$$\pi_2 = \frac{m_2}{m_1 + m_2} = 0.01215$$

Substituting this value of π_2 into Equation 2.195 and plotting the function yields the curves shown in Figure 2.31. By carefully determining where various branches of the

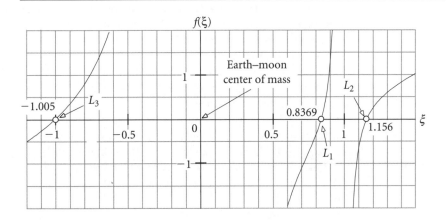

Figure 2.31 Graph of Equation 2.195 for earth–moon data ($\pi_2 = 0.01215$), showing the three real roots.

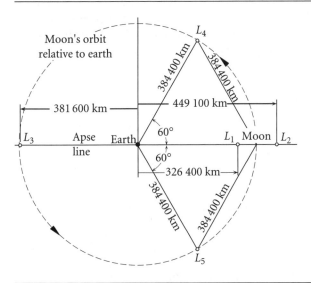

Figure 2.32 Location of the five Lagrange points of the earth–moon system. These points orbit the earth with the same period as the moon.

curve cross the ξ axis, we find the real roots, which are the three additional Lagrange points for the earth–moon system, all lying on the apse line:

$$\begin{aligned} L_1 &: x = 0.8369 r_{12} = 3.217 \times 10^5 \text{ km} \\ L_2 &: x = 1.156 r_{12} = 4.444 \times 10^5 \text{ km} \\ L_3 &: x = -1.005 r_{12} = -3.863 \times 10^5 \text{ km} \end{aligned} \quad (2.197)$$

The locations of the five Lagrange points for the earth–moon system are shown in Figure 2.32. For convenience, all of their positions are shown relative to the center of the earth, instead of the center of mass. As can be seen from Equation 2.168a, the center of mass of the earth–moon system is only 4670 km from the center of the earth. That is, it lies within the earth at 73 percent of its radius. Since the Lagrange points are fixed relative to the earth and moon, they follow circular orbits around the earth with the same period as the moon.

If an equilibrium point is stable, then a small mass occupying that point will tend to return to that point if nudged out of position. The perturbation results in a small oscillation (orbit) about the equilibrium point. Thus, objects can be placed in small orbits (called halo orbits) around stable equilibrium points without requiring much in the way of station keeping. On the other hand, if a body located at an unstable equilibrium point is only slightly perturbed, it will oscillate in a divergent fashion, drifting eventually completely away from that point. It turns out that the Lagrange points L_1, L_2 and L_3 on the apse line are unstable, whereas L_4 and L_5 – 60° ahead of and behind the moon in its orbit – are stable. However, L_4 and L_5 are destabilized by the influence of the sun's gravity, so that in actuality station keeping would be required to maintain position in the neighborhood of those points.

Solar observation spacecraft have been placed in halo orbits around the L_1 point of the sun–earth system. L_1 lies about 1.5 million kilometers from the earth (1/100 the distance to the sun) and well outside the earth's magnetosphere. Three such missions were the International Sun–Earth Explorer 3 (ISSE-3) launched in August 1978; the Solar and Heliocentric Observatory (SOHO) launched in December 1995; and the Advanced Composition Explorer (ACE) launched in August 1997.

2.12.2 JACOBI CONSTANT

Multiply Equation 2.183a by $\dot{x}$, Equation 2.183b by $\dot{y}$ and Equation 2.183c by $\dot{z}$ to obtain

$$\ddot{x}\dot{x} - 2\Omega\dot{x}\dot{y} - \Omega^2 x\dot{x} = -\frac{\mu_1}{r_1^3}(x\dot{x} + \pi_2 r_{12}\dot{x}) - \frac{\mu_2}{r_2^3}(x\dot{x} - \pi_1 r_{12}\dot{x})$$

$$\ddot{y}\dot{y} + 2\Omega\dot{x}\dot{y} - \Omega^2 y\dot{y} = -\frac{\mu_1}{r_1^3}y\dot{y} - \frac{\mu_2}{r_2^3}y\dot{y}$$

$$\ddot{z}\dot{z} = -\frac{\mu_1}{r_1^3}z\dot{z} - \frac{\mu_2}{r_2^3}z\dot{z}$$

Sum the left and right sides of these equations to get

$$\ddot{x}\dot{x}+\ddot{y}\dot{y}+\ddot{z}\dot{z}-\Omega^2(x\dot{x}+y\dot{y}) = -\left(\frac{\mu_1}{r_1^3}+\frac{\mu_2}{r_2^3}\right)(x\dot{x}+y\dot{y}+z\dot{z})+r_{12}\left(\frac{\pi_1\mu_2}{r_2^3}-\frac{\pi_2\mu_1}{r_1^3}\right)\dot{x}$$

or, rearranging terms,

$$\ddot{x}\dot{x} + \ddot{y}\dot{y} + \ddot{z}\dot{z} - \Omega^2(x\dot{x}+y\dot{y}) = -\frac{\mu_1}{r_1^3}(x\dot{x}+y\dot{y}+z\dot{z}+\pi_2 r_{12}\dot{x})$$

$$-\frac{\mu_2}{r_2^3}(x\dot{x}+y\dot{y}+z\dot{z}-\pi_1 r_{12}\dot{x}) \quad (2.198)$$

Note that

$$\ddot{x}\dot{x} + \ddot{y}\dot{y} + \ddot{z}\dot{z} = \frac{1}{2}\frac{d}{dt}(\dot{x}^2+\dot{y}^2+\dot{z}^2) = \frac{1}{2}\frac{dv^2}{dt} \quad (2.199)$$

2.12 Restricted three-body problem

where v is the speed of the secondary mass relative to the rotating frame. Similarly,

$$x\dot{x} + y\dot{y} = \frac{1}{2}\frac{d}{dt}(x^2 + y^2) \qquad (2.200)$$

From Equation 2.170 we obtain

$$r_1^2 = (x + \pi_2 r_{12})^2 + y^2 + z^2$$

Therefore

$$2r_1 \frac{dr_1}{dt} = 2(x + \pi_2 r_{12})\dot{x} + 2y\dot{y} + 2z\dot{z}$$

or

$$\frac{dr_1}{dt} = \frac{1}{r_1}(\pi_2 r_{12}\dot{x} + x\dot{x} + y\dot{y} + z\dot{z})$$

It follows that

$$\frac{d}{dt}\frac{1}{r_1} = -\frac{1}{r_1^2}\frac{dr_1}{dt} = -\frac{1}{r_1^3}(x\dot{x} + y\dot{y} + z\dot{z} + \pi_2 r_{12}\dot{x}) \qquad (2.201)$$

In a similar fashion, starting with Equation 2.171, we find

$$\frac{d}{dt}\frac{1}{r_2} = -\frac{1}{r_2^3}(x\dot{x} + y\dot{y} + z\dot{z} - \pi_1 r_{12}\dot{x}) \qquad (2.202)$$

Substituting Equations 2.199, 2.200, 2.201 and 2.202 into Equation 2.198 yields

$$\frac{1}{2}\frac{dv^2}{dt} - \frac{1}{2}\Omega^2 \frac{d}{dt}(x^2 + y^2) = \mu_1 \frac{d}{dt}\frac{1}{r_1} + \mu_2 \frac{d}{dt}\frac{1}{r_2}$$

Alternatively, upon rearranging terms

$$\frac{d}{dt}\left[\frac{1}{2}v^2 - \frac{1}{2}\Omega^2(x^2 + y^2) - \frac{\mu_1}{r_1} - \frac{\mu_2}{r_2}\right] = 0$$

which means the bracketed expression is a constant

$$\frac{1}{2}v^2 - \frac{1}{2}\Omega^2(x^2 + y^2) - \frac{\mu_1}{r_1} - \frac{\mu_2}{r_2} = C \qquad (2.203)$$

$v^2/2$ is the kinetic energy per unit mass relative to the rotating frame. $-\mu_1/r_1$ and $-\mu_2/r_2$ are the gravitational potential energies of the two primary masses. $-\Omega^2(x^2 + y^2)/2$ may be interpreted as the potential energy of the centrifugal force per unit mass $\Omega^2(x\hat{\mathbf{i}} + y\hat{\mathbf{j}})$ induced by the rotation of the reference frame. The constant C is known as the Jacobi constant, after the German mathematician Carl Jacobi (1804–1851), who discovered it in 1836. Jacobi's constant may be interpreted as the total energy of the secondary particle relative to the rotating frame. C is a constant of the motion of the secondary mass just like the energy and angular momentum are constants of the relative motion in the two-body problem.

Solving Equation 2.203 for v^2 yields

$$v^2 = \Omega^2(x^2 + y^2) + \frac{2\mu_1}{r_1} + \frac{2\mu_2}{r_2} + 2C \qquad (2.204)$$

If we restrict the motion of the secondary mass to lie in the plane of motion of the primary masses, then

$$r_1 = \sqrt{(x+\pi_2 r_{12})^2 + y^2} \quad r_2 = \sqrt{(x-\pi_1 r_{12})^2 + y^2} \tag{2.205}$$

For a given value of the Jacobi constant, v^2 is a function only of position in the rotating frame. Since v^2 cannot be negative, it must be true that

$$\Omega^2(x^2 + y^2) + \frac{2\mu_1}{r_1} + \frac{2\mu_2}{r_2} + 2C \geq 0 \tag{2.206}$$

Trajectories of the secondary body in regions where this inequality is violated are not allowed. The boundaries between forbidden and allowed regions of motion are found by setting $v^2 = 0$, i.e.,

$$\Omega^2(x^2 + y^2) + \frac{2\mu_1}{r_1} + \frac{2\mu_2}{r_2} + 2C = 0 \tag{2.207}$$

For a given value of the Jacobi constant the curves of zero velocity are determined by this equation. These boundaries cannot be crossed by a secondary mass (spacecraft) moving within an allowed region.

Since the first three terms on the left of Equation 2.207 are all positive, it follows that the zero velocity curves correspond to negative values of the Jacobi constant. Large negative values of C mean that the secondary body is far from the system center of mass ($x^2 + y^2$ is large) or that the body is close to one of the primary bodies (r_1 is small or r_2 is small).

Let us consider again the earth–moon system. From Equations 2.165, 2.166, 2.167, 2.181 and 2.196, together with Table A.2, we have

$$\begin{aligned} \mu_1 &= \mu_{\text{earth}} = 398\,600\text{ km}^3/\text{s}^2 \\ \mu_2 &= \mu_{\text{moon}} = 4903.02\text{ km}^3/\text{s}^2 \\ \Omega &= \sqrt{\frac{\mu_1 + \mu_2}{r_{12}^3}} = \sqrt{\frac{398\,600 + 4903}{384\,400^3}} \\ &= 2.66538 \times 10^{-6}\text{ rad/s} \end{aligned} \tag{2.208}$$

Substituting these values into Equation 2.207, we can plot the zero velocity curves for different values of Jacobi's constant. The curves bound regions in which the motion of a spacecraft is not allowed.

For $C = -1.8\text{ km}^2/\text{s}^2$, the allowable regions are circles surrounding the earth and the moon, as shown in Figure 2.33(a). A spacecraft launched from the earth with this value of C cannot reach the moon, to say nothing of escaping the earth–moon system.

Substituting the coordinates of the Lagrange points L_1, L_2 and L_3 into Equation 2.207, we obtain the successively larger values of the Jacobi constants C_1, C_2 and C_3 which are required to arrive at those points with zero velocity. These are shown along with the allowable regions in Figure 2.33. From part (c) of that figure we see that C_2 represents the minimum energy for a spacecraft to escape the earth–moon system via a narrow corridor around the moon. Increasing C widens that corridor and at C_3 escape becomes possible in the opposite direction from the moon. The last vestiges of

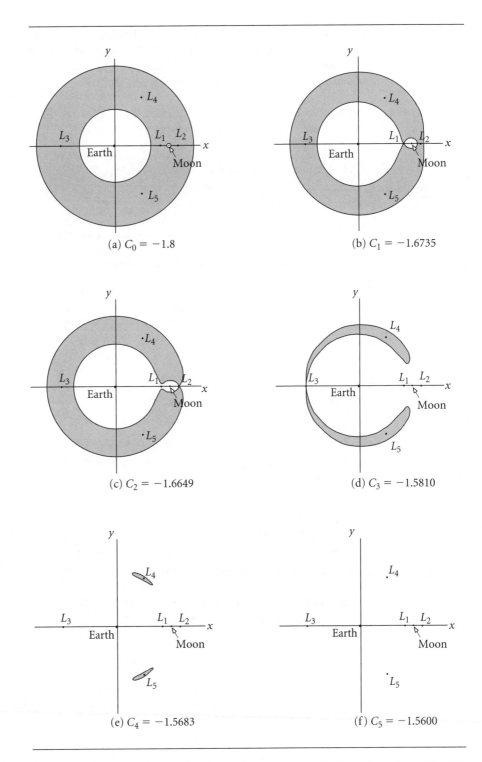

Figure 2.33 Forbidden regions (shaded) within the earth–moon system for increasing values of Jacobi's constant (km²/s²).

the forbidden regions surround L_4 and L_5. Further increase in Jacobi's constant makes the entire earth–moon system and beyond accessible to an earth-launched spacecraft.

For a given value of the Jacobi constant, the relative speed at any point within an allowable region can be found using Equation 2.204.

EXAMPLE 2.11

A spacecraft has a burnout velocity v_{bo} at a point on the earth–moon line with an altitude of 200 km. Find the value of v_{bo} for each of the scenarios depicted in Figure 2.33.

From Equations 2.168 and 2.196 we have

$$\pi_1 = \frac{m_1}{m_1 + m_2} = \frac{5.974 \times 10^{24}}{6.047 \times 10^{24}} = 0.9878 \quad \pi_2 = 1 - \pi_1 = 0.1215$$

$$x_1 = -\pi_1 r_{12} = -0.9878 \cdot 384\,400 = -4670.6\,\text{km}$$

Therefore, the coordinates of the burnout point are

$$x = 6578 - 4670.6 = 1907.3\,\text{km} \quad y = 0$$

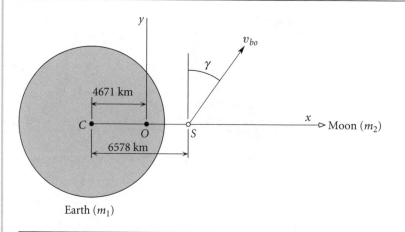

Figure 2.34 Spacecraft S burnout position and velocity relative to the rotating earth–moon frame.

Substituting these values along with the Jacobi constant into Equations 2.204 and 2.205 yields the burnout velocity v_{bo}. For the six Jacobi constants in Figure 2.33 we obtain

$$C_0: v_{bo} = 10.845\,\text{km/s}$$
$$C_1: v_{bo} = 10.857\,\text{km/s}$$
$$C_2: v_{bo} = 10.858\,\text{km/s}$$
$$C_3: v_{bo} = 10.866\,\text{km/s}$$
$$C_4: v_{bo} = 10.867\,\text{km/s}$$
$$C_5: v_{bo} = 10.868\,\text{km/s}$$

These velocities are not substantially different from the escape velocity (Equation 2.81) at 200 km altitude,

$$v_{esc} = \sqrt{\frac{2\mu}{r}} = \sqrt{\frac{2 \cdot 398\,600}{6578}} = 11.01 \text{ km/s}$$

It is remarkable that a change in v_{bo} on the order of only 10 m/s or less can have a significant influence on the regions of earth–moon space accessible to the spacecraft.

PROBLEMS

For man-made earth satellites use $\mu = 398\,600 \text{ km}^3/\text{s}^2$. $R_E = 6378$ km (Tables A.1 and A.2).

2.1 If $\mathbf{r}$, in meters, is given by $\mathbf{r} = 3t^4\hat{\mathbf{I}} + 2t^3\hat{\mathbf{J}} + 9t^2\hat{\mathbf{K}}$, where t is time in seconds, calculate $\dot{r}$ (where $r = \|\mathbf{r}\|$) and $\|\dot{\mathbf{r}}\|$ at $t = 2$ s.
{Ans.: $\dot{r} = 101.3$ m/s, $\|\dot{\mathbf{r}}\| = 105.3$ m/s}

2.2 Show that, in general, if $\hat{\mathbf{u}}_r = \mathbf{r}/r$, then $\hat{\mathbf{u}}_r \cdot d\hat{\mathbf{u}}_r/dt = 0$.

2.3 Two particles of identical mass m are acted on only by the gravitational force of one upon the other. If the distance d between the particles is constant, what is the angular velocity of the line joining them?
{Ans.: $\omega = \sqrt{2\,Gm/d^3}$}

2.4 Three particles of identical mass m are acted on only by their mutual gravitational attraction. They are located at the vertices of an equilateral triangle with sides of length d. Consider the motion of any one of the particles about the system center of mass and use Newton's second law to determine the angular velocity ω required for d to remain constant.
{Ans.: $\omega = \sqrt{3\,Gm/d^3}$}

2.5 A satellite is in a circular, 350 km orbit (i.e., it is 350 km above the earth's surface). Calculate
(a) the speed in km/s;
(b) the period.
{Ans.: (a) 7.697 km/s; (b) 91 min 32 s}

2.6 A spacecraft is in a circular orbit of the moon at an altitude of 80 km. Calculate its speed and its period.
{Ans.: 1.642 km/s; 1 hr 56 min}

2.7 It is desired to place a satellite in earth polar orbit such that successive ground tracks at the equator are spaced 3000 km apart. Determine the required altitude of the circular orbit.
{Ans.: 1440 km}

2.8 Find the minimum additional speed required to escape from GEO.
{Ans.: 1.274 km/s}

2.9 What velocity, relative to the earth, is required to escape the solar system on a parabolic path from the earth's orbit?
{12.34 km/s}

2.10 Calculate the area A swept out during the time $t = T/3$ since periapsis, where T is the period of the elliptical orbit.
{Ans.: $1.047ab$}

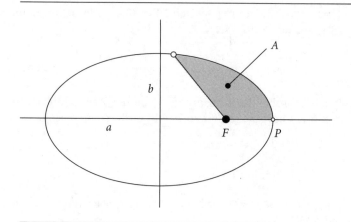

Figure P.2.10

2.11 Show that $v = \frac{\mu}{h}\sqrt{1 + 2e\cos\theta + e^2}$ for any orbit.

2.12 Determine the true anomaly θ of the point(s) on an elliptical orbit at which the speed equals the speed of a circular orbit with the same radius, i.e., $v_{\text{ellipse}} = v_{\text{circle}}$. {Ans.: $\theta = \cos^{-1}(-e)$, where e is the eccentricity of the ellipse}

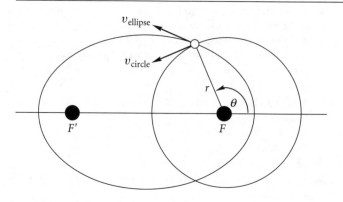

Figure P.2.12

2.13 Calculate the flight path angle at the locations found in Exercise 2.12. $\left\{\text{Ans.} : \gamma = \tan^{-1}\left(e/\sqrt{1-e^2}\right)\right\}$

2.14 An unmanned satellite orbits the earth with a perigee radius of 7000 km and an apogee radius of 70 000 km. Calculate
  (a) the eccentricity of the orbit;
  (b) the semimajor axis of the orbit (km);
  (c) the period of the orbit (hours);
  (d) the specific energy of the orbit (km²/s²);
  (e) the true anomaly at which the altitude is 1000 km (degrees);

(f) v_r and $v_\perp$ at the points found in part (e) (km/s);
(g) the speed at perigee and apogee (km/s).
{Partial ans.: (c) 20.88 hr; (e) 27.61°; (g) 10.18 km/s, 1.018 km/s}

2.15 A spacecraft is in a 250 km by 300 km low earth orbit. How long (in minutes) does it take to fly from perigee to apogee?
{Ans.: 45.00 min}

2.16 The altitude of a satellite in an elliptical orbit around the earth is 1600 km at apogee and 600 km at perigee. Determine
(a) the eccentricity of the orbit;
(b) the orbital speeds at perigee and apogee;
(c) the period of the orbit.
{Ans.: (a) 0.06686; (b) $v_P = 7.81$ km/s; (c) $v_A = 6.83$ km/s; (d) $T = 107.2$ min}

2.17 A satellite is placed into an earth orbit at perigee at an altitude of 1270 km with a speed of 9 km/s. Calculate the flight path angle γ and the altitude of the satellite at a true anomaly of 100°.
{Ans.: $\gamma = 31.1°$; $z = 6774$ km}

2.18 A satellite is launched into earth orbit at an altitude of 640 km with a speed of 9.2 km/s and a flight path angle of 10°. Calculate the true anomaly of the launch point and the period of the orbit.
{Ans.: $\theta = 29.8°$; $T = 4.46$ hr}

2.19 A satellite has perigee and apogee altitudes of 250 km and 42 000 km. Calculate the orbit period, eccentricity, and the maximum speed.
{Ans.: 12 hr 36 min, 0.759, 10.3 km/s}

2.20 A satellite is launched parallel to the earth's surface with a speed of 8 km/s at an altitude of 640 km. Calculate the apogee altitude and the period.
{Ans.: 2679 km, 1 hr 59 min 30 s}

2.21 A satellite in orbit around the earth has a perigee velocity of 8 km/s. Its period is 2 hours. Calculate its altitude at perigee.
{Ans.: 648 km}

2.22 A satellite in polar orbit around the earth comes within 150 km of the North Pole at its point of closest approach. If the satellite passes over the pole once every 90 minutes, calculate the eccentricity of its orbit.
{Ans.: 0.0187}

2.23 A hyperbolic earth departure trajectory has a perigee altitude of 300 km and a perigee speed of 15 km/s.
(a) Calculate the hyperbolic excess speed (km/s);
(b) Find the radius (km) when the true anomaly is 100°; {Ans.: 48 497 km}
(c) Find v_r and $v_\perp$ (km/s) when the true anomaly is 100°.

2.24 A meteoroid is first observed approaching the earth when it is 402 000 km from the center of the earth with a true anomaly of 150°. If the speed of the meteoroid at that time is 2.23 km/s, calculate
(a) the eccentricity of the trajectory;
(b) the altitude at closest approach;
(c) the speed at closest approach.
{Ans.: (a) 1.086; (b) 5088 km; (c) 8.516 km/s}

2.25 Calculate the radius r at which the speed on a hyperbolic trajectory is 1.1 times the hyperbolic excess speed. Express your result in terms of the periapse radius r_p and the eccentricity e.
{Ans.: $r = 9.524 r_p/(e - 1)$}

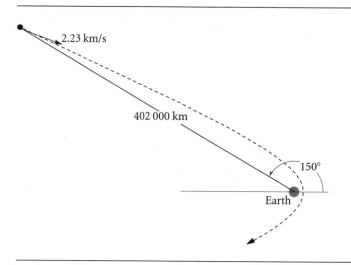

Figure P.2.24

2.26 A hyperbolic trajectory has an eccentricity $e = 3.0$ and an angular momentum $h = 105\,000 \text{ km}^2/\text{s}$. Without using the energy equation, calculate the hyperbolic excess speed.
{Ans.: 10.7 km/s}

2.27 The following position data for an earth orbiter is given:

Altitude = 1700 km at a true anomaly of 130°.
Altitude = 500 km at a true anomaly of 50°.

Calculate
(a) the eccentricity;
(b) the perigee altitude (km);
(c) the semimajor axis (km).
{Ans.: (c) 7547 km}

2.28 An earth satellite has a speed of 7 km/s and a flight path angle of 15° when its radius is 9000 km. Calculate
(a) the true anomaly (degrees);
(b) the eccentricity of the orbit.
{Ans.: (a) 83.35°; (b) 0.2785}

2.29 If, for an earth satellite, the specific angular momentum is $60\,000 \text{ km}^2/\text{s}$ and the specific energy is $-20 \text{ km}^2/\text{s}^2$, calculate the apogee and perigee altitudes.
{Ans.: 6637 km and 537.2 km}

2.30 A rocket launched from the surface of the earth has a speed of 8.85 km/s when powered flight ends at an altitude of 550 km. The flight path angle at this time is 6°. Determine
(a) the eccentricity of the trajectory;
(b) the period of the orbit.
{Ans.: (a) $e = 0.3742$; (b) $T = 187.4$ min}

2.31 A space vehicle has a velocity of 10 km/s in the direction shown when it is 10 000 km from the center of the earth. Calculate its true anomaly.
{Ans.: 51°}

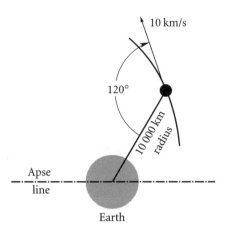

Figure P.2.31

2.32 A space vehicle has a velocity of 10 km/s and a flight path angle of 20° when it is 15 000 km from the center of the earth. Calculate its true anomaly.
{Ans.: 27.5°}

2.33 For a spacecraft trajectory around the earth, $r = 10\,000$ km when $\theta = 30°$, and $r = 30\,000$ km when $\theta = 105°$. Calculate the eccentricity.
{Ans.: 1.22}

2.34 A spacecraft in a 500 km altitude circular orbit is given a delta-v equal to one-half its orbital speed. Use the energy equation to calculate the hyperbolic excess velocity.
{Ans.: 3.806 km/s}

2.35 A satellite is in a circular orbit at an altitude of 320 km above the earth's surface. If an onboard rocket provides a delta-v of 500 m/s in the direction of the satellite's motion, calculate the altitude of the new orbit's apogee.
{Ans.: 2390 km}

2.36 A spacecraft is in a circular orbit of radius r and speed v around an unspecified planet. A rocket on the spacecraft is fired, instantaneously increasing the speed in the direction of motion by the amount $\Delta v = \alpha$, where $\alpha > 0$. Calculate the eccentricity of the new orbit.
{Ans.: $e = \alpha(\alpha + 2)$}

2.37 A satellite is in a circular earth orbit of altitude 400 km. Determine the new perigee and apogee altitudes if the satellite on-board engine
   (a) increases the speed of the satellite in the flight direction by 240 m/s;
   (b) gives the satellite a radial (outward) component of velocity of 240 m/s.
{Ans.: (a) $z_A = 1230$ km, $z_P = 400$ km; (b) $z_A = 621$ km, $z_P = 196$ km}

2.38 For the sun–earth system, find the distance of the L_1, L_2 and L_3 Lagrange points from the center of mass of the sun–earth system.
{Ans.: $x_1 = 151.101 \times 10^6$ km, $x_2 = 148.108 \times 10^6$ km, $x_3 = -149.600 \times 10^6$ km (opposite side of the sun)}

Chapter 3
Orbital position as a function of time

Chapter outline

3.1	Introduction	107
3.2	Time since periapsis	108
3.3	Circular orbits	108
3.4	Elliptical orbits	109
3.5	Parabolic trajectories	124
3.6	Hyperbolic trajectories	125
3.7	Universal variables	134
Problems		145

3.1 Introduction

In Chapter 2 we found the relationship between position and true anomaly for the two-body problem. The only place time appeared explicitly was in the expression for the period of an ellipse. Obtaining position as a function of time is a simple matter for circular orbits. For elliptical, parabolic and hyperbolic paths we are led to the various forms of Kepler's equation relating position to time. These transcendental equations must be solved iteratively using a procedure like Newton's method, which is presented and illustrated in the chapter.

The different forms of Kepler's equation are combined into a single universal Kepler's equation by introducing universal variables. Implementation of this appealing notion is accompanied by the introduction of an unfamiliar class of functions known as Stumpff functions. The universal variable formulation is required for the Lambert and Gauss orbit determination algorithms in Chapter 5.

The road map of Appendix B may aid in grasping how the material presented here depends on that of Chapter 2.

3.2 Time since periapsis

The orbit formula, $r = (h^2/\mu)/(1 + e\cos\theta)$, gives the position of body m_2 in its orbit around m_1 as a function of the true anomaly. For many practical reasons we need to be able to determine the position of m_2 as a function of time. For elliptical orbits, we have a formula for the period T (Equation 2.72), but we cannot yet calculate the time required to fly between any two true anomalies. The purpose of this section is to come up with the formulas that allow us to do that calculation.

The one equation we have which relates true anomaly directly to time is Equation 2.37, $h = r^2\dot\theta$, which can be written

$$\frac{d\theta}{dt} = \frac{h}{r^2}$$

Substituting $r = (h^2/\mu)/(1 + e\cos\theta)$, we find, after separating variables,

$$\frac{\mu^2}{h^3}dt = \frac{d\theta}{(1+e\cos\theta)^2}$$

Integrating both sides of this equation yields

$$\frac{\mu^2}{h^3}(t - t_p) = \int_0^\theta \frac{d\vartheta}{(1 + e\cos\vartheta)^2} \qquad (3.1)$$

in which the constant of integration t_p is the time at periapse passage, where by definition $\theta = 0$. t_p is the sixth constant of the motion that was missing in Chapter 2. The origin of time is arbitrary. It is convenient to measure time from periapse passage, so we will usually set $t_p = 0$. In that case we have

$$\frac{\mu^2}{h^3}t = \int_0^\theta \frac{d\vartheta}{(1 + e\cos\vartheta)^2} \qquad (3.2)$$

The integral on the right may be found in any standard mathematical handbook. See, for example, Beyer (1991), integrals 341, 366 and 372. The specific form of the integral depends on whether the value of the eccentricity e corresponds to a circle, ellipse, parabola or hyperbola.

3.3 Circular orbits

For a circle, $e = 0$, so the integral in Equation 3.2 is simply $\int_0^\theta d\vartheta$. Thus we have

$$t = \frac{h^3}{\mu^2}\theta$$

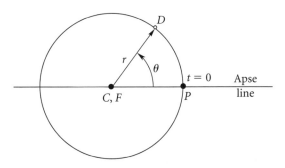

Figure 3.1 Time since periapsis is directly proportional to true anomaly in a circular orbit.

Recall that for a circle (Equation 2.52), $r = h^2/\mu$. Therefore $h^3 = r^{\frac{3}{2}}\mu^{\frac{3}{2}}$, so that

$$t = \frac{r^{\frac{3}{2}}}{\sqrt{\mu}}\theta$$

Finally, substituting the formula (Equation 2.54) for the period T of a circular orbit, $T = 2\pi r^{\frac{3}{2}}/\sqrt{\mu}$, yields

$$t = \frac{\theta}{2\pi}T$$

or

$$\theta = \frac{2\pi}{T}t$$

The reason that t is directly proportional to θ in a circular orbit is simply that the angular velocity $2\pi/T$ is constant. Therefore the time Δt to fly through a true anomaly of $\Delta\theta$ is $(\Delta\theta/2\pi)T$.

Because the circle is symmetric about any diameter, the apse line – and therefore the periapsis – can be chosen arbitrarily.

3.4 Elliptical orbits

For $0 < e < 1$, we find in integral tables that

$$\int_0^\theta \frac{d\vartheta}{(1+e\cos\vartheta)^2} = \frac{1}{(1-e^2)^{\frac{3}{2}}}\left[2\tan^{-1}\left(\sqrt{\frac{1-e}{1+e}}\tan\frac{\theta}{2}\right) - \frac{e\sqrt{1-e^2}\sin\theta}{1+e\cos\theta}\right]$$

Therefore, Equation 3.2 in this case becomes

$$\frac{\mu^2}{h^3}t = \frac{1}{(1-e^2)^{\frac{3}{2}}}\left[2\tan^{-1}\left(\sqrt{\frac{1-e}{1+e}}\tan\frac{\theta}{2}\right) - \frac{e\sqrt{1-e^2}\sin\theta}{1+e\cos\theta}\right]$$

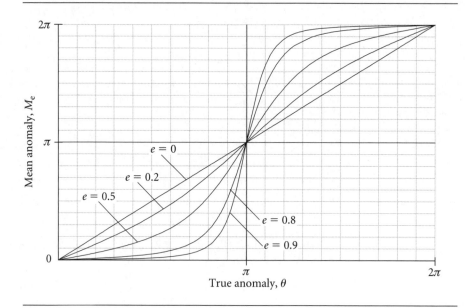

Figure 3.2 Mean anomaly versus true anomaly for ellipses of various eccentricities.

or

$$M_e = 2\tan^{-1}\left(\sqrt{\frac{1-e}{1+e}}\tan\frac{\theta}{2}\right) - \frac{e\sqrt{1-e^2}\sin\theta}{1+e\cos\theta} \quad (3.3)$$

where

$$M_e = \frac{\mu^2}{h^3}(1-e^2)^{\frac{3}{2}}t \quad (3.4)$$

M_e is called the mean anomaly. Equation 3.3 is plotted in Figure 3.2. Observe that for all values of the eccentricity e, M_e is a monotonically increasing function of the true anomaly θ.

From Equation 2.72, the formula for the period T of an elliptical orbit, we have $\mu^2(1-e^2)^{\frac{3}{2}}/h^3 = 2\pi/T$, so that the mean anomaly can be written much more simply as

$$M_e = \frac{2\pi}{T}t \quad (3.5)$$

The angular velocity of the position vector of an elliptical orbit is not constant, but since 2π radians are swept out per period T, the ratio $2\pi/T$ is the average angular velocity, which is given the symbol n and called the mean motion,

$$n = \frac{2\pi}{T} \quad (3.6)$$

In terms of the mean motion, Equation 3.5 can be written simpler still,

$$M_e = nt$$

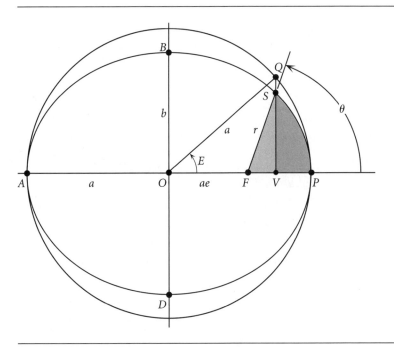

Figure 3.3 Ellipse and the circumscribed auxiliary circle.

The mean anomaly is the azimuth position (in radians) of a fictitious body moving around the ellipse at the constant angular speed n. For a circular orbit, the mean anomaly M_e and the true anomaly θ are identical.

It is convenient to simplify Equation 3.3 by introducing an auxiliary angle E called the eccentric anomaly, which is shown in Figure 3.3. This is done by circumscribing the ellipse with a concentric auxiliary circle having a radius equal to the semimajor axis a of the ellipse. Let S be that point on the ellipse whose true anomaly is θ. Through point S we pass a perpendicular to the apse line, intersecting the auxiliary circle at point Q and the apse line at point V. The angle between the apse line and the radius drawn from the center of the circle to Q on its circumference is the eccentric anomaly E. Observe that E lags θ from P to A, whereas it leads θ from A to P.

To find E as a function of θ, we first observe from Figure 3.3 that, in terms of the eccentric anomaly, $\overline{OV} = a \cos E$ whereas in terms of the true anomaly, $\overline{OV} = ae + r \cos \theta$. Thus

$$a \cos E = ae + r \cos \theta$$

Using Equation 2.62, $r = a(1-e^2)/(1+e\cos\theta)$, we can write this as

$$a \cos E = ae + \frac{a(1-e^2)\cos\theta}{1+e\cos\theta}$$

Simplifying the right-hand side, we get

$$\cos E = \frac{e + \cos \theta}{1 + e \cos \theta} \qquad (3.7a)$$

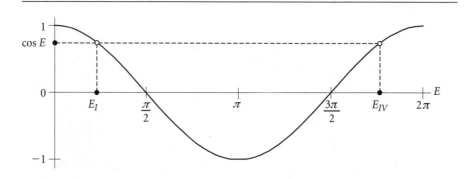

Figure 3.4 For $0 < \cos E < 1$, E can lie in the first or fourth quadrant. For $-1 < \cos E < 0$, E can lie in the second or third quadrant.

Solving this for $\cos\theta$ we obtain the inverse relation,

$$\cos\theta = \frac{e - \cos E}{e \cos E - 1} \qquad (3.7b)$$

Substituting Equation 3.7a into the trigonometric identity $\sin^2 E + \cos^2 E = 1$ and solving for $\sin E$ yields

$$\sin E = \frac{\sqrt{1-e^2}\sin\theta}{1 + e\cos\theta} \qquad (3.8)$$

Equation 3.7a would be fine for obtaining E from θ, except that, given a value of $\cos E$ between -1 and 1, there are *two* values of E between $0°$ and $360°$, as illustrated in Figure 3.4. The same comments hold for Equation 3.8. To resolve this quadrant ambiguity, we use the following trigonometric identity

$$\tan^2\frac{E}{2} = \frac{1 - \cos E}{1 + \cos E} \qquad (3.9)$$

From Equation 3.7a

$$1 - \cos E = \frac{1-\cos\theta}{1+e\cos\theta}(1-e) \quad \text{and} \quad 1 + \cos E = \frac{1+\cos\theta}{1+e\cos\theta}(1+e)$$

Therefore,

$$\tan^2\frac{E}{2} = \frac{1-e}{1+e}\frac{1-\cos\theta}{1+\cos\theta} = \frac{1-e}{1+e}\tan^2\frac{\theta}{2}$$

where the last step required applying the trig identity in Equation 3.9 to the term $(1 - \cos\theta)/(1 + \cos\theta)$. Finally, therefore, we obtain

$$\tan\frac{E}{2} = \sqrt{\frac{1-e}{1+e}}\tan\frac{\theta}{2} \qquad (3.10a)$$

or

$$E = 2\tan^{-1}\left(\sqrt{\frac{1-e}{1+e}}\tan\frac{\theta}{2}\right) \qquad (3.10b)$$

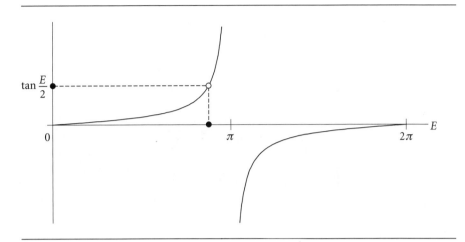

Figure 3.5 To any value of $\tan(E/2)$ there corresponds a unique value of E in the range 0 to 2π.

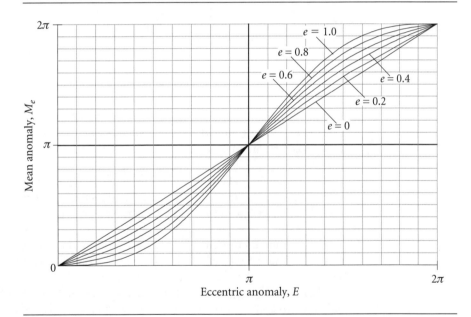

Figure 3.6 Plot of Kepler's equation for an elliptical orbit.

Observe from Figure 3.5 that for any value of $\tan(E/2)$, there is only one value of E between 0° and 360°. There is no quadrant ambiguity.

Substituting Equations 3.8 and 3.10b into Equation 3.3 yields Kepler's equation,

$$M_e = E - e \sin E \qquad (3.11)$$

This monotonically increasing relationship between mean anomaly and eccentric anomaly is plotted for several values of eccentricity in Figure 3.6.

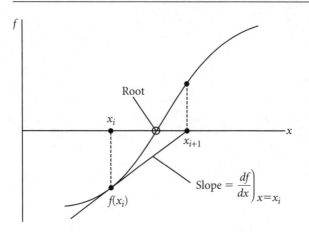

Figure 3.7 Newton's method for finding a root of $f(x)=0$.

Given the true anomaly θ, we calculate the eccentric anomaly E using Equations 3.10. Substituting E into Kepler's formula, Equation 3.11, yields the mean anomaly directly. From the mean anomaly and the period T we find the time (since periapsis) from Equation 3.5,

$$t = \frac{M_e}{2\pi} T \qquad (3.12)$$

On the other hand, if we are given the time, then Equation 3.12 yields the mean anomaly M_e. Substituting M_e into Kepler's equation we get the following expression for the eccentric anomaly,

$$E - e \sin E = M_e$$

We cannot solve this transcendental equation directly for E. A rough value of E might be read off Figure 3.6. However, an accurate solution requires an iterative, 'trial and error' procedure.

Newton's method, or one of its variants, is one of the more common and efficient ways of finding the root of a well-behaved function. To find a root of the equation $f(x)=0$ in Figure 3.7, we estimate it to be x_i, and evaluate the function $f(x)$ and its first derivative $f'(x)$ at that point. We then extend the tangent to the curve at $f(x_i)$ until it intersects the x axis at x_{i+1}, which becomes our updated estimate of the root. The intercept x_{i+1} is found by setting the slope of the tangent line equal to the slope of the curve at x_i, that is,

$$f'(x_i) = \frac{0 - f(x_i)}{x_{i+1} - x_i}$$

from which we obtain

$$x_{i+1} = x_i - \frac{f(x_i)}{f'(x_i)} \qquad (3.13)$$

The process is repeated, using x_{i+1} to estimate x_{i+2}, and so on, until the root has been found to the desired level of precision.

3.4 Elliptical orbits

To apply Newton's method to the solution of Kepler's equation, we form the function

$$f(E) = E - e \sin E - M_e$$

and seek the value of eccentric anomaly that makes $f(E) = 0$. Since

$$f'(E) = 1 - e \cos E$$

for this problem Equation 3.13 becomes

$$E_{i+1} = E_i - \frac{E_i - e \sin E_i - M_e}{1 - e \cos E_i} \qquad (3.14)$$

ALGORITHM 3.1

Solve Kepler's equation for the eccentric anomaly E given the eccentricity e and the mean anomaly M_e. See Appendix D.2 for the implementation of this algorithm in MATLAB®.

1. Choose an initial estimate of the root E as follows (Prussing and Conway, 1993). If $M_e < \pi$, then $E = M_e + e/2$. If $M_e > \pi$, then $E = M_e - e/2$. Remember that the angles E and M_e are in radians. (When using a hand-held calculator, be sure it is in radian mode.)
2. At any given step, having obtained E_i from the previous step, calculate $f(E_i) = E_i - e \sin E_i - M_e$ and $f'(E_i) = 1 - e \cos E_i$.
3. Calculate $\text{ratio}_i = f(E_i)/f'(E_i)$.
4. If $|\text{ratio}_i|$ exceeds the chosen tolerance (e.g., 10^{-8}), then calculate an updated value of E

$$E_{i+1} = E_i - \text{ratio}_i$$

Return to step 2.

5. If $|\text{ratio}_i|$ is less than the tolerance, then accept E_i as the solution to within the chosen accuracy.

EXAMPLE 3.1

A geocentric elliptical orbit has a perigee radius of 9600 km and an apogee radius of 21 000 km. Calculate the time to fly from perigee P to a true anomaly of 120°.

The eccentricity is readily obtained from the perigee and apogee radii by means of Equation 2.74,

$$e = \frac{r_a - r_p}{r_a + r_p} = \frac{21\,000 - 9600}{21\,000 + 9600} = 0.37255 \qquad (a)$$

We find the angular momentum using the orbit equation,

$$9600 = \frac{h^2}{398\,600} \frac{1}{1 + 0.37255 \cos(0)} \Rightarrow h = 72\,472 \text{ km}^2/\text{s}$$

With h and e, the period of the orbit is obtained from Equation 2.72,

$$T = \frac{2\pi}{\mu^2} \left(\frac{h}{\sqrt{1-e^2}}\right)^3 = \frac{2\pi}{398\,600^2} \left(\frac{72\,472}{\sqrt{1-0.37255^2}}\right)^3 = 18\,834 \text{ s} \qquad (b)$$

(Example 3.1 continued)

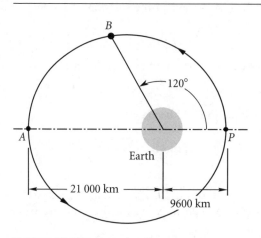

Figure 3.8 Geocentric elliptical orbit.

Equation 3.10a yields the eccentric anomaly from the true anomaly,

$$\tan\frac{E}{2} = \sqrt{\frac{1-e}{1+e}}\tan\frac{\theta}{2} = \sqrt{\frac{1-0.37255}{1+0.37255}}\tan\frac{120°}{2} = 1.1711 \Rightarrow E = 1.7281 \text{ rad}$$

Then Kepler's equation, Equation 3.11, is used to find the mean anomaly,

$$M_e = 1.7281 - 0.37255 \sin 1.7281 = 1.3601 \text{ rad}$$

Finally, the time follows from Equation 3.12,

$$t = \frac{M_e}{2\pi}T = \frac{1.3601}{2\pi}18\,834 = \underline{4077\,\text{s}} \quad (1.132\,\text{hr})$$

EXAMPLE 3.2

In the previous example, find the true anomaly at three hours after perigee passage.

Since the time (10 800 seconds) is greater than one-half the period, the true anomaly must be greater than 180°.

First, we use Equation 3.12 to calculate the mean anomaly for $t = 10\,800$ s:

$$M_e = 2\pi\frac{t}{T} = 2\pi\frac{10\,800}{18\,830} = 3.6029 \text{ rad} \quad \text{(a)}$$

Kepler's equation, $E - e\sin(E) = M_e$ (with all angles in radians), is then employed to find the eccentric anomaly. This transcendental equation will be solved using Algorithm 3.1 with an error tolerance of 10^{-6}. Since $M_e > \pi$, a good starting value for the iteration is $E_0 = M_e - e/2 = 3.4166$. Executing the algorithm yields the following steps:

Step 1:

$$E_0 = 3.4166$$

$$f(E_0) = -0.085124$$

$$f'(E_0) = 1.3585$$
$$\text{ratio} = -0.062658$$
$$|\text{ratio}| > 10^{-6}, \text{ so repeat.}$$

Step 2:
$$E_1 = 3.4166 - (-0.062658) = 3.4793$$
$$f(E_1) = -0.0002134$$
$$f'(E_1) = 1.3515$$
$$\text{ratio} = -1.5778 \times 10^{-4}$$
$$|\text{ratio}| > 10^{-6}, \text{ so repeat.}$$

Step 3:
$$E_2 = 3.4793 - (-1.5778 \times 10^{-4}) = 3.4794$$
$$f(E_2) = -1.5366 \times 10^{-9}$$
$$f'(E_2) = 1.3515$$
$$\text{ratio} = -1.137 \times 10^{-9}$$
$$|f(E_2)| < 10^{-6}, \text{ so accept } \underline{E = 3.4794} \text{ as the solution.}$$

Convergence to even more than the desired accuracy occurred after just two iterations. With this value of the eccentric anomaly, the true anomaly is found from Equation 3.10a

$$\tan\frac{\theta}{2} = \sqrt{\frac{1+e}{1-e}} \tan\frac{E}{2} = \sqrt{\frac{1+0.37255}{1-0.37255}} \tan\frac{3.4794}{2} = -8.6721 \Rightarrow \underline{\theta = 193.2°}$$

EXAMPLE 3.3

Let a satellite be in a 500 km by 5000 km orbit with its apse line parallel to the line from the earth to the sun, as shown below. Find the time that the satellite is in the earth's shadow if: (a) the apogee is towards the sun; (b) the perigee is towards the sun.

(a) If the apogee is towards the sun, as in Figure 3.9, then the satellite is in earth's shadow between points a and b on its orbit. These are two of the four points of intersection of the orbit with lines parallel to the earth–sun line which are a distance R_E from the center of the earth. The true anomaly of b is therefore given by $\sin\theta = R_E/r$, where r is the radial position of the satellite. It follows that the radius of b is

$$r = \frac{R_E}{\sin\theta} \qquad (a)$$

From Equation 2.62 we also have

$$r = \frac{a(1-e^2)}{1+e\cos\theta} \qquad (b)$$

(Example 3.3 continued)

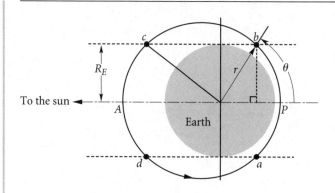

Figure 3.9 Satellite passing in and out of the earth's shadow.

Equating (a) and (b), collecting terms and simplifying yields an equation in θ,

$$e \cos\theta - (1-e^2)\frac{a}{R_E}\sin\theta + 1 = 0 \qquad (c)$$

From the data given in the problem statement, we obtain

$$e = \frac{r_a - r_p}{r_a + r_p} = \frac{(6378+5000)-(6378+500)}{(6378+5000)+(6378+500)} = 0.24649 \qquad (d)$$

$$a = \frac{r_p + r_a}{2} = \frac{(6378+500)+(6378+5000)}{2} = 9128\,\text{km} \qquad (e)$$

$$T = \frac{2\pi}{\sqrt{\mu}}a^{\frac{3}{2}} = \frac{2\pi}{\sqrt{398\,600}}(9128)^{\frac{3}{2}} = 8679.1\,\text{s}\ (2.4109\,\text{hr}) \qquad (f)$$

Substituting (d) and (e) together with $R_E = 6378$ km into (c) yields

$$0.24649\cos\theta - 1.3442\sin\theta = -1 \qquad (g)$$

This equation is of the form

$$a\cos\theta + b\sin\theta = c \qquad (h)$$

It has two roots, which are given by (see Problem 3.9)

$$\theta = \tan^{-1}\frac{b}{a} \pm \cos^{-1}\left[\frac{c}{a}\cos\left(\tan^{-1}\frac{b}{a}\right)\right] \qquad (i)$$

For the case at hand,

$$\theta = \tan^{-1}\frac{-1.3442}{0.24649} \pm \cos^{-1}\left[\frac{-1}{0.24649}\cos\left(\tan^{-1}\frac{-1.3442}{0.24649}\right)\right]$$

$$= -79.607° \pm 137.03°$$

That is

$$\begin{aligned}\theta_b &= 57.423°\\ \theta_c &= -216.64°\ (+143.36°)\end{aligned} \qquad (j)$$

For apogee towards the sun, the flight from perigee to point b will be in shadow. To find the time of flight from perigee to point b, we first compute the eccentric anomaly of b using Equation 3.10b:

$$E_b = 2\tan^{-1}\left(\sqrt{\frac{1-e}{1+e}}\tan\frac{\theta_b}{2}\right) = 2\tan^{-1}\left(\sqrt{\frac{1-0.24649}{1+0.24649}}\tan\frac{1.0022}{2}\right)$$

$$= 0.80521\text{ rad} \tag{k}$$

From this we find the mean anomaly using Kepler's equation,

$$M_e = E - e\sin E = 0.80521 - 0.24649\sin 0.80521 = 0.62749\text{ rad} \tag{l}$$

Finally, Equation (3.5) yields the time at b,

$$t_b = \frac{M_e}{2\pi}T = \frac{0.62749}{2\pi}8679.1 = 866.77\text{ s} \tag{m}$$

The total time in shadow, from a to b, during which the satellite passes through perigee, is

$$t = 2t_b = 1734\text{ s }(28.98\text{ min}) \tag{n}$$

(b) If the perigee is towards the sun, then the satellite is in shadow near apogee, from point c ($\theta_c = 143.36°$) to d on the orbit. Following the same procedure as above we obtain (see Problem 3.12),

$$E_c = 2.3364\text{ rad}$$
$$M_c = 2.1587\text{ rad} \tag{o}$$
$$t_c = 2981.8\text{ s}$$

The total time in shadow, from c to d, is

$$t = T - 2t_c = 8679.1 - 2\cdot 2891.8 = 2716\text{ s }(45.26\text{ min}) \tag{p}$$

The time is longer than that given by (n) since the satellite travels slower near apogee.

We have observed that there is no closed form solution for the eccentric anomaly E in Kepler's equation, $E - e\sin E = M_e$. However, there exist infinite series solutions. One of these, due to Lagrange (Battin, 1999), is a power series in the eccentricity e,

$$E = M_e + \sum_{n=1}^{\infty} a_n e^n \tag{3.15}$$

where the coefficients a_n are given by the somewhat intimidating expression

$$a_n = \frac{1}{2^{n-1}} \sum_{k=0}^{\text{floor}(n/2)} (-1)^k \frac{1}{(n-k)!k!}(n-2k)^{n-1}\sin[(n-2k)M] \tag{3.16}$$

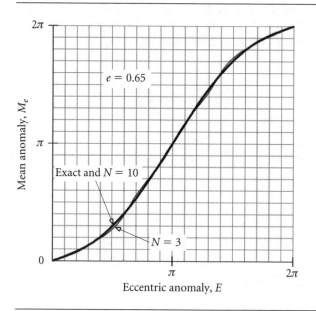

Figure 3.10 Comparison of the exact solution of Kepler's equation with the truncated Lagrange series solution ($N = 3$ and $N = 10$) for an eccentricity of 0.65.

Here, *floor(x)* means rounded to the next lowest integer [e.g., *floor*(0.5) = 0, *floor*(π) = 3]. If e is sufficiently small, then the Lagrange series converges. That means by including enough terms in the summation, we can obtain E to any desired degree of precision. Unfortunately, if e exceeds 0.662743419, the series diverges, which means taking more and more terms yields worse and worse results for some values of M.

The limiting value for the eccentricity was discovered by the French mathematician Pierre-Simon Laplace (1749–1827) and is called the Laplace limit.

In practice, we must truncate the Lagrange series to a finite number of terms N, so that

$$E = M_e + \sum_{n=1}^{N} a_n e^n \qquad (3.17)$$

For example, setting $N = 3$ and calculating each a_n by means of Equation 3.16 leads to

$$E = M_e + e \sin M_e + \frac{e^2}{2} \sin 2M_e + \frac{e^3}{8}(3 \sin 3M_e - \sin M_e) \qquad (3.18)$$

For small values of the eccentricity e this yields good agreement with the exact solution of Kepler's equation (plotted in Figure 3.6). However, as we approach the Laplace limit, the accuracy degrades unless more terms of the series are included. Figure 3.10 shows that for an eccentricity of 0.65, just below the Laplace limit, Equation 3.18 ($N = 3$) yields a solution which oscillates around the exact solution, but is fairly close to it everywhere. Setting $N = 10$ in Equation 3.17 produces a curve which, at the given scale, is indistinguishable from the exact solution. On the other hand, for an eccentricity of 0.90, far above the Laplace limit, Figure 3.11 reveals that Equation 3.18

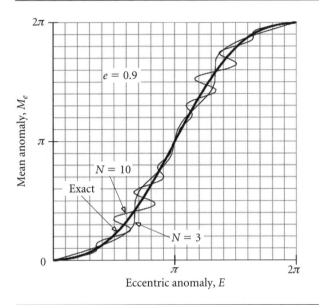

Figure 3.11 Comparison of the exact solution of Kepler's equation with the truncated Lagrange series solution ($N = 3$ and $N = 10$) for an eccentricity of 0.90.

is a poor approximation to the exact solution, and using $N = 10$ makes matters even worse.

Another infinite series for E (Battin, 1999) is given by

$$E = M_e + \sum_{n=1}^{\infty} \frac{2}{n} J_n(ne) \sin nM_e \qquad (3.19)$$

where the coefficients J_n are functions due to the German astronomer and mathematician Friedrich Bessel (1784–1846). These *Bessel functions of the first kind* are defined by the infinite series

$$J_n(x) = \sum_{k=0}^{\infty} \frac{(-1)^k}{k!(n+k)!} \left(\frac{x}{2}\right)^{n+2k} \qquad (3.20)$$

J_1 through J_5 are plotted in Figure 3.12. Clearly, they are oscillatory in appearance and tend towards zero with increasing x.

It turns out that, unlike the Lagrange series, the Bessel function series solution converges for all values of the eccentricity less than 1. Figure 3.13 shows how the truncated Bessel series solution

$$E = M_e + \sum_{n=1}^{N} \frac{2}{n} J_n(ne) \sin nM_e \qquad (3.21)$$

for $N = 3$ and $N = 10$ compares to the exact solution of Kepler's equation for the very large elliptical eccentricity of $e = 0.99$. It can be seen that the case $N = 3$ yields a poor

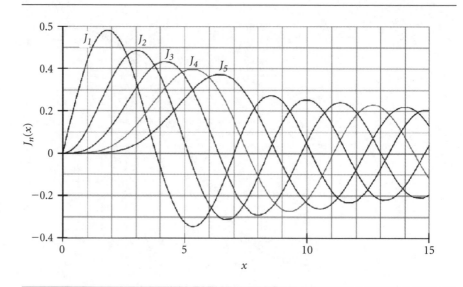

Figure 3.12 Bessel functions of the first kind.

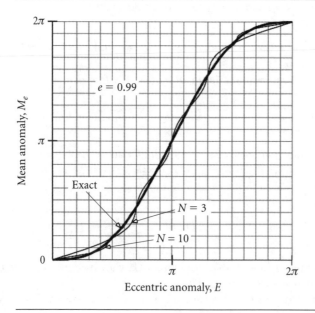

Figure 3.13 Comparison of the exact solution of Kepler's equation with the truncated Bessel series solution ($N = 3$ and $N = 10$) for an eccentricity of 0.99.

approximation for all but a few values of M_e. Increasing the number of terms in the series to $N = 10$ obviously improves the approximation, and adding even more terms will make the truncated series solution indistinguishable from the exact solution at the given scale.

Observe that we can combine Equations 3.7 and 2.62 as follows to obtain the orbit equation for the ellipse in terms of the eccentric anomaly:

$$r = \frac{a(1-e^2)}{1+e\cos\theta} = \frac{a(1-e^2)}{1+e\left(\dfrac{e-\cos E}{e\cos E - 1}\right)}$$

From this it is easy to see that

$$r = a(1 - e\cos E) \tag{3.22}$$

In Equation 2.76 we defined the true-anomaly-averaged radius $\bar{r}_\theta$ of an elliptical orbit. Alternatively, the time-averaged radius $\bar{r}_t$ of an elliptical orbit is defined as

$$\bar{r}_t = \frac{1}{T}\int_0^T r\,dt \tag{3.23}$$

According to Equations 3.11 and 3.12,

$$t = \frac{T}{2\pi}(E - e\sin E)$$

Therefore,

$$dt = \frac{T}{2\pi}(1 - e\cos E)dE$$

Upon using this relationship to change the variable of integration from t to E and substituting Equation 3.22, Equation 3.23 becomes

$$\bar{r}_t = \frac{1}{T}\int_0^{2\pi} [a(1-e\cos E)]\left[\frac{T}{2\pi}(1-e\cos E)\right]dE$$

$$= \frac{a}{2\pi}\int_0^{2\pi} (1 - e\cos E)^2\,dE$$

$$= \frac{a}{2\pi}\int_0^{2\pi} (1 - 2e\cos E + e^2\cos^2 E)dE$$

$$= \frac{a}{2\pi}(2\pi - 0 + e^2\pi)$$

so that

$$\bar{r}_t = a\left(1 + \frac{e^2}{2}\right) \quad \text{Time-averaged radius of an elliptical orbit.} \tag{3.24}$$

Comparing this result with Equation 2.77 reveals, as we should have expected (Why?), that $\bar{r}_t > \bar{r}_\theta$. In fact, combining Equations 2.77 and 3.24 yields

$$\bar{r}_\theta = a\sqrt{3 - 2\frac{\bar{r}_t}{a}} \tag{3.25}$$

3.5 PARABOLIC TRAJECTORIES

For the parabola ($e = 1$), Equation 3.2 becomes

$$\frac{\mu^2}{h^3} t = \int_0^\theta \frac{d\vartheta}{(1 + \cos\vartheta)^2} \tag{3.26}$$

In integral tables we find that

$$\int_0^\theta \frac{d\vartheta}{(1 + \cos\vartheta)^2} = \frac{1}{2} \tan\frac{\theta}{2} + \frac{1}{6} \tan^3 \frac{\theta}{2}$$

Therefore, Equation 3.26 may be written as

$$M_p = \frac{1}{2} \tan\frac{\theta}{2} + \frac{1}{6} \tan^3\frac{\theta}{2} \tag{3.27}$$

where

$$M_p = \frac{\mu^2 t}{h^3} \tag{3.28}$$

M_p is dimensionless, and it may be thought of as the 'mean anomaly' for the parabola. Equation 3.27 is plotted in Figure 3.14. Equation 3.27 is also known as Barker's equation.

Given the true anomaly θ, we find the time directly from Equations 3.27 and 3.28. If time is the given variable, then we must solve the cubic equation

$$\frac{1}{6}\left(\tan\frac{\theta}{2}\right)^3 + \frac{1}{2}\tan\frac{\theta}{2} - M_p = 0$$

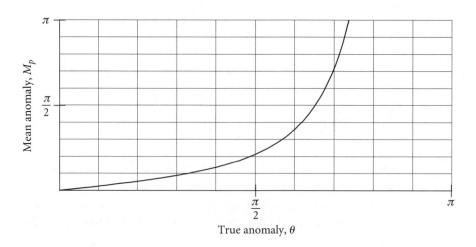

Figure 3.14 Graph of Equation 3.27.

which has but one real root, namely,

$$\tan\frac{\theta}{2} = \left[3M_p + \sqrt{(3M_p)^2 + 1}\right]^{\frac{1}{3}} - \left[(3M_p + \sqrt{(3M_p)^2 + 1})\right]^{-\frac{1}{3}} \qquad (3.29)$$

EXAMPLE 3.4

A geocentric parabola has a perigee velocity of 10 km/s. How far is the satellite from the center of the earth six hours after perigee passage?

Using Equation 2.80, we find the perigee radius,

$$r_p = \frac{2\mu}{v_p^2} = \frac{2 \cdot 398\,600}{10^2} = 7972 \text{ km}$$

so that the angular momentum is

$$h = r_p v_p = 7972 \cdot 10 = 79\,720 \text{ km}^2/\text{s}$$

Now we can calculate the parabolic mean anomaly using Equation 3.28,

$$M_p = \frac{\mu^2 t}{h^3} = \frac{398\,600^2 \cdot (6 \cdot 3600)}{79\,720^3} = 6.7737 \text{ rad}$$

so that $3M_p = 20.321$ rad. Equation 3.29 yields the true anomaly,

$$\tan\frac{\theta}{2} = \left[20.321 + \sqrt{20.321^2 + 1}\right]^{\frac{1}{3}} - \left[(20.321 + \sqrt{20.321^2 + 1})\right]^{-\frac{1}{3}}$$

$$= 3.1481 \Rightarrow \theta = 144.75°$$

Finally, we substitute the true anomaly into the orbit equation to find the radius,

$$r = \frac{79\,720^2}{398\,600} \frac{1}{1 + \cos(144.75°)} = \underline{86\,899 \text{ km}}$$

3.6 Hyperbolic trajectories

For the hyperbola ($e > 1$), integral tables reveal

$$\int_0^\theta \frac{d\vartheta}{(1 + e\cos\vartheta)^2}$$

$$= \frac{1}{e^2 - 1}\left[\frac{e\sin\theta}{1 + e\cos\theta} - \frac{1}{\sqrt{e^2 - 1}}\ln\left(\frac{\sqrt{e+1} + \sqrt{e-1}\tan(\theta/2)}{\sqrt{e+1} - \sqrt{e-1}\tan(\theta/2)}\right)\right]$$

so that Equation 3.1 becomes

$$\frac{\mu^2}{h^3}t = \frac{1}{e^2 - 1}\frac{e\sin\theta}{1 + e\cos\theta} - \frac{1}{(e^2 - 1)^{\frac{3}{2}}}\ln\left(\frac{\sqrt{e+1} + \sqrt{e-1}\tan(\theta/2)}{\sqrt{e+1} - \sqrt{e-1}\tan(\theta/2)}\right)$$

Multiplying both sides by $(e^2 - 1)^{\frac{3}{2}}$, we get

$$M_h = \frac{e\sqrt{e^2 - 1}\sin\theta}{1 + e\cos\theta} - \ln\left(\frac{\sqrt{e+1} + \sqrt{e-1}\tan(\theta/2)}{\sqrt{e+1} - \sqrt{e-1}\tan(\theta/2)}\right) \qquad (3.30)$$

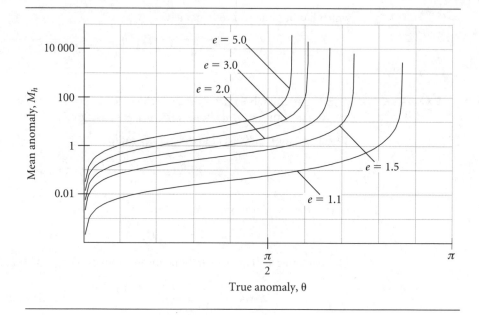

Figure 3.15 Plots of Equation 3.30 for several different eccentricities.

where

$$M_h = \frac{\mu^2}{h^3}(e^2 - 1)^{\frac{3}{2}} t \tag{3.31}$$

M_h is the hyperbolic mean anomaly. Equation 3.30 is plotted in Figure 3.15. Recall that $|\theta| < \cos^{-1}(-1/e)$.

We can simplify Equation 3.30 by introducing an auxiliary angle analogous to the eccentric anomaly E for the ellipse. Consider a point on a hyperbola whose polar coordinates are r and θ. Referring to Figure 3.16, let x be the distance of the point from the center C of the hyperbola, and let y be its distance above the apse line. The ratio y/b defines the hyperbolic sine of the dimensionless variable F that we will use as the hyperbolic eccentric anomaly. That is, we define F to be such that

$$\sinh F = \frac{y}{b} \tag{3.32}$$

In view of the equation of a hyperbola

$$\frac{x^2}{a^2} - \frac{y^2}{b^2} = 1$$

it is consistent with the definition of $\sinh F$ to define the hyperbolic cosine as

$$\cosh F = \frac{x}{a} \tag{3.33}$$

(It should be recalled that $\sinh x = (e^x - e^{-x})/2$ and $\cosh x = (e^x + e^{-x})/2$ and, therefore, that $\cosh^2 x - \sinh^2 x = 1$.)

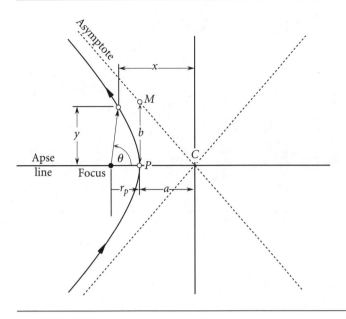

Figure 3.16 Hyperbola parameters.

From Figure 3.16 we see that $y = r \sin \theta$. Substituting this into Equation 3.32, along with $r = a(e^2 - 1)/(1 + e \cos \theta)$ (Equation 2.94) and $b = a\sqrt{e^2 - 1}$ (Equation 2.96), we get

$$\sinh F = \frac{1}{b} r \sin \theta = \frac{1}{a\sqrt{e^2 - 1}} \frac{a(e^2 - 1)}{1 + e \cos \theta} \sin \theta$$

so that

$$\sinh F = \frac{\sqrt{e^2 - 1} \sin \theta}{1 + e \cos \theta} \qquad (3.34)$$

This can be used to solve for F in terms of the true anomaly,

$$F = \sinh^{-1}\left(\frac{\sqrt{e^2 - 1} \sin \theta}{1 + e \cos \theta}\right) \qquad (3.35)$$

Using the formula $\sinh^{-1} x = \ln(x + \sqrt{x^2 + 1})$, we can, after simplifying the algebra, write Equation 3.35 as

$$F = \ln\left(\frac{\sin \theta \sqrt{e^2 - 1} + \cos \theta + e}{1 + e \cos \theta}\right)$$

Substituting the trigonometric identities

$$\sin \theta = \frac{2 \tan(\theta/2)}{1 + \tan^2(\theta/2)} \qquad \cos \theta = \frac{1 - \tan^2(\theta/2)}{1 + \tan^2(\theta/2)}$$

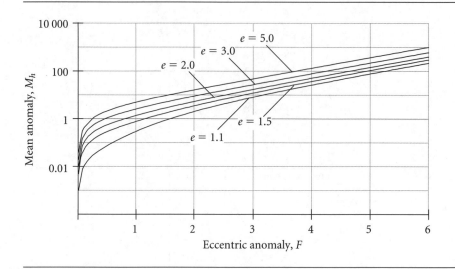

Figure 3.17 Plot of Kepler's equation for the hyperbola.

and doing some more algebra yields

$$F = \ln\left[\frac{1 + e + (e-1)\tan^2(\theta/2) + 2\tan(\theta/2)\sqrt{e^2-1}}{1 + e + (1-e)\tan^2(\theta/2)}\right]$$

Fortunately, but not too obviously, the numerator and the denominator in the brackets have a common factor, so that this expression for the hyperbolic eccentric anomaly reduces to

$$F = \ln\left[\frac{\sqrt{e+1} + \sqrt{e-1}\tan(\theta/2)}{\sqrt{e+1} - \sqrt{e-1}\tan(\theta/2)}\right] \quad (3.36)$$

Substituting Equations 3.34 and 3.36 into Equation 3.30 yields Kepler's equation for the hyperbola,

$$M_h = e \sinh F - F \quad (3.37)$$

This equation is plotted for several different eccentricities in Figure 3.17.

If we substitute the expression for $\sinh F$, Equation 3.34, into the hyperbolic trig identity $\cosh^2 F - \sinh^2 F = 1$, we get

$$\cosh^2 F = 1 + \left(\frac{\sqrt{e^2-1}\sin\theta}{1+e\cos\theta}\right)^2$$

A few steps of algebra lead to

$$\cosh^2 F = \left(\frac{\cos\theta + e}{1 + e\cos\theta}\right)^2$$

so that

$$\cosh F = \frac{\cos\theta + e}{1 + e\cos\theta} \quad (3.38a)$$

Solving this for $\cos\theta$, we obtain the inverse relation,

$$\cos\theta = \frac{\cosh F - e}{1 - e\cosh F} \quad (3.38b)$$

The hyperbolic tangent is found in terms of the hyperbolic sine and cosine by the formula

$$\tanh F = \frac{\sinh F}{\cosh F}$$

In mathematical handbooks we can find the hyperbolic trig identity,

$$\tanh\frac{F}{2} = \frac{\sinh F}{1 + \cosh F} \quad (3.39)$$

Substituting Equations 3.34 and 3.38a into this formula and simplifying yields

$$\tanh\frac{F}{2} = \sqrt{\frac{e-1}{e+1}}\frac{\sin\theta}{1+\cos\theta} \quad (3.40)$$

Interestingly enough, Equation 3.39 holds for ordinary trig functions, too; that is,

$$\tan\frac{\theta}{2} = \frac{\sin\theta}{1+\cos\theta}$$

Therefore, Equation 3.40 can be written

$$\tanh\frac{F}{2} = \sqrt{\frac{e-1}{e+1}}\tan\frac{\theta}{2} \quad (3.41a)$$

This is a somewhat simpler alternative to Equation 3.36 for computing eccentric anomaly from true anomaly, and it is a whole lot simpler to invert:

$$\tan\frac{\theta}{2} = \sqrt{\frac{e+1}{e-1}}\tanh\frac{F}{2} \quad (3.41b)$$

If time is the given quantity, then Equation 3.37 – a transcendental equation – must be solved for F by an iterative procedure, as was the case for the ellipse. To apply Newton's procedure to the solution of Kepler's equation for the hyperbola, we form the function

$$f(F) = e\sinh F - F - M_h$$

and seek the value of F that makes $f(F) = 0$. Since

$$f'(F) = e\cosh F - 1$$

Equation 3.13 becomes

$$F_{i+1} = F_i - \frac{e\sinh F_i - F_i - M_h}{e\cosh F_i - 1} \quad (3.42)$$

All quantities in this formula are dimensionless (radians, not degrees).

ALGORITHM 3.2

Solve Kepler's equation for the hyperbola for the hyperbolic eccentric anomaly F given the eccentricity e and the hyperbolic mean anomaly M_h. See Appendix D.3 for the implementation of this algorithm in MATLAB.

1. Choose an initial estimate of the root F.

   (a) For hand computations read a rough value of F_0 (no more than two significant figures) from Figure 3.17 in order to keep the number of iterations to a minimum.

   (b) In computer software let $F_0 = M_h$, an inelegant choice which may result in many iterations but will nevertheless rapidly converge on today's high speed desktop and laptop computers.

2. At any given step, having obtained F_i from the previous step, calculate $f(F_i) = e \sinh F_i - F_i - M_h$ and $f'(F_i) = e \cosh F_i - 1$.

3. Calculate $\text{ratio}_i = f(F_i)/f'(F_i)$.

4. If $|\text{ratio}_i|$ exceeds the chosen tolerance (e.g., 10^{-8}), then calculate an updated value of F,

$$F_{i+1} = F_i - \text{ratio}_i$$

Return to step 2.

5. If $|\text{ratio}_i|$ is less than the tolerance, then accept F_i as the solution to within the desired accuracy.

EXAMPLE 3.5

A geocentric trajectory has a perigee velocity of 15 km/s and a perigee altitude of 300 km. Find (a) the radius when the true anomaly is 100° and (b) the position and speed three hours later.

(a) The angular momentum is calculated from the given perigee data:

$$h = r_p v_p = (6378 + 300) \cdot 15 = 100\,170 \text{ km}^2/\text{s}$$

The eccentricity is found by evaluating the orbit equation, $r = (h^2/\mu)[1/(1 + e \cos \theta)]$, at perigee:

$$6378 + 300 = \frac{100\,170^2}{398\,600} \frac{1}{1+e} \Rightarrow e = 2.7696 \qquad (a)$$

Since $e > 1$ the trajectory is a hyperbola. Note that the true anomaly of the asymptote of the hyperbola is, from Equation 2.87,

$$\theta_\infty = \cos^{-1}\left(-\frac{1}{2.7696}\right) = 111.17°$$

Solving the orbit equation at $\theta = 100°$ yields

$$r = \frac{100\,170^2}{398\,600} \frac{1}{1 + 2.7696 \cos 100°} = \underline{48\,497 \text{ km}}$$

(b) The time since perigee passage at $\theta = 100°$ must be found next so that we can add the three hour time increment needed to find the final position of the satellite. Using Equation 3.41a to calculate the hyperbolic eccentric anomaly, we find

$$\tanh \frac{F}{2} = \sqrt{\frac{2.7696 - 1}{2.7696 + 1}} \tan \frac{100°}{2} = 0.81653 \Rightarrow F = 2.2927 \text{ rad}$$

Kepler's equation for the hyperbola then yields the mean anomaly,

$$M_h = e \sinh F - F = 2.7696 \sinh 2.2927 - 2.2927 = 11.279 \text{ rad}$$

Now we can obtain the time since perigee passage by means of Equation 3.31,

$$t = \frac{h^3}{\mu^2} \frac{1}{(e^2 - 1)^{\frac{3}{2}}} M_h = \frac{100\,170^3}{398\,600^2} \frac{1}{(2.7696^2 - 1)^{\frac{3}{2}}} 11.279 = \underline{4141 \text{ s}}$$

Three hours later the time since perigee passage is

$$t = 4141.4 + 3 \cdot 3600 = 14\,941 \text{ s } (4.15 \text{ hr})$$

The corresponding mean anomaly, from Equation 3.31, is

$$M_h = \frac{398\,600^2}{100\,170^3} (2.7696^2 - 1)^{\frac{3}{2}} 14\,941 = 40.690 \text{ rad} \quad \text{(b)}$$

We will use Algorithm 3.2 with an error tolerance of 10^{-6} to find the hyperbolic eccentric anomaly F. Referring to Figure 3.17, we see that for $M_h = 40.69$ and $e = 2.7696$, F lies between 3 and 4. Let us arbitrarily choose $F_0 = 3$ as our initial estimate of F. Executing the algorithm yields the following steps:

$$F_0 = 3$$

Step 1:

$$f(F_0) = -15.944494$$
$$f'(F_0) = 26.883397$$
$$\text{ratio} = -0.59309818$$
$$F_1 = 3 - (-0.59309818) = 3.5930982$$
$$|\text{ratio}| > 10^{-6}, \text{ so repeat.}$$

Step 2:

$$f(F_1) = 6.0114484$$
$$f'(F_1) = 49.370747$$
$$\text{ratio} = -0.12176134$$
$$F_2 = 3.5930982 - (-0.12176134) = 3.4713368$$
$$|\text{ratio}| > 10^{-6}, \text{ so repeat.}$$

(Example 3.5 continued)

Step 3:
$$f(F_2) = 0.35812370$$
$$f'(F_2) = 43.605527$$
$$\text{ratio} = 8.2128052 \times 10^{-3}$$
$$F_3 = 3.4713368 - (8.2128052 \times 10^{-3}) = 3.4631240$$
$|\text{ratio}| > 10^{-6}$, so repeat.

Step 4:
$$f(F_3) = 1.4973128 \times 10^{-3}$$
$$f'(F_3) = 43.241398$$
$$\text{ratio} = 3.4626836 \times 10^{-5}$$
$$F_4 = 3.4631240 - (3.4626836 \times 10^{-5}) = 3.4630894$$
$|\text{ratio}| > 10^{-6}$, so repeat.

Step 5:
$$f(F_4) = 2.6470781 \times 10^{-3}$$
$$f'(F_4) = 43.239869$$
$$\text{ratio} = 6.1218459 \times 10^{-10}$$
$$F_5 = 3.4630894 - (6.1218459 \times 10^{-10}) = 3.4630894$$
$|\text{ratio}| < 10^{-6}$, so accept $\underline{F = 3.4631}$ as the solution.

We substitute this value of F into Equation 3.41b to find the true anomaly,

$$\tan\frac{\theta}{2} = \sqrt{\frac{e+1}{e-1}}\tanh\frac{F}{2} = \sqrt{\frac{2.7696+1}{2.7696-1}}\tanh\frac{3.4631}{2} = 1.3708 \Rightarrow \theta = 107.78°$$

With the true anomaly, the orbital equation yields the radial coordinate at the final time

$$r = \frac{h^2}{\mu}\frac{1}{1+e\cos\theta} = \frac{100\,170^2}{398\,600}\frac{1}{1+2.7696\cos 107.78} = \underline{163\,180\text{ km}}$$

The velocity components are obtained from Equation 2.21,

$$v_\perp = \frac{h}{r} = \frac{100\,170}{163\,180} = 0.61386\text{ km/s}$$

and Equation 2.39,

$$v_r = \frac{\mu}{h}e\sin\theta = \frac{398\,600}{100\,170}2.7696\sin 107.78° = 10.494\text{ km/s}$$

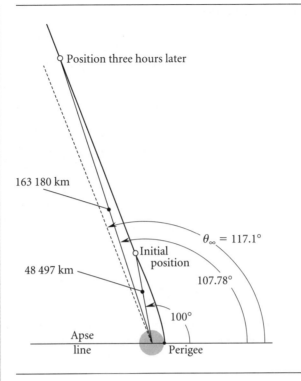

Figure 3.18 Given and computed data for Example 3.5.

Therefore, the speed of the spacecraft is

$$v = \sqrt{v_r^2 + v_\perp^2} = \sqrt{10.494^2 + 0.61386^2} = \underline{10.51 \text{ km/s}}$$

Note that the hyperbolic excess speed for this orbit is

$$v_\infty = \frac{\mu}{h} e \sin \theta_\infty = \frac{398\,600}{100\,170} \cdot 2.7696 \cdot \sin 111.7° = 10.277 \text{ km/s}$$

The results of this analysis are shown in Figure 3.18.

When determining orbital position as a function of time with the aid of Kepler's equation, it is convenient to have position r as a function of eccentric anomaly F. This is obtained by substituting Equation 3.38b into Equation 2.94,

$$r = \frac{a(e^2 - 1)}{1 + e \cos \theta} = \frac{a(e^2 - 1)}{1 + e \left(\dfrac{\cosh F - e}{1 - e \cos F} \right)}$$

This reduces to

$$r = a(e \cosh F - 1) \tag{3.43}$$

3.7 Universal variables

The equations for elliptical and hyperbolic trajectories are very similar, as can be seen from Table 3.1. Observe, for example, that the hyperbolic mean anomaly is obtained from that of the ellipse as follows:

$$M_h = \frac{\mu^2}{h^3}(e^2-1)^{\frac{3}{2}}t$$

$$= \frac{\mu^2}{h^3}\left[(-1)(1-e^2)\right]^{\frac{3}{2}}t$$

$$= \frac{\mu^2}{h^3}(-1)^{\frac{3}{2}}(1-e^2)^{\frac{3}{2}}t$$

$$= \frac{\mu^2}{h^3}(-i)(1-e^2)^{\frac{3}{2}}t$$

$$= -i\left[\frac{\mu^2}{h^3}(1-e^2)^{\frac{3}{2}}t\right]$$

$$= -iM_e$$

In fact, the formulas for the hyperbola can all be obtained from those of the ellipse by replacing the variables in the ellipse equations according to the following scheme, wherein '←' means 'replace by':

$$\begin{aligned} a &\leftarrow -a \\ b &\leftarrow ib \\ M_e &\leftarrow -iM_h \\ E &\leftarrow iF \end{aligned} \quad (i=\sqrt{-1})$$

Note in this regard that $\sin(iF) = i\sinh F$ and $\cos(iF) = \cosh F$. Relations among the circular and hyperbolic trig functions are found in mathematics handbooks, such as Beyer (1991).

In the universal variable approach, the semimajor axis of the hyperbola is considered to have a negative value, so that the energy equation (row 5 of Table 3.1) has the same form for any type of orbit, including the parabola, for which $a = \infty$. In this formulation, the semimajor axis of any orbit is found using (row 3),

$$a = \frac{h^2}{\mu}\frac{1}{1-e^2} \quad (3.44)$$

If the position r and velocity v are known at a given point on the path, then the energy equation (row 5) is convenient for finding the semimajor axis of any orbit,

$$a = \frac{1}{\frac{2}{r}-\frac{v^2}{\mu}} \quad (3.45)$$

Kepler's equation may also be written in terms of a universal variable, or universal 'anomaly' χ, that is valid for all orbits. See, for example, Battin (1999), Bond and Allman (1993) and Prussing and Conway (1993). If t_0 is the time when the universal

3.7 Universal variables

Table 3.1 Comparison of some of the orbital formulas for the ellipse and hyperbola

	Equation	Ellipse ($e < 1$)	Hyperbola ($e > 1$)
1.	Orbit equation (2.35)	$r = \dfrac{h^2}{\mu} \dfrac{1}{1 + e \cos\theta}$	same
2.	Conic equation in cartesian coordinates (2.69), (2.99)	$\dfrac{x^2}{a^2} + \dfrac{y^2}{b^2} = 1$	$\dfrac{x^2}{a^2} - \dfrac{y^2}{b^2} = 1$
3.	Semimajor axis (2.61), (2.93)	$a = \dfrac{h^2}{\mu} \dfrac{1}{1 - e^2}$	$a = \dfrac{h^2}{\mu} \dfrac{1}{e^2 - 1}$
4.	Semiminor axis (2.66), (2.96)	$b = a\sqrt{1 - e^2}$	$b = a\sqrt{e^2 - 1}$
5.	Energy equation (2.71), (2.101)	$\dfrac{v^2}{2} - \dfrac{\mu}{r} = -\dfrac{\mu}{2a}$	$\dfrac{v^2}{2} - \dfrac{\mu}{r} = \dfrac{\mu}{2a}$
6.	Mean anomaly (3.4), (3.31)	$M_e = \dfrac{\mu^2}{h^3}(1 - e^2)^{\frac{3}{2}} t$	$M_h = \dfrac{\mu^2}{h^3}(e^2 - 1)^{\frac{3}{2}} t$
7.	Kepler's equation (3.11), (3.37)	$M_e = E - e \sin E$	$M_h = e \sinh F - F$
8.	Orbit equation in terms of eccentric anomaly (3.22), (3.43)	$r = a(1 - e \cos E)$	$r = a(e \cosh F - 1)$

variable is zero, then the value of χ at time $t_0 + \Delta t$ is found by iterative solution of the universal Kepler's equation

$$\sqrt{\mu}\Delta t = \frac{r_0 v_{r0}}{\sqrt{\mu}} \chi^2 C(\alpha \chi^2) + (1 - \alpha r_0) \chi^3 S(\alpha \chi^2) + r_0 \chi \qquad (3.46)$$

in which r_0 and v_{r0} are the radius and radial velocity at $t = t_0$, and α is the reciprocal of the semimajor axis

$$\alpha = \frac{1}{a} \qquad (3.47)$$

$\alpha < 0$, $\alpha = 0$ and $\alpha > 0$ for hyperbolas, parabolas and ellipses, respectively. The units of χ are km$^{\frac{1}{2}}$ (so $\alpha \chi^2$ is dimensionless). The functions $C(z)$ and $S(z)$ belong to the class known as Stumpff functions, and they are defined by the infinite series,

$$S(z) = \sum_{k=0}^{\infty} (-1)^k \frac{z^k}{(2k+3)!} = \frac{1}{6} - \frac{z}{120} + \frac{z^2}{5040} - \frac{z^3}{362\,880} + \frac{z^4}{39\,916\,800}$$

$$- \frac{z^5}{6\,227\,020\,800} + \cdots \qquad (3.48\text{a})$$

$$C(z) = \sum_{k=0}^{\infty} (-1)^k \frac{z^k}{(2k+2)!} = \frac{1}{2} - \frac{z}{24} + \frac{z^2}{720} - \frac{z^3}{40\,320} + \frac{z^4}{3\,628\,800}$$

$$- \frac{z^5}{479\,001\,600} + \cdots \qquad (3.48\text{b})$$

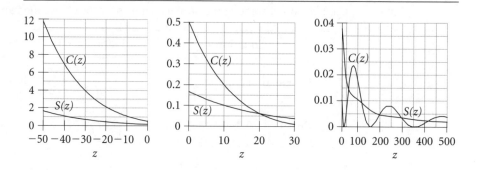

Figure 3.19 A plot of the Stumpff functions $C(z)$ and $S(z)$.

$C(z)$ and $S(z)$ are related to the circular and hyperbolic trig functions as follows:

$$S(z) = \begin{cases} \dfrac{\sqrt{z} - \sin\sqrt{z}}{(\sqrt{z})^3} & (z > 0) \\[6pt] \dfrac{\sinh\sqrt{-z} - \sqrt{-z}}{(\sqrt{-z})^3} & (z < 0) \\[6pt] \dfrac{1}{6} & (z = 0) \end{cases} \qquad (z = \alpha\chi^2) \qquad (3.49)$$

$$C(z) = \begin{cases} \dfrac{1 - \cos\sqrt{z}}{z} & (z > 0) \\[6pt] \dfrac{\cosh\sqrt{-z} - 1}{-z} & (z < 0) \\[6pt] \dfrac{1}{2} & (z = 0) \end{cases} \qquad (z = \alpha\chi^2) \qquad (3.50)$$

Clearly, $z < 0$, $z = 0$ and $z > 0$ for hyperbolas, parabolas and ellipses, respectively. It should be pointed out that if $C(z)$ and $S(z)$ are computed by the series expansions, Equations 3.48a and 3.48b, then the forms of $C(z)$ and $S(z)$, depending on the sign of z, are selected, so to speak, automatically. $C(z)$ and $S(z)$ behave as shown in Figure 3.19. Both $C(z)$ and $S(z)$ are non-negative functions of z. They increase without bound as z approaches $-\infty$ and tend towards zero for large positive values of z. As can be seen from Equation 3.50$_1$, for $z > 0$ $C(z) = 0$ when $\cos\sqrt{z} = 1$, that is, when $z = (2\pi)^2, (4\pi)^2, (6\pi)^2, \ldots$.

The price we pay for using the universal variable formulation is having to deal with the relatively unknown Stumpff functions. However, Equations 3.49 and 3.50 are easy to implement in both computer programs and programmable calculators. See Appendix D.4 for the implementation of these expressions in MATLAB.

To gain some insight into how Equation 3.46 represents the Kepler equations for all of the conic sections, let t_0 be the time at periapse passage and let us set $t_0 = 0$, as we have assumed previously. Then $\Delta t = t$, $v_{r0} = 0$ and r_0 equals r_p, the periapse radius. In that case Equation 3.46 reduces to

$$\sqrt{\mu}\,t = (1 - \alpha r_p)\chi^3 S(\alpha\chi^2) + r_p\chi \qquad (t = 0 \text{ at periapse passage}) \qquad (3.51)$$

Consider first the parabola. In that case $\alpha = 0$ and $S = S(0) = 1/6$, so that Equation 3.51 becomes a cubic polynomial in χ,

$$\sqrt{\mu}t = \frac{1}{6}\chi^3 + r_p\chi$$

Multiply this equation through by $(\sqrt{\mu}/h)^3$ to obtain

$$\frac{\mu^2}{h^3}t = \frac{1}{6}\left(\frac{\chi\sqrt{\mu}}{h}\right)^3 + r_p\chi\left(\frac{\sqrt{\mu}}{h}\right)^3$$

Since $r_p = h^2/2\mu$ for a parabola, we can write this as

$$\frac{\mu^2}{h^3}t = \frac{1}{6}\left(\frac{\sqrt{\mu}}{h}\chi\right)^3 + \frac{1}{2}\left(\frac{\sqrt{\mu}}{h}\chi\right) \quad (3.52)$$

Upon setting $\chi = h\tan(\theta/2)/\sqrt{\mu}$, Equation 3.52 becomes identical to Equation 3.27, the time versus true anomaly relation for the parabola.

Kepler's equation for the ellipse can be obtained by multiplying Equation 3.51 through by $\left(\sqrt{\mu(1-e^2)}/h\right)^3$:

$$\frac{\mu^2}{h^3}(1-e^2)^{\frac{3}{2}}t = \left(\chi\frac{\sqrt{\mu}}{h}\sqrt{1-e^2}\right)^3 (1 - \alpha r_p)S(z)$$

$$+ r_p\chi\left(\frac{\sqrt{\mu}}{h}\sqrt{1-e^2}\right)^3 \quad (z = \alpha\chi^2) \quad (3.53)$$

Recall that for the ellipse, $r_p = h^2/[\mu(1+e)]$ and $\alpha = 1/a = \mu(1-e^2)/h^2$. Using these two expressions in Equation 3.53, along with $S(z) = [\sqrt{\alpha}\chi - \sin(\sqrt{\alpha}\chi)]/\alpha^{\frac{3}{2}}\chi^3$ (from Equation 3.49$_1$), and working through the algebra ultimately leads to

$$M_e = \frac{\chi}{\sqrt{a}} - e\sin\left(\frac{\chi}{\sqrt{a}}\right)$$

Comparing this with Kepler's equation for an ellipse (Equation 3.11) reveals that the relationship between the universal variable χ and the eccentric anomaly E is $\chi = \sqrt{a}E$. Similarly, it can be shown for hyperbolic orbits that $\chi = \sqrt{-a}F$. In summary, the universal anomaly χ is related to the previously encountered anomalies as follows:

$$\chi = \begin{cases} \frac{h}{\sqrt{\mu}}\tan\frac{\theta}{2} & \text{parabola} \\ \sqrt{a}E & \text{ellipse} \\ \sqrt{-a}F & \text{hyperbola} \end{cases} \quad (t_0 = 0, \text{ at periapsis}) \quad (3.54)$$

When t_0 is the time at a point other than periapsis, so that Equation 3.46 applies, then Equations 3.54 become

$$\chi = \begin{cases} \frac{h}{\sqrt{\mu}}\left(\tan\frac{\theta}{2} - \tan\frac{\theta_0}{2}\right) & \text{parabola} \\ \sqrt{a}(E - E_0) & \text{ellipse} \\ \sqrt{-a}(F - F_0) & \text{hyperbola} \end{cases} \quad (3.55)$$

As before, we can use Newton's method to solve Equation 3.46 for the universal anomaly χ, given the time interval Δt. To do so, we form the function

$$f(\chi) = \frac{r_0 v_{r0}}{\sqrt{\mu}} \chi^2 C(z) + (1 - \alpha r_0)\chi^3 S(z) + r_0 \chi - \sqrt{\mu} \Delta t \quad (3.56)$$

and its derivative

$$\frac{df(\chi)}{d\chi} = 2\frac{r_0 v_{r0}}{\sqrt{\mu}} \chi C(z) + \frac{r_0 v_{r0}}{\sqrt{\mu}} \chi^2 \frac{dC(z)}{dz}\frac{dz}{d\chi}$$
$$+ 3(1 - \alpha r_0)\chi^2 S(z) + (1 - r_0\alpha)\chi^3 \frac{dS(z)}{dz}\frac{dz}{d\chi} + r_0 \quad (3.57)$$

where it is to be recalled that

$$z = \alpha \chi^2 \quad (3.58)$$

which means of course that

$$\frac{dz}{d\chi} = 2\alpha \chi \quad (3.59)$$

It turns out that

$$\frac{dS(z)}{dz} = \frac{1}{2z}[C(z) - 3S(z)]$$
$$\frac{dC(z)}{dz} = \frac{1}{2z}[1 - zS(z) - 2C(z)] \quad (3.60)$$

Substituting Equations 3.58, 3.59 and 3.60 into Equation 3.57 and simplifying the result yields

$$\frac{df(\chi)}{d\chi} = \frac{r_0 v_{r0}}{\sqrt{\mu}} \chi[1 - \alpha\chi^2 S(z)] + (1 - \alpha r_0)\chi^2 C(z) + r_0 \quad (3.61)$$

With Equations 3.56 and 3.61, Newton's algorithm (Equation 3.13) for the universal Kepler equation becomes

$$\chi_{i+1} = \chi_i - \frac{\frac{r_0 v_{r0}}{\sqrt{\mu}}\chi_i^2 C(z_i) + (1 - \alpha r_0)\chi_i^3 S(z_i) + r_0 \chi_i - \sqrt{\mu}\Delta t}{\frac{r_0 v_{r0}}{\sqrt{\mu}}\chi_i[1 - \alpha\chi_i^2 S(z_i)] + (1 - \alpha r_0)\chi_i^2 C(z_i) + r_0} \quad (z_i = \alpha\chi_i^2)$$

$$(3.62)$$

According to Chobotov (2002), a reasonable estimate for the starting value χ_0 is

$$\chi_0 = \sqrt{\mu}|\alpha|\Delta t \quad (3.63)$$

ALGORITHM 3.3

Solve the universal Kepler's equation for the universal anomaly χ given Δt, r_0, v_{r0} and α. See Appendix D.5 for an implementation of this procedure in MATLAB.

1. Use Equation 3.63 for an initial estimate of χ_0.
2. At any given step, having obtained χ_i from the previous step, calculate

$$f(\chi_i) = \frac{r_0 v_{r0}}{\sqrt{\mu}} \chi_i^2 C(z_i) + (1 - \alpha r_0)\chi_i^3 S(z_i) + r_0 \chi_i - \sqrt{\mu}\Delta t$$

and
$$f'(\chi_i) = \frac{r_0 v_{r0}}{\sqrt{\mu}} \chi_i [1 - \alpha \chi_i^2 S(z_i)] + (1 - \alpha r_0) \chi_i^2 C(z_i) + r_0$$

where $z_i = \alpha \chi_i^2$.

3. Calculate $\text{ratio}_i = f(\chi_i)/f'(\chi_i)$.
4. If $|\text{ratio}_i|$ exceeds the chosen tolerance (e.g., 10^{-8}), then calculate an updated value of χ,

$$\chi_{i+1} = \chi_i - \text{ratio}_i$$

Return to step 2.

5. If $|\text{ratio}_i|$ is less than the tolerance, then accept χ_i as the solution to within the desired accuracy.

EXAMPLE 3.6

An earth satellite has an initial true anomaly of $\theta_0 = 30°$, a radius of $r_0 = 10\,000$ km, and a speed of $v_0 = 10$ km/s. Use the universal Kepler's equation to find the change in universal anomaly χ after one hour and use that information to determine the true anomaly θ at that time.

Using the initial conditions, let us first determine the angular momentum and the eccentricity of the trajectory. From the orbit formula, Equation 2.35, we have

$$h = \sqrt{\mu r_0 (1 + e \cos \theta_0)} = \sqrt{398\,600 \cdot 10\,000 \cdot (1 + e \cos 30°)}$$
$$= 63\,135\sqrt{1 + 0.86602e} \tag{a}$$

This, together with the angular momentum formula, Equation 2.21, yields

$$v_{\perp 0} = \frac{h}{r_0} = \frac{63\,135\sqrt{1 + 0.86602e}}{10\,000} = 6.3135\sqrt{1 + 0.86602e}$$

Using the radial velocity relation, Equation 2.39, we find

$$v_{r0} = \frac{\mu}{h} e \sin \theta_0 = \frac{398\,600}{63\,135\sqrt{1 + 0.86602e}} e \sin 30° = 3.1567 \frac{e}{\sqrt{1 + 0.86602e}}$$

Since $v_{r0}^2 + v_{\perp 0}^2 = v_0^2$, it follows that

$$\left(3.1567 \frac{e}{\sqrt{1 + 0.86602e}}\right)^2 + \left(6.3135\sqrt{1 + 0.86602e}\right)^2 = 10^2$$

which simplifies to become $39.86e^2 - 17.563e - 60.14 = 0$. The only positive root of this quadratic equation is

$$e = 1.4682$$

Substituting this value of the eccentricity back into (a) yields the angular momentum

$$h = 95\,154 \text{ km}^2/\text{s}$$

(Example 3.6 continued)

The hyperbolic eccentric anomaly F_0 for the initial conditions may now be found from Equation 3.41a,

$$\tanh \frac{F_0}{2} = \sqrt{\frac{e-1}{e+1}} \tan \frac{\theta_0}{2} = \sqrt{\frac{1.4682-1}{1.4682+1}} \tan \frac{30°}{2} = 0.16670$$

Solving for F_0 yields

$$F_0 = 0.23448 \text{ rad} \qquad \text{(b)}$$

The initial radial speed (required in Equation 3.46) is obtained from Equation 2.39,

$$v_{r0} = \frac{\mu}{h} e \sin \theta_0 = \frac{398\,600}{95\,154} \cdot 1.4682 \cdot \sin 30° = 3.0752 \text{ km/s} \qquad \text{(c)}$$

We calculate the semimajor axis of the orbit by means of Equation 3.44,

$$a = \frac{h^2}{\mu} \frac{1}{1-e^2} = \frac{95\,154^2}{398\,600} \frac{1}{1 - 1.4682^2} = -19\,655 \text{ km}$$

The fact that the semimajor axis is negative means the orbit is a hyperbola. Equation 3.47 implies that

$$\alpha = \frac{1}{a} = \frac{1}{-19\,655} = -5.0878 \times 10^{-5} \text{ km}^{-1} \qquad \text{(d)}$$

We will use Algorithm 3.3 with an error tolerance of 10^{-6} to find the universal anomaly. From Equation 3.63, our initial estimate is

$$\chi_0 = \sqrt{398\,600} \cdot |-5.0878 \times 10^{-6}| \cdot 3600 = 115.6$$

Executing the algorithm yields the following steps:

$$\chi_0 = 115.6$$

Step 1:

$$f(\chi_0) = -370\,650.01$$
$$f'(\chi_0) = 26\,956.300$$
$$\text{ratio} = -13.750033$$
$$\chi_1 = 115.6 - (-13.750033) = 129.35003$$
$$|\text{ratio}| > 10^{-6}, \text{ so repeat.}$$

Step 2:

$$f(\chi_1) = 25\,729.002$$
$$f'(\chi_1) = 30\,776.401$$
$$\text{ratio} = 0.83599669$$
$$\chi_2 = 129.35003 - 0.83599669 = 128.51404$$
$$|\text{ratio}| > 10^{-6}, \text{ so repeat.}$$

Step 3:

$$f(\chi_2) = 102.83891$$
$$f'(\chi_2) = 30\,530.672$$
$$\text{ratio} = 3.3683800 \times 10^{-3}$$
$$\chi_3 = 128.51404 - 3.3683800 \times 10^{-3} = 128.51067$$

$|\text{ratio}| > 10^{-6}$, so repeat.

Step 4:

$$f(\chi_3) = 1.6614116 \times 10^{-3}$$
$$f'(\chi_3) = 30\,529.686$$
$$\text{ratio} = 5.4419545 \times 10^{-8}$$
$$\chi_4 = 128.51067 - 5.4419545 \times 10^{-8} = 128.51067$$

$|\text{ratio}| < 10^{-6}$

So we accept

$$\chi = 128.51 \text{ km}^{\frac{1}{2}}$$

as the solution after four iterations. Substituting this value of χ together with the semimajor axis [Equation (d)] into Equation 3.55$_3$ yields

$$F - F_0 = \frac{\chi}{\sqrt{-a}} = \frac{128.51}{\sqrt{-(-19\,655)}} = 0.91664$$

It follows from (b) that the hyperbolic eccentric anomaly after one hour is

$$F = 0.23448 + 0.91664 = 1.1511$$

Finally, we calculate the corresponding true anomaly using Equation 3.41b,

$$\tan\frac{\theta}{2} = \sqrt{\frac{e+1}{e-1}}\tanh\frac{F}{2} = \sqrt{\frac{1.4682+1}{1.4682-1}}\tanh\frac{1.1511}{2} = 1.1926$$

which means that after one hour

$$\theta = 100.04°$$

Recall from Section 2.11 that the position **r** and velocity **v** on a trajectory at any time t can be found in terms of the position $\mathbf{r}_0$ and velocity $\mathbf{v}_0$ at time t_0 by means of the Lagrange f and g coefficients and their first derivatives,

$$\mathbf{r} = f\mathbf{r}_0 + g\mathbf{v}_0 \tag{3.64}$$

$$\mathbf{v} = \dot{f}\mathbf{r}_0 + \dot{g}\mathbf{v}_0 \tag{3.65}$$

Equations 2.148 give f, g, $\dot{f}$ and $\dot{g}$ explicitly in terms of the change in true anomaly $\Delta\theta$ over the time interval $\Delta t = t - t_0$. The Lagrange coefficients can also be derived in terms of changes in the eccentric anomaly ΔE for elliptical orbits, ΔF for hyperbolas or $\Delta \tan(\theta/2)$ for parabolas. However, if we take advantage of the universal variable formulation, we can cover all of these cases with the same set of Lagrange coefficients. In terms of the universal anomaly χ and the Stumpff functions $C(z)$ and $S(z)$, the Lagrange coefficients are (Bond and Allman, 1996)

$$f = 1 - \frac{\chi^2}{r_0} C(\alpha\chi^2) \tag{3.66a}$$

$$g = \Delta t - \frac{1}{\sqrt{\mu}} \chi^3 S(\alpha\chi^2) \tag{3.66b}$$

$$\dot{f} = \frac{\sqrt{\mu}}{rr_0} \left[\alpha\chi^3 S(\alpha\chi^2) - \chi \right] \tag{3.66c}$$

$$\dot{g} = 1 - \frac{\chi^2}{r} C(\alpha\chi^2) \tag{3.66d}$$

The implementation of these four functions in MATLAB is found in Appendix D.6.

ALGORITHM 3.4

Given $\mathbf{r}_0$ and $\mathbf{v}_0$, find $\mathbf{r}$ and $\mathbf{v}$ at a time Δt later. See Appendix D.7 for an implementation of this procedure in MATLAB.

1. Use the initial conditions to find:

   (a) The magnitude of $\mathbf{r}_0$ and $\mathbf{v}_0$,

   $$r_0 = \sqrt{\mathbf{r}_0 \cdot \mathbf{r}_0} \qquad v_0 = \sqrt{\mathbf{v}_0 \cdot \mathbf{v}_0}$$

   (b) The radial component velocity of v_{r0} by projecting $\mathbf{v}_0$ onto the direction of $\mathbf{r}_0$,

   $$v_{r0} = \frac{\mathbf{r}_0 \cdot \mathbf{v}_0}{r_0}$$

   (c) The reciprocal α of the semimajor axis, using Equation 3.45

   $$\alpha = \frac{2}{r_0} - \frac{v_0^2}{\mu}$$

   The sign of α determines whether the trajectory is an ellipse ($\alpha > 0$), parabola ($\alpha = 0$) or hyperbola ($\alpha < 0$).

2. With r_0, v_{r0}, α and Δt, use Algorithm 3.3 to find the universal anomaly χ.
3. Substitute α, r_0, Δt and χ into Equations 3.66a and 3.66b to obtain f, g.
4. Use Equation 3.64 to compute $\mathbf{r}$ and, from that, its magnitude r.
5. Substitute α, r_0, r and χ into Equations 3.66c and 3.66d to obtain $\dot{f}$ and $\dot{g}$.
6. Use Equation 3.65 to compute $\mathbf{v}$.

3.7 Universal variables

EXAMPLE 3.7

An earth satellite moves in the *xy* plane of an inertial frame with origin at the earth's center. Relative to that frame, the position and velocity of the satellite at time t_0 are

$$\mathbf{r}_0 = 7000.0\hat{\mathbf{i}} - 12\,124\hat{\mathbf{j}}\ (\text{km}) \qquad \mathbf{v}_0 = 2.6679\hat{\mathbf{i}} + 4.6210\hat{\mathbf{j}}\ (\text{km/s}) \qquad (a)$$

Compute the position and velocity vectors of the satellite 60 minutes later using Algorithm 3.4.

Step 1:

$$r_0 = \sqrt{7000.0^2 + (-12\,124)^2} = 14\,000\ \text{km}$$

$$v_0 = \sqrt{2.6679^2 + 4.6210^2} = 5.3359\ \text{km/s}$$

$$v_{r0} = \frac{7000.0 \cdot 2.6679 + (-12\,124) \cdot 4.6210}{14\,000} = -2.6679\ \text{km/s}$$

$$\alpha = \frac{2}{14\,000} - \frac{5.3359^2}{398\,600} = 7.1429 \times 10^{-5}\ \text{km}^{-1}$$

The trajectory is an ellipse, because α is positive.

Step 2:

Using the results of Step 1, Algorithm 3.3 yields

$$\chi = 253.53\ \text{km}^{\frac{1}{2}}$$

which means

$$z = \alpha\chi^2 = 7.1429 \times 10^{-5} \cdot 253.53^2 = 4.5911$$

Step 3:

Substituting the above values of χ and z into Equations 3.66a and 3.66b we find

$$f = 1 - \frac{\chi^2}{r_0}C(\alpha\chi^2) = 1 - \frac{253.53^2}{14\,000}\overbrace{C(4.5911)}^{0.3357} = -0.54123$$

$$g = \Delta t - \frac{1}{\sqrt{\mu}}\chi^3 S(\alpha\chi^2) = 3600 - \frac{253.53^2}{\sqrt{398\,600}}\overbrace{S(4.5911)}^{0.13233} = 184.35\ \text{s}^{-1}$$

Step 4:

$$\mathbf{r} = f\mathbf{r}_0 + g\mathbf{v}_0$$

$$= (-0.54123)(7000.0\hat{\mathbf{i}} - 12.124\hat{\mathbf{j}}) + 184.35(2.6679\hat{\mathbf{i}} + 4.6210\hat{\mathbf{j}})$$

$$= \underline{-3296.8\hat{\mathbf{i}} + 7413.9\hat{\mathbf{j}}\ (\text{km})}$$

(Example 3.7 continued)

Therefore, the magnitude of **r** is

$$r = \sqrt{(-3296.8)^2 + 7413.9^2} = 8113.9 \text{ km}$$

Step 5:

$$\dot{f} = \frac{\sqrt{\mu}}{rr_0}\left[\alpha\chi^3 S(\alpha\chi^2) - \chi\right]$$

$$= \frac{\sqrt{398\,600}}{8113.9 \cdot 14\,000}\left[(7.1429 \times 10^5) \cdot 253.53^2 \cdot \overbrace{S(4.5911)}^{0.13233} - 253.53\right]$$

$$= -0.00055298 \text{ s}^{-1}$$

$$\dot{g} = 1 - \frac{\chi^2}{r}C(\alpha\chi^2) = 1 - \frac{253.53^2}{8113.9}\overbrace{C(4.5911)}^{0.3357} = -1.6593$$

Step 6:

$$\mathbf{v} = \dot{f}\mathbf{r}_0 + \dot{g}\mathbf{v}_0$$
$$= (-0.00055298)(7000.0\hat{\mathbf{i}} - 12.124\hat{\mathbf{j}}) + (-1.6593)\mathbf{v}_0(2.6679\hat{\mathbf{i}} + 4.6210\hat{\mathbf{j}})$$
$$= \underline{-8.2977\hat{\mathbf{i}} - 0.96309\hat{\mathbf{j}}} \text{ (km/s)}$$

The initial and final position and velocity vectors, as well as the trajectory, are accurately illustrated in Figure 3.20.

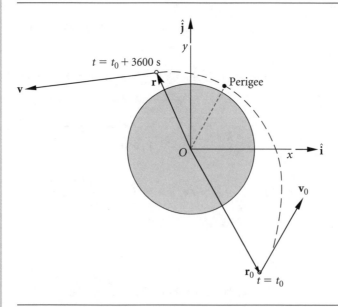

Figure 3.20 Initial and final points on a geocentric trajectory.

Problems

3.1 Use Newton's method to find, to eight significant figures, the positive roots of the equation $10e^{\sin x} = x^2 - 5x + 4$. In each case, starting with your initial guess, list each successive approximation until subsequent iterations produce changes only beyond eight significant figures. Recall that successive estimates of a root of the equation $f(x) = 0$ are obtained from the formula $x_{i+1} = x_i - f(x_i)/f'(x_i)$.

3.2 Use Newton's method to find, to eight significant figures, the first four non-negative roots of the equation $\tan(x) = \tanh(x)$. Starting with your initial guess, list each successive approximation until subsequent iterations produce changes only beyond eight significant figures.

3.3 A satellite is in earth orbit for which perigee altitude is 200 km and apogee altitude is 600 km. Find the time interval during which the satellite remains above an altitude of 400 km.
{Ans.: 47.15 min}

3.4 An earth-orbiting satellite has a perigee radius of 7000 km and an apogee radius of 10 000 km.
(a) What true anomaly $\Delta\theta$ is swept out between $t = 0.5$ hr and $t = 1.5$ hr after perigee passage?
(b) What area is swept out by the position vector during that time interval?
{Ans.: (a) 128.7°; (b) 1.03×10^8 km^2}

3.5 An earth-orbiting satellite has a period of 15.743 hours and a perigee radius of 12 756 km. At time $t = 10$ hours after perigee passage, determine
(a) the radius;
(b) the speed;
(c) the radial component of the velocity.
{Ans.: (a) 48 290 km; (b) 2.00 km/s; (c) −0.7210 km/s}

3.6 In terms of the eccentricity e and the period T, calculate
(a) the time required to fly from D to B through perigee;
(b) the time required to fly from B to D through apogee.
{Ans.: (a) $t_{DPB} = (1/2 - e/\pi)T$; (b) $t_{BAD} = (1/2 + e/\pi)T$}

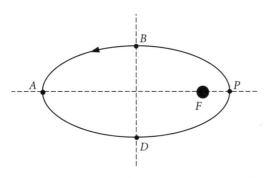

Figure P.3.6

3.7 If the eccentricity of the elliptical orbit is 0.3, calculate, in terms of the period T, the time required to fly from P to B.
{Ans.: $0.157T$}

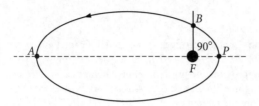

Figure P.3.7

3.8 A satellite in earth orbit has perigee and apogee radii of $r_p = 7000$ km and $r_a = 14\,000$ km, respectively. Find its true anomaly 30 minutes after passing true anomaly of 60°.
{Ans.: 127°}

3.9 Show that the solution to $a\cos\theta + b\sin\theta = c$, where a, b and c are given, is

$$\theta = \phi \pm \cos^{-1}\left(\frac{c}{a}\cos\phi\right)$$

where $\tan\phi = b/a$.

3.10 Calculate the time required to fly from P to B, in terms of the eccentricity e and the period T. B lies on the minor axis.
{Ans.: $(0.25 - 0.1592e)T$}

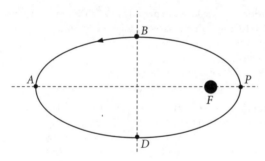

Figure P.3.10

3.11 If the eccentricity of the elliptical orbit is 0.5, calculate, in terms of the period T, the time required to fly from P to B.
{Ans.: $0.170T$}

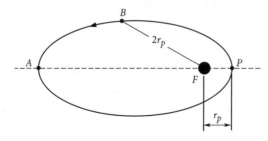

Figure P.3.11

3.12 Verify the results of part (b) of Example 3.3.

3.13 Calculate the time required for a spacecraft launched into a parabolic trajectory at a perigee altitude of 500 km to leave the earth's sphere of influence (see Table A.2).
{Ans.: 7 d 18 hr 34 min}

3.14 A spacecraft on a parabolic trajectory around the earth has a perigee radius of 7500 km.
(a) How long does it take to fly from $\theta = -90°$ to $\theta = +90°$?
(b) How far is the spacecraft from the center of the earth 24 hours after passing through perigee?
{Ans.: (a) 1.078 hr; (b) 230 200 km}

3.15 A spacecraft on a hyperbolic trajectory around the earth has a perigee radius of 7500 km and a perigee speed of $1.1 v_{esc}$.
(a) How long does it take to fly from $\theta = -90°$ to $\theta = +90°$?
(b) How far is the spacecraft from the center of the earth 24 hours after passing through perigee?
{Ans.: (a) 1.14 hr; (b) 456 000 km}

3.16 A trajectory has a perigee velocity of 11.5 km/s and a perigee altitude of 300 km. If at 6 AM the satellite is traveling towards the earth with a speed of 10 km/s, how far will it be from the earth's surface at 11 AM the same day?
{Ans.: 88 390 km}

3.17 An incoming object is sighted at an altitude of 37 000 km with a speed of 8 km/s and a flight path angle of $-65°$.
(a) Will it impact the earth or fly by?
(b) What is the time to impact or closest passage?
{Ans.: (b) 1 hr 24 min}

3.18 At a given instant the radial position of an earth-orbiting satellite is 7200 km, its radial speed is 1 km/s. If the semimajor axis is 10 000 km, use Algorithm 3.3 to find the universal anomaly 60 minutes later. Check your result using Equation 3.55.

3.19 At a given instant a space object has the following position and velocity vectors relative to an earth-centered inertial frame of reference:

$$\mathbf{r}_0 = 20\,000\hat{\mathbf{i}} - 105\,000\hat{\mathbf{j}} - 19\,000\hat{\mathbf{k}} \text{ (km)}$$

$$\mathbf{v}_0 = 0.9000\hat{\mathbf{i}} - 3.4000\hat{\mathbf{j}} - 1.5000\hat{\mathbf{k}} \text{ (km/s)}$$

Find **r** and **v** two hours later.
{Ans.: $\mathbf{r} = 26\,338\hat{\mathbf{i}} - 128\,750\hat{\mathbf{j}} - 29\,656\hat{\mathbf{k}}$ (km); $\mathbf{v} = 0.862800\hat{\mathbf{i}} - 3.2116\hat{\mathbf{j}} - 1.4613\hat{\mathbf{k}}$ (km/s)}

CHAPTER 4

Orbits in Three Dimensions

Chapter outline

4.1	Introduction	149
4.2	Geocentric right ascension–declination frame	150
4.3	State vector and the geocentric equatorial frame	154
4.4	Orbital elements and the state vector	158
4.5	Coordinate transformation	164
4.6	Transformation between geocentric equatorial and perifocal frames	172
4.7	Effects of the earth's oblateness	177
Problems		187

4.1 Introduction

The discussion of orbital mechanics up to now has been confined to two dimensions, i.e., to the plane of the orbits themselves. This chapter explores the means of describing orbits in three-dimensional space, which, of course, is the setting for real missions and orbital maneuvers. Our focus will be on the orbits of earth satellites, but the applications are to any two-body trajectories, including interplanetary missions to be discussed in Chapter 8.

We begin with a discussion of the ancient concept of the celestial sphere and the use of right ascension and declination to define the location of stars, planets and other celestial objects on the sphere. This leads to the establishment of the inertial geocentric equatorial frame of reference and the concept of state vector. The six components of this vector give the instantaneous position and velocity of an object relative to the inertial frame and define the characteristics of the orbit. Following that discussion is a presentation of the six classical orbital elements, which also uniquely define the shape and orientation of an orbit and the location of a body on it. We then show how to transform the state vector into orbital elements and vice versa, taking advantage of the perifocal frame introduced in Chapter 2.

The chapter concludes with a summary of two major perturbations of earth orbits due to the earth's non-spherical shape. These perturbations are exploited to place satellites in sun-synchronous and molniya orbits.

4.2 Geocentric right ascension–declination frame

The coordinate system used to describe earth orbits in three dimensions is defined in terms of earth's equatorial plane, the ecliptic plane, and the earth's axis of rotation. The ecliptic is the plane of the earth's orbit around the sun, as illustrated in Figure 4.1. The earth's axis of rotation, which passes through the North and South Poles, is not perpendicular to the ecliptic. It is tilted away by an angle known as the obliquity of the ecliptic, ε. For the earth ε is approximately 23.4°. Therefore, the earth's equatorial plane and the ecliptic intersect along a line, which is known as the vernal equinox line. On the calendar, 'vernal equinox' is the first day of spring in the northern hemisphere, when the noontime sun crosses the equator from south to north. The position of the sun at that instant defines the location of a point in the sky called the vernal equinox, for which the symbol γ is used. On the day of the vernal equinox, the number of hours of daylight and darkness is equal; hence, the word equinox. The other equinox occurs

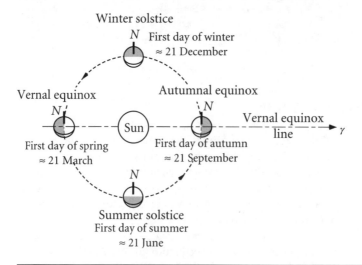

Figure 4.1 The earth's orbit around the sun, viewed from above the ecliptic plane, showing the change of seasons in the northern hemisphere.

precisely one-half year later, when the sun crosses back over the equator from north to south, thereby defining the first day of autumn. The vernal equinox lies today in the constellation Pisces, which is visible in the night sky during the fall. The direction of the vernal equinox line is from the earth towards γ, as shown in Figure 4.1.

For many practical purposes, the vernal equinox line may be considered fixed in space. However, it actually rotates slowly because the earth's tilted spin axis precesses westward around the normal to the ecliptic at the rate of about 1.4° per century. This slow precession is due primarily to the action of the sun and the moon on the non-spherical distribution of mass within the earth. Due to the centrifugal force of rotation about its own axis, the earth bulges very slightly outward at its equator. This effect is shown highly exaggerated in Figure 4.2. One of the bulging sides is closer to the sun than the other, so the force of the sun's gravity $\mathbf{f}_1$ on its mass is slightly larger than the force $\mathbf{f}_2$ on the opposite side, farthest from the sun. The forces $\mathbf{f}_1$ and $\mathbf{f}_2$, along with the dominant force $\mathbf{F}$ on the spherical mass, comprise the total force of the sun on the earth, holding in its solar orbit. Taken together, $\mathbf{f}_1$ and $\mathbf{f}_2$ produce a net clockwise moment (a vector into the page) about the center of the earth. That moment would rotate the earth's equator into alignment with the ecliptic if it were not for the fact that the earth has an angular momentum directed along its south-to-north polar axis due to its spin around that axis at an angular velocity ω_E of 360° per day. The effect of the moment is to rotate the angular momentum vector in the direction of the moment (into the page). The result is that the spin axis is forced to precess in a counterclockwise direction around the normal to the ecliptic, sweeping out a cone as illustrated in the figure. The moon exerts a torque on the earth for the same reason, and the combined effect of the sun and the moon is a precession of the spin axis, and hence γ, with a period of about 26 000 years. The moon's action also superimposes a small nutation on the precession. This causes the obliquity ε to vary with a maximum amplitude of 0.0025° over a period of 18.6 years.

Four thousand years ago, when the first recorded astronomical observations were being made, γ was located in the constellation Aries, the ram. The Greek letter γ is a descendent of the ancient Babylonian symbol resembling the head of a ram.

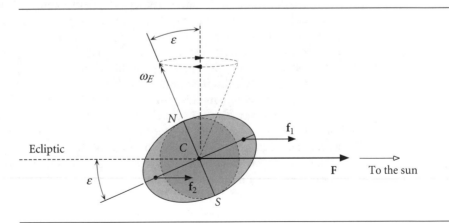

Figure 4.2 Secondary (perturbing) gravitational forces on the earth.

152 Chapter 4 Orbits in three dimensions

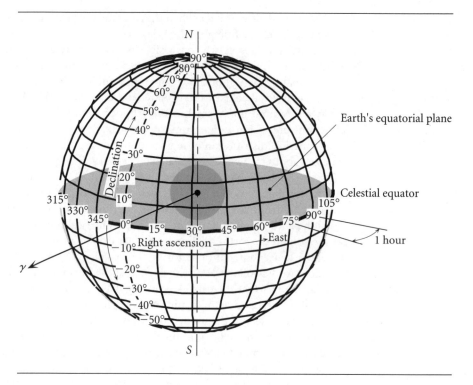

Figure 4.3 The celestial sphere, with grid lines of right ascension and declination.

To the human eye, objects in the night sky appear as points on a celestial sphere surrounding the earth, as illustrated in Figure 4.3. The north and south poles of this fixed sphere correspond to those of the earth rotating within it. Coordinates of latitude and longitude are used to locate points on the celestial sphere in much the same way as on the surface of the earth. The projection of the earth's equatorial plane outward onto the celestial sphere defines the celestial equator. The vernal equinox γ, which lies on the celestial equator, is the origin for measurement of longitude, which in astronomical parlance is called right ascension. Right ascension (RA or α) is measured along the celestial equator in degrees east from the vernal equinox. (Astronomers measure right ascension in hours instead of degrees, where 24 hours equals 360°.) Latitude on the celestial sphere is called declination. Declination (Dec or δ) is measured along a meridian in degrees, positive to the north of the equator and negative to the south. Figure 4.4 is a sky chart showing how the heavenly grid appears from a given point on the earth. Notice that the sun is located at the intersection of the equatorial and ecliptic planes, so this must be the first day of spring.

Stars are so far away from the earth that their positions relative to each other appear stationary on the celestial sphere. Planets, comets, satellites, etc., move upon the fixed backdrop of the stars. The coordinates of celestial bodies as a function of time is called an ephemeris, for example, the *Astronomical Almanac* (US Naval Observatory, 2004). Table 4.1 is an abbreviated ephemeris for the moon and for Venus. An ephemeris depends on the location of the vernal equinox at a given time

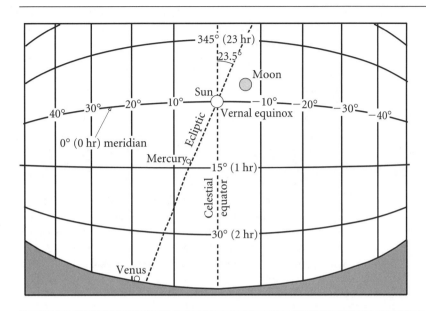

Figure 4.4 A view of the sky above the eastern horizon from 0° longitude on the equator at 9 am local time, 20 March, 2004. (Precession epoch AD 2000.)

Table 4.1 Venus and moon ephemeris for 0 hours universal time (Precession epoch: AD 2000)

Date	Venus RA	Venus Dec	Moon RA	Moon Dec
1 Jan 2004	21 hr 05.0 min	−18°36′	1 hr 44.9 min	+8°47′
1 Feb 2004	23 hr 28.0 min	−04°30′	4 hr 37.0 min	+24°11′
1 Mar 2004	01 hr 30.0 min	+10°26′	6 hr 04.0 min	+08°32′
1 Apr 2004	03 hr 37.6 min	+22°51′	9 hr 18.7 min	+21°08′
1 May 2004	05 hr 20.3 min	+27°44′	11 hr 28.8 min	+07°53′
1 Jun 2004	05 hr 25.9 min	+24°43′	14 hr 31.3 min	−14°48′
1 Jul 2004	04 hr 34.5 min	+17°48′	17 hr 09.0 min	−26°08′
1 Aug 2004	05 hr 37.4 min	+19°04′	21 hr 05.9 min	−21°49′
1 Sep 2004	07 hr 40.9 min	+19°16′	00 hr 17.0 min	−00°56′
1 Oct 2004	09 hr 56.5 min	+12°42′	02 hr 20.9 min	+14°35′
1 Nov 2004	12 hr 15.8 min	+00°01′	05 hr 26.7 min	+27°18′
1 Dec 2004	14 hr 34.3 min	−13°21′	07 hr 50.3 min	+26°14′
1 Jan 2005	17 hr 12.9 min	−22°15′	10 hr 49.4 min	+11°39′

or epoch, for we know that even the positions of the stars relative to the equinox change slowly with time. For example, Table 4.2 shows the celestial coordinates of the star Regulus at five epochs since AD 1700. Currently, the position of the vernal equinox in the year 2000 is used to define the standard grid of the celestial sphere.

Table 4.2 Variation of the coordinates of the star Regulus due to precession of the equinox

Precession epoch	RA	Dec
AD 1700	9 hr 52.2 min (148.05°)	+13°25′
AD 1800	9 hr 57.6 min (149.40°)	+12°56′
AD 1900	10 hr 3.0 min (150.75°)	+12°27′
AD 1950	10 hr 5.7 min (151.42°)	+12°13′
AD 2000	10 hr 8.4 min (152.10°)	+11°58′

In 2025, the position will be updated to that of the year 2050; in 2075 to that of the year 2100; and so on at 50 year intervals. Since observations are made relative to the actual orientation of the earth, these measurements must be transformed into the standardized celestial frame of reference. As Table 4.2 suggests, the adjustments will be small if the current epoch is within 25 years of the standard precession epoch.

4.3 STATE VECTOR AND THE GEOCENTRIC EQUATORIAL FRAME

At any given time, the state vector of a satellite comprises its velocity $\mathbf{v}$ and acceleration $\mathbf{a}$. Orbital mechanics is concerned with specifying or predicting state vectors over intervals of time. From Chapter 2, we know that the equation governing the state vector of a satellite traveling around the earth is, under the familiar assumptions,

$$\ddot{\mathbf{r}} = -\frac{\mu}{r^3}\mathbf{r} \tag{4.1}$$

$\mathbf{r}$ is the position vector of the satellite relative to the center of the earth. The components of $\mathbf{r}$ and, especially, those of its time derivatives $\dot{\mathbf{r}} = \mathbf{v}$ and $\ddot{\mathbf{r}} = \mathbf{a}$, must be measured in a non-rotating frame attached to the earth. A commonly used non-rotating right-handed cartesian coordinate system is the geocentric equatorial frame shown in Figure 4.5. The X axis points in the vernal equinox direction. The XY plane is the earth's equatorial plane, and the Z axis coincides with the earth's axis of rotation and points northward. The unit vectors $\hat{\mathbf{I}}, \hat{\mathbf{J}}$ and $\hat{\mathbf{K}}$ form a right-handed triad. The non-rotating geocentric equatorial frame serves as an inertial frame for the two-body earth satellite problem, as embodied in Equation 4.1. It is not truly an inertial frame, however, since the center of the earth is always accelerating towards a third body, the sun (to say nothing of the moon), a fact which we ignore in the two-body formulation.

In the geocentric equatorial frame the state vector is given in component form by

$$\mathbf{r} = X\hat{\mathbf{I}} + Y\hat{\mathbf{J}} + Z\hat{\mathbf{K}} \tag{4.2}$$

$$\mathbf{v} = v_X\hat{\mathbf{I}} + v_Y\hat{\mathbf{J}} + v_Z\hat{\mathbf{K}} \tag{4.3}$$

If r is the magnitude of the position vector, then

$$\mathbf{r} = r\hat{\mathbf{u}}_r \tag{4.4}$$

4.3 State vector and the geocentric equatorial frame

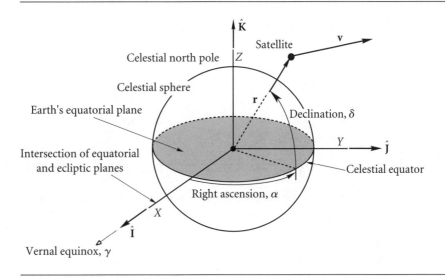

Figure 4.5 Geocentric equatorial frame.

From Figure 4.5 we see that the components of $\hat{\mathbf{u}}_r$ (the direction cosines of $\mathbf{r}$) are found in terms of the right ascension α and declination δ as follows,

$$\hat{\mathbf{u}}_r = \cos\delta \cos\alpha \hat{\mathbf{I}} + \cos\delta \sin\alpha \hat{\mathbf{J}} + \sin\delta \hat{\mathbf{K}} \tag{4.5}$$

Therefore, given the state vector, we can then compute the right ascension and declination. However, the right ascension and declination alone do not furnish $\mathbf{r}$. For that we need the distance r to obtain $\mathbf{r}$ from Equation 4.4.

EXAMPLE 4.1

If the position vector of the International Space Station is

$$\mathbf{r} = -5368\hat{\mathbf{I}} - 1784\hat{\mathbf{J}} + 3691\hat{\mathbf{K}} \text{ (km)}$$

what are its right ascension and declination?

The magnitude of $\mathbf{r}$ is

$$r = \sqrt{(-5368)^2 + (-1784)^2 + 3691^2} = 6754 \text{ km}$$

Hence,

$$\hat{\mathbf{u}}_r = \frac{\mathbf{r}}{r} = -0.7947\hat{\mathbf{I}} - 0.2642\hat{\mathbf{J}} + 0.5464\hat{\mathbf{K}} \tag{a}$$

From this and Equation 4.5 we see that $\sin\delta = 0.5464$ which means

$$\delta = \sin^{-1} 0.5464 = \underline{33.12°}$$

There is no quadrant ambiguity since, by definition, the declination lies between $-90°$ and $+90°$, which is precisely the range of the principal values of the arcsin function. It also follows that $\cos\delta$ cannot be negative.

(Example 4.1 continued)

From Equation 4.5 and Equation (a) just above we have

$$\cos \delta \cos \alpha = -0.7947 \tag{b}$$
$$\cos \delta \sin \alpha = -0.2642 \tag{c}$$

Therefore

$$\cos \alpha = \frac{-0.7947}{\cos 33.12°} = -0.9489$$

which implies

$$\alpha = \cos^{-1}(-0.9489) = 161.6° \text{ (second quadrant) or } 198.4° \text{ (third quadrant)}$$

From (c) we observe that $\sin \alpha$ is negative, which means α lies in the third quadrant,

$$\alpha = \underline{198.4°}$$

If we are provided with the state vector $\mathbf{r}_0$, $\mathbf{v}_0$ at a given instant, then we can determine the state vector at any other time in terms of the initial vector by means of the expressions

$$\begin{aligned} \mathbf{r} &= f\mathbf{r}_0 + g\mathbf{v}_0 \\ \mathbf{v} &= \dot{f}\mathbf{r}_0 + \dot{g}\mathbf{v}_0 \end{aligned} \tag{4.6}$$

where the Lagrange coefficients f and g and their time derivatives are given in Equation 3.66. Specifying the total of six components of $\mathbf{r}_0$ and $\mathbf{v}_0$ therefore completely determines the size, shape and orientation of the orbit.

EXAMPLE 4.2

At time t_0 the state vector of an earth satellite is

$$\mathbf{r}_0 = 1600\hat{\mathbf{I}} + 5310\hat{\mathbf{J}} + 3800\hat{\mathbf{K}} \text{ (km)} \tag{a}$$
$$\mathbf{v}_0 = -7.350\hat{\mathbf{I}} + 0.4600\hat{\mathbf{J}} + 2.470\hat{\mathbf{K}} \text{ (km/s)} \tag{b}$$

Determine the position and velocity 3200 seconds later and plot the orbit in three dimensions.

We will use the universal variable formulation and Algorithm 3.4, which was illustrated in detail in Example 3.7. Therefore, only the results of each step are presented here.

Step 1:

$$\alpha = 1.4613 \times 10^{-4} \text{ km}^{-1}. \text{ Since this is positive, the orbit is an ellipse.}$$

Step 2:

$$\chi = 294.42 \text{ km}^{\frac{1}{2}}.$$

Step 3:

$$f = -0.94843 \quad \text{and} \quad g = -354.89 \, \text{s}^{-1}.$$

Step 4:
$$\mathbf{r} = 1090.9\hat{\mathbf{I}} - 5199.4\hat{\mathbf{J}} - 4480.6\hat{\mathbf{K}}\,(\text{km}), \quad r = 6949.8\,\text{km}.$$

Step 5:
$$\dot{f} = 0.00045324\,\text{s}^{-1}, \quad \dot{g} = -0.88479.$$

Step 6:
$$\mathbf{v} = 7.2284\hat{\mathbf{I}} + 1.9997\hat{\mathbf{J}} - 0.46311\hat{\mathbf{K}}\,(\text{km/s})$$

To plot the orbit, we observe that one complete revolution means a change in the eccentric anomaly E of 2π radians. According to Equation 3.54$_2$, the corresponding change in the universal anomaly is

$$\chi = \sqrt{a}E = \sqrt{\frac{1}{\alpha}}E = \sqrt{\frac{1}{0.00014613}} \cdot 2\pi = 519.77\,\text{km}^{\frac{1}{2}}$$

Letting χ vary from 0 to 519.77 in small increments, we employ the Lagrange coefficient formulation (Equation 3.64 plus 3.66a and 3.66b) to compute

$$\mathbf{r} = \left[1 - \frac{\chi^2}{r_0}C(\alpha\chi^2)\right]\mathbf{r}_0 + \left[\Delta t - \frac{1}{\sqrt{\mu}}\chi^3 S(\alpha\chi^2)\right]\mathbf{v}_0$$

where Δt for a given value of χ is given by Equation 3.45. Using a computer to plot the points obtained in this fashion yields Figure 4.6, which also shows the state vectors at t_0 and $t_0 + 3200$ s.

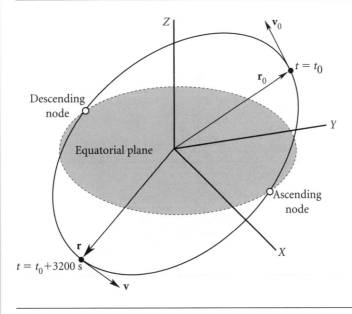

Figure 4.6   The orbit corresponding to the initial conditions given in Equations (a) and (b) of Example 4.2.

The previous example illustrates the fact that the six quantities or orbital elements comprising the state vector **r** and **v** completely determine the orbit. Other elements may be chosen. The classical orbital elements are introduced and related to the state vector in the next section.

4.4 Orbital elements and the state vector

To define an orbit in the plane requires two parameters: eccentricity and angular momentum. Other parameters, such as the semimajor axis, the specific energy, and (for an ellipse) the period, are obtained from these two. To locate a point on the orbit requires a third parameter, the true anomaly, which leads us to the time since perigee. Describing the orientation of an orbit in three dimensions requires three additional parameters, called the Euler angles, which are illustrated in Figure 4.7.

First, we locate the intersection of the orbital plane with the equatorial (XY) plane. That line is called the node line. The point on the node line where the orbit passes above the equatorial plane from below it is called the ascending node. The node line vector **N** extends outward from the origin through the ascending node. At the other end of the node line, where the orbit dives below the equatorial plane, is the descending node. The angle between the positive X axis and the node line is the first Euler angle Ω, the right ascension of the ascending node. Recall from Section 4.2 that right ascension is a positive number lying between $0°$ and $360°$.

The dihedral angle between the orbital plane and the equatorial plane is the inclination i, measured according to the right-hand rule, that is, counterclockwise around the node line vector from the equator to the orbit. The inclination is also the angle between the positive Z axis and the normal to the plane of the orbit. The two

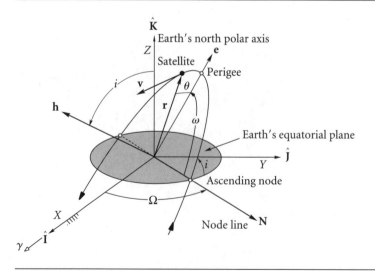

Figure 4.7 Geocentric equatorial frame and the orbital elements.

equivalent means of measuring i are indicated in Figure 4.7. Recall from Chapter 2 that the angular momentum vector **h** is normal to the plane of the orbit. Therefore, the inclination i is the angle between the positive Z axis and **h**. The inclination is a positive number between 0° and 180°.

It remains to locate the perigee of the orbit. Recall that perigee lies at the intersection of the eccentricity vector **e** with the orbital path. The third Euler angle ω, the argument of perigee, is the angle between the node line vector **N** and the eccentricity vector **e**, measured in the plane of the orbit. The argument of perigee is a positive number between 0° and 360°.

In summary, the six orbital elements are

h specific angular momentum

i inclination

Ω right ascension (RA) of the ascending node

e eccentricity

ω argument of perigee

θ true anomaly

The angular momentum h and true anomaly θ are frequently replaced by the semimajor axis a and the mean anomaly M, respectively.

Given the position **r** and velocity **v** of a satellite in the geocentric equatorial frame, how do we obtain the orbital elements? The step-by-step procedure is outlined in Algorithm 4.1. Note that each step incorporates results obtained in the previous steps.

ALGORITHM 4.1

Obtain orbital elements from the state vector. A MATLAB version of this procedure appears in Appendix D.8. Applying this algorithm to orbits around other planets or the sun amounts to defining the frame of reference and substituting the appropriate gravitational parameter μ.

1. Calculate the distance,

$$r = \sqrt{\mathbf{r} \cdot \mathbf{r}} = \sqrt{X^2 + Y^2 + Z^2}$$

2. Calculate the speed,

$$v = \sqrt{\mathbf{v} \cdot \mathbf{v}} = \sqrt{v_X^2 + v_Y^2 + v_Z^2}$$

3. Calculate the radial velocity,

$$v_r = \mathbf{r} \cdot \mathbf{v}/r = (Xv_X + Yv_Y + Zv_Z)/r$$

Note that if $v_r > 0$, the satellite is flying away from perigee. If $v_r < 0$, it is flying towards perigee.

4. Calculate the specific angular momentum,

$$\mathbf{h} = \mathbf{r} \times \mathbf{v} = \begin{vmatrix} \hat{\mathbf{I}} & \hat{\mathbf{J}} & \hat{\mathbf{K}} \\ X & Y & Z \\ v_X & v_Y & v_Z \end{vmatrix}$$

(Algorithm 4.1 continued)

5. Calculate the magnitude of the specific angular momentum,

$$h = \sqrt{\mathbf{h} \cdot \mathbf{h}}$$

the first orbital element.

6. Calculate the inclination,

$$i = \cos^{-1}\left(\frac{h_Z}{h}\right) \tag{4.7}$$

This is the second orbital element. Recall that i must lie between $0°$ and $180°$, so there is no quadrant ambiguity. If $90° < i \leq 180°$, the orbit is retrograde.

7. Calculate

$$\mathbf{N} = \hat{\mathbf{K}} \times \mathbf{h} = \begin{vmatrix} \hat{\mathbf{I}} & \hat{\mathbf{J}} & \hat{\mathbf{K}} \\ 0 & 0 & 1 \\ h_X & h_Y & h_Z \end{vmatrix} \tag{4.8}$$

This vector defines the node line.

8. Calculate the magnitude of $\mathbf{N}$,

$$N = \sqrt{\mathbf{N} \cdot \mathbf{N}}$$

9. Calculate the *RA* of the ascending node,

$$\Omega = \cos^{-1}(N_X/N)$$

the third orbital element. If $(N_X/N) > 0$, then Ω lies in either the first or fourth quadrant. If $(N_X/N) < 0$, then Ω lies in either the second or third quadrant. To place Ω in the proper quadrant, observe that the ascending node lies on the positive side of the vertical *XZ* plane $(0 \leq \Omega < 180°)$ if $N_Y > 0$. On the other hand, the ascending node lies on the negative side of the *XZ* plane $(180° \leq \Omega < 360°)$ if $N_Y < 0$. Therefore, $N_Y > 0$ implies that $0 < \Omega < 180°$, whereas $N_Y < 0$ implies that $180° < \Omega < 360°$. In summary,

$$\Omega = \begin{cases} \cos^{-1}\left(\dfrac{N_X}{N}\right) & (N_Y \geq 0) \\ 360° - \cos^{-1}\left(\dfrac{N_X}{N}\right) & (N_Y < 0) \end{cases} \tag{4.9}$$

10. Calculate the eccentricity vector. Starting with Equation 2.30,

$$\mathbf{e} = \frac{1}{\mu}\left[\mathbf{v} \times \mathbf{h} - \mu\frac{\mathbf{r}}{r}\right] = \frac{1}{\mu}\left[\mathbf{v} \times (\mathbf{r} \times \mathbf{v}) - \mu\frac{\mathbf{r}}{r}\right] = \frac{1}{\mu}\left[\overbrace{r v^2 - \mathbf{v}(\mathbf{r} \cdot \mathbf{v})}^{bac-cab\ rule} - \mu\frac{\mathbf{r}}{r}\right]$$

so that

$$\mathbf{e} = \frac{1}{\mu}\left[\left(v^2 - \frac{\mu}{r}\right)\mathbf{r} - r v_r \mathbf{v}\right] \tag{4.10}$$

11. Calculate the eccentricity,

$$e = \sqrt{\mathbf{e} \cdot \mathbf{e}}$$

the fourth orbital element. Substituting Equation 4.10 leads to a form depending only on the scalars obtained thus far,

$$e = \frac{1}{\mu}\sqrt{(2\mu - rv^2)rv_r^2 + (\mu - rv^2)^2} \qquad (4.11)$$

12. Calculate the argument of perigee,

$$\omega = \cos^{-1}(\mathbf{N} \cdot \mathbf{e}/Ne)$$

the fifth orbital element. If $\mathbf{N} \cdot \mathbf{e} > 0$, then ω lies in either the first or fourth quadrant. If $\mathbf{N} \cdot \mathbf{e} < 0$, then ω lies in either the second or third quadrant. To place ω in the proper quadrant, observe that perigee lies above the equatorial plane ($0 \leq \omega < 180°$) if $\mathbf{e}$ points up (in the positive Z direction), and perigee lies below the plane ($180° \leq \omega < 360°$) if $\mathbf{e}$ points down. Therefore, $e_Z \geq 0$ implies that $0 < \omega < 180°$, whereas $e_Z < 0$ implies that $180° < \omega < 360°$. To summarize,

$$\omega = \begin{cases} \cos^{-1}\left(\dfrac{\mathbf{N} \cdot \mathbf{e}}{Ne}\right) & (e_Z \geq 0) \\ 360° - \cos^{-1}\left(\dfrac{\mathbf{N} \cdot \mathbf{e}}{Ne}\right) & (e_Z < 0) \end{cases} \qquad (4.12)$$

13. Calculate the true anomaly,

$$\theta = \cos^{-1}\left(\frac{\mathbf{e} \cdot \mathbf{r}}{er}\right)$$

the sixth and final orbital element. If $\mathbf{e} \cdot \mathbf{r} > 0$, then θ lies in the first or fourth quadrant. If $\mathbf{e} \cdot \mathbf{r} < 0$, then θ lies in the second or third quadrant. To place θ in the proper quadrant, note that if the satellite is flying away from perigee ($\mathbf{r} \cdot \mathbf{v} \geq 0$), then $0 \leq \theta < 180°$, whereas if the satellite is flying towards perigee ($\mathbf{r} \cdot \mathbf{v} < 0$), then $180° \leq \theta < 360°$. Therefore, using the results of step 3 above

$$\theta = \begin{cases} \cos^{-1}\left(\dfrac{\mathbf{e} \cdot \mathbf{r}}{er}\right) & (v_r \geq 0) \\ 360° - \cos^{-1}\left(\dfrac{\mathbf{e} \cdot \mathbf{r}}{er}\right) & (v_r < 0) \end{cases} \qquad (4.13a)$$

Substituting Equation 4.10 yields an alternative form of this expression,

$$\theta = \begin{cases} \cos^{-1}\left[\dfrac{1}{e}\left(\dfrac{h^2}{\mu r} - 1\right)\right] & (v_r \geq 0) \\ 360° - \cos^{-1}\left[\dfrac{1}{e}\left(\dfrac{h^2}{\mu r} - 1\right)\right] & (v_r < 0) \end{cases} \qquad (4.13b)$$

The procedure described above for calculating the orbital elements is not unique.

EXAMPLE 4.3

Given the state vector,

$$\mathbf{r} = -6045\hat{\mathbf{I}} - 3490\hat{\mathbf{J}} + 2500\hat{\mathbf{K}} \text{ (km)}$$

$$\mathbf{v} = -3.457\hat{\mathbf{I}} + 6.618\hat{\mathbf{J}} + 2.533\hat{\mathbf{K}} \text{ (km/s)}$$

find the orbital elements h, i, Ω, e, ω and θ using Algorithm 4.1.

(Example 4.3 continued)

Step 1:
$$r = \sqrt{\mathbf{r} \cdot \mathbf{r}} = \sqrt{(-6045)^2 + (-3490)^2 + 2500^2} = 7414 \text{ km} \quad \text{(a)}$$

Step 2:
$$v = \sqrt{\mathbf{v} \cdot \mathbf{v}} = \sqrt{(-3.457)^2 + 6.618^2 + 2.533^2} = 7.884 \text{ km/s} \quad \text{(b)}$$

Step 3:
$$v_r = \frac{\mathbf{v} \cdot \mathbf{r}}{r} = \frac{(-3.457) \cdot (-6045) + 6.618 \cdot (-3490) + 2.533 \cdot 2500}{7414}$$
$$= 0.5575 \text{ km/s} \quad \text{(c)}$$

Since $v_r > 0$, the satellite is flying *away* from perigee.

Step 4:
$$\mathbf{h} = \mathbf{r} \times \mathbf{v} = \begin{vmatrix} \hat{\mathbf{I}} & \hat{\mathbf{J}} & \hat{\mathbf{K}} \\ -6045 & -3490 & 2500 \\ -3.457 & 6.618 & 2.533 \end{vmatrix} = -25\,380\hat{\mathbf{I}} + 6670\hat{\mathbf{J}} - 52\,070\hat{\mathbf{K}} \text{ (km}^2\text{/s)} \quad \text{(d)}$$

Step 5:
$$h = \sqrt{\mathbf{h} \cdot \mathbf{h}} = \sqrt{(-25\,380)^2 + 6670^2 + (-52\,070)^2} = \underline{58\,310 \text{ km}^2/\text{s}} \quad \text{(e)}$$

Step 6:
$$i = \cos^{-1} \frac{h_Z}{h} = \cos^{-1}\left(\frac{-52\,070}{58\,310}\right) = \underline{153.2°} \quad \text{(f)}$$

Since i is greater than 90°, this is a retrograde orbit.

Step 7:
$$\mathbf{N} = \hat{\mathbf{K}} \times \mathbf{h} = \begin{vmatrix} \hat{\mathbf{I}} & \hat{\mathbf{J}} & \hat{\mathbf{K}} \\ 0 & 0 & 1 \\ -25\,380 & 6670 & -52\,070 \end{vmatrix} = -6670\hat{\mathbf{I}} - 25\,380\hat{\mathbf{J}} \quad \text{(g)}$$

Step 8:
$$N = \sqrt{\mathbf{N} \cdot \mathbf{N}} = \sqrt{(-6670)^2 + (-25\,380)^2} = 26\,250 \quad \text{(h)}$$

Using (g) and (h), we compute the right ascension of the node.

Step 9:
$$\Omega = \cos^{-1} \frac{N_X}{N} = \cos^{-1}\left(\frac{-6670}{26\,250}\right) = 104.7° \text{ or } 255.3°$$

From (g) we know that $N_Y < 0$; therefore, Ω must lie in the third quadrant,

$$\underline{\Omega = 255.3°} \quad \text{(i)}$$

Step 10:

$$\mathbf{e} = \frac{1}{\mu}\left[\left(v^2 - \frac{\mu}{r}\right)\mathbf{r} - (\mathbf{r}\cdot\mathbf{v})\mathbf{v}\right]$$

$$= \frac{1}{398\,600}\left[\left(7.884^2 - \frac{398\,600}{7414}\right)(-6045\hat{\mathbf{I}} - 3490\hat{\mathbf{J}} + 2500\hat{\mathbf{K}})\right.$$

$$\left. -4133(-3.457\hat{\mathbf{I}} + 6.618\hat{\mathbf{J}} + 2.533\hat{\mathbf{K}})\right]$$

$$= -0.09160\hat{\mathbf{I}} - 0.1422\hat{\mathbf{J}} + 0.02644\hat{\mathbf{K}} \tag{j}$$

Step 11:

$$e = \sqrt{\mathbf{e}\cdot\mathbf{e}} = \sqrt{(-0.09160)^2 + (-0.1422)^2 + (0.02644)^2} = \underline{0.1712} \tag{k}$$

Clearly, the orbit is an ellipse.

Step 12:

$$\omega = \cos^{-1}\frac{\mathbf{N}\cdot\mathbf{e}}{Ne}$$

$$= \cos^{-1}\left[\frac{(-6670)(-0.09160) + (-25\,380)(-0.1422) + (0)(0.02644)}{(26\,250)(0.1712)}\right]$$

$$= 20.07° \text{ or } 339.9°$$

ω lies in the first quadrant if $e_Z > 0$, which is true in this case, as we see from (j). Therefore,

$$\underline{\omega = 20.07°} \tag{l}$$

Step 13:

$$\theta = \cos^{-1}\left(\frac{\mathbf{e}\cdot\mathbf{r}}{er}\right)$$

$$= \cos^{-1}\left[\frac{(-0.09160)(-6045) + (-0.1422)\cdot(-3490) + (0.02644)(2500)}{(0.1712)(7414)}\right]$$

$$= 28.45° \text{ or } 331.6°$$

From (c) we know that $v_r > 0$, which means $0 \leq \theta < 180°$. Therefore,

$$\underline{\theta = 28.45°}$$

Having found the orbital elements, we can go on to compute other parameters. The perigee and apogee radii are

$$r_p = \frac{h^2}{\mu}\frac{1}{1 + e\cos(0)} = \frac{58\,310^2}{398\,600}\frac{1}{1 + 0.1712} = 7284\text{ km}$$

$$r_a = \frac{h^2}{\mu}\frac{1}{1 + e\cos(180°)} = \frac{58\,310^2}{398\,600}\frac{1}{1 - 0.1712} = 10\,290\text{ km}$$

(Example 4.3 continued)

From these it follows that the semimajor axis of the ellipse is

$$a = \frac{1}{2}(r_p + r_a) = 8788 \text{ km}$$

This leads to the period,

$$T = \frac{2\pi}{\sqrt{\mu}} a^{\frac{3}{2}} = 2.278 \text{ hr}$$

The orbit is illustrated in Figure 4.8.

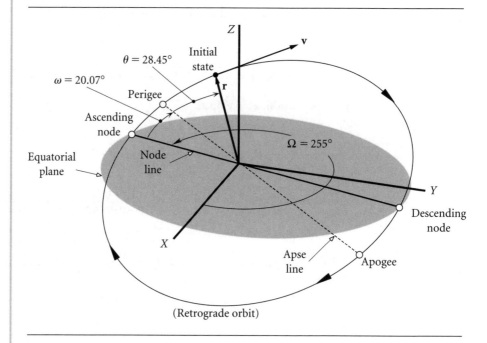

Figure 4.8 A plot of the orbit identified in Example 4.3.

We have seen how to obtain the orbital elements from the state vector. To arrive at the state vector, given the orbital elements, requires performing coordinate transformations, which are discussed in the next section.

4.5 Coordinate transformation

Figure 4.9 shows two cartesian coordinate systems: the unprimed system with axes xyz, and the primed system with axes $x'y'z'$. The orthogonal unit basis vectors for the unprimed system are $\hat{\mathbf{i}}, \hat{\mathbf{j}}$ and $\hat{\mathbf{k}}$. The fact they are unit vectors means

$$\hat{\mathbf{i}} \cdot \hat{\mathbf{i}} = \hat{\mathbf{j}} \cdot \hat{\mathbf{j}} = \hat{\mathbf{k}} \cdot \hat{\mathbf{k}} = 1 \qquad (4.14)$$

Since they are orthogonal,

$$\hat{\mathbf{i}} \cdot \hat{\mathbf{j}} = \hat{\mathbf{i}} \cdot \hat{\mathbf{k}} = \hat{\mathbf{j}} \cdot \hat{\mathbf{k}} = 0 \qquad (4.15)$$

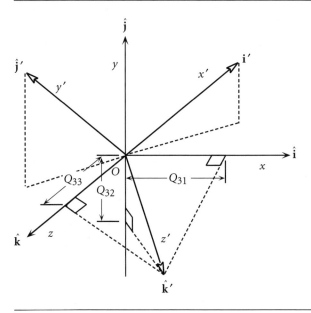

Figure 4.9 Two sets of cartesian reference axes, xyz and $x'y'z'$.

The orthonormal basis vectors $\hat{\mathbf{i}}', \hat{\mathbf{j}}'$ and $\hat{\mathbf{k}}'$ of the primed system share these same properties. That is,

$$\hat{\mathbf{i}}' \cdot \hat{\mathbf{i}}' = \hat{\mathbf{j}}' \cdot \hat{\mathbf{j}}' = \hat{\mathbf{k}}' \cdot \hat{\mathbf{k}}' = 1 \tag{4.16}$$

and

$$\hat{\mathbf{i}}' \cdot \hat{\mathbf{j}}' = \hat{\mathbf{i}}' \cdot \hat{\mathbf{k}}' = \hat{\mathbf{j}}' \cdot \hat{\mathbf{k}}' = 0 \tag{4.17}$$

We can express the unit vectors of the primed system in terms of their components in the unprimed system as follows

$$\begin{aligned} \hat{\mathbf{i}}' &= Q_{11}\hat{\mathbf{i}} + Q_{12}\hat{\mathbf{j}} + Q_{13}\hat{\mathbf{k}} \\ \hat{\mathbf{j}}' &= Q_{21}\hat{\mathbf{i}} + Q_{22}\hat{\mathbf{j}} + Q_{23}\hat{\mathbf{k}} \\ \hat{\mathbf{k}}' &= Q_{31}\hat{\mathbf{i}} + Q_{32}\hat{\mathbf{j}} + Q_{33}\hat{\mathbf{k}} \end{aligned} \tag{4.18}$$

The Qs in these expressions are just the direction cosines of $\hat{\mathbf{i}}', \hat{\mathbf{j}}'$ and $\hat{\mathbf{k}}'$. Figure 4.9 illustrates the components of $\hat{\mathbf{k}}'$, which are, of course, the projections of $\hat{\mathbf{k}}'$ onto the x, y and z axes. The unprimed unit vectors may be resolved into components along the primed system to obtain a set of equations similar to Equations 4.18:

$$\begin{aligned} \hat{\mathbf{i}} &= Q'_{11}\hat{\mathbf{i}}' + Q'_{12}\hat{\mathbf{j}}' + Q'_{13}\hat{\mathbf{k}}' \\ \hat{\mathbf{j}} &= Q'_{21}\hat{\mathbf{i}}' + Q'_{22}\hat{\mathbf{j}}' + Q'_{23}\hat{\mathbf{k}}' \\ \hat{\mathbf{k}} &= Q'_{31}\hat{\mathbf{i}}' + Q'_{32}\hat{\mathbf{j}}' + Q'_{33}\hat{\mathbf{k}}' \end{aligned} \tag{4.19}$$

However, $\hat{\mathbf{i}}' \cdot \hat{\mathbf{i}} = \hat{\mathbf{i}} \cdot \hat{\mathbf{i}}'$, so that, from Equations 4.18$_1$ and 4.19$_1$, we find $Q_{11} = Q'_{11}$. Likewise, $\hat{\mathbf{i}}' \cdot \hat{\mathbf{j}} = \hat{\mathbf{j}} \cdot \hat{\mathbf{i}}'$, which, according to Equations 4.18$_1$ and 4.19$_2$, means $Q_{12} = Q'_{21}$. Proceeding in this fashion, it is clear that the direction cosines in Equations 4.18 may be expressed in terms of those in Equations 4.19. That is, Equations 4.19 may be written

$$\hat{\mathbf{i}} = Q_{11}\hat{\mathbf{i}}' + Q_{21}\hat{\mathbf{j}}' + Q_{31}\hat{\mathbf{k}}'$$
$$\hat{\mathbf{j}} = Q_{12}\hat{\mathbf{i}}' + Q_{22}\hat{\mathbf{j}}' + Q_{32}\hat{\mathbf{k}}' \quad (4.20)$$
$$\hat{\mathbf{k}} = Q_{13}\hat{\mathbf{i}}' + Q_{23}\hat{\mathbf{j}}' + Q_{33}\hat{\mathbf{k}}'$$

Substituting Equations 4.20 into Equations 4.14 and making use of Equations 4.16 and 4.17, we get the three relations

$$\hat{\mathbf{i}} \cdot \hat{\mathbf{i}} = 1 \Rightarrow Q_{11}^2 + Q_{21}^2 + Q_{31}^2 = 1$$
$$\hat{\mathbf{j}} \cdot \hat{\mathbf{j}} = 1 \Rightarrow Q_{12}^2 + Q_{22}^2 + Q_{32}^2 = 1 \quad (4.21)$$
$$\hat{\mathbf{k}} \cdot \hat{\mathbf{k}} = 1 \Rightarrow Q_{13}^2 + Q_{23}^2 + Q_{33}^2 = 1$$

Substituting Equations 4.20 into Equations 4.15 and, again, making use of Equations 4.16 and 4.17, we obtain the three equations

$$\hat{\mathbf{i}} \cdot \hat{\mathbf{j}} = 0 \Rightarrow Q_{11}Q_{12} + Q_{21}Q_{22} + Q_{31}Q_{32} = 0$$
$$\hat{\mathbf{i}} \cdot \hat{\mathbf{k}} = 0 \Rightarrow Q_{11}Q_{13} + Q_{21}Q_{23} + Q_{31}Q_{33} = 0 \quad (4.22)$$
$$\hat{\mathbf{j}} \cdot \hat{\mathbf{k}} = 0 \Rightarrow Q_{12}Q_{13} + Q_{22}Q_{23} + Q_{32}Q_{33} = 0$$

Let [**Q**] represent the matrix of direction cosines of $\hat{\mathbf{i}}', \hat{\mathbf{j}}'$ and $\hat{\mathbf{k}}'$ relative to $\hat{\mathbf{i}}, \hat{\mathbf{j}}$ and $\hat{\mathbf{k}}$, as given by Equations 4.19. Then

$$[\mathbf{Q}] = \begin{bmatrix} Q_{11} & Q_{12} & Q_{13} \\ Q_{21} & Q_{22} & Q_{23} \\ Q_{31} & Q_{32} & Q_{33} \end{bmatrix} = \begin{bmatrix} \hat{\mathbf{i}}' \cdot \hat{\mathbf{i}} & \hat{\mathbf{i}}' \cdot \hat{\mathbf{j}} & \hat{\mathbf{i}}' \cdot \hat{\mathbf{k}} \\ \hat{\mathbf{j}}' \cdot \hat{\mathbf{i}} & \hat{\mathbf{j}}' \cdot \hat{\mathbf{j}} & \hat{\mathbf{j}}' \cdot \hat{\mathbf{k}} \\ \hat{\mathbf{k}}' \cdot \hat{\mathbf{i}} & \hat{\mathbf{k}}' \cdot \hat{\mathbf{j}} & \hat{\mathbf{k}}' \cdot \hat{\mathbf{k}} \end{bmatrix} \quad (4.23)$$

The transpose of the matrix [**Q**], denoted $[\mathbf{Q}]^T$, is obtained by interchanging the rows and columns of [**Q**]. Thus,

$$[\mathbf{Q}]^T = \begin{bmatrix} Q_{11} & Q_{21} & Q_{31} \\ Q_{12} & Q_{22} & Q_{32} \\ Q_{13} & Q_{23} & Q_{33} \end{bmatrix} = \begin{bmatrix} \hat{\mathbf{i}} \cdot \hat{\mathbf{i}}' & \hat{\mathbf{i}} \cdot \hat{\mathbf{j}}' & \hat{\mathbf{i}} \cdot \hat{\mathbf{k}}' \\ \hat{\mathbf{j}} \cdot \hat{\mathbf{i}}' & \hat{\mathbf{j}} \cdot \hat{\mathbf{j}}' & \hat{\mathbf{j}} \cdot \hat{\mathbf{k}}' \\ \hat{\mathbf{k}} \cdot \hat{\mathbf{i}}' & \hat{\mathbf{k}} \cdot \hat{\mathbf{j}}' & \hat{\mathbf{k}} \cdot \hat{\mathbf{k}}' \end{bmatrix} \quad (4.24)$$

Forming the product $[\mathbf{Q}]^T[\mathbf{Q}]$ we get

$$[\mathbf{Q}]^T[\mathbf{Q}] = \begin{bmatrix} Q_{11} & Q_{21} & Q_{31} \\ Q_{12} & Q_{22} & Q_{32} \\ Q_{13} & Q_{23} & Q_{33} \end{bmatrix} \begin{bmatrix} Q_{11} & Q_{12} & Q_{13} \\ Q_{21} & Q_{22} & Q_{23} \\ Q_{31} & Q_{32} & Q_{33} \end{bmatrix}$$

$$= \begin{bmatrix} Q_{11}^2 + Q_{21}^2 + Q_{31}^2 & Q_{11}Q_{12} + Q_{21}Q_{22} + Q_{31}Q_{32} & Q_{11}Q_{13} + Q_{21}Q_{23} + Q_{31}Q_{33} \\ Q_{12}Q_{11} + Q_{22}Q_{21} + Q_{32}Q_{31} & Q_{12}^2 + Q_{22}^2 + Q_{32}^2 & Q_{12}Q_{13} + Q_{22}Q_{23} + Q_{32}Q_{33} \\ Q_{13}Q_{11} + Q_{23}Q_{21} + Q_{33}Q_{31} & Q_{13}Q_{12} + Q_{23}Q_{22} + Q_{33}Q_{32} & Q_{13}^2 + Q_{23}^2 + Q_{33}^2 \end{bmatrix}$$

From this we obtain, with the aid of Equations 4.21 and 4.22,

$$[\mathbf{Q}]^T[\mathbf{Q}] = [\mathbf{1}] \tag{4.25}$$

where

$$[\mathbf{1}] = \begin{bmatrix} 1 & 0 & 0 \\ 0 & 1 & 0 \\ 0 & 0 & 1 \end{bmatrix}$$

[**1**] stands for the identity matrix or unit matrix.

In a similar fashion, we can substitute Equations 4.18 into Equations 4.16 and 4.17 and make use of Equations 4.14 and 4.15 to finally obtain

$$[\mathbf{Q}][\mathbf{Q}]^T = [\mathbf{1}] \tag{4.26}$$

Since [**Q**] satisfies Equations 4.25 and 4.26, it is called an orthogonal matrix.

Let **v** be a vector. It can be expressed in terms of its components along the unprimed system,

$$\mathbf{v} = v_x\hat{\mathbf{i}} + v_y\hat{\mathbf{j}} + v_z\hat{\mathbf{k}}$$

or along the primed system,

$$\mathbf{v} = v'_x\hat{\mathbf{i}}' + v'_y\hat{\mathbf{j}}' + v'_z\hat{\mathbf{k}}'$$

These two expressions for **v** are equivalent (**v** = **v**) since a vector is independent of the coordinate system used to describe it. Thus,

$$v'_x\hat{\mathbf{i}}' + v'_y\hat{\mathbf{j}}' + v'_z\hat{\mathbf{k}}' = v_x\hat{\mathbf{i}} + v_y\hat{\mathbf{j}} + v_z\hat{\mathbf{k}} \tag{4.27}$$

Substituting Equations 4.20 into the right-hand side of Equation 4.27 yields

$$v'_x\hat{\mathbf{i}}' + v'_y\hat{\mathbf{j}}' + v'_z\hat{\mathbf{k}}' = v_x(Q_{11}\hat{\mathbf{i}}' + Q_{21}\hat{\mathbf{j}}' + Q_{31}\hat{\mathbf{k}}') + v_y(Q_{12}\hat{\mathbf{i}}' + Q_{22}\hat{\mathbf{j}}' + Q_{32}\hat{\mathbf{k}}')$$
$$+ v_z(Q_{13}\hat{\mathbf{i}}' + Q_{23}\hat{\mathbf{j}}' + Q_{33}\hat{\mathbf{k}}')$$

Upon collecting terms on the right, we get

$$v'_x\hat{\mathbf{i}}' + v'_y\hat{\mathbf{j}}' + v'_z\hat{\mathbf{k}}' = (Q_{11}v_x + Q_{12}v_y + Q_{13}v_z)\hat{\mathbf{i}}' + (Q_{21}v_x + Q_{22}v_y + Q_{23}v_z)\hat{\mathbf{j}}'$$
$$+ (Q_{31}v_x + Q_{32}v_y + Q_{33}v_z)\hat{\mathbf{k}}'$$

Equating the components of like unit vectors on each side of the equals sign yields

$$\begin{aligned} v'_x &= Q_{11}v_x + Q_{12}v_y + Q_{13}v_z \\ v'_y &= Q_{21}v_x + Q_{22}v_y + Q_{23}v_z \\ v'_z &= Q_{31}v_x + Q_{32}v_y + Q_{33}v_z \end{aligned} \tag{4.28}$$

In matrix notation, this may be written

$$\{\mathbf{v}'\} = [\mathbf{Q}]\{\mathbf{v}\} \tag{4.29}$$

where

$$\{\mathbf{v}'\} = \begin{Bmatrix} v'_x \\ v'_y \\ v'_z \end{Bmatrix} \qquad \{\mathbf{v}\} = \begin{Bmatrix} v_x \\ v_y \\ v_z \end{Bmatrix} \qquad (4.30)$$

and $[\mathbf{Q}]$ is given by Equation 4.23. Equation 4.28 (or Equation 4.29) shows how to transform the components of the vector **v** in the unprimed system into its components in the primed system. The inverse transformation, from primed to unprimed, is found by multiplying Equation 4.29 through by $[\mathbf{Q}]^T$:

$$[\mathbf{Q}]^T\{\mathbf{v}'\} = [\mathbf{Q}]^T[\mathbf{Q}]\{\mathbf{v}\}$$

But, according to Equation 4.25, $[\mathbf{Q}][\mathbf{Q}]^T = [\mathbf{1}]$, so that

$$[\mathbf{Q}]^T\{\mathbf{v}'\} = [\mathbf{1}]\{\mathbf{v}\}$$

Since $[\mathbf{1}]\{\mathbf{v}\} = \{\mathbf{v}\}$, we obtain

$$\{\mathbf{v}\} = [\mathbf{Q}]^T\{\mathbf{v}'\} \qquad (4.31)$$

Therefore, to go from the primed system to the unprimed system use $[\mathbf{Q}]$, and in the reverse direction – from primed to unprimed – use $[\mathbf{Q}]^T$.

EXAMPLE 4.4

In Figure 4.10, the x' axis is defined by the line segment $O'P$. The $x'y'$ plane is defined by the intersecting line segments $O'P$ and $O'Q$. The z' axis is normal to the plane of $O'P$ and $O'Q$ and obtained by rotating $O'P$ towards $O'Q$ and using the right-hand rule. (a) Find the transformation matrix $[\mathbf{Q}]$. (b) If $\{\mathbf{v}\} = \lfloor 2 \quad 4 \quad 6 \rfloor^T$, find $\{\mathbf{v}'\}$. (c) If $\{\mathbf{v}'\} = \lfloor 2 \quad 4 \quad 0 \rfloor^T$, find $\{\mathbf{v}\}$.

(a) Resolve the directed line segments $\vec{O'P}$ and $\vec{O'Q}$ into components along the unprimed system:

$$\vec{O'P} = (-5 - 3)\hat{\mathbf{i}} + (5 - 1)\hat{\mathbf{j}} + (4 - 2)\hat{\mathbf{k}} = -8\hat{\mathbf{i}} + 4\hat{\mathbf{j}} + 2\hat{\mathbf{k}}$$

$$\vec{O'Q} = (-6 - 3)\hat{\mathbf{i}} + (3 - 1)\hat{\mathbf{j}} + (5 - 2)\hat{\mathbf{k}} = -9\hat{\mathbf{i}} + 2\hat{\mathbf{j}} + 3\hat{\mathbf{k}}$$

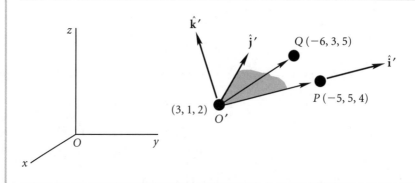

Figure 4.10 Defining a unit triad from the coordinates of three non-collinear points, O', P and Q.

Taking the cross product of $\vec{O'P}$ into $\vec{O'Q}$ yields a vector **Z'** which lies in the direction of the desired positive z' axis:

$$\mathbf{Z'} = \vec{O'P} \times \vec{O'Q} = 8\hat{i} + 6\hat{j} + 20\hat{k}$$

Taking the cross product of **Z'** into $\vec{O'P}$ then yields a vector **Y'** which points in the positive y' direction:

$$\mathbf{Y'} = \mathbf{Z} \times \vec{O'P} = -68\hat{i} - 176\hat{j} + 80\hat{k}$$

Normalizing the vectors $\vec{O'P}, \mathbf{Y'}$ and $\mathbf{Z'}$ produces the $\hat{i}', \hat{j}'$ and $\hat{k}'$ unit vectors, respectively. Thus

$$\hat{i}' = \frac{\vec{O'P}}{\|\vec{O'P}\|} = -0.8729\hat{i} + 0.4364\hat{j} + 0.2182\hat{k}$$

$$\hat{j}' = \frac{\mathbf{Y'}}{\|\mathbf{Y'}\|} = -0.3318\hat{i} - 0.8588\hat{j} + 0.3904\hat{k}$$

and

$$\hat{k}' = \frac{\mathbf{Z'}}{\|\mathbf{Z'}\|} = 0.3578\hat{i} + 0.2683\hat{j} + 0.8944\hat{k}$$

The components of $\hat{i}', \hat{j}'$ and $\hat{k}'$ are the rows of the orthogonal transformation matrix [**Q**]. Thus,

$$[\mathbf{Q}] = \begin{bmatrix} -0.8729 & 0.4364 & 0.2182 \\ -0.3318 & -0.8588 & 0.3904 \\ 0.3578 & 0.2683 & 0.8944 \end{bmatrix}$$

(b)

$$\{\mathbf{v'}\} = [\mathbf{Q}]\{\mathbf{v}\} = \begin{bmatrix} -0.8729 & 0.4364 & 0.2182 \\ -0.3318 & -0.8588 & 0.3904 \\ 0.3578 & 0.2683 & 0.8944 \end{bmatrix} \begin{Bmatrix} 2 \\ 4 \\ 6 \end{Bmatrix} = \begin{Bmatrix} 1.309 \\ -1.756 \\ 7.155 \end{Bmatrix}$$

(c)

$$\{\mathbf{v}\} = [\mathbf{Q}]^T\{\mathbf{v'}\} = \begin{bmatrix} -0.8729 & -0.3318 & 0.3578 \\ 0.4364 & -0.8588 & 0.2683 \\ 0.2182 & 0.3904 & 0.8944 \end{bmatrix} \begin{Bmatrix} 2 \\ 4 \\ 0 \end{Bmatrix} = \begin{Bmatrix} -3.073 \\ -2.562 \\ 1.998 \end{Bmatrix}$$

Let us consider the special case in which the coordinate transformation involves a rotation about only one of the coordinate axes, as shown in Figure 4.11. If the rotation is about the x axis, then according to Equations 4.18 and 4.23,

$$\hat{i}' = \hat{i}$$

$$\hat{j}' = (\hat{j}' \cdot \hat{i})\hat{i} + (\hat{j}' \cdot \hat{j})\hat{j} + (\hat{j}' \cdot \hat{k})\hat{k} = \cos\phi\hat{j} + \cos(90 - \phi)\hat{k} = \cos\phi\hat{j} + \sin(\phi)\hat{k}$$

$$\hat{k}' = (\hat{k}' \cdot \hat{j})\hat{j} + (\hat{k}' \cdot \hat{k})\hat{k} = \cos(90° + \phi)\hat{j} + \cos\phi\hat{k} = -\sin\phi\hat{j} + \cos\phi\hat{k}$$

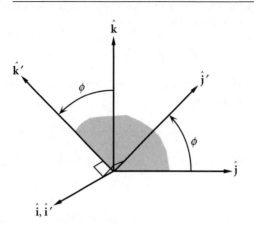

Figure 4.11 Rotation about the x axis.

or

$$\left\{\begin{array}{c}\hat{\mathbf{i}}' \\ \hat{\mathbf{j}}' \\ \hat{\mathbf{k}}'\end{array}\right\} = \begin{bmatrix} 1 & 0 & 0 \\ 0 & \cos\phi & \sin\phi \\ 0 & -\sin\phi & \cos\phi \end{bmatrix} \left\{\begin{array}{c}\hat{\mathbf{i}} \\ \hat{\mathbf{j}} \\ \hat{\mathbf{k}}\end{array}\right\}$$

The transformation from the xyz coordinate system to the $xy'z'$ system having a common x axis is given by the matrix coefficient of the unit vectors on the right. Since this is a rotation through the angle ϕ about the x axis, we denote this matrix by $[\mathbf{R}_1(\phi)]$, in which the subscript 1 stands for axis 1 (the x axis). Thus,

$$[\mathbf{R}_1(\phi)] = \begin{bmatrix} 1 & 0 & 0 \\ 0 & \cos\phi & \sin\phi \\ 0 & -\sin\phi & \cos\phi \end{bmatrix} \quad (4.32)$$

If the rotation is about the y axis, as shown in Figure 4.12, then Equation 4.18 yields

$$\hat{\mathbf{i}}' = (\hat{\mathbf{i}}' \cdot \hat{\mathbf{i}})\hat{\mathbf{i}} + (\hat{\mathbf{i}}' \cdot \hat{\mathbf{k}})\hat{\mathbf{k}} = \cos\phi\hat{\mathbf{i}} + \cos(\phi + 90°)\hat{\mathbf{k}} = \cos\phi\hat{\mathbf{i}} - \sin\phi\hat{\mathbf{k}}$$
$$\hat{\mathbf{j}}' = \hat{\mathbf{j}}$$
$$\hat{\mathbf{k}}' = (\hat{\mathbf{k}}' \cdot \hat{\mathbf{i}})\hat{\mathbf{i}} + (\hat{\mathbf{k}}' \cdot \hat{\mathbf{k}})\hat{\mathbf{k}} = \cos(90° - \phi)\hat{\mathbf{i}} + \cos\phi\hat{\mathbf{k}} = \sin\phi\hat{\mathbf{i}} + \cos\phi\hat{\mathbf{k}}$$

or, more compactly,

$$\left\{\begin{array}{c}\hat{\mathbf{i}}' \\ \hat{\mathbf{j}}' \\ \hat{\mathbf{k}}'\end{array}\right\} = \begin{bmatrix} \cos\phi & 0 & -\sin\phi \\ 0 & 1 & 0 \\ \sin\phi & 0 & \cos\phi \end{bmatrix} \left\{\begin{array}{c}\hat{\mathbf{i}} \\ \hat{\mathbf{j}} \\ \hat{\mathbf{k}}\end{array}\right\}$$

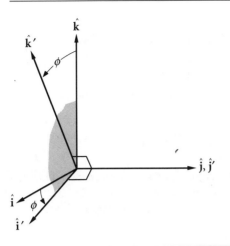

Figure 4.12 Rotation about the y axis.

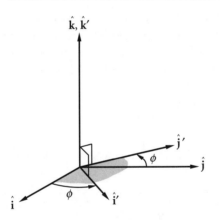

Figure 4.13 Rotation about the z axis.

We represent this transformation between two cartesian coordinate systems having a common y axis (axis 2) as $[\mathbf{R}_2(\phi)]$. Therefore,

$$[\mathbf{R}_2(\phi)] = \begin{bmatrix} \cos\phi & 0 & -\sin\phi \\ 0 & 1 & 0 \\ \sin\phi & 0 & \cos\phi \end{bmatrix} \quad (4.33)$$

Finally, if the rotation is about the z axis, as shown in Figure 4.13, then we have from Equation 4.18 that

$$\hat{\mathbf{i}}' = (\hat{\mathbf{i}}' \cdot \hat{\mathbf{i}})\hat{\mathbf{i}} + (\hat{\mathbf{i}}' \cdot \hat{\mathbf{j}})\hat{\mathbf{j}} = \cos\phi\hat{\mathbf{i}} + \cos(90° - \phi)\hat{\mathbf{j}} = \cos\phi\hat{\mathbf{i}} + \sin\phi\hat{\mathbf{j}}$$

$$\hat{\mathbf{j}}' = (\hat{\mathbf{j}}' \cdot \hat{\mathbf{i}})\hat{\mathbf{i}} + (\hat{\mathbf{j}}' \cdot \hat{\mathbf{j}})\hat{\mathbf{j}} = \cos(90° + \phi)\hat{\mathbf{i}} + \cos\phi\hat{\mathbf{j}} = -\sin\phi\hat{\mathbf{i}} + \cos\phi\hat{\mathbf{j}}$$

$$\hat{\mathbf{k}}' = \hat{\mathbf{k}}$$

or

$$\begin{Bmatrix} \hat{\mathbf{i}}' \\ \hat{\mathbf{j}}' \\ \hat{\mathbf{k}}' \end{Bmatrix} = \begin{bmatrix} \cos\phi & \sin\phi & 0 \\ -\sin\phi & \cos\phi & 0 \\ 0 & 0 & 1 \end{bmatrix} \begin{Bmatrix} \hat{\mathbf{i}} \\ \hat{\mathbf{j}} \\ \hat{\mathbf{k}} \end{Bmatrix}$$

In this case the rotation is around axis 3, the z axis, so

$$[\mathbf{R}_3(\phi)] = \begin{bmatrix} \cos\phi & \sin\phi & 0 \\ -\sin\phi & \cos\phi & 0 \\ 0 & 0 & 1 \end{bmatrix} \quad (4.34)$$

A transformation between two cartesian coordinate systems can be broken down into a sequence of two-dimensional rotations using the matrices $[\mathbf{R}_i(\phi)], i = 1, 2, 3$. We will use this to great advantage in the following sections.

4.6 TRANSFORMATION BETWEEN GEOCENTRIC EQUATORIAL AND PERIFOCAL FRAMES

The perifocal frame of reference for a given orbit was introduced in Section 2.10. Figure 4.14 illustrates the relationship between the perifocal and geocentric equatorial frames. Since the orbit lies in the $\bar{x}\bar{y}$ plane, the components of the state vector of a body relative to its perifocal reference are, according to Equations 2.109 and 2.115,

$$\mathbf{r} = \bar{x}\hat{\mathbf{p}} + \bar{y}\hat{\mathbf{q}} = \frac{h^2}{\mu}\frac{1}{1+e\cos\theta}(\cos\theta\hat{\mathbf{p}} + \sin\theta\hat{\mathbf{q}}) \quad (4.35)$$

$$\mathbf{v} = \dot{\bar{x}}\hat{\mathbf{p}} + \dot{\bar{y}}\hat{\mathbf{q}} = \frac{\mu}{h}[-\sin\theta\hat{\mathbf{p}} + (e+\cos\theta)\hat{\mathbf{q}}] \quad (4.36)$$

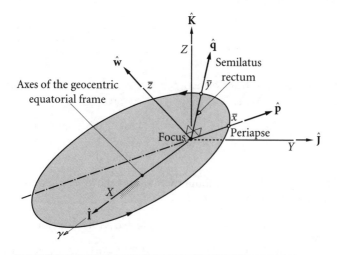

Figure 4.14 Perifocal ($\bar{x}\bar{y}\bar{z}$) and geocentric equatorial (XYZ) frames.

4.6 Transformation between geocentric equatorial and perifocal frames

In matrix notation these may be written

$$\{\mathbf{r}\}_{\bar{x}} = \frac{h^2}{\mu} \frac{1}{1 + e\cos\theta} \begin{Bmatrix} \cos\theta \\ \sin\theta \\ 0 \end{Bmatrix} \quad (4.37)$$

$$\{\mathbf{v}\}_{\bar{x}} = \frac{\mu}{h} \begin{Bmatrix} -\sin\theta \\ e + \cos\theta \\ 0 \end{Bmatrix} \quad (4.38)$$

The subscript $\bar{x}$ is shorthand for 'the $\bar{x}\bar{y}\bar{z}$ coordinate system' and is used to indicate that the components of these vectors are given in the perifocal frame, as opposed to, say, the geocentric equatorial frame (Equations 4.2 and 4.3).

The transformation from the geocentric equatorial frame into the perifocal frame may be accomplished by the sequence of three rotations illustrated in Figure 4.15. The first rotation, ①, is around the $\hat{\mathbf{K}}$ axis, through the right ascension Ω. It rotates the

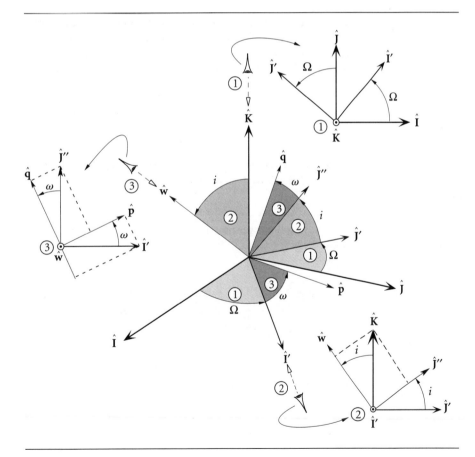

Figure 4.15 Sequence of three rotations transforming $\hat{\mathbf{I}}\hat{\mathbf{J}}\hat{\mathbf{K}}$ into $\hat{\mathbf{p}}\hat{\mathbf{q}}\hat{\mathbf{w}}$. The 'eye' viewing down an axis sees the illustrated rotation about that axis.

$\hat{\mathbf{I}}, \hat{\mathbf{J}}$ directions into the $\hat{\mathbf{I}}', \hat{\mathbf{J}}'$ directions. Viewed down the Z axis, this rotation appears as shown in the insert at the top of the figure. The orthogonal transformation matrix associated with this rotation is

$$[\mathbf{R}_3(\Omega)] = \begin{bmatrix} \cos\Omega & \sin\Omega & 0 \\ -\sin\Omega & \cos\Omega & 0 \\ 0 & 0 & 1 \end{bmatrix} \quad (4.39)$$

Recall that the subscript on $\mathbf{R}$ means that the rotation is around the '3' direction, in this case the $\hat{\mathbf{K}}$ axis.

The second rotation, ②, is around the node line ($\hat{\mathbf{I}}'$), through the angle i required to bring the XY plane parallel to the orbital plane. In other words, it rotates $\hat{\mathbf{K}}$ into alignment with $\hat{\mathbf{w}}$, and $\hat{\mathbf{J}}'$ simultaneously rotates into $\hat{\mathbf{J}}''$. The insert in the lower right of Figure 4.15 shows how this rotation appears when viewed from the $\hat{\mathbf{I}}'$ direction. The orthogonal transformation matrix for this rotation is

$$[\mathbf{R}_1(i)] = \begin{bmatrix} 1 & 0 & 0 \\ 0 & \cos i & \sin i \\ 0 & -\sin i & \cos i \end{bmatrix} \quad (4.40)$$

The third and final rotation, ③, is in the orbital plane and rotates the unit vectors $\hat{\mathbf{I}}'$ and $\hat{\mathbf{J}}''$ through the angle ω around the $\hat{\mathbf{w}}$ axis so that they become aligned with $\hat{\mathbf{p}}$ and $\hat{\mathbf{q}}$, respectively. This rotation appears from the $\hat{\mathbf{w}}$ direction as shown in the insert on the left of Figure 4.15. The orthogonal transformation matrix is seen to be

$$[\mathbf{R}_3(\omega)] = \begin{bmatrix} \cos\omega & \sin\omega & 0 \\ -\sin\omega & \cos\omega & 0 \\ 0 & 0 & 1 \end{bmatrix} \quad (4.41)$$

Finally, let us note that the transformation matrix $[\mathbf{Q}]_{X\bar{x}}$ from the geocentric equatorial frame into the perifocal frame is just the product of the three rotation matrices given by Equations 4.39, 4.40 and 4.41; i.e.,

$$[\mathbf{Q}]_{X\bar{x}} = [\mathbf{R}_3(\omega)][\mathbf{R}_1(i)][\mathbf{R}_3(\Omega)] \quad (4.42)$$

Substituting the three matrices on the right and carrying out the matrix multiplications yields

$[\mathbf{Q}]_{X\bar{x}}$

$$= \begin{bmatrix} \cos\Omega\cos\omega - \sin\Omega\sin\omega\cos i & \sin\Omega\cos\omega + \cos\Omega\cos i\sin\omega & \sin i\sin\omega \\ -\cos\Omega\sin\omega - \sin\Omega\cos i\cos\omega & -\sin\Omega\sin\omega + \cos\Omega\cos i\cos\omega & \sin i\cos\omega \\ \sin\Omega\sin i & -\cos\Omega\sin i & \cos i \end{bmatrix}$$
(4.43)

Remember, this is an orthogonal matrix, so that for the inverse transformation, from $\bar{x}\bar{y}\bar{z}$ to XYZ we have $[\mathbf{Q}]_{\bar{x}X} = ([\mathbf{Q}]_{X\bar{x}})^T$, or

$[\mathbf{Q}]_{\bar{x}X}$

$$= \begin{bmatrix} \cos\Omega\cos\omega - \sin\Omega\sin\omega\cos i & -\cos\Omega\sin\omega - \sin\Omega\cos i\cos\omega & \sin\Omega\sin i \\ \sin\Omega\cos\omega + \cos\Omega\cos i\sin\omega & -\sin\Omega\sin\omega + \cos\Omega\cos i\cos\omega & -\cos\Omega\sin i \\ \sin i\sin\omega & \sin i\cos\omega & \cos i \end{bmatrix}$$
(4.44)

4.6 Transformation between geocentric equatorial and perifocal frames

If the components of the state vector are given in the geocentric equatorial frame

$$\mathbf{r} = \{\mathbf{r}\}_X = \begin{Bmatrix} X \\ Y \\ Z \end{Bmatrix} \qquad \mathbf{v} = \{\mathbf{v}\}_X = \begin{Bmatrix} v_X \\ v_Y \\ v_Z \end{Bmatrix}$$

the components in the perifocal frame are found by carrying out the matrix multiplications

$$\{\mathbf{r}\}_{\bar{x}} = \begin{Bmatrix} \bar{x} \\ \bar{y} \\ 0 \end{Bmatrix} = [Q]_{X\bar{x}} \{\mathbf{r}\}_X \qquad \{\mathbf{v}\}_{\bar{x}} = \begin{Bmatrix} \dot{\bar{x}} \\ \dot{\bar{y}} \\ 0 \end{Bmatrix} = [Q]_{X\bar{x}} \{\mathbf{v}\}_X \qquad (4.45)$$

Likewise, the transformation from perifocal to geocentric equatorial components is

$$\{\mathbf{r}\}_X = [Q]_{\bar{x}X} \{\mathbf{r}\}_{\bar{x}} \qquad \{\mathbf{v}\}_X = [Q]_{\bar{x}X} \{\mathbf{v}\}_{\bar{x}} \qquad (4.46)$$

ALGORITHM 4.2

Given the orbital elements h, e, i, Ω, ω and θ, compute the position vectors $\mathbf{r}$ and $\mathbf{v}$ in the geocentric equatorial frame of reference. A MATLAB implementation of this procedure is listed in Appendix D.9. This algorithm can be applied to orbits around other planets or the sun.

1. Calculate position vector $\{\mathbf{r}\}_{\bar{x}}$ in perifocal coordinates using Equation 4.37.
2. Calculate velocity vector $\{\mathbf{v}\}_{\bar{x}}$ in perifocal coordinates using Equation 4.38.
3. Calculate the matrix $[Q]_{\bar{x}X}$ of the transformation from perifocal to geocentric equatorial coordinates using Equation 4.44.
4. Transform $\{\mathbf{r}\}_{\bar{x}}$ and $\{\mathbf{v}\}_{\bar{x}}$ into the geocentric frame by means of Equations 4.46.

EXAMPLE 4.5

For a given earth orbit, the elements are $h = 80\,000$ km^2/s, $e = 1.4$, $i = 30°$, $\Omega = 40°$, $\omega = 60°$ and $\theta = 30°$. Using Algorithm 4.2 find the state vectors $\mathbf{r}$ and $\mathbf{v}$ in the geocentric equatorial frame.

Step 1:

$$\{\mathbf{r}\}_{\bar{x}} = \frac{h^2}{\mu} \frac{1}{1 + e \cos \theta} \begin{Bmatrix} \cos \theta \\ \sin \theta \\ 0 \end{Bmatrix} = \frac{80\,000^2}{398\,600} \frac{1}{1 + 1.4 \cos 30°} \begin{Bmatrix} \cos 30° \\ \sin 30° \\ 0 \end{Bmatrix} = \begin{Bmatrix} 6285.0 \\ 3628.6 \\ 0 \end{Bmatrix} \text{ km}$$

Step 2:

$$\{\mathbf{v}\}_{\bar{x}} = \frac{\mu}{h} \begin{Bmatrix} -\sin \theta \\ e + \cos \theta \\ 0 \end{Bmatrix} = \frac{398\,600}{80\,000} \begin{Bmatrix} -\sin 30° \\ 1.4 + \cos 30° \\ 0 \end{Bmatrix} = \begin{Bmatrix} -2.4913 \\ 11.290 \\ 0 \end{Bmatrix} \text{ km/s}$$

Step 3:

$$[Q]_{X\bar{x}} = \begin{bmatrix} \cos \omega & \sin \omega & 0 \\ -\sin \omega & \cos \omega & 0 \\ 0 & 0 & 1 \end{bmatrix} \begin{bmatrix} 1 & 0 & 0 \\ 0 & \cos i & \sin i \\ 0 & -\sin i & \cos i \end{bmatrix} \begin{bmatrix} \cos \Omega & \sin \Omega & 0 \\ -\sin \Omega & \cos \Omega & 0 \\ 0 & 0 & 1 \end{bmatrix}$$

(Example 4.5 continued)

$$= \begin{bmatrix} \cos 60° & \sin 60° & 0 \\ -\sin 60° & \cos 60° & 0 \\ 0 & 0 & 1 \end{bmatrix} \begin{bmatrix} 1 & 0 & 0 \\ 0 & \cos 30° & \sin 30° \\ 0 & -\sin 30° & \cos 30° \end{bmatrix} \begin{bmatrix} \cos 40° & \sin 40° & 0 \\ -\sin 40° & \cos 40° & 0 \\ 0 & 0 & 1 \end{bmatrix}$$

$$= \begin{bmatrix} -0.099068 & 0.89593 & 0.43301 \\ -0.94175 & -0.22496 & 0.25 \\ 0.32139 & -0.38302 & 0.86603 \end{bmatrix}$$

This is the transformation matrix for $XYZ \to \bar{x}\bar{y}\bar{z}$. The transformation matrix for $\bar{x}\bar{y}\bar{z} \to XYZ$ is the transpose,

$$[Q]_{\bar{x}X} = \begin{bmatrix} -0.099068 & -0.94175 & 0.32139 \\ 0.89593 & -0.22496 & -0.38302 \\ 0.43301 & 0.25 & 0.86603 \end{bmatrix}$$

Step 4:

The geocentric equatorial position vector is

$$\{\mathbf{r}\}_X = [Q]_{\bar{x}X}\{\mathbf{r}\}_{\bar{x}}$$

$$= \begin{bmatrix} -0.099068 & -0.94175 & 0.32139 \\ 0.89593 & -0.22496 & -0.38302 \\ 0.43301 & 0.25 & 0.86603 \end{bmatrix} \begin{Bmatrix} 6285.0 \\ 3628.6 \\ 0 \end{Bmatrix} = \begin{Bmatrix} -4040 \\ 4815 \\ 3629 \end{Bmatrix} \text{ (km)} \quad \text{(a)}$$

whereas the geocentric equatorial velocity vector is

$$\{\mathbf{v}\}_X = [Q]_{\bar{x}X}\{\mathbf{v}\}_{\bar{x}}$$

$$= \begin{bmatrix} -0.099068 & -0.94175 & 0.32139 \\ 0.89593 & -0.22496 & -0.38302 \\ 0.43301 & 0.25 & 0.86603 \end{bmatrix} \begin{Bmatrix} -2.4913 \\ 11.290 \\ 0 \end{Bmatrix} = \begin{Bmatrix} -10.39 \\ -4.772 \\ 1.744 \end{Bmatrix} \text{ (km/s)}$$

The state vectors **r** and **v** are shown in Figure 4.16. By holding all of the orbital parameters except the true anomaly fixed and allowing θ to take on a range of values, we generate a sequence of position vectors $\mathbf{r}_{\bar{x}}$ from Equations 4.37. Each of these is projected into the geocentric equatorial frame as in (a), using repeatedly the same transformation matrix $[Q]_{\bar{x}X}$. By connecting the end points of all of the position vectors $\mathbf{r}_X$, we trace out the trajectory illustrated in Figure 4.16.

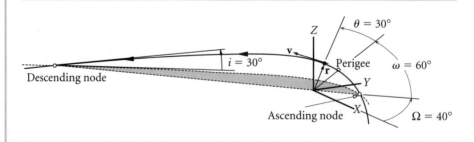

Figure 4.16 A portion of the hyperbolic trajectory of Example 4.5.

4.7 Effects of the earth's oblateness

The earth, like all of the planets with comparable or higher rotational rates, bulges out at the equator because of centrifugal force. The earth's equatorial radius is 21 km (13 miles) larger than the polar radius. This flattening at the poles is called oblateness, which is defined as follows

$$\text{oblateness} = \frac{\text{equatorial radius} - \text{polar radius}}{\text{equatorial radius}}$$

The earth is an oblate spheroid, lacking the perfect symmetry of a sphere. (A basketball can be made an oblate spheroid by sitting on it.) This lack of symmetry means that the force of gravity on an orbiting body is not directed towards the center of the earth. Whereas the gravitational field of a perfectly spherical planet depends only on the distance from its center, oblateness causes a variation also with latitude, that is, the angular distance from the equator (or pole). This is called a zonal variation. The dimensionless parameter which quantifies the major effects of oblateness on orbits is J_2, the second zonal harmonic. J_2 is not a universal constant. Each planet has its own value, as illustrated in Table 4.3, which lists variations of J_2 as well as oblateness.

The gravitational acceleration (force per unit mass) arising from an oblate planet is given by

$$\ddot{\mathbf{r}} = -\frac{\mu}{r^2}\hat{\mathbf{u}}_r + \mathbf{p}$$

The first term on the right is the familiar one (Equation 2.15) due to a spherical planet. The second term, $\mathbf{p}$, which is several orders of magnitude smaller than μ/r^2, is a perturbing acceleration due to the oblateness. This perturbing acceleration can be resolved into components,

$$\mathbf{p} = p_r\hat{\mathbf{u}}_r + p_\perp\hat{\mathbf{u}}_\perp + p_h\hat{\mathbf{h}}$$

where $\hat{\mathbf{u}}_r$, $\hat{\mathbf{u}}_\perp$ and $\hat{\mathbf{h}}$ are the radial, transverse and normal unit vectors attached to the satellite, as illustrated in Figure 4.17. $\hat{\mathbf{u}}_r$ points in the direction of the radial position

Table 4.3 Oblateness and second zonal harmonics

Planet	Oblateness	J_2
Mercury	0.000	60×10^{-6}
Venus	0.000	4.458×10^{-6}
Earth	0.003353	1.08263×10^{-3}
Mars	0.00648	1.96045×10^{-3}
Jupiter	0.06487	14.736×10^{-3}
Saturn	0.09796	16.298×10^{-3}
Uranus	0.02293	3.34343×10^{-3}
Neptune	0.01708	3.411×10^{-3}
Pluto	0.000	–
(Moon)	0.0012	202.7×10^{-6}

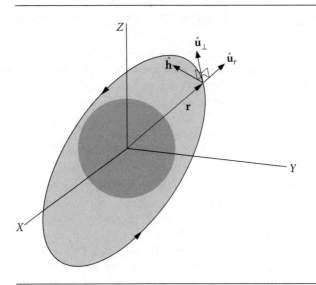

Figure 4.17 Unit vectors attached to an orbiting body.

vector $\mathbf{r}$, $\hat{\mathbf{h}}$ is the unit vector normal to the plane of the orbit and $\hat{\mathbf{u}}_\perp$ is perpendicular to $\mathbf{r}$, lying in the orbital plane and pointing in the direction of the motion.

The perturbation components p_r, $p_\perp$ and p_h are all directly proportional to J_2 and are functions of otherwise familiar orbital parameters as well as the planet radius R,

$$p_r = -\frac{\mu}{r^2}\frac{3}{2}J_2\left(\frac{R}{r}\right)^2 \left[1 - 3\sin^2 i \sin^2(\omega + \theta)\right]$$

$$p_\perp = -\frac{\mu}{r^2}\frac{3}{2}J_2\left(\frac{R}{r}\right)^2 \sin^2 i \sin[2(\omega + \theta)]$$

$$p_h = -\frac{\mu}{r^2}\frac{3}{2}J_2\left(\frac{R}{r}\right)^2 \sin 2i \sin(\omega + \theta)$$

These relations are derived by Prussing and Conway (1993), who also show how p_r, $p_\perp$ and p_h induce time rates of change in all of the orbital parameters. For example,

$$\dot{\Omega} = \frac{h}{\mu} \frac{\sin(\omega + \theta)}{\sin i (1 + e\cos\theta)} p_h$$

$$\dot{\omega} = -\frac{r\cos\theta}{eh}p_r + \frac{(2 + e\cos\theta)\sin\theta}{eh}p_\perp - \frac{r\sin(\omega + \theta)}{h\tan i}p_h$$

Clearly, the time variation of the right ascension Ω depends only on the component of the perturbing force normal to the (instantaneous) orbital plane, whereas the rate of change of the argument of perigee is influenced by all three perturbation components. Integrating $\dot{\Omega}$ over one complete orbit yields the average rate of change,

$$\dot{\Omega}_{avg} = \frac{1}{T}\int_0^T \dot{\Omega}\, dt$$

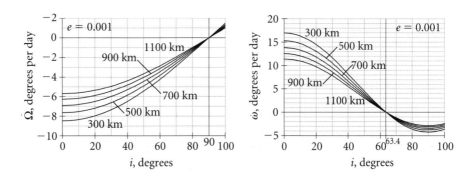

Figure 4.18 Regression of the node and advance of perigee for nearly circular orbits of altitudes 300 to 1100 km.

where T is the period. Carrying out the mathematical details leads to an expression for the average rate of precession of the node line, and hence, the orbital plane,

$$\dot{\Omega} = -\left[\frac{3}{2}\frac{\sqrt{\mu}J_2 R^2}{\left(1-e^2\right)^2 a^{\frac{7}{2}}}\right]\cos i \qquad (4.47)$$

where we have dropped the subscript avg. R and μ are the radius and gravitational parameter of the planet, a and e are the semimajor axis and eccentricity of the orbit, and i is the orbit's inclination. Observe that if $0 \leq i < 90°$, then $\dot{\Omega} < 0$. That is, for posigrade orbits, the node line drifts westward. Since the right ascension of the node continuously decreases, this phenomenon is called regression of the nodes. If $90° < i \leq 180°$, we see that $\dot{\Omega} > 0$. The node line of retrograde orbits therefore advances eastward. For polar orbits ($i = 90°$), the node line is stationary.

In a similar fashion the time rate of change of the argument of perigee is found to be

$$\dot{\omega} = -\left[\frac{3}{2}\frac{\sqrt{\mu}J_2 R^2}{\left(1-e^2\right)^2 a^{\frac{7}{2}}}\right]\left(\frac{5}{2}\sin^2 i - 2\right) \qquad (4.48)$$

This expression shows that if $0° \leq i < 63.4°$ or $116.6° < i \leq 180°$ then $\dot{\omega}$ is positive, which means the perigee advances in the direction of the motion of the satellite (hence, the name advance of perigee for this phenomenon). If $63.4° < i \leq 116.6°$, the perigee regresses, moving opposite to the direction of motion. $i = 63.4°$ and $i = 116.6°$ are the critical inclinations at which the apse line does not move.

Observe that the coefficient of the trigonometric terms in Equations 4.47 and 4.48 are identical.

Figure 4.18 is a plot of Equations 4.47 and 4.48 for several low-earth orbits. The effect of oblateness on both $\dot{\Omega}$ and $\dot{\omega}$ is greatest at low inclinations, for which the orbit is near the equatorial bulge for longer portions of each revolution. The effect decreases with increasing semimajor axis because the satellite becomes further from the bulge and its gravitational influence. Obviously, $\dot{\Omega} = \dot{\omega} = 0$ if $J_2 = 0$ (no equatorial bulge).

The time averaged rates of change for the inclination, eccentricity and semimajor axis are zero.

EXAMPLE 4.6

The space shuttle is in a 280 km by 400 km orbit with an inclination of 51.43°. Find the rates of node regression and perigee advance.

The perigee and apogee radii are

$$r_p = 6378 + 280 = 6658 \text{ km} \qquad r_a = 6378 + 400 = 6778 \text{ km}$$

Therefore the eccentricity and semimajor axis are

$$e = \frac{r_a - r_p}{r_a + r_p} = 0.008931$$

$$a = \frac{1}{2}(r_a + r_p) = 6718 \text{ km}$$

From Equation 4.47 we obtain the rate of node line regression:

$$\dot{\Omega} = -\left[\frac{3}{2} \frac{\sqrt{398\,600} \cdot 0.0010826 \cdot 6378^2}{(1 - 0.0089312^2)^2 \cdot 6718^{\frac{7}{2}}}\right] \cos 51.43°$$

$$= -1.6786 \times 10^{-6} \cdot \cos 51.43°$$

$$= -1.0465 \times 10^{-6} \text{ rad/s}$$

or

$$\underline{\dot{\Omega} = 5.181° \text{ per day to the west}}$$

From Equation 4.48,

$$\dot{\omega} = \overbrace{-1.6786 \times 10^{-6}}^{\text{same as in } \dot{\Omega}} \cdot \left(\frac{5}{2} \sin^2 51.43° - 2\right) = +7.9193 \times 10^{-7} \text{ rad/s}$$

or

$$\underline{\dot{\omega} = 3.920° \text{ per day in the flight direction}}$$

The effect of orbit inclination on node regression and advance of perigee is taken advantage of for two very important types of orbits. Sun-synchronous orbits are those whose orbital plane makes a constant angle α with the radial from the sun, as illustrated in Figure 4.19. For that to occur, the orbital plane must rotate in inertial space with the angular velocity of the earth in its orbit around the sun, which is 360° per 365.26 days, or 0.9856° per day. With the orbital plane precessing eastward at this rate, the ascending node will lie at a fixed local time. In the illustration it happens to be 3 pm. During every orbit, the satellite sees any given swath of the planet under nearly the same conditions of daylight or darkness day after day. The satellite also has a constant perspective on the sun. Sun-synchronous satellites, like the NOAA Polar-orbiting Operational Environmental Satellites (NOAA/POES) and those

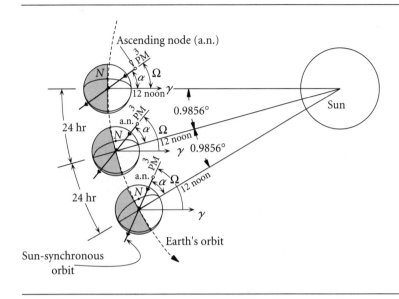

Figure 4.19 Sun-synchronous orbit.

of the Defense Meteorological Satellite Program (DMSP) are used for global weather coverage, while Landsat and the French SPOT series are intended for high-resolution earth observation.

EXAMPLE 4.7

A satellite is to be launched into a sun-synchronous circular orbit with period of 100 minutes. Determine the required altitude and inclination of its orbit.

We find the altitude z from the period relation for a circular orbit, Equation 2.54:

$$T = \frac{2\pi}{\sqrt{\mu}}(R_E + z)^{\frac{3}{2}} \Rightarrow 100 \cdot 60 = \frac{2\pi}{\sqrt{398\,600}}(6378 + z)^{\frac{3}{2}} \Rightarrow \underline{z = 758.63\,\text{km}}$$

For a sun-synchronous orbit, the ascending node must advance at the rate

$$\dot{\Omega} = \frac{2\pi\,\text{rad}}{365.26 \cdot 24 \cdot 3600\,\text{s}} = 1.991 \times 10^{-7}\,\text{rad/s}$$

Substituting this and the altitude into Equation 4.47, we obtain,

$$1.991 \times 10^{-7} = -\left[\frac{3}{2}\frac{\sqrt{398\,600} \cdot 0.00108263 \cdot 6378^2}{(1 - 0^2)^2 (6378 + 758.63)^{\frac{7}{2}}}\right]\cos i \Rightarrow \cos i = -0.14658$$

Thus, the inclination of the orbit is

$$i = \cos^{-1}(-0.14658) = 98.43°$$

This illustrates the fact that sun-synchronous orbits are very nearly polar orbits ($i = 90°$).

182 Chapter 4 *Orbits in three dimensions*

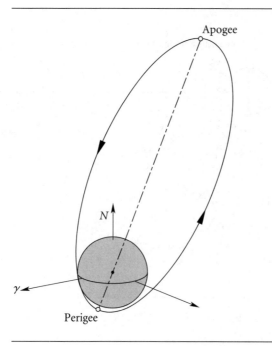

Figure 4.20 A typical Molniya orbit (to scale).

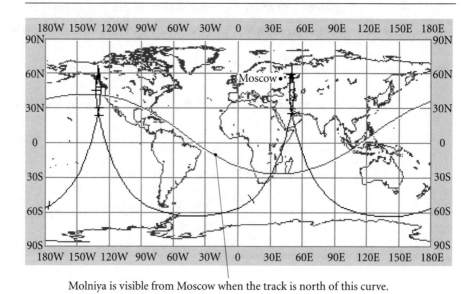

Molniya is visible from Moscow when the track is north of this curve.

Figure 4.21 Ground track of a Molniya satellite. Tick marks are one hour apart.

If a satellite is launched into an orbit with an inclination of 63.4° (prograde) or 116.6° (retrograde), then Equation 4.48 shows that the apse line will remain stationary. The Russian space program made this a key element in the design of the system of Molniya ('lightning') communications satellites. All of the Russian launch sites are above 45° latitude, the northernmost, Plesetsk, being located at 62.8°N. As we shall see in Chapter 6, launching a satellite into a geostationary orbit would involve a costly plane change maneuver. Furthermore, recall from Example 2.4 that a geostationary satellite cannot view effectively the far northern latitudes into which Russian territory extends.

The Molniya telecommunications satellites are launched from Plesetsk into 63° inclination orbits having a period of 12 hours. From Equation 2.73 we conclude that the apse line of these orbits is 53 000 km long. Perigee (typically 500 km altitude) lies in the southern hemisphere, while apogee is at an altitude of 40 000 km (25 000 miles) above the northern latitudes, farther out than the geostationary satellites. Figure 4.20 illustrates a typical Molniya orbit, and Figure 4.21 shows a ground track. A Molniya 'constellation' consists of eight satellites in planes separated by 45°. Each satellite is above 30° north latitude for over eight hours, coasting towards and away from apogee.

EXAMPLE 4.8

Determine the perigee and apogee for an earth satellite whose orbit satisfies all of the following conditions: it is sun-synchronous, its argument of perigee is constant, and its period is three hours.

The period determines the semimajor axis,

$$T = \frac{2\pi}{\sqrt{\mu}} a^{\frac{3}{2}} \Rightarrow 3 \cdot 3600 = \frac{2\pi}{\sqrt{398\,600}} a^{\frac{3}{2}} \Rightarrow a = 10\,560 \text{ km}$$

For the apse line to be stationary we know from Equation 4.48 that $i = 64.435°$ or $i = 116.57°$. But an inclination of less than 90° causes a westward regression of the node, whereas a sun-synchonous orbit requires an eastward advance, which $i = 116.57°$ provides. Substituting this, the semimajor axis and the $\dot{\Omega}$ in radians per second for a sun-synchronous orbit (cf. Example 4.7) into Equation 4.47, we get

$$1.991 \times 10^{-7} = -\frac{3}{2} \frac{\sqrt{398\,600} \cdot 0.0010826 \cdot 6378^2}{(1-e^2)^2 \cdot 10\,560^{\frac{7}{2}}} \cos 116.57° \Rightarrow e = 0.3466$$

Now we can find the angular momentum from the period expression (Equation 2.72)

$$T = \frac{2\pi}{\mu^2} \left(\frac{h}{\sqrt{1-e^2}}\right)^3 \Rightarrow 3 \cdot 3600 = \frac{2\pi}{398\,600^2} \left(\frac{h}{\sqrt{1-0.34655^2}}\right)^3$$

$$\Rightarrow h = 60\,850 \text{ km}^2/\text{s}$$

Finally, to obtain the perigee and apogee radii, we use the orbit formula:

$$z_p + 6378 = \frac{h^2}{\mu} \frac{1}{1+e} = \frac{60\,860^2}{398\,600} \frac{1}{1+0.34655} \Rightarrow \underline{z_p = 522.6 \text{ km}}$$

$$z_a + 6378 = \frac{h^2}{\mu} \frac{1}{1-e} \Rightarrow \underline{z_a = 7842 \text{ km}}$$

EXAMPLE 4.9

Given the following state vector of a satellite in geocentric equatorial coordinates,

$$\mathbf{r} = -3670\hat{\mathbf{I}} - 3870\hat{\mathbf{J}} + 4400\hat{\mathbf{K}} \text{ km}$$

$$\mathbf{v} = 4.7\hat{\mathbf{I}} - 7.4\hat{\mathbf{J}} + 1\hat{\mathbf{K}} \text{ km/s}$$

find the state vector four days (96 hours) later, assuming that there are no perturbations other than the influence of the earth's oblateness on Ω and ω.

Four days is a long enough time interval that we need to take into consideration not only the change in true anomaly but also the regression of the ascending node and the advance of perigee. First we must determine the orbital elements at the initial time using Algorithm 4.1, which yields

$$h = 58\,930 \text{ km}^2/\text{s}$$
$$i = 39.687°$$
$$e = 0.42607 \text{ (the orbit is an ellipse)}$$
$$\Omega_0 = 130.32°$$
$$\omega_0 = 42.373°$$
$$\theta_0 = 52.404°$$

We use Equation 2.61 to determine the semimajor axis,

$$a = \frac{h^2}{\mu}\frac{1}{1-e^2} = \frac{58\,930^2}{398\,600}\frac{1}{1-0.4261^2} = 10\,640 \text{ km}$$

so that, according to Equation 2.73, the period is

$$T = \frac{2\pi}{\sqrt{\mu}} a^{\frac{3}{2}} = 10\,928 \text{ s}$$

From this we obtain the mean motion

$$n = \frac{2\pi}{T} = 0.00057495 \text{ rad/s}$$

The initial value E_0 of eccentric anomaly is found from the true anomaly θ_0 using Equation 3.10a,

$$\tan\frac{E_0}{2} = \sqrt{\frac{1-e}{1+e}}\tan\frac{\theta_0}{2} = \sqrt{\frac{1-0.42607}{1+0.42607}}\tan\frac{52.404°}{2} \Rightarrow E_0 = 0.60520 \text{ rad}$$

With E_0, we use Kepler's equation to calculate the time t_0 since perigee at the initial epoch,

$$nt_0 = E_0 - e\sin E_0 \Rightarrow 0.00057495 t_0 = 0.60520 - 0.42607\sin 0.60520 \Rightarrow t_1 = 631.00 \text{ s}$$

Now we advance the time to t_f, that of the final epoch, given as 96 hours later. That is, $\Delta t = 345\,600$ s, so that

$$t_f = t_1 + \Delta t = 631.00 + 345\,600 = 346\,230 \text{ s}$$

4.7 Effects of the earth's oblateness

The number of periods n_P since passing perigee in the first orbit is

$$n_P = \frac{t_f}{T} = \frac{346\,230}{10\,928} = 31.682$$

From this we see that the final epoch occurs in the 32nd orbit, whereas t_0 was in orbit 1. Time since passing perigee in the 32nd orbit, which we will denote t_{32}, is

$$t_{32} = (31.682 - 31)\,T \Rightarrow t_{32} = 7455.7\,\text{s}$$

The mean anomaly corresponding to that time in the 32nd orbit is

$$M_{32} = nt_{32} = 0.00057495 \cdot 7455.7 = 4.2866\,\text{rad}$$

Kepler's equation yields the eccentric anomaly

$$E_{32} - e\sin E_{32} = M_{32} \Rightarrow E_{32} - 0.42607\sin E_{32} = 4.2866 \Rightarrow E_{32} = 3.9721\,\text{rad}$$
$$\text{(Algorithm 3.1)}$$

The true anomaly follows in the usual way,

$$\tan\frac{\theta_{32}}{2} = \sqrt{\frac{1+e}{1-e}}\tan\frac{E_{32}}{2} \Rightarrow \theta_{32} = 211.25°$$

At this point, we use the newly found true anomaly to calculate the state vector of the satellite in perifocal coordinates. Thus, from Equation 4.35

$$\mathbf{r}_{\bar{x}} = r\cos\theta_{32}\hat{\mathbf{p}} + r\sin\theta_{32}\hat{\mathbf{q}} = -11\,714\hat{\mathbf{p}} - 7108.8\hat{\mathbf{q}}\,(\text{km})$$

or, in matrix notation,

$$\{\mathbf{r}\}_{\bar{x}} = \begin{Bmatrix} -11\,714 \\ -7108.8 \\ 0 \end{Bmatrix}\,(\text{km})$$

Likewise, from Equation 4.36,

$$\mathbf{v}_{\bar{x}} = -\frac{\mu}{h}\sin\theta_{32}\hat{\mathbf{p}} + \frac{\mu}{h}(e+\cos\theta_{32})\,\hat{\mathbf{q}} = 3.5093\hat{\mathbf{p}} - 2.9007\hat{\mathbf{q}}\,(\text{km/s})$$

or

$$\{\mathbf{v}\}_{\bar{x}} = \begin{Bmatrix} 3.5093 \\ -2.9007 \\ 0 \end{Bmatrix}\,(\text{km/s})$$

Before we can project $\mathbf{r}_{\bar{x}}$ and $\mathbf{v}_{\bar{x}}$ into the geocentric equatorial frame, we must update the right ascension of the node and the argument of perigee. The regression rate of the ascending node is

$$\dot{\Omega} = -\left[\frac{3}{2}\frac{\sqrt{\mu}J_2R^2}{(1-e^2)^2 a^{\frac{7}{2}}}\right]\cos i = -\frac{3}{2}\frac{\sqrt{398\,600}\cdot 00108263 \cdot 6378^2}{(1-0.42607^2)^2 \cdot 10\,644^{\frac{7}{2}}}\cos 39.69°$$

$$= -3.8514\times 10^{-7}\,(\text{rad/s}) = -2.2067\times 10^{-5}\,°/\text{s}$$

Therefore, right ascension at epoch in the 32nd orbit is

$$\Omega_{32} = \Omega_0 + \dot{\Omega}\Delta t = 130.32 + (-2.2067\times 10^{-5})\cdot 345\,600 = 122.70°$$

(Example 4.9 continued)

Likewise, the perigee advance rate is

$$\dot{\omega} = -\left[\frac{3}{2}\frac{\sqrt{\mu}J_2 R^2}{(1-e^2)^2 a^{\frac{7}{2}}}\right]\left(\frac{5}{2}\sin^2 i - 2\right) = 4.9072 \times 10^{-7}\,\text{rad/s} = 2.8116 \times 10^{-5}\,°/\text{s}$$

which means the argument of perigee at epoch in the 32nd orbit is

$$\omega_{32} = \omega_0 + \dot{\omega}\Delta t = 42.373 + 2.8116 \times 10^{-5} \cdot 345\,600 = 52.090°$$

Substituting the updated values of Ω and ω, together with the inclination i, into Equation 4.43 yields the updated transformation matrix from geocentric equatorial to the perifocal frame,

$$[Q]_{X\bar{x}} = \begin{bmatrix} \cos\omega_{32} & \sin\omega_{32} & 0 \\ -\sin\omega_{32} & \cos\omega_{32} & 0 \\ 0 & 0 & 1 \end{bmatrix}\begin{bmatrix} 1 & 0 & 0 \\ 0 & \cos i & \sin i \\ 0 & -\sin i & \cos i \end{bmatrix}\begin{bmatrix} \cos\Omega_{32} & \sin\Omega_{32} & 0 \\ -\sin\Omega_{32} & \cos\Omega_{32} & 0 \\ 0 & 0 & 1 \end{bmatrix}$$

$$= \begin{bmatrix} \cos 52.09° & \sin 52.09° & 0 \\ -\sin 52.09° & \cos 52.09° & 0 \\ 0 & 0 & 1 \end{bmatrix}\begin{bmatrix} 1 & 0 & 0 \\ 0 & \cos 39.687° & \sin 39.687° \\ 0 & -\sin 39.687° & \cos 39.687° \end{bmatrix}$$

$$\times \begin{bmatrix} \cos 122.70° & \sin 122.70° & 0 \\ -\sin 122.70° & \cos 122.70° & 0 \\ 0 & 0 & 1 \end{bmatrix}$$

or

$$[Q]_{X\bar{x}} = \begin{bmatrix} -0.84285 & 0.18910 & 0.50383 \\ 0.028276 & -0.91937 & 0.39237 \\ 0.53741 & 0.34495 & 0.76955 \end{bmatrix}$$

For the inverse transformation, from perifocal to geocentric equatorial, we need the transpose of this matrix,

$$[Q]_{\bar{x}X} = \begin{bmatrix} -0.84285 & 0.18910 & 0.50383 \\ 0.028276 & -0.91937 & 0.39237 \\ 0.53741 & 0.34495 & 0.76955 \end{bmatrix}^T = \begin{bmatrix} -0.84285 & 0.028276 & 0.53741 \\ 0.18910 & -0.91937 & 0.34495 \\ 0.50383 & 0.39237 & 0.76955 \end{bmatrix}$$

Thus, according to Equations 4.46, the final state vector in the geocentric equatorial frame is

$$\{\mathbf{r}\}_X = [Q]_{\bar{x}X}\{\mathbf{r}\}_{\bar{x}}$$

$$= \begin{bmatrix} -0.84285 & 0.028276 & 0.53741 \\ 0.18910 & -0.91937 & 0.34495 \\ 0.50383 & 0.39237 & 0.76955 \end{bmatrix}\begin{Bmatrix} -11\,714 \\ -7108.8 \\ 0 \end{Bmatrix} = \begin{Bmatrix} 9672 \\ 4320 \\ -8691 \end{Bmatrix}\,(\text{km})$$

$$\{\mathbf{v}\}_X = [Q]_{\bar{x}X}\{\mathbf{v}\}_{\bar{x}}$$

$$= \begin{bmatrix} -0.84285 & 0.028276 & 0.53741 \\ 0.18910 & -0.91937 & 0.34495 \\ 0.50383 & 0.39237 & 0.76955 \end{bmatrix}\begin{Bmatrix} 3.5093 \\ -2.9007 \\ 0 \end{Bmatrix} = \begin{Bmatrix} -3.040 \\ 3.330 \\ 0.6299 \end{Bmatrix}\,(\text{km/s})$$

or, in vector notation,

$$\mathbf{r}_X = 9672\hat{\mathbf{I}} + 4320\hat{\mathbf{J}} - 8691\hat{\mathbf{K}} \text{ (km)}$$

$$\mathbf{v}_X = -3.040\hat{\mathbf{I}} + 3.330\hat{\mathbf{J}} + 0.6299\hat{\mathbf{K}} \text{(km/s)}$$

The two orbits are plotted in Figure 4.22.

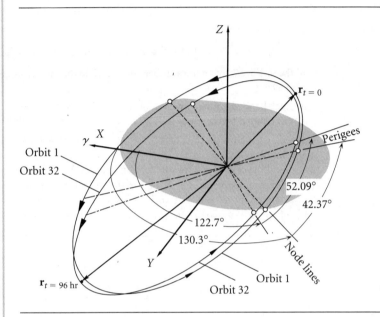

Figure 4.22   The initial and final position vectors.

PROBLEMS

4.1 Find the orbital elements of a geocentric satellite whose inertial position and velocity vectors in a geocentric equatorial frame are

$$\mathbf{r} = 2615\hat{\mathbf{I}} + 15\,881\hat{\mathbf{J}} + 3980\hat{\mathbf{K}} \text{ (km)}$$

$$\mathbf{v} = -2.767\hat{\mathbf{I}} - 0.7905\hat{\mathbf{J}} + 4.980\hat{\mathbf{K}} \text{ (km/s)}$$

{Ans.: $e = 0.3760$, $h = 95\,360$ km^2/s, $i = 63.95°$, $\Omega = 73.71°$, $\omega = 15.43°$, $\theta = 0.06764°$}

4.2 At a given instant the position $\mathbf{r}$ and velocity $\mathbf{v}$ of a satellite in the geocentric equatorial frame are $\mathbf{r} = 12\,670\hat{\mathbf{K}}$ (km) and $\mathbf{v} = -3.874\hat{\mathbf{J}} - 0.7905\hat{\mathbf{K}}$ (km/s). Find the orbital elements.
{Ans.: $h = 49\,080$ km^2/s, $e = 0.5319$, $\Omega = 90°$, $\omega = 259.5°$, $\theta = 190.5°$, $i = 90°$}

4.3 At time t_o the position $\mathbf{r}$ and velocity $\mathbf{v}$ of a satellite in the geocentric equatorial frame are $\mathbf{r} = 6472.7\hat{\mathbf{I}} - 7470.8\hat{\mathbf{J}} - 2469.8\hat{\mathbf{K}}$ (km) and $\mathbf{v} = 3.9914\hat{\mathbf{I}} + 2.7916\hat{\mathbf{J}} - 3.2948\hat{\mathbf{K}}$ (km/s). Find the orbital elements.
{Ans.: $h = 58\,461$ km^2/s, $e = 0.2465$, $\Omega = 110°$, $\omega = 75°$, $\theta = 130°$, $i = 35°$}

4.4 Given that, with respect to the geocentric equatorial frame,

$$\mathbf{r} = -6634.2\hat{\mathbf{I}} - 1261.8\hat{\mathbf{J}} - 5230.9\hat{\mathbf{K}} \text{ (km)}, \quad \mathbf{v} = 5.7644\hat{\mathbf{I}} - 7.2005\hat{\mathbf{J}} - 1.8106\hat{\mathbf{K}} \text{ (km/s)}$$

and the eccentricity vector is

$$\mathbf{e} = -0.40907\hat{\mathbf{I}} - 0.48751\hat{\mathbf{J}} - 0.63640\hat{\mathbf{K}} \text{ (dimensionless)}$$

calculate the true anomaly θ of the earth-orbiting satellite.
{Ans.: 330°}

4.5 Given that, relative to the geocentric equatorial frame,

$$\mathbf{r} = -6634.2\hat{\mathbf{I}} - 1261.8\hat{\mathbf{J}} - 5230.9\hat{\mathbf{K}} \text{ (km)}$$

the eccentricity vector is

$$\mathbf{e} = -0.40907\hat{\mathbf{I}} - 0.48751\hat{\mathbf{J}} - 0.63640\hat{\mathbf{K}} \text{ (dimensionless)}$$

and the satellite is flying towards perigee, calculate the inclination of the orbit.
{Ans.: 69.3°}

4.6 The right-handed, primed xyz system is defined by the three points A, B and C. The $x'y'$ plane is defined by the plane ABC. The x' axis runs from A through B. The z' axis is defined by the cross product of $\vec{AB}$ into $\vec{AC}$, so that the $+y'$ axis lies on the same side of the x' axis as point C.
(a) Find the orthogonal transformation matrix [$\mathbf{Q}$] relating the two coordinate bases.
(b) If the components of a vector $\mathbf{v}$ in the primed system are $\lfloor 2 \quad -1 \quad 3 \rfloor^T$, find the components of $\mathbf{v}$ in the unprimed system.
{Ans.: $\lfloor -1.307 \quad 2.390 \quad 2.565 \rfloor^T$}

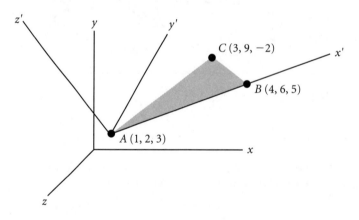

Figure P.4.6

4.7 The unit vectors in a uvw cartesian coordinate frame have the following components in the xyz frame

$$\hat{\mathbf{u}} = 0.26726\hat{\mathbf{i}} + 0.53452\hat{\mathbf{j}} + 0.80178\hat{\mathbf{k}}$$

$$\hat{\mathbf{v}} = -0.44376\hat{\mathbf{i}} + 0.80684\hat{\mathbf{j}} - 0.38997\hat{\mathbf{k}}$$

$$\hat{\mathbf{w}} = -0.85536\hat{\mathbf{i}} - 0.25158\hat{\mathbf{j}} + 0.45284\hat{\mathbf{k}}$$

If, in the xyz frame, $\mathbf{V} = -50\hat{\mathbf{i}} + 100\hat{\mathbf{j}} + 75\hat{\mathbf{k}}$, find the components of the vector $\mathbf{V}$ in the uvw frame.
{Ans.: $\mathbf{V} = 100.2\hat{\mathbf{u}} + 73.62\hat{\mathbf{v}} + 51.57\hat{\mathbf{w}}$}

4.8 Calculate the transformation matrix [**Q**] for the sequence of two rotations: $\alpha = 40°$ about the positive X axis, followed by $\beta = 25°$ about the positive y' axis. The result is that the XYZ axes are rotated into the $x''y'z''$ axes.
{Partial ans.: $Q_{11} = 0.9063 \quad Q_{12} = 0.2716 \quad Q_{13} = -0.3237$}

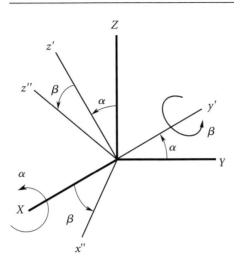

Figure P.4.8

4.9 At time t_o the position **r** and velocity **v** of a satellite in the geocentric equatorial frame are

$$\mathbf{r} = -5102\hat{\mathbf{I}} - 8228\hat{\mathbf{J}} - 2105\hat{\mathbf{K}} \text{ (km)}$$
$$\mathbf{v} = -4.348\hat{\mathbf{I}} + 3.478\hat{\mathbf{J}} - 2.846\hat{\mathbf{K}} \text{ (km/s)}$$

Find **r** and **v** at time $t_o + 50$ minutes. ($t_o \neq 0!$)
{Ans.: $\mathbf{r} = -4198\hat{\mathbf{I}} + 7856\hat{\mathbf{J}} - 3199\hat{\mathbf{K}}$ (km); $\mathbf{v} = 4.952\hat{\mathbf{I}} + 3.482\hat{\mathbf{J}} + 2.495\hat{\mathbf{K}}$ (km/s)}

4.10 For a spacecraft, the following orbital parameters are given: $e = 1.5$; perigee altitude $= 300$ km; $i = 35°$; $\Omega = 130°$; $\omega = 115°$. Calculate **r** and **v** at perigee relative to
(a) the perifocal reference frame, and
(b) the geocentric equatorial frame.
{Ans.: (a) $\mathbf{r} = 6678\hat{\mathbf{p}}$ (km), $\mathbf{v} = 12.22\hat{\mathbf{q}}$ (km/s)
(b) $\mathbf{r} = -1984\hat{\mathbf{I}} - 5348\hat{\mathbf{J}} + 3471\hat{\mathbf{K}}$ (km), $\mathbf{v} = 10.36\hat{\mathbf{I}} - 5.763\hat{\mathbf{J}} - 2.961\hat{\mathbf{K}}$ (km/s)}

4.11 For the spacecraft of Problem 4.10 calculate **r** and **v** at two hours past perigee relative to
(a) the perifocal reference frame, and
(b) the geocentric equatorial frame.
{Ans.: (a) $\mathbf{r} = -25\,010\hat{\mathbf{p}} + 48\,090\hat{\mathbf{q}}$ (km), $\mathbf{v} = -4.335\hat{\mathbf{p}} + 5.075\hat{\mathbf{q}}$ (km/s)
(b) $\mathbf{r} = 48\,200\hat{\mathbf{I}} - 2658\hat{\mathbf{J}} - 24\,660\hat{\mathbf{K}}$ (km), $\mathbf{v} = 5.590\hat{\mathbf{I}} + 1.078\hat{\mathbf{J}} - 3.484\hat{\mathbf{K}}$ (km/s)}

4.12 Calculate **r** and **v** for the satellite in Problem 4.3 at time $t_0 + 50$ minutes. ($t_o \neq 0!$)
{Ans.: $\mathbf{r} = 6864\hat{\mathbf{I}} + 5916\hat{\mathbf{J}} - 5933\hat{\mathbf{K}}$ (km), $\mathbf{v} = -3.564\hat{\mathbf{I}} + 3.905\hat{\mathbf{J}} + 1.410\hat{\mathbf{K}}$ (km/s)}

4.13 For a spacecraft, the following orbital parameters are given: $e = 1.2$; perigee altitude $= 200$ km; $i = 50°$; $\Omega = 75°$; $\omega = 80°$. Calculate **r** and **v** at perigee relative to
(a) the perifocal reference frame, and
(b) the geocentric equatorial frame.
{Ans.: (a) $\mathbf{r} = 6578\hat{\mathbf{p}}$ (km); $\mathbf{v} = 11.55\hat{\mathbf{q}}$ (km/s)
(b) $\mathbf{r} = -3726\hat{\mathbf{I}} + 2181\hat{\mathbf{J}} + 4962\hat{\mathbf{K}}$ (km), $\mathbf{v} = -4.188\hat{\mathbf{I}} - 10.65\hat{\mathbf{J}} + 1.536\hat{\mathbf{K}}$ (km/s)}

4.14 For the spacecraft of Exercise 4.13 calculate **r** and **v** at two hours past perigee relative to
(a) the perifocal reference frame, and
(b) the geocentric equatorial frame.
{Ans.: (a) $\mathbf{r} = -26\,340\hat{\mathbf{p}} + 37\,810\hat{\mathbf{q}}$ (km), $\mathbf{v} = -4.306\hat{\mathbf{p}} + 3.298\hat{\mathbf{q}}$ (km/s)
(b) $\mathbf{r} = 1207\hat{\mathbf{I}} - 43\,600\hat{\mathbf{J}} - 14\,840\hat{\mathbf{K}}$ (km), $\mathbf{v} = 1.243\hat{\mathbf{I}} - 4.4700\hat{\mathbf{J}} - 2.810\hat{\mathbf{K}}$ (km/s)}

4.15 Given that $e = 0.7$, $h = 75\,000$ km^2/s, and $\theta = 25°$, calculate the components of velocity in the geocentric equatorial frame if $[Q]_{X\bar{x}} = \begin{bmatrix} -0.83204 & -0.13114 & 0.53899 \\ 0.02741 & -0.98019 & -0.19617 \\ 0.55403 & -0.14845 & 0.81915 \end{bmatrix}$.

{Ans.: $\mathbf{v} = 2.103\hat{\mathbf{I}} - 8.073\hat{\mathbf{J}} - 2.885\hat{\mathbf{K}}$ (km/s)}

4.16 The apse line of the elliptical orbit lies in the XY plane of the geocentric equatorial frame, whose Z axis lies in the plane of the orbit. At B (for which $\theta = 140°$) the perifocal velocity vector is $\{\mathbf{v}\}_{\bar{x}} = \lfloor -3.208 \quad -0.8288 \quad 0 \rfloor^T$ (km/s). Calculate the geocentric equatorial components of the velocity at B.
{Ans.: $\{\mathbf{v}\}_X = \lfloor -1.604 \quad -2.778 \quad -0.8288 \rfloor^T$ (km/s)}

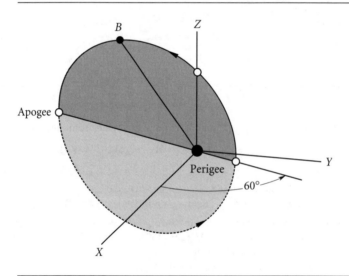

Figure P.4.16

4.17 A satellite in earth orbit has the following orbital parameters: $a = 7016$ km, $e = 0.05$, $i = 45°$, $\Omega = 0°$, $\omega = 20°$ and $\theta = 10°$. Find the position vector in the geocentric-equatorial frame.
{Ans.: $\mathbf{r} = 5776.4\hat{\mathbf{I}} + 2358.2\hat{\mathbf{J}} + 2358.2\hat{\mathbf{K}}$ (km)}

4.18 Calculate the orbital inclination required to place an earth satellite in a 500 km by 1000 km sun-synchronous orbit.
{Ans.: 98.37°}

4.19 The space shuttle is in a circular orbit of 180 km altitude and inclination 30°. What is the spacing, in kilometers, between successive ground tracks at the equator, including the effect of earth's oblateness?
{Ans.: 2511 km}

4.20 A satellite in a circular, sun-synchronous low earth orbit passes over the same point on the equator once each day, at 12 o'clock noon. Calculate the inclination, altitude and period of the orbit.
{This problem has more than one solution.}

4.21 The orbit of a satellite around an unspecified planet has an inclination of 40°, and its perigee advances at the rate of 7° per day. At what rate does the node line regress?
{Ans.: $\dot{\Omega} = 5.545°/\text{day}$}

4.22 At a given time, the position and velocity of an earth satellite in the geocentric equatorial frame are $\mathbf{r} = -2429.1\hat{\mathbf{I}} + 4555.1\hat{\mathbf{J}} + 4577.0\hat{\mathbf{K}}$ (km) and $\mathbf{v} = -4.7689\hat{\mathbf{I}} - 5.6113\hat{\mathbf{J}} + 3.0535\hat{\mathbf{K}}$ (km/s). Find $\mathbf{r}$ and $\mathbf{v}$ precisely 72 hours later, taking into consideration the node line regression and the advance of perigee.
{Ans.: $\mathbf{r} = 4596\hat{\mathbf{I}} + 5759\hat{\mathbf{J}} - 1266\hat{\mathbf{K}}$ km, $\mathbf{v} = -3.601\hat{\mathbf{I}} + 3.179\hat{\mathbf{J}} + 5.617\hat{\mathbf{K}}$ km/s}

CHAPTER 5

Preliminary orbit determination

Chapter outline

5.1	Introduction	193
5.2	Gibbs' method of orbit determination from three position vectors	194
5.3	Lambert's problem	202
5.4	Sidereal time	213
5.5	Topocentric coordinate system	218
5.6	Topocentric equatorial coordinate system	221
5.7	Topocentric horizon coordinate system	223
5.8	Orbit determination from angle and range measurements	228
5.9	Angles-only preliminary orbit determination	235
5.10	Gauss's method of preliminary orbit determination	236
Problems		250

5.1 Introduction

In this chapter we will consider some (by no means all) of the classical ways in which the orbit of a satellite can be determined from earth-bound observations. All of the methods presented here are based on the two-body equations of motion. As such, they must be considered preliminary orbit determination techniques because the actual orbit is influenced over time by other phenomena (perturbations), such as the gravitational force of the moon and sun, atmospheric drag, solar wind and the non-spherical shape and non-uniform mass distribution of the earth. We took a brief look at the dominant effects of the earth's oblateness in Section 4.7. To accurately propagate an orbit into the future from a set of initial observations requires taking the various perturbations, as well as instrumentation errors themselves, into account. More detailed considerations, including the means of updating the orbit on the basis of additional observations, are beyond our scope. Introductory discussions may be found elsewhere – see Bate, Mueller and White (1971), Boulet (1991), Prussing and Conway (1993) and Wiesel (1997), to name but a few.

We begin with the Gibbs method of predicting an orbit using three geocentric position vectors. This is followed by a presentation of Lambert's problem, in which an orbit is determined from two position vectors and the time between them. Both the Gibbs and Lambert procedures are based on the fact that two-body orbits lie in a plane. The Lambert problem is more complex and requires using the Lagrange f and g functions introduced in Chapter 2 as well as the universal variable formulation introduced in Chapter 3. The Lambert algorithm is employed in Chapter 8 to analyze interplanetary missions.

In preparation for explaining how satellites are tracked, the Julian day numbering scheme is introduced along with the notion of sidereal time. This is followed by a description of the topocentric coordinate systems and the relationships among topocentric right ascension/declension angles and azimuth/elevation angles. We then describe how orbits are determined from measuring the range and angular orientation of the line of sight together with their rates. The chapter concludes with a presentation of the Gauss method of angles-only orbit determination.

5.2 Gibbs' method of orbit determination from three position vectors

Suppose that from observations of a space object at the three successive times t_1, t_2 and t_3 ($t_1 < t_2 < t_3$) we have obtained the geocentric position vectors $\mathbf{r}_1, \mathbf{r}_2$ and $\mathbf{r}_3$. The problem is to determine the velocities $\mathbf{v}_1, \mathbf{v}_2$ and $\mathbf{v}_3$ at t_1, t_2 and t_3 assuming that the object is in a two-body orbit. The solution using purely vector analysis is due to J. W. Gibbs (1839–1903), an American scholar who is known primarily for his contributions to thermodynamics. Our explanation is based on that in Bate, Mueller and White (1971).

We know that the conservation of angular momentum requires that the position vectors of an orbiting body must lie in the same plane. In other words, the unit vector normal to the plane of $\mathbf{r}_2$ and $\mathbf{r}_3$ must be perpendicular to the unit vector in the direction of $\mathbf{r}_1$. Thus, if $\hat{\mathbf{u}}_{r1} = \mathbf{r}_1/r_1$ and $\hat{\mathbf{C}}_{23} = (\mathbf{r}_2 \times \mathbf{r}_3)/\|\mathbf{r}_2 \times \mathbf{r}_3\|$, then the dot product of these two unit vectors must vanish,

$$\hat{\mathbf{u}}_{r1} \cdot \hat{\mathbf{C}}_{23} = 0$$

Furthermore, as illustrated in Figure 5.1, the fact that $\mathbf{r}_1, \mathbf{r}_2$ and $\mathbf{r}_3$ lie in the same plane means we can apply scalar factors c_1 and c_3 to $\mathbf{r}_1$ and $\mathbf{r}_3$ so that $\mathbf{r}_2$ is the vector sum of $c_1\mathbf{r}_1$ and $c_3\mathbf{r}_3$

$$\mathbf{r}_2 = c_1\mathbf{r}_1 + c_3\mathbf{r}_3 \qquad (5.1)$$

The coefficients c_1 and c_3 are readily obtained from $\mathbf{r}_1, \mathbf{r}_2$ and $\mathbf{r}_3$ as we shall see in Section 5.10 (Equations 5.89 and 5.90).

5.2 Gibbs' method of orbit determination from three position vectors

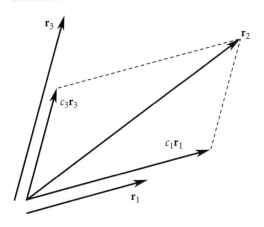

Figure 5.1 Any one of a set of three coplanar vectors (r_1, r_2, r_3) can be expressed as the vector sum of the other two.

To find the velocity **v** corresponding to any of the three given position vectors **r**, we start with Equation 2.30, which may be written

$$\mathbf{v} \times \mathbf{h} = \mu \left(\frac{\mathbf{r}}{r} + \mathbf{e} \right)$$

where **h** is the angular momentum and **e** is the eccentricity vector. To isolate the velocity, take the cross product of this equation with the angular momentum,

$$\mathbf{h} \times (\mathbf{v} \times \mathbf{h}) = \mu \left(\frac{\mathbf{h} \times \mathbf{r}}{r} + \mathbf{h} \times \mathbf{e} \right) \quad (5.2)$$

By means of the *bac – cab* rule (Equation 2.23), the left side becomes

$$\mathbf{h} \times (\mathbf{v} \times \mathbf{h}) = \mathbf{v}(\mathbf{h} \cdot \mathbf{h}) - \mathbf{h}(\mathbf{h} \cdot \mathbf{v})$$

But $\mathbf{h} \cdot \mathbf{h} = h^2$ and $\mathbf{v} \times \mathbf{h} = 0$, since **v** is perpendicular to **h**. Therefore

$$\mathbf{h} \times (\mathbf{v} \times \mathbf{h}) = h^2 \mathbf{v}$$

which means Equation 5.2 may be written

$$\mathbf{v} = \frac{\mu}{h^2} \left(\frac{\mathbf{h} \times \mathbf{r}}{r} + \mathbf{h} \times \mathbf{e} \right) \quad (5.3)$$

In Section 2.10 we introduced the perifocal coordinate system, in which the unit vector $\hat{\mathbf{p}}$ lies in the direction of the eccentricity vector **e** and $\hat{\mathbf{w}}$ is the unit vector normal to the orbital plane, in the direction of the angular momentum vector **h**. Thus, we can write

$$\mathbf{e} = e\hat{\mathbf{p}} \quad (5.4a)$$

$$\mathbf{h} = h\hat{\mathbf{w}} \quad (5.4b)$$

so that Equation 5.3 becomes

$$\mathbf{v} = \frac{\mu}{h^2}\left(\frac{h\hat{\mathbf{w}} \times \mathbf{r}}{r} + h\hat{\mathbf{w}} \times e\hat{\mathbf{p}}\right) = \frac{\mu}{h}\left[\frac{\hat{\mathbf{w}} \times \mathbf{r}}{r} + e(\hat{\mathbf{w}} \times \hat{\mathbf{p}})\right] \quad (5.5)$$

Since $\hat{\mathbf{p}}$, $\hat{\mathbf{q}}$ and $\hat{\mathbf{w}}$ form a right-handed triad of unit vectors, it follows that $\hat{\mathbf{p}} \times \hat{\mathbf{q}} = \hat{\mathbf{w}}$, $\hat{\mathbf{q}} \times \hat{\mathbf{w}} = \hat{\mathbf{p}}$ and

$$\hat{\mathbf{w}} \times \hat{\mathbf{p}} = \hat{\mathbf{q}} \quad (5.6)$$

Therefore, Equation 5.5 reduces to

$$\mathbf{v} = \frac{\mu}{h}\left(\frac{\hat{\mathbf{w}} \times \mathbf{r}}{r} + e\hat{\mathbf{q}}\right) \quad (5.7)$$

This is an important result, because if we can somehow use the position vectors $\mathbf{r}_1, \mathbf{r}_2$ and $\mathbf{r}_3$ to calculate $\hat{\mathbf{q}}, \hat{\mathbf{w}}, h$ and e, then the velocities $\mathbf{v}_1, \mathbf{v}_2$ and $\mathbf{v}_3$ will each be determined by this formula.

So far the only condition we have imposed on the three position vectors is that they are coplanar (Equation 5.1). To bring in the fact that they describe an orbit, let us take the dot product of Equation 5.1 with the eccentricity vector $\mathbf{e}$ to obtain the scalar equation

$$\mathbf{r}_2 \cdot \mathbf{e} = c_1 \mathbf{r}_1 \cdot \mathbf{e} + c_3 \mathbf{r}_3 \cdot \mathbf{e} \quad (5.8)$$

According to Equation 2.34 – the orbit equation – we have the following relation among h, e and each of the position vectors,

$$\mathbf{r}_1 \cdot \mathbf{e} = \frac{h^2}{\mu} - r_1 \qquad \mathbf{r}_2 \cdot \mathbf{e} = \frac{h^2}{\mu} - r_2 \qquad \mathbf{r}_3 \cdot \mathbf{e} = \frac{h^2}{\mu} - r_3 \quad (5.9)$$

Substituting these relations into Equation 5.8 yields

$$\frac{h^2}{\mu} - r_2 = c_1\left(\frac{h^2}{\mu} - r_1\right) + c_3\left(\frac{h^2}{\mu} - r_3\right) \quad (5.10)$$

To eliminate the unknown coefficients c_1 and c_2 from this expression, let us take the cross product of Equation 5.1 first with $\mathbf{r}_1$ and then $\mathbf{r}_3$. This results in two equations, both having $\mathbf{r}_3 \times \mathbf{r}_1$ on the right,

$$\mathbf{r}_2 \times \mathbf{r}_1 = c_3(\mathbf{r}_3 \times \mathbf{r}_1) \qquad \mathbf{r}_2 \times \mathbf{r}_3 = -c_1(\mathbf{r}_3 \times \mathbf{r}_1) \quad (5.11)$$

Now multiply Equation 5.10 through by the vector $\mathbf{r}_3 \times \mathbf{r}_1$ to obtain

$$\frac{h^2}{\mu}(\mathbf{r}_3 \times \mathbf{r}_1) - r_2(\mathbf{r}_3 \times \mathbf{r}_1) = c_1(\mathbf{r}_3 \times \mathbf{r}_1)\left(\frac{h^2}{\mu} - r_1\right) + c_3(\mathbf{r}_3 \times \mathbf{r}_1)\left(\frac{h^2}{\mu} - r_3\right)$$

Using Equations 5.11, this becomes

$$\frac{h^2}{\mu}(\mathbf{r}_3 \times \mathbf{r}_1) - r_2(\mathbf{r}_3 \times \mathbf{r}_1) = -(\mathbf{r}_2 \times \mathbf{r}_3)\left(\frac{h^2}{\mu} - r_1\right) + (\mathbf{r}_2 \times \mathbf{r}_1)\left(\frac{h^2}{\mu} - r_3\right)$$

5.2 Gibbs' method of orbit determination from three position vectors

Observe that c_1 and c_2 have been eliminated. Rearranging terms we get

$$\frac{h^2}{\mu}(\mathbf{r}_1 \times \mathbf{r}_2 + \mathbf{r}_2 \times \mathbf{r}_3 + \mathbf{r}_3 \times \mathbf{r}_1) = r_1(\mathbf{r}_2 \times \mathbf{r}_3) + r_2(\mathbf{r}_3 \times \mathbf{r}_1) + r_3(\mathbf{r}_1 \times \mathbf{r}_2) \quad (5.12)$$

This is an equation involving the given position vectors and the unknown angular momentum h. Let us introduce the following notation for the vectors on each side of Equation 5.12,

$$\mathbf{N} = r_1(\mathbf{r}_2 \times \mathbf{r}_3) + r_2(\mathbf{r}_3 \times \mathbf{r}_1) + r_3(\mathbf{r}_1 \times \mathbf{r}_2) \quad (5.13)$$

and

$$\mathbf{D} = \mathbf{r}_1 \times \mathbf{r}_2 + \mathbf{r}_2 \times \mathbf{r}_3 + \mathbf{r}_3 \times \mathbf{r}_1 \quad (5.14)$$

Then Equation 5.12 may be written more simply as

$$\mathbf{N} = \frac{h^2}{\mu}\mathbf{D}$$

from which we obtain

$$N = \frac{h^2}{\mu}D \quad (5.15)$$

where $N = \|\mathbf{N}\|$ and $D = \|\mathbf{D}\|$. It follows from Equation 5.15 that the angular momentum h is determined from $\mathbf{r}_1, \mathbf{r}_2$ and $\mathbf{r}_3$ by the formula

$$h = \sqrt{\mu \frac{N}{D}} \quad (5.16)$$

Since $\mathbf{r}_1, \mathbf{r}_2$ and $\mathbf{r}_3$ are coplanar, all of the cross products $\mathbf{r}_1 \times \mathbf{r}_2, \mathbf{r}_2 \times \mathbf{r}_3$ and $\mathbf{r}_3 \times \mathbf{r}_1$ lie in the same direction, namely, normal to the orbital plane. Therefore, it is clear from Equation 5.14 that $\mathbf{D}$ must be normal to the orbital plane. In the context of the perifocal frame, we use $\hat{\mathbf{w}}$ to denote the orbit unit normal. Therefore,

$$\hat{\mathbf{w}} = \frac{\mathbf{D}}{D} \quad (5.17)$$

So far we have found h and $\hat{\mathbf{w}}$ in terms of $\mathbf{r}_1, \mathbf{r}_2$ and $\mathbf{r}_3$. We need likewise to find an expression for $\hat{\mathbf{q}}$ to use in Equation 5.7. From Equations 5.4a, 5.6, and 5.17 it follows that

$$\hat{\mathbf{q}} = \hat{\mathbf{w}} \times \hat{\mathbf{p}} = \frac{1}{De}(\mathbf{D} \times \mathbf{e}) \quad (5.18)$$

Substituting Equation 5.14 we get

$$\hat{\mathbf{q}} = \frac{1}{De}[(\mathbf{r}_1 \times \mathbf{r}_2) \times \mathbf{e} + (\mathbf{r}_2 \times \mathbf{r}_3) \times \mathbf{e} + (\mathbf{r}_3 \times \mathbf{r}_1) \times \mathbf{e}] \quad (5.19)$$

We can apply the *bac − cab* rule to the right side by noting

$$(\mathbf{A} \times \mathbf{B}) \times \mathbf{C} = -\mathbf{C} \times (\mathbf{A} \times \mathbf{B}) = \mathbf{B}(\mathbf{A} \cdot \mathbf{C}) - \mathbf{A}(\mathbf{B} \cdot \mathbf{C})$$

Using this vector identity we obtain

$$(\mathbf{r}_2 \times \mathbf{r}_3) \times \mathbf{e} = \mathbf{r}_3(\mathbf{r}_2 \cdot \mathbf{e}) - \mathbf{r}_2(\mathbf{r}_3 \cdot \mathbf{e})$$
$$(\mathbf{r}_3 \times \mathbf{r}_1) \times \mathbf{e} = \mathbf{r}_1(\mathbf{r}_3 \cdot \mathbf{e}) - \mathbf{r}_3(\mathbf{r}_1 \cdot \mathbf{e})$$
$$(\mathbf{r}_1 \times \mathbf{r}_2) \times \mathbf{e} = \mathbf{r}_2(\mathbf{r}_1 \cdot \mathbf{e}) - \mathbf{r}_1(\mathbf{r}_2 \cdot \mathbf{e})$$

Once again employing Equations 5.9, these become

$$(\mathbf{r}_2 \times \mathbf{r}_3) \times \mathbf{e} = \mathbf{r}_3\left(\frac{h^2}{\mu} - r_2\right) - \mathbf{r}_2\left(\frac{h^2}{\mu} - r_3\right) = \frac{h^2}{\mu}(\mathbf{r}_3 - \mathbf{r}_2) + r_3\mathbf{r}_2 - r_2\mathbf{r}_3$$

$$(\mathbf{r}_3 \times \mathbf{r}_1) \times \mathbf{e} = \mathbf{r}_1\left(\frac{h^2}{\mu} - r_3\right) - \mathbf{r}_3\left(\frac{h^2}{\mu} - r_1\right) = \frac{h^2}{\mu}(\mathbf{r}_1 - \mathbf{r}_3) + r_1\mathbf{r}_3 - r_3\mathbf{r}_1$$

$$(\mathbf{r}_1 \times \mathbf{r}_2) \times \mathbf{e} = \mathbf{r}_2\left(\frac{h^2}{\mu} - r_1\right) - \mathbf{r}_1\left(\frac{h^2}{\mu} - r_2\right) = \frac{h^2}{\mu}(\mathbf{r}_2 - \mathbf{r}_1) + r_2\mathbf{r}_1 - r_1\mathbf{r}_2$$

Summing these three equations, collecting terms and substituting the result into Equation 5.19 yields

$$\hat{\mathbf{q}} = \frac{1}{De}\mathbf{S} \quad (5.20)$$

where

$$\mathbf{S} = \mathbf{r}_1(r_2 - r_3) + \mathbf{r}_2(r_3 - r_1) + \mathbf{r}_3(r_1 - r_2) \quad (5.21)$$

Finally, we substitute Equations 5.16, 5.17 and 5.20 into Equation 5.7 to obtain

$$\mathbf{v} = \frac{\mu}{h}\left(\frac{\hat{\mathbf{w}} \times \mathbf{r}}{r} + e\hat{\mathbf{q}}\right) = \frac{\mu}{\sqrt{\mu\frac{N}{D}}}\left[\frac{\frac{\mathbf{D}}{D} \times \mathbf{r}}{r} + e\left(\frac{1}{De}\mathbf{S}\right)\right]$$

Simplifying this expression for the velocity yields

$$\mathbf{v} = \sqrt{\frac{\mu}{ND}}\left(\frac{\mathbf{D} \times \mathbf{r}}{r} + \mathbf{S}\right) \quad (5.22)$$

All of the terms on the right depend only on the given position vectors $\mathbf{r}_1, \mathbf{r}_2$ and $\mathbf{r}_3$. The Gibbs procedure may be summarized in the following algorithm.

ALGORITHM 5.1

Gibbs' method of preliminary orbit determination. A MATLAB implementation of this procedure is found in Appendix D.10.

Given $\mathbf{r}_1, \mathbf{r}_2$ and $\mathbf{r}_3$, the steps are as follows.

1. Calculate r_1, r_2 and r_3.
2. Calculate $\mathbf{C}_{12} = \mathbf{r}_1 \times \mathbf{r}_2$, $\mathbf{C}_{23} = \mathbf{r}_2 \times \mathbf{r}_3$ and $\mathbf{C}_{31} = \mathbf{r}_3 \times \mathbf{r}_1$.
3. Verify that $\hat{\mathbf{u}}_{r1} \cdot \hat{\mathbf{C}}_{23} = 0$.
4. Calculate $\mathbf{N}, \mathbf{D}$ and $\mathbf{S}$ using Equations 5.13, 5.14 and 5.21, respectively.
5. Calculate $\mathbf{v}_2$ using Equation 5.22.
6. Use $\mathbf{r}_2$ and $\mathbf{v}_2$ to compute the orbital elements by means of Algorithm 4.1.

5.2 Gibbs' method of orbit determination from three position vectors

EXAMPLE 5.1

The geocentric position vectors of a space object at three successive times are

$$\mathbf{r}_1 = -294.32\hat{\mathbf{I}} + 4265.1\hat{\mathbf{J}} + 5986.7\hat{\mathbf{K}} \text{ (km)}$$

$$\mathbf{r}_2 = -1365.5\hat{\mathbf{I}} + 3637.6\hat{\mathbf{J}} + 6346.8\hat{\mathbf{K}} \text{ (km)}$$

$$\mathbf{r}_3 = -2940.3\hat{\mathbf{I}} + 2473.7\hat{\mathbf{J}} + 6555.8\hat{\mathbf{K}} \text{ (km)}$$

Determine the classical orbital elements using Gibbs' procedure.

Step 1:

$$r_1 = \sqrt{(-294.32)^2 + 4265.1^2 + 5986.7^2} = 7356.5 \text{ km}$$

$$r_2 = \sqrt{(-1365.5)^2 + 3637.6^2 + 6346.8^2} = 7441.7 \text{ km}$$

$$r_3 = \sqrt{(-2940.3)^2 + 2473.7^2 + 6555.8^2} = 7598.9 \text{ km}$$

Step 2:

$$\mathbf{C}_{12} = \begin{vmatrix} \hat{\mathbf{I}} & \hat{\mathbf{J}} & \hat{\mathbf{K}} \\ -294.32 & 4265.1 & 5986.7 \\ -1365.5 & 3637.6 & 6346.8 \end{vmatrix}$$

$$= (5.292\hat{\mathbf{I}} - 6.3066\hat{\mathbf{J}} + 4.7531\hat{\mathbf{K}}) \times 10^6 \text{ (km}^2\text{)}$$

$$\mathbf{C}_{23} = \begin{vmatrix} \hat{\mathbf{I}} & \hat{\mathbf{J}} & \hat{\mathbf{K}} \\ -1365.5 & 3637.6 & 6346.8 \\ -2940.3 & 2473.7 & 6555.8 \end{vmatrix}$$

$$= (8.1473\hat{\mathbf{I}} - 9.7095\hat{\mathbf{J}} + 7.3179\hat{\mathbf{K}}) \times 10^6 \text{ (km}^2\text{)}$$

$$\mathbf{C}_{31} = \begin{vmatrix} \hat{\mathbf{I}} & \hat{\mathbf{J}} & \hat{\mathbf{K}} \\ -2940.3 & 2473.7 & 6555.8 \\ -294.32 & 4265.1 & 5986.7 \end{vmatrix}$$

$$= (-1.3151\hat{\mathbf{I}} + 1.5673\hat{\mathbf{J}} - 1.1812\hat{\mathbf{K}}) \times 10^6 \text{ (km}^2\text{)}$$

Step 3:

$$\hat{\mathbf{C}}_{23} = \frac{\mathbf{C}_{23}}{\|\mathbf{C}_{23}\|} = \frac{8.1473\hat{\mathbf{I}} - 9.7095\hat{\mathbf{J}} + 7.3179\hat{\mathbf{K}}}{\sqrt{8.1473^2 + (-9.7095)^2 + 7.3179^2}}$$

$$= 0.55667\hat{\mathbf{I}} - 0.66341\hat{\mathbf{J}} + 0.5000\hat{\mathbf{K}}$$

Therefore

$$\hat{\mathbf{u}}_{r_1} \cdot \hat{\mathbf{C}}_{23} = \frac{-294.32\hat{\mathbf{I}} + 4265.1\hat{\mathbf{J}} + 5986.7\hat{\mathbf{K}}}{7356.5} \cdot (0.55667\hat{\mathbf{I}} - 0.66341\hat{\mathbf{J}} + 0.5000\hat{\mathbf{K}})$$

$$= 6.9200 \times 10^{-20}$$

This certainly is close enough to zero for our purposes. The three vectors $\mathbf{r}_1, \mathbf{r}_2$ and $\mathbf{r}_3$ are coplanar.

(Example 5.1 continued)

Step 4:

$$\mathbf{N} = r_1 \mathbf{C}_{23} + r_2 \mathbf{C}_{31} + r_3 \mathbf{C}_{12}$$

$$= 7356.5[(8.1473\hat{\mathbf{I}} - 9.7095\hat{\mathbf{J}} + 7.3179\hat{\mathbf{K}}) \times 10^6]$$
$$+ 7441.7[(-1.3151\hat{\mathbf{I}} + 1.5673\hat{\mathbf{J}} - 1.1812\hat{\mathbf{K}}) \times 10^6]$$
$$+ 7598.9[(5.292\hat{\mathbf{I}} - 6.3066\hat{\mathbf{J}} + 4.7531\hat{\mathbf{K}}) \times 10^6]$$

or

$$\mathbf{N} = (2.2807\hat{\mathbf{I}} - 2.7181\hat{\mathbf{J}} + 2.0486\hat{\mathbf{K}}) \times 10^9 \; (\text{km}^3)$$

so that

$$N = \sqrt{[2.2807^2 + (-2.7181)^2 + 2.0486^2] \times 10^{18}}$$
$$= 4.0971 \times 10^9 \; (\text{km}^3)$$

$$\mathbf{D} = \mathbf{C}_{12} + \mathbf{C}_{23} + \mathbf{C}_{31}$$
$$= [(5.292\hat{\mathbf{I}} - 6.3066\hat{\mathbf{J}} + 4.7531\hat{\mathbf{K}}) \times 10^6] + [(8.1473\hat{\mathbf{I}} - 9.7095\hat{\mathbf{J}}$$
$$+ 7.3179\hat{\mathbf{K}}) \times 10^6] + [(-1.3151\hat{\mathbf{I}} + 1.5673\hat{\mathbf{J}} - 1.1812\hat{\mathbf{K}}) \times 10^6]$$

or

$$\mathbf{D} = (2.8797\hat{\mathbf{I}} - 3.4319\hat{\mathbf{J}} + 2.5866\hat{\mathbf{K}}) \times 10^6 \; (\text{km}^2)$$

so that

$$D = \sqrt{[2.8797^2 + (-3.4319)^2 + 2.5866^2] \times 10^{12}}$$
$$= 5.1731 \times 10^5 \; (\text{km}^2)$$

Lastly,

$$\mathbf{S} = \mathbf{r}_1(r_2 - r_3) + \mathbf{r}_2(r_3 - r_1) + \mathbf{r}_3(r_1 - r_2)$$
$$= (-294.32\hat{\mathbf{I}} + 4265.1\hat{\mathbf{J}} + 5986.7\hat{\mathbf{K}})(7441.7 - 7598.9)$$
$$+ (-1365.5\hat{\mathbf{I}} + 3637.6\hat{\mathbf{J}} + 6346.8\hat{\mathbf{K}})(7598.9 - 7356.5)$$
$$+ (-2940.3\hat{\mathbf{I}} + 2473.7\hat{\mathbf{J}} + 6555.8\hat{\mathbf{K}})(7356.5 - 7441.7)$$

or

$$\mathbf{S} = -34\,213\hat{\mathbf{I}} + 533.51\hat{\mathbf{J}} + 38\,798\hat{\mathbf{K}} \; (\text{km}^2)$$

Step 5:

$$\mathbf{v}_2 = \sqrt{\frac{\mu}{ND}} \left(\frac{\mathbf{D} \times \mathbf{r}_2}{r_2} + \mathbf{S} \right)$$

$$= \sqrt{\frac{398\,600}{(4.0971 \times 10^9)(5.1731 \times 10^3)}}$$

$$\times \left[\frac{\begin{vmatrix} \hat{\mathbf{I}} & \hat{\mathbf{J}} & \hat{\mathbf{K}} \\ 2.8797 \times 10^6 & -3.4319 \times 10^6 & 2.5866 \times 10^6 \\ -1365.5 & 3637.6 & 6346.8 \end{vmatrix}}{7441.7} + \begin{pmatrix} -34\,213\hat{\mathbf{I}} + 533.51\hat{\mathbf{J}} \\ +38\,798\hat{\mathbf{K}} \end{pmatrix} \right]$$

or

$$\mathbf{v}_2 = -6.2171\hat{\mathbf{I}} - 4.0117\hat{\mathbf{J}} + 1.5989\hat{\mathbf{K}} \text{ (km/s)}$$

Step 6:

Using $\mathbf{r}_2$ and $\mathbf{v}_2$, Algorithm 4.1 yields the orbital elements:

$$a = 8000\,\text{km}$$
$$e = 0.1$$
$$i = 60°$$
$$\Omega = 40°$$
$$\omega = 30°$$
$$\theta = 50° \text{ (for position vector } \mathbf{r}_2\text{)}$$

The orbit is sketched in Figure 5.2.

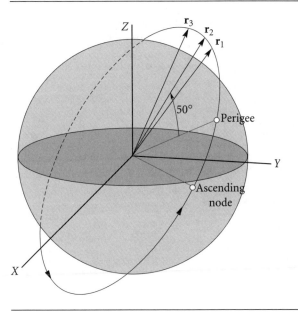

Figure 5.2 Sketch of the orbit of Example 5.1.

5.3 Lambert's problem

Suppose we know the position vectors $\mathbf{r}_1$ and $\mathbf{r}_2$ of two points P_1 and P_2 on the path of mass m around mass M, as illustrated in Figure 5.3. $\mathbf{r}_1$ and $\mathbf{r}_2$ determine the change in the true anomaly $\Delta\theta$, since

$$\cos\Delta\theta = \frac{\mathbf{r}_1 \cdot \mathbf{r}_2}{r_1 r_2} \tag{5.23}$$

where

$$r_1 = \sqrt{\mathbf{r}_1 \cdot \mathbf{r}_1} \qquad r_2 = \sqrt{\mathbf{r}_2 \cdot \mathbf{r}_2} \tag{5.24}$$

However, if $\cos\Delta\theta > 0$, then $\Delta\theta$ lies in either the first or fourth quadrant, whereas if $\cos\Delta\theta < 0$, then $\Delta\theta$ lies in the second or third quadrant. (Recall Figure 3.4.) The first step in resolving this quadrant ambiguity is to calculate the Z component of $\mathbf{r}_1 \times \mathbf{r}_2$,

$$(\mathbf{r}_1 \times \mathbf{r}_2)_Z = \hat{\mathbf{K}} \cdot (\mathbf{r}_1 \times \mathbf{r}_2) = \hat{\mathbf{K}} \cdot (r_1 r_2 \sin\Delta\theta \hat{\mathbf{w}}) = r_1 r_2 \sin\Delta\theta (\hat{\mathbf{K}} \cdot \hat{\mathbf{w}})$$

where $\hat{\mathbf{w}}$ is the unit normal to the orbital plane. Therefore, $\hat{\mathbf{K}} \cdot \hat{\mathbf{w}} = \cos i$, where i is the inclination of the orbit, so that

$$(\mathbf{r}_1 \times \mathbf{r}_2)_Z = r_1 r_2 \sin\Delta\theta \cos i \tag{5.25}$$

We use the sign of the scalar $(\mathbf{r}_1 \times \mathbf{r}_2)_Z$ to determine the correct quadrant for $\Delta\theta$.

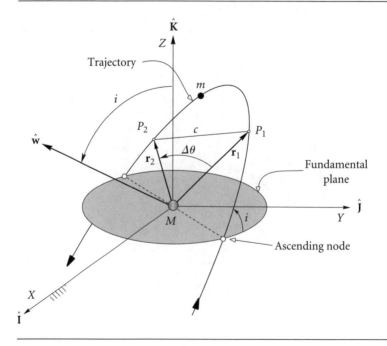

Figure 5.3 Lambert's problem.

5.3 Lambert's problem

There are two cases to consider: prograde trajectories ($0 < i < 90°$), and retrograde trajectories ($90° < i < 180°$).

For prograde trajectories (like the one illustrated in Figure 5.3), $\cos i > 0$, so that if $(\mathbf{r}_1 \times \mathbf{r}_2)_Z > 0$, then Equation 5.25 implies that $\sin \Delta\theta > 0$, which means $0° < \Delta\theta < 180°$. Since $\Delta\theta$ therefore lies in the first or second quadrant, it follows that $\Delta\theta$ is given by $\cos^{-1}(\mathbf{r}_1 \cdot \mathbf{r}_2/r_1 r_2)$. On the other hand, if $(\mathbf{r}_1 \times \mathbf{r}_2)_Z < 0$, Equation 5.25 implies that $\sin \Delta\theta < 0$, which means $180° < \Delta\theta < 360°$. In this case $\Delta\theta$ lies in the third or fourth quadrant and is given by $360° - \cos^{-1}(\mathbf{r}_1 \cdot \mathbf{r}_2/r_1 r_2)$. For retrograde trajectories, $\cos i < 0$. Thus, if $(\mathbf{r}_1 \times \mathbf{r}_2)_Z > 0$ then $\sin \Delta\theta < 0$, which places $\Delta\theta$ in the third or fourth quadrant. Similarly, if $(\mathbf{r}_1 \times \mathbf{r}_2)_Z > 0$, $\Delta\theta$ must lie in the first or second quadrant.

This logic can be expressed more concisely as follows:

$$\Delta\theta = \begin{cases} \cos^{-1}\left(\dfrac{\mathbf{r}_1 \cdot \mathbf{r}_2}{r_1 r_2}\right) & \text{if } (\mathbf{r}_1 \times \mathbf{r}_2)_Z \geq 0 \\ 360° - \cos^{-1}\left(\dfrac{\mathbf{r}_1 \cdot \mathbf{r}_2}{r_1 r_2}\right) & \text{if } (\mathbf{r}_1 \times \mathbf{r}_2)_Z < 0 \end{cases} \quad \text{prograde trajectory} \\ \cdots\cdots\cdots\cdots\cdots\cdots\cdots\cdots\cdots\cdots\cdots\cdots\cdots\cdots\cdots\cdots\cdots\cdots\cdots\cdots\cdots\cdots\cdots\cdots \\ \begin{cases} \cos^{-1}\left(\dfrac{\mathbf{r}_1 \cdot \mathbf{r}_2}{r_1 r_2}\right) & \text{if } (\mathbf{r}_1 \times \mathbf{r}_2)_Z < 0 \\ 360° - \cos^{-1}\left(\dfrac{\mathbf{r}_1 \cdot \mathbf{r}_2}{r_1 r_2}\right) & \text{if } (\mathbf{r}_1 \times \mathbf{r}_2)_Z \geq 0 \end{cases} \quad \text{retrograde trajectory}$$

(5.26)

J. H. Lambert (1728–1777) was a French-born German astronomer, physicist and mathematician. According to a theorem of Lambert, the transfer time Δt from P_1 to P_2 is independent of the orbit's eccentricity and depends only on the sum $r_1 + r_2$ of the magnitudes of the position vectors, the semimajor axis a and the length c of the chord joining P_1 and P_2. It is noteworthy that the period (of an ellipse) and the specific mechanical energy are also independent of the eccentricity (Equations 2.73, 2.70 and 2.100).

If we know the time of flight Δt from P_1 to P_2, then Lambert's problem is to find the trajectory joining P_1 and P_2. The trajectory is determined once we find $\mathbf{v}_1$, because, according to Equations 2.125 and 2.126, the position and velocity of any point on the path are determined by $\mathbf{r}_1$ and $\mathbf{v}_1$. That is, in terms of the notation in Figure 5.3,

$$\mathbf{r}_2 = f\mathbf{r}_1 + g\mathbf{v}_1 \qquad (5.27a)$$

$$\mathbf{v}_2 = \dot{f}\mathbf{r}_1 + \dot{g}\mathbf{v}_1 \qquad (5.27b)$$

Solving the first of these for $\mathbf{v}_1$ yields

$$\mathbf{v}_1 = \frac{1}{g}(\mathbf{r}_2 - f\mathbf{r}_1) \qquad (5.28)$$

Substitute this result into Equation 5.27b to get

$$\mathbf{v}_2 = \dot{f}\mathbf{r}_1 + \frac{\dot{g}}{g}(\mathbf{r}_2 - f\mathbf{r}_1) = \frac{\dot{g}}{g}\mathbf{r}_2 - \frac{f\dot{g} - \dot{f}g}{g}\mathbf{r}_1$$

But, according to Equation 2.129, $f\dot{g} - \dot{f}g = 1$. Hence,

$$\mathbf{v}_2 = \frac{1}{g}(\dot{g}\mathbf{r}_2 - \mathbf{r}_1) \tag{5.29}$$

By means of Algorithm 4.1 we can find the orbital elements from either $\mathbf{r}_1$ and $\mathbf{v}_1$ or $\mathbf{r}_2$ and $\mathbf{v}_2$.

Clearly, Lambert's problem is solved once we determine the Lagrange coefficients f, g and $\dot{g}$. We will follow the procedure presented by Bate, Mueller and White (1971) and Bond and Allman (1996).

The Lagrange f and g coefficients and their time derivatives are listed as functions of the change in true anomaly $\Delta\theta$ in Equations 2.148,

$$f = 1 - \frac{\mu r_2}{h^2}(1 - \cos\Delta\theta) \qquad g = \frac{r_1 r_2}{h}\sin\Delta\theta \tag{5.30a}$$

$$\dot{f} = \frac{\mu}{h}\frac{1-\cos\Delta\theta}{\sin\Delta\theta}\left[\frac{\mu}{h^2}(1-\cos\Delta\theta) - \frac{1}{r_1} - \frac{1}{r_2}\right] \qquad \dot{g} = 1 - \frac{\mu r_1}{h^2}(1 - \cos\Delta\theta) \tag{5.30b}$$

Equations 3.66 express these quantities in terms of the universal anomaly χ,

$$f = 1 - \frac{\chi^2}{r_1}C(z) \qquad g = \Delta t - \frac{1}{\sqrt{\mu}}\chi^3 S(z) \tag{5.31a}$$

$$\dot{f} = \frac{\sqrt{\mu}}{r_1 r_2}\chi[zS(z) - 1] \qquad \dot{g} = 1 - \frac{\chi^2}{r_2}C(z) \tag{5.31b}$$

where $z = \alpha\chi^2$. The f and g functions do not depend on the eccentricity, which would seem to make them an obvious choice for the solution of Lambert's problem. Ignoring for the time being that $z = \alpha\chi^2$, the unknowns on the right of the above sets of equations are h, χ and z, whereas $\Delta\theta$, Δt, r and r_0 are given.

While $\Delta\theta$ appears throughout Equations 5.30, the time interval Δt does not. However, Δt does appear in Equation 5.31a. A relationship between $\Delta\theta$ and Δt can therefore be found by equating the two expressions for g,

$$\frac{r_1 r_2}{h}\sin\Delta\theta = \Delta t - \frac{1}{\sqrt{\mu}}\chi^3 S(z) \tag{5.32}$$

To eliminate the unknown angular momentum h, equate the expressions for f in Equations 5.30a and 5.31a,

$$1 - \frac{\mu r_2}{h^2}(1 - \cos\Delta\theta) = 1 - \frac{\chi^2}{r_1}C(z)$$

Upon solving this for h we obtain

$$h = \sqrt{\frac{\mu r_1 r_2(1 - \cos\Delta\theta)}{\chi^2 C(z)}} \tag{5.33}$$

(Equating the two expressions for $\dot{g}$ leads to the same result.) Substituting Equation 5.33 into 5.32, simplifying and rearranging terms yields

$$\sqrt{\mu}\Delta t = \chi^3 S(z) + \chi\sqrt{C(z)}\left(\sin\Delta\theta\sqrt{\frac{r_1 r_2}{1 - \cos\Delta\theta}}\right) \tag{5.34}$$

The term in parentheses on the right is a constant comprised solely of the given data. Let us assign it the symbol A,

$$A = \sin \Delta\theta \sqrt{\frac{r_1 r_2}{1 - \cos \Delta\theta}} \tag{5.35}$$

Then Equation 5.34 assumes the simpler form

$$\sqrt{\mu}\Delta t = \chi^3 S(z) + A\chi\sqrt{C(z)} \tag{5.36}$$

The right side of this equation contains both of the unknown variables χ and z. We cannot use the fact that $z = \alpha \chi^2$ to reduce the unknowns to one since α is the reciprocal of the semimajor axis of the unknown orbit.

In order to find a relationship between z and χ which does not involve orbital parameters, we equate the expressions for $\dot{f}$ (Equations 5.30b and 5.31b) to obtain

$$\frac{\mu}{h} \frac{1 - \cos \Delta\theta}{\sin \Delta\theta} \left[\frac{\mu}{h^2}(1 - \cos \Delta\theta) - \frac{1}{r_1} - \frac{1}{r_2} \right] = \frac{\sqrt{\mu}}{r_1 r_2} \chi[zS(z) - 1]$$

Multiplying through by $r_1 r_2$ and substituting for the angular momentum using Equation 5.33 yields

$$\frac{\mu}{\sqrt{\frac{\mu r_1 r_2 (1 - \cos \Delta\theta)}{\chi^2 C(z)}}} \frac{1 - \cos \Delta\theta}{\sin \Delta\theta} \left[\frac{\mu}{\frac{\mu r_1 r_2 (1 - \cos \Delta\theta)}{\chi^2 C(z)}}(1 - \cos \Delta\theta) - r_1 - r_2 \right]$$

$$= \sqrt{\mu}\chi[zS(z) - 1]$$

Simplifying and dividing out common factors leads to

$$\frac{\sqrt{1 - \cos \Delta\theta}}{\sqrt{r_1 r_2} \sin \Delta\theta} \sqrt{C(z)}[\chi^2 C(z) - r_1 - r_2] = zS(z) - 1$$

We recognize the reciprocal of A on the left, so we can rearrange this expression to read as follows,

$$\chi^2 C(z) = r_1 + r_2 + A \frac{zS(z) - 1}{\sqrt{C(z)}}$$

The right-hand side depends exclusively on z. Let us call that function $y(z)$, so that

$$\chi = \sqrt{\frac{y(z)}{C(z)}} \tag{5.37}$$

where

$$y(z) = r_1 + r_2 + A \frac{zS(z) - 1}{\sqrt{C(z)}} \tag{5.38}$$

Equation 5.37 is the relation between χ and z that we were seeking. Substituting it back into Equation 5.36 yields

$$\sqrt{\mu}\Delta t = \left[\frac{y(z)}{C(z)}\right]^{\frac{3}{2}} S(z) + A\sqrt{y(z)} \tag{5.39}$$

We can use this equation to solve for z, given the time interval Δt. It must be done iteratively. Using Newton's method, we form the function

$$F(z) = \left[\frac{y(z)}{C(z)}\right]^{\frac{3}{2}} S(z) + A\sqrt{y(z)} - \sqrt{\mu}\Delta t \quad (5.40)$$

and its derivative

$$F'(z) = \frac{1}{2\sqrt{y(z)C(z)^5}} \left\{ [2C(z)S'(z) - 3C'(z)S(z)]y^2(z) \right.$$
$$\left. + \left[AC(z)^{\frac{5}{2}} + 3C(z)S(z)y(z)\right]y'(z) \right\} \quad (5.41)$$

in which $C'(z)$ and $S'(z)$ are the derivatives of the Stumpff functions, which are given by Equations 3.60. $y'(z)$ is obtained by differentiating $y(z)$ in Equation 5.38,

$$y'(z) = \frac{A}{2C(z)^{\frac{3}{2}}} \{[1 - zS(z)]C'(z) + 2[S(z) + zS'(z)]C(z)\}$$

If we substitute Equations 3.60 into this expression a much simpler form is obtained, namely

$$y'(z) = \frac{A}{4}\sqrt{C(z)} \quad (5.42)$$

This result can be worked out by using Equations 3.49 and 3.50 to express $C(z)$ and $S(z)$ in terms of the more familiar trig functions. Substituting Equation 5.42 along with Equations 3.60 into Equation 5.41 yields

$$F'(z) = \begin{cases} \left[\frac{y(z)}{C(z)}\right]^{\frac{3}{2}} \left\{\frac{1}{2z}\left[C(z) - \frac{3}{2}\frac{S(z)}{C(z)}\right] + \frac{3}{4}\frac{S(z)^2}{C(z)}\right\} \\ + \frac{A}{8}\left[3\frac{S(z)}{C(z)}\sqrt{y(z)} + A\sqrt{\frac{C(z)}{y(z)}}\right] & (z \neq 0) \\ \frac{\sqrt{2}}{40}y(0)^{\frac{3}{2}} + \frac{A}{8}\left[\sqrt{y(0)} + A\sqrt{\frac{1}{2y(0)}}\right] & (z = 0) \end{cases} \quad (5.43)$$

Evaluating $F'(z)$ at $z = 0$ must be done carefully (and is therefore shown as a special case), because of the z in the denominator within the curly brackets. To handle $z = 0$, we assume that z is very small (almost, but not quite zero) so that we can retain just the first two terms in the series expansions of $C(z)$ and $S(z)$ (Equations 3.47 and 3.48),

$$C(z) = \frac{1}{2} - \frac{z}{24} + \ldots \qquad S(z) = \frac{1}{6} - \frac{z}{120} + \ldots$$

5.3 Lambert's problem

Then we evaluate the term within the curly brackets as follows:

$$\frac{1}{2z}\left[C(z) - \frac{3}{2}\frac{S(z)}{C(z)}\right] \approx \frac{1}{2z}\left[\left(\frac{1}{2} - \frac{z}{24}\right) - \frac{3}{2}\frac{\left(\frac{1}{6} - \frac{z}{120}\right)}{\left(\frac{1}{2} - \frac{z}{24}\right)}\right]$$

$$= \frac{1}{2z}\left[\left(\frac{1}{2} - \frac{z}{24}\right) - 3\left(\frac{1}{6} - \frac{z}{120}\right)\left(1 - \frac{z}{12}\right)^{-1}\right]$$

$$\approx \frac{1}{2z}\left[\left(\frac{1}{2} - \frac{z}{24}\right) - 3\left(\frac{1}{6} - \frac{z}{120}\right)\left(1 + \frac{z}{12}\right)\right]$$

$$= \frac{1}{2z}\left(-\frac{7z}{120} + \frac{z^2}{480}\right)$$

$$= -\frac{7}{240} + \frac{z}{960}$$

In the third step we used the familiar binomial expansion theorem,

$$(a+b)^n = a^n + na^{n-1}b + \frac{n(n-1)}{2!}a^{n-2}b^2 + \frac{n(n-1)(n-2)}{3!}a^{n-3}b^3 + \ldots \quad (5.44)$$

to set $(1 - z/12)^{-1} \approx 1 + z/12$, which is true if z is close to zero. Thus, when z is actually zero,

$$\frac{1}{2z}\left[C(z) - \frac{3}{2}\frac{S(z)}{C(z)}\right] = -\frac{7}{240}$$

Evaluating the other terms in $F'(z)$ presents no difficulties.

$F(z)$ in Equation 5.40 and $F'(z)$ in Equation 5.43 are used in Newton's formula, Equation 3.13, for the iterative procedure,

$$z_{i+1} = z_i - \frac{F(z_i)}{F'(z_i)} \quad (5.45)$$

For choice of a starting value for z, recall that $z = (1/a)\chi^2$. According to Equation 3.54, $z = E^2$ for an ellipse and $z = -F^2$ for a hyperbola. Since we do not know what the orbit is, setting $z_0 = 0$ seems a reasonable, simple choice. Alternatively, one can plot or tabulate $F(z)$ and choose z_0 to be a point near where $F(z)$ changes sign.

Substituting Equations 5.37 and 5.39 into Equations 5.31 yields the Lagrange coefficients as functions of z alone:

$$f = 1 - \frac{\left[\sqrt{\frac{y(z)}{C(z)}}\right]^2}{r_1}C(z) = 1 - \frac{y(z)}{r_1} \quad (5.46a)$$

$$g = \frac{1}{\sqrt{\mu}}\left\{\left[\frac{y(z)}{C(z)}\right]^{\frac{3}{2}}S(z) + A\sqrt{y(z)}\right\} - \frac{1}{\sqrt{\mu}}\left[\frac{y(z)}{C(z)}\right]^{\frac{3}{2}}S(z) = A\sqrt{\frac{y(z)}{\mu}} \quad (5.46b)$$

208 Chapter 5 *Preliminary orbit determination*

$$\dot{f} = \frac{\sqrt{\mu}}{r_1 r_2}\sqrt{\frac{y(z)}{C(z)}}[zS(z) - 1] \tag{5.46c}$$

$$\dot{g} = 1 - \frac{\left[\sqrt{\frac{y(z)}{C(z)}}\right]^2}{r_2}C(z) = 1 - \frac{y(z)}{r_2} \tag{5.46d}$$

We are now in a position to present the solution of Lambert's problem in universal variables, following Bond and Allman (1996).

ALGORITHM 5.2

Solve Lambert's problem. A MATLAB implementation appears in Appendix D.12.

Given $\mathbf{r}_1, \mathbf{r}_2$ and Δt, the steps are as follows.

1. Calculate r_1 and r_2 using Equation 5.24.
2. Choose either a prograde or retrograde trajectory and calculate $\Delta\theta$ using Equation 5.26.
3. Calculate A in Equation 5.35.
4. By iteration, using Equations 5.40, 5.43 and 5.45, solve Equation 5.39 for z. The sign of z tells us whether the orbit is a hyperbola ($z < 0$), parabola ($z = 0$) or ellipse ($z > 0$).
5. Calculate y using Equation 5.38.
6. Calculate the Lagrange f, g and $\dot{g}$ functions using Equations 5.46.
7. Calculate $\mathbf{v}_1$ and $\mathbf{v}_2$ from Equations 5.28 and 5.29.
8. Use $\mathbf{r}_1$ and $\mathbf{v}_1$ (or $\mathbf{r}_2$ and $\mathbf{v}_2$) in Algorithm 4.1 to obtain the orbital elements.

EXAMPLE 5.2

The position of an earth satellite is first determined to be $\mathbf{r}_1 = 5000\hat{\mathbf{I}} + 10\,000\hat{\mathbf{J}} + 2100\hat{\mathbf{K}}$ (km). After one hour the position vector is $\mathbf{r}_2 = -14\,600\hat{\mathbf{I}} + 2500\hat{\mathbf{J}} + 7000\hat{\mathbf{K}}$ (km). Determine the orbital elements and find the perigee altitude and the time since perigee passage of the first sighting.

We first must execute the steps of Algorithm 5.2 in order to find $\mathbf{v}_1$ and $\mathbf{v}_2$.

Step 1:

$$r_1 = \sqrt{5000^2 + 10\,000^2 + 2100^2} = 11\,375\,\text{km}$$

$$r_2 = \sqrt{(-14\,600)^2 + 2500^2 + 7000^2} = 16\,383\,\text{km}$$

Step 2: assume a prograde trajectory:

$$\mathbf{r}_1 \times \mathbf{r}_2 = (64.75\hat{\mathbf{I}} - 65.66\hat{\mathbf{J}} + 158.5\hat{\mathbf{K}}) \times 10^6$$

$$\cos^{-1}\frac{\mathbf{r}_1 \cdot \mathbf{r}_2}{r_1 r_2} = 100.29°$$

Since the trajectory is prograde and the z component of $\mathbf{r}_1 \times \mathbf{r}_2$ is positive, it follows from Equation 5.26 that

$$\Delta\theta = 100.29°$$

5.3 Lambert's problem

Step 3:

$$A = \sin \Delta\theta \sqrt{\frac{r_1 r_2}{1 - \cos \Delta\theta}} = \sin 100.29° \sqrt{\frac{11\,375 \cdot 16\,383}{1 - \cos 100.29°}} = 12\,372 \text{ km}$$

Step 4:

Using this value of A and $\Delta t = 3600$ s, we can evaluate the functions $F(z)$ and $F'(z)$ given by Equations 5.40 and 5.43, respectively. Let us first plot $F(z)$ to get at least a rough idea of where it crosses the z axis. As can be seen from Figure 5.4, $F(z) = 0$ near $z = 1.5$. With $z_0 = 1.5$ as our initial estimate, we execute Newton's procedure, Equation 5.45,

$$z_{i+1} = z_i - \frac{F(z_i)}{F'(z_i)}$$

$$z_1 = 1.5 - \frac{-14\,476.4}{362\,642} = 1.53991$$

$$z_2 = 1.53991 - \frac{23.6274}{363\,828} = 1.53985$$

$$z_3 = 1.53985 - \frac{6.29457 \times 10^{-5}}{363\,826} = 1.53985$$

Thus, to five significant figures $z = 1.5398$. The fact that z is positive means the orbit is an ellipse.

Step 5:

$$y = r_1 + r_2 + A \frac{zS(z) - 1}{\sqrt{C(z)}} = 11\,375 + 16\,383 + 12\,372 \frac{1.5398 S(1.5398)}{\sqrt{C(1.5398)}} = 13\,523 \text{ km}$$

Step 6:

Equations 5.46 yield the Lagrange functions

$$f = 1 - \frac{y}{r_1} = 1 - \frac{13\,523}{11\,375} = -0.18877$$

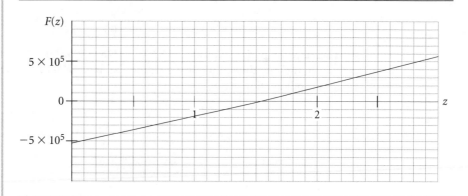

Figure 5.4 Graph of $F(z)$.

(*Example 5.2 continued*)

$$g = A\sqrt{\frac{y}{\mu}} = 12\,372\sqrt{\frac{13\,523}{398\,600}} = 2278.9\,\text{s}$$

$$\dot{g} = 1 - \frac{y}{r_2} = 1 - \frac{13\,523}{16\,383} = 0.17457$$

Step 7:

$$\mathbf{v}_1 = \frac{1}{g}(\mathbf{r}_2 - f\mathbf{r}_1) = \frac{1}{2278.9}[(-14\,600\hat{\mathbf{I}} + 2500\hat{\mathbf{J}} + 7000\hat{\mathbf{K}})$$
$$- (-0.18877)(5000\hat{\mathbf{I}} + 10\,000\hat{\mathbf{J}} + 2100\hat{\mathbf{K}})]$$

$$\mathbf{v}_1 = -5.9925\hat{\mathbf{I}} + 1.9254\hat{\mathbf{J}} + 3.2456\hat{\mathbf{K}} \,(\text{km})$$

$$\mathbf{v}_2 = \frac{1}{g}(\dot{g}\mathbf{r}_2 - \mathbf{r}_1) = \frac{1}{2278.9}[(0.17457)(-14\,600\hat{\mathbf{I}} + 2500\hat{\mathbf{J}} + 7000\hat{\mathbf{K}})$$
$$- (5000\hat{\mathbf{I}} + 10\,000\hat{\mathbf{J}} + 2100\hat{\mathbf{K}})]$$

$$\mathbf{v}_2 = -3.3125\hat{\mathbf{I}} - 4.1966\hat{\mathbf{J}} - 0.38529\hat{\mathbf{K}} \,(\text{km})$$

Step 8:

Using $\mathbf{r}_1$ and $\mathbf{v}_1$, Algorithm 4.1 yields the orbital elements:

$$h = 80\,470\,\text{km}^2/\text{s}$$
$$a = 20\,000\,\text{km}$$
$$e = 0.4335$$
$$\Omega = 44.60°$$
$$i = 30.19°$$
$$\omega = 30.71°$$
$$\theta_1 = 350.8°$$

This elliptical orbit is plotted in Figure 5.5. The perigee of the orbit is

$$r_p = \frac{h^2}{\mu}\frac{1}{1 + e\cos(0)} = \frac{80\,470^2}{398\,600}\frac{1}{1 + 0.4335} = 11\,330\,\text{km}$$

Therefore the perigee altitude is $11\,330 - 6378 = \underline{4952\,\text{km}}$.

To find the time of the first sighting, we first calculate the eccentric anomaly by means of Equation 3.10b,

$$E_1 = 2\tan^{-1}\left(\sqrt{\frac{1-e}{1+e}}\tan\frac{\theta}{2}\right) = 2\tan^{-1}\left(\sqrt{\frac{1-0.4335}{1+0.4335}}\tan\frac{350.8°}{2}\right)$$

$$= 2\tan^{-1}(-0.05041) = -0.1007\,\text{rad}$$

Then using Kepler's equation for the ellipse (Equation 3.11), the mean anomaly is found to be

$$M_{e_1} = E_1 - e\sin E_1 = -0.1007 - 0.4335\sin(-0.1007) = -0.05715\,\text{rad}$$

5.3 Lambert's problem

so that from Equation 3.4, the time since perigee passage is

$$t_1 = \frac{h^3}{\mu^2} \frac{1}{(1-e^2)^{\frac{3}{2}}} M_{e_1} = \frac{80\,470^3}{398\,600^2} \frac{1}{(1-0.4335^2)^{\frac{3}{2}}} (-0.05715) = -256.1 \text{ s}$$

The minus sign means there are 256.1 seconds until perigee encounter after the initial sighting.

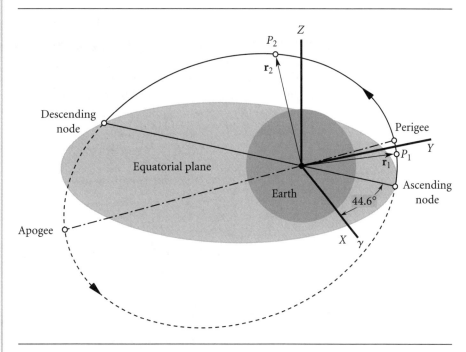

Figure 5.5   The solution of Lambert's problem.

EXAMPLE 5.3

A meteoroid is sighted at an altitude of 267 000 km. 13.5 hours later, after a change in true anomaly of 5°, the altitude is observed to be 140 000 km. Calculate the perigee altitude and the time to perigee after the second sighting.

We have

$$P_1: \quad r_1 = 6378 + 267\,000 = 273\,378 \text{ km}$$
$$P_2: \quad r_2 = 6378 + 140\,000 = 146\,378 \text{ km}$$
$$\Delta t = 13.5 \cdot 3600 = 48\,600 \text{ s}$$
$$\Delta \theta = 5°$$

Since r_1, r_2 and $\Delta \theta$ are given, we can skip to step 3 of Algorithm 5.2 and compute

$$A = 2.8263 \times 10^5 \text{ km}$$

(Example 5.3 continued)

Then, solving for z as in the previous example, we obtain

$$z = -0.17344$$

Since z is negative, the path of the meteoroid is a hyperbola.

With z available, we evaluate the Lagrange functions,

$$f = 0.95846$$
$$g = 47\,708 \text{ s} \quad \text{(a)}$$
$$\dot{g} = 0.92241$$

Step 7 requires the initial and final position vectors. Therefore, for the purposes of this problem let us define a geocentric coordinate system with the x axis aligned with $\mathbf{r}_1$ and the y axis at 90° thereto in the direction of the motion (see Figure 5.6). The z axis is therefore normal to the plane of the orbit. Then

$$\mathbf{r}_1 = r_1 \hat{\mathbf{i}} = 273\,378 \hat{\mathbf{i}} \text{ (km)}$$
$$\mathbf{r}_2 = r_2 \cos \Delta\theta \hat{\mathbf{i}} + r_2 \sin \Delta\theta \hat{\mathbf{j}} = 145\,820 \hat{\mathbf{i}} + 12\,758 \hat{\mathbf{j}} \text{ (km)} \quad \text{(b)}$$

With (a) and (b) we obtain the velocity at P_1,

$$\mathbf{v}_1 = \frac{1}{g}(\mathbf{r}_2 - f\mathbf{r}_1)$$
$$= \frac{1}{47\,708}[(145\,820\hat{\mathbf{i}} + 12\,758\hat{\mathbf{j}}) - 0.95846(273\,378\hat{\mathbf{i}})]$$
$$= -2.4356\hat{\mathbf{i}} - 0.26741\hat{\mathbf{j}} \text{ (km/s)}$$

Using $\mathbf{r}_1$ and $\mathbf{v}_1$, Algorithm 4.1 yields

$$h = 73\,105 \text{ km}^2/\text{s}$$
$$e = 1.0506$$
$$\theta_1 = 205.16°$$

The orbit is now determined except for its orientation in space, for which no information was provided. In the plane of the orbit, the trajectory is as shown in Figure 5.6.

The perigee radius is

$$r_p = \frac{h^2}{\mu} \frac{1}{1 + e\cos(0)} = 6538.2 \text{ km}$$

which means the perigee altitude is dangerously low for a large meteoroid,

$$z_p = 6538.2 - 6378 = \underline{160.2 \text{ km}} \text{ (100 miles)}$$

To find the time of flight from P_2 to perigee, we note that the true anomaly of P_2 is

$$\theta_2 = \theta_1 + 5° = 210.16°$$

The hyperbolic eccentric anomaly F_2 follows from Equation 3.42,

$$F_2 = 2\tanh^{-1}\left(\sqrt{\frac{e-1}{e+1}}\tan\frac{\theta_2}{2}\right) = -1.3347 \text{ rad}$$

From this we appeal to Kepler's equation (Equation 3.37) for the mean anomaly M_h,

$$M_{h_2} = e\sinh(F_2) - F_2 = -0.52265 \text{ rad}$$

Finally, Equation 3.31 yields the time

$$t_2 = \frac{M_{h_2} h^3}{\mu^2 (e^2 - 1)^{\frac{3}{2}}} = -38\,396 \text{ s}$$

The minus sign means that 38 396 seconds (a scant 10.6 hours) remain until the meteoroid passes through perigee.

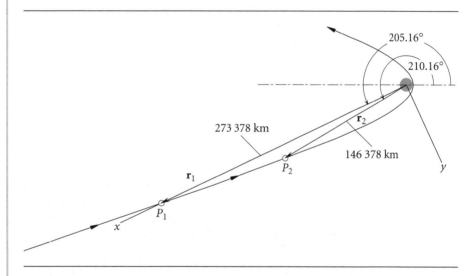

Figure 5.6 Solution of Lambert's problem for the incoming meteoroid.

5.4 SIDEREAL TIME

To deduce the orbit of a satellite or celestial body from observations requires, among other things, recording the time of each observation. The time we use in every day life, the time we set our clocks by, is solar time. It is reckoned by the motion of the sun across the sky. A solar day is the time required for the sun to return to the same position overhead, that is, to lie on the same meridian. A solar day – from high noon to high noon – comprises 24 hours. Universal time (UT) is determined by the sun's passage across the Greenwich meridian, which is zero degrees terrestrial longitude.

See Figure 1.9. At noon UT the sun lies on the Greenwich meridian. Local standard time, or civil time, is obtained from universal time by adding one hour for each time zone between Greenwich and the site, measured westward.

Sidereal time is measured by the rotation of the earth relative to the fixed stars (i.e., the celestial sphere, Figure 4.3). The time it takes for a distant star to return to its same position overhead, i.e., to lie on the same meridian, is one sidereal day (24 sidereal hours). As illustrated in Figure 4.19, the earth's orbit around the sun results in the sidereal day being slightly shorter than the solar day. One sidereal day is 23 hours and 56 minutes. To put it another way, the earth rotates $360°$ in one sidereal day whereas it rotates $360.986°$ in a solar day.

Local sidereal time θ of a site is the time elapsed since the local meridian of the site passed through the vernal equinox. The number of degrees (measured eastward) between the vernal equinox and the local meridian is the sidereal time multiplied by 15. To know the location of a point on the earth at any given instant relative to the geocentric equatorial frame requires knowing its local sidereal time. The local sidereal time of a site is found by first determining the Greenwich sidereal time θ_G (the sidereal time of the Greenwich meridian), and then adding the east longitude (or subtracting the west longitude) of the site. Algorithms for determining sidereal time rely on the notion of the Julian day (JD).

The Julian day number is the number of days since noon UT on 1 January 4713 BC. The origin of this time scale is placed in antiquity so that, except for prehistoric events, we do not have to deal with positive and negative dates. The Julian day count is uniform and continuous and does not involve leap years or different numbers of days in different months. The number of days between two events is found by simply subtracting the Julian day of one from that of the other. The Julian day begins at noon rather than at midnight so that astronomers observing the heavens at night would not have to deal with a change of date during their watch.

The Julian day numbering system is not to be confused with the Julian calendar which the Roman emperor Julius Caesar introduced in 46 BC. The Gregorian calendar, introduced in 1583, has largely supplanted the Julian calender and is in common civil use today throughout much of the world.

J_0 is the symbol for the Julian day number at 0 hr UT (which is half way into the Julian day). At any other UT, the Julian day is given by

$$JD = J_0 + \frac{UT}{24} \qquad (5.47)$$

Algorithms and tables for obtaining J_0 from the ordinary year (y), month (m) and day (d) exist in the literature and on the World Wide Web. One of the simplest formulas is found in Boulet (1991),

$$J_0 = 367y - \text{INT}\left\{\frac{7\left[y + \text{INT}\left(\frac{m+9}{12}\right)\right]}{4}\right\} + \text{INT}\left(\frac{275m}{9}\right) + d + 1\,721\,013.5$$

$$(5.48)$$

where y, m and d are integers lying in the following ranges

$$1901 \leq y \leq 2099$$
$$1 \leq m \leq 12$$
$$1 \leq d \leq 31$$

INT(x) means to retain only the integer portion of x, without rounding (or, in other words, round towards zero); that is, INT(-3.9) $=-3$ and INT(3.9) $=3$. Appendix D.12 lists a MATLAB implementation of Equation 5.48.

EXAMPLE 5.4

What is the Julian day number for 12 May 2004 at 14:45:30 UT?

In this case $y = 2004$, $m = 5$ and $d = 12$. Therefore, Equation 5.48 yields the Julian day number at 0 hr UT,

$$J_0 = 367 \cdot 2004 - \text{INT}\left\{\frac{7\left[2004 + \text{INT}\left(\frac{5+9}{12}\right)\right]}{4}\right\} + \text{INT}\left(\frac{275 \cdot 5}{9}\right)$$

$$+ 12 + 1\,721\,013.5$$

$$= 735\,468 - \text{INT}\left\{\frac{7\,[2004 + 1]}{4}\right\} + 152 + 12 + 1\,721\,013.5$$

$$= 735\,468 - 3508 + 152 + 12 + 1\,721\,013.5$$

or

$$J_0 = 2\,453\,137.5 \text{ days}$$

The universal time, in hours, is

$$UT = 14 + \frac{45}{60} + \frac{30}{3600} = 14.758 \text{ hr}$$

Therefore, from Equation 5.47 we obtain the Julian day number at the desired UT,

$$JD = 2\,453\,137.5 + \frac{14.758}{24} = \underline{2\,453\,138.115 \text{ days}}$$

EXAMPLE 5.5

Find the elapsed time between 4 October 1957 UT 19:26:24 and the date of the previous example.

Proceeding as in Example 5.4 we find that the Julian day number of the given event (the launch of the first man-made satellite, Sputnik I) is

$$JD_1 = 2\,436\,116.3100 \text{ days}$$

The Julian day of the previous example is

$$JD_2 = 2\,453\,138.1149 \text{ days}$$

Hence, the elapsed time is

$$\Delta JD = 2\,453\,138.1149 - 2\,436\,116.3100 = \underline{17\,021.805 \text{ days}} \text{ (46 years, 220 days)}$$

The current Julian epoch is defined to have been noon on 1 January 2000. This epoch is denoted J2000 and has the exact Julian day number 2 451 545.0. Since there are 365.25 days in a Julian year, a Julian century has 36 525 days. It follows that the time T_0 in Julian centuries between the Julian day J_0 and J2000 is

$$T_0 = \frac{J_0 - 2\,451\,545}{36\,525} \tag{5.49}$$

The Greenwich sidereal time θ_{G_0} at 0 hr UT may be found in terms of this dimensionless time (Seidelmann, 1992, Section 2.24). θ_{G_0} in degrees is given by the series

$$\theta_{G_0} = 100.4606184 + 36\,000.77004 T_0 + 0.000387933 T_0^2 - 2.583(10^{-8}) T_0^3 \text{ (degrees)} \tag{5.50}$$

This formula can yield a value outside of the range $0 \leq \theta_{G_0} \leq 360°$. If so, then the appropriate integer multiple of 360° must be added or subtracted to bring θ_{G_0} into that range.

Once θ_{G_0} has been determined, the Greenwich sidereal time θ_G at any other universal time are found using the relation

$$\theta_G = \theta_{G_0} + 360.98564724 \frac{UT}{24} \tag{5.51}$$

where UT is in hours. The coefficient of the second term on the right is the number of degrees the earth rotates in 24 hours (solar time).

Finally, the local sidereal time θ of a site is obtained by adding its east longitude Λ to the Greenwich sidereal time,

$$\theta = \theta_G + \Lambda \tag{5.52}$$

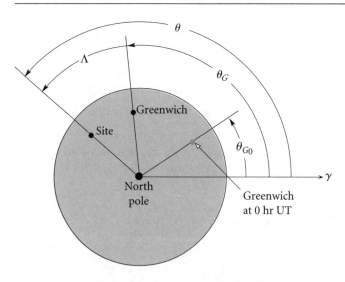

Figure 5.7 Schematic of the relationship among $\theta_{G_0}, \theta_G, \Lambda$ and θ.

5.4 Sidereal time

Here again it is possible for the computed value of θ to exceed 360°. If so, it must be reduced to within that limit by subtracting the appropriate integer multiple of 360°. Figure 5.7 illustrates the relationship among $\theta_{G_0}, \theta_G, \Lambda$ and θ.

ALGORITHM 5.3

Calculate the local sidereal time, given the date, the local time and the east longitude of the site. This is implemented in MATLAB in Appendix D.13.

1. Using the year, month and day, calculate J_0 using Equation 5.48.
2. Calculate T_0 by means of Equation 5.49.
3. Compute θ_{G_0} from Equation 5.50. If θ_{G_0} lies outside the range $0° \leq \theta_{G_0} \leq 360°$, then subtract the multiple of 360° required to place θ_{G_0} in that range.
4. Calculate θ_G using Equation 5.51.
5. Calculate the local sidereal time θ by means of Equation 5.52, adjusting the final value so it lies between 0 and 360°.

EXAMPLE 5.6

Use Algorithm 5.3 to find the local sidereal time (in degrees) of Tokyo, Japan, on 3 March 2004 at 4:30:00 UT. The east longitude of Tokyo is 139.80°. (This places Tokyo nine time zones ahead of Greenwich, so the local time is 1:30 in the afternoon.)

Step 1:

$$J_0 = 367 \cdot 2004 - \text{INT}\left\{\frac{7\left[2004 + \text{INT}\left(\frac{3+9}{12}\right)\right]}{4}\right\} + \text{INT}\left(\frac{275 \cdot 3}{9}\right)$$

$$+ 3 + 1\,721\,013.5$$

$$= 2\,453\,067.5 \text{ days}$$

Recall that the .5 means that we are half way into the Julian day, which began at noon UT of the previous day.

Step 2:

$$T_0 = \frac{2\,453\,067.5 - 2\,451\,545}{36\,525} = 0.041683778$$

Step 3:

$$\theta_{G_0} = 100.4606184 + 36\,000.77004(0.041683778)$$
$$+ 0.000387933(0.041683778)^2 - 2.583(10^{-8})(0.041683778)^3$$
$$= 1601.1087°$$

The right-hand side is too large. We must reduce θ_{G_0} to an angle which does not exceed 360°. To that end observe that

$$\text{INT}(1601.1087/360) = 4$$

(Example 5.6 continued)

Hence,
$$\theta_{G_0} = 1601.1087 - 4 \cdot 360 = 161.10873° \qquad (a)$$

Step 4:

The universal time of interest in this problem is
$$UT = 4 + \frac{30}{60} + \frac{0}{3600} = 4.5\,\text{hr}$$

Substitute this and (a) into Equation 5.51 to get the Greenwich sidereal time:
$$\theta_G = 161.10873 + 360.98564724 \frac{4.5}{24} = 228.79354°$$

Step 5:

Add the east longitude of Tokyo to this value to obtain the local sidereal time,
$$\theta = 228.79354 + 139.80 = 368.59°$$

To reduce this result into the range $0 \le \theta \le 360°$ we must subtract $360°$ to get
$$\theta = 368.59 - 360 = \underline{8.59°}\ (0.573\,\text{hr})$$

Observe that the right ascension of a celestial body lying on Tokyo's meridian is $8.59°$.

5.5 Topocentric coordinate system

A topocentric coordinate system is one which is centered at the observer's location on the surface of the earth. Consider an object B – a satellite or celestial body – and an observer O on the earth's surface, as illustrated in Figure 5.8. $\mathbf{r}$ is the position of the body B relative to the center of attraction C; $\mathbf{R}$ is the position vector of the observer relative to C; and $\boldsymbol{\varrho}$ is the position of the body B relative to the observer. $\mathbf{r}, \mathbf{R}$ and $\boldsymbol{\varrho}$ comprise the fundamental vector triangle. The relationship among these three vectors is

$$\mathbf{r} = \mathbf{R} + \boldsymbol{\varrho} \qquad (5.53)$$

As we know, the earth is not a sphere, but a slightly oblate spheroid. This ellipsoidal shape is exaggerated in Figure 5.8. The location of the observation site O is determined by specifying its east longitude Λ and latitude ϕ. East longitude Λ is measured positive eastward from the Greenwich meridian to the meridian through O. The angle between the vernal equinox direction (XZ plane) and the meridian of O is the local sidereal time θ. Likewise, θ_G is the Greenwich sidereal time. Once we know θ_G, then the local sidereal time is given by Equation 5.52.

Latitude ϕ is the angle between the equator and the normal $\hat{\mathbf{n}}$ to the earth's surface at O. Since the earth is not a perfect sphere, the position vector $\mathbf{R}$, directed from the center C of the earth to O, does not point in the direction of the normal except at the equator and the poles.

5.5 Topocentric coordinate system

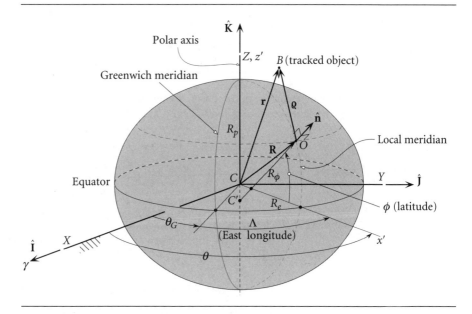

Figure 5.8 Oblate spheroidal earth (exaggerated).

The oblateness, or flattening f, was defined in Section 4.7,

$$f = \frac{R_e - R_p}{R_e}$$

where R_e is the equatorial radius and R_p is the polar radius. (Review from Table 4.3 that $f = 0.00335$ for the earth.) Figure 5.9 shows the ellipse of the meridian through O. Obviously, R_e and R_p are, respectively, the semimajor and semiminor axes of the ellipse. According to Equation 2.66,

$$R_p = R_a \left(1 - e^2\right)$$

It is easy to show from the above two relations that flattening and eccentricity are related as follows

$$e = \sqrt{2f - f^2} \qquad f = 1 - \sqrt{1 - e^2}$$

As illustrated in Figure 5.8 and again in Figure 5.9, the normal to the earth's surface at O intersects the polar axis at a point C' which lies below the center C of the earth (if O is in the northern hemisphere). The angle ϕ between the normal and the equator is called the geodetic latitude, as opposed to geocentric latitude ϕ', which is the angle between the equatorial plane and line joining O to the center of the earth. The distance from C to C' is $R_\phi e^2 \sin^2 \phi$, where R_ϕ, the distance from C' to O, is a function of latitude (Seidelmann, 1991, Section 4.22)

$$R_\phi = \frac{R_e}{\sqrt{1 - e^2 \sin^2 \phi}} = \frac{R_e}{\sqrt{1 - \left(2f - f^2\right) \sin^2 \phi}} \qquad (5.54)$$

Thus, the meridional coordinates of O are

$$x'_O = R_\phi \cos \phi$$
$$z'_O = (1 - e^2) R_\phi \sin \phi = (1 - f)^2 R_\phi \sin \phi$$

If the observation point O is at an elevation H above the ellipsoidal surface, then we must add $H \cos \phi$ to x'_O and $H \sin \phi$ to z'_O to obtain

$$x'_O = R_c \cos \phi \qquad z'_O = R_s \sin \phi \qquad (5.55a)$$

where

$$R_c = R_\phi + H \qquad R_s = (1 - f)^2 R_\phi + H \qquad (5.55b)$$

Observe that whereas R_c is the distance of O from point C' on the earth's axis, R_s is the distance from O to the intersection of the line OC' with the equatorial plane.

The geocentric equatorial coordinates of O are

$$X = x'_O \cos \theta \qquad Y = x'_O \sin \theta \qquad Z = z'_O$$

where θ is the local sidereal time given in Equation 5.52. Hence, the position vector **R** shown in Figure 5.8 is

$$\mathbf{R} = R_c \cos \phi \cos \theta \hat{\mathbf{I}} + R_c \cos \phi \sin \theta \hat{\mathbf{J}} + R_s \sin \phi \hat{\mathbf{K}}$$

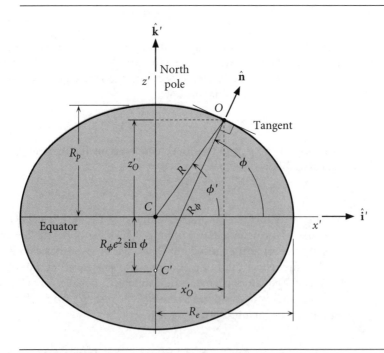

Figure 5.9 The relationship between geocentric latitude (ϕ') and geodetic latitude (ϕ).

Substituting Equation 5.54 and Equations 5.55b yields

$$\mathbf{R} = \left[\frac{R_e}{\sqrt{1-(2f-f^2)\sin^2\phi}} + H\right]\cos\phi(\cos\theta\hat{\mathbf{I}} + \sin\theta\hat{\mathbf{J}})$$

$$+ \left[\frac{R_e(1-f)^2}{\sqrt{1-(2f-f^2)\sin^2\phi}} + H\right]\sin\phi\hat{\mathbf{K}} \qquad (5.56)$$

In terms of the geocentric latitude ϕ'

$$\mathbf{R} = R_e\cos\phi'\cos\theta\hat{\mathbf{I}} + R_e\cos\phi'\sin\theta\hat{\mathbf{J}} + R_e\sin\phi'\hat{\mathbf{K}}$$

By equating these two expressions for $\mathbf{R}$ and setting $H = 0$ it is easy to show that at sea level geodetic latitude is related to geocentric latitude ϕ' as follows,

$$\tan\phi' = (1-f)^2 \tan\phi$$

5.6 Topocentric equatorial coordinate system

The topocentric equatorial coordinate system with origin at point O on the surface of the earth uses a non-rotating set of *xyz* axes through O which coincide with the *XYZ* axes of the geocentric equatorial frame, as illustrated in Figure 5.10. As can be

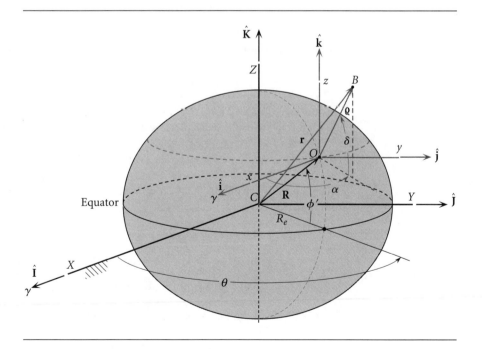

Figure 5.10 Topocentric equatorial coordinate system.

inferred from the figure, the relative position vector $\boldsymbol{\varrho}$ in terms of the topocentric right ascension and declination is

$$\boldsymbol{\varrho} = \varrho \cos\delta \cos\alpha \hat{\mathbf{I}} + \varrho \cos\delta \sin\alpha \hat{\mathbf{J}} + \varrho \sin\delta \hat{\mathbf{K}}$$

since at all times, $\hat{\mathbf{i}} = \hat{\mathbf{I}}, \hat{\mathbf{j}} = \hat{\mathbf{J}}$ and $\hat{\mathbf{k}} = \hat{\mathbf{K}}$ for this frame of reference. We can write $\boldsymbol{\varrho}$ as

$$\boldsymbol{\varrho} = \varrho \hat{\boldsymbol{\varrho}}$$

where ϱ is the slant range and $\hat{\boldsymbol{\varrho}}$ is the unit vector in the direction of $\boldsymbol{\varrho}$,

$$\hat{\boldsymbol{\varrho}} = \cos\delta \cos\alpha \hat{\mathbf{I}} + \cos\delta \sin\alpha \hat{\mathbf{J}} + \sin\delta \hat{\mathbf{K}} \tag{5.57}$$

Since the origins of the geocentric and topocentric systems do not coincide, the direction cosines of the position vectors $\mathbf{r}$ and $\boldsymbol{\varrho}$ will in general differ. In particular the topocentric right ascension and declination of an earth-orbiting body B will not be the same as the geocentric right ascension and declination. This is an example of parallax. On the other hand, if $\|\mathbf{r}\| \gg \|\mathbf{R}\|$ then the difference between the geocentric and topocentric position vectors, and hence the right ascension and declination, is negligible. This is true for the distant planets and stars.

EXAMPLE 5.7

At the instant when the Greenwich sidereal time is $\theta_G = 126.7°$, the geocentric equatorial position vector of the International Space Station is

$$\mathbf{r} = -5368\hat{\mathbf{I}} - 1784\hat{\mathbf{J}} + 3691\hat{\mathbf{K}} \text{ (km)}$$

Find the topocentric right ascension and declination at sea level ($H = 0$), latitude $\phi = 20°$ and east longitude $\Lambda = 60°$.

According to Equation 5.52, the local sidereal time at the observation site is

$$\theta = \theta_G + \Lambda = 126.7 + 60 = 186.7°$$

Substituting $R_e = 6378$ km, $f = 0.003353$ (Table 4.3), $\theta = 189.7°$ and $\phi = 20°$ into Equation 5.56 yields the geocentric position vector of the site:

$$\mathbf{R} = -5955\hat{\mathbf{I}} - 699.5\hat{\mathbf{J}} + 2168\hat{\mathbf{K}} \text{ (km)}$$

Having found $\mathbf{R}$, we obtain the position vector of the space station relative to the site from Equation 5.53:

$$\boldsymbol{\varrho} = \mathbf{r} - \mathbf{R}$$
$$= (-5368\hat{\mathbf{I}} - 1784\hat{\mathbf{J}} + 3691\hat{\mathbf{K}}) - (-5955\hat{\mathbf{I}} - 699.5\hat{\mathbf{J}} + 2168\hat{\mathbf{K}})$$
$$= 586.8\hat{\mathbf{I}} - 1084\hat{\mathbf{J}} + 1523\hat{\mathbf{K}} \text{ (km)}$$

The magnitude of this vector is $\varrho = 1960$ km, so that

$$\hat{\boldsymbol{\varrho}} = \frac{\boldsymbol{\varrho}}{\varrho} = 0.2994\hat{\mathbf{I}} - 0.5533\hat{\mathbf{J}} + 0.7773\hat{\mathbf{K}}$$

Comparing this equation with Equation 5.57 we see that

$$\cos\delta\cos\alpha = 0.2997$$
$$\cos\delta\sin\alpha = -0.5524$$
$$\sin\delta = 0.7778$$

From these we obtain the topocentric declension,

$$\delta = \sin^{-1} 0.7773 = \underline{51.01°} \qquad \text{(a)}$$

as well as

$$\sin\alpha = \frac{-0.5533}{\cos\delta} = -0.8795$$
$$\cos\alpha = \frac{0.2994}{\cos\delta} = 0.4759$$

Thus

$$\alpha = \cos^{-1}(0.4759) = 61.58° \text{ (first quadrant) or } 298.4° \text{(fourth quadrant)}$$

Since $\sin\alpha$ is negative, α must lie in the fourth quadrant, so that the right ascension is

$$\underline{\alpha = 298.4°} \qquad \text{(b)}$$

Compare (a) and (b) with the geocentric right ascension α_0 and declination δ_0, which were computed in Example 4.2,

$$\alpha_0 = 198.4° \qquad \delta_0 = 33.12°$$

5.7 Topocentric horizon coordinate system

The topocentric horizon system was introduced in Section 1.6 and is illustrated again in Figure 5.11. It is centered at the observation point O whose position vector is $\mathbf{R}$. The xy plane is the local horizon, which is the plane tangent to the ellipsoid at point O. The z axis is normal to this plane directed outward towards the zenith. The x axis is directed eastward and the y axis points north. Because the x axis points east, this may be referred to as an *ENZ* (East-North-Zenith) frame. In the *SEZ* topocentric reference frame the x axis points towards the south and the y axis towards the east. The *SEZ* frame is obtained from *ENZ* by a 90° clockwise rotation around the zenith. Therefore, the matrix of the transformation from *NEZ* to *SEZ* is $[\mathbf{R}_3(-90°)]$, where $[\mathbf{R}_3(\phi)]$ is found in Equation 4.33.

The position vector $\boldsymbol{\varrho}$ of a body B relative to the topocentric horizon system in Figure 5.11 is

$$\boldsymbol{\varrho} = \varrho\cos a\sin A\hat{\mathbf{i}} + \varrho\cos a\cos A\hat{\mathbf{j}} + \varrho\sin a\hat{\mathbf{k}}$$

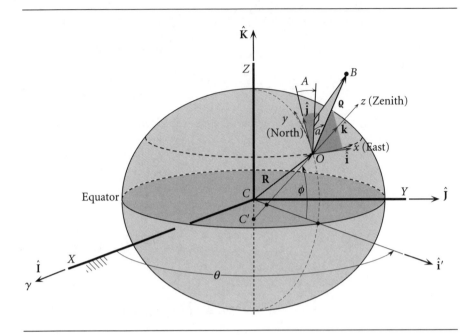

Figure 5.11 Topocentric horizon (*xyz*) coordinate system on the surface of the oblate earth.

in which ϱ is the range; A is the azimuth measured positive clockwise from due north ($0 \leq A \leq 360°$); and a is the elevation angle or altitude measured from the horizontal to the line of sight of the body B ($-90° \leq a \leq 90$). The unit vector $\hat{\varrho}$ in the line of sight direction is

$$\hat{\varrho} = \cos a \sin A \hat{i} + \cos a \cos A \hat{j} + \sin a \hat{k} \tag{5.58}$$

The transformation between geocentric equatorial and topocentric horizon systems is found by first determining the projections of the topocentric base vectors $\hat{i}\hat{j}\hat{k}$ onto those of the geocentric equatorial frame. From Figure 5.11 it is apparent that

$$\hat{k} = \cos\phi \hat{i}' + \sin\phi \hat{K}$$

and

$$\hat{i}' = \cos\theta \hat{I} + \sin\theta \hat{J}$$

where $\hat{i}'$ lies in the local meridional plane and is normal to the Z axis. Hence

$$\hat{k} = \cos\phi \cos\theta \hat{I} + \cos\phi \sin\theta \hat{J} + \sin\phi \hat{K} \tag{5.59}$$

The eastward-directed unit vector $\hat{i}$ may be found by taking the cross product of $\hat{K}$ into the unit normal $\hat{k}$,

$$\hat{i} = \frac{\hat{K} \times \hat{k}}{\left\|\hat{K} \times \hat{k}\right\|} = \frac{-\cos\phi \sin\theta \hat{I} + \cos\phi \cos\theta \hat{J}}{\sqrt{\cos^2\phi \left(\sin^2\theta + \cos^2\theta\right)}} = -\sin\theta \hat{I} + \cos\theta \hat{J} \tag{5.60}$$

5.7 Topocentric horizon coordinate system

Finally, crossing $\hat{\mathbf{k}}$ into $\hat{\mathbf{i}}$ yields $\hat{\mathbf{j}}$,

$$\hat{\mathbf{j}} = \hat{\mathbf{k}} \times \hat{\mathbf{i}} = \begin{vmatrix} \hat{\mathbf{I}} & \hat{\mathbf{J}} & \hat{\mathbf{K}} \\ \cos\phi\cos\theta & \cos\phi\sin\theta & \sin\phi \\ -\sin\theta & \cos\theta & 0 \end{vmatrix} = -\sin\phi\cos\theta\hat{\mathbf{I}} - \sin\phi\sin\theta\hat{\mathbf{J}} + \cos\phi\hat{\mathbf{K}} \quad (5.61)$$

Let us denote the matrix of the transformation from geocentric equatorial to topocentric horizon as $[\mathbf{Q}]_{Xx}$. Recall from Section 4.5 that the rows of this matrix comprise the direction cosines of $\hat{\mathbf{i}}, \hat{\mathbf{j}}$ and $\hat{\mathbf{k}}$, respectively. It follows from Equations 5.59 through 5.61 that

$$[\mathbf{Q}]_{Xx} = \begin{bmatrix} -\sin\theta & \cos\theta & 0 \\ -\sin\phi\cos\theta & -\sin\phi\sin\theta & \cos\phi \\ \cos\phi\cos\theta & \cos\phi\sin\theta & \sin\phi \end{bmatrix} \quad (5.62a)$$

The reverse transformation, from topocentric horizon to geocentric equatorial, is represented by the transpose of this matrix,

$$[\mathbf{Q}]_{xX} = \begin{bmatrix} -\sin\theta & -\sin\phi\cos\theta & \cos\phi\cos\theta \\ \cos\theta & -\sin\phi\sin\theta & \cos\phi\sin\theta \\ 0 & \cos\phi & \sin\phi \end{bmatrix} \quad (5.62b)$$

Observe that these matrices also represent the transformation between topocentric horizontal and topocentric equatorial frames because the unit basis vectors of the latter coincide with those of the geocentric equatorial coordinate system.

EXAMPLE 5.8

The east longitude and latitude of an observer near San Francisco are $\Lambda = 238°$ and $\phi = 38°$, respectively. The local sidereal time, in degrees, is $\theta = 215.1°$ (12 hr 42 min). At that time the planet Jupiter is observed by means of a telescope to be located at azimuth $A = 214.3°$ and angular elevation $a = 43°$. What are Jupiter's right ascension and declination in the topocentric equatorial system?

The given information allows us to formulate the matrix of the transformation from topocentric horizon to topocentric equatorial using Equation 5.62b,

$$[\mathbf{Q}]_{xX} = \begin{bmatrix} -\sin 215.1° & -\sin 38°\cos 215.1° & \cos 38°\cos 215.1° \\ \cos 215.1° & -\sin 38°\sin 215.1° & \cos 38°\sin 215.1° \\ 0 & \cos 38° & \sin 38° \end{bmatrix}$$

$$= \begin{bmatrix} 0.5750 & 0.5037 & -0.6447 \\ -0.8182 & 0.3540 & -0.4531 \\ 0 & 0.7880 & 0.6157 \end{bmatrix}$$

From Equation 5.58 we have

$$\hat{\boldsymbol{\varrho}} = \cos a \sin A \hat{\mathbf{i}} + \cos a \cos A \hat{\mathbf{j}} + \sin a \hat{\mathbf{k}}$$

$$= \cos 43° \sin 214.3° \hat{\mathbf{i}} + \cos 43° \cos 214.3° \hat{\mathbf{j}} + \sin 43° \hat{\mathbf{k}}$$

$$= -0.4121\hat{\mathbf{i}} - 0.6042\hat{\mathbf{j}} + 0.6820\hat{\mathbf{k}}$$

(Example 5.8 continued)

Therefore, in matrix notation the topocentric horizon components of $\hat{\boldsymbol{\rho}}$ are

$$\{\hat{\boldsymbol{\rho}}\}_x = \begin{Bmatrix} -0.4121 \\ -0.6042 \\ 0.6820 \end{Bmatrix}$$

We obtain the topocentric equatorial components $\{\hat{\boldsymbol{\rho}}\}_X$ by the matrix operation

$$\{\hat{\boldsymbol{\rho}}\}_X = [Q]_{xX}\{\hat{\boldsymbol{\rho}}\}_x = \begin{bmatrix} 0.5750 & 0.5037 & -0.6447 \\ -0.8182 & 0.3540 & -0.4531 \\ 0 & 0.7880 & 0.6157 \end{bmatrix} \begin{Bmatrix} -0.4121 \\ -0.6042 \\ 0.6820 \end{Bmatrix}$$

$$= \begin{Bmatrix} -0.9810 \\ -0.1857 \\ -0.05621 \end{Bmatrix}$$

so that

$$\hat{\boldsymbol{\rho}} = -0.9810\hat{\mathbf{I}} - 0.1857\hat{\mathbf{J}} - 0.05621\hat{\mathbf{K}}$$

Recall Equation 5.57,

$$\hat{\boldsymbol{\rho}} = \cos\delta\cos\alpha\hat{\mathbf{I}} + \cos\delta\sin\alpha\hat{\mathbf{J}} + \sin\delta\hat{\mathbf{K}}$$

Comparing the Z components of these two expressions, we see that

$$\sin\delta = -0.0562$$

which means the topocentric equatorial declension is

$$\delta = \sin^{-1}(-0.0562) = \underline{-3.222°}$$

Equating the X and Y components yields

$$\sin\alpha = \frac{-0.1857}{\cos\delta} = -0.1860$$

$$\cos\alpha = \frac{-0.9810}{\cos\delta} = -0.9825$$

Therefore,

$$\alpha = \cos^{-1}(-0.9825) = 169.3°\,(\text{second quadrant}) \text{ or } 190.7°\,(\text{fourth quadrant})$$

Since $\sin\alpha$ is negative, α is in the fourth quadrant, which means the topocentric equatorial right ascension is

$$\underline{\alpha = 190.7°}$$

Jupiter is sufficiently far away that we can ignore the radius of the earth in Equation 5.53. That is, to our level of precision, there is no distinction between the topocentric equatorial and geocentric equatorial systems:

$$\mathbf{r} \approx \boldsymbol{\rho}$$

Therefore the topocentric right ascension and declination computed above are the same as the geocentric equatorial values.

5.7 Topocentric horizon coordinate system

EXAMPLE 5.9

At a given time, the geocentric equatorial position vector of the International Space Station is

$$\mathbf{r} = -2032.4\hat{\mathbf{I}} + 4591.2\hat{\mathbf{J}} - 4544.8\hat{\mathbf{K}} \text{ (km)}$$

Determine the azimuth and elevation angle relative to a sea-level ($H = 0$) observer whose latitude is $\phi = -40°$ and local sidereal time is $\theta = 110°$.

Using Equation 5.56 we find the position vector of the observer to be

$$\mathbf{R} = -1673\hat{\mathbf{I}} + 4598\hat{\mathbf{J}} - 4078\hat{\mathbf{K}} \text{ (km)}$$

For the position vector of the space station relative to the observer we have (Equation 5.53)

$$\boldsymbol{\varrho} = \mathbf{r} - \mathbf{R}$$
$$= (-2032\hat{\mathbf{I}} + 4591\hat{\mathbf{J}} - 4545\hat{\mathbf{K}}) - (-1673\hat{\mathbf{I}} + 4598\hat{\mathbf{J}} - 4078\hat{\mathbf{K}})$$
$$= -359.0\hat{\mathbf{I}} - 6.342\hat{\mathbf{J}} - 466.9\hat{\mathbf{K}} \text{ (km)}$$

or, in matrix notation,

$$\{\boldsymbol{\varrho}\}_X = \begin{Bmatrix} -359.0 \\ -6.342 \\ -466.9 \end{Bmatrix} \text{ (km)}$$

To transform these geocentric equatorial components into the topocentric horizon system we need the transformation matrix $[Q]_{Xx}$, which is given by Equation 5.62a,

$$[Q]_{Xx} = \begin{bmatrix} -\sin\theta & \cos\theta & 0 \\ -\sin\phi\cos\theta & -\sin\phi\sin\theta & \cos\phi \\ \cos\phi\cos\theta & \cos\phi\sin\theta & \sin\phi \end{bmatrix}$$

$$= \begin{bmatrix} -\sin 110° & \cos 110° & 0 \\ -\sin(-40°)\cos 110° & -\sin(-40°)\sin 110° & \cos(-40°) \\ \cos(-40°)\cos 110° & \cos(-40°)\sin 110° & \sin(-40°) \end{bmatrix}$$

Thus,

$$\{\boldsymbol{\varrho}\}_x = [Q]_{Xx}\{\boldsymbol{\varrho}\}_X = \begin{bmatrix} -0.9397 & -0.3420 & 0 \\ -0.2198 & 0.6040 & 0.7660 \\ -0.2620 & 0.7198 & -0.6428 \end{bmatrix} \begin{Bmatrix} -359.0 \\ -6.342 \\ -466.9 \end{Bmatrix}$$

$$= \begin{Bmatrix} 339.5 \\ -282.6 \\ 389.6 \end{Bmatrix} \text{ (km)}$$

or, reverting to vector notation,

$$\boldsymbol{\varrho} = 339.5\hat{\mathbf{i}} - 282.6\hat{\mathbf{j}} + 389.6\hat{\mathbf{k}} \text{ (km)}$$

The magnitude of this vector is $\varrho = 589.0$ km. Hence, the unit vector in the direction of $\boldsymbol{\varrho}$ is

$$\hat{\boldsymbol{\varrho}} = \frac{\boldsymbol{\varrho}}{\varrho} = 0.5765\hat{\mathbf{i}} - 0.4787\hat{\mathbf{j}} + 0.6615\hat{\mathbf{k}}$$

(Example 5.9 continued)

Comparing this with Equation 5.58 we see that $\sin a = 0.6615$, so that the angular elevation is

$$a = \sin^{-1} 0.6615 = \underline{41.41°}$$

Furthermore

$$\sin A = \frac{0.5765}{\cos a} = 0.7687$$

$$\cos A = \frac{-0.4787}{\cos a} = -0.6397$$

It follows that

$$A = \cos^{-1}(-0.6397) = 129.8° \text{(second quadrant) or } 230.2° \text{(third quadrant)}$$

A must lie in the second quadrant because $\sin A > 0$. Thus the azimuth is

$$\underline{A = 129.8°}$$

5.8 Orbit determination from angle and range measurements

We know that an orbit around the earth is determined once the state vectors $\mathbf{r}$ and $\mathbf{v}$ in the inertial geocentric equatorial frame are provided at a given instant of time (epoch). Satellites are of course observed from the earth's surface and not from its center. Let us briefly consider how the state vector is determined from measurements by an earth-based tracking station.

The fundamental vector triangle formed by the topocentric position vector $\boldsymbol{\varrho}$ of a satellite relative to a tracking station, the position vector $\mathbf{R}$ of the station relative to the center of attraction C and the geocentric position vector $\mathbf{r}$ was illustrated in Figure 5.8 and is shown again schematically in Figure 5.12. The relationship among these three vectors is given by Equation 5.53, which can be written

$$\mathbf{r} = \mathbf{R} + \varrho \hat{\boldsymbol{\varrho}} \tag{5.63}$$

where the range ϱ is the distance of the body B from the tracking site and $\hat{\boldsymbol{\varrho}}$ is the unit vector containing the directional information about B. By differentiating Equation 5.63 with respect to time we obtain the velocity $\mathbf{v}$ and acceleration $\mathbf{a}$,

$$\mathbf{v} = \dot{\mathbf{r}} = \dot{\mathbf{R}} + \dot{\varrho}\hat{\boldsymbol{\varrho}} + \varrho\dot{\hat{\boldsymbol{\varrho}}} \tag{5.64}$$

$$\mathbf{a} = \ddot{\mathbf{r}} = \ddot{\mathbf{R}} + \ddot{\varrho}\hat{\boldsymbol{\varrho}} + 2\dot{\varrho}\dot{\hat{\boldsymbol{\varrho}}} + \varrho\ddot{\hat{\boldsymbol{\varrho}}} \tag{5.65}$$

The vectors in these equations must all be expressed in the common basis $(\hat{\mathbf{I}}\hat{\mathbf{J}}\hat{\mathbf{K}})$ of the inertial (non-rotating) geocentric equatorial frame.

Since $\mathbf{R}$ is a vector fixed in the earth, whose constant angular velocity is $\boldsymbol{\Omega} = \omega_E \hat{\mathbf{K}}$ (see Equation 2.57), it follows from Equations 1.24 and 1.25 that

$$\dot{\mathbf{R}} = \boldsymbol{\Omega} \times \mathbf{R} \tag{5.66}$$

5.8 Orbit determination from angle and range measurements

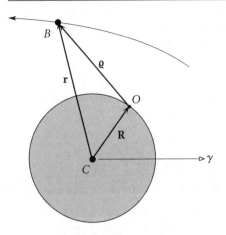

Figure 5.12 Earth-orbiting body B tracked by an observer O.

$$\ddot{\mathbf{R}} = \mathbf{\Omega} \times (\mathbf{\Omega} \times \mathbf{R}) \tag{5.67}$$

If L_X, L_Y and L_Z are the topocentric equatorial direction cosines, then the direction cosine vector $\hat{\boldsymbol{\rho}}$ is

$$\hat{\boldsymbol{\rho}} = L_X \hat{\mathbf{I}} + L_Y \hat{\mathbf{J}} + L_Z \hat{\mathbf{K}} \tag{5.68}$$

and its first and second derivatives are

$$\dot{\hat{\boldsymbol{\rho}}} = \dot{L}_X \hat{\mathbf{I}} + \dot{L}_Y \hat{\mathbf{J}} + \dot{L}_Z \hat{\mathbf{K}} \tag{5.69}$$

and

$$\ddot{\hat{\boldsymbol{\rho}}} = \ddot{L}_X \hat{\mathbf{I}} + \ddot{L}_Y \hat{\mathbf{J}} + \ddot{L}_Z \hat{\mathbf{K}} \tag{5.70}$$

Comparing Equations 5.57 and 5.68 reveals that the topocentric equatorial direction cosines in terms of the topocentric right ascension α and declension δ are

$$\begin{Bmatrix} L_X \\ L_Y \\ L_Z \end{Bmatrix} = \begin{Bmatrix} \cos \alpha \cos \delta \\ \sin \alpha \cos \delta \\ \sin \delta \end{Bmatrix} \tag{5.71}$$

Differentiating this equation twice yields

$$\begin{Bmatrix} \dot{L}_X \\ \dot{L}_Y \\ \dot{L}_Z \end{Bmatrix} = \begin{Bmatrix} -\dot{\alpha} \sin \alpha \cos \delta - \dot{\delta} \cos \alpha \sin \delta \\ \dot{\alpha} \cos \alpha \cos \delta - \dot{\delta} \sin \alpha \sin \delta \\ \dot{\delta} \cos \delta \end{Bmatrix} \tag{5.72}$$

and

$$\begin{Bmatrix} \ddot{L}_X \\ \ddot{L}_Y \\ \ddot{L}_Z \end{Bmatrix} = \begin{Bmatrix} -\ddot{\alpha} \sin \alpha \cos \delta - \ddot{\delta} \cos \alpha \sin \delta - (\dot{\alpha}^2 + \dot{\delta}^2) \cos \alpha \cos \delta + 2\dot{\alpha}\dot{\delta} \sin \alpha \sin \delta \\ \ddot{\alpha} \cos \alpha \cos \delta - \ddot{\delta} \sin \alpha \sin \delta - (\dot{\alpha}^2 + \dot{\delta}^2) \sin \alpha \cos \delta - 2\dot{\alpha}\dot{\delta} \cos \alpha \sin \delta \\ \ddot{\delta} \cos \delta - \dot{\delta}^2 \sin \delta \end{Bmatrix} \tag{5.73}$$

Equations 5.71 through 5.73 show how the direction cosines and their rates are obtained from the right ascension and declination and their rates.

In the topocentric horizon system, the relative position vector is written

$$\hat{\varrho} = l_x \hat{i} + l_y \hat{j} + l_z \hat{k} \tag{5.74}$$

where, according to Equation 5.58, the direction cosines l_x, l_y and l_z are found in terms of the azimuth A and elevation a as

$$\begin{Bmatrix} l_x \\ l_y \\ l_z \end{Bmatrix} = \begin{Bmatrix} \sin A \cos a \\ \cos A \cos a \\ \sin a \end{Bmatrix} \tag{5.75}$$

L_X, L_Y and L_Z are obtained from l_x, l_y and l_z by the coordinate transformation

$$\begin{Bmatrix} L_X \\ L_Y \\ L_Z \end{Bmatrix} = [Q]_{xX} \begin{Bmatrix} l_x \\ l_y \\ l_z \end{Bmatrix} \tag{5.76}$$

where $[Q]_{xX}$ is given by Equation 5.62b. Thus

$$\begin{Bmatrix} L_X \\ L_Y \\ L_Z \end{Bmatrix} = \begin{bmatrix} -\sin\theta & -\cos\theta\sin\phi & \cos\theta\cos\phi \\ \cos\theta & -\sin\theta\sin\phi & \sin\theta\cos\phi \\ 0 & \cos\phi & \sin\phi \end{bmatrix} \begin{Bmatrix} \sin A \cos a \\ \cos A \cos a \\ \sin a \end{Bmatrix} \tag{5.77}$$

Substituting Equation 5.71 we see that topocentric right ascension/declination and azimuth/elevation are related by

$$\begin{Bmatrix} \cos\alpha\cos\delta \\ \sin\alpha\cos\delta \\ \sin\delta \end{Bmatrix} = \begin{bmatrix} -\sin\theta & -\cos\theta\sin\phi & \cos\theta\cos\phi \\ \cos\theta & -\sin\theta\sin\phi & \sin\theta\cos\phi \\ 0 & \cos\phi & \sin\phi \end{bmatrix} \begin{Bmatrix} \sin A \cos a \\ \cos A \cos a \\ \sin a \end{Bmatrix}$$

Expanding the right-hand side and solving for $\sin\delta$, $\sin\alpha$ and $\cos\alpha$ we get

$$\sin\delta = \cos\phi\cos A \cos a + \sin\phi\sin a \tag{5.78a}$$

$$\sin\alpha = \frac{(\cos\phi\sin a - \cos A \cos a \sin\phi)\sin\theta + \cos\theta\sin A \cos a}{\cos\delta} \tag{5.78b}$$

$$\cos\alpha = \frac{(\cos\phi\sin a - \cos A \cos a \sin\phi)\cos\theta - \sin\theta\sin A \cos a}{\cos\delta} \tag{5.78c}$$

We can simplify Equations 5.78b and 5.78c by introducing the hour angle h,

$$h = \theta - \alpha \tag{5.79}$$

h is the angular distance between the object and the local meridian. If h is positive, the object is west of the meridian; if h is negative, the object is east of the meridian.

Using well-known trig identities we have

$$\sin(\theta - \alpha) = \sin\theta\cos\alpha - \cos\theta\sin\alpha \tag{5.80a}$$
$$\cos(\theta - \alpha) = \cos\theta\cos\alpha + \sin\theta\sin\alpha \tag{5.80b}$$

Substituting Equations 5.78b and 5.78c on the right of 5.80a and simplifying yields

$$\sin(h) = -\frac{\sin A \cos a}{\cos\delta} \tag{5.81}$$

5.8 Orbit determination from angle and range measurements

Likewise, Equation 5.80b leads to

$$\cos(h) = \frac{\cos\phi \sin a - \sin\phi \cos A \cos a}{\cos\delta} \qquad (5.82)$$

We calculate h from this equation, resolving quadrant ambiguity by checking the sign of $\sin(h)$. That is,

$$h = \cos^{-1}\left(\frac{\cos\phi \sin a - \sin\phi \cos A \cos a}{\cos\delta}\right)$$

if $\sin(h)$ is positive. Otherwise, we must subtract h from 360°. Since both the elevation angle a and the declension δ lie between −90° and +90°, neither $\cos a$ nor $\cos\delta$ can be negative. It follows from Equation 5.81 that the sign of $\sin(h)$ depends only on that of $\sin A$.

To summarize, given the topocentric azimuth A and altitude a of the target together with the sidereal time θ and latitude ϕ of the tracking station, we compute the topocentric declension δ and right ascension α as follows,

$$\delta = \sin^{-1}(\cos\phi \cos A \cos a + \sin\phi \sin a) \qquad (5.83a)$$

$$h = \begin{cases} 2\pi - \cos^{-1}\left(\dfrac{\cos\phi \sin a - \sin\phi \cos A \cos a}{\cos\delta}\right) & 0° < A < 180° \\ \cos^{-1}\left(\dfrac{\cos\phi \sin a - \sin\phi \cos A \cos a}{\cos\delta}\right) & 180° \leq A \leq 360° \end{cases} \qquad (5.83b)$$

$$\alpha = \theta - h \qquad (5.83c)$$

If A and a are provided as functions of time, then α and δ are found as functions of time by means of Equations 5.83. The rates $\dot{\alpha}$, $\ddot{\alpha}$, $\dot{\delta}$ and $\ddot{\delta}$ are determined by differentiating $\alpha(t)$ and $\delta(t)$ and substituting the results into Equations 5.68 through 5.73 to calculate the direction cosine vector $\hat{\mathbf{\varrho}}$ and its rates $\dot{\hat{\mathbf{\varrho}}}$ and $\ddot{\hat{\mathbf{\varrho}}}$.

It is a relatively simple matter to find $\dot{\alpha}$ and $\dot{\delta}$ in terms of $\dot{A}$ and $\dot{a}$. Differentiating Equation 5.78a with respect to time yields

$$\dot{\delta} = \frac{1}{\cos\delta}[-\dot{A}\cos\phi \sin A \cos a + \dot{a}(\sin\phi \cos a - \cos\phi \cos A \sin a)] \qquad (5.84)$$

Differentiating Equation 5.81, we get

$$\dot{h}\cos(h) = -\frac{1}{\cos^2\delta}[(\dot{A}\cos A \cos a - \dot{a}\sin A \sin a)\cos\delta + \dot{\delta}\sin A \cos a \sin\delta]$$

Substituting Equation 5.82 and simplifying leads to

$$\dot{h} = -\frac{\dot{A}\cos A \cos a - \dot{a}\sin A \sin a + \dot{\delta}\sin A \cos a \tan\delta}{\cos\phi \sin a - \sin\phi \cos A \cos a}$$

But $\dot{h} = \dot{\theta} - \dot{\alpha} = \omega_E - \dot{\alpha}$, so that, finally,

$$\dot{\alpha} = \omega_E + \frac{\dot{A}\cos A \cos a - \dot{a}\sin A \sin a + \dot{\delta}\sin A \cos a \tan\delta}{\cos\phi \sin a - \sin\phi \cos A \cos a} \qquad (5.85)$$

ALGORITHM 5.4.

Given the range ϱ, azimuth A, angular elevation a together with the rates $\dot{\varrho}$, $\dot{A}$ and $\dot{a}$ relative to an earth-based tracking station, calculate the state vectors $\mathbf{r}$ and $\mathbf{v}$ in the geocentric equatorial frame. A MATLAB script of this procedure appears in Appendix D.14.

1. Using the altitude H, latitude ϕ and local sidereal time θ of the site, calculate its geocentric position vector $\mathbf{R}$ from Equation 5.56:

$$\mathbf{R} = \left[\frac{R_e}{\sqrt{1 - (2f - f^2)\sin^2\phi}} + H \right] \cos\phi \left(\cos\theta \hat{\mathbf{I}} + \sin\theta \hat{\mathbf{J}} \right)$$

$$+ \left[\frac{R_e(1 - f)^2}{\sqrt{1 - (2f - f^2)\sin^2\phi}} + H \right] \sin\phi \hat{\mathbf{K}}$$

where f is the earth's flattening factor.

2. Calculate the topocentric declination δ using Equation 5.83a.
3. Calculate the topocentric right ascension α from Equations 5.83b and 5.83c.
4. Calculate the direction cosine unit vector $\hat{\boldsymbol{\varrho}}$ from Equations 5.68 and 5.71,

$$\hat{\boldsymbol{\varrho}} = \cos\delta(\cos\alpha \hat{\mathbf{I}} + \sin\alpha \hat{\mathbf{J}}) + \sin\delta \hat{\mathbf{K}}$$

5. Calculate the geocentric position vector $\mathbf{r}$ from Equation 5.63,

$$\mathbf{r} = \mathbf{R} + \varrho \hat{\boldsymbol{\varrho}}$$

6. Calculate the inertial velocity $\dot{\mathbf{R}}$ of the site from Equation 5.66.
7. Calculate the declination rate $\dot{\delta}$ using Equation 5.84.
8. Calculate the right ascension rate $\dot{\alpha}$ by means of Equation 5.85.
9. Calculate the direction cosine rate vector $\dot{\hat{\boldsymbol{\varrho}}}$ from Equations 5.69 and 5.72:

$$\dot{\hat{\boldsymbol{\varrho}}} = (-\dot{\alpha}\sin\alpha\cos\delta - \dot{\delta}\cos\alpha\sin\delta)\hat{\mathbf{I}} + (\dot{\alpha}\cos\alpha\cos\delta - \dot{\delta}\sin\alpha\sin\delta)\hat{\mathbf{J}} + \dot{\delta}\cos\delta\hat{\mathbf{K}}$$

10. Calculate the geocentric velocity vector $\mathbf{v}$ from Equation 5.64:

$$\mathbf{v} = \dot{\mathbf{R}} + \dot{\varrho}\hat{\boldsymbol{\varrho}} + \varrho\dot{\hat{\boldsymbol{\varrho}}}$$

EXAMPLE 5.10

At $\theta = 300°$ local sidereal time a sea-level ($H = 0$) tracking station at latitude $\phi = 60°$ detects a space object and obtains the following data:

$$\begin{aligned}
\text{Slant range}: &\quad \varrho = 2551 \text{ km} \\
\text{Azimuth}: &\quad A = 90° \\
\text{Elevation}: &\quad a = 30°
\end{aligned}$$

5.8 Orbit determination from angle and range measurements

Range rate: $\dot{\varrho} = 0$
Azimuth rate: $\dot{A} = 1.973 \times 10^{-3}$ rad/s (0.1130°/s)
Elevation rate: $\dot{a} = 9.864 \times 10^{-4}$ rad/s (0.05651°/s)

What are the orbital elements of the object?

We must first employ Algorithm 5.4 to obtain the state vectors **r** and **v** in order to compute the orbital elements by means of Algorithm 4.1.

Step 1:

The equatorial radius of the earth is $R_e = 6378$ km and the flattening factor is $f = 0.003353$. It follows from Equation 5.56 that the position vector of the observer is

$$\mathbf{R} = 1598\hat{\mathbf{I}} - 2769\hat{\mathbf{J}} + 5500\hat{\mathbf{K}} \text{ (km)}$$

Step 2:

$$\delta = \sin^{-1}(\cos\phi \cos A \cos a + \sin\phi \sin a)$$
$$= \sin^{-1}(\cos 60° \cos 90° \cos 30° + \sin 60° \sin 30°)$$
$$= 25.66°$$

Step 3:

Since the given azimuth lies between 0° and 180°, Equation 5.83b yields

$$h = 360° - \cos^{-1}\left(\frac{\cos\phi \sin a - \sin\phi \cos A \cos a}{\cos\delta}\right)$$
$$= 360° - \cos^{-1}\left(\frac{\cos 60° \sin 30° - \sin 60° \cos 90° \cos 30°}{\cos 25.66°}\right)$$
$$= 360° - 73.90° = 286.1°$$

Therefore, the right ascension is

$$\alpha = \theta - h = 300° - 286.1° = 13.90°$$

Step 4:

$$\hat{\boldsymbol{\varrho}} = \cos 25.66(\cos 13.90°\hat{\mathbf{I}} + \sin 13.90°\hat{\mathbf{J}}) + \sin\delta\hat{\mathbf{K}} = 0.8750\hat{\mathbf{I}} + 0.2165\hat{\mathbf{J}} + 0.4330\hat{\mathbf{K}}$$

Step 5:

$$\mathbf{r} = \mathbf{R} + \varrho\hat{\boldsymbol{\varrho}} = (1598\hat{\mathbf{I}} - 2769\hat{\mathbf{J}} + 5500\hat{\mathbf{K}}) + 2551(0.8750\hat{\mathbf{I}} + 0.2165\hat{\mathbf{J}} + 0.4330\hat{\mathbf{K}})$$
$$\mathbf{r} = 3831\hat{\mathbf{I}} - 2216\hat{\mathbf{J}} + 6605\hat{\mathbf{K}} \text{ (km)}$$

Step 6:

Recalling from Equation 2.57 that the angular velocity ω_E of the earth is 72.92×10^{-6} rad/s,

$$\dot{\mathbf{R}} = \boldsymbol{\Omega} \times \mathbf{R} = (72.92 \times 10^{-6}\hat{\mathbf{K}}) \times (1598\hat{\mathbf{I}} - 2769\hat{\mathbf{J}} + 5500\hat{\mathbf{K}})$$
$$= 0.2019\hat{\mathbf{I}} + 0.1166\hat{\mathbf{J}} \text{ (km/s)}$$

(Example 5.10 continued)

Step 7:

$$\dot{\delta} = \frac{1}{\cos\delta}\left[-\dot{A}\cos\phi\sin A\cos a + \dot{a}(\sin\phi\cos a - \cos\phi\cos A\sin a)\right]$$

$$= \frac{1}{\cos 25.66°}[-1.973 \times 10^{-3} \cdot \cos 60° \sin 90° \cos 30° + 9.864$$
$$\times 10^{-4}(\sin 60° \cos 30° - \cos 60° \cos 90° \sin 30°)]$$

$$\dot{\delta} = -1.2696 \times 10^{-4} \text{ (rad/s)}$$

Step 8:

$$\dot{\alpha} - \omega_E = \frac{\dot{A}\cos A\cos a - \dot{a}\sin A\sin a + \dot{\delta}\sin A\cos a\tan\delta}{\cos\phi\sin a - \sin\phi\cos A\cos a}$$

$$= \frac{\begin{array}{c}1.973 \times 10^{-3}\cos 90°\cos 30° - 9.864 \times 10^{-4}\sin 90°\sin 30°\\ +(-1.2696 \times 10^{-4})\sin 90°\cos 30°\tan 25.66°\end{array}}{\cos 60°\sin 30° - \sin 60°\cos 90°\cos 30°}$$

$$= -0.002184$$

$$\dot{\alpha} = 72.92 \times 10^{-6} - 0.002184 = -0.002111 \text{ (rad/s)}$$

Step 9:

$$\dot{\hat{\varrho}} = (-\dot{\alpha}\sin\alpha\cos\delta - \dot{\delta}\cos\alpha\sin\delta)\hat{I} + (\dot{\alpha}\cos\alpha\cos\delta - \dot{\delta}\sin\alpha\sin\delta)\hat{J} + \dot{\delta}\cos\delta\hat{K}$$

$$= \left[-(-0.002111)\sin 13.90°\cos 25.66° - (-0.1270)\cos 13.90°\sin 25.66°\right]\hat{I}$$
$$+ \left[(-0.002111)\cos 13.90°\cos 25.66° - (-0.1270)\sin 13.90°\sin 25.66°\right]\hat{J}$$
$$+ \left[-0.1270\cos 25.66°\right]\hat{K}$$

$$\dot{\hat{\varrho}} = (0.5104\hat{I} - 1.834\hat{J} - 0.1144\hat{K})(10^{-3}) \text{ (rad/s)}$$

Step 10:

$$\mathbf{v} = \dot{\mathbf{R}} + \dot{\varrho}\hat{\varrho} + \varrho\dot{\hat{\varrho}}$$

$$= (0.2019\hat{I} + 0.1166\hat{J}) + 0 \cdot (0.8750\hat{I} + 0.2165\hat{J} + 0.4330\hat{K})$$
$$+ 2551(0.5104 \times 10^{-3}\hat{I} - 1.834 \times 10^{-3}\hat{J} - 0.1144 \times 10^{-3}\hat{K})$$

$$\mathbf{v} = 1.504\hat{I} - 4.562\hat{J} - 0.2920\hat{K} \text{ (km/s)}$$

Using the position and velocity vectors from steps 5 and 10, the reader can verify that Algorithm 4.1 yields the following orbital elements of the tracked object

$$a = 5170 \text{ km}$$
$$i = 113.4°$$
$$\Omega = 109.8°$$
$$e = 0.6195$$
$$\omega = 309.8°$$
$$\theta = 165.3°$$

This is a highly elliptical orbit with a semimajor axis less than the earth's radius, so the object will impact the earth (at a true anomaly of 216°).

5.9 Angles-only preliminary orbit determination 235

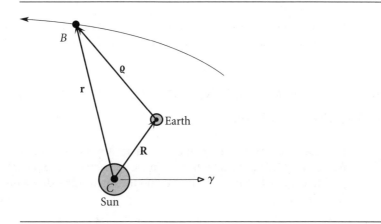

Figure 5.13 An object B orbiting the sun and tracked from earth.

For objects orbiting the sun (planets, asteroids, comets and man-made interplanetary probes), the fundamental vector triangle is as illustrated in Figure 5.13. The tracking station is on the earth but, of course, the sun rather than the earth is the center of attraction. The procedure for finding the heliocentric state vector **r** and **v** is similar to that outlined above. Because of the vast distances involved, the observer can usually be imagined to reside at the center of the earth. Dealing with **R** is different in this case. The daily position of the sun relative to the earth ($-$**R** in Figure 5.13) may be found in ephemerides, such as *Astronomical Almanac* (US Naval Observatory, 2004). A discussion of interplanetary trajectories appears in Chapter 8 of this text.

5.9 ANGLES-ONLY PRELIMINARY ORBIT DETERMINATION

To determine an orbit requires specifying six independent quantities. These can be the six classical orbital elements or the total of six components the state vector, **r** and **v**, at a given instant. To determine an orbit solely from observations therefore requires six independent measurements. In the previous section we assumed the tracking station was able to measure simultaneously the six quantities: range and range rate; azimuth and azimuth rate; plus elevation and elevation rate. This data leads directly to the state vector and, hence, to a complete determination of the orbit. In the absence of range and range rate measuring capability, as with a telescope, we must rely on measurements of just the two angles, azimuth and elevation, to determine the orbit. A minimum of three observations of azimuth and elevation is therefore required to accumulate the six quantities we need to predict the orbit. We shall henceforth assume that the angular measurements are converted to topocentric right ascension α and declination δ, as described in the previous section.

We shall consider the classical method of angles-only orbit determination due to Carl Friedrich Gauss (1777–1855), a German mathematician who many consider was one of the greatest mathematicians ever. This method requires gathering angular information over closely spaced intervals of time and yields a preliminary orbit determination based on those initial observations. We follow Boulet (1991).

5.10 GAUSS'S METHOD OF PRELIMINARY ORBIT DETERMINATION

Suppose we have three observations of an orbiting body at times t_1, t_2 and t_3, as shown in Figure 5.14. At each time the geocentric position vector $\mathbf{r}$ is related to the observer's position vector $\mathbf{R}$, the slant range ϱ and the topocentric direction cosine vector $\hat{\boldsymbol{\varrho}}$ by Equation 5.63,

$$\mathbf{r}_1 = \mathbf{R}_1 + \varrho_1 \hat{\boldsymbol{\varrho}}_1 \qquad (5.86a)$$

$$\mathbf{r}_2 = \mathbf{R}_2 + \varrho_2 \hat{\boldsymbol{\varrho}}_2 \qquad (5.86b)$$

$$\mathbf{r}_3 = \mathbf{R}_3 + \varrho_3 \hat{\boldsymbol{\varrho}}_3 \qquad (5.86c)$$

The positions $\mathbf{R}_1$, $\mathbf{R}_2$ and $\mathbf{R}_3$ of the observer O are known from the location of the tracking station and the time of the observations. $\hat{\boldsymbol{\varrho}}_1$, $\hat{\boldsymbol{\varrho}}_2$ and $\hat{\boldsymbol{\varrho}}_3$ are obtained by measuring the right ascension α and declination δ of the body at each of the three times (recall Equation 5.57). Equations 5.86 are three vector equations, and therefore nine scalar equations, in 12 unknowns: the three components of each of the three vectors $\mathbf{r}_1$, $\mathbf{r}_2$ and $\mathbf{r}_3$, plus the three slant ranges ϱ_1, ϱ_2 and ϱ_3.

An additional three equations are obtained by recalling from Chapter 2 that the conservation of angular momentum requires the vectors $\mathbf{r}_1$, $\mathbf{r}_2$ and $\mathbf{r}_3$ to lie in the

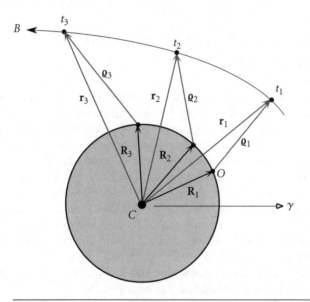

Figure 5.14 Center of attraction C, observer O and tracked body B.

same plane. As in our discussion of the Gibbs method in Section 5.2, that means $\mathbf{r}_2$ is a linear combination $\mathbf{r}_1$ and $\mathbf{r}_3$:

$$\mathbf{r}_2 = c_1\mathbf{r}_1 + c_3\mathbf{r}_3 \tag{5.87}$$

Adding this equation to those in 5.86 introduces two new unknowns c_1 and c_3. At this point we therefore have 12 scalar equations in 14 unknowns.

Another consequence of the two-body equation of motion (Equation 2.15) is that the state vectors $\mathbf{r}$ and $\mathbf{v}$ of the orbiting body can be expressed in terms of the state vector at any given time by means of the Lagrange coefficients, Equations 2.125 and 2.126. For the case at hand that means we can express the position vectors $\mathbf{r}_1$ and $\mathbf{r}_3$ in terms of the position $\mathbf{r}_2$ and velocity $\mathbf{v}_2$ at the intermediate time t_2 as follows,

$$\mathbf{r}_1 = f_1\mathbf{r}_2 + g_1\mathbf{v}_2 \tag{5.88a}$$

$$\mathbf{r}_3 = f_3\mathbf{r}_2 + g_3\mathbf{v}_2 \tag{5.88b}$$

where f_1 and g_1 are the Lagrange coefficients evaluated at t_1 while f_3 and g_3 are those same functions evaluated at time t_3. If the time intervals between the three observations are sufficiently small then Equations 2.163 reveal that f and g depend approximately only on the distance from the center of attraction at the initial time. For the case at hand that means the coefficients in Equations 5.88 depend only on r_2. Hence, Equations 5.88 add six scalar equations to our previous list of 12 while adding to the list of 14 unknowns only four: the three components of $\mathbf{v}_2$ and the radius r_2. We have arrived at 18 equations in 18 unknowns, so the problem is well posed and we can proceed with the solution. The ultimate objective is to determine the state vectors $\mathbf{r}_2, \mathbf{v}_2$ at the intermediate time t_2.

Let us start out by solving for c_1 and c_3 in Equation 5.87. First take the cross product of each term in that equation with $\mathbf{r}_3$,

$$\mathbf{r}_2 \times \mathbf{r}_3 = c_1(\mathbf{r}_1 \times \mathbf{r}_3) + c_3(\mathbf{r}_3 \times \mathbf{r}_3)$$

Since $\mathbf{r}_3 \times \mathbf{r}_3 = \mathbf{0}$, this reduces to

$$\mathbf{r}_2 \times \mathbf{r}_3 = c_1(\mathbf{r}_1 \times \mathbf{r}_3)$$

Taking the dot product of this result with $\mathbf{r}_1 \times \mathbf{r}_3$ and solving for c_1 yields

$$c_1 = \frac{(\mathbf{r}_2 \times \mathbf{r}_3) \cdot (\mathbf{r}_1 \times \mathbf{r}_3)}{\|\mathbf{r}_1 \times \mathbf{r}_3\|^2} \tag{5.89}$$

In a similar fashion, by forming the dot product of Equation 5.87 with $\mathbf{r}_1$, we are led to

$$c_3 = \frac{(\mathbf{r}_2 \times \mathbf{r}_1) \cdot (\mathbf{r}_3 \times \mathbf{r}_1)}{\|\mathbf{r}_1 \times \mathbf{r}_3\|^2} \tag{5.90}$$

Let us next use Equations 5.88 to eliminate $\mathbf{r}_1$ and $\mathbf{r}_3$ from the expressions for c_1 and c_3. First of all,

$$\mathbf{r}_1 \times \mathbf{r}_3 = (f_1\mathbf{r}_2 + g_1\mathbf{v}_2) \times (f_3\mathbf{r}_2 + g_3\mathbf{v}_2) = f_1g_3(\mathbf{r}_2 \times \mathbf{v}_2) + f_3g_1(\mathbf{v}_2 \times \mathbf{r}_2)$$

But $\mathbf{r}_2 \times \mathbf{v}_2 = \mathbf{h}$, where $\mathbf{h}$ is the constant angular momentum of the orbit (Equation 2.18). It follows that

$$\mathbf{r}_1 \times \mathbf{r}_3 = (f_1g_3 - f_3g_1)\mathbf{h} \tag{5.91}$$

and, of course,
$$\mathbf{r}_3 \times \mathbf{r}_1 = -(f_1 g_3 - f_3 g_1)\mathbf{h} \tag{5.92}$$

Therefore
$$\|\mathbf{r}_1 \times \mathbf{r}_3\|^2 = (f_1 g_3 - f_3 g_1)^2 h^2 \tag{5.93}$$

Similarly
$$\mathbf{r}_2 \times \mathbf{r}_3 = \mathbf{r}_2 \times (f_3 \mathbf{r}_2 + g_3 \mathbf{v}_2) = g_3 \mathbf{h} \tag{5.94}$$

and
$$\mathbf{r}_2 \times \mathbf{r}_1 = \mathbf{r}_2 \times (f_1 \mathbf{r}_2 + g_1 \mathbf{v}_2) = g_1 \mathbf{h} \tag{5.95}$$

Substituting Equations 5.91, 5.93 and 5.94 into Equation 5.89 yields
$$c_1 = \frac{g_3 \mathbf{h} \cdot (f_1 g_3 - f_3 g_1)\mathbf{h}}{(f_1 g_3 - f_3 g_1)^2 h^2} = \frac{g_3 (f_1 g_3 - f_3 g_1) h^2}{(f_1 g_3 - f_3 g_1)^2 h^2}$$

or
$$c_1 = \frac{g_3}{f_1 g_3 - f_3 g_1} \tag{5.96}$$

Likewise, substituting Equations 5.92, 5.93 and 5.95 into Equation 5.90 leads to
$$c_3 = -\frac{g_1}{f_1 g_3 - f_3 g_1} \tag{5.97}$$

The coefficients in Equation 5.87 are now expressed solely in terms of the Lagrange functions, and so far no approximations have been made. However, we will have to make some approximations in order to proceed.

We must approximate c_1 and c_3 under the assumption that the times between observations of the orbiting body are small. To that end, let us introduce the notation
$$\begin{aligned} \tau_1 &= t_1 - t_2 \\ \tau_3 &= t_3 - t_2 \end{aligned} \tag{5.98}$$

τ_1 and τ_3 are the time intervals between the successive measurements of $\hat{\boldsymbol{\varrho}}_1, \hat{\boldsymbol{\varrho}}_2$ and $\hat{\boldsymbol{\varrho}}_3$. If the time intervals τ_1 and τ_3 are small enough, we can retain just the first two terms of the series expressions for the Lagrange coefficients f and g in Equations 2.163, thereby obtaining the approximations

$$f_1 \approx 1 - \frac{1}{2}\frac{\mu}{r_2^3}\tau_1^2 \tag{5.99a}$$

$$f_3 \approx 1 - \frac{1}{2}\frac{\mu}{r_2^3}\tau_3^2 \tag{5.99b}$$

and
$$g_1 \approx \tau_1 - \frac{1}{6}\frac{\mu}{r_2^3}\tau_1^3 \tag{5.100a}$$

$$g_3 \approx \tau_3 - \frac{1}{6}\frac{\mu}{r_2^3}\tau_3^3 \tag{5.100b}$$

5.10 Gauss's method of preliminary orbit determination

We want to exclude all terms in f and g beyond the first two so that only the unknown r_2 appears in Equations 5.99 and 5.100. One can see from Equations 2.163 that the higher order terms include the unknown $\mathbf{v}_2$ as well.

Using Equations 5.99 and 5.100 we can calculate the denominator in Equations 5.96 and 5.97,

$$f_1 g_3 - f_3 g_1 = \left(1 - \frac{1}{2}\frac{\mu}{r_2^3}\tau_1^2\right)\left(\tau_3 - \frac{1}{6}\frac{\mu}{r_2^3}\tau_3^3\right) - \left(1 - \frac{1}{2}\frac{\mu}{r_2^3}\tau_3^2\right)\left(\tau_1 - \frac{1}{6}\frac{\mu}{r_2^3}\tau_1^3\right)$$

Expanding the right side and collecting terms yields

$$f_1 g_3 - f_3 g_1 = (\tau_3 - \tau_1) - \frac{1}{6}\frac{\mu}{r_2^3}(\tau_3 - \tau_1)^3 + \frac{1}{12}\frac{\mu^2}{r_2^6}(\tau_1^2\tau_3^3 - \tau_1^3\tau_3^2)$$

Retaining terms of at most third order in the time intervals τ_1 and τ_3, and setting

$$\tau = \tau_3 - \tau_1 \tag{5.101}$$

reduces this expression to

$$f_1 g_3 - f_3 g_1 \approx \tau - \frac{1}{6}\frac{\mu}{r_2^3}\tau^3 \tag{5.102}$$

From Equation 5.98 observe that τ is just the time interval between the first and last observations. Substituting Equations 5.100b and 5.102 into Equation 5.96, we get

$$c_1 \approx \frac{\tau_3 - \frac{1}{6}\frac{\mu}{r_2^3}\tau_3^3}{\tau - \frac{1}{6}\frac{\mu}{r_2^3}\tau^3} = \frac{\tau_3}{\tau}\left(1 - \frac{1}{6}\frac{\mu}{r_2^3}\tau_3^2\right)\cdot\left(1 - \frac{1}{6}\frac{\mu}{r_2^3}\tau^2\right)^{-1} \tag{5.103}$$

We can use the binomial theorem to simplify (linearize) the last term on the right. Setting $a = 1$, $b = -\frac{1}{6}\frac{\mu}{r_2^3}\tau^2$ and $n = -1$ in Equation 5.44, and neglecting terms of higher order than 2 in τ, yields

$$\left(1 - \frac{1}{6}\frac{\mu}{r_2^3}\tau^2\right)^{-1} \approx 1 + \frac{1}{6}\frac{\mu}{r_2^3}\tau^2$$

Hence Equation 5.103 becomes

$$c_1 \approx \frac{\tau_3}{\tau}\left[1 + \frac{1}{6}\frac{\mu}{r_2^3}(\tau^2 - \tau_3^2)\right] \tag{5.104}$$

where only second order terms in the time have been retained. In precisely the same way it can be shown that

$$c_3 \approx -\frac{\tau_1}{\tau}\left[1 + \frac{1}{6}\frac{\mu}{r_2^3}(\tau^2 - \tau_1^2)\right] \tag{5.105}$$

Finally, we have managed to obtain approximate formulas for the coefficients in Equation 5.87 in terms of just the time intervals between observations and the as yet unknown distance r_2 from the center of attraction at the central time t_2.

The next stage of the solution is to seek formulas for the slant ranges ϱ_1, ϱ_2 and ϱ_3 in terms of c_1 and c_3. To that end, substitute Equations 5.86 into Equation 5.87 to get

$$\mathbf{R}_2 + \varrho_2\hat{\mathbf{\varrho}}_2 = c_1(\mathbf{R}_1 + \varrho_1\hat{\mathbf{\varrho}}_1) + c_3(\mathbf{R}_3 + \varrho_3\hat{\mathbf{\varrho}}_3)$$

which we rearrange into the form

$$c_1 \varrho_1 \hat{\boldsymbol{\varrho}}_1 - \varrho_2 \hat{\boldsymbol{\varrho}}_2 + c_3 \varrho_3 \hat{\boldsymbol{\varrho}}_3 = -c_1 \mathbf{R}_1 + \mathbf{R}_2 - c_3 \mathbf{R}_3 \quad (5.106)$$

Let us isolate the slant ranges ϱ_1, ϱ_2 and ϱ_3 in turn by taking the dot product of this equation with appropriate vectors. To isolate ϱ_1 take the dot product of each term in this equation with $\hat{\boldsymbol{\varrho}}_2 \times \hat{\boldsymbol{\varrho}}_3$, which gives

$$c_1 \varrho_1 \hat{\boldsymbol{\varrho}}_1 \cdot (\hat{\boldsymbol{\varrho}}_2 \times \hat{\boldsymbol{\varrho}}_3) - \varrho_2 \hat{\boldsymbol{\varrho}}_2 \cdot (\hat{\boldsymbol{\varrho}}_2 \times \hat{\boldsymbol{\varrho}}_3) + c_3 \varrho_3 \hat{\boldsymbol{\varrho}}_3 \cdot (\hat{\boldsymbol{\varrho}}_2 \times \hat{\boldsymbol{\varrho}}_3)$$
$$= -c_1 \mathbf{R}_1 \cdot (\hat{\boldsymbol{\varrho}}_2 \times \hat{\boldsymbol{\varrho}}_3) + \mathbf{R}_2 \cdot (\hat{\boldsymbol{\varrho}}_2 \times \hat{\boldsymbol{\varrho}}_3) - c_3 \mathbf{R}_3 \cdot (\hat{\boldsymbol{\varrho}}_2 \times \hat{\boldsymbol{\varrho}}_3)$$

Since $\hat{\boldsymbol{\varrho}}_2 \cdot (\hat{\boldsymbol{\varrho}}_2 \times \hat{\boldsymbol{\varrho}}_3) = \hat{\boldsymbol{\varrho}}_3 \cdot (\hat{\boldsymbol{\varrho}}_2 \times \hat{\boldsymbol{\varrho}}_3) = 0$, this reduces to

$$c_1 \varrho_1 \hat{\boldsymbol{\varrho}}_1 \cdot (\hat{\boldsymbol{\varrho}}_2 \times \hat{\boldsymbol{\varrho}}_3) = (-c_1 \mathbf{R}_1 + \mathbf{R}_2 - c_3 \mathbf{R}_3) \cdot (\hat{\boldsymbol{\varrho}}_2 \times \hat{\boldsymbol{\varrho}}_3) \quad (5.107)$$

Let D_0 represent the scalar triple product of $\hat{\boldsymbol{\varrho}}_1, \hat{\boldsymbol{\varrho}}_2$ and $\hat{\boldsymbol{\varrho}}_3$,

$$D_0 = \hat{\boldsymbol{\varrho}}_1 \cdot (\hat{\boldsymbol{\varrho}}_2 \times \hat{\boldsymbol{\varrho}}_3) \quad (5.108)$$

We will assume that D_0 is not zero, which means that $\hat{\boldsymbol{\varrho}}_1, \hat{\boldsymbol{\varrho}}_2$ and $\hat{\boldsymbol{\varrho}}_3$ do not lie in the same plane. Then we can solve Equation 5.107 for ϱ_1 to get

$$\varrho_1 = \frac{1}{D_0} \left(-D_{11} + \frac{1}{c_1} D_{21} - \frac{c_3}{c_1} D_{31} \right) \quad (5.109a)$$

where the Ds stand for the scalar triple products

$$D_{11} = \mathbf{R}_1 \cdot (\hat{\boldsymbol{\varrho}}_2 \times \hat{\boldsymbol{\varrho}}_3) \quad D_{21} = \mathbf{R}_2 \cdot (\hat{\boldsymbol{\varrho}}_2 \times \hat{\boldsymbol{\varrho}}_3) \quad D_{31} = \mathbf{R}_3 \cdot (\hat{\boldsymbol{\varrho}}_2 \times \hat{\boldsymbol{\varrho}}_3) \quad (5.109b)$$

In a similar fashion, by taking the dot product of Equation 5.106 with $\hat{\boldsymbol{\varrho}}_1 \times \hat{\boldsymbol{\varrho}}_3$ and then $\hat{\boldsymbol{\varrho}}_1 \times \hat{\boldsymbol{\varrho}}_2$ we obtain ϱ_2 and ϱ_3,

$$\varrho_2 = \frac{1}{D_0} (-c_1 D_{12} + D_{22} - c_3 D_{32}) \quad (5.110a)$$

where

$$D_{12} = \mathbf{R}_1 \cdot (\hat{\boldsymbol{\varrho}}_1 \times \hat{\boldsymbol{\varrho}}_3) \quad D_{22} = \mathbf{R}_2 \cdot (\hat{\boldsymbol{\varrho}}_1 \times \hat{\boldsymbol{\varrho}}_3) \quad D_{32} = \mathbf{R}_3 \cdot (\hat{\boldsymbol{\varrho}}_1 \times \hat{\boldsymbol{\varrho}}_3) \quad (5.110b)$$

and

$$\varrho_3 = \frac{1}{D_0} \left(-\frac{c_1}{c_3} D_{13} + \frac{1}{c_3} D_{23} - D_{33} \right) \quad (5.111a)$$

where

$$D_{13} = \mathbf{R}_1 \cdot (\hat{\boldsymbol{\varrho}}_1 \times \hat{\boldsymbol{\varrho}}_2) \quad D_{23} = \mathbf{R}_2 \cdot (\hat{\boldsymbol{\varrho}}_1 \times \hat{\boldsymbol{\varrho}}_2) \quad D_{33} = \mathbf{R}_3 \cdot (\hat{\boldsymbol{\varrho}}_1 \times \hat{\boldsymbol{\varrho}}_2) \quad (5.111b)$$

To obtain these results we used the fact that $\hat{\boldsymbol{\varrho}}_2 \cdot (\hat{\boldsymbol{\varrho}}_1 \times \hat{\boldsymbol{\varrho}}_3) = -D_0$ and $\hat{\boldsymbol{\varrho}}_3 \cdot (\hat{\boldsymbol{\varrho}}_1 \times \hat{\boldsymbol{\varrho}}_2) = D_0$ (Equation 2.32).

Substituting Equations 5.104 and 5.105 into Equation 5.110a yields the approximate slant range ϱ_2,

$$\varrho_2 = A + \frac{\mu B}{r_2^3} \quad (5.112a)$$

5.10 Gauss's method of preliminary orbit determination

where

$$A = \frac{1}{D_0}\left(-D_{12}\frac{\tau_3}{\tau} + D_{22} + D_{32}\frac{\tau_1}{\tau}\right) \tag{5.112b}$$

$$B = \frac{1}{6D_0}\left[D_{12}(\tau_3^2 - \tau^2)\frac{\tau_3}{\tau} + D_{32}(\tau^2 - \tau_1^2)\frac{\tau_1}{\tau}\right] \tag{5.112c}$$

On the other hand, making the same substitutions into Equations 5.109 and 5.111 leads to the following approximate expressions for the slant ranges ϱ_1 and ϱ_3,

$$\varrho_1 = \frac{1}{D_0}\left[\frac{6\left(D_{31}\frac{\tau_1}{\tau_3} + D_{21}\frac{\tau}{\tau_3}\right)r_2^3 + \mu D_{31}(\tau^2 - \tau_1^2)\frac{\tau_1}{\tau_3}}{6r_2^3 + \mu(\tau^2 - \tau_3^2)} - D_{11}\right] \tag{5.113}$$

$$\varrho_3 = \frac{1}{D_0}\left[\frac{6\left(D_{13}\frac{\tau_3}{\tau_1} - D_{23}\frac{\tau}{\tau_1}\right)r_2^3 + \mu D_{13}(\tau^2 - \tau_3^2)\frac{\tau_3}{\tau_1}}{6r_2^3 + \mu(\tau^2 - \tau_3^2)} - D_{33}\right] \tag{5.114}$$

Equation 5.112a is a relation between the slant range ϱ_2 and the geocentric radius r_2. Another expression relating these two variables is obtained from Equation 5.86b,

$$\mathbf{r}_2 \cdot \mathbf{r}_2 = (\mathbf{R}_2 + \varrho_2\hat{\boldsymbol{\varrho}}_2) \cdot (\mathbf{R}_2 + \varrho_2\hat{\boldsymbol{\varrho}}_2)$$

or

$$r_2^2 = \varrho_2^2 + 2E\varrho_2 + R_2^2 \tag{5.115a}$$

where

$$E = \mathbf{R}_2 \cdot \hat{\boldsymbol{\varrho}}_2 \tag{5.115b}$$

Substituting Equation 5.112a into 5.115a gives

$$r_2^2 = \left(A + \frac{\mu B}{r_2^3}\right)^2 + 2C\left(A + \frac{\mu B}{r_2^3}\right) + R_2^2$$

Expanding and rearranging terms leads to an eighth order polynomial,

$$x^8 + ax^6 + bx^3 + c = 0 \tag{5.116}$$

where $x = r_2$ and the coefficients are

$$a = -(A^2 + 2AE + R_2^2) \quad b = -2\mu B(A + E) \quad c = -\mu^2 B^2 \tag{5.117}$$

We solve Equation 5.116 for r_2 and substitute the result into Equations 5.112 through 5.114 to obtain the slant ranges ϱ_1, ϱ_2 and ϱ_3. Then Equations 5.86 yield the position vectors $\mathbf{r}_1, \mathbf{r}_2$ and $\mathbf{r}_3$. Recall that finding $\mathbf{r}_2$ was one of our objectives.

To attain the other objective, the velocity $\mathbf{v}_2$, we first solve Equation 5.88a for $\mathbf{r}_2$

$$\mathbf{r}_2 = \frac{1}{f_1}\mathbf{r}_1 - \frac{g_1}{f_1}\mathbf{v}_2$$

Substitute this result into Equation 5.88b to get

$$\mathbf{r}_3 = \frac{f_3}{f_1}\mathbf{r}_1 + \left(\frac{f_1 g_3 - f_3 g_1}{f_1}\right)\mathbf{v}_2$$

Solving this for $\mathbf{v}_2$ yields

$$\mathbf{v}_2 = \frac{1}{f_1 g_3 - f_3 g_1}(-f_3\mathbf{r}_1 + f_1\mathbf{r}_3) \qquad (5.118)$$

in which the approximate Lagrange functions appearing in Equations 5.99 and 5.100 are used.

The approximate values we have found for $\mathbf{r}_2$ and $\mathbf{v}_2$ are used as the starting point for iteratively improving the accuracy of the computed $\mathbf{r}_2$ and $\mathbf{v}_2$ until convergence is achieved. The entire step-by-step procedure is summarized in Algorithms 5.5 and 5.6 presented below. See also Appendix D.15.

ALGORITHM 5.5

Gauss's method of preliminary orbit determination. Given the direction cosine vectors $\hat{\boldsymbol{\varrho}}_1$, $\hat{\boldsymbol{\varrho}}_2$ and $\hat{\boldsymbol{\varrho}}_3$ and the observer's position vectors $\mathbf{R}_1$, $\mathbf{R}_2$ and $\mathbf{R}_3$ at the times t_1, t_2 and t_3, proceed as follows.

1. Calculate the time intervals τ_1, τ_3 and τ using Equations 5.98 and 5.101.
2. Calculate the cross products $\mathbf{p}_1 = \hat{\boldsymbol{\varrho}}_2 \times \hat{\boldsymbol{\varrho}}_3$, $\mathbf{p}_2 = \hat{\boldsymbol{\varrho}}_1 \times \hat{\boldsymbol{\varrho}}_3$ and $\mathbf{p}_3 = \hat{\boldsymbol{\varrho}}_1 \times \hat{\boldsymbol{\varrho}}_2$.
3. Calculate $D_0 = \hat{\boldsymbol{\varrho}}_1 \cdot \mathbf{p}_1$ (Equation 5.108).
4. From Equations 5.109b, 5.110b and 5.111b compute the six scalar quantities

$$D_{11} = \mathbf{R}_1 \cdot \mathbf{p}_1 \qquad D_{12} = \mathbf{R}_1 \cdot \mathbf{p}_2 \qquad D_{13} = \mathbf{R}_1 \cdot \mathbf{p}_3$$
$$D_{21} = \mathbf{R}_2 \cdot \mathbf{p}_1 \qquad D_{22} = \mathbf{R}_2 \cdot \mathbf{p}_2 \qquad D_{23} = \mathbf{R}_2 \cdot \mathbf{p}_3$$
$$D_{31} = \mathbf{R}_3 \cdot \mathbf{p}_1 \qquad D_{32} = \mathbf{R}_3 \cdot \mathbf{p}_2 \qquad D_{33} = \mathbf{R}_3 \cdot \mathbf{p}_3$$

5. Calculate A and B using Equations 5.112b and 5.112c.
6. Calculate E, using Equation 5.115b, and $R_2^2 = \mathbf{R}_2 \cdot \mathbf{R}_2$.
7. Calculate a, b and c from Equation 5.117.
8. Find the roots of Equation 5.116 and select the most reasonable one as r_2. Newton's method can be used, in which case Equation 3.13 becomes

$$x_{i+1} = x_i - \frac{x_i^8 + ax_i^6 + bx_i^3 + c}{8x_i^7 + 6ax_i^5 + 3bx_i^2} \qquad (5.119)$$

One must first print or graph the function $F = x^8 + ax^6 + bx^3 + c$ for $x > 0$ and choose as an initial estimate a value of x near the point where F changes sign. If there is more than one physically reasonable root, then each one must be used and the resulting orbit checked against knowledge that may already be available about the general nature of the orbit. Alternatively, the analysis can be repeated using additional sets of observations.

9. Calculate ϱ_1, ϱ_2 and ϱ_3 using Equations 5.113, 5.112a and 5.114.
10. Use Equations 5.86 to calculate $\mathbf{r}_1$, $\mathbf{r}_2$ and $\mathbf{r}_3$.

11. Calculate the Lagrange coefficients f_1, g_1, f_3 and g_3 from Equations 5.99 and 5.100.
12. Calculate $\mathbf{v}_2$ using Equation 5.118.
13. (a) Use $\mathbf{r}_2$ and $\mathbf{v}_2$ from steps 10 and 12 to obtain the orbital elements from Algorithm 4.1. (b) Alternatively, proceed to Algorithm 5.6 to improve the preliminary estimate of the orbit.

ALGORITHM 5.6

Iterative improvement of the orbit determined by Algorithm 5.5.

Use the values of $\mathbf{r}_2$ and $\mathbf{v}_2$ obtained from Algorithm 5.5 to compute the 'exact' values of the f and g functions from their universal formulation, as follows:

1. Calculate the magnitude of $\mathbf{r}_2$ ($r_2 = \sqrt{\mathbf{r}_2 \cdot \mathbf{r}_2}$) and v_2 ($v_2 = \sqrt{\mathbf{v}_2 \cdot \mathbf{v}_2}$).
2. Calculate α, the reciprocal of the semimajor axis: $\alpha = 2/r_2 - v_2^2/\mu$.
3. Calculate the radial component of $\mathbf{v}_2$, $v_{r2} = \mathbf{v}_2 \cdot \mathbf{r}_2/r_2$.
4. Use Algorithm 3.3 to solve the universal Kepler's equation (Equation 3.46) for the universal variables χ_1 and χ_3 at times t_1 and t_3, respectively:

$$\sqrt{\mu}\tau_1 = \frac{r_2 v_{r2}}{\sqrt{\mu}} \chi_1^2 C(\alpha \chi_1^2) + (1 - \alpha r_2)\chi_1^3 S(\alpha \chi_1^2) + r_2 \chi_1$$

$$\sqrt{\mu}\tau_3 = \frac{r_2 v_{r2}}{\sqrt{\mu}} \chi_3^2 C(\alpha \chi_3^2) + (1 - \alpha r_2)\chi_3^3 S(\alpha \chi_3^2) + r_2 \chi_3$$

5. Use χ_1 and χ_3 to calculate f_1, g_1, f_3 and g_3 from Equations 3.66:

$$f_1 = 1 - \frac{\chi_1^2}{r_2}C(\alpha\chi_1^2) \qquad g_1 = \tau_1 - \frac{1}{\sqrt{\mu}}\chi_1^3 S(\alpha\chi_1^2)$$

$$f_3 = 1 - \frac{\chi_3^2}{r_2}C(\alpha\chi_3^2) \qquad g_3 = \tau_3 - \frac{1}{\sqrt{\mu}}\chi_3^3 S(\alpha\chi_3^2)$$

6. Use these values of f_1, g_1, f_3 and g_3 to calculate c_1 and c_3 from Equations 5.96 and 5.97.
7. Use c_1 and c_3 to calculate updated values of ϱ_1, ϱ_2 and ϱ_3 from Equations 5.109 through 5.111.
8. Calculate updated $\mathbf{r}_1$, $\mathbf{r}_2$ and $\mathbf{r}_3$ from Equations 5.86.
9. Calculate updated $\mathbf{v}_2$ using Equation 5.118 and the f and g values computed in step 5.
10. Go back to step 1 and repeat until, to the desired degree of precision, there is no further change in ϱ_1, ϱ_2 and ϱ_3.
11. Use $\mathbf{r}_2$ and $\mathbf{v}_2$ to compute the orbital elements by means of Algorithm 4.1.

EXAMPLE 5.11

A tracking station is located at $\phi = 40°$ north latitude at an altitude of $H = 1$ km. Three observations of an earth satellite yield the values for the topocentric right ascension and declination listed in the following table, which also shows the local sidereal time θ of the observation site.

(Example 5.11 continued)

Use the Gauss Algorithm 5.5 to estimate the state vector at the second observation time. Recall that $\mu = 398\,600 \text{ km}^3/\text{s}^2$.

Table 5.1   Data for Example 5.11

Observation	Time (seconds)	Right ascension, α (degrees)	Declination, δ (degrees)	Local sidereal time, θ (degrees)
1	0	43.537	−8.7833	44.506
2	118.10	54.420	−12.074	45.000
3	237.58	64.318	−15.105	45.499

Recalling that the equatorial radius of the earth is $R_e = 6378$ km and the flattening factor is $f = 0.003353$, we substitute $\phi = 40°$, $H = 1$ km and the given values of θ into Equation 5.56 to obtain the inertial position vector of the tracking station at each of the three observation times:

$$\mathbf{R}_1 = 3489.8\hat{\mathbf{I}} + 3430.2\hat{\mathbf{J}} + 4078.5\hat{\mathbf{K}} \text{ (km)}$$

$$\mathbf{R}_2 = 3460.1\hat{\mathbf{I}} + 3460.1\hat{\mathbf{J}} + 4078.5\hat{\mathbf{K}} \text{ (km)}$$

$$\mathbf{R}_3 = 3429.9\hat{\mathbf{I}} + 3490.1\hat{\mathbf{J}} + 4078.5\hat{\mathbf{K}} \text{ (km)}$$

Using Equation 5.57 we compute the direction cosine vectors at each of the three observation times from the right ascension and declination data:

$$\hat{\boldsymbol{\rho}}_1 = \cos(-8.7833°)\cos 43.537°\hat{\mathbf{I}} + \cos(-8.7833°)\sin 43.537°\hat{\mathbf{J}} + \sin(-8.7833°)\hat{\mathbf{K}}$$
$$= 0.71643\hat{\mathbf{I}} + 0.68074\hat{\mathbf{J}} - 0.15270\hat{\mathbf{K}}$$

$$\hat{\boldsymbol{\rho}}_2 = \cos(-12.074°)\cos 54.420°\hat{\mathbf{I}} + \cos(-12.074°)\sin 54.420°\hat{\mathbf{J}} + \sin(-12.074°)\hat{\mathbf{K}}$$
$$= 0.56897\hat{\mathbf{I}} + 0.79531\hat{\mathbf{J}} - 0.20917\hat{\mathbf{K}}$$

$$\hat{\boldsymbol{\rho}}_3 = \cos(-15.105°)\cos 64.318°\hat{\mathbf{I}} + \cos(-15.105°)\sin 64.318°\hat{\mathbf{J}} + \sin(-15.105°)\hat{\mathbf{K}}$$
$$= 0.41841\hat{\mathbf{I}} + 0.87007\hat{\mathbf{J}} - 0.26059\hat{\mathbf{K}}$$

We can now proceed with Algorithm 5.5.

Step 1:

$$\tau_1 = 0 - 118.10 = -118.10 \text{ s}$$
$$\tau_3 = 237.58 - 118.10 = 119.47 \text{ s}$$
$$\tau = 119.47 - (-118.1) = 237.58 \text{ s}$$

Step 2:

$$\mathbf{p}_1 = \hat{\boldsymbol{\rho}}_2 \times \hat{\boldsymbol{\rho}}_3 = -0.025258\hat{\mathbf{I}} + 0.060753\hat{\mathbf{J}} + 0.16229\hat{\mathbf{K}}$$
$$\mathbf{p}_2 = \hat{\boldsymbol{\rho}}_1 \times \hat{\boldsymbol{\rho}}_3 = -0.044538\hat{\mathbf{I}} + 0.12281\hat{\mathbf{J}} + 0.33853\hat{\mathbf{K}}$$
$$\mathbf{p}_3 = \hat{\boldsymbol{\rho}}_1 \times \hat{\boldsymbol{\rho}}_2 = -0.020950\hat{\mathbf{I}} + 0.062977\hat{\mathbf{J}} + 0.18246\hat{\mathbf{K}}$$

5.10 Gauss's method of preliminary orbit determination

Step 3:
$$D_0 = \hat{\boldsymbol{\rho}}_1 \cdot \mathbf{p}_1 = -0.0015198$$

Step 4:

$D_{11} = \mathbf{R}_1 \cdot \mathbf{p}_1 = 782.15 \text{ km}$ $D_{12} = \mathbf{R}_1 \cdot \mathbf{p}_2 = 1646.5 \text{ km}$ $D_{13} = \mathbf{R}_1 \cdot \mathbf{p}_3 = 887.10 \text{ km}$

$D_{21} = \mathbf{R}_2 \cdot \mathbf{p}_1 = 784.72 \text{ km}$ $D_{22} = \mathbf{R}_2 \cdot \mathbf{p}_2 = 1651.5 \text{ km}$ $D_{23} = \mathbf{R}_2 \cdot \mathbf{p}_3 = 889.60 \text{ km}$

$D_{31} = \mathbf{R}_3 \cdot \mathbf{p}_1 = 787.31 \text{ km}$ $D_{32} = \mathbf{R}_3 \cdot \mathbf{p}_2 = 1656.6 \text{ km}$ $D_{33} = \mathbf{R}_3 \cdot \mathbf{p}_3 = 892.13 \text{ km}$

Step 5:

$$A = \frac{1}{-0.0015198}\left[-1646.5\frac{119.47}{237.58} + 1651.5 + 1656.6\frac{(-118.10)}{237.58}\right] = -6.6858 \text{ km}$$

$$B = \frac{1}{6(-0.0015198)}\left\{1646.5(119.47^2 - 237.58^2)\frac{119.47}{237.58}\right.$$
$$\left. + 1656.6[237.58^2 - (-118.10)^2]\frac{(-118.10)}{237.58}\right\}$$
$$= 7.6667 \times 10^9 \text{ km} \cdot \text{s}^2$$

Step 6:
$$E = \mathbf{R}_2 \cdot \hat{\boldsymbol{\rho}}_2 = 3875.8 \text{ km}$$
$$R_2^2 = \mathbf{R}_2 \cdot \mathbf{R}_2 = 4.058 \times 10^7 \text{ km}^2$$

Step 7:
$$a = -[(-6.6858)^2 + 2(-6.6858)(3875.8) + 4.058 \times 10^7] = -4.0528 \times 10^7 \text{ km}^2$$
$$b = -2(389\,600)(7.6667 \times 10^9)(-6.6858 + 3875.8) = -2.3597 \times 10^{19} \text{ km}^5$$
$$c = -(398\,600)^2(7.6667 \times 10^9)^2 = -9.3387 \times 10^{30} \text{ km}^8$$

Step 8:
$$F(x) = x^8 - 4.0528 \times 10^7 x^6 - 2.3597 \times 10^{19} x^3 - 9.3387 \times 10^{30} = 0$$

The graph of $F(x)$ in Figure 5.15 shows that it changes sign near $x = 9000$ km. Let us use that as the starting value in Newton's method for finding the roots of $F(x)$. For the case at hand, Equation 5.119 is

$$x_{i+1} = x_i - \frac{x_i^8 - 4.0528 \times 10^7 x_i^6 - 2.3622 \times 10^{19} x_i^3 - 9.3186 \times 10^{30}}{8x_i^7 - 2.4317 \times 10^8 x_i^5 - 7.0866 \times 10^{19} x_i^2}$$

Stepping through Newton's iterative procedure yields

$$x_0 = 9000$$
$$x_1 = 9000 - (-276.93) = 9276.9$$
$$x_2 = 9276.9 - 34.526 = 9242.4$$

(Example 5.11 continued)

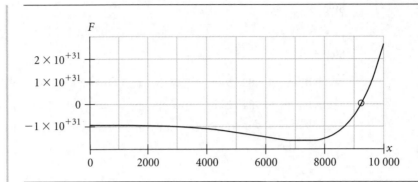

Figure 5.15 Graph of the polynomial in Equation (f).

$$x_3 = 9242.4 - 0.63428 = 9241.8$$
$$x_4 = 9241.8 - 0.00021048 = 9241.8$$

Thus, after four steps we converge to

$$r_2 = 9241.8 \text{ km}$$

The other roots are either negative or complex and are therefore physically unacceptable.

Step 9:

$$\varrho_1 = \frac{1}{-0.0015198} \times \left\{ \frac{6\left[787.31\frac{(-118.10)}{119.47} + 784.72\frac{237.58}{119.47}\right]9241.8^3 + 398\,600 \cdot 787.31[237.58^2 - (-118.10)^2]\frac{-118.10}{119.47}}{6 \cdot 9241.8^3 + 398\,600(237.58^2 - 119.47^2)} - 782.15 \right\}$$

$$= 3639.1 \text{ km}$$

$$\varrho_2 = -6.6858 + \frac{398\,600 \cdot 7.6667 \times 10^9}{9241.8^3} = 3864.8 \text{ km}$$

$$\varrho_3 = \frac{1}{-0.0015198} \times \left[\frac{6\left(887.10\frac{119.47}{-118.10} - 889.60\frac{237.58}{-118.10}\right)9241.8^3 + 398\,600 \cdot 887.10(237.58^2 - 119.47^2)\frac{119.47}{-118.10}}{6 \cdot 9241.8^3 + 398\,600(237.58^2 - 119.47^2)} - 892.13 \right]$$

$$= 4156.9 \text{ km}$$

5.10 Gauss's method of preliminary orbit determination 247

Step 10:

$$\mathbf{r}_1 = (3489.8\hat{\mathbf{I}} + 3430.2\hat{\mathbf{J}} + 4078.5\hat{\mathbf{K}}) + 3639.1(0.71643\hat{\mathbf{I}} + 0.68074\hat{\mathbf{J}} - 0.15270\hat{\mathbf{K}})$$
$$= 6096.9\hat{\mathbf{I}} + 5907.5\hat{\mathbf{J}} + 3522.9\hat{\mathbf{K}} \text{ (km)}$$

$$\mathbf{r}_2 = (3460.1\hat{\mathbf{I}} + 3460.1\hat{\mathbf{J}} + 4078.5\hat{\mathbf{K}}) + 3864.8(0.56897\hat{\mathbf{I}} + 0.79531\hat{\mathbf{J}} - 0.20917\hat{\mathbf{K}})$$
$$= 5659.1\hat{\mathbf{I}} + 6533.8\hat{\mathbf{J}} + 3270.1\hat{\mathbf{K}} \text{ (km)}$$

$$\mathbf{r}_3 = (3429.9\hat{\mathbf{I}} + 3490.1\hat{\mathbf{J}} + 4078.5\hat{\mathbf{K}}) + 4156.9(0.41841\hat{\mathbf{I}} + 0.87007\hat{\mathbf{J}} - 0.26059\hat{\mathbf{K}})$$
$$= 5169.1\hat{\mathbf{I}} + 7107.0\hat{\mathbf{J}} + 2995.3\hat{\mathbf{K}} \text{ (km)}$$

Step 11:

$$f_1 \approx 1 - \frac{1}{2}\frac{398\,600}{9241.8^3}(-118.10)^2 = 0.99648$$

$$f_3 \approx 1 - \frac{1}{2}\frac{398\,600}{9241.8^3}(119.47)^2 = 0.99640$$

$$g_1 \approx -118.10 - \frac{1}{6}\frac{398\,600}{9241.8^3}(-118.10)^3 = -117.97$$

$$g_3 \approx 119.47 - \frac{1}{6}\frac{398\,600}{9241.8^3}(119.47)^3 = 119.33$$

Step 12:

$$\mathbf{v}_2 = \frac{-0.99640(6096.9\hat{\mathbf{I}} + 5907.5\hat{\mathbf{J}} + 3522.9\hat{\mathbf{K}}) + 0.99648(5169.1\hat{\mathbf{I}} + 7107.0\hat{\mathbf{J}} + 2995.3\hat{\mathbf{K}})}{0.99648 \cdot 119.33 - 0.99640(-117.97)}$$

$$= -3.9080\hat{\mathbf{I}} + 5.0573\hat{\mathbf{J}} - 2.2222\hat{\mathbf{K}} \text{ (km/s)}$$

In summary, the state vector at time t_2 is, approximately,

$$\mathbf{r}_2 = 5659.1\hat{\mathbf{I}} + 6533.8\hat{\mathbf{J}} + 3270.1\hat{\mathbf{K}} \text{ (km)}$$
$$\mathbf{v}_2 = -3.9080\hat{\mathbf{I}} + 5.0573\hat{\mathbf{J}} - 2.2222\hat{\mathbf{K}} \text{ (km/s)}$$

EXAMPLE 5.12

Starting with the state vector determined in Example 5.11, use Algorithm 5.6 to improve the vector to five significant figures.

Step 1:

$$r_2 = \|\mathbf{r}_2\| = \sqrt{5659.1^2 + 6533.8^2 + 3270.1^2} = 9241.8 \text{ km}$$

$$v_2 = \|\mathbf{v}_2\| = \sqrt{(-3.9080)^2 + 5.0573 + (-2.2222)^2} = 6.7666 \text{ km/s}$$

Step 2:

$$\alpha = \frac{2}{r_2} - \frac{v_2^2}{\mu} = \frac{2}{9241.8} - \frac{6.7666^2}{398\,600} = 1.0154 \times 10^{-4} \text{ km}^{-1}$$

248 Chapter 5 Preliminary orbit determination

(Example 5.12 continued)

Step 3:

$$v_{r2} = \frac{\mathbf{v}_2 \cdot \mathbf{r}_2}{r_2} = \frac{(-3.9080) \cdot 5659.1 + 5.0573 \cdot 6533.8 + (-2.2222) \cdot 3270.1}{9241.8}$$

$$= 0.39611 \text{ km/s}$$

Step 4:

The universal Kepler's equation at times t_1 and t_3, respectively, becomes

$$\sqrt{398\,600}\,\tau_1 = \frac{9241.8 \cdot 0.39611}{\sqrt{398\,600}} \chi_1^2 C(1.0154 \times 10^{-4} \chi_1^2)$$

$$+ (1 - 1.0154 \times 10^{-4} \cdot 9241.8) \chi_1^3 S(1.0154 \times 10^{-4} \chi_1^2) + 9241.8 \chi_1$$

$$\sqrt{398\,600}\,\tau_3 = \frac{9241.8 \cdot 0.39611}{\sqrt{398\,600}} \chi_3^2 C(1.0154 \times 10^{-4} \chi_3^2)$$

$$+ (1 - 1.0154 \times 10^{-4} \cdot 9241.8) \chi_3^3 S(1.0154 \times 10^{-4} \chi_3^2) + 9241.8 \chi_3$$

or

$$631.35\tau_1 = 5.7983 \chi_1^2 C(1.0154 \times 10^{-4} \chi_1^2) + 0.061594 \chi_1^3 S(1.0154 \times 10^{-4} \chi_1^2)$$

$$+ 9241.8 \chi_1$$

$$631.35\tau_3 = 5.7983 \chi_3^2 C(1.0154 \times 10^{-4} \chi_3^2) + 0.061594 \chi_3^3 S(1.0154 \times 10^{-4} \chi_3^2)$$

$$+ 9241.8 \chi_3$$

Applying Algorithm 3.3 to each of these equations yields

$$\chi_1 = -8.0882 \sqrt{\text{km}}$$

$$\chi_3 = 8.1404 \sqrt{\text{km}}$$

Step 5:

$$f_1 = 1 - \frac{\chi_1^2}{r_2} C(\alpha \chi_1^2) = 1 - \frac{(-8.0882)^2}{9241.8} \cdot \overbrace{C[1.0154 \times 10^{-4}(-8.0882)^2]}^{0.49972} = 0.99646$$

$$g_1 = \tau_1 - \frac{1}{\sqrt{\mu}} \chi_1^3 S(\alpha \chi_1^2) = -118.1 - \frac{1}{\sqrt{398\,600}} (-8.0882)^3$$

$$\times \overbrace{S[1.0154 \times 10^{-4}(-8.0882)^2]}^{0.16661} = -117.96 \text{ s}$$

and

$$f_3 = 1 - \frac{\chi_3^2}{r_2} C(\alpha \chi_3^2) = 1 - \frac{8.1404^2}{9241.8} \cdot \overbrace{C[1.0154 \times 10^{-4} \cdot 8.1404^2]}^{0.49972} = 0.99642$$

$$g_3 = \tau_3 - \frac{1}{\sqrt{\mu}}\chi_3^3 S(\alpha\chi_3^2) = -118.1 - \frac{1}{\sqrt{398\,600}}8.1404^3$$

$$\times \overbrace{S[1.0154 \times 10^{-4}(-8.0882)^2]}^{0.16661} = 119.33$$

It turns out that the procedure converges more rapidly if the Lagrange coefficients are set equal to the average of those computed for the current step and those computed for the previous step. Thus, we set

$$f_1 = \frac{0.99648 + 0.99646}{2} = 0.99647$$

$$g_1 = \frac{-117.97 + (-117.96)}{2} = -117.96\,\text{s}$$

$$f_3 = \frac{0.99642 + 0.99641}{2} = 0.99641$$

$$g_3 = \frac{119.3 + 119.3}{2} = 119.3\,\text{s}$$

Step 6:

$$c_1 = \frac{119.3}{(0.99647)(119.3) - (0.99641)(-117.96\,\text{s})} = 0.50467$$

$$c_3 = -\frac{-117.96}{(0.99647)(119.3) - (0.99641)(-117.96)} = 0.49890$$

Step 7:

$$\varrho_1 = \frac{1}{-0.0015198}\left(-782.15 + \frac{1}{0.50467}784.72 - \frac{0.49890}{0.50467}787.31\right) = 3650.7\,\text{km}$$

$$\varrho_2 = \frac{1}{-0.0015198}(-0.50467 \cdot 1646.5 + 1651.5 - 0.49890 \cdot 1656.6) = 3877.2\,\text{km}$$

$$\varrho_3 = \frac{1}{-0.0015198}\left(-\frac{0.50467}{0.49890}887.10 + \frac{1}{0.49890}889.60 - 892.13\right) = 4186.2\,\text{km}$$

Step 8:

$$\mathbf{r}_1 = (3489.8\hat{\mathbf{I}} + 3430.2\hat{\mathbf{J}} + 4078.5\hat{\mathbf{K}}) + 3650.7(0.71643\hat{\mathbf{I}} + 0.68074\hat{\mathbf{J}} - 0.15270\hat{\mathbf{K}})$$

$$= 6105.3\hat{\mathbf{I}} + 5915.4\hat{\mathbf{J}} + 3521.1\hat{\mathbf{K}}\,(\text{km})$$

$$\mathbf{r}_2 = (3460.1\hat{\mathbf{I}} + 3460.1\hat{\mathbf{J}} + 4078.5\hat{\mathbf{K}}) + 3877.2(0.56897\hat{\mathbf{I}} + 0.79531\hat{\mathbf{J}} - 0.20917\hat{\mathbf{K}})$$

$$= 5662.1\hat{\mathbf{I}} + 6543.7\hat{\mathbf{J}} + 3267.5\hat{\mathbf{K}}\,(\text{km})$$

$$\mathbf{r}_3 = (3429.9\hat{\mathbf{I}} + 3490.1\hat{\mathbf{J}} + 4078.5\hat{\mathbf{K}}) + 4186.2(0.41841\hat{\mathbf{I}} + 0.87007\hat{\mathbf{J}} - 0.26059\hat{\mathbf{K}})$$

$$= 5181.4\hat{\mathbf{I}} + 7132.4\hat{\mathbf{J}} + 2987.6\hat{\mathbf{K}}\,(\text{km})$$

(Example 5.12 continued)

Step 9:

$$\mathbf{v}_2 = \frac{1}{0.99647 \cdot 119.3 - 0.99641(-117.96)} \times [-0.99641(6105.3\hat{\mathbf{I}} + 5915.4\hat{\mathbf{J}} + 3521.1\hat{\mathbf{K}}) + 0.99647(5181.4\hat{\mathbf{I}} + 7132.4\hat{\mathbf{J}} + 2987.6\hat{\mathbf{K}})]$$

$$= -3.8918\hat{\mathbf{I}} + 5.1307\hat{\mathbf{J}} - 2.2472\hat{\mathbf{K}} \text{ (km/s)}$$

This completes the first iteration.

The updated position $\mathbf{r}_2$ and velocity $\mathbf{v}_2$ are used to repeat the procedure beginning at step 1. The results of the first and subsequent iterations are shown in Table 5.2. Convergence to five significant figures in the slant ranges ϱ_1, ϱ_2 and ϱ_3 occurs in four steps, at which point the state vector is

$$\mathbf{r}_2 = 5662.1\hat{\mathbf{I}} + 6538.0\hat{\mathbf{J}} + 3269.0\hat{\mathbf{K}} \text{ (km)}$$
$$\mathbf{v}_2 = -3.8856\hat{\mathbf{I}} + 5.1214\hat{\mathbf{J}} - 2.2433\hat{\mathbf{K}} \text{ (km/s)}$$

Table 5.2 Key results at each step of the iterative procedure

Step	χ_1	χ_3	f_1	g_1	f_3	g_3	ϱ_1	ϱ_2	ϱ_3
1	−8.0882	8.1404	0.99647	−117.97	0.99641	119.33	3650.7	3877.2	4186.2
2	−8.0818	8.1282	0.99647	−117.96	0.99642	119.33	3643.8	3869.9	4178.3
3	−8.0871	8.1337	0.99647	−117.96	0.99642	119.33	3644.0	3870.1	4178.6
4	−8.0869	8.1336	0.99647	−117.96	0.99642	119.33	3644.0	3870.1	4178.6

Using Algorithm 4.1 we find that the orbital elements are

$$a = 10\,000 \text{ km} \quad (h = 62\,818 \text{ km}^2/\text{s})$$
$$e = 0.1000$$
$$i = 30°$$
$$\Omega = 270°$$
$$\omega = 90°$$
$$\theta = 45.01°$$

PROBLEMS

5.1 The geocentric equatorial position vectors of a satellite at three separate times are

$$\mathbf{r}_1 = 5887\hat{\mathbf{I}} - 3520\hat{\mathbf{J}} - 1204\hat{\mathbf{K}} \text{ (km)}$$
$$\mathbf{r}_2 = 5572\hat{\mathbf{I}} - 3457\hat{\mathbf{J}} - 2376\hat{\mathbf{K}} \text{ (km)}$$
$$\mathbf{r}_3 = 5088\hat{\mathbf{I}} - 3289\hat{\mathbf{J}} - 3480\hat{\mathbf{K}} \text{ (km)}$$

Use Gibbs' method to find $\mathbf{v}_2$.
{Partial ans.: $v_2 = 7.59$ km/s}

5.2 Calculate the orbital elements and perigee altitude of the space object in the previous problem.
{Partial ans.: $z_p = 567$ km}

5.3 At a given instant the altitude of an earth satellite is 600 km. Fifteen minutes later the altitude is 300 km and the true anomaly has increased by 60°. Find the perigee altitude.
{Ans.: $z_p = 298$ km}

5.4 At a given instant, the geocentric equatorial position vector of an earth satellite is

$$\mathbf{r}_1 = -3600\hat{\mathbf{I}} + 3600\hat{\mathbf{J}} + 5100\hat{\mathbf{K}} \text{ (km)}$$

Thirty minutes later the position is

$$\mathbf{r}_2 = -5500\hat{\mathbf{I}} - 6240\hat{\mathbf{J}} - 520\hat{\mathbf{K}} \text{ (km)}$$

Calculate $\mathbf{v}_1$ and $\mathbf{v}_2$.
{Partial ans.: $v_1 = 7.711$ km/s, $v_2 = 6.670$ km/s}

5.5 Compute the orbital elements and perigee altitude for the previous problem.
{Partial ans.: $z_p = 648$ km}

5.6 At a given instant, the geocentric equatorial position vector of an earth satellite is

$$\mathbf{r}_1 = 5644\hat{\mathbf{I}} - 2830\hat{\mathbf{J}} - 4170\hat{\mathbf{K}} \text{ (km)}$$

Twenty minutes later the position is

$$\mathbf{r}_2 = -2240\hat{\mathbf{I}} + 7320\hat{\mathbf{J}} - 4980\hat{\mathbf{K}} \text{ (km)}$$

Calculate $\mathbf{v}_1$ and $\mathbf{v}_2$.
{Partial ans.: $v_1 = 10.84$ km/s, $v_2 = 9.970$ km/s}

5.7 Compute the orbital elements and perigee altitude for the previous problem.
{Partial ans.: $z_p = 224$ km}

5.8 Calculate the Julian day number (JD) for the following epochs:
(a) 5:30 UT on August 14, 1914.
(b) 14:00 UT on April 18, 1946.
(c) 0:00 UT on September 1, 2010.
(d) 12:00 UT on October 16, 2007.
(e) Noon today, your local time.
{Ans.: (a) 2 420 358.729; (b) 2 431 929.083; (c) 2 455 440.500; (d) 2 454 390.000}

5.9 Calculate the number of days from 12:00 UT on your date of birth to 12:00 UT on today's date.

5.10 Calculate the local sidereal time (in degrees) at:
(a) Stockholm, Sweden (east longitude 18°03′) at 12:00 UT on 1 January 2008.
(b) Melbourne, Australia (east longitude 144°58′) at 10:00 UT on 21 December 2007.
(c) Los Angeles, California (west longitude 118°15′) at 20:00 UT on 4 July 2005.
(d) Rio de Janeiro, Brazil (west longitude 43°06′) at 3:00 UT on 15 February 2006.
(e) Vladivostok, Russia (east longitude 131°56′) at 8:00 UT on 21 March 2006.
(f) At noon today, your local time and place.
{Ans.: (a) 298.6°, (b) 24.6°, (c) 104.7°, (d) 146.9°, (e) 70.6°}

5.11 Relative to a tracking station whose local sidereal time is 117° and latitude is +51°, the azimuth and elevation angle of a satellite are 27.5156° and 67.5556°, respectively. Calculate the topocentric right ascension and declination of the satellite.
{Ans.: $\alpha = 145.3°$, $\delta = 68.24°$}

5.12 A sea-level tracking station at whose local sidereal time is 40° and latitude is 35° makes the following observations of a space object:

Azimuth:	36.0°
Azimuth rate:	0.590°/s
Elevation:	36.6°
Elevation rate:	−0.263°/s
Range:	988 km
Range rate:	4.86 km/s

What is the state vector of the object?
{Partial ans.: $r = 7003.3$ km, $v = 10.922$ km/s}

5.13 Calculate the orbital elements of the satellite in the previous problem.
{Partial ans.: $e = 1.1$, $i = 40°$}

5.14 A tracking station at latitude −20° and elevation 500 m makes the following observations of a satellite at the given times.

Time (min)	Local sidereal time (degrees)	Azimuth (degrees)	Elevation angle (degrees)	Range (km)
0	60.0	165.932	8.81952	1212.48
2	60.5014	145.970	44.2734	410.596
4	61.0027	2.40973	20.7594	726.464

Use the Gibbs method to calculate the state vector of the satellite at the central observation time.
{Partial ans.: $r_2 = 6684$ km, $v_2 = 7.7239$ km/s}

5.15 Calculate the orbital elements of the satellite in the previous problem.
{Partial ans.: $e = 0.001$, $i = 95°$}

5.16 A sea-level tracking station at latitude +29° makes the following observations of a satellite at the given times.

Time (min)	Local sidereal time (degrees)	Topocentric right ascension (degrees)	Topocentric declination (degrees)
0.0	0	0	51.5110
1.0	0.250684	65.9279	27.9911
2.0	0.501369	79.8500	14.6609

Use the Gauss method without iterative improvement to estimate the state vector of the satellite at the middle observation time.
{Partial ans.: $r = 6700.9$ km, $v = 8.0757$ km/s}

5.17 Refine the estimate in the previous problem using iterative improvement.
{Partial ans.: $r = 6701.5$ km, $v = 8.0881$ km/s}

5.18 Calculate the orbital elements from the state vector obtained in the previous problem.
{Partial ans.: $e = 0.10$, $i = 30°$}

5.19 A sea-level tracking station at latitude $+29°$ makes the following observations of a satellite at the given times.

Time (min)	Local sidereal time (degrees)	Topocentric right ascension (degrees)	Topocentric declination (degrees)
0.0	90	15.0394	20.7487
1.0	90.2507	25.7539	30.1410
2.0	90.5014	48.6055	43.8910

Use the Gauss method without iterative improvement to estimate the state vector of the satellite.
{Partial ans.: $r = 6999.1$ km, $v = 7.5541$ km/s}

5.20 Refine the estimate in the previous problem using iterative improvement.
{Partial ans.: $r = 7000.0$ km, $v = 7.5638$ km/s}

5.21 Calculate the orbital elements from the state vector obtained in the previous problem.
{Partial ans.: $e = 0.0048$, $i = 31°$}

5.22 The position vector $\mathbf{R}$ of a tracking station and the direction cosine vector $\hat{\boldsymbol{\rho}}$ of a satellite relative to the tracking station at three times are as follows:

$$t_1 = 0 \text{ min}$$
$$\mathbf{R}_1 = -1825.96\hat{\mathbf{I}} + 3583.66\hat{\mathbf{J}} + 4933.54\hat{\mathbf{K}} \text{ (km)}$$
$$\hat{\boldsymbol{\rho}}_1 = -0.301687\hat{\mathbf{I}} + 0.200673\hat{\mathbf{J}} + 0.932049\hat{\mathbf{K}}$$

$$t_2 = 1 \text{ min}$$
$$\mathbf{R}_2 = -1816.30\hat{\mathbf{I}} + 3575.63\hat{\mathbf{J}} + 4933.54\hat{\mathbf{K}} \text{ (km)}$$
$$\hat{\boldsymbol{\rho}}_2 = -0.793090\hat{\mathbf{I}} - 0.210324\hat{\mathbf{J}} + 0.571640\hat{\mathbf{K}}$$

$$t_3 = 2 \text{ min}$$
$$\mathbf{R}_3 = -1857.25\hat{\mathbf{I}} + 3567.54\hat{\mathbf{J}} + 4933.54\hat{\mathbf{K}} \text{ (km)}$$
$$\hat{\boldsymbol{\rho}}_3 = -0.873085\hat{\mathbf{I}} - 0.362969\hat{\mathbf{J}} + 0.325539\hat{\mathbf{K}}$$

Use the Gauss method without iterative improvement to estimate the state vector of the satellite at the central observation time.
{Partial ans.: $r = 6742.3$ km, $v = 7.6799$ km/s}

5.23 Refine the estimate in the previous problem using iterative improvement.
{Partial ans.: $r = 6743.0$ km, $v = 7.6922$ km/s}

5.24 Calculate the orbital elements from the state vector obtained in the previous problem.
{Partial ans.: $e = 0.001$, $i = 52°$}

5.25 A tracking station at latitude 60°N and 500 m elevation obtains the following data:

Time (min)	Local sidereal time (degrees)	Topocentric right ascension (degrees)	Topocentric declination (degrees)
0.0	150	157.783	24.2403
5.0	151.253	159.221	27.2993
10.0	152.507	160.526	29.8982

Use the Gauss method without iterative improvement to estimate the state vector of the satellite.
{Partial ans.: $r = 25\,132$ km, $v = 6.0588$ km/s}

5.26 Refine the estimate in the previous problem using iterative improvement.
{Partial ans.: $r = 25\,169$ km, $v = 6.0671$ km/s}

5.27 Calculate the orbital elements from the state vector obtained in the previous problem.
{Partial ans.: $e = 1.09$, $i = 63°$}

5.28 The position vector **R** of a tracking station and the direction cosine vector $\hat{\varrho}$ of a satellite relative to the tracking station at three times are as follows:

$$t_1 = 0 \text{ min}$$
$$\mathbf{R}_1 = 5582.84\hat{\mathbf{I}} + 3073.90\hat{\mathbf{K}} \text{ (km)}$$
$$\hat{\varrho}_1 = 0.846428\hat{\mathbf{I}} + 0.532504\hat{\mathbf{K}}$$

$$t_2 = 5 \text{ min}$$
$$\mathbf{R}_2 = 5581.50\hat{\mathbf{I}} + 122.122\hat{\mathbf{J}} + 3073.90\hat{\mathbf{K}} \text{ (km)}$$
$$\hat{\varrho}_2 = 0.749290\hat{\mathbf{I}} + 0.463023\hat{\mathbf{J}} + 0.473470\hat{\mathbf{K}}$$

$$t_3 = 10 \text{ min}$$
$$\mathbf{R}_3 = 5577.50\hat{\mathbf{I}} + 244.186\hat{\mathbf{J}} + 3073.90\hat{\mathbf{K}} \text{ (km)}$$
$$\hat{\varrho}_3 = 0.529447\hat{\mathbf{I}} + 0.777163\hat{\mathbf{J}} + 0.340152\hat{\mathbf{K}}$$

Use the Gauss method without iterative improvement to estimate the state vector of the satellite.
{Partial ans.: $r = 9729.6$ km, $v = 6.0234$ km/s}

5.29 Refine the estimate in the previous problem using iterative improvement.
{Partial ans.: $r = 9759.8$ km, $v = 6.0713$ km/s}

5.30 Calculate the orbital elements from the state vector obtained in the previous problem.
{Partial ans.: $e = 0.1$, $i = 30°$}

CHAPTER 6

ORBITAL MANEUVERS

CHAPTER OUTLINE

6.1	Introduction	255
6.2	Impulsive maneuvers	256
6.3	Hohmann transfer	257
6.4	Bi-elliptic Hohmann transfer	264
6.5	Phasing maneuvers	268
6.6	Non-Hohmann transfers with a common apse line	273
6.7	Apse line rotation	279
6.8	Chase maneuvers	285
6.9	Plane change maneuvers	290
Problems		304

6.1 INTRODUCTION

Orbital maneuvers transfer a spacecraft from one orbit to another. Orbital changes can be dramatic, such as the transfer from a low-earth parking orbit to an interplanetary trajectory. They can also be quite small, as in the final stages of the rendezvous of one spacecraft with another. Changing orbits requires the firing of onboard rocket engines. We will be concerned solely with impulsive maneuvers in which the rockets fire in relatively short bursts to produce the required velocity change (delta-v).

We start with the classical, energy-efficient Hohmann transfer maneuver, and generalize it to the bi-elliptic Hohmann transfer to see if even more efficiency can be obtained. The phasing maneuver, a form of Hohmann transfer, is considered next.

This is followed by a study of non-Hohmann transfer maneuvers with and without rotation of the apse line. We then analyze chase maneuvers, which involves solving Lambert's problem as explained in Chapter 5. The energy-demanding chase maneuvers may be impractical for low-earth orbits, but they are necessary for interplanetary missions, as we shall see in Chapter 8.

Up to this point, all of the maneuvers are transfers between coplanar orbits. The chapter ends with an introduction to plane change maneuvers and an explanation of why they require such large delta-vs compared to coplanar maneuvers.

6.2 IMPULSIVE MANEUVERS

Orbital maneuvers transfer a spacecraft from one orbit to another. Orbital changes can be dramatic, such as the transfer from a low-earth parking orbit to an interplanetary trajectory. They can also be quite small, as in the final stages of the rendezvous of one spacecraft with another. Impulsive maneuvers are those in which brief firings of onboard rocket motors change the magnitude and direction of the velocity vector instantaneously. During an impulsive maneuver, the position of the spacecraft is considered to be fixed; only the velocity changes. The impulsive maneuver is an idealization by means of which we can avoid having to solve the equations of motion (Equation 2.6) with the rocket thrust included. The idealization is satisfactory for those cases in which the position of the spacecraft changes only slightly during the time that the maneuvering rockets fire. This is true for high-thrust rockets with burn times short compared with the coasting time of the vehicle.

Each impulsive maneuver results in a change $\Delta \mathbf{v}$ in the velocity of the spacecraft. $\Delta \mathbf{v}$ can represent a change in the magnitude ('pumping maneuver') or the direction ('cranking maneuver') of the velocity vector, or both. The magnitude Δv of the velocity increment is related to Δm, the mass of propellant consumed, by the formula (see Equation 11.30)

$$\frac{\Delta m}{m} = 1 - e^{-\frac{\Delta v}{I_{sp}g_o}} \tag{6.1}$$

where m is the mass of the spacecraft before the burn, g_o is the sea-level standard acceleration of gravity, and I_{sp} is the specific impulse of the propellants. Specific impulse is defined as follows:

$$I_{sp} = \frac{\text{thrust}}{\text{sea-level weight rate of fuel consumption}}$$

Specific impulse has units of seconds, and it is a measure of the performance of a rocket propulsion system. I_{sp} for some common propellant combinations are shown in Table 6.1. Figure 6.1 is a graph of Equation 6.1 for a range of specific impulses. Note that for Δvs on the order of 1 km/s or higher, the required propellant exceeds 25 percent of the spacecraft mass prior to the burn.

There are no refueling stations in space, so a mission's delta-v schedule must be carefully planned to minimize the propellant mass carried aloft in favor of payload.

Table 6.1 Some typical specific impulses

Propellant	I_{sp} (seconds)
Cold gas	50
Monopropellant hydrazine	230
Solid propellant	290
Nitric acid/monomethylhydrazine	310
Liquid oxygen/liquid hydrogen	455

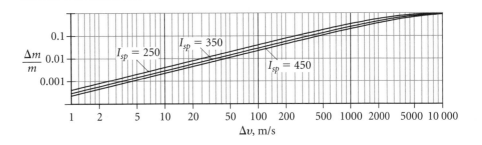

Figure 6.1 Propellant mass fraction versus Δv for typical specific impulses.

6.3 Hohmann transfer

The Hohmann transfer (Hohmann, 1925) is the most energy efficient two-impulse maneuver for transferring between two coplanar circular orbits sharing a common focus. The Hohmann transfer is an elliptical orbit tangent to both circles at its apse line, as illustrated in Figure 6.2. The periapse and apoapse of the transfer ellipse are the radii of the inner and outer circles, respectively. Obviously, only one-half of the ellipse is flown during the maneuver, which can occur in either direction, from the inner to the outer circle, or vice versa.

It may be helpful in sorting out orbit transfer strategies to use the fact that the energy of an orbit depends only on its semimajor axis a. Recall that for an ellipse (Equation 2.70), the specific energy is negative,

$$\varepsilon = -\frac{\mu}{2a}$$

Increasing the energy requires reducing its magnitude, in order to make ε less negative. Therefore, the larger the semimajor axis is, the more the energy the orbit has. In Figure 6.2, the energies increase as we move from the inner to the outer circle.

Starting at A on the inner circle, a velocity increment Δv_A in the direction of flight is required to boost the vehicle onto the higher-energy elliptical trajectory. After coasting from A to B, another forward velocity increment Δv_B places the vehicle on the still higher-energy, outer circular orbit. Without the latter delta-v burn, the spacecraft would, of course, remain on the Hohmann transfer ellipse and return to A. The total energy expenditure is reflected in the total delta-v requirement, $\Delta v_{\text{total}} = \Delta v_A + \Delta v_B$.

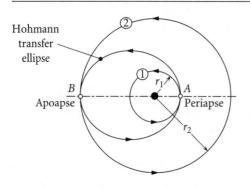

Figure 6.2 Hohmann transfer.

The same total delta-v is required if the transfer begins at B on the outer circular orbit. Since moving to the lower-energy inner circle requires lowering the energy of the spacecraft, the Δvs must be accomplished by retrofires. That is, the thrust of the maneuvering rocket is directed opposite to the flight direction in order to act as a brake on the motion. Since Δv represents the same propellant expenditure regardless of the direction the thruster is aimed, when summing up Δvs, we are concerned only with their magnitudes.

EXAMPLE 6.1

A spacecraft is in a 480 km by 800 km earth orbit (orbit 1 in Figure 6.3). Find (a) the Δv required at perigee A to place the spacecraft in a 480 km by 16 000 km transfer orbit (orbit 2); and (b) the Δv (apogee kick) required at B of the transfer orbit to establish a circular orbit of 16 000 km altitude (orbit 3).

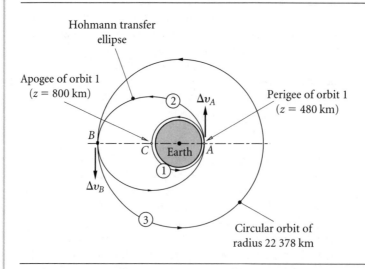

Figure 6.3 Hohmann transfer between two earth orbits.

(a) First, let us establish the primary orbital parameters of the original orbit 1. The perigee and apogee radii are

$$r_A = R_E + z_A = 6378 + 480 = 6858 \text{ km}$$
$$r_C = R_E + z_C = 6378 + 800 = 7178 \text{ km}$$

Therefore, the eccentricity of orbit 1 is

$$e_1 = \frac{r_C - r_A}{r_C + r_A} = 0.022799$$

Applying the orbit equation at perigee of orbit 1, we calculate the angular momentum,

$$r_A = \frac{h_1^2}{\mu} \frac{1}{1 + e_1 \cos(0)} \Rightarrow h_1 = 52\,876 \text{ km}^2/\text{s}$$

With the angular momentum, we can calculate the speed at A on orbit 1,

$$v_A)_1 = \frac{h_1}{r_A} = 7.7102 \text{ km/s} \tag{a}$$

Moving to the transfer orbit 2, we proceed in a similar fashion to get

$$r_B = R_E + z_B = 6378 + 16\,000 = 22\,378 \text{ km}$$
$$e_2 = \frac{r_B - r_A}{r_B + r_A} = 0.53085$$
$$r_A = \frac{h_2^2}{\mu} \frac{1}{1 + e_2 \cos(0)} \Rightarrow h_2 = 64\,690 \text{ km}$$

Thus, the speed at A on orbit 2 is

$$v_A)_2 = \frac{h_2}{r_A} = \frac{64\,690}{6858} = 9.4327 \text{ km/s} \tag{b}$$

The required forward velocity increment at A is now obtained from (a) and (b) as

$$\Delta v_A = v_A)_2 - v_A)_1 = \underline{1.7225 \text{ km/s}}$$

(b) We use the angular momentum formula to find the speed at B on orbit 2,

$$v_B)_2 = \frac{h_2}{r_B} = \frac{64\,690}{22\,378} = 2.8908 \text{ km/s} \tag{c}$$

Orbit 3 is circular, so its constant orbital speed is obtained from Equation 2.53,

$$v_B)_3 = \sqrt{\frac{398\,600}{22\,378}} = 4.2204 \text{ km/s} \tag{d}$$

(Example 6.1 continued)

Thus, the delta-v requirement at B to climb from orbit 2 to orbit 3 is

$$\Delta v_B = v_B)_3 - v_B)_2 = 4.2204 - 2.8908 = \underline{1.3297 \text{ km/s}}$$

Observe that the total delta-v requirement for this Hohmann transfer is

$$\Delta v_{\text{total}} = |\Delta v_A| + |\Delta v_B| = 1.7225 + 1.3297 = 3.0522 \text{ km/s}$$

In the previous example the initial orbit of the Hohmann transfer sequence was an ellipse, rather than a circle. Since no real orbit is perfectly circular, we must generalize the notion of a Hohmann transfer to include two impulsive transfers between elliptical orbits that are coaxial, i.e., share the same apse line, as shown in Figure 6.4. The transfer ellipse must be tangent to both the initial and target ellipses 1 and 2. As can be seen, there are two such transfer orbits, 3 and 3'. It is not immediately obvious which of the two requires the lowest energy expenditure.

To find out which is the best transfer orbit in general, we must calculate the individual total delta-v requirement for orbits 3 and 3'. This requires finding the velocities at A, A', B and B' for each pair of orbits having those points in common. To do so, recall from Equation 2.74 that for an ellipse,

$$e = \frac{r_a - r_p}{r_a + r_p}$$

where r_p and r_a are the radii to periapse and apoapse, respectively. Evaluating the orbit equation at periapse

$$r_p = \frac{h^2}{\mu} \frac{1}{1+e} = \frac{h^2}{\mu} \frac{1}{1 + \frac{r_a - r_p}{r_a + r_p}}$$

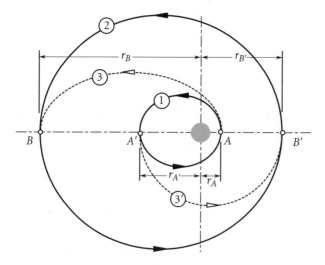

Figure 6.4 Hohmann transfers between coaxial elliptical orbits. In this illustration, $r_{A'}/r_0 = 3$, $r_B/r_0 = 8$ and $r_{B'}/r_0 = 4$.

6.3 Hohmann transfer

yields the angular momentum in terms of the periapse and apoapse radii,

$$h = \sqrt{2\mu}\sqrt{\frac{r_a r_p}{r_a + r_p}} \tag{6.2}$$

Equation 6.2 is used to evaluate the angular momentum of each of the four orbits in Figure 6.4:

$$h_1 = \sqrt{2\mu}\sqrt{\frac{r_A r_{A'}}{r_A + r_{A'}}} \qquad h_3 = \sqrt{2\mu}\sqrt{\frac{r_A r_B}{r_A + r_B}}$$

$$h_2 = \sqrt{2\mu}\sqrt{\frac{r_B r_{B'}}{r_B + r_{B'}}} \qquad h_{3'} = \sqrt{2\mu}\sqrt{\frac{r_{A'} r_{B'}}{r_{A'} + r_{B'}}}$$

From these we obtain the velocities,

$$v_A)_1 = \frac{h_1}{r_A} \qquad v_A)_3 = \frac{h_3}{r_A}$$

$$v_B)_2 = \frac{h_2}{r_B} \qquad v_B)_3 = \frac{h_3}{r_B}$$

$$v_{A'})_1 = \frac{h_1}{r_{A'}} \qquad v_{A'})_{3'} = \frac{h_{3'}}{r_{A'}}$$

$$v_{B'})_2 = \frac{h_2}{r_{B'}} \qquad v_{B'})_{3'} = \frac{h_{3'}}{r_{B'}}$$

These lead to the delta-vs

$$\Delta v_A = |v_A)_3 - v_A)_1| \qquad \Delta v_B = |v_B)_2 - v_B)_3|$$

$$\Delta v_{A'} = |v_{A'})_{3'} - v_{A'})_1| \qquad \Delta v_{B'} = |v_{B'})_2 - v_{B'})_{3'}|$$

and, finally, to the total delta-v requirement for the two possible transfer trajectories,

$$\Delta v_{\text{total}})_3 = \Delta v_A + \Delta v_B \qquad \Delta v_{\text{total}})_{3'} = \Delta v_{A'} + \Delta v_{B'}$$

If $\Delta v_{\text{total}})_{3'}/\Delta v_{\text{total}})_3 > 1$, then orbit 3 is the most efficient. On the other hand, if $\Delta v_{\text{total}})_{3'}/\Delta v_{\text{total}})_3 < 1$, then orbit 3' is more efficient than orbit 3.

Three contour plots of $\Delta v_{\text{total}})_{3'}/\Delta v_{\text{total}})_3$ are shown in Figure 6.5, for three different shapes of the inner orbit 1 of Figure 6.4. Figure 6.5(a) is for $r_{A'}/r_A = 3$, which is the situation represented in Figure 6.4, in which point A is the periapse of the initial ellipse. In Figure 6.5(b) $r_{A'}/r_A = 1$, which means the starting ellipse is a circle. Finally, in Figure 6.5(c) $r_{A'}/r_A = 1/3$, which corresponds to an initial orbit of the same shape as orbit 1 in Figure 6.4, but with point A being the apoapse instead of periapse.

Figure 6.5(a), for which $r_{A'} > r_A$, implies that if point A is the periapse of orbit 1, then transfer orbit 3 is the most efficient. Figure 6.5(c), for which $r_{A'} < r_A$, shows that if point A' is the periapse of orbit 1, then transfer orbit 3' is the most efficient. Together, these results lead us to conclude that it is most efficient for the transfer orbit to begin at the periapse on the inner orbit 1, where its kinetic energy is greatest, regardless of shape of the outer target orbit. If the starting orbit is a circle, then Figure 6.5(b) shows that transfer orbit 3' is most efficient if $r_{B'} > r_B$. That is, from an inner circular orbit, the transfer ellipse should terminate at apoapse of the outer target ellipse, where the speed is slowest.

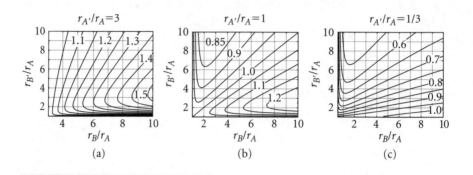

Figure 6.5 Contour plots of $(\Delta v_{\text{total}})_{3'}/(\Delta v_{\text{total}})_3$ for different relative sizes of the ellipses in Figure 6.4. Note that $r_B > r_{A'}$ and $r_{B'} > r_A$.

If the Hohmann transfer is in the reverse direction, i.e., to a lower-energy inner orbit, the above analysis still applies, since the same total delta-v is required whether the Hohmann transfer runs forwards or backwards. Thus, from an outer circle or ellipse to an inner ellipse, the most energy-efficient transfer ellipse terminates at periapse of the inner target orbit. If the inner orbit is a circle, the transfer ellipse should start at apoapse of the outer ellipse.

We close this section with an illustration of the careful planning required for one spacecraft to rendezvous with another at the end of a Hohmann transfer.

EXAMPLE 6.2

A spacecraft returning from a lunar mission approaches earth on a hyperbolic trajectory. At its closest approach A it is at an altitude of 5000 km, traveling at 10 km/s. At A retrorockets are fired to lower the spacecraft into a 500 km altitude circular orbit, where it is to rendezvous with a space station. Find the location of the space station at retrofire so that rendezvous will occur at B.

The time of flight from A to B is one-half the period T_2 of the elliptical transfer orbit 2. While the spacecraft coasts from A to B, the space station coasts through the angle ϕ_{CB} from C to B. Hence, this mission has to be carefully planned and executed, going all the way back to lunar departure, so that the two vehicles meet at B.

To calculate the period T_2, we must first obtain the primary orbital parameters, eccentricity and angular momentum. The apogee and perigee of orbit 2, the transfer ellipse, are

$$r_A = 5000 + 6378 = 11\,378 \text{ km}$$
$$r_B = 500 + 6378 = 6878 \text{ km}$$

Therefore, the eccentricity is

$$e_2 = \frac{11\,378 - 6878}{11\,378 + 6878} = 0.24649$$

Evaluating the orbit equation at perigee yields the angular momentum,

$$r_B = \frac{h_2^2}{\mu}\frac{1}{1+e_2} \Rightarrow 6878 = \frac{h_2^2}{398\,600}\frac{1}{1+0.24649} \Rightarrow h_2 = 58\,458 \text{ km}^2/\text{s}$$

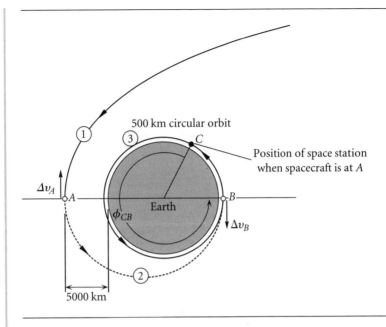

Figure 6.6 Relative position of spacecraft and space station at beginning of the transfer ellipse.

Now we can use Equation 2.72 to find the period of the transfer ellipse,

$$T_2 = \frac{2\pi}{\mu^2}\left(\frac{h_2}{\sqrt{1-e_2^2}}\right)^3 = \frac{2\pi}{398\,600^2}\left(\frac{58\,458}{\sqrt{1-0.24649^2}}\right)^3 = 8679.1 \text{ s} \quad \text{(a)}$$

The period of circular orbit 3 is, according to Equation 2.54,

$$T_3 = \frac{2\pi}{\sqrt{\mu}} r_B^{\frac{3}{2}} = \frac{2\pi}{\sqrt{398\,600}} 6878^{\frac{3}{2}} = 5676.8 \text{ s} \quad \text{(b)}$$

The time of flight from C to B on orbit 3 must equal the time of flight from A to B on orbit 2.

$$\Delta t_{CB} = \frac{1}{2}T_2 = \frac{1}{2} \cdot 8679.1 = 4339.5 \text{ s}$$

Since orbit 3 is a circle, its angular velocity, unlike an ellipse, is constant. Therefore, we can write

$$\frac{\phi_{CB}}{\Delta t_{CB}} = \frac{360°}{T_3} \Rightarrow \phi_{CB} = \frac{4339.5}{5676.8} \cdot 360 = \underline{275.2°}$$

(The student should verify that the total delta-v required to lower the spacecraft from the hyperbola into the parking orbit is 6.415 km/s. A glance at Figure 6.1 reveals the tremendous amount of propellant this would require.)

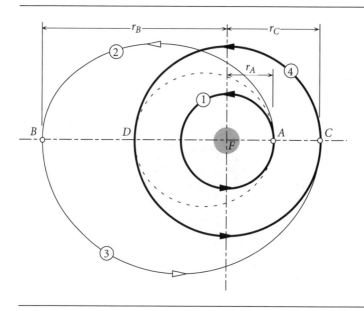

Figure 6.7 Bi-elliptic transfer from inner orbit 1 to outer orbit 4.

6.4 BI-ELLIPTIC HOHMANN TRANSFER

A Hohmann transfer from circular orbit 1 to circular orbit 4 in Figure 6.7 is the dotted ellipse lying inside the outer circle, outside the inner circle, and tangent to both. The bi-elliptical Hohmann transfer uses two coaxial semi-ellipses, 2 and 3, which extend beyond the outer target orbit. Each of the two ellipses is tangent to one of the circular orbits, and they are tangent to each other at B, which is the apoapse of both. The idea is to place B sufficiently far from the focus that the Δv_B will be very small. In fact, as r_B approaches infinity, Δv_B approaches zero. For the bi-elliptical scheme to be more energy efficient than the Hohmann transfer, it must be true that

$$\Delta v_{\text{total}})_{\text{bi-elliptical}} < \Delta v_{\text{total}})_{\text{Hohmann}} \tag{6.3}$$

Delta-v analyses of the Hohmann and bi-elliptical transfers lead to the following results,

$$\Delta v)_{\text{Hohmann}} = \left[\frac{1}{\sqrt{\alpha}} - \frac{\sqrt{2}(1-\alpha)}{\sqrt{\alpha(1+\alpha)}} - 1 \right] \sqrt{\frac{\mu}{r_A}}$$

$$\Delta v)_{\text{bi-elliptical}} = \left[\sqrt{\frac{2(\alpha+\beta)}{\alpha\beta}} - \frac{1+\sqrt{\alpha}}{\sqrt{\alpha}} - \sqrt{\frac{2}{\beta(1+\beta)}}(1-\beta) \right] \sqrt{\frac{\mu}{r_A}} \tag{6.4a}$$

where

$$\alpha = \frac{r_C}{r_A} \qquad \beta = \frac{r_B}{r_A} \tag{6.4b}$$

6.4 Bi-elliptic Hohmann transfer

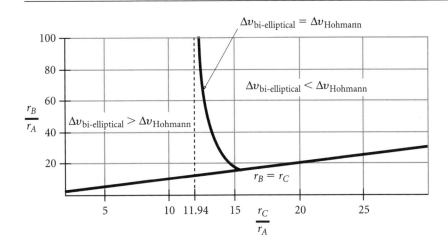

Figure 6.8 Orbits for which the bi-elliptical transfer is either less efficient or more efficient than the Hohmann transfer.

Plotting the difference between Hohmann and bi-elliptical Δv_{total} as a function of α and β reveals the regions in which the difference is positive, negative and zero. These are shown in Figure 6.8.

From the figure we see that if the radius r_C of the outer circular target orbit is less than about 11.9 times that of the inner one (r_A), the standard Hohmann maneuver is the more energy efficient. If the ratio exceeds about 15, then the bi-elliptical strategy is better in that regard. Between those two ratios, large values of the apoapse radius r_B favor the bi-elliptical transfer, while smaller values favor the Hohmann transfer.

Small gains in energy efficiency may be more than offset by the much longer flight times around the bi-elliptical trajectories as compared with the time of flight on the single semi-ellipse of the Hohmann transfer.

EXAMPLE 6.3

Find the total delta-v requirement for a bi-elliptical Hohmann transfer from a geocentric circular orbit of 7000 km radius to one of 105 000 km radius. Let the apogee of the first ellipse be 210 000 km. Compare the delta-v schedule and total flight time with that for an ordinary single Hohmann transfer ellipse.

Since

$$r_A = 7000 \text{ km} \qquad r_B = 210\,000 \text{ km} \qquad r_C = r_D = 105\,000 \text{ km}$$

we have $r_B/r_A = 30$ and $r_C/r_A = 15$, so that from Figure 6.8 it is apparent right away that the bi-elliptic transfer will be the more energy efficient.

To do the delta-v analysis requires analyzing each of the five orbits.

Orbit 1:

Since this is a circular orbit, we have, simply,

$$v_A)_1 = \sqrt{\frac{\mu}{r_A}} = \sqrt{\frac{398\,600}{7000}} = 7.546 \text{ km/s} \qquad (a)$$

(Example 6.3 continued)

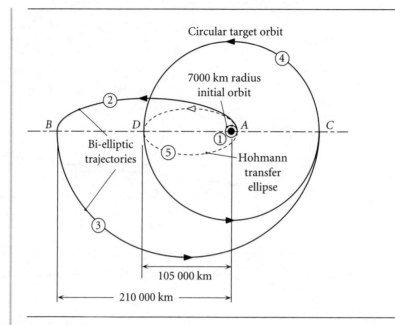

Figure 6.9 Bi-elliptic transfer.

Orbit 2:

For this transfer ellipse, Equation 6.2 yields

$$h_2 = \sqrt{2\mu}\sqrt{\frac{r_A r_B}{r_A + r_B}} = \sqrt{2 \cdot 398\,600}\sqrt{\frac{7000 \cdot 210\,000}{7000 + 210\,000}} = 73\,487\ \text{km}^2/\text{s}$$

Therefore,

$$v_A)_2 = \frac{h_2}{r_A} = \frac{73\,487}{7000} = 10.498\ \text{km/s} \tag{b}$$

$$v_B)_2 = \frac{h_2}{r_B} = \frac{73\,487}{210\,000} = 0.34994\ \text{km/s} \tag{c}$$

Orbit 3:

For the second transfer ellipse, we have

$$h_3 = \sqrt{2 \cdot 398\,600}\sqrt{\frac{105\,000 \cdot 210\,000}{105\,000 + 210\,000}} = 236\,230\ \text{km}^2/\text{s}$$

From this we obtain

$$v_B)_3 = \frac{h_3}{r_B} = \frac{236\,230}{210\,000} = 1.1249\ \text{km/s} \tag{d}$$

$$v_C)_3 = \frac{h_3}{r_C} = \frac{236\,230}{105\,000} = 2.2498\ \text{km/s} \tag{e}$$

Orbit 4:

The target orbit, like orbit 1, is a circle, which means

$$v_C)_4 = v_D)_4 = \sqrt{\frac{398\,600}{105\,000}} = 1.9484 \text{ km/s} \qquad \text{(f)}$$

For the bi-elliptical maneuver, the total delta-v is, therefore,

$$\Delta v_{\text{total}})_{\text{bi-elliptical}} = \Delta v_A + \Delta v_B + \Delta v_C$$
$$= |v_A)_2 - v_A)_1| + |v_B)_3 - v_B)_2| + |v_C)_4 - v_C)_3|$$
$$= |10.498 - 7.546| + |1.1249 - 0.34994| + |1.9484 - 2.2498|$$
$$= 2.9521 + 0.77496 + 0.30142$$

or,

$$\Delta v_{\text{total}})_{\text{bi-elliptical}} = \underline{4.0285 \text{ km/s}} \qquad \text{(g)}$$

The semimajor axes of transfer orbits 2 and 3 are

$$a_2 = \frac{1}{2}(7000 + 210\,000) = 108\,500 \text{ km}$$

$$a_3 = \frac{1}{2}(105\,000 + 210\,000) = 157\,500 \text{ km}$$

With this information and the period formula, Equation 2.73, the time of flight for the two semi-ellipses of the bi-elliptical transfer is found to be

$$t_{\text{bi-elliptical}} = \frac{1}{2}\left(\frac{2\pi}{\sqrt{\mu}}a_2^{\frac{3}{2}} + \frac{2\pi}{\sqrt{\mu}}a_3^{\frac{3}{2}}\right) = 488\,870 \text{ s} = \underline{5.66 \text{ days}} \qquad \text{(h)}$$

For the Hohmann transfer ellipse 5,

$$h_5 = \sqrt{2 \cdot 398\,600}\sqrt{\frac{7000 \cdot 105\,000}{7000 + 105\,000}} = 72\,330 \text{ km}^2/\text{s}$$

Hence,

$$v_A)_5 = \frac{h_5}{r_A} = \frac{72\,330}{7000} = 10.333 \text{ km/s} \qquad \text{(i)}$$

$$v_D)_5 = \frac{h_5}{r_D} = \frac{72\,330}{105\,000} = 0.68886 \text{ km/s} \qquad \text{(j)}$$

It follows that

$$\Delta v_{\text{total}})_{\text{Hohmann}} = |v_A)_5 - v_A)_1| + |v_D)_5 - v_D)_1|$$
$$= (10.333 - 7.546) + (1.9484 - 0.68886)$$
$$= 2.7868 + 1.2595$$

(Example 6.3 continued)

or

$$(\Delta v_{\text{total}})_{\text{Hohmann}} = 4.0463 \text{ km/s} \tag{k}$$

This is only slightly (0.44 percent) larger than that of the bi-elliptical transfer. Since the semimajor axis of the Hohmann semi-ellipse is

$$a_5 = \frac{1}{2}(7000 + 105\,000) = 56\,000 \text{ km}$$

the time of flight from A to D is

$$t_{\text{Hohmann}} = \frac{1}{2}\left(\frac{2\pi}{\sqrt{\mu}}a_5^{\frac{3}{2}}\right) = 65\,942 \text{ s} = \underline{0.763 \text{ days}} \tag{l}$$

The time of flight of the bi-elliptical maneuver is over seven times longer than that of the Hohmann transfer.

6.5 Phasing maneuvers

A phasing maneuver is a two-impulse Hohmann transfer from and back to the same orbit, as illustrated in Figure 6.10. The Hohmann transfer ellipse is the phasing orbit with a period selected to return the spacecraft to the main orbit within a specified time. Phasing maneuvers are used to change the position of a spacecraft in its orbit. If two spacecraft, destined to rendezvous, are at different locations in the same orbit, then one of them may perform a phasing maneuver in order to catch the other one. Communications and weather satellites in geostationary earth orbit use phasing maneuvers to move to new locations above the equator. In that case, the rendezvous

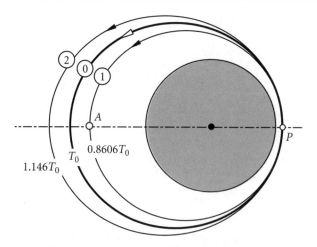

Figure 6.10 Main orbit (0) and two phasing orbits, faster (1) and slower (2). T_0 is the period of the main orbit.

is with an empty point in space rather than with a physical target. In Figure 6.10, phasing orbit 1 might be used to return to *P* in less than one period of the main orbit. This would be appropriate if the target is ahead of the chasing vehicle. Note that a retrofire is required to enter orbit 1 at *P*. That is, it is necessary to slow the spacecraft down in order to speed it up, relative to the main orbit. If the chaser is ahead of the target, then phasing orbit 2 with its longer period might be appropriate. A forward fire of the thruster boosts the spacecraft's speed in order to slow it down.

Once the period *T* of the phasing orbit is established, then Equation 2.73 should be used to determine the semimajor axis of the phasing ellipse,

$$a = \left(\frac{T\sqrt{\mu}}{2\pi}\right)^{\frac{2}{3}} \quad (6.5)$$

With the semimajor axis established, the radius of point *A* opposite to *P* is obtained from the fact that $2a = r_P + r_A$. It is then apparent whether *P* is periapse or apoapse, so that Equation 2.74 can be used to calculate the eccentricity of the phasing orbit. The orbit equation, Equation 2.35, may then be applied at either *P* or *A* to obtain the angular momentum, whereupon the phasing orbit is characterized completely.

EXAMPLE 6.4

Spacecraft at *A* and *B* are in the same orbit (1). At the instant shown, the chaser vehicle at *A* executes a phasing maneuver so as to catch the target spacecraft back at *A* after just one revolution of the chaser's phasing orbit (2). What is the required total delta-v?

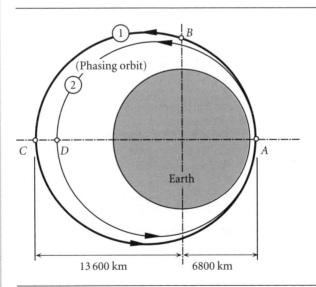

Figure 6.11 Phasing maneuver.

From the figure,

$$r_A = 6800 \text{ km} \qquad r_C = 13\,600 \text{ km}$$

(Example 6.4 continued)

Orbit 1:

The eccentricity of orbit 1 is

$$e_1 = \frac{r_C - r_A}{r_C + r_A} = 0.33333$$

Evaluating the orbit equation at A, we find

$$r_A = \frac{h_1^2}{\mu} \frac{1}{1 + e_1 \cos(0)} \Rightarrow 6800 = \frac{h_1^2}{398\,600} \frac{1}{1 + 0.3333} \Rightarrow h_1 = 60\,116\,\text{km}^2/\text{s}$$

The period is found using Equation 2.72,

$$T_1 = \frac{2\pi}{\mu^2} \left(\frac{h_1}{\sqrt{1-e_1^2}}\right)^3 = \frac{2\pi}{398\,600^2} \left(\frac{60\,116}{\sqrt{1-0.33333^2}}\right)^3 = 10\,252\,\text{s}$$

Since A is perigee, there is no radial velocity component there. The speed, directed entirely in the transverse direction, is found from the angular momentum formula,

$$v_{A1} = \frac{h_1}{r_A} = \frac{60\,116}{6800} = 8.8406\,\text{km/s}$$

The phasing orbit must have a period T_2 equal to the time it takes the target vehicle at B to coast around to point A on orbit 1. We can determine the flight time by calculating the time Δt_{AB} from A to B and subtracting that result from the period T_1 of orbit 1. At B the true anomaly is $\theta_A = 90°$. Therefore, according to Equation 3.10a,

$$\tan\frac{E_B}{2} = \sqrt{\frac{1-e_1}{1+e_1}} \tan\frac{\theta_B}{2} = \sqrt{\frac{1-0.33333}{1+0.33333}} \tan\frac{90°}{2}$$

$$= 0.70711 \Rightarrow E_B = 1.2310\,\text{rad}$$

Then, from Kepler's equation (Equations 3.5 and 3.11), we get

$$\Delta t_{AB} = \frac{T_1}{2\pi}(E_B - e_1 \sin E_B) = \frac{10\,252}{2\pi}(1.231 - 0.33333 \cdot \sin 1.231) = 1495.7\,\text{s}$$

Thus, the time of flight of the target spacecraft from B to A is

$$\Delta t_{BA} = T_1 - \Delta t_{AB} = 10\,252 - 1495.7 = 8756.3\,\text{s}$$

Orbit 2:

The period of orbit 2 must equal Δt_{BA} so that the chaser will arrive at A when the target does. That is,

$$T_2 = 8756.3\,\text{s}$$

This, together with the period formula, Equation 2.73, yields the semimajor axis of orbit 2,

$$T_2 = \frac{2\pi}{\sqrt{\mu}} a_2^{\frac{3}{2}} \Rightarrow 8756.2 = \frac{2\pi}{\sqrt{398\,600}} a_2^{\frac{3}{2}} \Rightarrow a_2 = 9182.1\,\text{km} \quad \text{(a)}$$

Since $2a_2 = r_A + r_D$, we find

$$r_D = 2a_2 - r_A = 2 \cdot 9182.1 - 6800 = 11\,564\,\text{km}$$

Therefore, point A is indeed the perigee of orbit 2, the eccentricity of which can now be determined:

$$e_2 = \frac{r_D - r_A}{r_D + r_A} = 0.25943$$

Evaluating the orbit equation at point A of orbit 2 yields its angular momentum,

$$r_A = \frac{h_2^2}{\mu} \frac{1}{1 + e_2 \cos(0)} \Rightarrow 6800 = \frac{h_2^2}{398\,600} \frac{1}{1 + 0.25943} \Rightarrow h_2 = 58\,426\,\text{km}^2/\text{s}$$

Finally, we can calculate the speed at perigee of orbit 2,

$$v_{A_2} = \frac{h_2}{r_A} = \frac{58\,426}{6800} = 8.5921\,\text{km/s}$$

At the beginning of the phasing maneuver,

$$\Delta v_A = v_{A_2} - v_{A_1} = 8.5921 - 8.8406 = -0.24851\,\text{km/s}$$

At the end of the phasing maneuver,

$$\Delta v_A = v_{A_1} - v_{A_2} = 8.8406 - 8.5921 = 0.24851\,\text{km/s}$$

The total delta-v, therefore, is

$$\Delta v_{\text{total}} = |-0.24851| + |0.24851| = \underline{0.4970\,\text{km/s}}$$

EXAMPLE 6.5

It is desired to shift the longitude of a GEO satellite 12° westward in three revolutions of its phasing orbit. Calculate the delta-v requirement.

This problem is illustrated in Figure 6.12. It may be recalled from Equations 2.57, 2.58 and 2.59 that the angular velocity of the earth, the radius to GEO and the speed in GEO are, respectively,

$$\omega_E = \omega_{\text{GEO}} = 72.922 \times 10^{-6}\,\text{rad/s}$$
$$r_{\text{GEO}} = 42\,164\,\text{km} \quad \text{(a)}$$
$$v_{\text{GEO}} = 3.0747\,\text{km/s}$$

(Example 6.5 continued)

Let $\Delta\Lambda$ be the change in longitude in radians. Then the period T_2 of the phasing orbit can be obtained from the following formula,

$$\omega_E(3T_2) = 3 \cdot 2\pi + \Delta\Lambda \qquad (b)$$

which states that after three circuits of the phasing orbit, the original position of the satellite will be $\Delta\Lambda$ radians east of P. In other words, the satellite will end up $\Delta\Lambda$ radians west of its original position in GEO, as desired. From (b) we obtain,

$$T_2 = \frac{1}{3}\frac{\Delta\Lambda + 6\pi}{\omega_E} = \frac{1}{3} \cdot \frac{12° \cdot \frac{\pi}{180°} + 6\pi}{72.922 \times 10^{-6}} = 87\,121\text{ s}$$

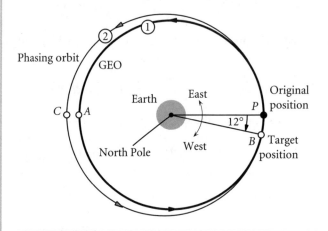

Figure 6.12 GEO repositioning.

Note that the period of GEO is

$$T_{GEO} = \frac{2\pi}{\omega_{GEO}} = 86\,163\text{ s}$$

The satellite in its slower phasing orbit appears to drift westward at the rate

$$\dot{\Lambda} = \frac{\Delta\Lambda}{3T_2} = 8.0133 \times 10^{-7}\text{ rad/s} = 3.9669°/\text{day}$$

Having the period, we can use Equation 6.5 to obtain the semimajor axis of orbit 2,

$$a = \left(\frac{T\sqrt{\mu}}{2\pi}\right)^{\frac{2}{3}} = \left(\frac{87\,121\sqrt{398\,600}}{2\pi}\right)^{\frac{2}{3}} = 42\,476\text{ km}$$

From this we find the radial coordinate of C,

$$2a_2 = r_P + r_C \implies r_C = 2 \cdot 42\,476 - 42\,164 = 42\,787\text{ km}$$

Now we can find the eccentricity of orbit 2,

$$e_2 = \frac{r_C - r_A}{r_C + r_A} = \frac{42\,787 - 42\,164}{42\,787 + 42\,164} = 0.0073395$$

and the angular momentum follows from applying the orbit equation at P (or C) of orbit 2:

$$r_P = \frac{h_2^2}{\mu} \frac{1}{1 + e_2 \cos(0)} \Rightarrow 42\,164 = \frac{h_2^2}{398\,600} \frac{1}{1 + 0.0073395} \Rightarrow h_2 = 130\,120\,\text{km}^2/\text{s}$$

At P the speed in orbit 2 is

$$v_{P_2} = \frac{130\,120}{42\,164} = 3.0859\,\text{km/s}$$

Therefore, at the beginning of the phasing orbit,

$$\Delta v = v_{P_2} - v_{\text{GEO}} = 3.0859 - 3.0747 = 0.01126\,\text{km/s}$$

at the end of the phasing maneuver,

$$\Delta v = v_{\text{GEO}} - v_{P_2} = 3.0747 - 3.08597 = -0.01126\,\text{km/s}$$

Therefore,

$$\Delta v_{\text{total}} = |0.01126| + |-0.01126| = \underline{0.022525\,\text{km/s}}$$

6.6 NON-HOHMANN TRANSFERS WITH A COMMON APSE LINE

Figure 6.13 illustrates a transfer between two coaxial elliptical orbits in which the transfer trajectory shares the apse line but is not necessarily tangent to either the initial or target orbit. The problem is to determine whether there exists such a trajectory joining points A and B, and, if so, to find the total delta-v requirement.

r_A and r_B are given, as are the true anomalies θ_A and θ_B. Because of the common apse line assumption, θ_A and θ_B are the true anomalies of points A and B on the transfer orbit as well. Applying the orbit equation to A and B on orbit 3 yields

$$r_A = \frac{h_3^2}{\mu} \frac{1}{1 + e_3 \cos \theta_A}$$

$$r_B = \frac{h_3^2}{\mu} \frac{1}{1 + e_3 \cos \theta_B}$$

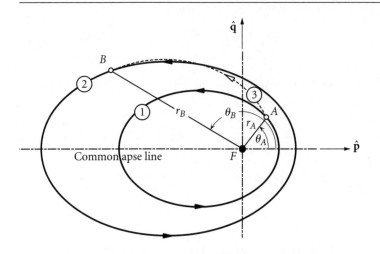

Figure 6.13 Non-Hohmann transfer (3) between two coaxial elliptical orbits.

Solving these two equations for e_3 and h_3, we get

$$e_3 = \frac{r_B - r_A}{r_A \cos\theta_A - r_B \cos\theta_B}$$

$$h_3 = \sqrt{\mu r_A r_B} \sqrt{\frac{\cos\theta_A - \cos\theta_B}{r_A \cos\theta_A - r_B \cos\theta_B}} \quad (6.6)$$

With these, the transfer orbit is determined and velocity may be found at any true anomaly. For a Hohmann transfer, in which $\theta_A = 0$ and $\theta_B = \pi$, Equations 6.6 become

$$e_3 = \frac{r_B - r_A}{r_B + r_A} \qquad h_3 = \sqrt{2\mu}\sqrt{\frac{r_A r_B}{r_A + r_B}} \quad \text{(Hohmann transfer)} \quad (6.7)$$

When a delta-v calculation is done at a point which is not on the apse line, care must be taken to include the change in direction as well as the magnitude of the velocity vector. Figure 6.14 shows a point where an impulsive maneuver changes the velocity vector from $\mathbf{v}_1$ on orbit 1 to $\mathbf{v}_2$ on orbit 2. The difference in length of the two vectors shows the change in the speed, and the difference in the flight path angles indicates the change in the direction. It is important to observe that the Δv we seek is the magnitude of the change in the velocity vector, not the change in its magnitude (speed). That is,

$$\Delta v = \|\mathbf{v}_2 - \mathbf{v}_1\| \quad (6.8)$$

Only if $\mathbf{v}_1$ and $\mathbf{v}_2$ are parallel, as in Hohmann transfers, is it true that $\Delta v = \|\mathbf{v}_2\| - \|\mathbf{v}_1\|$.

From Figure 6.14 and the law of cosines, we find that

$$\Delta v = \sqrt{v_1^2 + v_2^2 - 2v_1 v_2 \cos\Delta\gamma} \quad (6.9)$$

where $v_1 = \|\mathbf{v}_1\|$, $v_2 = \|\mathbf{v}_2\|$ and $\Delta\gamma = \gamma_2 - \gamma_1$.

6.6 Non-Hohmann transfers with a common apse line 275

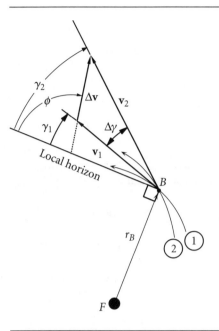

Figure 6.14 Vector diagram of the change in velocity and flight path angle at the intersection of two orbits.

The direction of $\Delta \mathbf{v}$ shows the required alignment of the thruster that produces the impulse. The orientation of $\Delta \mathbf{v}$ relative to the local horizon is found by replacing v_r and $v_\perp$ in Equation 2.41 by Δv_r and $\Delta v_\perp$, so that

$$\tan \phi = \frac{\Delta v_r}{\Delta v_\perp} \qquad (6.10)$$

where ϕ is the angle from the local horizon to the $\Delta \mathbf{v}$ vector.

Finally, recall the formula for specific mechanical energy of an orbit, Equation 2.47,

$$\varepsilon = \frac{\mathbf{v} \cdot \mathbf{v}}{2} - \frac{\mu}{r} \quad (v^2 = \mathbf{v} \cdot \mathbf{v})$$

An impulsive maneuver results in a change of orbit and, therefore, a change in the specific energy ε. If the expenditure of propellant Δm is negligible compared to the initial mass m_1 of the vehicle, then $\Delta \varepsilon = \varepsilon_2 - \varepsilon_1$. For the situation illustrated in Figure 6.14,

$$\varepsilon_1 = \frac{v_1^2}{2} - \frac{\mu}{r_B}$$

and

$$\varepsilon_2 = \frac{(\mathbf{v}_1 + \Delta \mathbf{v}) \cdot (\mathbf{v}_1 + \Delta \mathbf{v})}{2} - \frac{\mu}{r_B} = \frac{v_1^2 + 2\mathbf{v}_1 \cdot \Delta \mathbf{v} + \Delta v^2}{2} - \frac{\mu}{r_B}$$

Hence

$$\Delta\varepsilon = \mathbf{v}_1 \cdot \Delta\mathbf{v} + \frac{\Delta v^2}{2}$$

From Figure 6.14 it is apparent that $\mathbf{v}_1 \cdot \Delta\mathbf{v} = v_1 \Delta v \cos \Delta\gamma$, so that

$$\Delta\varepsilon = v_1 \Delta v \cos \Delta\gamma + \frac{\Delta v^2}{2} = v_1 \Delta v \left(\cos \Delta\gamma + \frac{1}{2} \frac{\Delta v}{v_1} \right)$$

For consistency with our assumption that $\Delta m \ll m_1$, it must be true (recall Figure 6.1) that $\Delta v \ll v_1$. It follows that

$$\Delta\varepsilon \approx v_1 \Delta v \cos \Delta\gamma \qquad (6.11)$$

This shows that, for a given Δv, the change in specific energy is larger the faster the spacecraft is moving (unless, of course, the change in flight path angle is 90°). The larger the $\Delta\varepsilon$ associated with a given Δv, the more efficient the maneuver. As we know, a spacecraft has its greatest speed at periapsis.

EXAMPLE 6.6

A geocentric satellite in orbit 1 of Figure 6.15 executes a delta-v maneuver at A which places it on orbit 2, for re-entry at D. Calculate Δv at A and its direction relative to the local horizon.

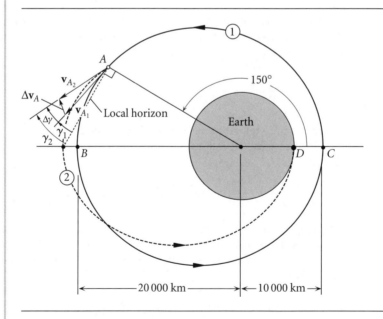

Figure 6.15 Non-Hohmann transfer with a common apse line.

From the figure we see that

$$r_B = 20\,000 \text{ km} \qquad r_C = 10\,000 \text{ km} \qquad r_D = 6378 \text{ km}$$

Orbit 1:

The eccentricity is

$$e_1 = \frac{r_B - r_C}{r_B + r_C} = 0.33333$$

The angular momentum is obtained from the orbit equation, noting that point C is perigee:

$$r_C = \frac{h_1^2}{\mu} \frac{1}{1 + e_1 \cos(0)} \Rightarrow 10\,000 = \frac{h_1^2}{398\,600} \frac{1}{1 + 0.33333} \Rightarrow h_1 = 72\,902 \text{ km}^2/\text{s}$$

With the angular momentum and the eccentricity, we can use the orbit equation to find the radial coordinate of point A,

$$r_A = \frac{72\,902^2}{398\,600} \frac{1}{1 + 0.33333 \cdot \cos 150°} = 18\,744 \text{ km}$$

Equations 2.21 and 2.38 yield the transverse and radial components of velocity at A on orbit 1,

$$v_{\perp_A})_1 = \frac{h_1}{r_A} = 3.8893 \text{ km/s} \quad \text{(a)}$$

$$v_{r_A})_1 = \frac{\mu}{h_1} e_1 \sin 150° = 0.91127 \text{ km/s}$$

From these we find the speed at A

$$v_A)_1 = \sqrt{v_{\perp_A})_1^2 + v_{r_A})_1^2} = 3.9946 \text{ km/s}$$

and the flight path angle,

$$\gamma_1 = \tan^{-1} \frac{v_{r_A})_1}{v_{\perp_A})_1} = \tan^{-1} \frac{0.91127}{3.8893} = 13.187°$$

Orbit 2:

The radius and true anomaly of points A and D on orbit 2 are known. Applying the orbit equation at A, we get

$$18\,744 = \frac{h_2^2}{398\,600} \frac{1}{1 + e_2 \cos 150°} \Rightarrow h_2^2 = 7.4715 \times 10^9 - 6.4705 \times 10^9 e_2 \quad \text{(b)}$$

Likewise, at point D, which is perigee of orbit 2,

$$6378 = \frac{h_2^2}{398\,600} \frac{1}{1 + e_2} \Rightarrow h_2^2 = 2.5423 \times 10^9 + 2.5423 \times 10^9 e_2 \quad \text{(c)}$$

Equating the expressions for h_2^2 in (b) and (c), and solving for e_2, yields

$$e_2 = 0.54692$$

(Example 6.6 continued)

whereupon either (b) or (c) may be used to find

$$h_2 = 62\,711 \text{ km}^2/\text{s}$$

Now we can calculate the radial and perpendicular components of velocity on orbit 2 at point A:

$$v_{\perp_A})_2 = \frac{h_2}{r_A} = 3.3456 \text{ km/s}$$

$$v_{r_A})_2 = \frac{\mu}{h_2}e_2 \sin 150° = 1.7381 \text{ km/s} \qquad (d)$$

Hence, the speed and flight path angle at A on orbit 2 are

$$v_A)_2 = \sqrt{v_{\perp_A})_2^2 + v_{r_A})_2^2} = 3.7702 \text{ km/s}$$

$$\gamma_2 = \tan^{-1}\frac{v_{r_A})_2}{v_{\perp_A})_2} = \tan^{-1}\frac{1.7381}{3.3456} = 27.453°$$

The change in the flight path angle as a result of the impulsive maneuver is

$$\Delta\gamma = \gamma_2 - \gamma_1 = 27.453° - 13.187° = 14.266°$$

With this we can use Equation 6.9 to finally obtain Δv_A,

$$\Delta v_A = \sqrt{v_A)_1^2 + v_A)_2^2 - 2v_A)_1 v_A)_2 \cos \Delta\gamma}$$

$$= \sqrt{3.9946^2 + 3.7702^2 - 2 \cdot 3.9946 \cdot 3.7702 \cdot \cos 14.266}$$

$$\Delta v_A = 0.9896 \text{ km/s} \qquad (e)$$

Note that Δv_A is the magnitude of the change in velocity vector $\Delta \mathbf{v}_A$ at A. That is not the same as the change in the magnitude of the velocity (i.e., the change in speed), which is $v_A)_2 - v_A)_1 = 3.9946 - 3.7702 = 0.2244$ km/s.

To find the orientation of $\Delta\mathbf{v}_A$, we use Equation 6.10,

$$\tan\phi = \frac{\Delta v_r)_A}{\Delta v_\perp)_A} = \frac{v_{r_A})_2 - v_{r_A})_1}{v_{\perp_A})_2 - v_{\perp_A})_1} = \frac{1.7381 - 0.9113}{3.3456 - 3.8893} = -1.5207$$

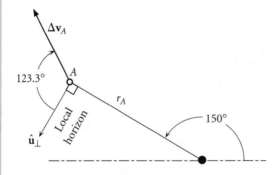

Figure 6.16 Orientation of Δv_A to the local horizon.

so that

$$\phi = 123.3°$$

This angle is illustrated in Figure 6.16. Prior to firing, the spacecraft would have to be rotated so that the centerline of the rocket motor coincides with the line of action of $\Delta \mathbf{v}_A$, with the nozzle aimed in the opposite direction.

6.7 APSE LINE ROTATION

Figure 6.17 shows two intersecting orbits which have a common focus, but their apse lines are not collinear. A Hohmann transfer between them is clearly impossible. The opportunity for transfer from one orbit to the other by a single impulsive maneuver occurs where they intersect, at points I and J in this case. As can be seen from the figure, the rotation η of the apse line is the difference between the true anomalies of the point of intersection, measured from periapse of each orbit. That is,

$$\eta = \theta_1 - \theta_2 \tag{6.12}$$

We will consider two cases of apse line rotation.

The first case is that in which the apse line rotation η is given as well as the orbital parameters e and h of both orbits. The problem is then to find the true anomaly of I and J relative to both orbits. The radius of the point of intersection I is given by either of the following

$$r_I)_1 = \frac{h_1^2}{\mu} \frac{1}{1 + e_1 \cos \theta_1} \qquad r_I)_2 = \frac{h_2^2}{\mu} \frac{1}{1 + e_2 \cos \theta_2}$$

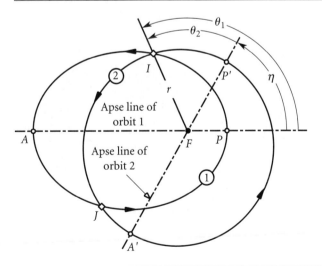

Figure 6.17 Two intersecting orbits whose apse lines do not coincide.

Since $r_I)_1 = r_I)_2$, we can equate these two expressions and rearrange terms to get

$$e_1 h_2^2 \cos\theta_1 - e_2 h_1^2 \cos\theta_2 = h_1^2 - h_2^2$$

Setting $\theta_2 = \theta_1 - \eta$ and using the trig identity $\cos(\theta_1 - \eta) = \cos\theta_1 \cos\eta + \sin\theta_1 \sin\eta$ leads to an equation for θ_1

$$a \cos\theta_1 + b \sin\theta_1 = c \qquad (6.13a)$$

where

$$a = e_1 h_2^2 - e_2 h_1^2 \cos\eta \qquad b = -e_2 h_1^2 \sin\eta \qquad c = h_1^2 - h_2^2 \qquad (6.13b)$$

Equation 6.13a has two roots (see Problem 3.9), corresponding to the two points of intersection I and J of the two orbits:

$$\theta_1 = \phi \pm \cos^{-1}\left(\frac{c}{a}\cos\phi\right) \qquad (6.14a)$$

where

$$\phi = \tan^{-1}\frac{b}{a} \qquad (6.14b)$$

Having found θ_1 we obtain θ_2 from Equation 6.12. Δv for the impulsive maneuver may then be computed as illustrated in the following example.

EXAMPLE 6.7

An earth satellite is in an 8000 km by 16 000 km radius orbit (orbit 1 of Figure 6.18). Calculate the delta-v and the true anomaly θ_1 required to obtain a 7000 km by 21 000 km radius orbit (orbit 2) whose apse line is rotated 25° counterclockwise. Indicate the orientation ϕ of $\Delta\mathbf{v}$ to the local horizon.

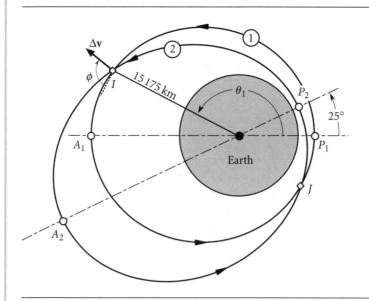

Figure 6.18 $\Delta\mathbf{v}$ produces a rotation of the apse line.

6.7 Apse line rotation

The eccentricities of the two orbits are

$$e_1 = \frac{r_{A_1} - r_{P_1}}{r_{A_1} + r_{P_1}} = \frac{16\,000 - 8000}{16\,000 + 8000} = 0.33333$$

$$e_2 = \frac{r_{A_2} - r_{P_2}}{r_{A_2} + r_{P_2}} = \frac{21\,000 - 7000}{21\,000 + 7000} = 0.5 \quad \text{(a)}$$

The orbit equation yields the angular momenta

$$r_{P_1} = \frac{h_1^2}{\mu} \frac{1}{1 + e_1 \cos(0)} \Rightarrow 8000 = \frac{h_1^2}{398\,600} \frac{1}{1 + 0.33333} \Rightarrow h_1 = 65\,205\,\text{km}^2/\text{s}$$

$$r_{P_2} = \frac{h_2^2}{\mu} \frac{1}{1 + e_2 \cos(0)} \Rightarrow 7000 = \frac{h_2^2}{398\,600} \frac{1}{1 + 0.5} \Rightarrow h_2 = 64\,694 \quad \text{(b)}$$

Using these orbital parameters and the fact that $\eta = 25°$, we calculate the terms in Equations 6.13b,

$$a = e_1 h_2^2 - e_2 h_1^2 \cos\eta = 0.3333 \cdot 64\,694^2 - 0.5 \cdot 65\,205^2 \cdot \cos 25°$$
$$= -5.3159 \times 10^8 \,\text{km}^4/\text{s}^2$$

$$b = -e_2 h_1^2 \sin\eta = -0.5 \cdot 65\,205^2 \sin 25° = -8.9843 \times 10^8 \,\text{km}^4/\text{s}^2$$

$$c = h_1^2 - h_2^2 = 65\,205^2 - 64\,694^2 = 6.6433 \times 10^7 \,\text{km}^4/\text{s}^2$$

Then Equations 6.14 yield

$$\phi = \tan^{-1} \frac{-8.9843 \times 10^8}{-5.3159 \times 10^8} = 59.39°$$

$$\theta_1 = 59.39° \pm \cos^{-1}\left(\frac{6.6433 \times 10^7}{-5.3159 \times 10^8} \cos 59.39°\right) = 59.39° \pm 93.65°$$

Thus, the true anomaly of point I, the point of interest, is

$$\theta_1 = 153.04° \quad \text{(c)}$$

(For point J, $\theta_1 = 325.74°$.)

With the true anomaly available, we can evaluate the radial coordinate of the maneuver point,

$$r = \frac{h_1^2}{\mu} \frac{1}{1 + e_1 \cos 153.04°} = 15\,175\,\text{km}$$

The velocity components and flight path angle for orbit 1 at point I are

$$v_{\perp_1} = \frac{h_1}{r} = \frac{65\,205}{15\,175} = 4.2968\,\text{km/s}$$

$$v_{r_1} = \frac{\mu}{h_1} e_1 \sin 153.04° = \frac{398\,600}{65\,205} \cdot 0.33333 \cdot \sin 153.04° = 0.92393\,\text{km/s}$$

$$\gamma_1 = \tan^{-1} \frac{v_{r_1}}{v_{\perp_1}} = 12.135°$$

(Example 6.7 continued)

The speed of the satellite in orbit 1 is, therefore,

$$v_1 = \sqrt{v_{r_1}^2 + v_{\perp_1}^2} = 4.3950 \text{ km/s}$$

Likewise, for orbit 2,

$$v_{\perp_2} = \frac{h_2}{r} = \frac{64\,694}{15\,175} = 4.2631 \text{ km/s}$$

$$v_{r_2} = \frac{\mu}{h_2} e_2 \sin(153.04° - 25°) = \frac{398\,600}{64\,694} \cdot 0.5 \cdot \sin 128.04° = 2.4264 \text{ km/s}$$

$$\gamma_2 = \tan^{-1} \frac{v_{r_2}}{v_{\perp_2}} = 29.647°$$

$$v_2 = \sqrt{v_{r_2}^2 + v_{\perp_2}^2} = 4.9053 \text{ km/s}$$

Equation 6.9 is used to find Δv,

$$\Delta v = \sqrt{v_1^2 + v_2^2 - 2v_1 v_2 \cos(\gamma_2 - \gamma_1)}$$

$$= \sqrt{4.3950^2 + 4.9053^2 - 2 \cdot 4.3950 \cdot 4.9053 \cos(29.647° - 12.135°)}$$

$$\underline{\Delta v = 1.503 \text{ km/s}}$$

The angle ϕ which the vector $\Delta \mathbf{v}$ makes with the local horizon is given by Equation 6.10,

$$\phi = \tan^{-1} \frac{\Delta v_r}{\Delta v_\perp} = \tan^{-1} \frac{v_{r_2} - v_{r_1}}{v_{\perp_2} - v_{\perp_1}} = \tan^{-1} \frac{2.4264 - 0.92393}{4.2631 - 4.2968} = \underline{91.28°}$$

The second case of apse line rotation is that in which the impulsive maneuver takes place at a given true anomaly θ_1 on orbit 1. The problem is to determine the angle of rotation η and the eccentricity e_2 of the new orbit.

The impulsive maneuver creates a change in the radial and transverse velocity components at point I of orbit 1. From the angular momentum formula, $h = rv_\perp$, we obtain the angular momentum of orbit 2,

$$h_2 = r(v_\perp + \Delta v_\perp) = h_1 + r\Delta v_\perp \tag{6.15}$$

The formula for radial velocity, $v_r = (\mu/h)e \sin \theta$, applied to orbit 2 at point I, where

$$v_{r_2} = v_{r_1} + \Delta v_r \quad \text{and} \quad \theta_2 = \theta_1 - \eta, \quad \text{yields}$$

$$v_{r_1} + \Delta v_r = \frac{\mu}{h_2} e_2 \sin \theta_2$$

6.7 Apse line rotation

Substituting Equation 6.15 into this expression and solving for $\sin\theta_2$ leads to

$$\sin\theta_2 = \frac{1}{e_2}\frac{(h_1+r\Delta v_\perp)(\mu e_1 \sin\theta_1 + h_1\Delta v_r)}{\mu h_1} \qquad (6.16)$$

From the orbit equation, we have at point I

$$r = \frac{h_1^2}{\mu}\frac{1}{1+e_1\cos\theta_1} \quad \text{(orbit 1)}$$

$$r = \frac{h_2^2}{\mu}\frac{1}{1+e_2\cos\theta_2} \quad \text{(orbit 2)}$$

Equating these two expressions for r, substituting Equation 6.15, and solving for $\cos\theta_2$, yields

$$\cos\theta_2 = \frac{1}{e_2}\frac{(h_1+r\Delta v_\perp)^2 e_1 \cos\theta_1 + (2h_1+r\Delta v_\perp)r\Delta v_\perp}{h_1^2} \qquad (6.17)$$

Finally, substituting Equations 6.16 and 6.17 into the trig identity $\tan\theta_2 = \sin\theta_2/\cos\theta_2$, we obtain

$$\tan\theta_2 = \frac{h_1}{\mu}\frac{(h_1+r\Delta v_\perp)(\mu e_1 \sin\theta_1 + h_1\Delta v_r)}{(h_1+r\Delta v_\perp)^2 e_1 \cos\theta_1 + (2h_1+r\Delta v_\perp)r\Delta v_\perp} \qquad (6.18a)$$

Equation 6.18a can be simplified a bit by replacing $\mu e_1 \sin\theta_1$ with $h_1 v_{r_1}$ and h_1 with $rv_{\perp_1}$, so that

$$\tan\theta_2 = \frac{(v_{\perp_1}+\Delta v_\perp)(v_{r_1}+\Delta v_r)}{(v_{\perp_1}+\Delta v_\perp)^2 e_1 \cos\theta_1 + (2v_{\perp_1}+\Delta v_\perp)\Delta v_\perp}\frac{v_{\perp_1}^2}{(\mu/r)} \qquad (6.18b)$$

Equations 6.18 show how the apse line rotation, $\eta = \theta_1 - \theta_2$, is completely determined by the components of $\Delta\mathbf{v}$ imparted at the true anomaly θ_1.

After solving Equation 6.18 (a or b), we substitute θ_2 into Equation 6.16 or 6.17 to calculate the eccentricity e_2 of orbit 2. Therefore, with h_2 from Equation 6.15, the rotated orbit 2 is completely specified.

If the impulsive maneuver takes place at the periapse of orbit 1, so that $\theta_1 = v_r = 0$, and if it is also true that $\Delta v_\perp = 0$, then Equation 6.18b yields

$$\tan\eta = -\frac{rv_{\perp_1}}{\mu e_1}\Delta v_r \quad \text{(with radial impulse at periapse)}$$

Thus, if the velocity vector is given an outward radial component at periapse, then $\eta < 0$, which means the apse line of the resulting orbit is rotated clockwise relative to the original one. That makes sense, since having acquired $v_r > 0$ means the spacecraft is now flying away from its new periapse. Likewise, applying an inward radial velocity component at periapse rotates the apse line counterclockwise.

EXAMPLE 6.8

An earth satellite in orbit 1 of Figure 6.19 undergoes the indicated delta-v maneuver at its perigee. Determine the rotation η of its apse line.

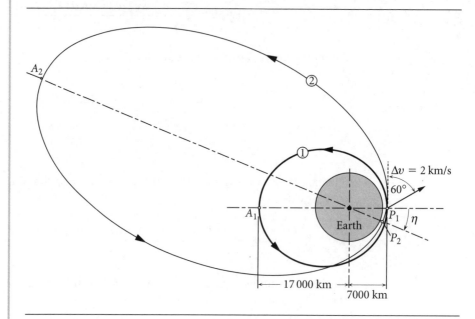

Figure 6.19 Apse line rotation maneuver.

From the figure

$$r_{A_1} = 17\,000\,\text{km} \qquad r_{P_1} = 7000\,\text{km}$$

The eccentricity of orbit 1 is

$$e_1 = \frac{r_{A_1} - r_{P_1}}{r_{A_1} + r_{P_1}} = 0.41667 \qquad (a)$$

As usual, we use the orbit equation to find the angular momentum,

$$r_{P_1} = \frac{h_1^2}{\mu} \frac{1}{1 + e_1 \cos(0)} \Rightarrow 7000 = \frac{h_1^2}{398\,600} \frac{1}{1 + 0.41667} \Rightarrow h_1 = 62\,871\,\text{km}^2/\text{s}$$

At the maneuver point P_1, the angular momentum formula and the fact that P_1 is perigee of orbit 1 ($\theta_1 = 0$) imply that

$$v_{\perp_1} = \frac{h_1}{r_{P_1}} = \frac{62\,871}{7000} = 8.9816\,\text{km/s}$$

$$v_{r_1} = 0 \qquad (b)$$

From Figure 6.18 it is clear that

$$\Delta v_\perp = \Delta v \cos 60° = 1\,\text{km/s}$$
$$\Delta v_r = \Delta v \sin 60° = 1.7321\,\text{km/s} \qquad (c)$$

To compute θ_2, we use Equation 6.18b together with (a), (b) and (c):

$$\tan \theta_2 = \frac{(v_{\perp_1} + \Delta v_\perp)(v_{r_1} + \Delta v_r)}{(v_{\perp_1} + \Delta v_\perp)^2 e_1 \cos \theta_1 + (2v_{\perp_1} + \Delta v_\perp)\Delta v_\perp} \frac{v_{\perp_1}^2}{(\mu/r_{P_1})}$$

$$= \frac{(8.9816 + 1)(0 + 1.7321)}{(8.9816 + 1)^2 \cdot 0.41667 \cdot \cos(0) + (2 \cdot 8.9816 + 1) \cdot 1} \cdot \frac{8.9816^2}{(398\,600/7000)}$$

$$= 0.4050$$

It follows that $\theta_2 = 22.047°$, so that Equation 6.12 yields

$$\eta = -22.05°$$

which means the rotation of the apse line is clockwise, as indicated in Figure 6.19.

From Equation 6.17 we obtain the eccentricity of orbit 2,

$$e_2 = \frac{(h_1 + r_{P_1}\Delta v_\perp)^2 e_1 \cos \theta_1 + (2h_1 + r_{P_1}\Delta v_\perp)r_{P_1}\Delta v_\perp}{h_1^2 \cos \theta_2}$$

$$= \frac{(62\,871 + 7000 \cdot 1)^2 \cdot 0.41667 \cdot \cos(0) + (2 \cdot 62\,871 + 7000 \cdot 1) \cdot 7000 \cdot 1}{62\,871^2 \cdot \cos 22.047°}$$

$$= 0.808830$$

With this and the angular momentum we find using the orbit equation that the perigee and apogee radii of orbit 2 are

$$r_{P_2} = \frac{h_2^2}{\mu} \frac{1}{1 + e_2} = \frac{69\,871^2}{398\,600} \frac{1}{1 + 0.808830} = 6771.1 \text{ km}$$

$$r_{A_2} = \frac{69\,871^2}{398\,600} \frac{1}{1 - 0.808830} = 64\,069 \text{ km}$$

6.8 CHASE MANEUVERS

Whereas Hohmann transfers and phasing maneuvers are leisurely, energy-efficient procedures that require some preconditions (e.g., coaxial elliptical, orbits) in order to work, a chase or intercept trajectory is one which answers the question, 'How do I get from point A to point B in space in a given amount of time?' The nature of the orbit lies in the answer to the question rather than being prescribed at the outset. Intercept trajectories near a planet are likely to require delta-vs beyond the capabilities of today's technology, so they are largely of theoretical rather than practical interest. We might refer to them as 'star wars maneuvers.' Chase trajectories can be found as solutions to Lambert's problem (Section 5.3).

Example 6.9

Spacecraft B and C are both in the geocentric elliptical orbit (1) shown in Figure 6.20, from which it can be seen that the true anomalies are $\theta_B = 45°$ and $\theta_C = 150°$. At the instant shown, spacecraft B executes a delta-v maneuver, embarking upon a trajectory (2) which will intercept vehicle C in precisely one hour. Find the orbital parameters (e and h) of the intercept trajectory and the total delta-v required for the chase maneuver.

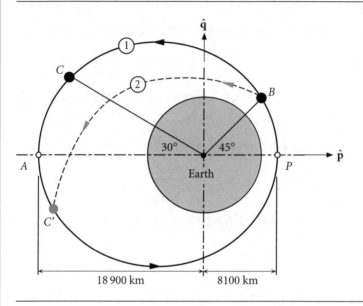

Figure 6.20 Intercept trajectory (2) required for B to catch C in one hour.

First, we must determine the parameters of orbit 1 in the usual way. The eccentricity is found using the orbit's perigee and apogee, shown in Figure 6.20,

$$e_1 = \frac{18\,900 - 8100}{18\,900 + 8100} = 0.4000$$

From the orbit equation,

$$r_P = \frac{h_1^2}{\mu} \frac{1}{1 + e_1 \cos(0)} \Rightarrow 8100 = \frac{h_1^2}{398\,600} \frac{1}{1 + 0.4} \Rightarrow h_1 = 67\,232 \text{ km}^2/\text{s}$$

Using Equation 2.72 yields the period,

$$T_1 = \frac{2\pi}{\mu^2}\left(\frac{h_1}{\sqrt{1 - e_1^2}}\right)^3 = \frac{2\pi}{398\,600^2}\left(\frac{67\,232}{\sqrt{1 - 0.4^2}}\right)^3 = 15\,610 \text{ s}$$

In perifocal coordinates (Equation 2.109) the position vector of B is

$$\mathbf{r}_B = \frac{h_1^2}{\mu} \frac{1}{1 + e_1 \cos\theta_B} (\cos\theta_B \hat{\mathbf{p}} + \sin\theta_B \hat{\mathbf{q}})$$

$$= \frac{67\,232^2}{398\,600} \frac{1}{1 + 0.4 \cos 45°} (\cos 45° \hat{\mathbf{p}} + \sin 45° \hat{\mathbf{q}})$$

or

$$\mathbf{r}_B = 6250.6 \hat{\mathbf{p}} + 6250.6 \hat{\mathbf{q}} \text{ (km)} \quad (a)$$

Likewise, according to Equation 2.115, the velocity at B on orbit 1 is

$$\mathbf{v}_{B_1} = \frac{\mu}{h}[-\sin\theta_B \hat{\mathbf{p}} + (e + \cos\theta_B)\hat{\mathbf{q}}] = \frac{398\,600}{67\,232}[-\sin 45° \hat{\mathbf{p}} + (0.4 + \cos 45°)\hat{\mathbf{q}}]$$

so that

$$\mathbf{v}_{B_1} = -4.1922 \hat{\mathbf{p}} + 6.5637 \hat{\mathbf{q}} \text{ (km/s)} \quad (b)$$

Now we need to move spacecraft C along orbit 1 to the position C' that it will occupy one hour later (Δt), when it will presumably be met by spacecraft B. To do that, we must first calculate the time since perigee passage at C. Since we know the true anomaly, the eccentric anomaly follows from Equation 3.10a,

$$\tan\frac{E_C}{2} = \sqrt{\frac{1-e_1}{1+e_1}} \tan\frac{\theta_C}{2} = \sqrt{\frac{1-0.4}{1+0.4}} \tan\frac{150°}{2} = 2.4432 \Rightarrow E_C = 2.3646 \text{ rad}$$

Substituting this value into Kepler's equation (Equations 3.5 and 3.11) yields the time since perigee passage,

$$t_C = \frac{T_1}{2\pi}(E_C - e_1 \sin E_C) = \frac{15\,610}{2\pi}(2.3646 - 0.4 \cdot \sin 2.3646) = 5178 \text{ s}$$

One hour later ($\Delta t = 3600$ s), the spacecraft will be in intercept position at C',

$$t_{C'} = t_C + \Delta t = 5178 + 3600 = 8778 \text{ s}$$

The corresponding mean anomaly is

$$M_e)_{C'} = 2\pi \frac{t_{C'}}{T_1} = 2\pi \frac{8778}{15\,610} = 3.5331 \text{ rad}$$

With this value of the mean anomaly, Kepler's equation becomes

$$E_{C'} - e_1 \sin E_{C'} = 3.5331$$

Applying Algorithm 4.1 to the solution of this equation we get

$$E_{C'} = 3.4223 \text{ rad}$$

Substituting this result into Equation 3.10a yields the true anomaly at C',

$$\tan\frac{\theta_{C'}}{2} = \sqrt{\frac{1+0.4}{1-0.4}} \tan\frac{3.4223}{2} = -10.811 \Rightarrow \theta_{C'} = 190.57°$$

(Example 6.9 continued)

We are now able to calculate the perifocal position and velocity vectors at C' on orbit 1:

$$\mathbf{r}_{C'} = \frac{67\,232^2}{398\,600} \frac{1}{1 + 0.4 \cos 190.57°} (\cos 190.57° \hat{\mathbf{p}} + \sin 4190.57° \hat{\mathbf{q}})$$

$$= -18\,372\hat{\mathbf{p}} - 3428.1\hat{\mathbf{q}} \text{ (km)}$$

$$\mathbf{v}_{C'_1} = \frac{398\,600}{67\,232} [-\sin 190.57° \hat{\mathbf{p}} + (0.4 + \cos 190.57°)\hat{\mathbf{q}}]$$

$$= 1.0875\hat{\mathbf{p}} - 3.4566\hat{\mathbf{q}} \text{ (km/s)} \tag{c}$$

The intercept trajectory connecting points B and C' are found by solving Lambert's problem. Substituting $\mathbf{r}_B$ and $\mathbf{r}_{C'}$ along with $\Delta t = 3600$ s into Algorithm 5.2 yields

$$\mathbf{v}_{B_2} = -8.1349\hat{\mathbf{p}} + 4.0506\hat{\mathbf{q}} \text{ (km/s)} \tag{d}$$

$$\mathbf{v}_{C'_2} = -3.4745\hat{\mathbf{p}} - 4.7943\hat{\mathbf{q}} \text{ (km/s)} \tag{e}$$

These velocities are most easily obtained by running the following MATLAB script, which executes Algorithm 5.2 by means of the function M-file lambert.m (Appendix D.11).

```
clear
global mu
    deg = pi/180;
     mu = 398600;
      e = 0.4;
      h = 67232;
 theta1 = 45*deg;
 theta2 = 190.57*deg;
delta_t = 3600;
rB = h^2/mu/(1 + e*cos(theta1))...
                *[cos(theta1),sin(theta1),0];
rC_prime = h^2/mu/(1 + e*cos(theta2))...
                *[cos(theta2),sin(theta2),0];
string = 'pro';
[vB2 vC_prime_2] = lambert(rB, rC_prime,...
                    delta_t, string)
```

From (b) and (d) we find

$$\Delta \mathbf{v}_B = \mathbf{v}_{B_2} - \mathbf{v}_{B_1} = -3.9426\hat{\mathbf{p}} - 2.5132\hat{\mathbf{q}} \text{ (km/s)}$$

whereas (c) and (e) yield

$$\Delta \mathbf{v}_{C'} = \mathbf{v}_{C'_1} - \mathbf{v}_{C'_2} = 4.5620\hat{\mathbf{p}} + 1.3376\hat{\mathbf{q}} \text{ (km/s)}$$

The anticipated, extremely large, delta-v requirement for this chase maneuver is the sum of the magnitudes of these two vectors,

$$\Delta v = ||\Delta \mathbf{v}_B|| + ||\Delta \mathbf{v}_{C'}|| = 4.6755 + 4.7540 = \underline{9.430 \text{ km/s}}$$

We know that orbit 2 is an ellipse. To pin it down a bit more, we can use $\mathbf{r}_B$ and $\mathbf{v}_{B_2}$ to obtain the orbital elements from Algorithm 4.1, which yields

$$\underline{h_2 = 76\,167 \text{ km}^2/\text{s}}$$

$$\underline{e_2 = 0.8500}$$

$$\underline{a_2 = 52\,449 \text{ km}}$$

$$\underline{\theta_{B_2} = 319.52°}$$

These may be found quickly by running the following MATLAB script, in which the M-function coe_from_sv.m is Algorithm 4.1 (see Appendix D.8):

```
clear
global mu
 mu = 398600;
 rB = [6250.6 6250.6 0];
 vB2 = [-8.1349 4.0506 0];
 orbital_elements = coe_from_sv(rB, vB2)
```

The details of the intercept trajectory and the delta-v maneuvers are shown in Figure 6.21. A far less dramatic though more leisurely (and realistic) way for B to catch up with C would be to use a phasing maneuver.

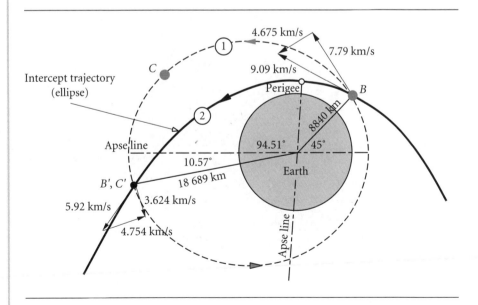

Figure 6.21 Details of the large elliptical orbit, a portion of which serves as the intercept trajectory.

6.9 PLANE CHANGE MANEUVERS

Orbits having a common focus F need not, and generally do not, lie in a common plane. Figure 6.22 shows two such orbits and their line of intersection BD. A and P denote the apoapses and periapses. Since the common focus lies in every orbital plane, it must lie on the line of intersection of any two orbits. For a spacecraft in orbit 1 to change its plane to that of orbit 2 by means of a single delta-v maneuver (cranking maneuver), it must do so when it is on the line of intersection of the orbital planes. Those two opportunities occur only at points B and D in Figure 6.22(a).

A view down the line of intersection, from B towards D, is shown in Figure 6.22(b). Here we can see in true view the dihedral angle δ between the two planes. The transverse component of velocity $\mathbf{v}_\perp$ at B is evident in this perspective, whereas the radial component $\mathbf{v}_r$, lying as it does on the line of intersection, is normal to the view plane (thus appearing as a dot). It is apparent that changing the plane of orbit 1 requires simply rotating $\mathbf{v}_\perp$ around the intersection line, through the dihedral angle. If the magnitudes of $\mathbf{v}_\perp$ and $\mathbf{v}_r$ remain unchanged in the process, then we have a rigid body rotation of the orbit. That is, except for its new orientation in space, the orbit remains unchanged. If the magnitudes of $\mathbf{v}_r$ and $\mathbf{v}_\perp$ change in the process, then the rotated orbit acquires a new size and shape.

To find the delta-v associated with a plane change, let $\mathbf{v}_1$ be the velocity before and $\mathbf{v}_2$ the velocity after the impulsive maneuver. Then

$$\mathbf{v}_1 = v_{r_1}\hat{\mathbf{u}}_r + v_{\perp_1}\hat{\mathbf{u}}_{\perp_1}$$
$$\mathbf{v}_2 = v_{r_2}\hat{\mathbf{u}}_r + v_{\perp_2}\hat{\mathbf{u}}_{\perp_2}$$

where $\hat{\mathbf{u}}_r$ is the radial unit vector directed along the line of intersection of the two orbital planes. $\hat{\mathbf{u}}_r$ does not change during the maneuver. As we know, the transverse

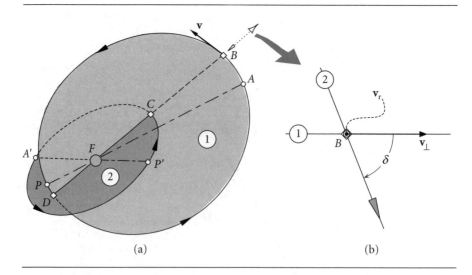

Figure 6.22 (a) Two non-coplanar orbits about F. (b) A view down the line of intersection of the two orbital planes.

6.9 Plane change maneuvers

unit vector $\hat{\mathbf{u}}_\perp$ is perpendicular to $\hat{\mathbf{u}}_r$ and lies in the orbital plane. Therefore it rotates through the dihedral angle δ from its initial orientation $\hat{\mathbf{u}}_{\perp_1}$ to its final orientation $\hat{\mathbf{u}}_{\perp_2}$.

The change $\Delta \mathbf{v}$ in the velocity vector is

$$\Delta \mathbf{v} = \mathbf{v}_2 - \mathbf{v}_1 = (v_{r_2} - v_{r_1})\hat{\mathbf{u}}_r + v_{\perp_2}\hat{\mathbf{u}}_{\perp_2} - v_{\perp_1}\hat{\mathbf{u}}_{\perp_1}$$

Δv is found by taking the dot product of $\Delta \mathbf{v}$ with itself,

$$\Delta v^2 = \Delta \mathbf{v} \cdot \Delta \mathbf{v}$$
$$= \left[(v_{r_2} - v_{r_1})\hat{\mathbf{u}}_r + v_{\perp_2}\hat{\mathbf{u}}_{\perp_2} - v_{\perp_1}\hat{\mathbf{u}}_{\perp_1} \right] \cdot \left[(v_{r_2} - v_{r_1})\hat{\mathbf{u}}_r + v_{\perp_2}\hat{\mathbf{u}}_{\perp_2} - v_{\perp_1}\hat{\mathbf{u}}_{\perp_1} \right]$$

Carrying out the dot products while noting that $\hat{\mathbf{u}}_r \cdot \hat{\mathbf{u}}_r = \hat{\mathbf{u}}_{\perp_1} \cdot \hat{\mathbf{u}}_{\perp_1} = \hat{\mathbf{u}}_{\perp_2} \cdot \hat{\mathbf{u}}_{\perp_2} = 1$ and $\hat{\mathbf{u}}_r \cdot \hat{\mathbf{u}}_{\perp_1} = \hat{\mathbf{u}}_r \cdot \hat{\mathbf{u}}_{\perp_2} = 0$, yields

$$\Delta v^2 = (v_{r_2} - v_{r_1})^2 + v_{\perp_1}^2 + v_{\perp_2}^2 - 2v_{\perp_1}v_{\perp_2}(\hat{\mathbf{u}}_{\perp_1} \cdot \hat{\mathbf{u}}_{\perp_2})$$

But $\hat{\mathbf{u}}_{\perp_1} \cdot \hat{\mathbf{u}}_{\perp_2} = \cos \delta$, so that we finally obtain a general formula for Δv with plane change,

$$\Delta v = \sqrt{(v_{r_2} - v_{r_1})^2 + v_{\perp_1}^2 + v_{\perp_2}^2 - 2v_{\perp_1}v_{\perp_2}\cos\delta} \quad (6.19)$$

From the definition of the flight path angle (cf. Figure 2.11),

$$v_{r_1} = v_1 \sin \gamma_1 \qquad v_{\perp_1} = v_1 \cos \gamma_1$$
$$v_{r_2} = v_2 \sin \gamma_2 \qquad v_{\perp_2} = v_2 \cos \gamma_2$$

Substituting these relations into Equation 6.19, expanding and collecting terms, and using the trig identities

$$\sin^2 \gamma_1 + \cos^2 \gamma_1 = \sin^2 \gamma_2 + \cos^2 \gamma_2 = 1$$
$$\cos(\gamma_2 - \gamma_1) = \cos \gamma_2 \cos \gamma_1 + \sin \gamma_2 \sin \gamma_1$$

leads to another version of the same equation,

$$\Delta v = \sqrt{v_1^2 + v_2^2 - 2v_1v_2[\cos \Delta\gamma - \cos \gamma_2 \cos \gamma_1(1 - \cos\delta)]} \quad (6.20)$$

where $\Delta \gamma = \gamma_2 - \gamma_1$. If there is no plane change ($\delta = 0$), then $\cos \delta = 1$ and Equation 6.20 reduces to

$$\Delta v = \sqrt{v_1^2 + v_2^2 - 2v_1v_2 \cos \Delta\gamma}$$

which is the cosine law we have been using to compute Δv in coplanar maneuvers. Therefore, Equation 6.19 contains Equation 6.9 as a special case.

To keep Δv at a minimum, it is clear from Equation 6.19 that the radial velocity should remain unchanged during a plane change maneuver. For the same reason, it is apparent that the maneuver should occur where $v_\perp$ is smallest, which is at apoapse. Figure 6.23 illustrates a plane change maneuver at apoapse. In this case $v_{r_1} = v_{r_2} = 0$, so that $v_{\perp_1} = v_1$ and $v_{\perp_2} = v_2$, thereby reducing Equation 6.19 to

$$\Delta v = \sqrt{v_1^2 + v_2^2 - 2v_1v_2 \cos\delta} \quad \text{Plane change at apoapse (or periapse).} \quad (6.21)$$

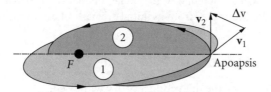

Figure 6.23 Plane change at apoapse.

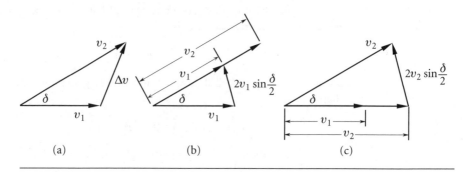

Figure 6.24 Plane changes at apoapse or periapse. (a) Speed change accompanied by plane change. (b) Plane change followed by speed change. (c) Speed change followed by plane change.

Equation 6.21 is for a speed change accompanied by a plane change, as illustrated in Figure 6.24(a). Using the trig identity

$$\cos\delta = 1 - 2\sin^2\frac{\delta}{2}$$

we can rewrite Equation 6.21 as follows for a plane change together with a speed change at apoapse or periapse,

$$\Delta v_\mathrm{I} = \sqrt{(v_2 - v_1)^2 + 4v_1 v_2 \sin^2\frac{\delta}{2}} \qquad (6.22)$$

If there is no change in the speed, so that $v_2 = v_1$, Equation 6.22 yields

$$\Delta v_\delta = 2v \sin\frac{\delta}{2} \qquad (6.23)$$

The subscript δ reminds us that this is the delta-v for a pure rotation of the velocity vector through the angle δ.

Another plane-change strategy, illustrated in Figure 6.24(b), is to rotate the velocity vector and then change its magnitude. In that case, the delta-v is

$$\Delta v_\mathrm{II} = 2v_1 \sin\frac{\delta}{2} + |v_2 - v_1|$$

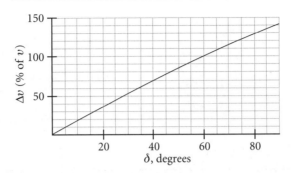

Figure 6.25 Δv required to rotate the velocity vector through an angle δ.

Yet another possibility is to change the speed first, and then rotate the velocity vector (Figure 6.24(c)). Then

$$\Delta v_{\text{III}} = |v_2 - v_1| + 2v_2 \sin \frac{\delta}{2}$$

It is easy to show that

$$\Delta v_{\text{II}} = \sqrt{\Delta v_{\text{I}}^2 + 4v_1|v_2 - v_1| \sin \frac{\delta}{2} \left(1 - \sin \frac{\delta}{2}\right)} > \Delta v_{\text{I}}$$

$$\Delta v_{\text{III}} = \sqrt{\Delta v_{\text{I}}^2 + 4v_2|v_2 - v_1| \sin \frac{\delta}{2} \left(1 - \sin \frac{\delta}{2}\right)} > \Delta v_{\text{I}}$$

It follows that plane change accompanied by speed change is the most efficient of the above three maneuvers.

Equation 6.23, the delta-v formula for pure rotation of the velocity vector, is plotted in Figure 6.25, which shows why significant plane changes are so costly in terms of propellant expenditure. For example, a plane change of just 24° requires a delta-v equal to that needed for an escape trajectory (41.4 percent). A 60° plane change requires a delta-v equal to the speed of the spacecraft itself, which in earth orbit operations is about 7.5 km/s. This would require placing in orbit, a spacecraft about the size of that which launched the satellite into orbit. Of course, this launch-vehicle sized satellite would itself have to be launched atop a vehicle of monstrous proportions. The space shuttle is capable of a plane change in orbit of only about 3°, a maneuver which would exhaust its entire fuel capacity. Orbit plane adjustments are therefore made during the powered ascent phase when the energy is available to do so.

For some missions, however, plane changes must occur in orbit. A common example is the maneuvering of GEO satellites into position. These must orbit the earth in the equatorial plane, but it is impossible to throw a satellite directly into an equatorial orbit from a launch site which is not on the equator. That is not difficult to understand when we realize that the plane of the orbit must contain the center of the earth (the focus) as well as the point at which the satellite is inserted into orbit, as illustrated in Figure 6.26. So if the insertion point is anywhere but on the equator, the

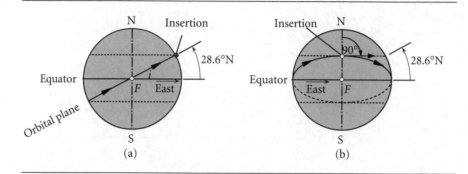

Figure 6.26 Two views of the orbit of a satellite launched directly east at 28.6° north latitude. (a) Edge-on view of the orbital plane. (b) View towards insertion point meridian.

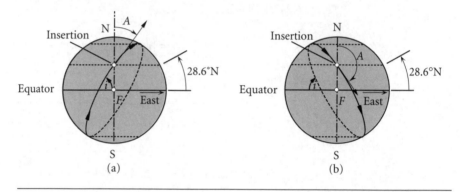

Figure 6.27 (a) Northeasterly launch (0° < A < 270) from a latitude of 28.6°N. (b) Southeasterly launch (90° < A < 270).

plane of the orbit will be tilted away from the earth's axis. As we know from Chapter 4, the angle between the equatorial plane and the plane of the orbiting satellite is called the inclination i.

Launching a satellite due east takes full advantage of the earth's rotational velocity, which is about 0.5 km/s at the equator and diminishes towards the poles. Figure 6.26 shows a spacecraft launched due east into low earth orbit at a latitude ϕ of 28.6° north, which is the latitude of Kennedy Space Flight Center (KSC). As can be seen from the figure, the inclination of the orbit will be 28.6°. One-fourth of the way around the earth the satellite will cross the equator. Halfway around the earth it reaches its southernmost latitude, $\phi = 28.6°$ south. It then heads north, crossing over the equator at the three-quarters point and returning after one complete revolution to $\phi = 28.6°$ north.

Launch azimuth A is the flight direction at insertion, measured clockwise from north on the local meridian. Thus $A = 90°$ is due east. If the launch direction is not directly eastward, then the orbit will have an inclination greater than the launch latitude, as illustrated in Figure 6.27 for $\phi = 28.6°$N. Northeasterly ($0 < A < 90°$) or southeasterly ($90° < A < 180°$) launches take only partial advantage of the earth's

6.9 Plane change maneuvers 295

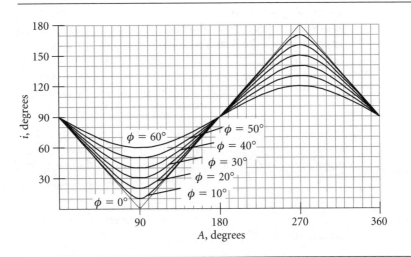

Figure 6.28 Orbit inclination i versus launch azimuth A for several latitudes ϕ.

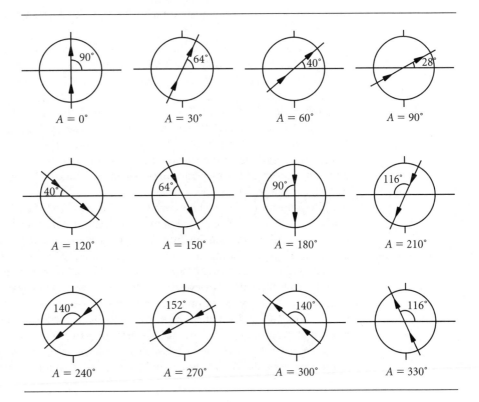

Figure 6.29 Variation of orbit inclinations with launch azimuth at $\phi = 28°$. Note the retrograde orbits for $A > 180°$.

rotational speed and both produce an inclination i greater than the launch latitude but less than 90°. Since these orbits have an eastward velocity component, they are called prograde orbits. Launches to the west produce retrograde orbits with inclinations between 90° and 180°. Launches directly north or directly south result in polar orbits.

Spherical trigonometry is required to obtain the relationship between orbital inclination i, launch platform latitude ϕ, and launch azimuth A. It turns out that

$$\cos i = \cos \phi \sin A \qquad (6.24)$$

From this we verify, for example, that $i = \phi$ when $A = 90°$, as pointed out above. A plot of this relation is presented in Figure 6.28, while Figure 6.29 illustrates the orientation of orbits for a range of launch azimuths at $\phi = 28°$.

EXAMPLE 6.10

Determine the required launch azimuth for the sun-synchronous satellite of Example 4.7 if it is launched from Vandenburgh AFB on the California coast (latitude = 34.5°N).

In Example 4.7 the inclination of the sun-synchronous orbit was determined to be 98.43°. Equation 6.24 is used to calculate the launch azimuth,

$$\sin A = \frac{\cos i}{\cos l} = \frac{\cos 98.43°}{\cos 34.5°} = -0.1779$$

From this, $A = 190.2°$, a launch to the south or $A = 349.8°$, a launch to the north.

Figure 6.30 shows the effect that the choice of launch azimuth has on the orbit. It does not change the fact that the orbit is retrograde; it simply determines whether the ascending node will be in the same hemisphere as the launch site or on the opposite side of the earth. Actually, a launch to the north from Vandenburgh is not an option because of the safety hazard to the populated land lying below the ascent trajectory. Launches to the south, over open water, are not a hazard. Working this problem for Kennedy Space Center (latitude 28.6°N) yields nearly the same values of A. Since safety considerations on the Florida east coast limit launch azimuths to between 35° and 120°, polar and sun-synchronous satellites cannot be launched from the US eastern test range.

The projection of a satellite's orbit onto the earth's surface is called its ground track. Because the satellite reaches a maximum and minimum latitude ('amplitude') during each orbit while passing over the equator twice, on a Mercator projection the ground track of a satellite in low-earth orbit resembles a sine curve. If the earth did not rotate, there would be just one sinusoid-like track, traced over and over again as the satellite orbits the earth. However, the earth rotates eastward beneath the satellite orbit at 15.04° per hour, so the ground track advances westward at that rate. Figure 6.31 shows about two and a half orbits of a satellite, with the beginning and end of this portion of the ground track labeled. The distance between two successive crossings of the equator is measured to be 23.2°, which is the amount of earth rotation in one orbit of the spacecraft. Therefore, the ground track reveals that the period of the satellite is

$$T = \frac{23.2°}{15.04°/\text{hr}} = 1.54 \, \text{hr} = 92.6 \, \text{min}$$

This is a typical low earth orbital period.

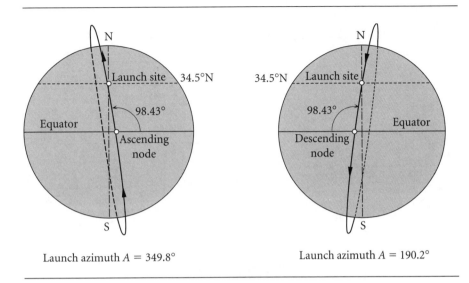

Figure 6.30 Effect of launch azimuth on the position of the orbit.

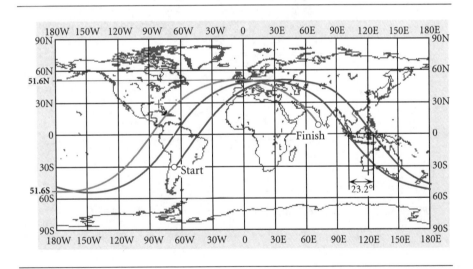

Figure 6.31 Ground track of a satellite.

EXAMPLE 6.11

Find the delta-v required to transfer a satellite from a circular, 300 km altitude low-earth orbit of 28° inclination to a geostationary equatorial orbit. Circularize and change the inclination at altitude. Compare that delta-v requirement with the one in which the plane change is done in the low-earth orbit.

Figure 6.32 shows the 28° inclined low-earth parking orbit (1), the coplanar Hohmann transfer ellipse (2), and the coplanar GEO orbit (3). From the figure we see that

$$r_B = 6678 \text{ km} \quad r_c = 42\,164 \text{ km}$$

(Example 6.11 continued)

Orbit 1:

For this circular orbit the speed at B is

$$v_{B_1} = \sqrt{\frac{\mu}{r_B}} = \sqrt{\frac{398\,600}{6678}} = 7.7258 \text{ km/s}$$

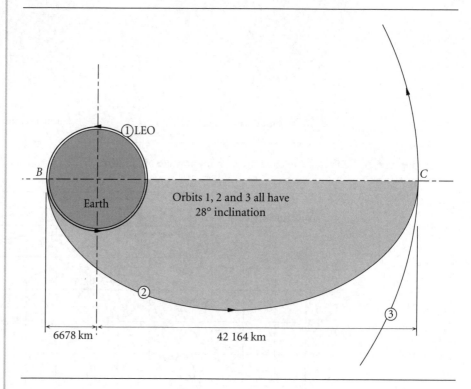

Figure 6.32 Transfer from LEO to GEO in an orbit of 28° inclination.

Orbit 2:

The eccentricity of the transfer orbit is

$$e_2 = \frac{r_C - r_B}{r_C + r_B} = 0.72655$$

Let us evaluate the orbit equation at B to find the angular momentum of the Hohmann transfer orbit 2,

$$r_B = \frac{h_2^2}{\mu} \frac{1}{1 + e_2 \cos(0)} \quad \Rightarrow \quad 6678 = \frac{h_2^2}{398\,600} \frac{1}{1 + 0.72655} \quad \Rightarrow \quad h_2 = 67\,792 \text{ km/s}$$

The velocities at perigee and apogee of orbit 2 are, from the angular momentum formula,

$$v_{B_2} = \frac{h_2}{r_B} = 10.152 \text{ km/s} \qquad v_{C_2} = \frac{h_2}{r_C} = 1.6078 \text{ km/s}$$

At this point we can calculate Δv_B,

$$\Delta v_B = v_{B_2} - v_{B_1} = 10.152 - 7.7258 = 2.4258 \text{ km/s} \tag{a}$$

Orbit 3:

For this orbit, which is circular, the speed at C is

$$v_{C_3} = \sqrt{\frac{\mu}{r_C}} = 3.0747 \text{ km/s}$$

so that

$$\Delta v_C = v_{C_3} - v_{C_2} = 3.0747 - 1.6078 = 1.4668 \text{ km/s} \tag{b}$$

We can now calculate the total delta-v for the Hohmann transfer:

$$\Delta v_{\text{Hohmann}} = \Delta v_B + \Delta v_C = 2.4258 + 1.4668 = 3.8926 \text{ km/s}$$

This places the satellite in a circular orbit of the correct radius, but the wrong inclination. The velocity vector at C must be rotated into the plane of the equator, as illustrated in Figure 6.33. According to Equation 6.30, the delta-v required to rotate that velocity through the change in inclination of 28° is

$$\Delta v_{iC} = 2v_{C_3} \sin \frac{\Delta i}{2} = 2 \cdot 3.0747 \cdot \sin \frac{28°}{2} = 1.4877 \text{ km/s}$$

Therefore, the total delta-v requirement is

$$\Delta v_{\text{total}} = \Delta v_{\text{Hohmann}} + \Delta v_{iC} = \underline{5.3803 \text{ km/s}}$$

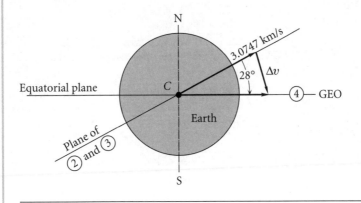

Figure 6.33 Plane change maneuver required after the Hohmann transfer.

(*Example 6.11 continued*)

Suppose we make the plane change at LEO instead of at GEO. To rotate the velocity vector $\mathbf{v}_{B_1}$ through 28° requires

$$\Delta v_{B_i} = 2v_{B_1} \sin \frac{\Delta i}{2} = 2 \cdot 7.7258 \cdot \sin \frac{28°}{2} = 3.7381 \text{ km/s}$$

This, together with (a) and (b), yields the total delta-v schedule for insertion into GEO:

$$\Delta v_{\text{total}} = \Delta v_{B_i} + \Delta v_B + \Delta v_C = 3.7381 + 2.4258 + 1.4668 = \underline{7.6307 \text{ km/s}}$$

This is a 42 percent increase over the total delta-v with plane change at GEO. Clearly, it is best to do plane change maneuvers at the largest possible distance (apoapse) from the primary attractor, where the velocities are smallest.

EXAMPLE 6.12

Suppose in the previous example that part of the plane change, Δi, takes place at B, the perigee of the Hohmann transfer ellipse, and the remainder, $28° - \Delta i$, occurs at the apogee C. What is the value of Δi which results in the minimum Δv_{total}?

We found in Example 6.11 that if $\Delta i = 0$, then $\Delta v_{\text{total}} = 5.3803$ km/s, whereas $\Delta i = 28°$ made $\Delta v_{\text{total}} = 7.6307$ km/s. Here we are to determine if there is a value of Δi between 0° and 28° that yields a Δv_{total} which is smaller than either of those two.

In this case a plane change occurs at both B and C. Recall that the most efficient strategy is to combine the plane change with the speed change, so that the delta-vs at those points are (Equation 6.21)

$$\Delta v_B = \sqrt{v_{B_1}^2 + v_{B_2}^2 - 2v_{B_1} v_{B_2} \cos \Delta i}$$

$$= \sqrt{7.7258^2 + 10.152^2 - 2 \cdot 7.7258 \cdot 10.152 \cdot \cos \Delta i}$$

$$= \sqrt{162.74 - 156.86 \cos \Delta i}$$

and

$$\Delta v_C = \sqrt{v_{C_2}^2 + v_{C_3}^2 - 2v_{C_2} v_{C_3} \cos(28° - \Delta i)}$$

$$= \sqrt{1.6078^2 + 3.0747^2 - 2 \cdot 1.6078 \cdot 3.0747 \cdot \cos(28° - \Delta i)}$$

$$= \sqrt{12.039 - 9.8871 \cos(28° - \Delta i)}$$

Thus,

$$\Delta v_{\text{total}} = \Delta v_B + \Delta v_C$$

$$= \sqrt{162.74 - 156.86 \cos \Delta i} + \sqrt{12.039 - 9.8871 \cos(28° - \Delta i)} \quad \text{(a)}$$

6.9 Plane change maneuvers

To determine if there is a Δi which minimizes Δv_{total}, we take its derivative with respect to Δi and set it equal to zero:

$$\frac{d\Delta v_{\text{total}}}{d\Delta i} = \frac{78.43 \sin \Delta i}{\sqrt{162.74 - 156.86 \cos \Delta i}} - \frac{4.9435 \sin(28° - \Delta i)}{\sqrt{12.039 - 9.8871 \cos(28° - \Delta i)}} = 0$$

This is a transcendental equation which must be solved iteratively. The solution, as the reader may verify, is

$$\Delta i = 2.1751° \quad \text{(b)}$$

That is, an inclination change of 2.1751° should occur in low-earth orbit, while the rest of the plane change, 25.825°, is done at GEO. Substituting (b) into (a) yields

$$\underline{\Delta v_{\text{total}} = 4.2207 \text{ km/s}}$$

This is 21 percent less than the smallest Δv_{total} computed in Example 6.11.

EXAMPLE 6.13

A spacecraft is in a 500 km by 10 000 km altitude geocentric orbit which intersects the equatorial plane at a true anomaly of 120° (see Figure 6.34). If the inclination to the equatorial plane is 15°, what is the minimum velocity increment required to make this an equatorial orbit?

The orbital parameters are

$$e = \frac{r_A - r_P}{r_A + r_P} = \frac{(6378 + 10\,000) - (6378 + 500)}{(6378 + 10\,000) + (6378 + 500)} = 0.4085$$

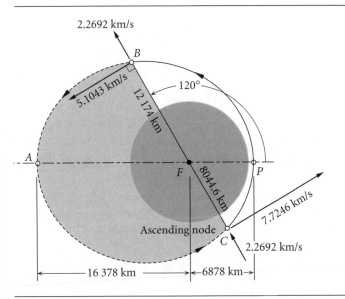

Figure 6.34 An orbit which intersects the equatorial plane along line BC. The equatorial plane makes an angle of 15° with the plane of the page.

(Example 6.13 continued)

$$r_P = \frac{h^2}{\mu}\frac{1}{1+e\cos(0)} \Rightarrow 6878 = \frac{h^2}{398\,600}\frac{1}{1+0.4085} \Rightarrow h = 62\,141 \text{ km/s}$$

The radial coordinate and velocity components at points B and C, on the line of intersection with the equatorial plane, are

$$r_B = \frac{h^2}{\mu}\frac{1}{1+e\cos\theta_B} = \frac{62\,141^2}{398\,600}\frac{1}{1+0.4085\cdot\cos 120°} = 12\,174 \text{ km}$$

$$v_{\perp_B} = \frac{h}{r_B} = \frac{62\,141}{12\,174} = 5.1043 \text{ km/s}$$

$$v_{r_B} = \frac{\mu}{h}e\sin\theta_B = \frac{398\,600}{62\,141}\cdot 0.4085 \cdot \sin 120° = 2.2692 \text{ km/s}$$

and

$$r_C = \frac{h^2}{\mu}\frac{1}{1+e\cos\theta_C} = \frac{62\,141^2}{398\,600}\frac{1}{1+0.4085\cdot\cos 300°} = 8044.6 \text{ km}$$

$$v_{\perp_C} = \frac{h}{r_C} = \frac{62\,141}{8044.6} = 7.7246 \text{ km/s}$$

$$v_{r_C} = \frac{\mu}{h}e\sin\theta_C = \frac{398\,600}{62\,141}\cdot 0.4085 \cdot \sin 300° = -2.2692 \text{ km/s}$$

All we wish to do here is rotate the plane of the orbit rigidly around the node line BC. The impulsive maneuver must occur at either B or C. Equation 6.19 applies, and since the radial and perpendicular velocity components remain fixed, it reduces to

$$\Delta v = v_\perp \sqrt{2(1-\cos\delta)}$$

where $\delta = 15°$. For the minimum Δv, the maneuver must be done where $v_\perp$ is smallest, which is at B, the point farthest from the center of attraction F. Thus,

$$\Delta v = 5.1043\sqrt{2(1-\cos 15°)} = \underline{1.3325 \text{ km/s}}$$

EXAMPLE 6.14

Orbit 1 has angular momentum h and eccentricity e. The direction of motion is shown. Calculate the Δv required to rotate the orbit 90° about its latus rectum BC without changing h and e. The required direction of motion in orbit 2 is shown in Figure 6.35.

By symmetry, the required maneuver may occur at either B or C, and it involves a rigid body rotation of the ellipse, so that v_r and $v_\perp$ remain unaltered. Because of the directions of motion shown, the true anomalies of B on the two orbits are

$$\theta_{B_1} = -90° \qquad \theta_{B_2} = +90°$$

The radial coordinate of B is

$$r_B = \frac{h^2}{\mu}\frac{1}{1+e\cos(\pm 90)} = \frac{h^2}{\mu}$$

6.9 Plane change maneuvers

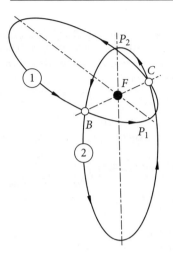

Figure 6.35 Identical ellipses intersecting at 90° along their common latus rectum, *BC*.

For the velocity components at *B*, we have

$$v_{\perp_B})_1 = v_{\perp_B})_2 = \frac{h}{r_B} = \frac{\mu}{h}$$

$$v_{r_B})_1 = \frac{\mu}{h}e\sin(\theta_{B_1}) = -\frac{\mu e}{h} \qquad v_{r_B})_2 = \frac{\mu}{h}e\sin(\theta_{B_2}) = \frac{\mu e}{h}$$

Substituting these into Equation 6.19, yields

$$\Delta v_B = \sqrt{\left[v_{r_B})_2 - v_{r_B})_1\right]^2 + v_{\perp_B})_1^2 + v_{\perp_B})_2^2 - 2v_{\perp_B})_1 v_{\perp_B})_2 \cos 90°}$$

$$= \sqrt{\left[\frac{\mu e}{h} - \left(-\frac{\mu e}{h}\right)\right]^2 + \left(\frac{\mu}{h}\right)^2 + \left(\frac{\mu}{h}\right)^2 - 2\left(\frac{\mu}{h}\right)\left(\frac{\mu}{h}\right) \cdot 0}$$

$$= \sqrt{4\frac{\mu^2}{h^2}e^2 + 2\frac{\mu^2}{h^2}}$$

so that

$$\Delta v_B = \frac{\sqrt{2}\mu}{h}\sqrt{1 + 2e^2} \qquad (a)$$

If the motion on ellipse 2 were opposite to that shown in Figure 6.35, then the radial velocity components at *B* (and *C*) would be in the same rather than in the opposite direction on both ellipses, so that instead of (a) we would find a smaller velocity increment,

$$\Delta v_B = \frac{\sqrt{2}\mu}{h}$$

Problems

6.1 The shuttle orbiter has a mass of 125 000 kg. The two orbital maneuvering engines produce a combined (non-throttleable) thrust of 53.4 kN. The orbiter is in a 300 km circular orbit. A delta-v maneuver transfers the spacecraft to a coplanar 250 km by 300 km elliptical orbit. Neglecting propellant loss and using elementary physics (linear impulse equals change in linear momentum, distance equals speed times time), estimate
(a) the time required for the Δv burn, and
(b) the distance traveled by the orbiter during the burn.
(c) Calculate the ratio of your answer for (b) to the circumference of the initial circular orbit.
{Ans.: (a) $\Delta t = 34$ s; (b) 263 km; (c) 0.0063}

6.2 A satellite traveling at 8.2 km/s at perigee fires a retrorocket at perigee altitude of 480 km. What delta-v is necessary to reach a minimum altitude of 100 miles during the next orbit?
{Ans.: −66.8 m/s}

6.3 A spacecraft is in a 300 km circular earth orbit. Calculate
(a) the total delta-v required for a Hohmann transfer to a 3000 km coplanar circular earth orbit, and
(b) the transfer orbit time.
{Ans.: (a) 1.198 km/s; (b) 59 min 39 s}

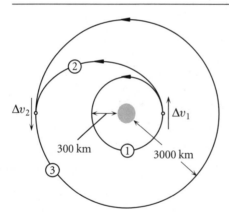

Figure P.6.3

6.4 A spacecraft S is in a geocentric hyperbolic trajectory with a perigee radius of 7000 km and a perigee speed of $1.3v_{esc}$. At perigee, the spacecraft releases a projectile B with a speed of 7.1 km/s parallel to the spacecraft's velocity. How far d from the earth's surface is S at the instant B impacts the earth? Neglect the atmosphere.
{Ans.: $d = 8978$ km}

6.5 Assuming the orbits of earth and Mars are circular and coplanar, calculate
(a) the time required for a Hohmann transfer from earth to Mars, and
(b) the initial position of Mars (α) in its orbit relative to earth for interception to occur. Radius of earth orbit $= 1.496 \times 10^8$ km. Radius of Mars orbit $= 2.279 \times 10^8$ km. $\mu_{sun} = 1.327 \times 10^{11}$ km^3/s^2.
{Ans.: (a) 259 days; (b) $\alpha = 44.3°$}

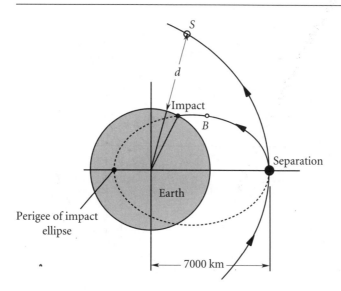

Figure P.6.4

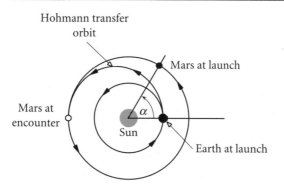

Figure P.6.5

6.6 Two geocentric elliptical orbits have common apse lines and their perigees are on the same side of the earth. The first orbit has a perigee radius of $r_p = 7000$ km and $e = 0.3$, whereas for the second orbit $r_p = 32\,000$ km and $e = 0.5$.
  (a) Find the minimum total delta-v and the time of flight for a transfer from the perigee of the inner orbit to the apogee of the outer orbit.
  (b) Do part (a) for a transfer from the apogee of the inner orbit to the perigee of the outer orbit.
  {Ans.: (a) $\Delta v_{total} = 2.388$ km/s, TOF $= 16.2$ hr; (b) $\Delta v_{total} = 3.611$ km/s, TOF $= 4.66$ hr}

6.7 A spacecraft is in a 500 km altitude circular earth orbit. Neglecting the atmosphere, find the delta-v required at A in order to impact the earth at
  (a) point B
  (b) point C.
  {Ans.: (a) 192 m/s; (b) 7.61 km/s}

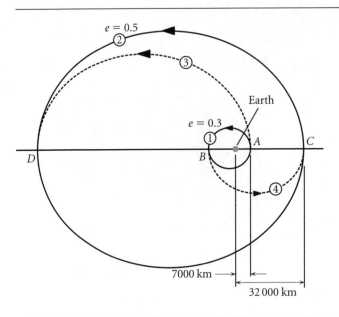

Figure P.6.6

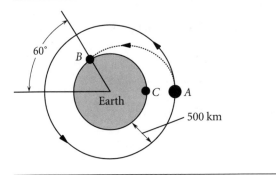

Figure P.6.7

6.8 A spacecraft is in a 200 km circular earth orbit. At $t=0$, it fires a projectile in the direction opposite to the spacecraft's motion. Thirty minutes after leaving the spacecraft, the projectile impacts the earth. What delta-v was imparted to the projectile? Neglect the atmosphere.
{Ans.: $\Delta v = 77.2$ m/s}

6.9 The space shuttle was launched on a 15-day mission. There were four orbits after injection, all of them at 39° inclination.
    Orbit 1: 302 by 296 km
    Orbit 2 (day 11): 291 by 259 km
    Orbit 3 (day 12): 259 km circular
    Orbit 4 (day 13): 255 by 194 km

Calculate the total delta-v, which should be as small as possible, assuming Hohmann transfers.
{Ans.: $\Delta v_{total} = 43.5$ m/s}

6.10 A space vehicle in a circular orbit at an altitude of 500 km above the earth executes a Hohmann transfer to a 1000 km circular orbit. Calculate the total delta-v requirement.
{Ans.: 0.2624 km/s}

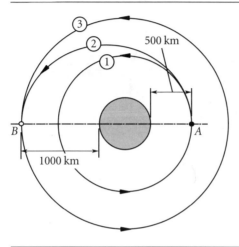

Figure P.6.10

6.11 Calculate the total delta-v required for a Hohmann transfer from a circular orbit of radius r to a circular orbit of radius $12r$.
{Ans.: $0.5342\sqrt{\mu/r}$}

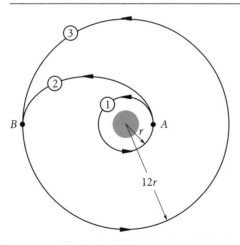

Figure P.6.11

6.12 A spacecraft in circular orbit 1 of radius r leaves for infinity on parabolic trajectory 2 and returns from infinity on a parabolic trajectory 3 to a circular orbit 4 of radius $12r$. Find the total delta-v required for this non-Hohmann orbit change maneuver.
{Ans.: $0.5338\sqrt{\mu/r}$}

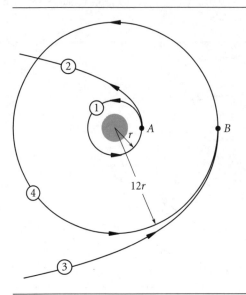

Figure P.6.12

6.13 Calculate the total delta-v required for a Hohmann transfer from the smaller circular orbit to the larger one.
{Ans.: $0.394v_1$, where v_1 is the speed in orbit 1}

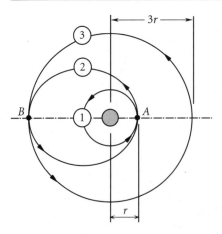

Figure P.6.13

6.14 A spacecraft is in a 300 km circular earth orbit. Calculate
(a) the total delta-v required for the bi-elliptical transfer to a 3000 km altitude coplanar circular orbit shown, and
(b) the total transfer time.
{Ans.: (a) 2.039 km/s; (b) 2.86 hr}

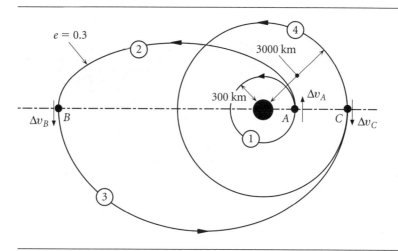

Figure P.6.14

6.15
(a) With a single delta-v maneuver, the earth orbit of a satellite is to be changed from a circle of radius 15 000 km to a coplanar ellipse with perigee altitude of 500 km and apogee radius of 22 000 km. Calculate the magnitude of the required delta-v and the change in the flight path angle $\Delta\gamma$.
(b) What is the minimum total delta-v if the orbit change is accomplished instead by a Hohmann transfer?
{Ans.: (a) $||\Delta\mathbf{v}|| = 2.77$ km/s, $\Delta\gamma = 31.51°$; (b) $\Delta v_{Hohmann} = 1.362$ km/s}

6.16 An earth satellite has a perigee altitude of 1270 km and a perigee speed of 9 km/s. It is required to change its orbital eccentricity to 0.4, without rotating the apse line, by a delta-v maneuver at $\theta = 100°$. Calculate the magnitude of the required $\Delta\mathbf{v}$ and the change in flight path angle $\Delta\gamma$.
{Ans.: $||\Delta\mathbf{v}|| = 0.915$ km/s; $\Delta\gamma = -8.18°$}

6.17 At point A on its earth orbit, the radius, speed and flight path angle of a satellite are $r_A = 12\,756$ km, $v_A = 6.5992$ km/s and $\gamma_A = 20°$. At point B, at which the true anomaly is 150°, an impulsive maneuver causes $\Delta v_\perp = +0.75820$ km/s and $\Delta v_r = 0$.
(a) What is the time of flight from A to B?
(b) What is the rotation of the apse line as a result of this maneuver?
{Ans.: (a) 2.045 hr; (b) 43.39° counterclockwise}

6.18 A satellite is in elliptical orbit 1. Calculate the true anomaly θ (relative to the apse line of orbit 1) of an impulsive maneuver which rotates the apse line at an angle η counterclockwise but leaves the eccentricity and the angular momentum unchanged.
{Ans.: $\theta = \eta/2$}

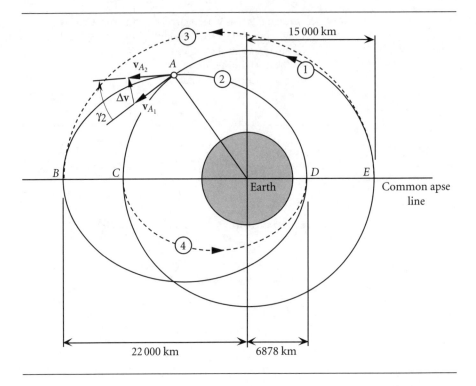

Figure P.6.15

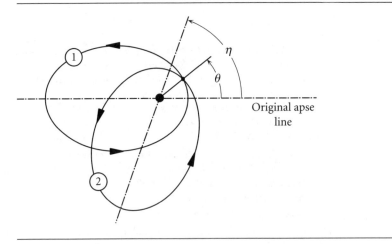

Figure P.6.18

6.19 A satellite in orbit 1 undergoes a delta-v maneuver at perigee P_1 such that the new orbit 2 has the same eccentricity e, but its apse line is rotated 90° clockwise from the original one. Calculate the specific angular momentum of orbit 2 in terms of that of orbit 1 and the eccentricity e.
{Ans.: $h_2 = h_1/\sqrt{1+e}$}

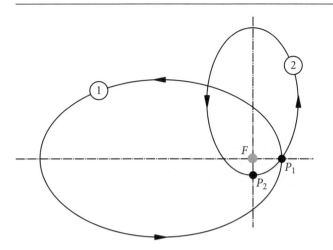

Figure P.6.19

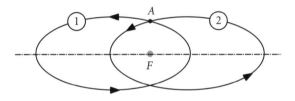

Figure P.6.20

6.20 Calculate the delta-v required at A in orbit 1 for a single impulsive maneuver to rotate the apse line 180° counterclockwise (to become orbit 2), but keep the eccentricity e and the angular momentum $\mathbf{h}$ the same.
{Ans.: $\Delta v = 2\mu e/h$}

6.21 The space station and spacecraft A and B are all in the same circular earth orbit of 350 km altitude. Spacecraft A is 600 km behind the space station and spacecraft B is 600 km ahead of the space station. At the same instant, both spacecraft apply a $\Delta v_\perp$ so as to arrive at the space station in one revolution of their phasing orbits.
(a) Calculate the times required for each spacecraft to reach the space station.
(b) Calculate the total delta-v requirement for each spacecraft.
{Ans.: (a) spacecraft A: 90.2 min; spacecraft B: 92.8 min; (b) $\Delta v_A = 73.9$ m/s; $\Delta v_B = 71.5$ m/s}

6.22 Satellites A and B are in the same circular orbit of radius r. B is 180° ahead of A. Calculate the semimajor axis of a phasing orbit in which A will rendezvous with B after just one revolution in the phasing orbit.
{Ans.: $a = 0.63r$}

6.23 Two spacecraft are in the same elliptical earth orbit with perigee radius 8000 km and apogee radius 13 000 km. Spacecraft 1 is at perigee and spacecraft 2 is 30° ahead. Calculate the total delta-v required for spacecraft 1 to intercept and rendezvous with spacecraft 2 when spacecraft 2 has traveled 60°.
{Ans.: $\Delta v_{\text{total}} = 6.24$ km/s}

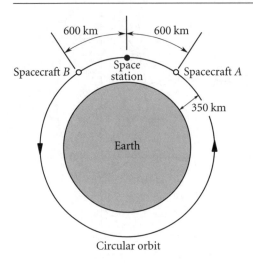

Figure P.6.21

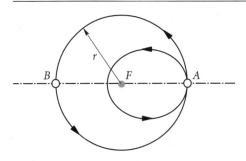

Figure P.6.22

6.24 An earth satellite has the following orbital elements: $a = 15\,000$ km, $e = 0.5$, $W = 45°$, $w = 30°$, $i = 10°$. What minimum delta-v is required to reduce the inclination to zero?
{Ans.: 0.588 km/s}

6.25 With a single impulsive maneuver, an earth satellite changes from a 400 km circular orbit inclined at 60° to an elliptical orbit of eccentricity $e = 0.5$ with an inclination of 40°. Calculate the minimum required delta-v.
{Ans.: 3.41 km/s}

6.26 An earth satellite is in an elliptical orbit of eccentricity 0.3 and angular momentum 60 000 km²/s. Find the delta-v required for a 90° change in inclination at apogee (no change in speed).
{Ans.: 6.58 km/s}

6.27 A spacecraft is in a circular, equatorial orbit (1) of radius r_o about a planet. At point B it impulsively transfers to polar orbit (2), whose eccentricity is 0.25 and whose perigee is directly over the North Pole. Calculate the minimum delta-v required at B for this maneuver.
{Ans.: $1.436\sqrt{\mu/r_o}$}

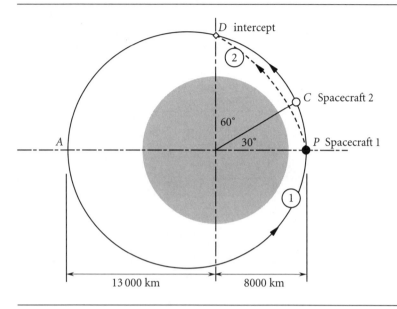

Figure P.6.23

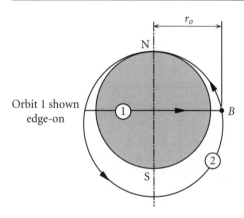

Figure P.6.27

6.28 A spacecraft is in a 300 km circular parking orbit. It is desired to increase the altitude to 600 km and change the inclination by 20°. Find the total delta-v required if
   (a) the plane change is made after insertion into the 600 km orbit (so that there are a total of three delta-v burns);
   (b) the plane change and insertion into the 600 km orbit are accomplished simultaneously (so that the total number of delta-v burns is two);
   (c) the plane change is made upon departing the lower orbit (so that the total number of delta-v burns is two).
   {Ans.: (a) 2.793 km/s; (b) 2.696 km/s; (c) 2.783 km/s}

6.29 At time $t=0$, manned spacecraft a and unmanned spacecraft b are at the positions shown in circular earth orbits 1 and 2, respectively. For assigned values of $\theta_0^{(a)}$ and $\theta_0^{(b)}$, design a series of impulsive maneuvers by means of which spacecraft a transfers from orbit 1 to orbit 2 so as to rendezvous with spacecraft b (i.e., occupy the same position in space). The total time and total delta-v required for the transfer should be as small as possible. Consider earth's gravity only.

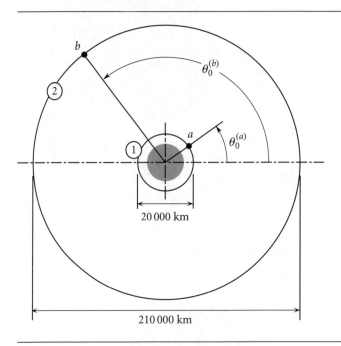

Figure P.6.29

6.30 What must the launch azimuth be if the satellite in Example 4.8 is launched from
  (a) Kennedy Space Center (latitude $= 28.5°$N);
  (b) Vandenburgh AFB (latitude $= 34.5°$N);
  (c) Kourou, French Guiana (latitude $5.5°$N).
{Ans.: (a) 329.4°; (b) 327.1°; (c) 333.3°}

CHAPTER 7

RELATIVE MOTION AND RENDEZVOUS

CHAPTER OUTLINE

7.1	Introduction	315
7.2	Relative motion in orbit	316
7.3	Linearization of the equations of relative motion in orbit	322
7.4	Clohessy–Wiltshire equations	324
7.5	Two-impulse rendezvous maneuvers	330
7.6	Relative motion in close-proximity circular orbits	338
Problems		340

7.1 INTRODUCTION

Up to now we have mostly referenced the motion of orbiting objects to a non-rotating coordinate system fixed to the center of attraction (e.g., the center of the earth). This platform served as an inertial frame of reference, in which Newton's second law can be written

$$\mathbf{F}_{net} = m\mathbf{a}_{absolute}$$

An exception to this rule was the discussion of the restricted three-body problem at the end of Chapter 2, in which we made use of the relative motion equations developed in Chapter 1. In a rendezvous maneuver, two orbiting vehicles observe one another from each of their own free-falling, rotating, clearly non-inertial frames of reference. To base impulsive maneuvers on observations made from a moving platform requires transforming relative velocity and acceleration measurements into an inertial frame.

Otherwise, the true thrusting forces cannot be sorted out from the fictitious 'inertial forces' that appear in Newton's law when it is written incorrectly as

$$\mathbf{F}_{net} = m\mathbf{a}_{rel}$$

The purpose of this chapter is to use relative motion analysis to gain some familiarity with the problem of maneuvering one spacecraft relative to another, especially when they are in close proximity.

7.2 Relative motion in orbit

A rendezvous maneuver usually involves a target vehicle, which is passive and non-maneuvering, and a chase vehicle which is active and performs the maneuvers required to bring itself alongside the target. An obvious example is the space shuttle, the chaser, rendezvousing with the international space station, the target. The position vector of the target in the geocentric equatorial frame is $\mathbf{r}_0$. This outward radial is sometimes called 'r-bar'. The moving frame of reference has its origin at the target, as illustrated in Figure 7.1. The x axis is directed along $\mathbf{r}_0$, the outward radial to the target. The y axis is perpendicular to $\mathbf{r}_0$ and points in the direction of the target satellite's local horizon. The x and y axes therefore lie in the target's orbital plane, and the z axis is normal to that plane.

The angular velocity of the xyz axes attached to the target is just the angular velocity of the position vector $\mathbf{r}_0$, and it is obtained from the fact that

$$\mathbf{h} = \mathbf{r}_0 \times \mathbf{v}_0 = (r_0 v_{0\perp})\hat{\mathbf{k}} = (r_0^2 \Omega)\hat{\mathbf{k}} = r_0^2 \mathbf{\Omega}$$

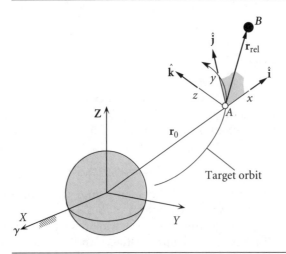

Figure 7.1 Co-moving reference frame attached to A, from which body B is observed.

which means that

$$\boldsymbol{\Omega} = \frac{\mathbf{r}_0 \times \mathbf{v}_0}{r_0^2} \tag{7.1}$$

To find the angular acceleration $\dot{\boldsymbol{\Omega}}$, we take the derivative of $\boldsymbol{\Omega}$ in Equation 7.1 to obtain

$$\dot{\boldsymbol{\Omega}} = \frac{1}{r_0^2}(\dot{\mathbf{r}}_0 \times \mathbf{v}_0 + \mathbf{r}_0 \times \dot{\mathbf{v}}_0) - \frac{2}{r_0^3}\dot{r}_0(\mathbf{r}_0 \times \mathbf{v}_0) \tag{7.2}$$

But

$$\dot{\mathbf{r}}_0 \times \mathbf{v}_0 = \mathbf{v}_0 \times \mathbf{v}_0 = \mathbf{0} \tag{7.3}$$

According to Equation 2.15, the acceleration $\dot{\mathbf{v}}_0$ of the target satellite is given by

$$\dot{\mathbf{v}}_0 = -\frac{\mu}{r_0^3}\mathbf{r}_0$$

Hence,

$$\mathbf{r}_0 \times \dot{\mathbf{v}}_0 = \mathbf{r}_0 \times \left(-\frac{\mu}{r_0^3}\mathbf{r}_0\right) = -\frac{\mu}{r_0^3}(\mathbf{r}_0 \times \mathbf{r}_0) = \mathbf{0} \tag{7.4}$$

Substituting Equations 7.1, 7.3 and 7.4 into Equation 7.2 yields

$$\dot{\boldsymbol{\Omega}} = -\frac{2}{r_0}\dot{r}_0\boldsymbol{\Omega}$$

Finally, recalling from Equation 2.25a that $\dot{r}_0 = \mathbf{v}_0 \cdot \mathbf{r}_0/r_0$, we obtain

$$\dot{\boldsymbol{\Omega}} = -\frac{2(\mathbf{r}_0 \cdot \mathbf{v}_0)}{r_0^2}\boldsymbol{\Omega} \tag{7.5}$$

Equations 7.1 and 7.5 are the means of determining the angular velocity and acceleration of the co-moving frame for use in the relative velocity and acceleration formulas, Equations 1.38 and 1.42.

EXAMPLE 7.1

Spacecraft A is in an elliptical earth orbit having the following parameters:

$$h = 52\,059 \text{ km}^2/\text{s}, \ e = 0.025724, \ i = 60°, \ \Omega = 40°, \ \omega = 30°, \theta = 40° \quad \text{(a)}$$

Spacecraft B is likewise in an orbit with these parameters:

$$h = 52\,362 \text{ km}^2/\text{s}, \ e = 0.0072696, \ i = 50°, \ \Omega = 40°, \ \omega = 120°, \ \theta = 40° \quad \text{(b)}$$

Calculate the velocity $\mathbf{v}_{\text{rel}})_{xyz}$ and acceleration $\mathbf{a}_{\text{rel}})_{xyz}$ of spacecraft B relative to spacecraft A, measured along the xyz axes of the co-moving coordinate system of spacecraft A, as defined in Figure 7.1.

(Example 7.1 continued)

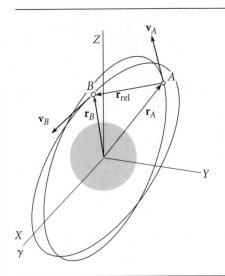

Figure 7.2 Spacecraft A and B in slightly different orbits.

From the orbital elements in (a) and (b) we can use Algorithm 4.2 to find the position and velocity of the spacecraft relative to the geocentric equatorial reference frame. Omitting those calculations, we find, for spacecraft A,

$$\mathbf{r}_A = -266.74\hat{\mathbf{I}} + 3865.4\hat{\mathbf{J}} + 5425.7\hat{\mathbf{K}} \text{ (km)} \tag{a}$$

$$\mathbf{v}_A = -6.4842\hat{\mathbf{I}} - 3.6201\hat{\mathbf{J}} + 2.4159\hat{\mathbf{K}} \text{ (km/s)} \tag{b}$$

and for spacecraft B,

$$\mathbf{r}_B = -5890.0\hat{\mathbf{I}} - 2979.4\hat{\mathbf{J}} + 1792.0\hat{\mathbf{K}} \text{ (km)} \tag{c}$$

$$\mathbf{v}_B = 0.93594\hat{\mathbf{I}} - 5.2409\hat{\mathbf{J}} - 5.5016\hat{\mathbf{K}} \text{ (km/s)} \tag{d}$$

According to Equation 2.15, the accelerations of the two spacecraft are

$$\mathbf{a}_A = -\mu \frac{\mathbf{r}_A}{\|\mathbf{r}_A\|^3} = 0.00035876\hat{\mathbf{I}} - 0.0051989\hat{\mathbf{J}} - 0.0072975\hat{\mathbf{K}} \text{ (km/s}^2) \tag{e}$$

$$\mathbf{a}_B = -\mu \frac{\mathbf{r}_B}{\|\mathbf{r}_B\|^3} = 0.0073377\hat{\mathbf{I}} + 0.0037117\hat{\mathbf{J}} - 0.0022325\hat{\mathbf{K}} \text{ (km/s}^2) \tag{f}$$

The unit vector $\hat{\mathbf{i}}$ along the x axis of spacecraft A's rigid, co-moving frame of reference is

$$\hat{\mathbf{i}} = \frac{\mathbf{r}_A}{r_A} = -0.040008\hat{\mathbf{I}} + 0.57977\hat{\mathbf{J}} + 0.81380\hat{\mathbf{K}} \tag{g}$$

Since the z axis is in the direction of $\mathbf{h}_A$, and

$$\mathbf{h}_A = \mathbf{r}_A \times \mathbf{v}_A = \begin{vmatrix} \hat{\mathbf{I}} & \hat{\mathbf{J}} & \hat{\mathbf{K}} \\ -266.74 & 3865.4 & 5425.7 \\ -6.4842 & -3.6201 & 2.4159 \end{vmatrix}$$

$$= 28\,980\hat{\mathbf{I}} - 34\,537\hat{\mathbf{J}} + 26\,030\hat{\mathbf{K}} \; (\text{km}/\text{s}^2)$$

we obtain

$$\hat{\mathbf{k}} = \frac{\mathbf{h}_A}{h_A} = 0.55667\hat{\mathbf{I}} - 0.66341\hat{\mathbf{J}} + 0.5000\hat{\mathbf{K}} \tag{h}$$

Finally, $\hat{\mathbf{j}} = \hat{\mathbf{k}} \times \hat{\mathbf{i}}$, so that

$$\hat{\mathbf{j}} = -0.82977\hat{\mathbf{I}} - 0.47302\hat{\mathbf{J}} + 0.29620\hat{\mathbf{K}} \tag{i}$$

The angular velocity $\mathbf{\Omega}$ of the xyz frame attached to spacecraft A is given by Equation 7.1,

$$\mathbf{\Omega} = \frac{28\,980\hat{\mathbf{I}} - 34\,537\hat{\mathbf{J}} + 26\,030\hat{\mathbf{K}}}{6667.1^2}$$

$$= 0.00065196\hat{\mathbf{I}} - 0.00077698\hat{\mathbf{J}} + 0.00058559\hat{\mathbf{K}} \; (\text{rad}/\text{s}) \tag{j}$$

We find the angular acceleration $\dot{\mathbf{\Omega}}$ using Equation 7.5,

$$\dot{\mathbf{\Omega}} = -\frac{2(844.41)}{6667.1^2}(0.00065196\hat{\mathbf{I}} - 0.00077698\hat{\mathbf{J}} + 0.00058559\hat{\mathbf{K}})$$

$$= -2.4763(10^{-8})\hat{\mathbf{I}} + 2.9512(10^{-8})\hat{\mathbf{J}} - 2.2242(10^{-8})\hat{\mathbf{K}} \; (\text{rad}/\text{s}^2) \tag{k}$$

According to Equation 1.38, the relative velocity relation is

$$\mathbf{v}_B = \mathbf{v}_A + \mathbf{\Omega} \times \mathbf{r}_{\text{rel}} + \mathbf{v}_{\text{rel}} \tag{l}$$

where $\mathbf{r}_{\text{rel}}$ and $\mathbf{v}_{\text{rel}}$ are the position and velocity of B as measured relative to the moving xyz frame attached to A. From (a) and (b), we have

$$\mathbf{r}_{\text{rel}} = \mathbf{r}_B - \mathbf{r}_A = -5623.3\hat{\mathbf{I}} - 6844.8\hat{\mathbf{J}} - 3633.7\hat{\mathbf{K}} \; (\text{km}) \tag{m}$$

Substituting this, together with (b), (d) and (j) into (l), we get

$$0.93594\hat{\mathbf{I}} - 5.2409\hat{\mathbf{J}} - 5.5016\mathbf{K} = (-6.4842\hat{\mathbf{I}} - 3.6201\hat{\mathbf{J}} + 2.4159\hat{\mathbf{K}})$$

$$+ \begin{vmatrix} \hat{\mathbf{I}} & \hat{\mathbf{J}} & \hat{\mathbf{K}} \\ 0.00065196 & -0.00077698 & 0.00058559 \\ -5623.3 & -6844.8 & -3633.7 \end{vmatrix} + \mathbf{v}_{\text{rel}}$$

Solving for $\mathbf{v}_{\text{rel}}$ yields

$$\mathbf{v}_{\text{rel}} = 0.58865\hat{\mathbf{I}} - 0.69692\hat{\mathbf{J}} + 0.91414\hat{\mathbf{K}} \; (\text{km}/\text{s}) \tag{n}$$

The relative acceleration formula, Equation 1.42, is

$$\mathbf{a}_B = \mathbf{a}_A + \dot{\mathbf{\Omega}} \times \mathbf{r}_{\text{rel}} + \mathbf{\Omega} \times (\mathbf{\Omega} \times \mathbf{r}_{\text{rel}}) + 2\mathbf{\Omega} \times \mathbf{v}_{\text{rel}} + \mathbf{a}_{\text{rel}} \tag{o}$$

(Example 7.1 continued)

Substituting (e), (f), (j), (k), (m), and (n) into (o), we get

$$0.0073377\hat{\mathbf{I}} + 0.0037117\hat{\mathbf{J}} - 0.0022325\hat{\mathbf{K}}$$

$$= 0.00035876\hat{\mathbf{I}} - 0.0051989\hat{\mathbf{J}} - 0.0072975\mathbf{K}$$

$$+ \begin{vmatrix} \hat{\mathbf{I}} & \hat{\mathbf{J}} & \hat{\mathbf{K}} \\ -2.4770(10^{-8}) & 2.9520(10^{-8}) & -2.2248(10^{-8}) \\ -5623.3 & -6844.8 & -3633.7 \end{vmatrix}$$

$$+ (0.00065196\hat{\mathbf{I}} - 0.00077698\hat{\mathbf{J}} + 0.00058559\hat{\mathbf{K}})$$

$$\times \begin{vmatrix} \hat{\mathbf{I}} & \hat{\mathbf{J}} & \hat{\mathbf{K}} \\ 0.00065196 & -0.00077698 & 0.00058559 \\ -5623.3 & -6844.8 & -3633.7 \end{vmatrix}$$

$$+ 2 \begin{vmatrix} \hat{\mathbf{I}} & \hat{\mathbf{J}} & \hat{\mathbf{K}} \\ 0.00065196 & -0.00077698 & 0.00058559 \\ 0.58865 & -0.69692 & 0.91414 \end{vmatrix} + \mathbf{a}_{\text{rel}}$$

Carrying out the cross products, combining terms and solving for $\mathbf{a}_{\text{rel}}$ yields

$$\mathbf{a}_{\text{rel}} = 0.00043984\hat{\mathbf{I}} - 0.00038019\hat{\mathbf{J}} + 0.000017988\hat{\mathbf{K}} \ (\text{km/s}^2) \quad (p)$$

From (g), (h), and (i), we see that the orthogonal transformation matrix $[\mathbf{Q}]_{Xx}$ from the inertial XYZ frame into the co-moving xyz frame is

$$[\mathbf{Q}]_{Xx} = \begin{bmatrix} -0.040008 & 0.57977 & 0.81380 \\ -0.82977 & -0.47302 & 0.29620 \\ 0.55667 & -0.66341 & 0.5000 \end{bmatrix}$$

To get the components of $\mathbf{r}_{\text{rel}}$, $\mathbf{v}_{\text{rel}}$, and $\mathbf{a}_{\text{rel}}$ along the axes of the co-moving xyz frame of spacecraft A, we multiply each of their expressions as components in the XYZ frame [(m), (n) and (p), respectively] by $[\mathbf{Q}]_{Xx}$ as follows:

$$\mathbf{r}_{\text{rel}})_{xyz} = \begin{bmatrix} -0.040008 & 0.57977 & 0.81380 \\ -0.82977 & -0.47302 & 0.29620 \\ 0.55667 & -0.66341 & 0.5000 \end{bmatrix} \begin{Bmatrix} -5623.3 \\ -6844.8 \\ -3633.7 \end{Bmatrix} = \begin{Bmatrix} -6700.5 \\ 6827.4 \\ -406.22 \end{Bmatrix} \ (\text{km})$$

$$\mathbf{v}_{\text{rel}})_{xyz} = \begin{bmatrix} -0.040008 & 0.57977 & 0.81380 \\ -0.82977 & -0.47302 & 0.29620 \\ 0.55667 & -0.66341 & 0.5000 \end{bmatrix} \begin{Bmatrix} 0.58865 \\ -0.69692 \\ 0.91414 \end{Bmatrix} = \begin{Bmatrix} 0.31632 \\ 0.11199 \\ 1.2471 \end{Bmatrix} \ (\text{km/s})$$

$$\mathbf{a}_{\text{rel}})_{xyz} = \begin{bmatrix} -0.040008 & 0.57977 & 0.81380 \\ -0.82977 & -0.47302 & 0.29620 \\ 0.55667 & -0.66341 & 0.5000 \end{bmatrix} \begin{Bmatrix} 0.00043984 \\ -0.00038019 \\ 0.000017988 \end{Bmatrix}$$

$$= \begin{Bmatrix} -0.00022338 \\ -0.00017980 \\ 0.00050607 \end{Bmatrix} \ (\text{km/s}^2)$$

7.2 Relative motion in orbit

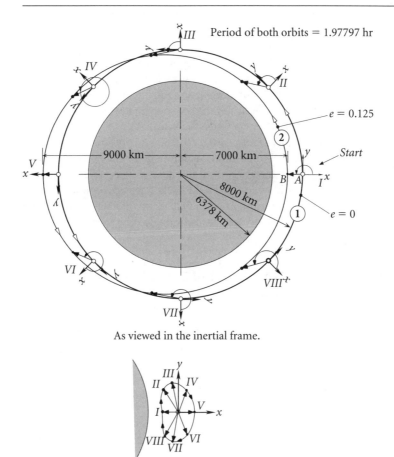

Figure 7.3 The spacecraft B in elliptical orbit 2 appears to orbit the observer A in circular orbit 1.

The motion of one spacecraft relative to another in orbit may be hard to visualize at first. Figure 7.3 is offered as an assist. Orbit 1 is circular and orbit 2 is an ellipse with eccentricity 0.125. Both orbits were chosen to have the same semimajor axis length, so they both have the same period. A co-moving frame is shown attached to the observers A in circular orbit 1. At epoch I the spacecraft B in elliptical orbit 2 is directly below the observers. In other words, A must draw an arrow in the negative local x direction to determine the position vector of B in the lower orbit. The figure shows eight different epochs $(I, II, III, \ldots)$, equally spaced around the circular orbit, at which observers A construct the position vector pointing from them to B in the elliptical orbit. Of course, A's frame is rotating, because its x axis must always be directed away from the earth. Observers A cannot sense this rotation and record the set of observations in their (to them) fixed xy coordinate system, as shown at the bottom of the figure. Coasting at a uniform speed along his circular orbit, A sees the other vehicle orbiting them clockwise in a sort of bean-shaped path. The distance between the two spacecraft in this case never becomes so great that the earth intervenes.

If A declared theirs to be an inertial frame of reference, they would be faced with the task of explaining the physical origin of the force holding B in its bean-shaped orbit. Of course, there is no such force. The apparent path is due to the actual, combined motion of both spacecraft in their free fall towards the earth. When B is below A (having a negative x coordinate), conservation of angular momentum demands that B move faster than A, thereby speeding up in A's positive y direction until the orbits cross ($x=0$). When B's x coordinate becomes positive, i.e., B is above A, the laws of momentum dictate that B slow down, which it does, progressing in A's negative y direction until the next crossing of the orbits. B then falls below and begins to pick up speed. The process repeats over and over. From inertial space, the process is the motion of two satellites on intersecting orbits, appearing not at all like the orbiting motion seen by the moving observers A.

7.3 Linearization of the equations of relative motion in orbit

Figure 7.4 shows two spacecraft in earth orbit. The inertial position vector of the target vehicle A is $\mathbf{r}_0$, and that of the chase vehicle B is $\mathbf{r}$. The position vector of the chase vehicle relative to the target is $\delta\mathbf{r}$, so that

$$\mathbf{r} = \mathbf{r}_0 + \delta\mathbf{r} \tag{7.6}$$

The symbol δ is used to represent the fact that the relative position vector has a magnitude which is very small compared to the magnitude of $\mathbf{r}_0$ (and $\mathbf{r}$); i.e.,

$$\frac{\delta r}{r_0} \ll 1 \tag{7.7}$$

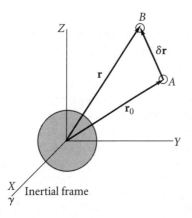

Figure 7.4 Position of chaser B relative to the target A.

7.3 Linearization of the equations of relative motion in orbit

where $\delta r = \|\delta \mathbf{r}\|$ and $r_0 = \|\mathbf{r}_0\|$. This is true if the two vehicles are in close proximity to each other, as is the case in a rendezvous maneuver. Our purpose in this section is to seek the equations of motion of the chase vehicle relative to the target.

The equation of motion of the chase vehicle B is

$$\ddot{\mathbf{r}} = -\mu \frac{\mathbf{r}}{r^3} \qquad (7.8)$$

where $r = \|\mathbf{r}\|$. Substituting Equation 7.6 into Equation 7.8 yields the equation of motion of the chaser relative to the target,

$$\delta \ddot{\mathbf{r}} = -\ddot{\mathbf{r}}_0 - \mu \frac{\mathbf{r}_0 + \delta \mathbf{r}}{r^3} \qquad (7.9)$$

We will simplify this equation by making use of the fact that $\|\delta \mathbf{r}\|$ is very small, as expressed in Equation 7.7. First, note that

$$r^2 = \mathbf{r} \cdot \mathbf{r} = (\mathbf{r}_0 + \delta \mathbf{r}) \cdot (\mathbf{r}_0 + \delta \mathbf{r}) = \mathbf{r}_0 \cdot \mathbf{r}_0 + 2\mathbf{r}_0 \cdot \delta \mathbf{r} + \delta \mathbf{r} \cdot \delta \mathbf{r}$$

Since $\mathbf{r}_0 \cdot \mathbf{r}_0 = r_0^2$ and $\delta \mathbf{r} \cdot \delta \mathbf{r} = \delta r^2$, we can factor out r_0^2 on the right to obtain

$$r^2 = r_0^2 \left[1 + \frac{2\mathbf{r}_0 \cdot \delta \mathbf{r}}{r_0^2} + \left(\frac{\delta r}{r_0} \right)^2 \right]$$

By virtue of Equation 7.7, we can neglect the last term in the brackets, so that

$$r^2 = r_0^2 \left(1 + \frac{2\mathbf{r}_0 \cdot \delta \mathbf{r}}{r_0^2} \right) \qquad (7.10)$$

In fact, we will neglect all powers of $\delta r/r_0$ greater than unity, wherever they appear. Since $r^{-3} = (r^2)^{-3/2}$, it follows from Equation 7.10 that

$$r^{-3} = r_0^{-3} \left(1 + \frac{2\mathbf{r}_0 \cdot \delta \mathbf{r}}{r_0^2} \right)^{-\frac{3}{2}} \qquad (7.11)$$

Using the binomial theorem (Equation 5.52) and neglecting terms of higher order than 1 in $\delta r/r_0$, we obtain

$$\left(1 + \frac{2\mathbf{r}_0 \cdot \delta \mathbf{r}}{r_0^2} \right)^{-\frac{3}{2}} = 1 + \left(-\frac{3}{2} \right) \left(\frac{2\mathbf{r}_0 \cdot \delta \mathbf{r}}{r_0^2} \right)$$

Therefore, Equation 7.11 becomes

$$r^{-3} = r_0^{-3} \left(1 - \frac{3}{r_0^2} \mathbf{r}_0 \cdot \delta \mathbf{r} \right)$$

which can be written

$$\frac{1}{r^3} = \frac{1}{r_0^3} - \frac{3}{r_0^5} \mathbf{r}_0 \cdot \delta \mathbf{r} \qquad (7.12)$$

Substituting Equation 7.12 into Equation 7.9 (the equation of motion), we get

$$\delta \ddot{\mathbf{r}} = -\ddot{\mathbf{r}}_0 - \mu \left(\frac{1}{r_0^3} - \frac{3}{r_0^5} \mathbf{r}_0 \cdot \delta \mathbf{r} \right) (\mathbf{r}_0 + \delta \mathbf{r})$$

$$= -\ddot{\mathbf{r}}_0 - \mu \left[\frac{\mathbf{r}_0 + \delta \mathbf{r}}{r_0^3} - \frac{3}{r_0^5} (\mathbf{r}_0 \cdot \delta \mathbf{r})(\mathbf{r}_0 + \delta \mathbf{r}) \right]$$

$$= -\ddot{\mathbf{r}}_0 - \mu \left[\frac{\mathbf{r}_0}{r_0^3} + \frac{\delta \mathbf{r}}{r_0^3} - \frac{3}{r_0^5} (\mathbf{r}_0 \cdot \delta \mathbf{r}) \mathbf{r}_0 + \overbrace{\text{terms of higher order than 1 in } \delta \mathbf{r}}^{\text{neglect}} \right]$$

That is,

$$\delta \ddot{\mathbf{r}} = -\ddot{\mathbf{r}}_0 - \mu \frac{\mathbf{r}_0}{r_0^3} - \frac{\mu}{r_0^3} \left[\delta \mathbf{r} - \frac{3}{r_0^2} (\mathbf{r}_0 \cdot \delta \mathbf{r}) \mathbf{r}_0 \right] \quad (7.13)$$

But the equation of motion of the target vehicle is

$$\ddot{\mathbf{r}}_0 = -\mu \frac{\mathbf{r}_0}{r_0^3}$$

Substituting this into Equation 7.13 finally yields

$$\delta \ddot{\mathbf{r}} = -\frac{\mu}{r_0^3} \left[\delta \mathbf{r} - \frac{3}{r_0^2} (\mathbf{r}_0 \cdot \delta \mathbf{r}) \mathbf{r}_0 \right] \quad (7.14)$$

This is the linearized version of Equation 7.8, the equation which governs the motion of the chaser with respect to the target. The expression is linear because $\delta \mathbf{r}$ appears only in the numerator and only to the first power throughout. We achieved this by dropping a lot of terms that are insignificant when Equation 7.7 is valid.

7.4 Clohessy–Wiltshire equations

Let us attach a moving frame of reference xyz to the target vehicle A, as shown in Figure 7.5. This is similar to Figure 7.1, the difference being that $\delta \mathbf{r}$ is restricted by Equation 7.7. The origin of the moving system is at A. The x axis lies along $\mathbf{r}_0$, so that

$$\hat{\mathbf{i}} = \frac{\mathbf{r}_0}{r_0} \quad (7.15)$$

The y axis is in the direction of the local horizon, and the z axis is normal to the orbital plane of A, such that $\hat{\mathbf{k}} = \hat{\mathbf{i}} \times \hat{\mathbf{j}}$. The inertial angular velocity of the moving frame of reference is $\mathbf{\Omega}$, and the inertial angular acceleration is $\dot{\mathbf{\Omega}}$.

According to the relative acceleration formula (Equation 1.42), we have

$$\ddot{\mathbf{r}} = \ddot{\mathbf{r}}_0 + \dot{\mathbf{\Omega}} \times \delta \mathbf{r} + \mathbf{\Omega} \times (\mathbf{\Omega} \times \delta \mathbf{r}) + 2\mathbf{\Omega} \times \delta \mathbf{v}_{\text{rel}} + \delta \mathbf{a}_{\text{rel}} \quad (7.16)$$

where, in terms of their components in the moving frame, the relative position, velocity and acceleration are given by

$$\delta \mathbf{r} = \delta x \hat{\mathbf{i}} + \delta y \hat{\mathbf{j}} + \delta z \hat{\mathbf{k}} \quad (7.17a)$$

7.4 Clohessy–Wiltshire equations

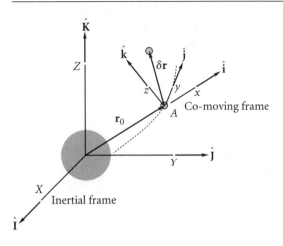

Figure 7.5 Co-moving Clohessy–Wiltshire frame.

$$\delta \mathbf{v}_{\mathrm{rel}} = \delta \dot{x}\hat{\mathbf{i}} + \delta \dot{y}\hat{\mathbf{j}} + \delta \dot{z}\hat{\mathbf{k}} \quad (7.17\mathrm{b})$$

$$\delta \mathbf{a}_{\mathrm{rel}} = \delta \ddot{x}\hat{\mathbf{i}} + \delta \ddot{y}\hat{\mathbf{j}} + \delta \ddot{z}\hat{\mathbf{k}} \quad (7.17\mathrm{c})$$

For simplicity, we assume at this point that the orbit of the target vehicle A is circular. (Note that for a space station in low-earth orbit, this is a very good assumption.) Then $\dot{\boldsymbol{\Omega}} = \mathbf{0}$. Substituting this together with Equation 7.6 into Equation 7.16 yields

$$\delta \ddot{\mathbf{r}} = \boldsymbol{\Omega} \times (\boldsymbol{\Omega} \times \delta \mathbf{r}) + 2\boldsymbol{\Omega} \times \delta \mathbf{v}_{\mathrm{rel}} + \delta \mathbf{a}_{\mathrm{rel}}$$

Applying the *bac − cab* rule to the first term on the right-hand side, we get

$$\delta \ddot{\mathbf{r}} = \boldsymbol{\Omega}(\boldsymbol{\Omega} \cdot \delta \mathbf{r}) - \Omega^2 \delta \mathbf{r} + 2\boldsymbol{\Omega} \times \delta \mathbf{v}_{\mathrm{rel}} + \delta \mathbf{a}_{\mathrm{rel}} \quad (7.18)$$

Since the orbit of A is circular, we may write the angular velocity as

$$\boldsymbol{\Omega} = n\hat{\mathbf{k}} \quad (7.19)$$

where n, the mean motion, is constant. Thus,

$$\boldsymbol{\Omega} \cdot \delta \mathbf{r} = n\hat{\mathbf{k}} \cdot (\delta x\hat{\mathbf{i}} + \delta y\hat{\mathbf{j}} + \delta z\hat{\mathbf{k}}) = n\delta z \quad (7.20)$$

and

$$\boldsymbol{\Omega} \times \delta \mathbf{v}_{\mathrm{rel}} = n\hat{\mathbf{k}} \times (\delta \dot{x}\hat{\mathbf{i}} + \delta \dot{y}\hat{\mathbf{j}} + \delta \dot{z}\hat{\mathbf{k}}) = -n\delta \dot{y}\hat{\mathbf{i}} + n\delta \dot{x}\hat{\mathbf{j}} \quad (7.21)$$

Substituting Equations 7.19, 7.20 and 7.21, along with Equations 7.17, into Equation 7.18 yields

$$\delta \ddot{\mathbf{r}} = n\hat{\mathbf{k}}(n\delta z) - n^2(\delta x\hat{\mathbf{i}} + \delta y\hat{\mathbf{j}} + \delta z\hat{\mathbf{k}}) + 2(-n\delta \dot{y}\hat{\mathbf{i}} + n\delta \dot{x}\hat{\mathbf{j}}) + \delta \ddot{x}\hat{\mathbf{i}} + \delta \ddot{y}\hat{\mathbf{j}} + \delta \ddot{z}\hat{\mathbf{k}}$$

Finally, collecting terms leads to

$$\delta \ddot{\mathbf{r}} = (-n^2\delta x - 2n\delta \dot{y} + \delta \ddot{x})\hat{\mathbf{i}} + (-n^2\delta y + 2n\delta \dot{x} + \delta \ddot{y})\hat{\mathbf{j}} + \delta \ddot{z}\hat{\mathbf{k}} \quad (7.22)$$

This expression gives the components of the chaser's absolute relative acceleration vector in terms of quantities measured in the moving reference.

Since the orbit of A is circular, the mean motion is found as

$$n = \frac{v}{r_0} = \frac{1}{r_0}\sqrt{\frac{\mu}{r_0}} = \sqrt{\frac{\mu}{r_0^3}}$$

Therefore,

$$\frac{\mu}{r_0^3} = n^2 \quad (7.23)$$

Recalling Equations 7.15 and 7.17a, we also note that

$$\mathbf{r}_0 \cdot \delta\mathbf{r} = (r_0\hat{\mathbf{i}}) \cdot (\delta x\hat{\mathbf{i}} + \delta y\hat{\mathbf{j}} + \delta z\hat{\mathbf{k}}) = r_0 \delta x \quad (7.24)$$

Substituting Equations 7.17a, 7.23 and 7.24 into Equation 7.14 (the equation of motion) yields

$$\delta\ddot{\mathbf{r}} = -n^2\left[\delta x\hat{\mathbf{i}} + \delta y\hat{\mathbf{j}} + \delta z\hat{\mathbf{k}} - \frac{3}{r_0^2}(r_0\delta x)r_0\hat{\mathbf{i}}\right] = 2n^2\delta x\hat{\mathbf{i}} - n^2\delta y\hat{\mathbf{j}} - n^2\delta z\hat{\mathbf{k}} \quad (7.25)$$

Combining Equation 7.22 (a kinematic relationship) and Equation 7.25 (the equation of motion), we obtain

$$(-n^2\delta x - 2n\delta\dot{y} + \delta\ddot{x})\hat{\mathbf{i}} + (-n^2\delta y + 2n\delta\dot{x} + \delta\ddot{y})\hat{\mathbf{j}} + \delta\ddot{z}\hat{\mathbf{k}} = 2n^2\delta x\hat{\mathbf{i}} - n^2\delta y\hat{\mathbf{j}} - n^2\delta z\hat{\mathbf{k}}$$

Upon collecting terms to the left-hand side, we get

$$(\delta\ddot{x} - 3n^2\delta x - 2n\delta\dot{y})\hat{\mathbf{i}} + (\delta\ddot{y} + 2n\delta\dot{x})\hat{\mathbf{j}} + (\delta\ddot{z} + n^2\delta z)\hat{\mathbf{k}} = 0$$

That is,

$$\delta\ddot{x} - 3n^2\delta x - 2n\delta\dot{y} = 0 \quad (7.26a)$$

$$\delta\ddot{y} + 2n\delta\dot{x} = 0 \quad (7.26b)$$

$$\delta\ddot{z} + n^2\delta z = 0 \quad (7.26c)$$

These are the Clohessy–Wiltshire (CW) equations. When using these equations we will refer to the moving frame of reference in which they were derived as the Clohessy–Wiltshire frame (or CW frame). Equations 7.26 are a set of coupled, second order differential equations with constant coefficients. The initial conditions are

$$\text{At } t = 0 \quad \delta x = \delta x_0 \quad \delta y = \delta y_0 \quad \delta z = \delta z_0$$
$$\delta\dot{x} = \delta\dot{x}_0 \quad \delta\dot{y} = \delta\dot{y}_0 \quad \delta\dot{z} = \delta\dot{z}_0 \quad (7.27)$$

From Equation 7.26b,

$$\frac{d}{dt}(\delta\dot{y} + 2n\delta x) = 0$$

which means

$$\delta\dot{y} + 2n\delta x = \text{const}$$

7.4 Clohessy–Wiltshire equations

We find the constant by evaluating the left-hand side at $t=0$. Therefore,

$$\delta\dot{y} + 2n\delta x = \delta\dot{y}_0 + 2n\delta x_0$$

so that

$$\delta\dot{y} = \delta\dot{y}_0 + 2n(\delta x_0 - \delta x) \qquad (7.28)$$

Substituting this result into Equation 7.26a yields

$$\delta\ddot{x} - 3n^2\delta x - 2n[\delta\dot{y}_0 + 2n(\delta x_0 - \delta x)] = 0$$

which, upon rearrangement, becomes,

$$\delta\ddot{x} + n^2\delta x = 2n\delta\dot{y}_0 + 4n^2\delta x_0 \qquad (7.29)$$

The solution of this differential equation is

$$\delta x = \overbrace{A\sin nt + B\cos nt}^{\text{complementary solution}} + \overbrace{\frac{1}{n^2}(2n\delta\dot{y}_0 + 4n^2\delta x_0)}^{\text{particular solution}}$$

or

$$\delta x = A\sin nt + B\cos nt + \frac{2}{n}\delta\dot{y}_0 + 4\delta x_0 \qquad (7.30)$$

Differentiating this equation once with respect to time, we obtain

$$\delta\dot{x} = nA\cos nt - nB\sin nt \qquad (7.31)$$

Evaluating Equation 7.30 at $t=0$ we find

$$\delta x_0 = B + \frac{2}{n}\delta\dot{y}_0 + 4\delta x_0 \quad \Rightarrow \quad B = -3\delta x_0 - 2\frac{\delta\dot{y}_0}{n}$$

Evaluating Equation 7.31 at $t=0$ yields

$$\delta\dot{x}_0 = nA \quad \Rightarrow \quad A = \frac{\delta\dot{x}_0}{n}$$

Substituting these values of A and B back into Equation 7.30 leads to

$$\delta x = \frac{\delta\dot{x}_0}{n}\sin nt + \left(-3\delta x_0 - 2\frac{\delta\dot{y}_0}{n}\right)\cos nt + \frac{2}{n}\delta\dot{y}_0 + 4\delta x_0$$

which, upon combining terms, becomes

$$\delta x = (4 - 3\cos nt)\delta x_0 + \frac{\sin nt}{n}\delta\dot{x}_0 + \frac{2}{n}(1 - \cos nt)\delta\dot{y}_0 \qquad (7.32)$$

Therefore,

$$\delta\dot{x} = 3n\sin nt\,\delta x_0 + \cos nt\,\delta\dot{x}_0 + 2\sin nt\,\delta\dot{y}_0 \qquad (7.33)$$

Substituting Equation 7.32 into Equation 7.28 yields

$$\delta \dot{y} = \delta \dot{y}_0 + 2n\left[\delta x_0 - (4 - 3\cos nt)\delta x_0 - \frac{\sin nt}{n}\delta \dot{x}_0 - \frac{2}{n}(1 - \cos nt)\delta \dot{y}_0\right]$$

which simplifies to become

$$\delta \dot{y} = 6n(\cos nt - 1)\delta x_0 - 2\sin nt\, \delta \dot{x}_0 + (4\cos nt - 3)\delta \dot{y}_0 \qquad (7.34)$$

Integrating this expression with respect to time, we find

$$\delta y = 6n\left(\frac{1}{n}\sin nt - t\right)\delta x_0 + \frac{2}{n}\cos nt\, \delta \dot{x}_0 + \left(\frac{4}{n}\sin nt - 3t\right)\delta \dot{y}_0 + C \qquad (7.35)$$

Evaluating δy at $t = 0$ yields

$$\delta y_0 = \frac{2}{n}\delta \dot{x}_0 + C \quad \Rightarrow \quad C = \delta y_0 - \frac{2}{n}\delta \dot{x}_0$$

Substituting this value for C into Equation 7.35, we get

$$\delta y = 6(\sin nt - nt)\delta x_0 + \delta y_0 + \frac{2}{n}(\cos nt - 1)\delta \dot{x}_0 + \left(\frac{4}{n}\sin nt - 3t\right)\delta \dot{y}_0 \qquad (7.36)$$

Finally, the solution of Equation 7.26c is

$$\delta z = D\cos nt + E\sin nt \qquad (7.37)$$

so that

$$\delta \dot{z} = -nD\sin nt + nE\cos nt \qquad (7.38)$$

We evaluate these two expressions at $t = 0$ to obtain the constants of integration:

$$\delta z_0 = D$$
$$\delta \dot{z}_0 = nE$$

Putting these values of D and E back into Equations 7.36 and 7.38 yields

$$\delta z = \cos nt\, \delta z_0 + \frac{1}{n}\sin nt\, \delta \dot{z}_0 \qquad (7.39)$$

$$\delta \dot{z} = -n\sin nt\, \delta z_0 + \cos nt\, \delta \dot{z}_0 \qquad (7.40)$$

Now that we have finished solving the Clohessy–Wiltshire equations, let us change our notation a bit and denote the x, y and z components of relative velocity in the moving frame as δu, δv and δw, respectively. That is,

$$\delta u = \delta \dot{x} \quad \delta v = \delta \dot{y} \quad \delta w = \delta \dot{z}$$

The initial conditions on the relative velocity components are then written

$$\delta u_0 = \delta \dot{x}_0 \quad \delta v_0 = \delta \dot{y}_0 \quad \delta w_0 = \delta \dot{z}_0$$

7.4 Clohessy–Wiltshire equations

Using this notation we write Equations 7.32, 7.33, 7.34, 7.36, 7.39 and 7.40 as

$$\delta x = (4 - 3\cos nt)\delta x_0 + \frac{\sin nt}{n}\delta u_0 + \frac{2}{n}(1 - \cos nt)\delta v_0$$

$$\delta y = 6(\sin nt - nt)\delta x_0 + \delta y_0 + \frac{2}{n}(\cos nt - 1)\delta u_0 + \frac{1}{n}(4\sin nt - 3nt)\delta v_0$$

$$\delta z = \cos nt\, \delta z_0 + \frac{1}{n}\sin nt\, \delta w_0 \tag{7.41}$$

$$\delta u = 3n\sin nt\, \delta x_0 + \cos nt\, \delta u_0 + 2\sin nt\, \delta v_0$$

$$\delta v = 6n(\cos nt - 1)\delta x_0 - 2\sin nt\, \delta u_0 + (4\cos nt - 3)\delta v_0$$

$$\delta w = -n\sin nt\, \delta z_0 + \cos nt\, \delta w_0$$

Let us introduce matrix notation to define the relative position and velocity vectors

$$\{\delta \mathbf{r}(t)\} = \begin{Bmatrix} \delta x(t) \\ \delta y(t) \\ \delta z(t) \end{Bmatrix} \quad \{\delta \mathbf{v}(t)\} = \begin{Bmatrix} \delta u(t) \\ \delta v(t) \\ \delta w(t) \end{Bmatrix}$$

and their initial values (at $t=0$)

$$\{\delta \mathbf{r}_0\} = \begin{Bmatrix} \delta x_0 \\ \delta y_0 \\ \delta z_0 \end{Bmatrix} \quad \{\delta \mathbf{v}_0\} = \begin{Bmatrix} \delta u_0 \\ \delta v_0 \\ \delta w_0 \end{Bmatrix}$$

Observe that we have dropped the subscript rel introduced in Equations 7.17 because it is superfluous in rendezvous analysis, where all kinematic quantities are relative to the Clohessy–Wiltshire frame. In matrix notation Equations 7.41 appear more compactly as

$$\{\delta \mathbf{r}(t)\} = [\mathbf{\Phi}_{rr}(t)]\{\delta \mathbf{r}_0\} + [\mathbf{\Phi}_{rv}(t)]\{\delta \mathbf{v}_0\} \tag{7.42a}$$

$$\{\delta \mathbf{v}(t)\} = [\mathbf{\Phi}_{vr}(t)]\{\delta \mathbf{r}_0\} + [\mathbf{\Phi}_{vv}(t)]\{\delta \mathbf{v}_0\} \tag{7.42b}$$

where the Clohessy–Wiltshire matrices are

$$[\mathbf{\Phi}_{rr}(t)] = \begin{bmatrix} 4 - 3\cos nt & 0 & 0 \\ 6(\sin nt - nt) & 1 & 0 \\ 0 & 0 & \cos nt \end{bmatrix}$$

$$[\mathbf{\Phi}_{rv}(t)] = \begin{bmatrix} \frac{1}{n}\sin nt & \frac{2}{n}(1 - \cos nt) & 0 \\ \frac{2}{n}(\cos nt - 1) & \frac{1}{n}(4\sin nt - 3nt) & 0 \\ 0 & 0 & \frac{1}{n}\sin nt \end{bmatrix} \tag{7.43}$$

$$[\mathbf{\Phi}_{vr}(t)] = \begin{bmatrix} 3n\sin nt & 0 & 0 \\ 6n(\cos nt - 1) & 0 & 0 \\ 0 & 0 & -n\sin nt \end{bmatrix}$$

$$[\mathbf{\Phi}_{vv}(t)] = \begin{bmatrix} \cos nt & 2\sin nt & 0 \\ -2\sin nt & 4\cos nt - 3 & 0 \\ 0 & 0 & \cos nt \end{bmatrix}$$

7.5 TWO-IMPULSE RENDEZVOUS MANEUVERS

Figure 7.6 illustrates the rendezvous problem. At time $t = 0^-$ (the instant preceding $t = 0$), the position $\delta \mathbf{r}_0$ and velocity $\delta \mathbf{v}_0^-$ of the chase vehicle B relative to the target A are known. At $t = 0$ an impulsive maneuver instantaneously changes the relative velocity to $\delta \mathbf{v}_0^+$ at $t = 0^+$ (the instant after $t = 0$). The components of $\delta \mathbf{v}_0^+$ are shown in Figure 7.6. We must determine the values of δu_0^+, δv_0^+, δw_0^+, at the beginning of the rendezvous trajectory, so that B will arrive at the target in a specified time t_f. The delta-v required to place B on the rendezvous trajectory is

$$\{\Delta \mathbf{v}_0\} = \{\delta \mathbf{v}_0^+\} - \{\delta \mathbf{v}_0^-\} = \begin{Bmatrix} \delta u_0^+ \\ \delta v_0^+ \\ \delta w_0^+ \end{Bmatrix} - \begin{Bmatrix} \delta u_0^- \\ \delta v_0^- \\ \delta w_0^- \end{Bmatrix} \qquad (7.44)$$

At time t_f, B arrives at A, at the origin of the co-moving frame, which means $\{\delta \mathbf{r}_f\} = \{\delta \mathbf{r}(t_f)\} = \{\mathbf{0}\}$. Evaluating Equation 7.42a at t_f, we find

$$\{\mathbf{0}\} = [\mathbf{\Phi}_{rr}(t_f)]\{\delta \mathbf{r}_0\} + [\mathbf{\Phi}_{rv}(t_f)]\{\delta \mathbf{v}_0^+\} \qquad (7.45)$$

Solving this for $\{\delta \mathbf{v}_0^+\}$ yields

$$\{\delta \mathbf{v}_0^+\} = -[\mathbf{\Phi}_{rv}(t_f)]^{-1}[\mathbf{\Phi}_{rr}(t_f)]\{\delta \mathbf{r}_0\} \qquad (7.46)$$

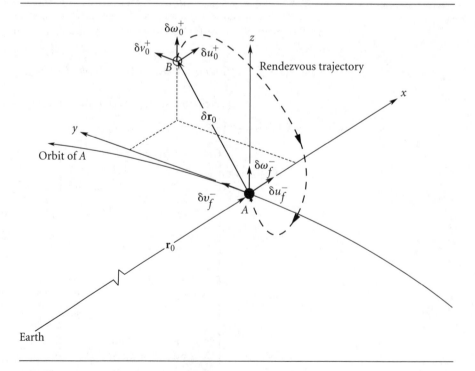

Figure 7.6 Rendezvous with a target A in the neighborhood of the chase vehicle B.

where $[\mathbf{\Phi}_{rv}(t_f)]^{-1}$ is the matrix inverse of $[\mathbf{\Phi}_{rv}(t_f)]$. We know the velocity $\delta\mathbf{v}_0^+$ at the beginning of the rendezvous path substituting Equation 7.46 into Equation 7.42b we obtain the velocity $\delta\mathbf{v}_f^-$ at which B arrives at the target A, when $t = t_f^-$:

$$\{\delta\mathbf{v}_f^-\} = [\mathbf{\Phi}_{vr}(t_f)]\{\delta\mathbf{r}_0\} + [\mathbf{\Phi}_{vv}(t_f)]\{\delta\mathbf{v}_0^+\}$$
$$= [\mathbf{\Phi}_{vr}(t_f)]\{\delta\mathbf{r}_0\} + [\mathbf{\Phi}_{vv}(t_f)](-[\mathbf{\Phi}_{rv}(t_f)]^{-1}[\mathbf{\Phi}_{rr}(t_f)]\{\delta\mathbf{r}_0\})$$

Simplifying, we get

$$\{\delta\mathbf{v}_f^-\} = ([\mathbf{\Phi}_{vr}(t_f)] - [\mathbf{\Phi}_{vv}(t_f)][\mathbf{\Phi}_{rv}(t_f)]^{-1}[\mathbf{\Phi}_{rr}(t_f)])\{\delta\mathbf{r}_0\} \quad (7.47)$$

Obviously, an impulsive delta-v maneuver is required at $t = t_f$ to bring vehicle B to rest relative to A ($\delta\mathbf{v}_f^+ = \mathbf{0}$):

$$\{\Delta\mathbf{v}_f\} = \{\delta\mathbf{v}_f^+\} - \{\delta\mathbf{v}_f^-\} = \{\mathbf{0}\} - \{\delta\mathbf{v}_f^-\} = -\{\delta\mathbf{v}_f^-\} \quad (7.48)$$

Note that in Equations 7.44 and 7.48 we are using the difference between relative velocities to calculate delta-v, which is the difference in absolute velocities. To show that this is valid, use Equation 1.38, to write

$$\begin{aligned} \mathbf{v}^- &= \mathbf{v}_0^- + \mathbf{\Omega}^- \times \mathbf{r}_{rel}^- + \mathbf{v}_{rel}^- \\ \mathbf{v}^+ &= \mathbf{v}_0^+ + \mathbf{\Omega}^+ \times \mathbf{r}_{rel}^+ + \mathbf{v}_{rel}^+ \end{aligned} \quad (7.49)$$

Since the target is passive, the impulsive maneuver has no effect on its state of motion, which means $\mathbf{v}_0^+ = \mathbf{v}_0^-$ and $\mathbf{\Omega}^+ = \mathbf{\Omega}^-$. Furthermore, by definition of an impulsive maneuver, there is no change in the position, i.e., $\mathbf{r}_{rel}^+ = \mathbf{r}_{rel}^-$. It follows from Equation 7.49 that

$$\mathbf{v}^+ - \mathbf{v}^- = \mathbf{v}_{rel}^+ - \mathbf{v}_{rel}^- \quad \text{or} \quad \Delta\mathbf{v} = \Delta\mathbf{v}_{rel}$$

EXAMPLE 7.2

A space station and spacecraft are in orbits with the following parameters:

	Space station	Spacecraft
Perigee × apogee (altitude)	300 km circular	318.50 × 515.51 km
Period (computed using above data)	1.508 hr	1.548 hr
True anomaly, θ	60°	349.65°
Inclination, i	40°	40.130°
RA, Ω	20°	19.819°
Argument of perigee, ω	0° (arbitrary)	70.662°

Compute the total delta-v required for an eight-hour, two-impulse rendezvous trajectory.

We use the given data in Algorithm 4.1 to obtain the state vectors of the two spacecraft in the geocentric equatorial frame.

Space station:

$$\mathbf{r}_0 = 1622.39\hat{\mathbf{I}} + 5305.10\hat{\mathbf{J}} + 3717.44\hat{\mathbf{K}} \text{ (km)}$$

$$\mathbf{v}_0 = -7.29977\hat{\mathbf{I}} + 0.492357\hat{\mathbf{J}} + 2.48318\hat{\mathbf{K}} \text{ (km/s)}$$

(Example 7.2 continued)

Spacecraft:

$$\mathbf{r} = 1612.75\hat{\mathbf{I}} + 5310.19\hat{\mathbf{J}} + 3750.33\hat{\mathbf{K}} \text{ (km)}$$

$$\mathbf{v} = -7.35321\hat{\mathbf{I}} + 0.463856\hat{\mathbf{J}} + 2.46920\hat{\mathbf{K}} \text{ (km/s)}$$

The space station reference frame unit vectors (at this instant) are, by definition:

$$\hat{\mathbf{i}} = \frac{\mathbf{r}_0}{\|\mathbf{r}_0\|} = 0.242945\hat{\mathbf{I}} + 0.794415\hat{\mathbf{J}} + 0.556670\hat{\mathbf{K}}$$

$$\hat{\mathbf{j}} = \frac{\mathbf{v}_0}{\|\mathbf{v}_0\|} = -0.944799\hat{\mathbf{I}} + 0.063725\hat{\mathbf{J}} + 0.321394\hat{\mathbf{K}}$$

$$\hat{\mathbf{k}} = \hat{\mathbf{i}} \times \hat{\mathbf{j}} = 0.219846\hat{\mathbf{I}} - 0.604023\hat{\mathbf{J}} + 0.766044\hat{\mathbf{K}}$$

Therefore, the transformation matrix from the geocentric equatorial frame into space station frame is (at this instant)

$$[Q]_{Xx} = \begin{bmatrix} 0.242945 & 0.794415 & 0.556670 \\ -0.944799 & 0.063725 & 0.321394 \\ 0.219846 & -0.604023 & 0.766044 \end{bmatrix}$$

The position vector of the spacecraft relative to the space station (in the geocentric equatorial frame) is

$$\delta\mathbf{r} = \mathbf{r} - \mathbf{r}_0 = -9.63980\hat{\mathbf{I}} + 5.08240\hat{\mathbf{J}} + 32.8821\hat{\mathbf{K}} \text{ (km)}$$

The relative velocity is given by the formula (Equation 1.38)

$$\delta\mathbf{v} = \mathbf{v} - \mathbf{v}_0 - \mathbf{\Omega}_{\text{space station}} \times \delta\mathbf{r}$$

where $\mathbf{\Omega}_{\text{space station}} = n\hat{\mathbf{k}}$ and n, the mean motion of the space station, is

$$n = \frac{v_0}{r_0} = \frac{7.72627}{6678} = 0.00115697 \text{ rad/s} \qquad (a)$$

Thus

$$\delta\mathbf{v} = -7.35321\hat{\mathbf{I}} + 0.463856\hat{\mathbf{J}} + 2.46920\hat{\mathbf{K}} - (-7.29977\hat{\mathbf{I}} + 0.492357\hat{\mathbf{J}} + 2.48318\hat{\mathbf{K}})$$

$$- (0.00115697) \begin{vmatrix} \hat{\mathbf{I}} & \hat{\mathbf{J}} & \hat{\mathbf{K}} \\ 0.219846 & -0.604023 & 0.766044 \\ -9.63980 & 5.08240 & 32.8821 \end{vmatrix}$$

so that

$$\delta\mathbf{v} = -0.024854\hat{\mathbf{I}} - 0.01159370\hat{\mathbf{J}} - 0.00853577\hat{\mathbf{K}} \text{ (km/s)}$$

In space station coordinates, the relative position vector $\delta\mathbf{r}_0$ at the beginning of the rendezvous maneuver is

$$\{\delta\mathbf{r}_0\} = [Q]_{Xx}\{\delta\mathbf{r}\} = \begin{bmatrix} 0.242945 & 0.794415 & 0.556670 \\ -0.944799 & 0.063725 & 0.321394 \\ 0.219846 & -0.604023 & 0.766044 \end{bmatrix} \begin{Bmatrix} -9.63980 \\ 5.08240 \\ 32.8821 \end{Bmatrix}$$

$$= \begin{Bmatrix} 20 \\ 20 \\ 20 \end{Bmatrix} \text{ (km)} \qquad (b)$$

7.5 Two-impulse rendezvous maneuvers

Likewise, the relative velocity $\delta \mathbf{v}_0^-$ just *before* launch into the rendezvous trajectory is

$$\{\delta \mathbf{v}_0^-\} = [\mathbf{Q}]_{Xx}\{\delta \mathbf{v}\} = \begin{bmatrix} 0.242945 & 0.794415 & 0.556670 \\ -0.944799 & 0.063725 & 0.321394 \\ 0.219846 & -0.604023 & 0.766044 \end{bmatrix} \begin{Bmatrix} -0.024854 \\ -0.0115937 \\ -0.00853578 \end{Bmatrix}$$

$$= \begin{Bmatrix} -0.02000 \\ 0.02000 \\ -0.005000 \end{Bmatrix} \text{(km/s)}$$

The Clohessy–Wiltshire matrices, for $t = t_f = 8\,\text{hr} = 28\,800\,\text{s}$ and $n = 0.00115697$ rad/s [from (a)], are

$$[\mathbf{\Phi}_{rr}] = \begin{bmatrix} 4 - 3\cos nt & 0 & 0 \\ 6(\sin nt - nt) & 1 & 0 \\ 0 & 0 & \cos nt \end{bmatrix} = \begin{bmatrix} 4.98383 & 0 & 0 \\ -194.257 & 1.000 & 0 \\ 0 & 0 & -0.327942 \end{bmatrix}$$

$$[\mathbf{\Phi}_{rv}] = \begin{bmatrix} \frac{1}{n}\sin nt & \frac{2}{n}(1-\cos nt) & 0 \\ \frac{2}{n}(\cos nt - 1) & \frac{1}{n}(4\sin nt - 3nt) & 0 \\ 0 & 0 & \frac{1}{n}\sin nt \end{bmatrix}$$

$$= \begin{bmatrix} 816.525 & 2295.54 & 0 \\ -2295.54 & -83\,133.9 & 0 \\ 0 & 0 & 816.525 \end{bmatrix}$$

$$[\mathbf{\Phi}_{vr}] = \begin{bmatrix} 3n\sin nt & 0 & 0 \\ 6n(\cos nt - 1) & 0 & 0 \\ 0 & 0 & -n\sin nt \end{bmatrix} = \begin{bmatrix} 0.00327897 & 0 & 0 \\ -0.00921837 & 0 & 0 \\ 0 & 0 & -0.00109299 \end{bmatrix}$$

$$[\mathbf{\Phi}_{vv}] = \begin{bmatrix} \cos nt & 2\sin nt & 0 \\ -2\sin nt & 4\cos nt - 3 & 0 \\ 0 & 0 & \cos nt \end{bmatrix}$$

$$= \begin{bmatrix} -0.327942 & 1.88940 & 0 \\ -1.88940 & -4.31177 & 0 \\ 0 & 0 & -0.327942 \end{bmatrix}$$

From Equation 7.46 and (b) we find $\delta \mathbf{v}_0^+$:

$$\begin{Bmatrix} \delta u_0^+ \\ \delta v_0^+ \\ \delta w_0^+ \end{Bmatrix} = -\begin{bmatrix} 816.525 & 2295.54 & 0 \\ -2295.54 & -83\,133.9 & 0 \\ 0 & 0 & 816.525 \end{bmatrix}^{-1}$$

$$\times \begin{bmatrix} 4.98383 & 0 & 0 \\ -194.257 & 1.000 & 0 \\ 0 & 0 & -0.327942 \end{bmatrix} \begin{Bmatrix} 20 \\ 20 \\ 20 \end{Bmatrix}$$

(Example 7.2 continued)

$$= -\begin{bmatrix} 816.525 & 2295.54 & 0 \\ -2295.54 & -83\,133.9 & 0 \\ 0 & 0 & 816.525 \end{bmatrix}^{-1} \begin{Bmatrix} 99.6765 \\ -3865.14 \\ -6.55884 \end{Bmatrix}$$

$$= \begin{Bmatrix} 0.00936084 \\ -0.0467514 \\ 0.00803263 \end{Bmatrix} \text{ (km/s)} \qquad (c)$$

From Equation 7.42b, evaluated at $t = t_f$, we have

$$\{\delta \mathbf{v}_f\} = [\mathbf{\Phi}_{vr}(t_f)]\{\delta \mathbf{r}_0\} + [\mathbf{\Phi}_{vv}(t_f)]\{\delta \mathbf{v}_0^+\}$$

Substituting (b) and (c),

$$\begin{Bmatrix} \delta u_f^- \\ \delta v_f^- \\ \delta w_f^- \end{Bmatrix} = \begin{bmatrix} 0.00327897 & 0 & 0 \\ -0.00921837 & 0 & 0 \\ 0 & 0 & -0.00109299 \end{bmatrix} \begin{Bmatrix} 20 \\ 20 \\ 20 \end{Bmatrix}$$

$$+ \begin{bmatrix} -0.327942 & 1.88940 & 0 \\ -1.88940 & -4.31177 & 0 \\ 0 & 0 & -0.327942 \end{bmatrix} \begin{Bmatrix} 0.00936084 \\ -0.0467514 \\ 0.00803263 \end{Bmatrix}$$

$$\begin{Bmatrix} \delta u_f^- \\ \delta v_f^- \\ \delta w_f^- \end{Bmatrix} = \begin{Bmatrix} -0.0258223 \\ -0.000472444 \\ -0.0222449 \end{Bmatrix} \text{ (km/s)} \qquad (d)$$

Delta-v at the beginning of the rendezvous maneuver is found as

$$\{\Delta \mathbf{v}_0\} = \{\delta \mathbf{v}_0^+\} - \{\delta \mathbf{v}_0^-\} = \begin{Bmatrix} 0.00936084 \\ -0.0467514 \\ 0.00803263 \end{Bmatrix} - \begin{Bmatrix} -0.02 \\ 0.02 \\ -0.005 \end{Bmatrix} = \begin{Bmatrix} 0.0293608 \\ -0.0667514 \\ 0.0130326 \end{Bmatrix}$$

Delta-v at the conclusion of the maneuver is

$$\{\Delta \mathbf{v}_f\} = \{\delta \mathbf{v}_f^+\} - \{\delta \mathbf{v}_f^-\} = \begin{Bmatrix} 0 \\ 0 \\ 0 \end{Bmatrix} - \begin{Bmatrix} -0.0258223 \\ -0.000472444 \\ -0.0222449 \end{Bmatrix} = \begin{Bmatrix} 0.0258223 \\ 0.000472444 \\ 0.0222449 \end{Bmatrix} \text{ (km/s)}$$

The total delta-v requirement is

$$\Delta v_{\text{total}} = \|\Delta \mathbf{v}_0\| + \|\Delta \mathbf{v}_f\| = 0.0740787 + 0.03559465 = 0.109673 \text{ km/s} = \underline{109.7 \text{ m/s}}$$

7.5 Two-impulse rendezvous maneuvers

From Equation 7.42a, we have, for $0 < t < t_f$,

$$\begin{Bmatrix} \delta x(t) \\ \delta y(t) \\ \delta z(t) \end{Bmatrix} = \begin{bmatrix} 4 - 3\cos nt & 0 & 0 \\ 6(\sin nt - nt) & 1 & 0 \\ 0 & 0 & \cos nt \end{bmatrix} \begin{Bmatrix} 20 \\ 20 \\ 20 \end{Bmatrix}$$

$$+ \begin{bmatrix} \frac{1}{n}\sin nt & \frac{2}{n}(1 - \cos nt) & 0 \\ \frac{2}{n}(\cos nt - 1) & \frac{1}{n}(4\sin nt - 3nt) & 0 \\ 0 & 0 & \frac{1}{n}\sin nt \end{bmatrix} \begin{Bmatrix} 0.00936084 \\ -0.0467514 \\ 0.00803263 \end{Bmatrix}$$

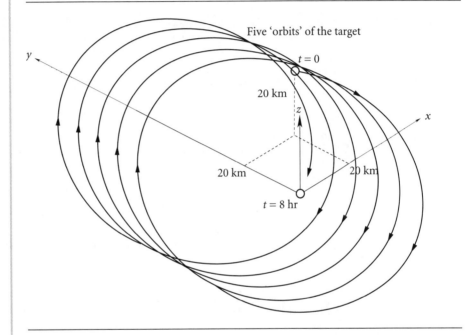

Figure 7.7 Rendezvous trajectory of the chase vehicle relative to the target.

Substituting n from (a), we obtain the relative position vector as a function of time. It is plotted in Figure 7.7.

EXAMPLE 7.3

A target and a chase vehicle are in the same 300 km circular earth orbit. The chaser is 2 km behind the target when the chaser initiates a two-impulse rendezvous maneuver so as to rendezvous with the target in 1.49 hours. Find the total delta-v requirement.

For the circular orbit

$$v = \sqrt{\frac{\mu}{r}} = \sqrt{\frac{398\,600}{6378 + 300}} = 7.726 \text{ km/s} \quad \text{(a)}$$

so that the mean motion is

$$n = \frac{v}{r} = \frac{7.726}{6678} = 0.0011569 \text{ rad/s} \quad \text{(b)}$$

(Example 7.3 continued)

For this mean motion and the rendezvous trajectory time $t = 1.49\,\text{hr} = 5364\,\text{s}$, the Clohessy–Wiltshire matrices are

$$[\Phi_{rr}] = \begin{bmatrix} 1.0090 & 0 & 0 \\ -37.699 & 1 & 0 \\ 0 & 0 & 0.99700 \end{bmatrix}$$

$$[\Phi_{rv}] = \begin{bmatrix} -66.946 & 5.1928 & 0 \\ -5.1928 & -16\,360 & 0 \\ 0 & 0 & -66.946 \end{bmatrix}$$

$$[\Phi_{vr}] = \begin{bmatrix} -2.6881 \times 10^{-4} & 0 & 0 \\ -2.0851 \times 10^{-5} & 0 & 0 \\ 0 & 0 & 8.9603 \times 10^{-5} \end{bmatrix} \quad (c)$$

$$[\Phi_{vv}] = \begin{bmatrix} 0.99700 & -0.15490 & 0 \\ 0.15490 & 0.98798 & 0 \\ 0 & 0 & 0.99700 \end{bmatrix}$$

The initial and final positions of the chaser in the CW frame are

$$\{\delta\mathbf{r}_0\} = \begin{Bmatrix} 0 \\ -2 \\ 0 \end{Bmatrix}\,(\text{km}) \qquad \{\delta\mathbf{r}_f\} = \begin{Bmatrix} 0 \\ 0 \\ 0 \end{Bmatrix} \quad (d)$$

Thus, solving the first CW equation, $\{\delta\mathbf{r}_f\} = [\Phi_{rr}]\{\delta\mathbf{r}_0\} + [\Phi_{rv}]\{\delta\mathbf{v}_0^+\}$, for $\{\delta\mathbf{v}_0^+\}$, we get

$$\{\delta\mathbf{v}_0^+\} = -[\Phi_{rv}]^{-1}[\Phi_{rr}]\{\delta\mathbf{r}_0\} = -\begin{bmatrix} -0.014937 & -4.7412 \times 10^{-6} & 0 \\ 4.7412 \times 10^{-6} & -6.1124 \times 10^{-5} & 0 \\ 0 & 0 & -0.014937 \end{bmatrix}$$

$$\times \begin{bmatrix} 1.0090 & 0 & 0 \\ -37.699 & 1 & 0 \\ 0 & 0 & 0.99700 \end{bmatrix} \begin{Bmatrix} 0 \\ -2 \\ 0 \end{Bmatrix}$$

$$\{\delta\mathbf{v}_0^+\} = \begin{Bmatrix} -9.4824 \times 10^{-6} \\ -1.2225 \times 10^{-4} \\ 0 \end{Bmatrix}\,(\text{km/s}) \quad (e)$$

Therefore, the second CW equation, $\{\delta\mathbf{v}_f^-\} = [\Phi_{vr}]\{\delta\mathbf{r}_0\} + [\Phi_{vv}]\{\delta\mathbf{v}_0^+\}$, yields

$$\{\delta\mathbf{v}_f^-\} = \begin{bmatrix} -2.6881 \times 10^{-4} & 0 & 0 \\ -2.0851 \times 10^{-5} & 0 & 0 \\ 0 & 0 & 8.9603 \times 10^{-5} \end{bmatrix} \begin{Bmatrix} 0 \\ -2 \\ 0 \end{Bmatrix}$$

$$+ \begin{bmatrix} 0.99700 & -0.15490 & 0 \\ 0.15490 & 0.98798 & 0 \\ 0 & 0 & 0.99700 \end{bmatrix} \begin{Bmatrix} -9.4824 \times 10^{-6} \\ -1.2225 \times 10^{-4} \\ 0 \end{Bmatrix}$$

$$\{\delta \mathbf{v}_f^-\} = \begin{Bmatrix} 9.4824 \times 10^{-6} \\ -1.2225 \times 10^{-4} \\ 0 \end{Bmatrix} \text{ (km/s)} \quad \text{(f)}$$

Since the chaser is in the same circular orbit as the target, its relative velocity is initially zero, i.e., $\{\delta \mathbf{v}_0^-\} = \{\mathbf{0}\}$. (See also Equation 7.58 at the end of the next section.) Thus,

$$\{\Delta \mathbf{v}_0\} = \{\delta \mathbf{v}_0^+\} - \{\delta \mathbf{v}_0^-\} = \begin{Bmatrix} -9.4824 \times 10^{-6} \\ -1.2225 \times 10^{-4} \\ 0 \end{Bmatrix} - \begin{Bmatrix} 0 \\ 0 \\ 0 \end{Bmatrix} = \begin{Bmatrix} -9.4824 \times 10^{-6} \\ -1.2225 \times 10^{-4} \\ 0 \end{Bmatrix} \text{ (km/s)}$$

which implies

$$\|\Delta \mathbf{v}_0\| = 0.1226 \text{ m/s} \quad \text{(g)}$$

At the end of the rendezvous maneuver, $\{\delta \mathbf{v}_f^+\} = \{\mathbf{0}\}$, so that

$$\{\Delta \mathbf{v}_f\} = \{\delta \mathbf{v}_f^+\} - \{\delta \mathbf{v}_f^-\} = \begin{Bmatrix} 0 \\ 0 \\ 0 \end{Bmatrix} - \begin{Bmatrix} 9.4824 \times 10^{-6} \\ -1.2225 \times 10^{-4} \\ 0 \end{Bmatrix} = \begin{Bmatrix} -9.4824 \times 10^{-6} \\ 1.2225 \times 10^{-4} \\ 0 \end{Bmatrix} \text{ (km/s)}$$

Therefore

$$\|\Delta \mathbf{v}_f\| = 0.1226 \text{ m/s} \quad \text{(h)}$$

The total delta-v required is

$$\Delta v_{\text{total}} = \|\Delta \mathbf{v}_0\| + \|\Delta \mathbf{v}_f\| = 0.2452 \text{ m/s} \quad \text{(i)}$$

Observe that in this case the motion takes place entirely in the plane of the target orbit. There is no motion normal to the plane (in the z direction). The coplanar rendezvous trajectory relative to the CW frame is sketched in Figure 7.8.

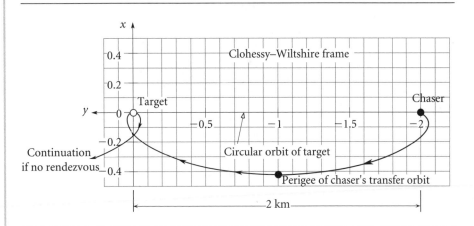

Figure 7.8 Motion of the chaser relative to the target.

7.6 Relative motion in close-proximity circular orbits

Figure 7.9 shows two spacecraft in coplanar circular orbits. Let us calculate the velocity $\delta\mathbf{v}$ of the chase vehicle B relative to the target A when they are in close proximity. 'Close proximity' means that

$$\frac{\delta r}{r_0} \ll 1$$

To solve this problem, we must use the relative velocity equation,

$$\mathbf{v}_B = \mathbf{v}_A + \mathbf{\Omega} \times \delta\mathbf{r} + \delta\mathbf{v} \qquad (7.50)$$

where $\mathbf{\Omega}$ is the angular velocity of the CW frame attached to A,

$$\mathbf{\Omega} = n\hat{\mathbf{k}}$$

n is the mean motion of the target vehicle,

$$n = \frac{v_A}{r_0} \qquad (7.51)$$

where, by virtue of the circular orbit,

$$v_A = \sqrt{\frac{\mu}{r_0}} \qquad (7.52)$$

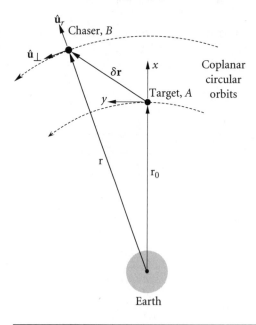

Figure 7.9 Two spacecraft in close proximity.

7.6 Relative motion in close-proximity circular orbits

Solving Equation 7.50 for the relative velocity $\delta \mathbf{v}$ yields

$$\delta \mathbf{v} = \mathbf{v}_B - \mathbf{v}_A - (n\hat{\mathbf{k}}) \times \delta \mathbf{r} \qquad (7.53)$$

Since the chase orbit is circular, we have for the first term on the right-hand side of Equation 7.53

$$\mathbf{v}_B = \sqrt{\frac{\mu}{r}} \hat{\mathbf{u}}_\perp = \sqrt{\frac{\mu}{r}} (\hat{\mathbf{k}} \times \hat{\mathbf{u}}_r) = \sqrt{\mu}\,\hat{\mathbf{k}} \times \left(\frac{1}{\sqrt{r}}\frac{\mathbf{r}}{r}\right) \qquad (7.54)$$

Since, as is apparent from Figure 7.9, $\mathbf{r} = \mathbf{r}_0 + \delta \mathbf{r}$, we can write this expression for $\mathbf{v}_B$ as follows:

$$\mathbf{v}_B = \sqrt{\mu}\,\hat{\mathbf{k}} \times r^{-\frac{3}{2}}(\mathbf{r}_0 + \delta \mathbf{r}) \qquad (7.55)$$

Now

$$r^{-\frac{3}{2}} = (r^2)^{-\frac{3}{4}} = \left[\overbrace{r_0^2\left(1 + \frac{2\mathbf{r}_0 \cdot \delta \mathbf{r}}{r_0^2}\right)}^{\text{See Equation 7.10}}\right]^{-\frac{3}{4}} = r_0^{-\frac{3}{2}}\left(1 + \frac{2\mathbf{r}_0 \cdot \delta \mathbf{r}}{r_0^2}\right)^{-\frac{3}{4}} \qquad (7.56)$$

Using the binomial theorem (cf. Equation 5.44), and retaining terms at most linear in $\delta \mathbf{r}$, we get

$$\left(1 + \frac{2\mathbf{r}_0 \cdot \delta \mathbf{r}}{r_0^2}\right)^{-\frac{3}{4}} = 1 - \frac{3}{2}\frac{\mathbf{r}_0 \cdot \delta \mathbf{r}}{r_0^2}$$

Substituting this into Equation 7.56 leads to

$$r^{-\frac{3}{2}} = r_0^{-\frac{3}{2}} - \frac{3}{2}\frac{\mathbf{r}_0 \cdot \delta \mathbf{r}}{r_0^{\frac{7}{2}}}$$

Upon substituting this result into Equation 7.55, we get

$$\mathbf{v}_B = \sqrt{\mu}\,\hat{\mathbf{k}} \times (\mathbf{r}_0 + \delta \mathbf{r})\left(r_0^{-\frac{3}{2}} - \frac{3}{2}\frac{\mathbf{r}_0 \cdot \delta \mathbf{r}}{r_0^{\frac{7}{2}}}\right)$$

Retaining terms at most linear in $\delta \mathbf{r}$, we can write this as

$$\mathbf{v}_B = \hat{\mathbf{k}} \times \left\{\sqrt{\frac{\mu}{r_0}}\frac{\mathbf{r}_0}{r_0} + \frac{\sqrt{\mu/r_0}}{r_0}\delta \mathbf{r} - \frac{3}{2}\frac{\sqrt{\mu r_0}}{r_0}\left[\left(\frac{\mathbf{r}_0}{r_0}\right) \cdot \delta \mathbf{r}\right]\frac{\mathbf{r}_0}{r_0}\right\}$$

Using Equations 7.51 and 7.52, together with the facts that $\delta \mathbf{r} = \delta x \hat{\mathbf{i}} + \delta y \hat{\mathbf{j}}$ and $\mathbf{r}_0/r_0 = \hat{\mathbf{i}}$, this reduces to

$$\mathbf{v}_B = \hat{\mathbf{k}} \times \left\{v_A \hat{\mathbf{i}} + \frac{v_A}{r_0}(\delta x \hat{\mathbf{i}} + \delta y \hat{\mathbf{j}}) - \frac{3}{2}\frac{v_A}{r_0}[\hat{\mathbf{i}} \cdot (\delta x \hat{\mathbf{i}} + \delta y \hat{\mathbf{j}})]\hat{\mathbf{i}}\right\}$$

$$= v_A \hat{\mathbf{j}} + (-n\delta y \hat{\mathbf{i}} + n\delta x \hat{\mathbf{j}}) - \frac{3}{2}n\delta x \hat{\mathbf{j}}$$

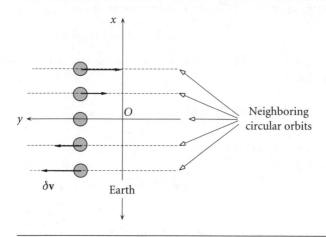

Figure 7.10 Circular orbits, with relative velocity directions, in the vicinity of the Clohessy–Wiltshire frame.

so that
$$\mathbf{v}_B = -n\delta y \hat{\mathbf{i}} + (v_A - \frac{1}{2}n\delta x)\hat{\mathbf{j}} \quad (7.57)$$

This is the absolute velocity of the chaser resolved into components in the target's Clohessy–Wiltshire frame.

Substituting Equation 7.57 into 7.53 and using the fact that $\mathbf{v}_A = v_A \hat{\mathbf{j}}$ yields

$$\delta\mathbf{v} = [-n\delta y \hat{\mathbf{i}} + (v_A - \frac{1}{2}n\delta x)\hat{\mathbf{j}}] - (v_A\hat{\mathbf{j}}) - (n\hat{\mathbf{k}}) \times (\delta x \hat{\mathbf{i}} + \delta y \hat{\mathbf{j}})$$

$$= -n\delta y \hat{\mathbf{i}} + v_A\hat{\mathbf{j}} - \frac{1}{2}n\delta x \hat{\mathbf{j}} - v_A\hat{\mathbf{j}} - n\delta x \hat{\mathbf{j}} + n\delta y \hat{\mathbf{i}}$$

so that
$$\delta\mathbf{v} = -\frac{3}{2}n\delta x \hat{\mathbf{j}} \quad (7.58)$$

This is the velocity of the chaser as measured in the moving reference frame of the neighboring target. Keep in mind that circular orbits were assumed at the outset.

In the Clohessy–Wiltshire frame, neighboring coplanar circular orbits appear to be straight lines parallel to the y axis, which is the orbit of the origin. Figure 7.10 illustrates this point, showing also the linear velocity variation according to Equation 7.58.

Problems

7.1 Two manned spacecraft, A and B (see figure), are in circular, polar ($i = 90°$) orbits around the earth. A's orbital altitude is 300 km; B's is 250 km. At the instant shown (A over the equator, B over the North Pole), calculate
  (a) the position,
  (b) velocity, and
  (c) the acceleration of B relative to A.

A's y axis points always in the flight direction, and its x axis is directed radially outward at all times.
{Ans.: (a) $\mathbf{r}_{\text{rel}})_{xyz} = -6678\hat{\mathbf{i}} + 6628\hat{\mathbf{j}}$ km; (b) $\mathbf{v}_{\text{rel}})_{xyz} = -0.08693\hat{\mathbf{i}}$ km/s; (c) $\mathbf{a}_{\text{rel}})_{xyz} = -1.140 \times 10^{-6}\hat{\mathbf{j}}$ km/s^2}

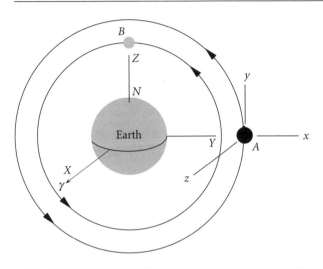

Figure P.7.1

7.2 Spacecraft A and B are in coplanar, circular geocentric orbits. The orbital radii are shown in the figure. When B is directly below A, as shown, calculate B's acceleration relative to A.
{Ans.: $(\mathbf{a}_{\text{rel}})_{xyz} = -0.268\hat{\mathbf{i}}$ (m/s^2)}

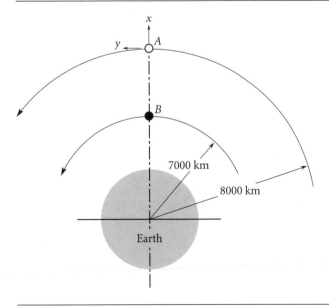

Figure P.7.2

7.3 Use the order of magnitude analysis in this chapter as a guide to answer the following questions.
  (a) If $\mathbf{r} = \mathbf{r}_0 + \delta\mathbf{r}$, express $\sqrt{r}$ (where $r = \sqrt{\mathbf{r} \cdot \mathbf{r}}$) to the first order in $\delta\mathbf{r}$ (i.e., to the first order in the components of $\delta\mathbf{r} = \delta x\hat{\mathbf{i}} + \delta y\hat{\mathbf{j}} + \delta z\hat{\mathbf{k}}$). In other words, find $O(\delta\mathbf{r})$, such that $\sqrt{r} = \sqrt{r_0} + O(\delta\mathbf{r})$, where $O(\delta\mathbf{r})$ is linear in $\delta\mathbf{r}$.
  (b) For the special case $\mathbf{r}_0 = 3\hat{\mathbf{i}} + 4\hat{\mathbf{j}} + 5\hat{\mathbf{k}}$ and $\delta\mathbf{r} = 0.01\hat{\mathbf{i}} - 0.01\hat{\mathbf{j}} + 0.03\hat{\mathbf{k}}$, calculate $\sqrt{r} - \sqrt{r_0}$ and compare that result with $O(\delta\mathbf{r})$.
  (c) Repeat part (b) using $\delta\mathbf{r} = \hat{\mathbf{i}} - \hat{\mathbf{j}} + 3\hat{\mathbf{k}}$ and compare the results.

{Ans.: (a) $O(\delta\mathbf{r}) = \mathbf{r}_0 \cdot \delta\mathbf{r}/\left(2r_0^{\frac{3}{2}}\right)$; (b) $O(\delta\mathbf{r})/\left(\sqrt{r} - \sqrt{r_0}\right) = 0.998$; (c) $O(\delta\mathbf{r})/\left(\sqrt{r} - \sqrt{r_0}\right) = 0.903$}

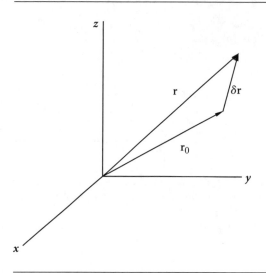

Figure P.7.3

7.4 Write the expression $r = \dfrac{a(1 - e^2)}{1 + e\cos\theta}$ as a linear function of e, valid for small values of $e (e \ll 1)$.

7.5 Given $\ddot{x} + 9x = 10$, with the initial conditions $x = 5$ and $\dot{x} = -3$ at $t = 0$, find x and $\dot{x}$ at $t = 1.2$.
{Ans.: $x(1.2) = -1.934$, $\dot{x}(1.2) = 7.853$}

7.6 Given that

$$\ddot{x} + 10x + 2\dot{y} = 0$$
$$\ddot{y} + 3\dot{x} = 0$$

with initial conditions $x(0) = 1$, $y(0) = 2$, $\dot{x}(0) = -3$ and $\dot{y}(0) = 4$, find x and y at $t = 5$.
{Ans.: $x(5) = -6.460$, $y(5) = 97.31$}

7.7 A space station is in a 90-minute period earth orbit. At $t = 0$, a satellite has the following position and velocity components relative to a Clohessy–Wiltshire frame attached to the

space station: $\{\delta \mathbf{r}\} = \lfloor 1 \quad 0 \quad 0 \rfloor^T$ (km), $\{\delta \mathbf{v}\} = \lfloor 0 \quad 10 \quad 0 \rfloor^T$ (m/s). How far is the satellite from the space station 15 minutes later?
{Ans.: 11.2 km}

7.8 A space station is in a circular earth orbit of radius 6600 km. An approaching spacecraft executes a delta-v burn when its position vector relative to the space station is $\{\delta \mathbf{r}_0\} = \lfloor 1 \quad 1 \quad 1 \rfloor^T$ (km). Just before the burn the relative velocity of the spacecraft was $\{\delta \mathbf{v}_0^-\} = \lfloor 0 \quad 0 \quad 5 \rfloor^T$ (m/s). Calculate the total delta-v required for the space shuttle to rendezvous with the station in one-third period of the space station orbit.
{Ans.: 6.21 m/s}

7.9 A space station is in circular orbit 2 of radius r_0. A spacecraft is in coplanar circular orbit 1 of radius $r_0 + \delta r$. At $t = 0$ the spacecraft executes an impulsive maneuver to rendezvous with the space station at time t_f = one-half the period T_0 of the space station. For a Hohmann transfer orbit ($\delta u_0^+ = 0$), find
(a) the initial position of the spacecraft relative to the space station, and
(b) the relative velocity of the spacecraft when it arrives at the target.
   Sketch the rendezvous trajectory relative to the target.
{Ans.: (a) $\{\delta \mathbf{r}_0\} = \lfloor \delta r \quad 3\pi/(4\delta r) \quad 0 \rfloor^T$, (b) $\{\delta \mathbf{v}_f^-\} = \lfloor 0 \quad \pi \delta r/(2T_0) \quad 0 \rfloor^T$}

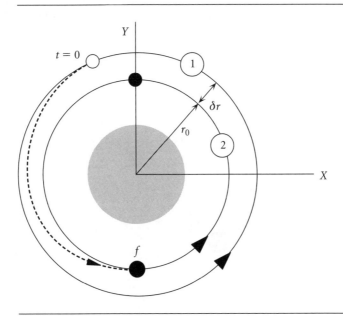

Figure P.7.9

7.10 Assuming a Hohmann transfer ($\delta u_0^+ = 0$), calculate the total delta-v required for rendezvous if $\{\delta \mathbf{r}_0\} = \lfloor 0 \quad \delta y_0 \quad 0 \rfloor^T$, $\{\delta \mathbf{v}_0^-\} = \lfloor 0 \quad 0 \quad 0 \rfloor^T$ and t_f = the period of the circular target orbit. Sketch the rendezvous trajectory relative to the target.
{Ans.: $\Delta v_{tot} = 2\delta y_0/(3T)$}

7.11 Spacecraft A and B are in the same circular earth orbit with a period of 2 hours. B is 6 km ahead of A. At $t=0$, B applies an in-track delta-v (retrofire) of 3 m/s. Using a Clohessy–Wiltshire frame attached to A, determine the distance between A and B at $t=30$ minutes and the velocity of B relative to A.
{Ans.: $\|\delta\mathbf{r}\|=10.9$ km, $\|\delta\mathbf{v}\|=10.8$ m/s}

7.12 A GEO satellite strikes some orbiting debris and is found 2 hours afterwards to have drifted to the position $\{\delta\mathbf{r}\}=\lfloor-10\ \ 10\ \ 0\rfloor^T$ km relative to its original location. At that time the only slightly damaged satellite initiates a two-impulse maneuver to return to its original location in 6 hours. Find the total delta-v for this maneuver.
{Ans.: 3.5 m/s}

7.13 A space station is in a 245 km circular earth orbit inclined at 30°. The right ascension of its node line is 40°. Meanwhile, a space shuttle has been launched into a 280 km by 250 km orbit inclined at 30.1°, with a nodal right ascension of 40° and argument of perigee equal to 60°. When the shuttle's true anomaly is 40°, the space station is 99° beyond its node line. At that instant, the space shuttle executes a delta-v burn to rendezvous with the space station in (precisely) t_f hours, where t_f is selected by you or assigned by the instructor. Calculate the total delta-v required and sketch the projection of the rendezvous trajectory on the xy plane of the space station coordinates.

7.14 The space station is in a circular earth orbit of radius 6600 km. The space shuttle is also in a circular orbit in the same plane as the space station's. At the instant that the position of the shuttle relative to the space station, in Clohessy–Wiltshire coordinates, is (5 km, 0, 0). What is the relative velocity $\delta\mathbf{v}$ of the space shuttle in meters/s?
{Ans.: 8.83 m/s}

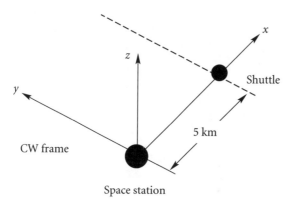

Figure P.7.14

7.15 The Clohessy–Wiltshire coordinates and velocities of a spacecraft upon entering a rendezvous trajectory with the target vehicle are shown. The spacecraft orbits are coplanar. Calculate the distance d of the spacecraft from the target when $t=\pi/2n$, where n is the mean motion of the target's circular orbit.
{Ans.: $0.900\delta r$}

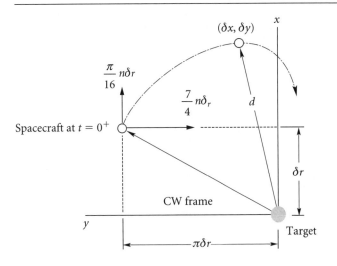

Figure P.7.15

7.16 The target T is in a circular earth orbit with mean motion n. The chaser C is directly above T in a slightly larger circular orbit having the same plane as T's. What relative initial velocity $\delta\mathbf{v}_0^+$ is required so that C arrives at the target T at time t_f = one-half the target's period? {Ans.: $\delta\mathbf{v}_0^+ = -0.589 n \delta x_0 \hat{\mathbf{i}} - 1.75 n \delta x_0 \hat{\mathbf{j}}$}

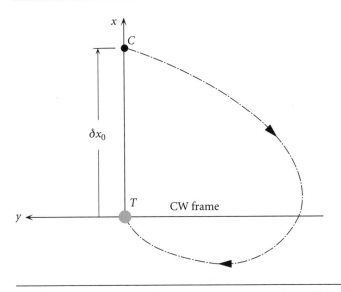

Figure P.7.16

7.17 The space shuttle and the International Space Station are in coplanar circular orbits. The space station has an orbital radius r and a mean motion n. The shuttle's radius is

$r - d (d \ll r)$. If a two-impulse rendezvous maneuver with $t_f = \pi/(4n)$ is initiated with zero relative velocity in the x direction ($\delta \dot{x}_0^+ = 0$), calculate the initial relative y coordinate of the shuttle.

{Ans.: $\delta y_0 = -1.98\,d$}

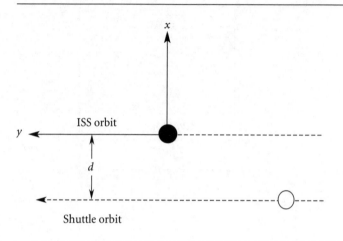

Figure P.7.17

CHAPTER 8

Interplanetary trajectories

Chapter outline

8.1	Introduction	347
8.2	Interplanetary Hohmann transfers	348
8.3	Rendezvous opportunities	349
8.4	Sphere of influence	354
8.5	Method of patched conics	359
8.6	Planetary departure	360
8.7	Sensitivity analysis	366
8.8	Planetary rendezvous	368
8.9	Planetary flyby	375
8.10	Planetary ephemeris	387
8.11	Non-Hohmann interplanetary trajectories	391
Problems		398

8.1 Introduction

In this chapter we consider some basic aspects of planning interplanetary missions. We begin by considering Hohmann transfers, which are the easiest to analyze and the most energy efficient. The orbits of the planets involved must lie in the same plane and the planets must be positioned just right for a Hohmann transfer to be used. The time between such opportunities is derived. The method of patched conics is employed to divide the mission into three parts: the hyperbolic departure trajectory relative to the home planet; the cruise ellipse relative to the sun; and the hyperbolic arrival trajectory, relative to the target planet.

The use of patched conics is justified by calculating the radius of a planet's sphere of influence and showing how small it is on the scale of the solar system. Matching the velocity of the spacecraft at the home planet's sphere of influence to that required to initiate the outbound cruise phase and then specifying the periapse radius of the departure hyperbola determines the delta-v requirement at departure. The sensitivity of the target radius to the burnout conditions is discussed. Matching the velocities at the target planet's sphere of influence and specifying the periapse of the arrival hyperbola yields the delta-v at the target for a planetary rendezvous or the direction of the outbound hyperbola for a planetary flyby. Flyby maneuvers are discussed, including the effect of leading and trailing side flybys, and some noteworthy examples of the use of gravity assist maneuvers are presented.

The chapter concludes with an analysis of the situation in which the planets' orbits are not coplanar and the transfer ellipse is tangent to neither orbit. This is akin to the chase maneuver in Chapter 6 and requires the solution of Lambert's problem using Algorithm 5.2.

8.2 Interplanetary Hohmann transfers

As can be seen from Table A.1, the orbits of most of the planets in the solar system lie very close to the earth's orbital plane (the ecliptic plane). The innermost planet, Mercury, and the outermost planet, Pluto, differ most in inclination (7° and 17°, respectively). The orbital planes of the other planets lie within 3.5° of the ecliptic. It is also evident from Table A.1 that most of the planetary orbits have small eccentricities, the exceptions once again being Mercury and Pluto. To simplify the beginning of our study of interplanetary trajectories, we will assume that all of the planets' orbits are circular and coplanar. Later on, in Section 8.10, we will relax this assumption.

The most energy efficient way for a spacecraft to transfer from one planet's orbit to another is to use a Hohmann transfer ellipse (Section 6.2). Consider Figure 8.1, which shows a Hohmann transfer from an inner planet 1 to an outer planet 2. The departure point D is at the periapse (perihelion) of the transfer ellipse and the arrival point is at the apoapse (aphelion). The circular orbital speed of planet 1 relative to the sun is given by Equation 2.53,

$$V_1 = \sqrt{\frac{\mu_{sun}}{R_1}} \qquad (8.1)$$

The specific angular momentum h of the transfer ellipse relative to the sun is found from Equation 6.2, so that the velocity of the space vehicle on the transfer ellipse at the departure point D is

$$V_D^{(v)} = \frac{h}{R_1} = \sqrt{2\mu_{sun}} \sqrt{\frac{R_2}{R_1(R_1 + R_2)}} \qquad (8.2)$$

This is greater than the speed of the planet. Therefore the required delta-v at D is

$$\Delta V_D = V_D^{(v)} - V_1 = \sqrt{\frac{\mu_{sun}}{R_1}} \left(\sqrt{\frac{2R_2}{R_1 + R_2}} - 1 \right) \qquad (8.3)$$

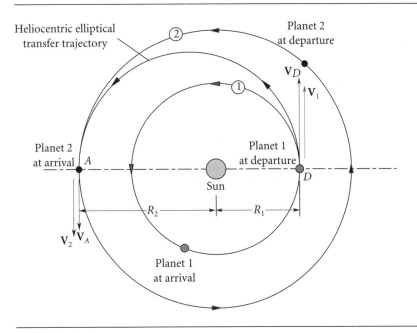

Figure 8.1 Hohmann transfer from inner planet 1 to outer planet 2.

Likewise, the delta-v at the arrival point A is

$$\Delta V_A = V_2 - V_A^{(v)} = \sqrt{\frac{\mu_{\text{sun}}}{R_2}}\left(1 - \sqrt{\frac{2R_1}{R_1 + R_2}}\right) \quad (8.4)$$

This velocity increment, like that at point D, is positive since planet 2 is traveling faster than the spacecraft at point A.

For a mission from an outer planet to an inner planet, as illustrated in Figure 8.2, the delta-vs computed using Equations 8.3 and 8.4 will both be negative instead of positive. That is because the departure point and arrival point are now at aphelion and perihelion, respectively, of the transfer ellipse. The speed of the spacecraft must be reduced for it to drop into the lower-energy transfer ellipse at the departure point D, and it must be reduced again at point A in order to arrive in the lower-energy circular orbit of planet 2.

8.3 Rendezvous opportunities

The purpose of an interplanetary mission is for the spacecraft not only to intercept a planet's orbit but also to rendezvous with the planet when it gets there. For rendezvous to occur at the end of a Hohmann transfer, the location of planet 2 in its orbit at the time of the spacecraft's departure from planet 1 must be such that planet 2 arrives at the apse line of the transfer ellipse at the same time the spacecraft does. Phasing

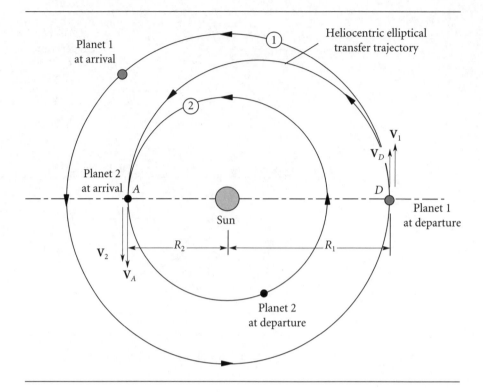

Figure 8.2 Hohmann transfer from outer planet 1 to inner planet 2.

maneuvers (Section 6.7) are clearly not practical, especially for manned missions, due to the large periods of the heliocentric orbits.

Consider planet 1 and planet 2 in circular orbits around the sun, as shown in Figure 8.3. Since the orbits are circular, we can choose a common horizontal apse line from which to measure the true anomaly θ. The true anomalies of planets 1 and 2, respectively, are

$$\theta_1 = \theta_{1_0} + n_1 t \tag{8.5}$$

$$\theta_2 = \theta_{2_0} + n_2 t \tag{8.6}$$

where n_1 and n_2 are the mean motions (angular velocities) of the planets and θ_{1_0} and θ_{2_0} are their true anomalies at time $t = 0$. The phase angle between the position vectors of the two planets is defined as

$$\phi = \theta_2 - \theta_1 \tag{8.7}$$

ϕ is the angular position of planet 2 relative to planet 1. Substituting Equations 8.5 and 8.6 into 8.7 we get

$$\phi = \phi_0 + (n_2 - n_1)t \tag{8.8}$$

ϕ_0 is the phase angle at time zero. $n_2 - n_1$ is the orbital angular velocity of planet 2 relative to planet 1. If the orbit of planet 1 lies inside that of planet 2, as in Figure 8.3(a), then $n_1 > n_2$. Therefore, the relative angular velocity $n_2 - n_1$ is negative, which means

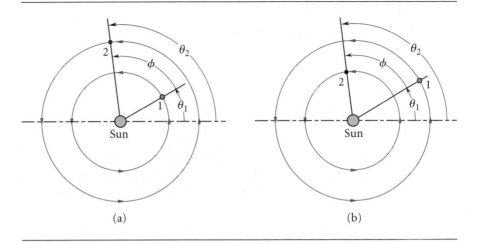

Figure 8.3 Planets in circular orbits around the sun. (a) Planet 2 outside the orbit of planet 1. (b) Planet 2 inside the orbit of planet 1.

planet 2 moves clockwise relative to planet 1. On the other hand, if planet 1 is outside of planet 2 then $n_2 - n_1$ is positive, so that the relative motion is counterclockwise.

The phase angle obviously varies linearly with time according to Equation 8.8. If the phase angle is ϕ_0 at $t = 0$, how long will it take to become ϕ_0 again? The answer: when the position vector of planet 2 rotates through 2π radians relative to planet 1. The time required for the phase angle to return to its initial value is called the synodic period, which is denoted T_{syn}. For the case shown in Figure 8.3(a) in which the relative motion is clockwise, T_{syn} is the time required for ϕ to change from ϕ_0 to $\phi_0 - 2\pi$. From Equation 8.8 we have

$$\phi_0 - 2\pi = \phi_0 + (n_2 - n_1)T_{syn}$$

so that

$$T_{syn} = \frac{2\pi}{n_1 - n_2} \quad (n_1 > n_2)$$

For the situation illustrated in Figure 8.3(b) ($n_2 > n_1$), T_{syn} is the time required for ϕ to go from ϕ_0 to $\phi_0 + 2\pi$, in which case Equation 8.8 yields

$$T_{syn} = \frac{2\pi}{n_2 - n_1} \quad (n_2 > n_1)$$

Both cases are covered by writing

$$T_{syn} = \frac{2\pi}{|n_1 - n_2|} \tag{8.9}$$

Recalling Equation 3.6, we can write $n_1 = 2\pi/T_1$ and $n_2 = 2\pi/T_2$. Thus, in terms of the orbital periods of the two planets,

$$T_{syn} = \frac{T_1 T_2}{|T_1 - T_2|} \tag{8.10}$$

Observe that T_{syn} is the orbital period of planet 2 relative to planet 1.

EXAMPLE 8.1

Calculate the synodic period of Mars relative to the earth

In Table A.1 we find the orbital periods of earth and Mars:

$$T_{earth} = 365.26 \text{ days (1 year)}$$
$$T_{Mars} = 1 \text{ year } 321.73 \text{ days} = 687.99 \text{ days}$$

Hence,

$$T_{syn} = \frac{T_{earth} T_{Mars}}{|T_{earth} - T_{Mars}|} = \frac{365.26 \times 687.99}{|365.26 - 687.99|} = \underline{777.9 \text{ days}}$$

These are earth days (1 day = 24 hours). Therefore it takes 2.13 years for a given configuration of Mars relative to the earth to occur again.

Figure 8.4 depicts a mission from planet 1 to planet 2. Following a heliocentric Hohmann transfer, the spacecraft intercepts and rendezvous with planet 2. Later it returns to planet 1 by means of another Hohmann transfer. The major axis of the heliocentric transfer ellipse is the sum of the radii of the two planets' orbits, $R_1 + R_2$. The time t_{12} required for the transfer is one-half the period of the ellipse. Hence, according to Equation 2.73,

$$t_{12} = \frac{\pi}{\sqrt{\mu_{sun}}} \left(\frac{R_1 + R_2}{2} \right)^{3/2} \quad (8.11)$$

During the time it takes the spacecraft to fly from orbit 1 to orbit 2, through an angle of π radians, planet 2 must move around its circular orbit and end up at a point directly opposite planet 1's position when the spacecraft departed. Since planet 2's angular velocity is n_2, the angular distance traveled by the planet during the spacecraft's trip is $n_2 t_{12}$. Hence, as can be seen from Figure 8.4(a), the initial phase angle ϕ_0 between the two planets is

$$\phi_0 = \pi - n_2 t_{12} \quad (8.12)$$

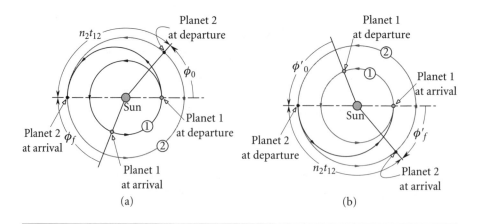

Figure 8.4 Round-trip mission, with layover, to planet 2. (a) Departure and rendezvous with planet 2. (b) Return and rendezvous with planet 1.

8.3 Rendezvous opportunities

When the spacecraft arrives at planet 2, the phase angle will be ϕ_f, which is found using Equations 8.8 and 8.12:

$$\phi_f = \phi_0 + (n_2 - n_1)t_{12} = (\pi - n_2 t_{12}) + (n_2 - n_1)t_{12}$$
$$\phi_f = \pi - n_1 t_{12} \tag{8.13}$$

For the situation illustrated in Figure 8.4, planet 2 ends up being *behind* planet 1 by an amount equal to the magnitude of ϕ_f.

At the start of the return trip, illustrated in Figure 8.4(b), planet 2 must be ϕ'_0 radians ahead of planet 2. Since the spacecraft flies the same Hohmann transfer trajectory back to planet 1, the time of flight is t_{12}, the same as the outbound leg. Therefore, the distance traveled by planet 1 during the return trip is the same as the outbound leg, which means

$$\phi'_0 = -\phi_f \tag{8.14}$$

In any case, the phase angle at the beginning of the return trip must be the negative of the phase angle at arrival from planet 1. The time required for the phase angle to reach its proper value is called the wait time, t_{wait}. Setting time equal to zero at the instant we arrive at planet 2, Equation 8.8 becomes

$$\phi = \phi_f + (n_2 - n_1)t$$

ϕ becomes $-\phi_f$ after the time t_{wait}. That is

$$-\phi_f = \phi_f + (n_2 - n_1)t_{wait}$$

or

$$t_{wait} = \frac{-2\phi_f}{n_2 - n_1} \tag{8.15}$$

where ϕ_f is given by Equation 8.13. Equation 8.15 may yield a negative result, which means the desired phase relation occurred in the past. Therefore we must add or subtract an integral multiple of 2π to the numerator in order to get a positive value for t_{wait}. Specifically, if $N = 0, 1, 2, \ldots$, then

$$t_{wait} = \frac{-2\phi_f - 2\pi N}{n_2 - n_1} \quad (n_1 > n_2) \tag{8.16}$$

$$t_{wait} = \frac{-2\phi_f + 2\pi N}{n_2 - n_1} \quad (n_1 < n_2) \tag{8.17}$$

where N is chosen to make t_{wait} positive. t_{wait} would probably be the smallest positive number thus obtained.

EXAMPLE 8.2

Calculate the minimum wait time for initiating a return trip from Mars to earth.

From Tables A.1 and A.2 we have

$$R_{earth} = 149.6 \times 10^6 \text{ km}$$
$$R_{Mars} = 227.9 \times 10^6 \text{ km}$$
$$\mu_{sun} = 132.71 \times 10^9 \text{ km}^3/\text{s}^2$$

(Example 8.2 continued)

According to Equation 8.11, the time of flight from earth to Mars is

$$t_{12} = \frac{\pi}{\sqrt{\mu_{sun}}} \left(\frac{R_{earth} + R_{Mars}}{2} \right)^{3/2}$$

$$= \frac{\pi}{\sqrt{132.71 \times 10^9}} \left(\frac{149.6 \times 10^6 + 227.9 \times 10^6}{2} \right)^{3/2} = 2.2362 \times 10^7 \text{ s}$$

or

$$t_{12} = 258.82 \text{ days}$$

From Equation 3.6 and the orbital periods of earth and Mars (see Example 8.1 above) we obtain the mean motions of the earth and Mars.

$$n_{earth} = \frac{2\pi}{365.26} = 0.017202 \text{ rad/day}$$

$$n_{Mars} = \frac{2\pi}{687.99} = 0.0091327 \text{ rad/day}$$

The phase angle between earth and Mars when the spacecraft reaches Mars is given by Equation 8.13.

$$\phi_f = \pi - n_{earth} t_{12} = \pi - 0.017202 \cdot 258.82 = -1.3107 \text{ (rad)}$$

Since $n_{earth} > n_{Mars}$, we choose Equation 8.16 to find the wait time:

$$t_{wait} = \frac{-2\phi_f - 2\pi N}{n_{Mars} - n_{earth}} = \frac{-2(-1.3107) - 2\pi N}{0.0091327 - 0.017202} = 778.65N - 324.85 \text{ (days)}$$

$N = 0$ yields a negative value, which we cannot accept. Setting $N = 1$, we get

$$t_{wait} = 453.8 \text{ days}$$

This is the minimum wait time. Obviously, we could set $N = 2, 3, \ldots$ to obtain longer wait times.

In order for a spacecraft to depart on a mission to Mars by means of a Hohmann (minimum energy) transfer, the phase angle between earth and Mars must be that given by Equation 8.12. Using the results of Example 8.2, we find it to be

$$\phi_0 = \pi - n_{Mars} t_{12} = \pi - 0.0091327 \cdot 258.82 = 0.7778 \text{ rad} = 44.57°$$

This opportunity occurs once every synodic period, which we found to be 2.13 years in Example 8.1. In Example 8.2 we found that the time to fly to Mars is 258.8 days, followed by a wait time of 453.8 days, followed by a return trip time of 258.8 days. Hence, the minimum total time for a manned Mars mission is

$$t_{total} = 258.8 + 453.8 + 253.8 = 971.4 \text{ days} = 2.66 \text{ years}$$

8.4 SPHERE OF INFLUENCE

The sun, of course, is the dominant celestial body in the solar system. It is over 1000 times more massive than the largest planet, Jupiter, and has a mass of over 300 000

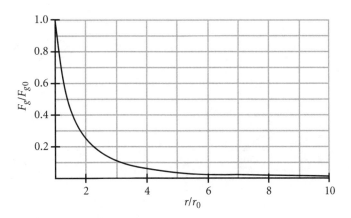

Figure 8.5 Decrease of gravitational force with distance from a planet's surface.

earths. The sun's gravitational pull holds all of the planets in its grasp according to Newton's law of gravity, Equation 2.6. However, near a given planet the influence of its own gravity exceeds that of the sun. For example, at its surface the earth's gravitational force is over 1600 times greater than the sun's. The inverse-square nature of the law of gravity means that the force of gravity F_g drops off rapidly with distance r from the center of attraction. If F_{g_0} is the gravitational force at the surface of a planet with radius r_0, then Figure 8.5 shows how rapidly the force diminishes with distance. At ten body radii, the force is 1 percent of its value at the surface. Eventually, the force of the sun's gravitational field overwhelms that of the planet.

In order to estimate the radius of a planet's gravitational sphere of influence, consider the three-body system comprising a planet p of mass m_p, the sun s of mass m_s and a space vehicle v of mass m_v illustrated in Figure 8.6. The position vectors of the planet and spacecraft relative to an inertial frame centered at the sun are $\mathbf{R}$ and $\mathbf{R}_v$, respectively. The position vector of the space vehicle relative to the planet is $\mathbf{r}$. (Throughout this chapter we will use upper case letters to represent position, velocity and acceleration measured relative to the sun and lower case letters when they are measured relative to a planet.) The gravitational force exerted on the vehicle by the planet is denoted $\mathbf{F}_p^{(v)}$, and that exerted by the sun is $\mathbf{F}_s^{(v)}$. Likewise, the forces on the planet are $\mathbf{F}_s^{(p)}$ and $\mathbf{F}_v^{(p)}$, whereas on the sun we have $\mathbf{F}_v^{(s)}$ and $\mathbf{F}_p^{(s)}$. According to Newton's law of gravitation (Equation 2.6), these forces are

$$\mathbf{F}_p^{(v)} = -\frac{Gm_v m_p}{r^3}\mathbf{r} \quad (8.18a)$$

$$\mathbf{F}_s^{(v)} = -\frac{Gm_v m_s}{R_v^3}\mathbf{R}_v \quad (8.18b)$$

$$\mathbf{F}_s^{(p)} = -\frac{Gm_p m_s}{R^3}\mathbf{R} \quad (8.18c)$$

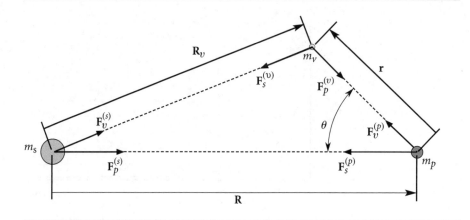

Figure 8.6 Relative position and gravitational force vectors among the three bodies.

Observe that

$$\mathbf{R}_v = \mathbf{R} + \mathbf{r} \tag{8.19}$$

From Figure 8.6 and the law of cosines we see that the magnitude of $\mathbf{R}_v$ is

$$R_v = (R^2 + r^2 - 2Rr\cos\theta)^{\frac{1}{2}} = R\left[1 - 2\frac{r}{R}\cos\theta + \left(\frac{r}{R}\right)^2\right]^{\frac{1}{2}} \tag{8.20}$$

We expect that within the planet's sphere of influence, $r/R \ll 1$. In that case, the terms involving r/R in Equation 8.20 can be neglected, so that, approximately,

$$R_v = R \tag{8.21}$$

The equation of motion of the spacecraft relative to the sun-centered inertial frame is

$$m_v \ddot{\mathbf{R}}_v = \mathbf{F}_s^{(v)} + \mathbf{F}_p^{(v)}$$

Solving for $\ddot{\mathbf{R}}_v$ and substituting the gravitational forces given by Equations 8.18a and 8.18b, we get

$$\ddot{\mathbf{R}}_v = \frac{1}{m_v}\left(-\frac{Gm_v m_s}{R_v^3}\mathbf{R}_v\right) + \frac{1}{m_v}\left(-\frac{Gm_v m_p}{r^3}\mathbf{r}\right) = -\frac{Gm_s}{R_v^3}\mathbf{R}_v - \frac{Gm_p}{r^3}\mathbf{r} \tag{8.22}$$

Let us write this as

$$\ddot{\mathbf{R}}_v = \mathbf{A}_s + \mathbf{P}_p \tag{8.23}$$

where

$$\mathbf{A}_s = -\frac{Gm_s}{R_v^3}\mathbf{R}_v \qquad \mathbf{P}_p = -\frac{Gm_p}{r^3}\mathbf{r} \tag{8.24}$$

8.4 Sphere of influence

$\mathbf{A}_s$ is the primary gravitational acceleration of the vehicle due to the sun, whereas $\mathbf{P}_p$ is the secondary or perturbing acceleration due to the planet. The magnitudes of $\mathbf{A}_s$ and $\mathbf{P}_p$ are

$$A_s = \frac{Gm_s}{R^2} \qquad P_p = \frac{Gm_p}{r^2} \tag{8.25}$$

where we made use of the approximation given by Equation 8.21. The ratio of the perturbing acceleration to the primary acceleration is, therefore,

$$\frac{P_p}{A_s} = \frac{\frac{Gm_p}{r^2}}{\frac{Gm_s}{R^2}} = \frac{m_p}{m_s}\left(\frac{R}{r}\right)^2 \tag{8.26}$$

The equation of motion of the planet relative to the inertial frame is

$$m_p \ddot{\mathbf{R}} = \mathbf{F}_v^{(p)} + \mathbf{F}_s^{(p)}$$

Solving for $\ddot{\mathbf{R}}$, noting that $\mathbf{F}_v^{(p)} = -\mathbf{F}_p^{(v)}$, and using Equations 8.18b and 8.18c, yields

$$\ddot{\mathbf{R}} = \frac{1}{m_p}\left(\frac{Gm_v m_p}{r^3}\mathbf{r}\right) + \frac{1}{m_p}\left(-\frac{Gm_p m_s}{R^3}\mathbf{R}\right) = \frac{Gm_v}{r^3}\mathbf{r} - \frac{Gm_s}{R^3}\mathbf{R} \tag{8.27}$$

Subtracting Equation 8.27 from 8.22 and collecting terms, we find

$$\ddot{\mathbf{R}}_v - \ddot{\mathbf{R}} = -\frac{Gm_p}{r^3}\mathbf{r}\left(1 + \frac{m_v}{m_p}\right) - \frac{Gm_s}{R_v^3}\left[\mathbf{R}_v - \left(\frac{R_v}{R}\right)^3 \mathbf{R}\right]$$

Recalling Equation 8.19, we can write this as

$$\ddot{\mathbf{r}} = -\frac{Gm_p}{r^3}\mathbf{r}\left(1 + \frac{m_v}{m_p}\right) - \frac{Gm_s}{R_v^3}\left\{\mathbf{r} + \left[1 - \left(\frac{R_v}{R}\right)^3\right]\mathbf{R}\right\} \tag{8.28}$$

This is the equation of motion of the vehicle relative to the planet. By using Equation 8.21 and the fact that $m_v \ll m_p$, we can write this in approximate form as

$$\ddot{\mathbf{r}} = \mathbf{a}_p + \mathbf{p}_s \tag{8.29}$$

where

$$\mathbf{a}_p = -\frac{Gm_p}{r^3}\mathbf{r} \qquad \mathbf{p}_s = -\frac{Gm_s}{R^3}\mathbf{r} \tag{8.30}$$

In this case $\mathbf{a}_p$ is the primary gravitational acceleration of the vehicle due to the planet, and $\mathbf{p}_s$ is the perturbation caused by the sun. The magnitudes of these vectors are

$$a_p = \frac{Gm_p}{r^2} \qquad p_s = \frac{Gm_s}{R^3}r \tag{8.31}$$

The ratio of the perturbing acceleration to the primary acceleration is

$$\frac{p_s}{a_p} = \frac{Gm_s \frac{r}{R^3}}{\frac{Gm_p}{r^2}} = \frac{m_s}{m_p}\left(\frac{r}{R}\right)^3 \tag{8.32}$$

For motion relative to the planet, the ratio p_s/a_p is a measure of the deviation of the vehicle's orbit from the Keplerian orbit arising from the planet acting by itself ($p_s/a_p = 0$). Likewise, P_p/A_s is a measure of the planet's influence on the orbit of the vehicle relative to the sun. If

$$\frac{p_s}{a_p} < \frac{P_p}{A_s} \tag{8.33}$$

then the perturbing effect of the sun on the vehicle's orbit around the planet is less than the perturbing effect of the planet on the vehicle's orbit around the sun. We say that the vehicle is therefore within the planet's sphere of influence. Substituting Equations 8.26 and 8.32 into 8.33 yields

$$\frac{m_s}{m_p}\left(\frac{r}{R}\right)^3 < \frac{m_p}{m_s}\left(\frac{R}{r}\right)^2$$

which means

$$\left(\frac{r}{R}\right)^5 < \left(\frac{m_p}{m_s}\right)^2$$

or

$$\frac{r}{R} < \left(\frac{m_p}{m_s}\right)^{\frac{2}{5}}$$

Let r_{SOI} be the radius of the sphere of influence. Within the planet's sphere of influence, defined by

$$\frac{r_{SOI}}{R} = \left(\frac{m_p}{m_s}\right)^{\frac{2}{5}} \tag{8.34}$$

the motion of the spacecraft is determined by its equations of motion relative to the planet (Equation 8.28). Outside of the sphere of influence, the path of the spacecraft is computed relative to the sun (Equation 8.22).

The sphere of influence radius presented in Equation 8.34 is not an exact quantity. It is simply a reasonable estimate of the distance beyond which the sun's gravitational attraction dominates that of a planet. The spheres of influence of all of the planets and the earth's moon are listed in Table A.2.

EXAMPLE 8.3

Calculate the radius of the earth's sphere of influence.

In Table A.1 we find

$$m_{earth} = 5.974 \times 10^{24} \text{ kg}$$
$$m_{sun} = 1.989 \times 10^{30} \text{ kg}$$
$$R_{earth} = 149.6 \times 10^6 \text{ km}$$

Substituting this data into Equation 8.34 yields

$$r_{SOI} = 149.6 \times 10^6 \left(\frac{5.974 \times 10^{24}}{1.989 \times 10^{24}}\right)^{\frac{2}{5}} = 925 \times 10^6 \text{ km}$$

Since the radius of the earth is 6378 km,

$$r_{SOI} = 145 \text{ earth radii}$$

Relative to the earth, its sphere of influence is very large. However, relative to the sun it is tiny, as illustrated in Figure 8.7.

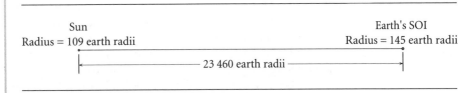

Figure 8.7 The earth's sphere of influence and the sun, drawn to scale.

8.5 METHOD OF PATCHED CONICS

'Conics' refers to the fact that two-body or Keplerian orbits are conic sections with the focus at the attracting body. To study an interplanetary trajectory we assume that when the spacecraft is outside the sphere of influence of a planet it follows an unperturbed Keplerian orbit around the sun. Because interplanetary distances are so vast, for heliocentric orbits we may neglect the size of the spheres of influence and consider them, like the planets they surround, to be just points in space coinciding with the planetary centers. Within each planetary sphere of influence, the spacecraft travels an unperturbed Keplerian path about the planet. While the sphere of influence appears as a mere speck on the scale of the solar system, from the point of view of the planet it is very large indeed and may be considered to lie at infinity.

To analyze a mission from planet 1 to planet 2 using the method of patched conics, we first determine the heliocentric trajectory – such as the Hohmann transfer ellipse discussed in Section 8.2 – that will intersect the desired positions of the two planets in their orbits. This trajectory takes the spacecraft from the sphere of influence of planet 1 to that of planet 2. At the spheres of influence, the heliocentric velocities of the transfer orbit are computed relative to the planet to establish the velocities 'at infinity' which are then used to determine planetocentric departure trajectory at planet 1 and arrival trajectory at planet 2. In this way we 'patch' together the three conics, one centered at the sun and the other two centered at the planets in question.

Whereas the method of patched conics is remarkably accurate for interplanetary trajectories, such is not the case for lunar rendezvous and return trajectories. The orbit of the moon is determined primarily by the earth, whose sphere of influence extends well beyond the moon's 384 400 km orbital radius. To apply patched conics to lunar trajectories we ignore the sun and consider the motion of a spacecraft as influenced by just the earth and moon, as in the restricted three-body problem discussed in Section 2.12. The size of the moon's sphere of influence is found using Equation 8.34, with the earth playing the role of the sun:

$$r_{SOI} = R \left(\frac{m_{moon}}{m_{earth}} \right)^{\frac{2}{5}}$$

where R is the radius of the moon's orbit. Thus, using Table A.1,

$$r_{SOI} = 384\,400 \left(\frac{73.48 \times 10^{21}}{5974 \times 10^{21}} \right)^{\frac{2}{5}} = 66\,200 \text{ km}$$

as recorded in Table A.2. The moon's sphere of influence extends out to over one-sixth of the distance to the earth. We can hardly consider it to be a mere speck relative to the earth. Another complication is the fact that the earth and the moon are somewhat comparable in mass, so that their center of mass lies almost three-quarters of an earth radius from the center of the earth. The motion of the moon cannot be accurately described as rotating around the center of the earth.

Complications such as these place the analysis of cislunar trajectories beyond our scope. Extensions of the patched conic technique to such orbits may be found in Bate, Mueller and White (1971), Kaplan (1976) and Battin (1999).

8.6 PLANETARY DEPARTURE

In order to escape the gravitational pull of a planet, the spacecraft must travel a hyperbolic trajectory relative to the planet, arriving at its sphere of influence with a relative velocity $\mathbf{v}_\infty$ (hyperbolic excess velocity) greater than zero. On a parabolic trajectory, according to Equation 2.80, the spacecraft will arrive at the sphere of influence ($r = \infty$) with a relative speed of zero. In that case the spacecraft remains in the same orbit as the planet and does not embark upon a heliocentric elliptical path.

Figure 8.8 shows a spacecraft departing on a Hohmann trajectory from planet 1 towards a target planet 2 which is farther away from the sun (as in Figure 8.1). At the sphere of influence crossing, the heliocentric velocity $\mathbf{V}_D^{(v)}$ of the spacecraft is parallel to the asymptote of the departure hyperbola as well as to the planet's heliocentric velocity vector $\mathbf{V}_1$. $\mathbf{V}_D^{(v)}$ and $\mathbf{V}_1$ must be parallel and in the same direction for a Hohmann transfer such that ΔV_D in Equation 8.3 is positive. Clearly, ΔV_D is the hyperbolic excess speed of the departure hyperbola,

$$v_\infty = \sqrt{\frac{\mu_{sun}}{R_1}} \left(\sqrt{\frac{2R_2}{R_1 + R_2}} - 1 \right) \quad (8.35)$$

It would be well at this point for the reader to review Section 2.9 on hyperbolic trajectories and compare Figures 8.8 and 2.23. Recall that point C is the center of the hyperbola.

A space vehicle is ordinarily launched into an interplanetary trajectory from a circular parking orbit. The radius of this parking orbit equals the periapse radius r_p of the departure hyperbola. According to Equation 2.40, the periapse radius is given by

$$r_p = \frac{h^2}{\mu_1} \frac{1}{1+e} \quad (8.36)$$

where h is the angular momentum of the departure hyperbola (relative to the planet), e is the eccentricity of the hyperbola and μ_1 is the planet's gravitational parameter.

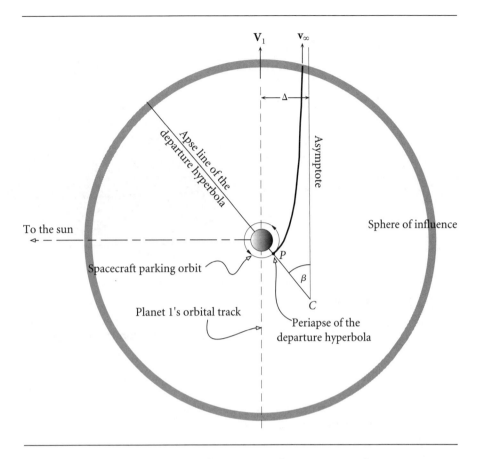

Figure 8.8 Departure of a spacecraft on a mission from an inner planet to an outer planet.

The hyperbolic excess speed is found in Equation 2.105, from which we obtain

$$h = \frac{\mu_1 \sqrt{e^2 - 1}}{v_\infty} \qquad (8.37)$$

Substituting this expression for the angular momentum into Equation 8.36 and solving for the eccentricity yields

$$e = 1 + \frac{r_p v_\infty^2}{\mu_1} \qquad (8.38)$$

We place this result back into Equation 8.37 to obtain the following expression for the angular momentum:

$$h = r_p \sqrt{v_\infty^2 + \frac{2\mu_1}{r_p}} \qquad (8.39)$$

Since the hyperbolic excess speed is specified by the mission requirements (Equation 8.35), choosing a departure periapse r_p yields the parameters e and h of the

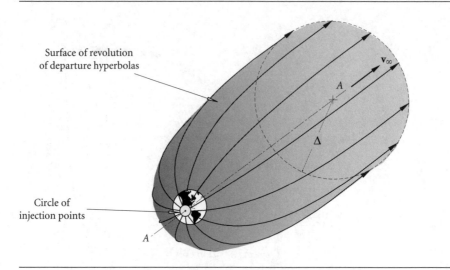

Figure 8.9 Locus of possible departure trajectories for a given v_∞ and r_p.

departure hyperbola. From the angular momentum we get the periapse speed,

$$v_p = \frac{h}{r_p} = \sqrt{v_\infty^2 + \frac{2\mu_1}{r_p}} \qquad (8.40)$$

which can also be found from an energy approach using Equation 2.103. With Equation 8.40 and the speed of the circular parking orbit (Equation 2.53),

$$v_c = \sqrt{\frac{\mu_1}{r_p}} \qquad (8.41)$$

we can calculate the delta-v required to put the vehicle onto the hyperbolic departure trajectory,

$$\Delta v = v_p - v_c = v_c \left(\sqrt{2 + \left(\frac{v_\infty}{v_c}\right)^2} - 1 \right) \qquad (8.42)$$

The location of periapse, where the delta-v maneuver must occur, is found using Equations 2.89 and 8.38,

$$\beta = \cos^{-1}\left(\frac{1}{e}\right) = \cos^{-1}\left(\frac{1}{1 + \frac{r_p v_\infty^2}{\mu_1}}\right) \qquad (8.43)$$

β gives the orientation of the apse line of the hyperbola to the planet's heliocentric velocity vector.

It should be pointed out that the only requirement on the orientation of the plane of the departure hyperbola is that it contains the center of mass of the planet as well as the relative velocity vector $\mathbf{v}_\infty$. Therefore, as shown in Figure 8.11, the hyperbola can be rotated about a line A–A which passes through the planet's center of mass and

8.6 Planetary departure

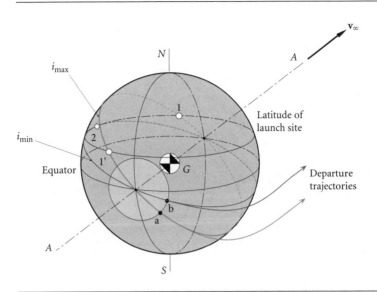

Figure 8.10 Parking orbits and departure trajectories for a launch site at a given latitude.

is parallel to $\mathbf{v}_\infty$ (or $\mathbf{V}_1$, which of course is parallel to $\mathbf{v}_\infty$ for Hohmann transfers). Rotating the hyperbola in this way sweeps out a surface of revolution on which lie all possible departure hyperbolas. The periapse of the hyperbola traces out a circle which, for the specified periapse radius r_p, is the locus of all possible points of injection into a departure trajectory towards the target planet. This circle is the base of a cone with vertex at the center of the planet. From Figure 3.23 we can determine that its radius is $r_p \sin \beta$, where β is given just above in Equation 8.43.

The plane of the parking orbit, or direct ascent trajectory, must contain the line A–A and the launch site at the time of launch. The possible inclinations of a prograde orbit range from a minimum of $i_{\min}$, where $i_{\min}$ is the latitude of the launch site, to $i_{\max}$, which cannot exceed 90°. Launch site safety considerations may place additional limits on that range. For example, orbits originating from the Kennedy Space Center in Florida, USA, (latitude 28.5°) are limited to inclinations between 28.5° and 52.5°. For the scenario illustrated in Figure 8.12 the location of the launch site limits access to just the departure trajectories having periapses lying between a and b. The figure shows that there are two times per day – when the planet rotates the launch site through positions 1 and 1' – that a spacecraft can be launched into a parking orbit. These times are closer together (the launch window is smaller) the lower the inclination of the parking orbit.

Once a spacecraft is established in its parking orbit, then an opportunity for launch into the departure trajectory occurs each orbital circuit.

If the mission is to send a spacecraft from an outer planet to an inner planet, as in Figure 8.2, then the spacecraft's heliocentric speed $V_D^{(v)}$ at departure must be less than that of the planet. That means the spacecraft must emerge from the back side of the sphere of influence with its relative velocity vector $\mathbf{v}_\infty$ directed opposite to $\mathbf{V}_1$, as shown in Figure 8.11. Figures 8.9 and 8.10 apply to this situation as well.

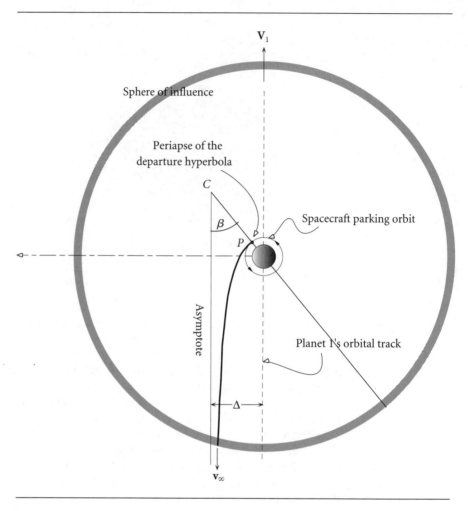

Figure 8.11 Departure of a spacecraft on a trajectory from an outer planet to an inner planet.

EXAMPLE 8.4

A spacecraft is launched on a mission to Mars starting from a 300 km circular parking orbit. Calculate (a) the delta-v required; (b) the location of perigee of the departure hyperbola; (c) the amount of propellant required as a percentage of the spacecraft mass before the delta-v burn, assuming a specific impulse of 300 seconds.

From Tables A.1 and A.2 we obtain the gravitational parameters for the sun and the earth,

$$\mu_{sun} = 1.327 \times 10^{11} \text{ km}^3/\text{s}^2$$

$$\mu_{earth} = 398\,600 \text{ km}^3/\text{s}^2$$

and the orbital radii of the earth and Mars,

$$R_{earth} = 149.6 \times 10^6 \text{ km}$$

$$R_{Mars} = 227.9 \times 10^6 \text{ km}$$

(a) According to Equation 8.35, the hyperbolic excess speed is

$$v_\infty = \sqrt{\frac{\mu_{sun}}{R_{earth}}}\left(\sqrt{\frac{2R_{Mars}}{R_{earth}+R_{Mars}}}-1\right)$$

$$= \sqrt{\frac{1.327\times 10^{11}}{149.6\times 10^6}}\left(\sqrt{\frac{2(227.9\times 10^6)}{149.6\times 10^6+227.9\times 10^6}}-1\right)$$

from which

$$v_\infty = 2.943\,\text{km/s}$$

The speed of the spacecraft in its 300 km circular parking orbit is given by Equation 8.41,

$$v_c = \sqrt{\frac{\mu_{earth}}{r_{earth}+300}} = \sqrt{\frac{398\,600}{6678}} = 7.726\,\text{km/s}$$

Finally, we use Equation 8.42 to calculate the delta-v required to step up to the departure hyperbola:

$$\Delta v = v_p - v_c = v_c\left(\sqrt{2+\left(\frac{v_\infty}{v_c}\right)^2}-1\right) = 7.726\left(\sqrt{2+\left(\frac{2.943}{7.726}\right)^2}-1\right)$$

$$\Delta v = 3.590\,\text{km/s}$$

(b) Perigee of the departure hyperbola, relative to the earth's orbital velocity vector, is found using Equation 8.43,

$$\beta = \cos^{-1}\left(\frac{1}{1+\frac{r_p v_\infty^2}{\mu_{earth}}}\right) = \cos^{-1}\left(\frac{1}{1+\frac{6678\cdot 2.943^2}{368\,600}}\right)$$

$$\beta = 29.16°$$

Figure 8.12 shows that the perigee can be located on either the sunlit or dark side of the earth. It is likely that the parking orbit would be a prograde orbit (west to east), which would place the burnout point on the dark side.

(c) From Equation 6.1 we have

$$\frac{\Delta m}{m} = 1 - e^{-\frac{\Delta v}{I_{sp}g_o}}$$

Substituting $\Delta v = 3.590$ km/s, $I_{sp} = 300$ s and $g_o = 9.81\times 10^{-3}$ km/s^2, this yields

$$\frac{\Delta m}{m} = 0.705$$

That is, prior to the delta-v maneuver, over 70 percent of the spacecraft mass must be propellant.

(Example 8.4 continued)

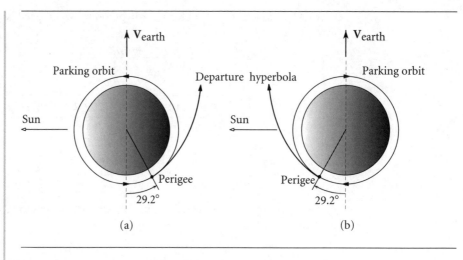

Figure 8.12 Departure trajectory to Mars initiated from (a) the dark side and (b) the sunlit side of the earth.

8.7 Sensitivity analysis

The initial maneuvers required to place a spacecraft on an interplanetary trajectory occur well within the sphere of influence of the departure planet. Since the sphere of influence is just a point on the scale of the solar system, one may ask what effects small errors in position and velocity at the maneuver point have on the trajectory. Assuming the mission is from an inner to an outer planet, let us consider the effect which small changes in the burnout velocity v_p and radius r_p have on the target radius R_2 of the heliocentric Hohmann transfer ellipse (see Figures 8.1 and 8.8).

R_2 is the radius of aphelion, so we use Equation 2.60 to obtain

$$R_2 = \frac{h^2}{\mu_{sun}} \frac{1}{1-e}$$

Substituting $h = R_1 V_D^{(v)}$ and $e = (R_2 - R_1)/(R_2 + R_1)$, and solving for R_2, yields

$$R_2 = \frac{R_1^2 [V_D^{(v)}]^2}{2\mu_{sun} - R_1 [V_D^{(v)}]^2} \quad (8.44)$$

(This expression holds as well for a mission from an outer to inner planet.) The change δR_2 in R_2 due to a small variation $\delta V_D^{(v)}$ of $V_D^{(v)}$ is

$$\delta R_2 = \frac{dR_2}{dV_D^{(v)}} \delta V_D^{(v)} = \frac{4R_1^2 \mu_{sun}}{\left\{2\mu_{sun} - R_1 [V_D^{(v)}]^2\right\}^2} V_D^{(v)} \delta V_D^{(v)}$$

Dividing this equation by Equation 8.44 leads to

$$\frac{\delta R_2}{R_2} = \frac{2}{1 - \frac{R_1 [V_D^{(v)}]^2}{2\mu_{sun}}} \frac{\delta V_D^{(v)}}{V_D^{(v)}} \quad (8.45)$$

The departure speed $V_D^{(v)}$ of the space vehicle is the sum of the planet's speed V_1 and excess speed v_∞:

$$V_D^{(v)} = V_1 + v_\infty$$

We can solve Equation 8.40 for v_∞,

$$v_\infty = \sqrt{v_p^2 - \frac{2\mu_1}{r_p}}$$

Hence

$$V_D^{(v)} = V_1 + \sqrt{v_p^2 - \frac{2\mu_1}{r_p}} \tag{8.46}$$

The change in $V_D^{(v)}$ due to variations δr_p and δv_p of the burnout position (periapse) r_p and speed v_p is given by

$$\delta V_D^{(v)} = \frac{\partial V_D^{(v)}}{\partial r_p} \delta r_p + \frac{\partial V_D^{(v)}}{\partial v_p} \delta v_p \tag{8.47}$$

From Equation 8.46 we obtain

$$\frac{\partial V_D^{(v)}}{\partial r_p} = \frac{\mu_1}{v_\infty r_p^2} \qquad \frac{\partial V_D^{(v)}}{\partial v_p} = \frac{v_p}{v_\infty}$$

Therefore

$$\delta V_D^{(v)} = \frac{\mu_1}{v_\infty r_p^2} \delta r_p + \frac{v_p}{v_\infty} \delta v_p$$

Once again making use of Equation 8.40, this can be written as follows

$$\frac{\delta V_D^{(v)}}{V_D^{(v)}} = \frac{\mu_1}{V_D^{(v)} v_\infty r_p} \frac{\delta r_p}{r_p} + \frac{v_\infty + \frac{2\mu_1}{r_p}}{V_D^{(v)}} \frac{\delta v_p}{v_p} \tag{8.48}$$

Substituting this into Equation 8.45 finally yields the desired result, an expression for the variation of R_2 due to variations in r_p and v_p:

$$\frac{\delta R_2}{R_2} = \frac{2}{1 - \frac{R_1 [V_D^{(v)}]^2}{2\mu_{\text{sun}}}} \left(\frac{\mu_1}{V_D^{(v)} v_\infty r_p} \frac{\delta r_p}{r_p} + \frac{v_\infty + \frac{2\mu_1}{r_p}}{V_D^{(v)}} \frac{\delta v_p}{v_p} \right) \tag{8.49}$$

Consider a mission from earth to Mars, starting from a 300 km parking orbit. We have

$$\mu_{\text{sun}} = 1.327 \times 10^{11} \text{ km}^3/\text{s}^2$$

$$\mu_{pl_1} = \mu_{\text{earth}} = 398\,600 \text{ km}^3/\text{s}^2$$

$$R_1 = 149.6 \times 10^6 \text{ km}$$

$$R_2 = 227.9 \times 10^6 \text{ km}$$

$$r_p = 6678 \text{ km}$$

In addition, from Equations 8.1 and 8.2,

$$V_1 = V_{earth} = \sqrt{\frac{\mu_{sun}}{R_1}} = \sqrt{\frac{1.327 \times 10^{11}}{149.6 \times 10^6}} = 29.78 \text{ km/s}$$

$$V_D^{(v)} = \sqrt{2\mu_{sun}} \sqrt{\frac{R_2}{R_1(R_1 + R_2)}}$$

$$= \sqrt{2 \cdot 1.327 \times 10^{11}} \sqrt{\frac{227.9 \times 10^6}{149.6 \times 10^6 (149.6 \times 10^6 + 227.9 \times 10^6)}} = 32.73 \text{ km/s}$$

Therefore

$$v_\infty = V_D^{(v)} - V_{earth} = 2.943 \text{ km/s}$$

and, from Equation 8.40,

$$v_p = \sqrt{v_\infty^2 + \frac{2\mu_{earth}}{r_p}} = \sqrt{2.943^2 + \frac{2 \cdot 398\,600}{6678}} = 11.32 \text{ km/s}$$

Substituting these values into Equation 8.49 yields

$$\frac{\delta R_2}{R_2} = 3.127 \frac{\delta r_p}{r_p} + 6.708 \frac{\delta v_p}{v_p}$$

This expression shows that a 0.01 percent variation (1.1 m/s) in the burnout speed v_p changes the target radius R_2 by 0.067 percent or 153 000 km! Likewise, an error of 0.01 percent (0.67 km) in burnout radius r_p produces an error of over 70 000 km. Thus small errors which are likely to occur in the launch phase of the mission must be corrected by midcourse maneuvers during the coasting flight along the elliptical transfer trajectory.

8.8 Planetary rendezvous

A spacecraft arrives at the sphere of influence of the target planet with a hyperbolic excess velocity $\mathbf{v}_\infty$ relative to the planet. In the case illustrated in Figure 8.1, a mission from an inner planet 1 to an outer planet 2 (e.g., earth to Mars), the spacecraft's heliocentric approach velocity $\mathbf{V}_A^{(v)}$ is smaller in magnitude than that of the planet, $\mathbf{V}_2$. Therefore, it crosses the forward portion of the sphere of influence, as shown in Figure 8.13. For a Hohmann transfer, $\mathbf{V}_A^{(v)}$ and $\mathbf{V}_2$ are parallel, so the magnitude of the hyperbolic excess velocity is, simply,

$$v_\infty = V_2 - V_A^{(v)} \qquad (8.50)$$

If the mission is as illustrated in Figure 8.2, from an outer planet to an inner one (e.g., earth to Venus), then $V_A^{(v)}$ is greater than V_2, and the spacecraft must cross the *rear* portion of the sphere of influence, as shown in Figure 8.14. In that case

$$v_\infty = V_A^{(v)} - V_2 \qquad (8.51)$$

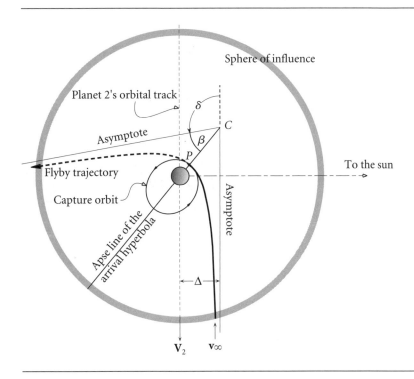

Figure 8.13 Spacecraft approach trajectory for a Hohmann transfer to an outer planet from an inner one. P is the periapse of the approach hyperbola.

What happens after crossing the sphere of influence depends on the nature of the mission. If the goal is to impact the planet (or its atmosphere), the aiming radius Δ of the approach hyperbola must be such that hyperbola's periapse r_p equals essentially the radius of the planet. If the intent is to go into orbit around the planet, then Δ must be chosen so that the delta-v burn at periapse will occur at the correct altitude above the planet. If there is no impact with the planet and no drop into a capture orbit around the planet, then the spacecraft will simply continue past periapse on a flyby trajectory, exiting the sphere of influence with the same relative speed v_∞ it entered, but with the velocity vector rotated through the turn angle δ, given by Equation 2.90,

$$\delta = 2\sin^{-1}\left(\frac{1}{e}\right) \quad (8.52)$$

With the hyperbolic excess speed v_∞ and the periapse radius r_p specified, the eccentricity of the approach hyperbola is found from Equation 8.38,

$$e = 1 + \frac{r_p v_\infty^2}{\mu_2} \quad (8.53)$$

where μ_2 is the gravitational parameter of planet 2. Hence, the turn angle is

$$\delta = 2\sin^{-1}\left(\frac{1}{1 + \frac{r_p v_\infty^2}{\mu_2}}\right) \quad (8.54)$$

370 Chapter 8 *Interplanetary trajectories*

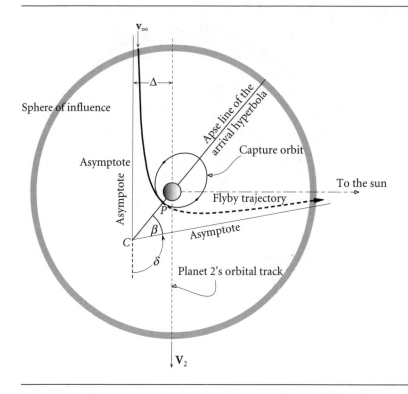

Figure 8.14 Spacecraft approach trajectory for a Hohmann transfer to an inner planet from an outer one. *P* is the periapse of the approach hyperbola.

We can combine Equations 2.93 and 2.97 to obtain the following expression for the aiming radius,

$$\Delta = \frac{h^2}{\mu_2} \frac{1}{\sqrt{e^2 - 1}} \qquad (8.55)$$

The angular momentum of the approach hyperbola relative to the planet is found using Equation 8.39,

$$h = r_p \sqrt{v_\infty^2 + \frac{2\mu_2}{r_p}} \qquad (8.56)$$

Substituting Equations 8.53 and 8.56 into 8.55 yields the aiming radius in terms of the periapse radius and the hyperbolic excess speed,

$$\Delta = r_p \sqrt{1 + \frac{2\mu_2}{r_p v_\infty^2}} \qquad (8.57)$$

Just as we observed when discussing departure trajectories, the approach hyperbola does not lie in a unique plane. We can rotate the hyperbolas illustrated in Figures 8.11 and 8.12 about a line *A–A* parallel to $\mathbf{v}_\infty$ and passing through the target planet's center

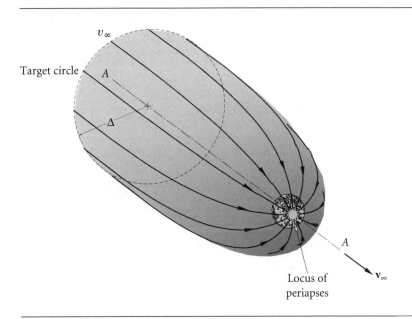

Figure 8.15 Locus of approach hyperbolas to the target planet.

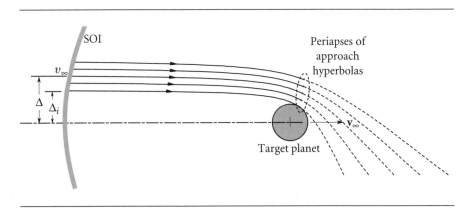

Figure 8.16 Family of approach hyperbolas having the same v_∞ but different Δ.

of mass, as shown in Figure 8.15. The approach hyperbolas in that figure terminate at the circle of periapses. Figure 8.16 is a plane through the solid of revolution revealing the shape of hyperbolas having a common v_∞ but varying Δ.

Let us suppose that the purpose of the mission is to enter an elliptical orbit of eccentricity e around the planet. This will require a delta-v maneuver at periapse P (Figures 8.13 and 8.14), which is also periapse of the ellipse. The speed in the hyperbolic trajectory at periapse is given by Equation 8.40

$$v_p)_{\text{hyp}} = \sqrt{v_\infty^2 + \frac{2\mu_2}{r_p}} \qquad (8.58)$$

The velocity at periapse of the capture orbit is found by setting $h = r_p v_p$ in Equation 2.40 and solving for v_p:

$$v_p)_{\text{capture}} = \sqrt{\frac{\mu_2(1+e)}{r_p}} \tag{8.59}$$

Hence, the required delta-v is

$$\Delta v = v_p)_{\text{hyp}} - v_p)_{\text{capture}} = \sqrt{v_\infty^2 + \frac{2\mu_2}{r_p}} - \sqrt{\frac{\mu_2(1+e)}{r_p}} \tag{8.60}$$

For a given v_∞, Δv clearly depends upon the choice of periapse radius r_p and capture orbit eccentricity e. Requiring the maneuver point to remain the periapse of the capture orbit means that Δv is maximum for a circular capture orbit and decreases with increasing eccentricity until $\Delta v = 0$, which, of course, means no capture (flyby).

In order to determine optimal capture radius, let us write Equation 8.60 in non-dimensional form as

$$\frac{\Delta v}{v_\infty} = \sqrt{1 + \frac{2}{\xi}} - \sqrt{\frac{1+e}{\xi}} \tag{8.61}$$

where

$$\xi = \frac{r_p v_\infty^2}{\mu_2} \tag{8.62}$$

The first and second derivatives of $\Delta v / v_\infty$ with respect to ξ are

$$\frac{d}{d\xi}\frac{\Delta v}{v_\infty} = \left(-\frac{1}{\sqrt{\xi+2}} + \frac{\sqrt{1+e}}{2}\right)\frac{1}{\xi^{\frac{3}{2}}} \tag{8.63}$$

$$\frac{d^2}{d\xi^2}\frac{\Delta v}{v_\infty} = \left[\frac{2\xi+3}{(\xi+2)^{\frac{3}{2}}} - \frac{3}{4}\sqrt{1+e}\right]\frac{1}{\xi^{\frac{5}{2}}} \tag{8.64}$$

Setting the first derivative equal to zero and solving for ξ yields

$$\xi = 2\frac{1-e}{1+e} \tag{8.65}$$

Substituting this value of ξ into Equation 8.64, we get

$$\frac{d^2}{d\xi^2}\frac{\Delta v}{v_\infty} = \frac{\sqrt{2}}{64}\frac{(1+e)^3}{(1-e)^{\frac{3}{2}}} \tag{8.66}$$

This expression is positive for elliptical orbits ($0 \leq e < 1$), which means that when ξ is given by Equation 8.65 Δv is a minimum. Therefore, from Equation 8.62, the optimal periapse radius as far as fuel expenditure is concerned is

$$r_p = \frac{2\mu_2}{v_\infty^2}\frac{1-e}{1+e} \tag{8.67}$$

8.8 Planetary rendezvous

We can combine Equations 2.40 and 2.60 to get

$$\frac{1-e}{1+e} = \frac{r_p}{r_a} \tag{8.68}$$

where r_a is the apoapse radius. Thus, Equation 8.67 implies

$$r_a = \frac{2\mu_2}{v_\infty^2} \tag{8.69}$$

That is, the apoapse of this capture ellipse is independent of the eccentricity and equals the radius of the optimal circular orbit.

Substituting Equation 8.65 back into Equation 8.61 yields the minimum Δv

$$\Delta v = v_\infty \sqrt{\frac{1-e}{2}} \tag{8.70}$$

Finally, placing the optimal r_p into Equation 8.57 leads to an expression for the aiming radius required for minimum Δv,

$$\Delta = 2\sqrt{2}\frac{\sqrt{1-e}}{1+e}\frac{\mu_2}{v_\infty^2} = \sqrt{\frac{2}{1-e}}r_p \tag{8.71}$$

Clearly, the optimal Δv (and periapse height) are reduced for highly eccentric elliptical capture orbits ($e \to 1$). However, it should be pointed out that the use of optimal Δv may have to be sacrificed in favor of a variety of other mission requirements.

EXAMPLE 8.5

After a Hohmann transfer from earth, calculate the minimum delta-v required to place a spacecraft in Mars orbit with a period of seven hours. Also calculate the periapse radius, the aiming radius and the angle between periapse and Mars' velocity vector.

The following data is required from Tables A.1 and A.2:

$$\mu_{sun} = 1.327 \times 10^{11} \text{ km}^3/\text{s}^2$$

$$\mu_{Mars} = 42\,830 \text{ km}^3/\text{s}^2$$

$$R_{earth} = 149.6 \times 10^6 \text{ km}$$

$$R_{Mars} = 227.9 \times 10^6 \text{ km}$$

$$r_{Mars} = 3396 \text{ km}$$

The hyperbolic excess speed is found using Equation 8.4,

$$v_\infty = \Delta V_A = \sqrt{\frac{\mu_{sun}}{R_{Mars}}}\left(1 - \sqrt{\frac{2R_{earth}}{R_{earth} + R_{Mars}}}\right)$$

$$= \sqrt{\frac{1.327 \times 10^{11}}{227.9 \times 10^6}}\left(1 - \sqrt{\frac{2 \cdot 149.6 \times 10^6}{149.6 \times 10^6 + 227.9 \times 10^6}}\right)$$

$$v_\infty = 2.648 \text{ km/s}$$

(Example 8.5 continued) We can use Equation 2.73 to express the semimajor axis a of the capture orbit in terms of its period T,

$$a = \left(\frac{T\sqrt{\mu_{\text{Mars}}}}{2\pi}\right)^{\frac{2}{3}}$$

Substituting $T = 7 \cdot 3600$ s yields

$$a = \left(\frac{25\,200\sqrt{42\,830}}{2\pi}\right)^{\frac{2}{3}} = 8832 \text{ km}$$

From Equation 2.63 we obtain

$$a = \frac{r_p}{1-e}$$

Upon substituting the optimal periapse radius, Equation 8.67, this becomes

$$a = \frac{2\mu_{\text{Mars}}}{v_\infty^2} \frac{1}{1+e}$$

from which

$$e = \frac{2\mu_{\text{Mars}}}{av_\infty^2} - 1 = \frac{2 \cdot 42\,830}{8832 \cdot 2.648^2} - 1 = 0.3833$$

Thus, using Equation 8.70, we find

$$\Delta v = v_\infty \sqrt{\frac{1-e}{2}} = 2.648\sqrt{\frac{1-0.3833}{2}} = \underline{1.470 \text{ km/s}}$$

From Equations 8.66 and 8.71 we obtain the periapse radius

$$r_p = \frac{2\mu_{\text{Mars}}}{v_\infty^2} \frac{1-e}{1+e} = \frac{2 \cdot 42\,830}{2.648^2} \frac{1-0.3833}{1+0.3833} = \underline{5447 \text{ km}}$$

and the aiming radius

$$\Delta = r_p\sqrt{\frac{2}{1-e}} = 5447\sqrt{\frac{2}{1-0.3833}} = \underline{9809 \text{ km}}$$

Finally, using Equation 8.43, we get the angle to periapse

$$\beta = \cos^{-1}\left(\frac{1}{1 + \frac{r_p v_\infty^2}{\mu_{\text{Mars}}}}\right) = \cos^{-1}\left(\frac{1}{1 + \frac{5447 \cdot 2.648^2}{42\,830}}\right) = \underline{58.09°}$$

Mars, the approach hyperbola, and the capture orbit are shown to scale in Figure 8.17. The approach could also be made from the dark side of the planet instead of the sunlit side. The approach hyperbola and capture ellipse would be the mirror image of that shown, as is the case in Figure 8.12.

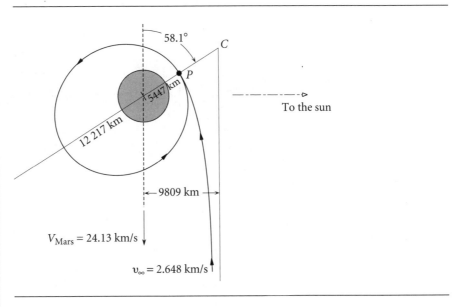

Figure 8.17 An optimal approach to a Mars capture orbit with a seven hour period. $r_{Mars} = 3396$ km.

8.9 PLANETARY FLYBY

A spacecraft which enters a planet's sphere of influence and does not impact the planet or go into orbit around it will continue in its hyperbolic trajectory through periapse P and exit the sphere of influence. Figure 8.18 shows a hyperbolic flyby trajectory along with the asymptotes and apse line of the hyperbola. It is a leading-side flyby because the periapse is on the side of the planet facing into the direction of motion. Likewise, Figure 8.19 illustrates a trailing-side flyby. At the inbound crossing point, the heliocentric velocity $\mathbf{V}_1^{(v)}$ of the spacecraft equals the planet's heliocentric velocity $\mathbf{V}$ plus the hyperbolic excess velocity $\mathbf{v}_\infty)_1$ of the spacecraft (relative to the planet),

$$\mathbf{V}_1^{(v)} = \mathbf{V} + \mathbf{v}_{\infty_1} \tag{8.72}$$

Similarly, at the outbound crossing we have

$$\mathbf{V}_2^{(v)} = \mathbf{V} + \mathbf{v}_{\infty_2} \tag{8.73}$$

The change $\Delta \mathbf{V}^{(v)}$ in the spacecraft's heliocentric velocity is

$$\Delta \mathbf{V}^{(v)} = \mathbf{V}_2^{(v)} - \mathbf{V}_1^{(v)} = (\mathbf{V} + \mathbf{v}_{\infty_2}) - (\mathbf{V} + \mathbf{v}_{\infty_1})$$

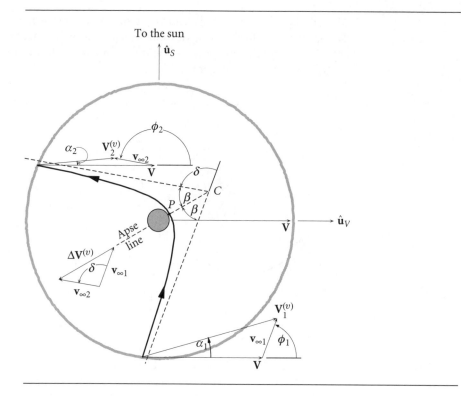

Figure 8.18 Leading-side planetary flyby.

which means

$$\Delta \mathbf{V}^{(v)} = \mathbf{v}_{\infty_2} - \mathbf{v}_{\infty_1} = \Delta \mathbf{v}_\infty \qquad (8.74)$$

The excess velocities $\mathbf{v}_{\infty_1}$ and $\mathbf{v}_{\infty_2}$ lie along the asymptotes of the hyperbola and are therefore inclined at the same angle β to the apse line (see Figure 2.23), with $\mathbf{v}_{\infty_1}$ pointing towards and $\mathbf{v}_{\infty_2}$ pointing away from the center C. They both have the same magnitude v_∞, with $\mathbf{v}_{\infty_2}$ having rotated relative to $\mathbf{v}_{\infty_1}$ by the turn angle δ. Hence, $\Delta \mathbf{v}_\infty$ – and therefore $\Delta \mathbf{V}^{(v)}$ – is a vector which lies along the apse line and always points away from periapse, as illustrated in Figures 8.18 and 8.19. From those figures it can be seen that, in a leading-side flyby, the component of $\Delta \mathbf{V}^{(v)}$ in the direction of the planet's velocity is negative, whereas for the trailing-side flyby it is positive. This means that a leading-side flyby results in a decrease in the spacecraft's heliocentric speed. On the other hand, a trailing-side flyby increases that speed.

In order to analyze a flyby problem, we proceed as follows. First, let $\hat{\mathbf{u}}_V$ be the unit vector in the direction of the planet's heliocentric velocity $\mathbf{V}$ and let $\hat{\mathbf{u}}_S$ be the unit vector pointing from the planet to the sun. At the inbound crossing of the sphere of influence, the heliocentric velocity $\mathbf{V}_1^{(v)}$ of the spacecraft is

$$\mathbf{V}_1^{(v)} = [V_1^{(v)}]_V \hat{\mathbf{u}}_V + [V_1^{(v)}]_S \hat{\mathbf{u}}_S \qquad (8.75)$$

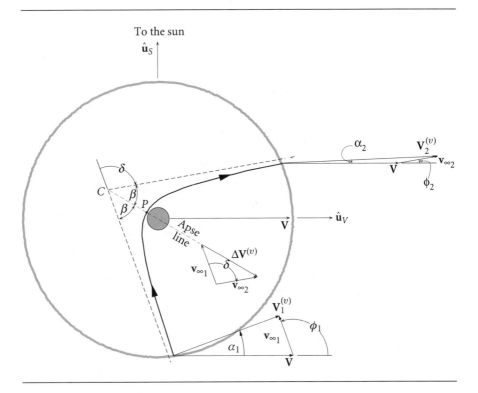

Figure 8.19 Trailing-side planetary flyby.

where the scalar components of $\mathbf{V}_1^{(v)}$ are

$$[V_1^{(v)}]_V = V_1^{(v)} \cos \alpha_1 \qquad [V_1^{(v)}]_S = V_1^{(v)} \sin \alpha_1 \qquad (8.76)$$

α_1 is the angle between $\mathbf{V}_1^{(v)}$ and $\mathbf{V}$. All angles are measured positive counterclockwise. Referring to Figure 2.11, we see that the magnitude of α_1 is the flight path angle γ of the spacecraft's heliocentric trajectory when it encounters the planet's sphere of influence (a mere speck) at the planet's distance R from the sun. Furthermore,

$$[V_1^{(v)}]_V = V_{\perp_1} \qquad [V_1^{(v)}]_S = -V_{r_1} \qquad (8.77)$$

$V_{\perp_1}$ and V_{r_1} are furnished by Equations 2.38 and 2.39

$$V_{\perp_1} = \frac{\mu_{\text{sun}}}{h_1} \frac{1}{1 + e_1 \cos \theta_1} \qquad V_{r_1} = \frac{\mu_{\text{sun}}}{h_1} e_1 \sin \theta_1 \qquad (8.78)$$

in which e_1, h_1 and θ_1 are the eccentricity, angular momentum and true anomaly of the heliocentric approach trajectory.

The velocity of the planet relative to the sun is

$$\mathbf{V} = V \hat{\mathbf{u}}_V \qquad (8.79)$$

where $V = \sqrt{\mu_{sun}/R}$. At the inbound crossing of the planet's sphere of influence, the hyperbolic excess velocity of the spacecraft is obtained from Equation 8.72

$$\mathbf{v}_{\infty_1} = \mathbf{V}_1^{(v)} - \mathbf{V}$$

Using this we find

$$\mathbf{v}_{\infty_1} = (v_{\infty_1})_V \hat{\mathbf{u}}_V + (v_{\infty_1})_S \hat{\mathbf{u}}_S \qquad (8.80)$$

where the scalar components of $\mathbf{v}_{\infty_1}$ are

$$(v_{\infty_1})_V = V_1^{(v)} \cos\alpha_1 - V \qquad (v_{\infty_1})_S = V_1^{(v)} \sin\alpha_1 \qquad (8.81)$$

v_∞ is the magnitude of $\mathbf{v}_{\infty_1}$,

$$v_\infty = \sqrt{\mathbf{v}_{\infty_1} \cdot \mathbf{v}_{\infty_1}} = \sqrt{\left[V_1^{(v)}\right]^2 + V^2 - 2V_1^{(v)} V \cos\alpha_1} \qquad (8.82)$$

At this point v_∞ is known, so that upon specifying the periapse radius r_p we can compute the angular momentum and eccentricity of the flyby hyperbola (relative to the planet), using Equations 8.38 and 8.39:

$$h = r_p \sqrt{v_\infty^2 + \frac{2\mu}{r_p}} \qquad e = 1 + \frac{r_p v_\infty^2}{\mu} \qquad (8.83)$$

where μ is the gravitational parameter of the planet.

The angle between $\mathbf{v}_{\infty_1}$ and the planet's heliocentric velocity is ϕ_1. It is found using the components of $\mathbf{v}_{\infty_1}$ in Equation 8.81,

$$\phi_1 = \tan^{-1}\frac{(v_{\infty_1})_S}{(v_{\infty_1})_V} = \tan^{-1}\frac{V_1^{(v)} \sin\alpha_1}{V_1^{(v)} \cos\alpha_1 - V} \qquad (8.84)$$

At the outbound crossing the angle between $\mathbf{v}_{\infty_2}$ and $\mathbf{V}$ is ϕ_2, where

$$\phi_2 = \phi_1 + \delta \qquad (8.85)$$

For the leading-side flyby in Figure 8.18, the turn angle is δ positive (counterclockwise) whereas in Figure 8.19 it is negative. Since the magnitude of $\mathbf{v}_{\infty_2}$ is v_∞, we can express $\mathbf{v}_{\infty_2}$ in components as

$$\mathbf{v}_{\infty_2} = v_\infty \cos\phi_2 \hat{\mathbf{u}}_V + v_\infty \sin\phi_2 \hat{\mathbf{u}}_S \qquad (8.86)$$

Therefore, the heliocentric velocity of the spacecraft at the outbound crossing is

$$\mathbf{V}_2^{(v)} = \mathbf{V} + \mathbf{v}_{\infty_2} = [V_2^{(v)}]_V \hat{\mathbf{u}}_V + [V_2^{(v)}]_S \hat{\mathbf{u}}_S \qquad (8.87)$$

where the components of $\mathbf{V}_2^{(v)}$ are

$$[V_2^{(v)}]_V = V + v_\infty \cos\phi_2 \qquad [V_2^{(v)}]_S = v_\infty \sin\phi_2 \qquad (8.88)$$

From this we obtain the radial and transverse heliocentric velocity components,

$$V_{\perp_2} = [V_2^{(v)}]_V \qquad V_{r_2} = -[V_2^{(v)}]_S \qquad (8.89)$$

Finally, we obtain the three elements e_2, h_2 and θ_2 of the new heliocentric departure trajectory by means of Equation 2.21,

$$h_2 = RV_{\perp_2} \tag{8.90}$$

Equation 2.35,

$$R = \frac{h_2^2}{\mu_{sun}} \frac{1}{1 + e_2 \cos \theta_2} \tag{8.91}$$

and Equation 2.39,

$$V_{r_2} = \frac{\mu_{sun}}{h_2} e_2 \sin \theta_2 \tag{8.92}$$

Notice that the flyby is considered to be an impulsive maneuver during which the heliocentric radius of the spacecraft, which is confined within the planet's sphere of influence, remains fixed at R. The heliocentric velocity analysis is similar to that described in Section 6.7.

EXAMPLE 8.6

A spacecraft departs earth with a velocity perpendicular to the sun line on a flyby mission to Venus. Encounter occurs at a true anomaly in the approach trajectory of $-30°$. Periapse altitude is to be 300 km. (a) For an approach from the dark side of the planet, show that the post-flyby orbit is as illustrated in Figure 8.20. (b) For an approach from the sunlit side of the planet, show that the post-flyby orbit is as illustrated in Figure 8.21.

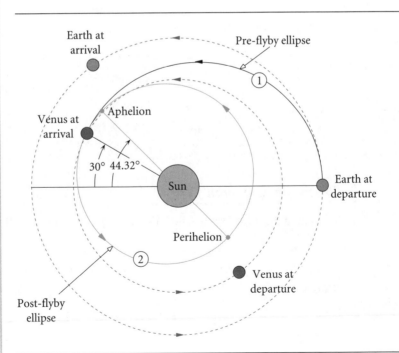

Figure 8.20 Spacecraft orbits before and after a flyby of Venus, approaching from the dark side.

(Example 8.6 continued)

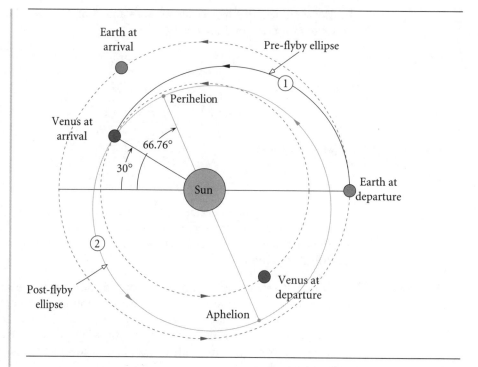

Figure 8.21 Spacecraft orbits before and after a flyby of Venus, approaching from the sunlit side.

The following data is found in Tables A.1 and A.2:

$$\mu_{sun} = 1.3271 \times 10^{11} \text{ km}^3/\text{s}^2$$
$$\mu_{Venus} = 324\,900 \text{ km}^3/\text{s}^2$$
$$R_{earth} = 149.6 \times 10^6 \text{ km}$$
$$R_{Venus} = 108.2 \times 10^6 \text{ km}$$
$$r_{Venus} = 6052 \text{ km}$$

Pre-flyby ellipse (orbit 1)

Evaluating the orbit formula, Equation 2.35, at perihelion of orbit 1 yields

$$R_{earth} = \frac{h_1^2}{\mu_{sun}} \frac{1}{1 - e_1}$$

Thus

$$h_1^2 = \mu_{sun} R_{earth}(1 - e_1) \tag{a}$$

At intercept

$$R_{Venus} = \frac{h_1^2}{\mu_{sun}} \frac{1}{1 + e_1 \cos(\theta_1)}$$

Substituting Equation (a) and $\theta_1 = -30°$ and solving the resulting expression for e_1 leads to

$$e_1 = \frac{R_{earth} - R_{Venus}}{R_{earth} + R_{Venus}\cos(\theta_1)} = \frac{149.6 \times 10^6 - 108.2 \times 10^6}{149.6 \times 10^6 + 108.2 \times 10^6 \cos(-30°)} = 0.1702$$

With this result, Equation (a) yields

$$h_1 = \sqrt{1.327 \times 10^{11} \cdot 149.6 \times 10^6 (1 - 0.1702)} = 4.059 \times 10^9 \, \text{km}^2/\text{s}$$

Now we can use Equations 8.78 to calculate the radial and transverse components of the spacecraft's heliocentric velocity at the inbound crossing of Venus's sphere of influence:

$$V_{\perp_1} = \frac{h_1}{R_{Venus}} = \frac{4.059 \times 10^9}{108.2 \times 10^6} = 37.51 \, \text{km/s}$$

$$V_{r_1} = \frac{\mu_{sun}}{h_1} e_1 \sin(\theta_1) = \frac{1.327 \times 10^{11}}{4.059 \times 10^9} \cdot 0.1702 \cdot \sin(-30°) = -2.782 \, \text{km/s}$$

The flight path angle, from Equation 2.41, is

$$\gamma_1 = \tan^{-1}\frac{V_{r_1}}{V_{\perp_1}} = \tan^{-1}\left(\frac{-2.782}{37.51}\right) = -4.241°$$

The negative sign is consistent with the fact that the spacecraft is flying towards perihelion of the pre-flyby elliptical trajectory (orbit 1).

The speed of the space vehicle at the inbound crossing is

$$V_1^{(v)} = \sqrt{V_{r_1}^2 + V_{\perp_1}^2} = \sqrt{(-2.782)^2 + 37.51^2} = 37.62 \, \text{km/s} \quad (b)$$

Flyby hyperbola

From Equations 8.75 and 8.77 we obtain

$$\mathbf{V}_1^{(v)} = 37.51\hat{\mathbf{u}}_V + 2.782\hat{\mathbf{u}}_S \, (\text{km/s})$$

The velocity of Venus in its presumed circular orbit around the sun is

$$\mathbf{V} = \sqrt{\frac{\mu_{sun}}{R_{Venus}}}\hat{\mathbf{u}}_V = \sqrt{\frac{1.327 \times 10^{11}}{108.2 \times 10^6}}\hat{\mathbf{u}}_V = 35.02\hat{\mathbf{u}}_V \, (\text{km/s}) \quad (c)$$

Hence

$$\mathbf{v}_{\infty_1} = \mathbf{V}_1^{(v)} - \mathbf{V} = (37.51\hat{\mathbf{u}}_V + 2.782\hat{\mathbf{u}}_S) - 35.02\hat{\mathbf{u}}_V = 2.490\hat{\mathbf{u}}_V + 2.782\hat{\mathbf{u}}_S \, (\text{km/s}) \quad (d)$$

It follows that

$$v_\infty = \sqrt{\mathbf{v}_{\infty_1} \cdot \mathbf{v}_{\infty_1}} = 3.733 \, \text{km/s}$$

The periapse radius is

$$r_p = r_{Venus} + 300 = 6352 \, \text{km}$$

(Example 8.6 continued)

Equations 8.38 and 8.39 are used to compute the angular momentum and eccentricity of the planetocentric hyperbola:

$$h = 6352\sqrt{v_\infty^2 + \frac{2\mu_{\text{Venus}}}{6352}} = 6352\sqrt{3.733^2 + \frac{2 \cdot 324\,900}{6352}} = 68\,480 \text{ km}^2/\text{s}$$

$$e = 1 + \frac{r_p v_\infty^2}{\mu_{\text{Venus}}} = 1 + \frac{6352 \cdot 3.733^2}{324\,900} = 1.272$$

The turn angle and true anomaly of the asymptote are

$$\delta = 2\sin^{-1}\left(\frac{1}{e}\right) = 2\sin^{-1}\left(\frac{1}{1.272}\right) = 103.6°$$

$$\theta_\infty = \cos^{-1}\left(-\frac{1}{e}\right) = \cos^{-1}\left(-\frac{1}{1.272}\right) = 141.8°$$

From Equations 2.40, 2.93 and 2.97, the aiming radius is

$$\Delta = r_p\sqrt{\frac{e+1}{e-1}} = 6352\sqrt{\frac{1.272+1}{1.272-1}} = 18\,340 \text{ km} \qquad \text{(e)}$$

Finally, from Equation (d) we obtain the angle between $\mathbf{v}_{\infty_1}$ and $\mathbf{V}$,

$$\phi_1 = \tan^{-1}\frac{2.782}{2.490} = 48.17° \qquad \text{(f)}$$

There are two flyby approaches, as shown in Figure 8.22. In the dark side approach, the turn angle is counterclockwise ($+102.9°$) whereas for the sunlit side approach it is clockwise ($-102.9°$).

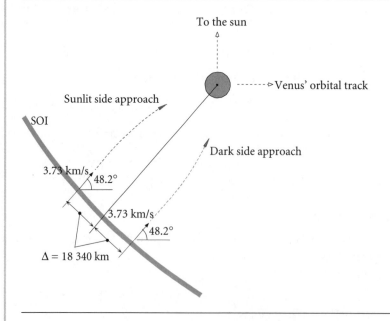

Figure 8.22 Initiation of a sunlit side approach and dark side approach at the inbound crossing.

Dark side approach
According to Equation 8.85, the angle between $\mathbf{v}_\infty$ and $\mathbf{V}_{\text{Venus}}$ at the outbound crossing is

$$\phi_2 = \phi_1 + \delta = 48.17° + 103.6° = 151.8°$$

Hence, by Equation 8.86,

$$\mathbf{v}_{\infty_2} = 3.733(\cos 151.8° \hat{\mathbf{u}}_V + \sin 151.8° \hat{\mathbf{u}}_S) = -3.289 \hat{\mathbf{u}}_V + 1.766 \hat{\mathbf{u}}_S \text{ (km/s)}$$

Using this and Equation (c) above, we compute the spacecraft's heliocentric velocity at the outbound crossing:

$$\mathbf{V}_2^{(v)} = \mathbf{V} + \mathbf{v}_{\infty_2} = 31.73 \hat{\mathbf{u}}_V + 1.766 \hat{\mathbf{u}}_S \text{ (km/s)}$$

It follows from Equation 8.89 that

$$V_{\perp_2} = 31.73 \text{ km/s} \qquad V_{r_2} = -1.766 \text{ km/s} \qquad (g)$$

The speed of the spacecraft at the outbound crossing is

$$V_2^{(v)} = \sqrt{V_{r_2}^2 + V_{\perp_2}^2} = \sqrt{(-1.766)^2 + 31.73^2} = 31.78 \text{ km/s}$$

This is 5.83 km/s less than the inbound speed.

Post-flyby ellipse (orbit 2) for the dark side approach
For the heliocentric post flyby trajectory, labeled orbit 2 in Figure 8.20, the angular momentum is found using Equation 8.90

$$h_2 = R_{\text{Venus}} V_{\perp_2} = (108.2 \times 10^6) \cdot 31.73 = 3.434 \times 10^9 \text{ (km}^2\text{/s)} \qquad (h)$$

From Equation 8.91,

$$e \cos \theta_2 = \frac{h_2^2}{\mu_{\text{sun}} R_{\text{Venus}}} - 1 = \frac{(3.434 \times 10^6)^2}{1.327 \times 10^{11} \cdot 108.2 \times 10^6} - 1 = -0.1790 \qquad (i)$$

and from Equation 8.92

$$e \sin \theta_2 = \frac{V_{r_2} h_2}{\mu_{\text{sun}}} = \frac{-1.766 \cdot 3.434 \times 10^9}{1.327 \times 10^{11}} = -0.04569 \qquad (j)$$

Thus

$$\tan \theta_2 = \frac{e \sin \theta_2}{e \cos \theta_2} = \frac{-0.04569}{-0.1790} = 0.2553 \qquad (k)$$

which means

$$\theta_2 = 14.32° \quad \text{or} \quad 194.32° \qquad (l)$$

But θ_2 must lie in the *third* quadrant since, according to Equations (i) and (j), both the sine and cosine are negative. Hence,

$$\theta_2 = 194.32° \qquad (m)$$

384 Chapter 8 *Interplanetary trajectories*

With this value of θ_2, we can use either Equation (i) or (j) to calculate the eccentricity,

$$e_2 = 0.1847 \tag{n}$$

Perihelion of the departure orbit lies 194.32° clockwise from the encounter point (so that aphelion is 14.32° therefrom), as illustrated in Figure 8.20. The perihelion radius is given by Equation 2.40,

$$R_{\text{perihelion}} = \frac{h_2^2}{\mu_{\text{sun}}} \frac{1}{1+e_2} = \frac{(3.434 \times 10^9)^2}{1.327 \times 10^{11}} \frac{1}{1+0.1847} = 74.98 \times 10^6 \text{ km}$$

which is well within the orbit of Venus.

Sunlit side approach
In this case the angle between $\mathbf{v}_\infty$ and $\mathbf{V}_{\text{Venus}}$ at the outbound crossing is

$$\phi_2 = \phi_1 - \delta = 48.17° - 103.6° = -55.44°$$

Therefore,

$$\mathbf{v}_{\infty_2} = 3.733[\cos(-55.44°)\hat{\mathbf{u}}_V + \sin(-55.44°)\hat{\mathbf{u}}_S] = 2.118\hat{\mathbf{u}}_V - 3.074\hat{\mathbf{u}}_S \text{ (km/s)}$$

The spacecraft's heliocentric velocity at the outbound crossing is

$$\mathbf{V}_2^{(v)} = \mathbf{V}_{\text{Venus}} + \mathbf{v}_{\infty_2} = 37.14\hat{\mathbf{u}}_V - 3.074\hat{\mathbf{u}}_S \text{ (km/s)}$$

which means

$$V_{\perp_2} = 37.14 \text{ km/s} \qquad V_{r_2} = 3.074 \text{ km/s}$$

The speed of the spacecraft at the outbound crossing is

$$V_2^{(v)} = \sqrt{3.074^2 + V_{\perp_2}^2} = \sqrt{3.050^2 + 37.14^2} = 37.27 \text{ km/s}$$

This speed is just 0.348 km/s less than the inbound crossing speed. The relatively small speed change is due to the fact that the apse line of this hyperbola is nearly perpendicular to Venus's orbital track, as shown in Figure 8.23. Nevertheless, the periapses of both hyperbolas are on the leading side of the planet.

Post-flyby ellipse (orbit 2) for the sunlit side approach
To determine the heliocentric post-flyby trajectory, labeled orbit 2 in Figure 8.21, we repeat steps (h) through (n) above:

$$h_2 = R_{\text{Venus}} V_{\perp_2} = (108.2 \times 10^6) \cdot 37.14 = 4.019 \times 10^9 \text{ (km}^2/\text{s)}$$

$$e \cos \theta_2 = \frac{h_2^2}{\mu_{\text{sun}} R_{\text{Venus}}} - 1 = \frac{(4.019 \times 10^9)^2}{1.327 \times 10^{11} \cdot 108.2 \times 10^6} - 1 = 0.1246 \tag{o}$$

$$e \sin \theta_2 = \frac{V_{r_2} h_2}{\mu_{\text{sun}}} = \frac{3.074 \cdot 4.019 \times 10^9}{1.327 \times 10^{11}} = 0.09309 \tag{p}$$

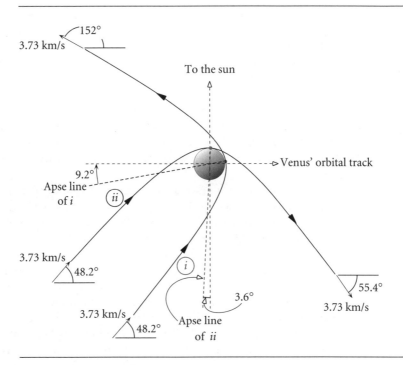

Figure 8.23 Hyperbolic flyby trajectories for (i) the dark side approach and (ii) the sunlit side approach.

$$\tan\theta_2 = \frac{e\sin\theta_2}{e\cos\theta_2} = \frac{0.09309}{0.1246} = 0.7469$$

$$\theta_2 = 36.08° \quad \text{or} \quad 216.08°$$

θ_2 must lie in the first quadrant since both the sine and cosine are positive. Hence,

$$\theta_2 = 36.76° \tag{q}$$

With this value of θ_2, we can use either Equation (o) or (p) to calculate the eccentricity,

$$e_2 = 0.1556$$

Perihelion of the departure orbit lies 36.76° clockwise from the encounter point as illustrated in Figure 8.21. The perihelion radius is

$$R_{\text{perihelion}} = \frac{h_2^2}{\mu_{\text{sun}}}\frac{1}{1+e_2} = \frac{(4.019\times 10^9)^2}{1.327\times 10^{11}}\frac{1}{1+0.1556} = 105.3\times 10^6 \text{ km}$$

which is just within the orbit of Venus. Aphelion lies between the orbits of earth and Venus.

Gravity assist maneuvers are used to add momentum to a spacecraft over and above that available from a spacecraft's on-board propulsion system. A sequence of flybys of planets can impart the delta-v needed to reach regions of the solar system that would be inaccessible using only existing propulsion technology. The technique can also reduce the flight time. Interplanetary missions using gravity assist flybys must be carefully designed in order to take advantage of the relative positions of planets.

The 260 kg spacecraft Pioneer 11, launched in April 1973, used a December 1974 flyby of Jupiter to gain the momentum required to carry it to the first ever flyby encounter with Saturn on 1 September 1979.

Following its September 1977 launch, Voyager 1 likewise used a flyby of Jupiter (March 1979) to reach Saturn in November 1980. In August 1977 Voyager 2 was launched on its 'grand tour' of the outer planets and beyond. This involved gravity assist flybys of Jupiter (July 1979), Saturn (August 1981), Uranus (January 1986) and Neptune (August 1989), after which the spacecraft departed at an angle of 30° to the ecliptic.

With a mass nine times that of Pioneer 11, the dual-spin Galileo spacecraft departed on 18 October 1989 for an extensive international exploration of Jupiter and its satellites lasting until September 2003. Galileo used gravity assist flybys of Venus (February 1990), earth (December 1990) and earth again (December 1992) before arriving at Jupiter in December 1995.

The international Cassini mission to Saturn also made extensive use of gravity assist flyby maneuvers. The Cassini spacecraft was launched on 15 October 1997 from Cape Canaveral, Florida, and arrived at Saturn nearly seven years later, on 1 July 2004. The mission involved four flybys, as illustrated in Figure 8.24. A little over eight months after launch, on 26 April 1998, Cassini flew by Venus at a periapse altitude of 284 km and received a speed boost of about 7 km/s. This placed the spacecraft in an orbit which sent it just outside the orbit of Mars (but well away from the planet) and returned it to Venus on 24 June 1999 for a second flyby, this time at an altitude of 600 km. The result was a trajectory that vectored Cassini toward the earth for an 18 August 1999 flyby at an altitude of 1171 km. The 5.5 km/s speed boost at earth sent the spacecraft toward Jupiter for its next flyby maneuver. This occurred on 30 December 2000 at a distance of 9.7 million km from Jupiter, boosting Cassini's speed by about 2 km/s and adjusting its trajectory so as to rendezvous with Saturn about three and a half years later.

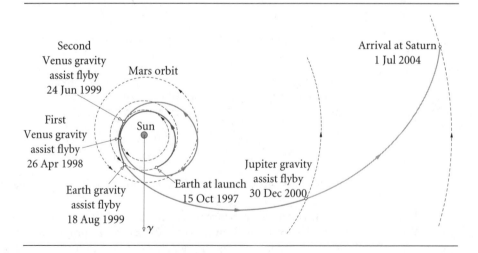

Figure 8.24 Cassini seven-year mission to Saturn.

8.10 Planetary ephemeris

The state vector **R**, **V** of a planet is defined relative to the heliocentric ecliptic frame of reference illustrated in Figure 8.25. This is very similar to the geocentric equatorial frame of Figure 4.5. The sun replaces the earth as the center of attraction, and the plane of the ecliptic replaces the earth's equatorial plane. The vernal equinox continues to define the inertial X axis.

In order to design realistic interplanetary missions we must be able to determine the state vector of a planet at any given time. Table 8.1 provides the orbital elements of the planets and their rates of change per century with respect to the J2000 epoch (1 January 2000, 12 hr UT). The table, covering the years 1800 to 2050, is sufficiently accurate for our needs. From the orbital elements we can infer the state vector using Algorithm 4.2.

In order to interpret Table 8.1, observe the following:

1 astronomical unit (1 AU) is 1.49597871×10^8 km, the average distance between the earth and the sun.

1 arcsecond (1″) is 1/3600 of a degree.

a is the semimajor axis.

e is the eccentricity.

i is the inclination to the ecliptic plane.

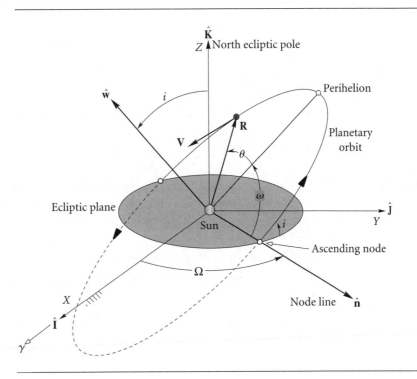

Figure 8.25   Planetary orbit in the heliocentric ecliptic frame.

Table 8.1 Planetary orbital elements and their centennial rates. From Standish et al. (1992). Used with permission

	a, AU $\dot{a}$, AU/Cy	e $\dot{e}$, 1/Cy	i, deg $\dot{i}$, "/Cy	Ω, deg $\dot{\Omega}$, "/Cy	$\tilde{\omega}$, deg $\dot{\tilde{\omega}}$, "/Cy	L, deg $\dot{L}$, "/Cy
Mercury	0.38709893 0.00000066	0.20563069 0.00002527	7.00487 −23.51	48.33167 −446.30	77.45645 573.57	252.25084 538 101 628.29
Venus	0.72333199 0.00000092	0.00677323 −0.00004938	3.39471 −2.86	76.68069 −996.89	131.53298 −108.80	181.97973 210 664 136.06
Earth	1.00000011 −0.00000005	0.01671022 −0.00003804	0.00005 −46.94	−11.26064 −18228.25	102.94719 1198.28	100.46435 129 597 740.63
Mars	1.52366231 −0.00007221	0.09341233 0.00011902	1.85061 −25.47	49.57854 −1020.19	336.04084 1560.78	355.45332 68 905 103.78
Jupiter	5.20336301 0.00060737	0.04839266 −0.00012880	1.30530 −4.15	100.55615 1217.17	14.75385 839.93	34.40438 10 925 078.35
Saturn	9.53707032 −0.00301530	0.05415060 −0.00036762	2.48446 6.11	113.71504 −1591.05	92.43194 −1948.89	49.94432 4 401 052.95
Uranus	19.19126393 0.00152025	0.04716771 −0.00019150	0.76986 −2.09	74.22988 −1681.4	170.96424 1312.56	313.23218 1 542 547.79
Neptune	30.06896348 −0.00125196	0.00858587 0.00002514	1.76917 −3.64	131.72169 −151.25	44.97135 −844.43	304.88003 786 449.21
Pluto	39.48168677 −0.00076912	0.24880766 0.00006465	17.14175 11.07	110.30347 −37.33	224.06676 −132.25	238.92881 522 747.90

Ω is the right ascension of the ascending node (relative to the J2000 vernal equinox).

$\tilde{\omega}$, the longitude of perihelion, is defined as $\tilde{\omega} = \omega + \Omega$, where ω is the argument of perihelion.

L, the mean longitude, is defined as $L = \tilde{\omega} + M$, where M is the mean anomaly.

$\dot{a}, \dot{e}, \dot{\Omega}$, etc., are the rates of change of the above orbital elements per Julian century. 1 century (Cy) equals 36 525 days.

ALGORITHM 8.1

Determine the state vector of a planet at a given date and time. All angular calculations must be adjusted so that they lie in the range 0° to 360°. Recall that the gravitational parameter of the sun is $\mu = 1.327 \times 10^{11}$ km^3/s^2. This procedure is implemented in MATLAB in Appendix D.17.

1. Use Equations 5.47 and 5.48 to calculate the Julian day number JD.

2. Calculate T_0, the number of Julian centuries between J2000 and the date in question

$$T_0 = \frac{JD - 2\,451\,545}{36\,525} \quad (8.104a)$$

3. If Q is any one of the six planetary orbital elements listed in Table 8.1, then calculate its value at JD by means of the formula

$$Q = Q_0 + \dot{Q}T_0 \quad (8.104b)$$

where Q_0 is the value listed for J2000 and $\dot{Q}$ is the tabulated rate. All angular quantities must be adjusted to lie in the range 0° to 360°.

4. Use the semimajor axis a and the eccentricity e to calculate the angular momentum h at JD from Equation 2.61

$$h = \sqrt{\mu a(1 - e^2)}$$

5. Obtain the argument of perihelion ω and mean anomaly M at JD from the results of step 3 by means of the definitions

$$\omega = \tilde{\omega} - \Omega$$
$$M = L - \tilde{\omega}$$

6. Substitute the eccentricity e and the mean anomaly M at JD into Kepler's equation (Equation 3.11) and calculate the eccentric anomaly E.
7. Calculate the true anomaly θ using Equation 3.10.
8. Use h, e, Ω, i, ω and θ to obtain the heliocentric position vector **R** and velocity **V** by means of Algorithm 4.2, with the heliocentric ecliptic frame replacing the geocentric equatorial frame.

EXAMPLE 8.7

Find the distance between the earth and Mars at 12 hr UT on 27 August 2003. Use Algorithm 8.1.

Step 1:

According to Equation 5.56, the Julian day number J_0 for midnight (0 hr UT) of this date is

$$J_0 = 367 \cdot 2003 - \text{INT}\left\{\frac{7\left[2003 + \text{INT}\left(\frac{8+9}{12}\right)\right]}{4}\right\}$$

$$+ \text{INT}\left(\frac{275 \cdot 8}{9}\right) + 27 + 1\,721\,013.5$$

$$= 735\,101 - 3507 + 244 + 27 + 1\,721\,013.5$$

$$= 2\,452\,878.5$$

At $UT = 12$, the Julian day number is

$$JD = 2\,452\,878.5 + \frac{12}{24} = 2\,452\,879.0$$

Step 2:

The number of Julian centuries between J2000 and this date is

$$T_0 = \frac{JD - 2\,451\,545}{36\,525} = \frac{2\,452\,879 - 2\,451\,545}{36\,525} = 0.036523 \text{ Cy}$$

(Example 8.7 continued)

Step 3:

Table 8.1 and Equation 8.104 yield the orbital elements of earth and Mars at 12 hr UT on 27 August 2003.

	a, km	e	i, deg	Ω, deg	$\tilde{\omega}$, deg	L, deg
Earth	1.4960×10^8	0.016709	0.00042622	348.55	102.96	335.27
Mars	2.2794×10^8	0.093417	1.8504	49.568	336.06	334.51

Step 4:
$$h_{\text{earth}} = 4.4451 \times 10^9 \text{ km}^2/\text{s}$$
$$h_{\text{Mars}} = 5.4760 \times 10^9 \text{ km}^2/\text{s}$$

Step 5:
$$\omega_{\text{earth}} = (\tilde{\omega} - \Omega)_{\text{earth}} = 102.96 - 348.55 = -245.59°(114.1°)$$
$$\omega_{\text{Mars}} = (\tilde{\omega} - \Omega)_{\text{Mars}} = 336.06 - 49.568 = 286.49°$$

$$M_{\text{earth}} = (L - \tilde{\omega})_{\text{earth}} = 335.27 - 102.96 = 232.31°$$
$$M_{\text{Mars}} = (L - \tilde{\omega})_{\text{Mars}} = 334.51 - 336.06 = -1.55°(358.45°)$$

Step 6:
$$E_{\text{earth}} - 0.016709 \sin E_{\text{earth}} = 232.31°(\pi/180) \Rightarrow E_{\text{earth}} = 231.56°$$
$$E_{\text{Mars}} - 0.093417 \sin E_{\text{Mars}} = 358.45°(\pi/180) \Rightarrow E_{\text{Mars}} = 358.30°$$

Step 7:
$$\theta_{\text{earth}} = 2 \tan^{-1}\left(\sqrt{\frac{1 - 0.016709}{1 + 0.016709}} \tan \frac{231.56°}{2}\right) = -129.19° \Rightarrow \theta_{\text{earth}} = 230.81°$$

$$\theta_{\text{Mars}} = 2 \tan^{-1}\left(\sqrt{\frac{1 - 0.093417}{1 + 0.093417}} \tan \frac{358.30°}{2}\right) = -1.8669° \Rightarrow \theta_{\text{Mars}} = 358.13°$$

Step 8:

From Algorithm 4.2,

$$\mathbf{R}_{\text{earth}} = (135.59\hat{\mathbf{I}} - 66.803\hat{\mathbf{J}} - 0.00028691\hat{\mathbf{K}}) \times 10^6 \text{ (km)}$$
$$\mathbf{V}_{\text{earth}} = 12.680\hat{\mathbf{I}} + 26.61\hat{\mathbf{J}} - 0.00021273\hat{\mathbf{K}} \text{ (km/s)}$$

$$\mathbf{R}_{\text{Mars}} = (185.95\hat{\mathbf{I}} - 89.916\hat{\mathbf{J}} - 6.4566\hat{\mathbf{K}}) \times 10^6 \text{ (km)}$$
$$\mathbf{V}_{\text{Mars}} = 11.474\hat{\mathbf{I}} + 23.884\hat{\mathbf{J}} + 0.21826\hat{\mathbf{K}} \text{ (km/s)}$$

The distance d between the two planets is therefore

$$d = \|\mathbf{R}_{\text{Mars}} - \mathbf{R}_{\text{earth}}\|$$
$$= \sqrt{(185.95 - 135.59)^2 + [-89.916 - (-66.803)]^2 + (-6.4566 - 0.00028691)^2} \times 10^6$$

or

$$d = 55.79 \times 10^6 \text{ km}$$

The positions of earth and Mars are illustrated in Figure 8.26. It is a rare event for Mars to be in opposition (lined up with earth on the same side of the sun) when Mars is at or near perihelion. The two planets had not been this close in recorded history.

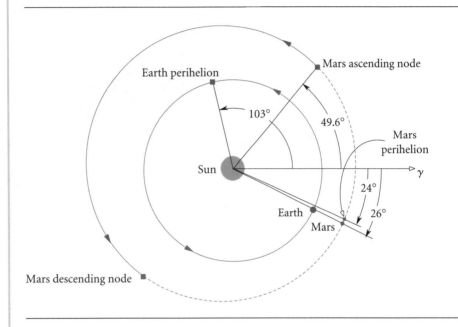

Figure 8.26 Earth and Mars on 27 August 2003. Angles shown are heliocentric latitude, measured in the plane of the ecliptic counterclockwise from the vernal equinox of J2000.

8.11 NON-HOHMANN INTERPLANETARY TRAJECTORIES

To implement a systematic patched conic procedure for three-dimensional trajectories, we will use vector notation and the procedures described in Sections 4.4 and 4.6 (Algorithms 4.1 and 4.2), together with the solution of Lambert's problem presented in Section 5.3 (Algorithm 5.2). The mission is to send a spacecraft from planet 1 to planet 2 in a specified time t_{12}. As previously in this chapter, we break the mission down into three parts: the departure phase, the cruise phase and the arrival phase. We start with the cruise phase.

The frame of reference that we use is the heliocentric ecliptic frame shown in Figure 8.27. The first step is to obtain the state vector of planet 1 at departure (time t) and the state vector of planet 2 at arrival (time $t + t_{12}$). That is accomplished by means of Algorithm 8.1.

The next step is to determine the spacecraft's transfer trajectory from planet 1 to planet 2. We first observe that, according to the patched conic procedure, the

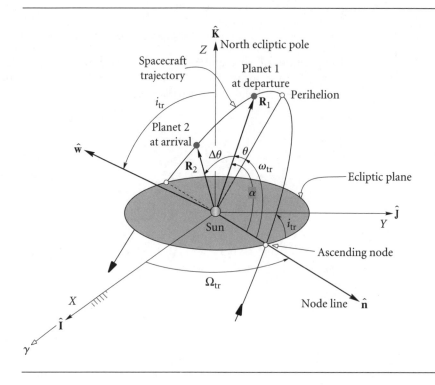

Figure 8.27 Heliocentric orbital elements of a three-dimensional transfer trajectory from planet 1 to planet 2.

heliocentric position vector of the spacecraft at time t is that of planet 1 ($\mathbf{R}_1$) and at time $t + t_{12}$ its position vector is that of planet 2 ($\mathbf{R}_2$). With $\mathbf{R}_1, \mathbf{R}_2$ and the time of flight t_{12} we can use Algorithm 5.2 (Lambert's problem) to obtain the spacecraft's departure and arrival velocities $\mathbf{V}_D^{(v)}$ and $\mathbf{V}_A^{(v)}$ relative to the sun. Either of the state vectors $\mathbf{R}_1, \mathbf{V}_D^{(v)}$ or $\mathbf{R}_2, \mathbf{V}_A^{(v)}$ can be used to obtain the transfer trajectory's six orbital elements by means of Algorithm 4.1.

The spacecraft's hyperbolic excess velocity upon exiting the sphere of influence of planet 1 is

$$\mathbf{v}_\infty)_{\text{Departure}} = \mathbf{V}_D^{(v)} - \mathbf{V}_1 \quad (8.102\text{a})$$

and its excess speed is

$$v_\infty)_{\text{Departure}} = \left\| \mathbf{V}_D^{(v)} - \mathbf{V}_1 \right\| \quad (8.102\text{b})$$

Likewise, at the sphere of influence crossing at planet 2,

$$\mathbf{v}_\infty)_{\text{Arrival}} = \mathbf{V}_A^{(v)} - \mathbf{V}_2 \quad (8.103\text{a})$$

$$v_\infty)_{\text{Arrival}} = \left\| \mathbf{V}_A^{(v)} - \mathbf{V}_2 \right\| \quad (8.103\text{b})$$

8.11 Non-Hohmann interplanetary trajectories

ALGORITHM 8.2

Given the departure and arrival dates (and, therefore, the time of flight), determine the trajectory for a mission from planet 1 to planet 2. This procedure is implemented in MATLAB in Appendix D.19.

1. Use Algorithm 8.1 to determine the state vector $\mathbf{R}_1$, $\mathbf{V}_1$ of planet 1 at departure and the state vector $\mathbf{R}_2$, $\mathbf{V}_2$ of planet 2 at arrival.
2. Use $\mathbf{R}_1$, $\mathbf{R}_2$ and the time of flight in Algorithm 5.2 to find the spacecraft velocity $\mathbf{V}_D^{(v)}$ at departure from planet 1's sphere of influence and its velocity $\mathbf{V}_A^{(v)}$ upon arrival at planet 2's sphere of influence.
3. Calculate the hyperbolic excess velocities at departure and arrival using Equations 8.102 and 8.103.

EXAMPLE 8.8

A spacecraft departs earth's sphere of influence on 7 November 1996 (0 hr UT) on a prograde coasting flight to Mars, arriving at Mars' sphere of influence on 12 September 1997 (0 hr UT). Use Algorithm 8.2 to determine the trajectory and then compute the hyperbolic excess velocities at departure and arrival.

Step 1:

Algorithm 8.1 yields the state vectors for earth and Mars:

$\mathbf{R}_{earth} = 1.0500 \times 10^8 \hat{\mathbf{I}} + 1.0466 \times 10^8 \hat{\mathbf{J}}$
$\quad + 988.33 \hat{\mathbf{K}}$ (km) $\quad\quad (R_{earth} = 1.482 \times 10^8 \text{ km})$

$\mathbf{V}_{earth} = -21.516 \hat{\mathbf{I}} + 20.987 \hat{\mathbf{J}} + 0.00013228 \hat{\mathbf{K}}$ (km/s) $\quad (V_{earth} = 30.06 \text{ km/s})$

$\mathbf{R}_{Mars} = -2.0833 \times 10^7 \hat{\mathbf{I}} - 2.1840 \times 10^8 \hat{\mathbf{J}}$
$\quad - 4.0629 \times 10^6 \hat{\mathbf{K}}$ (km) $\quad\quad (R_{Mars} = 2.194 \times 10^8 \text{ km})$

$\mathbf{V}_{Mars} = 25.047 \hat{\mathbf{I}} - 0.22029 \hat{\mathbf{J}} - 0.62062 \hat{\mathbf{K}}$ (km/s) $\quad (V_{Mars} = 25.05 \text{ km/s})$

Step 2:

The position vector $\mathbf{R}_1$ of the spacecraft at crossing the earth's sphere of influence is just that of the earth,

$$\mathbf{R}_1 = \mathbf{R}_{earth} = 1.0500 \times 10^8 \hat{\mathbf{I}} + 1.0466 \times 10^8 \hat{\mathbf{J}} + 988.33 \hat{\mathbf{K}} \text{ (km)}$$

Upon arrival at Mars' sphere of influence the spacecraft's position vector is

$$\mathbf{R}_2 = \mathbf{R}_{Mars} = -2.0833 \times 10^7 \hat{\mathbf{I}} - 2.1840 \times 10^8 \hat{\mathbf{J}} - 4.0629 \times 10^6 \hat{\mathbf{K}} \text{ (km)}$$

According to Equations 5.47 and 5.48

$$JD_{Departure} = 2\,450\,394.5$$
$$JD_{Arrival} = 2\,450\,703.5$$

Hence, the time of flight is

$$t_{12} = 2\,450\,703.5 - 2\,450\,394.5 = 309 \text{ days}$$

(Example 8.8 continued)

Entering $\mathbf{R}_1$, $\mathbf{R}_2$ and t_{12} into Algorithm 5.2 yields

$$\mathbf{V}_D^{(v)} = -24.427\hat{\mathbf{I}} + 21.781\hat{\mathbf{J}} + 0.94803\hat{\mathbf{K}} \text{ (km/s)} \qquad \left[V_D^{(v)} = 32.741 \text{ km/s}\right]$$

$$\mathbf{V}_A^{(v)} = 22.158\hat{\mathbf{I}} - 0.19668\hat{\mathbf{J}} - 0.45785\hat{\mathbf{K}} \text{ (km/s)} \qquad \left[V_A^{(v)} = 22.164 \text{ km/s}\right]$$

Using the state vector $\mathbf{R}_1$, $\mathbf{V}_D^{(v)}$ we employ Algorithm 4.1 to find the orbital elements of the transfer trajectory.

$$h = 4.8456 \times 10^6 \text{ km}^2/\text{s}$$
$$e = 0.20579$$
$$\Omega = 44.895°$$
$$i = 1.6621°$$
$$\omega = 19.969°$$
$$\theta_1 = 340.04°$$
$$a = 1.8474 \times 10^8 \text{ km}$$

Step 3:

At departure the hyperbolic excess velocity is

$$\mathbf{v}_\infty)_{\text{Departure}} = \mathbf{V}_D^{(v)} - \mathbf{V}_{\text{earth}} = -2.913\hat{\mathbf{I}} + 0.7958\hat{\mathbf{J}} + 0.9480\hat{\mathbf{K}} \text{ (km/s)}$$

Therefore, the hyperbolic excess speed is

$$v_\infty)_{\text{Departure}} = \left\|\mathbf{v}_\infty)_{\text{Departure}}\right\| = 3.1651 \text{ km/s} \qquad \text{(a)}$$

Likewise, at arrival

$$\mathbf{v}_\infty)_{\text{Arrival}} = \mathbf{V}_A^{(v)} - \mathbf{V}_{\text{Mars}} = -2.8804\hat{\mathbf{I}} + 0.023976\hat{\mathbf{J}} + 0.16277\hat{\mathbf{K}} \text{ (km/s)}$$

so that

$$v_\infty)_{\text{Arrival}} = \left\|\mathbf{v}_\infty)_{\text{Arrival}}\right\| = 2.8851 \text{ km/s} \qquad \text{(b)}$$

For the previous example, Figure 8.28 shows the orbits of earth, Mars and the spacecraft from directly above the ecliptic plane. Dotted lines indicate the portions of an orbit which are below the plane. λ is the heliocentric longitude measured counterclockwise from the vernal equinox of J2000. Also shown are the position of Mars at departure and the position of earth at arrival.

The transfer orbit resembles that of the Mars Global Surveyor, which departed earth on 7 November 1996 and arrived at Mars 309 days later, on 12 September 1997.

EXAMPLE 8.9

In Example 8.8, calculate the delta-v required to launch the spacecraft onto its cruise trajectory from a 180 km circular parking orbit. Sketch the departure trajectory.

Recall that

$$r_{\text{earth}} = 6378 \text{ km}$$
$$\mu_{\text{earth}} = 398\,600 \text{ km}^3/\text{s}^2$$

8.11 Non-Hohmann interplanetary trajectories

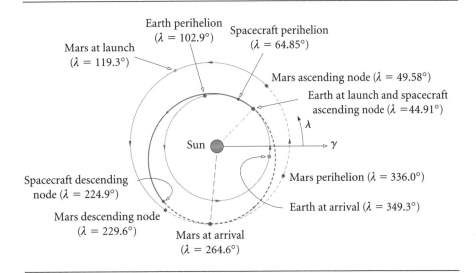

Figure 8.28 The transfer trajectory, together with the orbits of earth and Mars, as viewed from directly above the plane of the ecliptic.

The radius to periapse of the departure hyperbola is the radius of the earth plus the altitude of the parking orbit,

$$r_p = 6378 + 180 = 6558 \text{ km}$$

Substituting this and Equation (a) from Example 8.8 into Equation 8.40 we get the speed of the spacecraft at periapse of the departure hyperbola,

$$v_p = \sqrt{[(v_\infty)_{\text{Departure}}]^2 + \frac{2\mu_{\text{earth}}}{r_p}} = \sqrt{3.1651^2 + \frac{2 \cdot 398\,600}{6558}} = 11.47 \text{ km/s}$$

The speed of the spacecraft in its circular parking orbit is

$$v_0 = \sqrt{\frac{\mu_{\text{earth}}}{r_p}} = \sqrt{\frac{398\,600}{6558}} = 7.796 \text{ km/s}$$

Hence, the delta-v requirement is

$$\Delta v = v_p - v_0 = 3.674 \text{ km/s}$$

The eccentricity of the hyperbola is given by Equation 8.38,

$$e = 1 + \frac{r_p v_\infty^2}{\mu_{\text{earth}}} = 1 + \frac{6558 \cdot 3.1651^2}{398\,600} = 1.165$$

If we assume that the spacecraft is launched from a parking orbit of 28° inclination, then the departure appears as shown in the three-dimensional sketch in Figure 8.29.

(Example 8.9 continued)

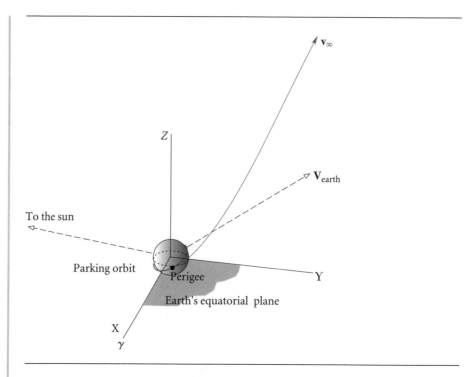

Figure 8.29   The departure hyperbola, assumed to be at 28° inclination to earth's equator.

EXAMPLE 8.10

In Example 8.8, calculate the delta-v required to place the spacecraft in an elliptical capture orbit around Mars with a periapse altitude of 300 km and a period of 48 hours. Sketch the approach hyperbola.

From Tables A.1 and A.2 we know that

$$r_{Mars} = 3380 \text{ km}$$
$$\mu_{Mars} = 42\,830 \text{ km}^3/\text{s}^2$$

The radius to periapse of the arrival hyperbola is the radius of Mars plus the periapse of the elliptical capture orbit,

$$r_p = 3380 + 300 = 3680 \text{ km}$$

According to Equation 8.40 and Equation (b) of Example 8.8, the speed of the spacecraft at periapse of the arrival hyperbola is

$$v_p)_{hyp} = \sqrt{[(v_\infty)_{Arrival}]^2 + \frac{2\mu_{Mars}}{r_p}} = \sqrt{2.8851^2 + \frac{2 \cdot 42\,830}{3680}} = 5.621 \text{ km/s}$$

To find the speed $v_p)_{ell}$ at periapse of the capture ellipse, we use the required period (48 hours) to determine the ellipse's semimajor axis, using Equation 2.73

$$a_{ell} = \left(\frac{T\sqrt{\mu_{Mars}}}{2\pi}\right)^{\frac{3}{2}} = \left(\frac{48 \cdot 3600 \cdot \sqrt{42\,830}}{2\pi}\right)^{\frac{3}{2}} = 31\,880 \text{ km}$$

8.11 Non-Hohmann interplanetary trajectories

From Equation 2.63 we obtain

$$e_{\text{ell}} = 1 - \frac{r_p}{a_{\text{ell}}} = 1 - \frac{3680}{31\,880} = 0.8846$$

Then Equation 8.59 yields

$$v_p)_{\text{ell}} = \sqrt{\frac{\mu_{\text{Mars}}}{r_p}(1 + e_{\text{ell}})} = \sqrt{\frac{42\,830}{3680}(1 + 0.8846)} = 4.683 \text{ km/s}$$

Hence, the delta-v requirement is

$$\Delta v = v_p)_{\text{hyp}} - v_p)_{\text{ell}} = \underline{0.9382 \text{ km/s}}$$

The eccentricity of the approach hyperbola is given by Equation 8.38,

$$e = 1 + \frac{r_p v_\infty^2}{\mu_{\text{Mars}}} = 1 + \frac{3680 \cdot 2.8851^2}{42\,830} = 1.715$$

Assuming that the capture ellipse is a polar orbit of Mars, then the approach hyperbola is as illustrated in Figure 8.30. Note that Mars' equatorial plane is inclined 25° to the plane of its orbit around the sun. Furthermore, the vernal equinox of Mars lies at an angle of 85° from that of the earth.

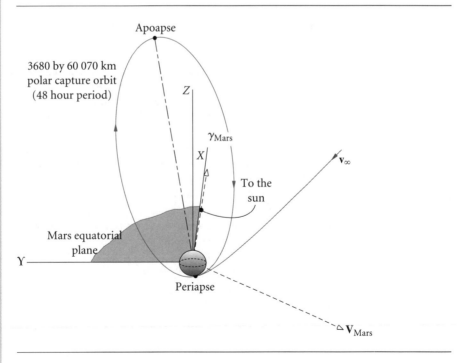

Figure 8.30   The approach hyperbola and capture ellipse.

Problems

8.1 On 6 February 2006, when the earth is 147.4×10^6 km from the sun, a spacecraft parked in a 200 km altitude circular earth orbit is to be launched directly into an elliptical orbit around the sun with perihelion of 120×10^6 km and aphelion equal to the earth's distance from the sun on the launch date. Calculate the delta-v required and v_∞ of the departure hyperbola.
{Ans.: $v_\infty = 30$ km/s, $\Delta v = 3.34$ km/s}

8.2 Estimate the total delta-v requirement for a Hohmann transfer from earth to Mercury, assuming a 150 km circular parking orbit at earth and a 150 km circular capture orbit at Mercury. Furthermore, assume that the planets have coplanar circular orbits with radii equal to the semimajor axes listed in Table A.1.
{Ans.: 15 km/s}

8.3 Calculate the radius of the spheres of influence of Mercury, Venus, Mars and Jupiter.
{Ans.: See Table A.2}

8.4 Calculate the radius of the spheres of influence of Saturn, Uranus, Neptune and Pluto.
{Ans.: See Table A.2}

8.5 Suppose a spacecraft approaches Jupiter on a Hohmann transfer ellipse from earth. If the spacecraft flies by Jupiter at an altitude of 200 000 km on the sunlit side of the planet, determine the orbital elements of the post-flyby trajectory and the delta-v imparted to the spacecraft by Jupiter's gravity. Assume that all of the orbits lie in the same (ecliptic) plane.
{Ans.: $\Delta V = 10.6$ km/s, $a = 4.79 \times 10^6$ km, $e = 0.8453$}

8.6 Use Table 8.1 to verify that the orbital elements for earth and Mars presented in Example 8.7.

8.7 Use Table 8.1 to determine the day of the year 2005 when the earth is farthest from the sun.
{Ans.: 4 July}

8.8 On 1 December 2005 a spacecraft leaves a 180 km altitude circular orbit around the earth on a mission to Venus. It arrives at Venus 121 days later on 1 April 2006, entering a 300 km by 9000 km capture ellipse around the planet. Calculate the total delta-v requirement for this mission.
{Ans.: 6.75 km/s}

8.9 On 15 August 2005 a spacecraft in a 190 km, 52° inclination circular parking orbit around the earth departs on a mission to Mars, arriving at the red planet on 15 March 2006, whereupon retro rockets place it into a highly elliptic orbit with a periapse of 300 km and a period of 35 hours. Determine the total delta-v required for this mission.
{Ans.: 4.86 km/s}

8.10 Calculate the propellant mass required to launch a 2000 kg spacecraft from a 180 km circular earth orbit on a Hohmann transfer trajectory to Saturn. Calculate the time required for the mission and compare it to that of Cassini. Assume the propulsion system has a specific impulse of 300 s.
{Ans.: 6.03 y; 21 810 kg}

CHAPTER 9

RIGID-BODY DYNAMICS

CHAPTER OUTLINE

9.1	Introduction	399
9.2	Kinematics	400
9.3	Equations of translational motion	408
9.4	Equations of rotational motion	410
9.5	Moments of inertia	414
	9.5.1 Parallel axis theorem	428
9.6	Euler's equations	435
9.7	Kinetic energy	441
9.8	The spinning top	443
9.9	Euler angles	448
9.10	Yaw, pitch and roll angles	459
Problems		463

9.1 Introduction

Just as Chapter 1 provides a foundation for the development of the equations of orbital mechanics, this chapter serves as a basis for developing the equations of satellite attitude dynamics. Chapter 1 deals with particles, whereas here we are concerned with rigid bodies. Those familiar with rigid body dynamics can move on to the next chapter, perhaps returning from time to time to review concepts.

The kinematics of rigid bodies is presented first. The subject depends on a theorem of the French mathematician Michel Chasles (1793–1880). Chasles' theorem states that the motion of a rigid body can be described by the displacement of any point of the body (the base point) plus a rotation about a unique axis through that point. The magnitude of the rotation does not depend on the base point. Thus, at any instant a rigid body in a general state of motion has an angular velocity vector whose direction is that of the instantaneous axis of rotation. Describing the rotational component of the motion a rigid body in three dimensions requires taking advantage of the vector nature of angular velocity and knowing how to take the time derivative of moving vectors, which is explained in Chapter 1. Several examples illustrate how this is done.

We then move on to study the interaction between the motion of a rigid body and the forces acting on it. Describing the translational component of the motion requires simply concentrating all of the mass at a point, the center of mass, and applying the methods of particle mechanics to determine its motion. Indeed, our study of the two-body problem up to this point has focused on the motion of their centers of mass without regard to the rotational aspect. Analyzing the rotational dynamics requires computing the body's angular momentum, and that in turn requires accounting for how the mass is distributed throughout the body. The mass distribution is described by the six components of the moment of inertia tensor.

Writing the equations of rotational motion relative to coordinate axes embedded in the rigid body and aligned with the principal axes of inertia yields the non-linear Euler equations of motion, which are applied to a study of the dynamics of a spinning top (or one-axis gyro).

The expression for the kinetic energy of a rigid body is derived because it will be needed in the following chapter.

The chapter concludes with a description of two sets of three angles commonly employed to specify the orientation of a body in three-dimensional space. One of these are the Euler angles, which are the same as the right ascension of the node (Ω), argument of periapse (ω) and inclination (i) introduced in Chapter 4 to orient orbits in space. The other set comprises the yaw, pitch and roll angles, which are suitable for describing the orientation of an airplane. Both the Euler angles and yaw–pitch–roll angles will be employed in Chapter 10.

9.2 KINEMATICS

Figure 9.1 shows a moving rigid body and its instantaneous axis of rotation, which defines the direction of the absolute angular velocity vector $\boldsymbol{\omega}$. The *XYZ* axes are a fixed, inertial frame of reference. The position vectors $\mathbf{R}_A$ and $\mathbf{R}_B$ of two points on the rigid body are measured in the inertial frame. The vector $\mathbf{R}_{B/A}$ drawn from point A to point B is the position vector of B relative to A. Since the body is rigid, $\mathbf{R}_{B/A}$ has a constant magnitude even though its direction is continuously changing. Clearly,

$$\mathbf{R}_B = \mathbf{R}_A + \mathbf{R}_{B/A}$$

Differentiating this equation through with respect to time, we get

$$\dot{\mathbf{R}}_B = \dot{\mathbf{R}}_A + \frac{d\mathbf{R}_{B/A}}{dt} \qquad (9.1)$$

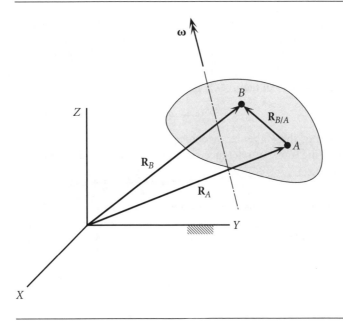

Figure 9.1 Rigid body and its instantaneous axis of rotation.

$\dot{\mathbf{R}}_A$ and $\dot{\mathbf{R}}_B$ are the absolute velocities $\mathbf{v}_A$ and $\mathbf{v}_B$ of points A and B. Because the magnitude of $\mathbf{R}_{B/A}$ does not change, its time derivative is given by Equation 1.24,

$$\frac{d\mathbf{R}_{B/A}}{dt} = \boldsymbol{\omega} \times \mathbf{R}_{B/A}$$

Thus, Equation 9.1 becomes

$$\mathbf{v}_B = \mathbf{v}_A + \boldsymbol{\omega} \times \mathbf{R}_{B/A} \tag{9.2}$$

Taking the time derivative of Equation 9.1 yields

$$\ddot{\mathbf{R}}_B = \ddot{\mathbf{R}}_A + \frac{d^2\mathbf{R}_{B/A}}{dt^2} \tag{9.3}$$

$\ddot{\mathbf{R}}_A$ and $\ddot{\mathbf{R}}_B$ are the absolute accelerations $\mathbf{a}_A$ and $\mathbf{a}_B$ of the two points of the rigid body, while from Equation 1.25 we have

$$\frac{d^2\mathbf{R}_{B/A}}{dt^2} = \boldsymbol{\alpha} \times \mathbf{R}_{B/A} + \boldsymbol{\omega} \times (\boldsymbol{\omega} \times \mathbf{R}_{B/A})$$

in which $\boldsymbol{\alpha}$ is the angular acceleration, $\boldsymbol{\alpha} = d\boldsymbol{\omega}/dt$. Therefore, Equation 9.3 can be written

$$\mathbf{a}_B = \mathbf{a}_A + \boldsymbol{\alpha} \times \mathbf{R}_{B/A} + \boldsymbol{\omega} \times (\boldsymbol{\omega} \times \mathbf{R}_{B/A}) \tag{9.4}$$

Equations 9.2 and 9.4 are the relative velocity and acceleration formulas. Note that all quantities in these expressions are measured in the same inertial frame of reference.

Chapter 9 Rigid-body dynamics

When the rigid body under consideration is connected to and moving relative to another rigid body, computation of its inertial angular velocity **ω** and angular acceleration **α** must be done with care. The key is to remember that angular velocity is a vector. It may be found as the vector sum of a sequence of angular velocities, each measured relative to another, starting with one measured relative to an absolute frame, as illustrated in Figure 9.2. In that case, the absolute angular velocity **ω** of body 4 is

$$\boldsymbol{\omega} = \boldsymbol{\omega}_1 + \boldsymbol{\omega}_{2/1} + \boldsymbol{\omega}_{3/2} + \boldsymbol{\omega}_{4/3} \tag{9.5}$$

Each of these angular velocities is resolved into components along the axes of the moving frame of reference *xyz* shown in Figure 9.2, so that

$$\boldsymbol{\omega} = \omega_x \hat{\mathbf{i}} + \omega_y \hat{\mathbf{j}} + \omega_z \hat{\mathbf{k}} \tag{9.6}$$

The moving frame is chosen for convenience of the analysis, and its inertial angular velocity is denoted **Ω**, as discussed in Section 1.5. According to Equation 1.30, the absolute angular acceleration **α** is obtained from Equation 9.6 by means of the following calculation,

$$\boldsymbol{\alpha} = \left(\frac{d\boldsymbol{\omega}}{dt}\right)_{rel} + \boldsymbol{\Omega} \times \boldsymbol{\omega} \tag{9.7}$$

where

$$\left(\frac{d\boldsymbol{\omega}}{dt}\right)_{rel} = \frac{d\omega_x}{dt}\hat{\mathbf{i}} + \frac{d\omega_y}{dt}\hat{\mathbf{j}} + \frac{d\omega_z}{dt}\hat{\mathbf{k}} \tag{9.8}$$

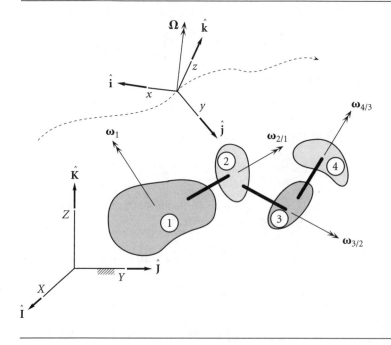

Figure 9.2 Angular velocity is the vector sum of the relative angular velocities starting with $\boldsymbol{\omega}_1$, measured relative to the inertial frame.

EXAMPLE 9.1

An airplane flies at constant speed v while simultaneously undergoing a constant yaw rate ω_{yaw} about a vertical axis and describing a circular loop in the vertical plane with a radius ϱ. The constant propeller spin rate is ω_{spin} relative to the airframe. Find the velocity and acceleration of the tip P of the propeller relative to the hub H, when P is directly above H. The propeller radius is l.

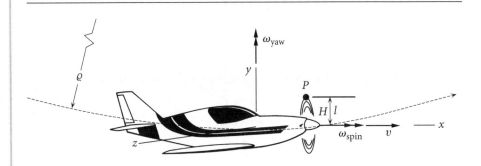

Figure 9.3 Airplane with attached *xyz* body frame.

The *xyz* axes are rigidly attached to the airplane. The x axis is aligned with the propeller's spin axis. The y axis is vertical, and the z axis is in the spanwise direction, so that *xyz* forms a right-handed triad. Although the *xyz* frame is not inertial, we can imagine it to instantaneously coincide with an inertial frame.

The absolute angular velocity of the airplane has two components, the yaw and the counterclockwise pitch angular velocity v/ϱ of its rotation in the circular loop,

$$\boldsymbol{\omega}_{\text{airplane}} = \omega_{\text{yaw}}\hat{\mathbf{j}} + \omega_{\text{pitch}}\hat{\mathbf{k}} = \omega_{\text{yaw}}\hat{\mathbf{j}} + \frac{v}{\varrho}\hat{\mathbf{k}}$$

The angular velocity of the body-fixed moving frame is that of the airplane, $\boldsymbol{\Omega} = \boldsymbol{\omega}_{\text{airplane}}$, so that

$$\boldsymbol{\Omega} = \omega_{\text{yaw}}\hat{\mathbf{j}} + \frac{v}{\varrho}\hat{\mathbf{k}}$$

The absolute angular velocity of the propeller is that of the airplane *plus* the angular velocity propeller relative to the airplane,

$$\boldsymbol{\omega}_{\text{prop}} = \boldsymbol{\omega}_{\text{airplane}} + \omega_{\text{spin}}\hat{\mathbf{i}}$$

which means

$$\boldsymbol{\omega}_{\text{prop}} = \omega_{\text{spin}}\hat{\mathbf{i}} + \omega_{\text{yaw}}\hat{\mathbf{j}} + \frac{v}{\varrho}\hat{\mathbf{k}} \qquad (a)$$

From Equation 9.2, the velocity of point P on the propeller relative to H on the hub, $\mathbf{v}_{P/H}$, is given by

$$\mathbf{v}_{P/H} = \mathbf{v}_P - \mathbf{v}_H = \boldsymbol{\omega}_{\text{prop}} \times \mathbf{r}_{P/H}$$

where $\mathbf{r}_{P/H}$ is the position vector of P relative to H. Thus, using (a),

$$\mathbf{v}_{P/H} = \left(\omega_{\text{spin}}\hat{\mathbf{i}} + \omega_{\text{yaw}}\hat{\mathbf{j}} + \frac{v}{\varrho}\hat{\mathbf{k}}\right) \times (l\hat{\mathbf{j}})$$

(Example 9.1 continued)

from which
$$\mathbf{v}_{P/H} = -\frac{v}{\varrho}l\hat{\mathbf{i}} + \omega_{\text{spin}}l\hat{\mathbf{k}}$$

The absolute angular acceleration of the propeller is found from Equation 9.7,

$$\boldsymbol{\alpha}_{\text{prop}} = \left(\frac{d\boldsymbol{\omega}_{\text{prop}}}{dt}\right)_{\text{rel}} + \boldsymbol{\Omega} \times \boldsymbol{\omega}_{\text{prop}} = \left(\frac{d\omega_{\text{spin}}}{dt}\hat{\mathbf{i}} + \frac{d\omega_{\text{yaw}}}{dt}\hat{\mathbf{j}} + \frac{d(v/\varrho)}{dt}\hat{\mathbf{k}}\right)$$
$$+ \left(\omega_{\text{yaw}}\hat{\mathbf{j}} + \frac{v}{\varrho}\hat{\mathbf{k}}\right) \times \left(\omega_{\text{spin}}\hat{\mathbf{i}} + \omega_{\text{yaw}}\hat{\mathbf{j}} + \frac{v}{\varrho}\hat{\mathbf{k}}\right)$$

Since ω_{spin}, ω_{yaw}, v and ϱ are all constant, this reduces to

$$\boldsymbol{\alpha}_{\text{prop}} = \left(\omega_{\text{yaw}}\hat{\mathbf{j}} + \frac{v}{\varrho}\hat{\mathbf{k}}\right) \times \left(\omega_{\text{spin}}\hat{\mathbf{i}} + \omega_{\text{yaw}}\hat{\mathbf{j}} + \frac{v}{\varrho}\hat{\mathbf{k}}\right) \tag{b}$$

Carrying out the cross product yields

$$\boldsymbol{\alpha}_{\text{prop}} = \frac{v}{\varrho}\omega_{\text{spin}}\hat{\mathbf{j}} - \omega_{\text{yaw}}\omega_{\text{spin}}\hat{\mathbf{k}} \tag{c}$$

From Equation 9.4, the acceleration of P relative to H, $\mathbf{a}_{P/H}$, is given by

$$\mathbf{a}_{P/H} = \mathbf{a}_P - \mathbf{a}_H = \boldsymbol{\alpha}_{\text{prop}} \times \mathbf{r}_{P/H} + \boldsymbol{\omega}_{\text{prop}} \times (\boldsymbol{\omega}_{\text{prop}} \times \mathbf{r}_{P/H})$$

Substituting (a) and (c) into this expression yields

$$\mathbf{a}_{P/H} = \left(\frac{v}{\varrho}\omega_{\text{spin}}\hat{\mathbf{j}} - \omega_{\text{yaw}}\omega_{\text{spin}}\hat{\mathbf{k}}\right) \times (l\hat{\mathbf{j}}) + \left(\omega_{\text{spin}}\hat{\mathbf{i}} + \omega_{\text{yaw}}\hat{\mathbf{j}} + \frac{v}{\varrho}\hat{\mathbf{k}}\right)$$
$$\times \left[\left(\omega_{\text{spin}}\hat{\mathbf{i}} + \omega_{\text{yaw}}\hat{\mathbf{j}} + \frac{v}{\varrho}\hat{\mathbf{k}}\right) \times \mathbf{r}_{P/H}\right]$$

From this we find

$$\mathbf{a}_{P/H} = \left(\omega_{\text{yaw}}\omega_{\text{spin}}l\hat{\mathbf{i}}\right) + \left(\omega_{\text{spin}}\hat{\mathbf{i}} + \omega_{\text{yaw}}\hat{\mathbf{j}} + \frac{v}{\varrho}\hat{\mathbf{k}}\right) \times \left(-\frac{v}{\varrho}l\hat{\mathbf{i}} + \omega_{\text{spin}}l\hat{\mathbf{k}}\right)$$
$$= \left(\omega_{\text{yaw}}\omega_{\text{spin}}l\hat{\mathbf{i}}\right) + \left[\omega_{\text{yaw}}\omega_{\text{spin}}l\hat{\mathbf{i}} - \left(\frac{v^2}{\varrho^2} + \omega_{\text{spin}}^2\right)l\hat{\mathbf{j}} + \omega_{\text{yaw}}\frac{v}{\varrho}l\hat{\mathbf{k}}\right]$$

so that finally,

$$\mathbf{a}_{P/H} = 2\omega_{\text{yaw}}\omega_{\text{spin}}l\hat{\mathbf{i}} - \left(\frac{v^2}{\varrho^2} + \omega_{\text{spin}}^2\right)l\hat{\mathbf{j}} + \omega_{\text{yaw}}\frac{v}{\varrho}l\hat{\mathbf{k}}$$

EXAMPLE 9.2

The satellite is rotating about the z axis at a constant rate N. The xyz axes are attached to the spacecraft, and the z axis has a fixed orientation in inertial space. The solar panels rotate at a constant rate $\dot{\theta}$ in the direction shown. Calculate the absolute velocity and acceleration of point A on the panel relative to point O which lies at the center of the spacecraft and on the centerline of the panels.

9.2 Kinematics

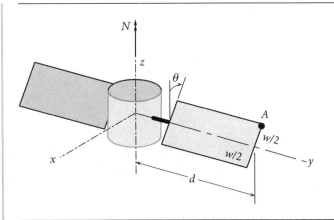

Figure 9.4 Rotating solar panel on a rotating satellite.

The position vector of A relative to O is

$$\mathbf{r}_{A/O} = -\frac{w}{2}\sin\theta\hat{\mathbf{i}} + d\hat{\mathbf{j}} + \frac{w}{2}\cos\theta\hat{\mathbf{k}} \quad (a)$$

The absolute angular velocity of the panel is the absolute angular velocity of the spacecraft plus the angular velocity of the panel relative to the spacecraft,

$$\boldsymbol{\omega}_{\text{panel}} = -\dot{\theta}\hat{\mathbf{j}} + N\hat{\mathbf{k}} \quad (b)$$

According to Equation 9.2, the velocity of A relative to O is

$$\mathbf{v}_{A/O} = \mathbf{v}_A - \mathbf{v}_O = \boldsymbol{\omega}_{\text{panel}} \times \mathbf{r}_{A/O} = \begin{vmatrix} \hat{\mathbf{i}} & \hat{\mathbf{j}} & \hat{\mathbf{k}} \\ 0 & -\dot{\theta} & N \\ -\dfrac{w}{2}\sin\theta & d & \dfrac{w}{2}\cos\theta \end{vmatrix}$$

from which

$$\mathbf{v}_{A/O} = -\left(\frac{w}{2}\dot{\theta}\cos\theta + Nd\right)\hat{\mathbf{i}} - \frac{w}{2}N\sin\theta\hat{\mathbf{j}} - \frac{w}{2}\dot{\theta}\sin\theta\hat{\mathbf{k}}$$

Since the moving xyz frame is attached to the body of the spacecraft, its angular velocity is

$$\boldsymbol{\Omega} = N\hat{\mathbf{k}}$$

The absolute angular acceleration of the panel is obtained from Equation 9.7,

$$\boldsymbol{\alpha}_{\text{panel}} = \left.\frac{d\boldsymbol{\omega}_{\text{panel}}}{dt}\right)_{\text{rel}} + \boldsymbol{\Omega} \times \boldsymbol{\omega}_{\text{panel}}$$

$$= \left(\frac{d(-\dot{\theta})}{dt}\hat{\mathbf{j}} + \frac{dN}{dt}\hat{\mathbf{k}}\right) + (N\hat{\mathbf{k}}) \times \left(-\dot{\theta}\hat{\mathbf{j}} + N\hat{\mathbf{k}}\right)$$

Since N and $\dot{\theta}$ are constants, this reduces to

$$\boldsymbol{\alpha}_{\text{panel}} = \dot{\theta}N\hat{\mathbf{i}} \quad (c)$$

(Example 9.2 continued)

The acceleration of A relative to O is found using Equation 9.4,

$$\mathbf{a}_{A/O} = \mathbf{a}_A - \mathbf{a}_O = \boldsymbol{\alpha}_{\text{panel}} \times \mathbf{r}_{A/O} + \boldsymbol{\omega}_{\text{panel}} \times \left(\boldsymbol{\omega}_{\text{panel}} \times \mathbf{r}_{A/O}\right)$$

$$= \begin{vmatrix} \hat{\mathbf{i}} & \hat{\mathbf{j}} & \hat{\mathbf{k}} \\ \dot{\theta}N & 0 & 0 \\ -\dfrac{w}{2}\sin\theta & d & \dfrac{w}{2}\cos\theta \end{vmatrix} + (-\dot{\theta}\hat{\mathbf{j}} + N\hat{\mathbf{k}}) \times \begin{vmatrix} \hat{\mathbf{i}} & \hat{\mathbf{j}} & \hat{\mathbf{k}} \\ 0 & -\dot{\theta} & N \\ -\dfrac{w}{2}\sin\theta & d & \dfrac{w}{2}\cos\theta \end{vmatrix}$$

$$= \left(-\dfrac{w}{2}N\dot{\theta}\cos\theta\,\hat{\mathbf{j}} + N\dot{\theta}d\,\hat{\mathbf{k}}\right) + \begin{vmatrix} \hat{\mathbf{i}} & \hat{\mathbf{j}} & \hat{\mathbf{k}} \\ 0 & -\dot{\theta} & N \\ -\dfrac{w}{2}\dot{\theta}\cos\theta - Nd & -N\dfrac{w}{2}\sin\theta & -\dfrac{w}{2}\dot{\theta}\sin\theta \end{vmatrix}$$

which leads to

$$\mathbf{a}_{A/O} = \dfrac{w}{2}(N^2 + \dot{\theta}^2)\sin\theta\,\hat{\mathbf{i}} - N(Nd + w\dot{\theta}\cos\theta)\hat{\mathbf{j}} - \dfrac{w}{2}\dot{\theta}^2\cos\theta\,\hat{\mathbf{k}}$$

EXAMPLE 9.3

The gyro rotor shown has a constant spin rate ω_{spin} around axis $b\text{–}a$ in the direction shown. The XYZ axes are fixed. The xyz axes are attached to the gimbal ring, whose angle θ with the vertical is increasing at the constant rate $\dot{\theta}$ in the direction shown. The assembly is forced to precess at the constant rate N around the vertical, as shown. Calculate the absolute angular velocity and acceleration of the rotor in the position shown, expressing the results in both the XYZ and the xyz frames of reference.

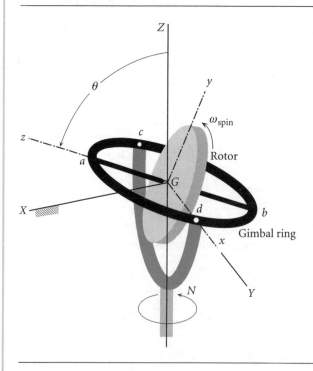

Figure 9.5 Rotating, precessing, nutating gyro.

We will need the instantaneous relationship between the unit vectors of the inertial XYZ axes and the co-moving xyz frame, which by inspection are

$$\hat{\mathbf{I}} = -\cos\theta\hat{\mathbf{j}} + \sin\theta\hat{\mathbf{k}}$$
$$\hat{\mathbf{J}} = \hat{\mathbf{i}} \qquad (a)$$
$$\hat{\mathbf{K}} = \sin\theta\hat{\mathbf{j}} + \cos\theta\hat{\mathbf{k}}$$

so that the matrix of the transformation from xyz to XYZ is

$$[Q]_{xX} = \begin{bmatrix} 0 & -\cos\theta & \sin\theta \\ 1 & 0 & 0 \\ 0 & \sin\theta & \cos\theta \end{bmatrix} \qquad (b)$$

The absolute angular velocity of the gimbal ring is that of the base plus the angular velocity of the gimbal relative to the base,

$$\boldsymbol{\omega}_{gimbal} = N\hat{\mathbf{K}} + \dot{\theta}\hat{\mathbf{i}} = N(\sin\theta\hat{\mathbf{j}} + \cos\theta\hat{\mathbf{k}}) + \dot{\theta}\hat{\mathbf{i}} = \dot{\theta}\hat{\mathbf{i}} + N\sin\theta\hat{\mathbf{j}} + N\cos\theta\hat{\mathbf{k}} \qquad (c)$$

where we made use of (a)$_3$. Since the moving xyz frame is attached to the gimbal,

$$\boldsymbol{\Omega} = \boldsymbol{\omega}_{gimbal}, \text{ so that}$$
$$\boldsymbol{\Omega} = \dot{\theta}\hat{\mathbf{i}} + N\sin\theta\hat{\mathbf{j}} + N\cos\theta\hat{\mathbf{k}} \qquad (d)$$

The absolute angular velocity of the rotor is its spin relative to the gimbal, plus the angular velocity of the gimbal,

$$\boldsymbol{\omega}_{rotor} = \boldsymbol{\omega}_{gimbal} + \omega_{spin}\hat{\mathbf{k}} \qquad (e)$$

From (c) it follows that

$$\boldsymbol{\omega}_{rotor} = \dot{\theta}\hat{\mathbf{i}} + N\sin\theta\hat{\mathbf{j}} + (N\cos\theta + \omega_{spin})\hat{\mathbf{k}} \qquad (f)$$

Because $\hat{\mathbf{i}}, \hat{\mathbf{j}}$ and $\hat{\mathbf{k}}$ move with the gimbal, this expression is valid for any time, not just the instant shown in Figure 9.5. Alternatively, applying the vector transformation

$$\{\boldsymbol{\omega}_{rotor}\}_{XYZ} = [Q]_{xX}\{\boldsymbol{\omega}_{rotor}\}_{xyz} \qquad (g)$$

we obtain the angular velocity of the rotor in the inertial frame, but only at the instant shown in the figure, i.e., when the x axis aligns with the Y axis

$$\begin{Bmatrix} \omega_X \\ \omega_Y \\ \omega_Z \end{Bmatrix} = \begin{bmatrix} 0 & -\cos\theta & \sin\theta \\ 1 & 0 & 0 \\ 0 & \sin\theta & \cos\theta \end{bmatrix} \begin{Bmatrix} \dot{\theta} \\ N\sin\theta \\ N\cos\theta + \omega_{spin} \end{Bmatrix}$$

$$= \begin{Bmatrix} -N\sin\theta\cos\theta + N\sin\theta\cos\theta + \omega_{spin}\sin\theta \\ \dot{\theta} \\ N\sin^2\theta + N\cos^2\theta + \omega_{spin}\cos\theta \end{Bmatrix}$$

or

$$\boldsymbol{\omega}_{rotor} = \omega_{spin}\sin\theta\hat{\mathbf{I}} + \dot{\theta}\hat{\mathbf{J}} + (N + \omega_{spin}\cos\theta)\hat{\mathbf{K}} \qquad (h)$$

(Example 9.3 continued)

The angular acceleration of the rotor is obtained from Equation 9.7, recalling that N, $\dot{\theta}$, and ω_{spin} are independent of time:

$$\boldsymbol{\alpha}_{\text{rotor}} = \left(\frac{d\boldsymbol{\omega}_{\text{rotor}}}{dt}\right)_{\text{rel}} + \boldsymbol{\Omega} \times \boldsymbol{\omega}_{\text{rotor}} = \left[\frac{d(\dot{\theta})}{dt}\hat{\mathbf{i}} + \frac{d(N\sin\theta)}{dt}\hat{\mathbf{j}} + \frac{d(N\cos\theta + \omega_{\text{spin}})}{dt}\hat{\mathbf{k}}\right]$$

$$+ \begin{vmatrix} \hat{\mathbf{i}} & \hat{\mathbf{j}} & \hat{\mathbf{k}} \\ \dot{\theta} & N\sin\theta & N\cos\theta \\ \dot{\theta} & N\sin\theta & N\cos\theta + \omega_{\text{spin}} \end{vmatrix}$$

$$= (N\dot{\theta}\cos\theta\hat{\mathbf{j}} - N\dot{\theta}\sin\theta\hat{\mathbf{k}}) + [\hat{\mathbf{i}}(N\omega_{\text{spin}}\sin\theta) - \hat{\mathbf{j}}(\omega_{\text{spin}}\dot{\theta}) + \hat{\mathbf{k}}(0)]$$

Upon simplification, this becomes

$$\boldsymbol{\alpha}_{\text{rotor}} = N\omega_{\text{spin}}\sin\theta\hat{\mathbf{i}} + \dot{\theta}(N\cos\theta - \omega_{\text{spin}})\hat{\mathbf{j}} - N\dot{\theta}\sin\theta\hat{\mathbf{k}} \qquad \text{(i)}$$

This expression, like (f), is valid at any time.

The components of $\boldsymbol{\alpha}_{\text{rotor}}$ along the XYZ axes are found in the same way as for $\boldsymbol{\omega}_{\text{rotor}}$,

$$\{\boldsymbol{\alpha}_{\text{rotor}}\}_{XYZ} = [\mathbf{Q}]_{xX}\{\boldsymbol{\alpha}_{\text{rotor}}\}_{xyz}$$

which means

$$\begin{Bmatrix} \alpha_X \\ \alpha_Y \\ \alpha_Z \end{Bmatrix} = \begin{bmatrix} 0 & -\cos\theta & \sin\theta \\ 1 & 0 & 0 \\ 0 & \sin\theta & \cos\theta \end{bmatrix} \begin{Bmatrix} N\omega_{\text{spin}}\sin\theta \\ \dot{\theta}(N\cos\theta - \omega_{\text{spin}}) \\ -N\dot{\theta}\sin\theta \end{Bmatrix}$$

$$= \begin{Bmatrix} -N\dot{\theta}\cos^2\theta + \dot{\theta}\omega_{\text{spin}}\cos\theta - N\dot{\theta}\sin^2\theta \\ N\omega_{\text{spin}}\sin\theta \\ N\dot{\theta}\sin\theta\cos\theta - \dot{\theta}\omega_{\text{spin}}\sin\theta - N\dot{\theta}\sin\theta\cos\theta \end{Bmatrix}$$

or

$$\boldsymbol{\alpha}_{\text{rotor}} = \dot{\theta}(\omega_{\text{spin}}\cos\theta - N)\hat{\mathbf{I}} + N\omega_{\text{spin}}\sin\theta\hat{\mathbf{J}} - \dot{\theta}\omega_{\text{spin}}\sin\theta\hat{\mathbf{K}} \qquad \text{(j)}$$

Note carefully that (j) is not simply the time derivative of (h). Equations (h) and (j) are valid only at the instant that the xyz and XYZ axes have the alignments shown in Figure 9.4.

9.3 Equations of translational motion

Figure 9.6 again shows an arbitrary, continuous, three-dimensional body of mass m. 'Continuous' means that as we zoom in on a point it remains surrounded by a continuous distribution of matter having the infinitesimal mass dm in the limit. The point never ends up in a void. In particular, we ignore the actual atomic and molecular microstructure in favor of this continuum hypothesis, as it is called. Molecular microstructure does not bear upon the overall dynamics of a finite body. We will use G to denote the center of mass. Position vectors of points relative to the origin of the inertial frame will be designated by capital letters. Thus, the position of the center of

9.3 Equations of translational motion

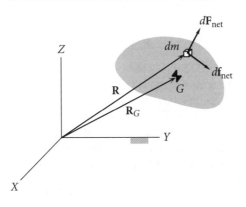

Figure 9.6 Forces on the mass element dm of a continuous medium.

mass is $\mathbf{R}_G$, defined as

$$m\mathbf{R}_G = \int_m \mathbf{R}\, dm \tag{9.9}$$

$\mathbf{R}$ is the position of a mass element dm within the continuum. Each element of mass is acted upon by a net external force $d\mathbf{F}_{net}$ and a net internal force $d\mathbf{f}_{net}$. The external force comes from direct contact with other objects and from action at a distance, such as gravitational attraction. The internal forces are those exerted from within the body by neighboring particles. These are the forces which hold the body together. For each mass element, Newton's second law, Equation 1.10, is written

$$d\mathbf{F}_{net} + d\mathbf{f}_{net} = dm\ddot{\mathbf{R}} \tag{9.10}$$

Writing this equation for the infinite number of mass elements of which the body is composed and then summing them all together leads to the integral,

$$\int d\mathbf{F}_{net} + \int d\mathbf{f}_{net} = \int_m \ddot{\mathbf{R}}\, dm$$

Because the internal forces occur in action–reaction pairs, $\int d\mathbf{f}_{net} = \mathbf{0}$. (External forces on the body are those without an internal reactant; the reactant lies outside the body and, hence, outside our purview.) Thus

$$\mathbf{F}_{net} = \int_m \ddot{\mathbf{R}}\, dm \tag{9.11}$$

where $\mathbf{F}_{net}$ is the resultant external force on the body, $\mathbf{F}_{net} = \int d\mathbf{F}_{net}$. From Equation 9.9

$$\int_m \ddot{\mathbf{R}}\, dm = m\ddot{\mathbf{R}}_G$$

where $\ddot{\mathbf{R}}_G = \mathbf{a}_G$, the absolute acceleration of the center of mass. Therefore, Equation 9.11 can be written

$$\mathbf{F}_{net} = m\ddot{\mathbf{R}}_G \tag{9.12}$$

We are therefore reminded that the motion of the center of mass of a body is determined solely by the resultant of the external forces acting on it. So far our study of orbiting bodies has focused exclusively on the motion of their centers of mass. In this chapter we will turn our attention to rotational motion around the center of mass. To simplify things, we will ultimately assume that the body is not only continuous, but that it is also rigid. That means all points of the body remain a fixed distance from each other and there is no flexing, bending or twisting deformation.

9.4 Equations of rotational motion

Our development of the rotational dynamics equations does not require at the outset that the body under consideration be rigid. It may be a solid, liquid or gas.

Point P in the Figure 9.7 is arbitrary; it need not be fixed in space nor attached to a point on the body. Then the moment about P of the forces on mass element dm (cf. Figure 9.6) is

$$d\mathbf{M}_P = \mathbf{r} \times d\mathbf{F}_{net} + \mathbf{r} \times d\mathbf{f}_{net}$$

where $\mathbf{r}$ is the position vector of the mass element dm relative to the point P. Writing the right-hand side as $\mathbf{r} \times (d\mathbf{F}_{net} + d\mathbf{f}_{net})$, substituting Equation 9.10, and integrating

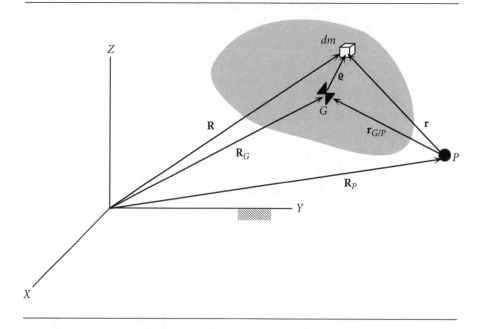

Figure 9.7 Position vectors of a mass element in a continuum from several key reference points.

over all of the mass elements of the body yields

$$\mathbf{M}_{P_{net}} = \int_m \mathbf{r} \times \ddot{\mathbf{R}} \, dm \tag{9.13}$$

where $\ddot{\mathbf{R}}$ is the absolute acceleration of dm relative to the inertial frame and

$$\mathbf{M}_{P_{net}} = \int \mathbf{r} \times d\mathbf{F}_{net} + \int \mathbf{r} \times d\mathbf{f}_{net}$$

But $\int \mathbf{r} \times d\mathbf{f}_{net} = 0$ because the internal forces occur in action–reaction pairs. Thus,

$$\mathbf{M}_{P_{net}} = \int \mathbf{r} \times d\mathbf{F}_{net}$$

which means the net moment includes only the moment of all of the external forces on the body.

Observe that

$$\mathbf{r} \times \ddot{\mathbf{R}} = \frac{d}{dt}(\mathbf{r} \times \dot{\mathbf{R}}) - \dot{\mathbf{r}} \times \dot{\mathbf{R}} \tag{9.14}$$

Since $\mathbf{r} = \mathbf{R} - \mathbf{R}_P$, where $\mathbf{R}_P$ is the absolute position vector of P, it is true that

$$\dot{\mathbf{r}} \times \dot{\mathbf{R}} = (\dot{\mathbf{R}} - \dot{\mathbf{R}}_P) \times \dot{\mathbf{R}} = -\dot{\mathbf{R}}_P \times \dot{\mathbf{R}} \tag{9.15}$$

Substituting Equation 9.15 into Equation 9.14, then moving that result into Equation 9.13 yields

$$\mathbf{M}_{P_{net}} = \frac{d}{dt} \int_m \mathbf{r} \times \dot{\mathbf{R}} \, dm + \dot{\mathbf{R}}_P \times \int_m \dot{\mathbf{R}} \, dm \tag{9.16}$$

Now, $\mathbf{r} \times \dot{\mathbf{R}} \, dm$ is the absolute angular momentum of mass element dm about P. The angular momentum of the entire body is the integral of this cross product over all of its mass elements. That is, the absolute angular momentum of the body relative to point P is

$$\mathbf{H}_P = \int_m \mathbf{r} \times \dot{\mathbf{R}} \, dm \tag{9.17}$$

Observing from Figure 9.7 that $\mathbf{r} = \mathbf{r}_{G/P} + \boldsymbol{\varrho}$, we can write Equation 9.17 as

$$\mathbf{H}_P = \int_m (\mathbf{r}_{G/P} + \boldsymbol{\varrho}) \times \dot{\mathbf{R}} \, dm = \mathbf{r}_{G/P} \times \int_m \dot{\mathbf{R}} \, dm + \int_m \boldsymbol{\varrho} \times \dot{\mathbf{R}} \, dm \tag{9.18}$$

The last term is the absolute angular momentum relative to the center of mass G,

$$\mathbf{H}_G = \int \boldsymbol{\varrho} \times \dot{\mathbf{R}} \, dm \tag{9.19}$$

Furthermore, by the definition of center of mass, Equation 9.9,

$$\int_m \dot{\mathbf{R}} \, dm = m\dot{\mathbf{R}}_G \tag{9.20}$$

Equations 9.19 and 9.20 allow us to write Equation 9.18 as

$$\mathbf{H}_P = \mathbf{H}_G + \mathbf{r}_{G/P} \times m\mathbf{v}_G \tag{9.21}$$

This useful relationship shows how to obtain the absolute angular momentum about any point P once $\mathbf{H}_G$ is known.

For calculating the angular momentum about the center of mass, Equation 9.19 can be cast in a much more useful form by making the substitution (cf. Figure 9.7) $\mathbf{R} = \mathbf{R}_G + \boldsymbol{\varrho}$, so that

$$\mathbf{H}_G = \int_m \boldsymbol{\varrho} \times (\dot{\mathbf{R}}_G + \dot{\boldsymbol{\varrho}}) dm = \int_m \boldsymbol{\varrho} \times \dot{\mathbf{R}}_G dm + \int_m \boldsymbol{\varrho} \times \dot{\boldsymbol{\varrho}} dm$$

In the two integrals on the right, the variable is $\boldsymbol{\varrho}$. $\dot{\mathbf{R}}_G$ is fixed and can therefore be factored out of the first integral to obtain

$$\mathbf{H}_G = \left(\int_m \boldsymbol{\varrho} dm\right) \times \dot{\mathbf{R}}_G + \int_m \boldsymbol{\varrho} \times \dot{\boldsymbol{\varrho}} dm$$

By definition of the center of mass, $\int_m \boldsymbol{\varrho} dm = \mathbf{0}$ (the position vector of the center of mass relative to itself is zero), which means

$$\mathbf{H}_G = \int_m \boldsymbol{\varrho} \times \dot{\boldsymbol{\varrho}} dm \qquad (9.22)$$

Since $\boldsymbol{\varrho}$ and $\dot{\boldsymbol{\varrho}}$ are the position and velocity relative to the center of mass G, $\int_m \boldsymbol{\varrho} \times \dot{\boldsymbol{\varrho}} dm$ is the total moment about the center of mass of the linear momentum relative to the center of mass, $\mathbf{H}_{G_{\text{rel}}}$. In other words,

$$\mathbf{H}_G = \mathbf{H}_{G_{\text{rel}}} \qquad (9.23)$$

This is a rather surprising fact, hidden in Equation 9.19 and true in general for no other point of the body.

Another useful angular momentum formula, similar to Equation 9.21, may be found by substituting $\mathbf{R} = \mathbf{R}_P + \mathbf{r}$ into Equation 9.17,

$$\mathbf{H}_P = \int_m \mathbf{r} \times (\dot{\mathbf{R}}_P + \dot{\mathbf{r}}) dm = \left(\int_m \mathbf{r} dm\right) \times \dot{\mathbf{R}}_P + \int_m \mathbf{r} \times \dot{\mathbf{r}} dm \qquad (9.24)$$

The term on the far right is the net moment of relative linear momentum about P,

$$\mathbf{H}_{P_{\text{rel}}} = \int_m \mathbf{r} \times \dot{\mathbf{r}} dm \qquad (9.25)$$

Also, $\int_m \mathbf{r} dm = m \mathbf{r}_{G/P}$, where $\mathbf{r}_{G/P}$ is the position of the center of mass relative to P. Thus, Equation 9.24 can be written

$$\mathbf{H}_P = \mathbf{H}_{P_{\text{rel}}} + \mathbf{r}_{G/P} \times m \mathbf{v}_P \qquad (9.26)$$

Finally, substituting this into Equation 9.21, solving for $\mathbf{H}_{P_{\text{rel}}}$, and noting that $\mathbf{v}_G - \mathbf{v}_P = \mathbf{v}_{G/P}$ yields

$$\mathbf{H}_{P_{\text{rel}}} = \mathbf{H}_G + \mathbf{r}_{G/P} \times m \mathbf{v}_{G/P} \qquad (9.27)$$

This expression is useful when the absolute velocity $\mathbf{v}_G$ of the center of mass, which is required in Equation 9.21, is not available.

So far we have written down some formulas for calculating the angular momentum about an arbitrary point in space and about the center of mass of the body itself. Let us now return to the problem of relating angular momentum to the applied torque. Substituting Equation 9.17 into 9.16 and noting that by definition of the center of mass,

$$\int_m \dot{\mathbf{R}} dm = m\dot{\mathbf{R}}_G$$

we obtain

$$\mathbf{M}_{P_{net}} = \dot{\mathbf{H}}_P + \dot{\mathbf{R}}_P \times m\dot{\mathbf{R}}_G$$

Thus, for an arbitrary point P,

$$\mathbf{M}_{P_{net}} = \dot{\mathbf{H}}_P + \mathbf{v}_P \times m\mathbf{v}_G \tag{9.28}$$

where $\mathbf{v}_P$ and $\mathbf{v}_G$ are the absolute velocities of points P and G, respectively. This expression is applicable to two important special cases.

If the point P is at rest in inertial space ($\mathbf{v}_P = \mathbf{0}$), then Equation 9.28 reduces to

$$\mathbf{M}_{P_{net}} = \dot{\mathbf{H}}_P \tag{9.29}$$

This equation holds as well if $\mathbf{v}_P$ and $\mathbf{v}_G$ are parallel, e.g., if P is the point of contact of a wheel rolling while slipping in the plane. Note that the validity of Equation 9.29 depends neither on the body's being rigid nor on its being in pure rotation about P. If point P is chosen to be the center of mass, then, since $\mathbf{v}_G \times \mathbf{v}_G = \mathbf{0}$, Equation 9.28 becomes

$$\mathbf{M}_{G_{net}} = \dot{\mathbf{H}}_G \tag{9.30}$$

This equation is valid for any state of motion.

If Equation 9.30 is integrated over a time interval, then we obtain the angular impulse–momentum principle,

$$\int_{t_1}^{t_2} \mathbf{M}_{G_{net}} dt = \mathbf{H}_{G_2} - \mathbf{H}_{G_1} \tag{9.31}$$

A similar expression follows from Equation 9.29. $\int \mathbf{M} dt$ is the angular impulse. If the net angular impulse is zero, then $\Delta \mathbf{H} = \mathbf{0}$, which is a statement of the conservation of angular momentum. Keep in mind that the angular impulse–momentum principle is not valid for just any reference point.

Additional versions of Equations 9.29 and 9.30 can be obtained which may prove useful in special circumstances. For example, substituting the expression for $\mathbf{H}_P$ (Equation 9.21) into Equation 9.28 yields

$$\mathbf{M}_{P_{net}} = \left[\dot{\mathbf{H}}_G + \frac{d}{dt}(\mathbf{r}_{G/P} \times m\mathbf{v}_G) \right] + \mathbf{v}_P \times m\mathbf{v}_G$$

$$= \dot{\mathbf{H}}_G + \frac{d}{dt}[(\mathbf{r}_G - \mathbf{r}_P) \times m\mathbf{v}_G] + \mathbf{v}_P \times m\mathbf{v}_G$$

$$= \dot{\mathbf{H}}_G + (\mathbf{v}_G - \mathbf{v}_P) \times m\mathbf{v}_G + \mathbf{r}_{G/P} \times m\mathbf{a}_G + \mathbf{v}_P \times m\mathbf{v}_G$$

or, finally,

$$\mathbf{M}_{P_{\text{net}}} = \dot{\mathbf{H}}_G + \mathbf{r}_{G/P} \times m\mathbf{a}_G \qquad (9.32)$$

This expression is useful when it is convenient to compute the net moment about a point other than the center of mass. Alternatively, by simply differentiating Equation 9.27 we get

$$\dot{\mathbf{H}}_{P_{\text{rel}}} = \dot{\mathbf{H}}_G + \overbrace{\mathbf{v}_{G/P} \times m\mathbf{v}_{G/P}}^{=0} + \mathbf{r}_{G/P} \times m\mathbf{a}_{G/P}$$

Solving for $\dot{\mathbf{H}}_G$, invoking Equation 9.30, and using the fact that $\mathbf{a}_{P/G} = -\mathbf{a}_{G/P}$ leads to

$$\mathbf{M}_{G_{\text{net}}} = \dot{\mathbf{H}}_{P_{\text{rel}}} + \mathbf{r}_{G/P} \times m\mathbf{a}_{P/G} \qquad (9.33)$$

Finally, if the body is rigid, the magnitude of the position vector $\boldsymbol{\varrho}$ of any point relative to the center of mass does not change with time. Therefore, Equation 1.24 requires that $\dot{\boldsymbol{\varrho}} = \boldsymbol{\omega} \times \boldsymbol{\varrho}$, leading us to conclude

$$\mathbf{H}_G = \int_m \boldsymbol{\varrho} \times (\boldsymbol{\omega} \times \boldsymbol{\varrho}) dm \qquad (9.34)$$

Again, the absolute angular momentum about the center of mass depends only on the absolute angular velocity and not on the absolute translational velocity of any point of the body.

No such simplification of Equation 9.17 exists for an arbitrary reference point P. However, if the point P is fixed in inertial space and the rigid body is rotating about P, then the magnitude of the position vector $\mathbf{r}$ from P to any point of the body is constant. It follows from Equation 1.24 that $\dot{\mathbf{r}} = \boldsymbol{\omega} \times \mathbf{r}$. According to Figure 9.7,

$$\mathbf{R} = \mathbf{R}_p + \mathbf{r}$$

Differentiating with respect to time gives

$$\dot{\mathbf{R}} = \dot{\mathbf{R}}_p + \dot{\mathbf{r}} = 0 + \boldsymbol{\omega} \times \mathbf{r} = \boldsymbol{\omega} \times \mathbf{r}$$

Substituting this into Equation 9.17 yields the formula for angular momentum in this special case,

$$\mathbf{H}_P = \int_m \mathbf{r} \times (\boldsymbol{\omega} \times \mathbf{r}) dm \qquad (9.35)$$

Although Equations 9.34 and 9.35 are mathematically identical, one must keep in mind the notation of Figure 9.7. Equation 9.35 applies only if the rigid body is in pure rotation about a stationary point in inertial space, whereas Equation 9.34 applies unconditionally to any situation.

9.5 MOMENTS OF INERTIA

To use Equation 9.29 or 9.30 to solve problems, the vectors within them have to be resolved into components. To find the components of angular momentum, we must

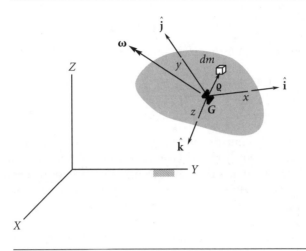

Figure 9.8 Co-moving *xyz* frame used to compute the moments of inertia.

appeal to its definition. We will focus on the formula for angular momentum of a rigid body about its center of mass, Equation 9.34, because the expression for fixed-point rotation (Equation 9.35) is mathematically the same. The integrand of Equation 9.34 can be rewritten using the *bac − cab* vector identity presented in Equation 2.23,

$$\varrho \times (\omega \times \varrho) = \omega \varrho^2 - \varrho(\omega \cdot \varrho) \tag{9.36}$$

Let the origin of a co-moving *xyz* coordinate system be attached to G, as shown in Figure 9.8. The unit vectors of this frame are $\hat{\mathbf{i}}$, $\hat{\mathbf{j}}$ and $\hat{\mathbf{k}}$. The vectors ϱ and ω can be resolved into components in the *xyz* directions to get $\varrho = x\hat{\mathbf{i}} + y\hat{\mathbf{j}} + z\hat{\mathbf{k}}$ and $\omega = \omega_x\hat{\mathbf{i}} + \omega_y\hat{\mathbf{j}} + \omega_z\hat{\mathbf{k}}$. Substituting these vector expressions into the right side of Equation 9.36 yields

$$\varrho \times (\omega \times \varrho) = (\omega_x\hat{\mathbf{i}} + \omega_y\hat{\mathbf{j}} + \omega_z\hat{\mathbf{k}})(x^2 + y^2 + z^2) - (x\hat{\mathbf{i}} + y\hat{\mathbf{j}} + z\hat{\mathbf{k}})(\omega_x x + \omega_y y + \omega_z z)$$

Expanding the right side and collecting terms having the unit vectors $\hat{\mathbf{i}}$, $\hat{\mathbf{j}}$ and $\hat{\mathbf{k}}$ in common, we get

$$\varrho \times (\omega \times \varrho) = [(y^2 + z^2)\omega_x - xy\omega_y - xz\omega_z]\hat{\mathbf{i}}$$
$$+ [-yx\omega_x + (x^2 + z^2)\omega_y - yz\omega_z]\hat{\mathbf{j}}$$
$$+ [-zx\omega_x - zy\omega_y + (x^2 + y^2)\omega_z]\hat{\mathbf{k}} \tag{9.37}$$

We put this result into the integrand of Equation 9.34 to obtain

$$\mathbf{H}_G = H_x\hat{\mathbf{i}} + H_y\hat{\mathbf{j}} + H_z\hat{\mathbf{k}} \tag{9.38}$$

where

$$\begin{Bmatrix} H_x \\ H_y \\ H_z \end{Bmatrix} = \begin{bmatrix} I_x & I_{xy} & I_{xz} \\ I_{yx} & I_y & I_{yz} \\ I_{zx} & I_{zy} & I_z \end{bmatrix} \begin{Bmatrix} \omega_x \\ \omega_y \\ \omega_z \end{Bmatrix} \tag{9.39a}$$

or, in matrix notation,

$$\{H\} = [I]\{\omega\} \quad (9.39b)$$

The components of the moment of inertia matrix $[I]$ about the center of mass are

$$\begin{array}{lll} I_x = \int(y^2+z^2)dm & I_{xy} = -\int xy\,dm & I_{xz} = -\int xz\,dm \\ I_{yx} = I_{xy} & I_y = \int(x^2+z^2)dm & I_{yz} = -\int yz\,dm \\ I_{zx} = I_{xz} & I_{zy} = I_{yz} & I_z = \int(x^2+y^2)dm \end{array} \quad (9.40)$$

$[I]$ is clearly a symmetric matrix: $[I]^T = [I]$. Observe that, whereas the products of inertia I_{xy}, I_{xz} and I_{yz} can be positive, negative or zero, the moments of inertia I_x, I_y and I_z are always positive (never zero or negative) for bodies of finite dimensions. For this reason, $[I]$ is a positive-definite matrix. Keep in mind that Equations 9.38 and 9.39 are valid as well for axes attached to a fixed point P about which the body is rotating.

The moments of inertia reflect how the mass of a rigid body is distributed. They manifest a body's rotational inertia, its resistance to being set into rotary motion or stopped once rotation is under way. It is not an object's mass alone but how that mass is distributed which determines how the body will respond to applied torques.

It is easy to show that the following statements are true:

If the xy plane is a plane of symmetry of the body, then $I_{xz} = I_{yz} = 0$.

If the xz plane is a plane of symmetry of the body, then $I_{xy} = I_{yz} = 0$.

If the yz plane is a plane of symmetry of the body, then $I_{xy} = I_{xz} = 0$.

Obviously, if the body has just two planes of symmetry relative to the xyz frame of reference, then all three products of inertia vanish, and $[I]$ becomes a diagonal matrix,

$$[I] = \begin{bmatrix} A & 0 & 0 \\ 0 & B & 0 \\ 0 & 0 & C \end{bmatrix} \quad (9.41)$$

where A, B and C are the principal moments of inertia (all positive), and the xyz axes are the principal axes of inertia. In this case, relative to either the center of mass or a fixed point of rotation, we have

$$H_x = A\omega_x \quad H_y = B\omega_y \quad H_z = C\omega_z \quad (9.42)$$

In general, the angular velocity $\boldsymbol{\omega}$ and the angular momentum $\mathbf{H}$ are not parallel. However, if (for example) $\boldsymbol{\omega} = \omega\hat{\mathbf{i}}$, then according to Equations 9.42, $\{H\} = A\{\omega\}$. In other words, if the angular velocity points in a principal direction, so does the angular momentum. In that case the two vectors are indeed parallel.

Each of the three principal moments of inertia can be expressed as follows:

$$A = mk_x^2 \quad B = mk_y^2 \quad C = mk_z^2 \quad (9.43)$$

where m is the mass of the body and k_x, k_y and k_z are the three radii of gyration. One may imagine the mass of a body to be concentrated around a principal axis at a distance equal to the radius of gyration.

The moments of inertia for several common shapes are listed in Figure 9.9. By symmetry, their products of inertia vanish for the coordinate axes used. Formulas

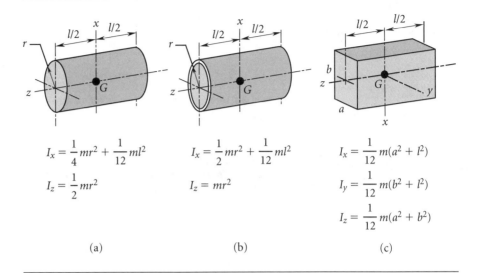

$$I_x = \frac{1}{4}mr^2 + \frac{1}{12}ml^2 \qquad I_x = \frac{1}{2}mr^2 + \frac{1}{12}ml^2 \qquad I_x = \frac{1}{12}m(a^2 + l^2)$$

$$I_z = \frac{1}{2}mr^2 \qquad\qquad I_z = mr^2 \qquad\qquad I_y = \frac{1}{12}m(b^2 + l^2)$$

$$I_z = \frac{1}{12}m(a^2 + b^2)$$

(a) (b) (c)

Figure 9.9 Moments of inertia for three common homogeneous solids of mass m. (a) Solid circular cylinder. (b) Circular cylindrical shell. (c) Rectangular parallelepiped.

for other solid geometries can be found in engineering handbooks and in dynamics textbooks.

For a mass concentrated at a point, the moments of inertia in Equation 9.40 are just the mass times the integrand evaluated at the point. That is, the moment of inertia matrix $[\mathbf{I}_m]$ of a point mass m is given by

$$[\mathbf{I}_m] = \begin{bmatrix} m(y^2 + z^2) & -mxy & -mxz \\ -mxy & m(x^2 + z^2) & -myz \\ -mxz & -myz & m(x^2 + y^2) \end{bmatrix} \qquad (9.44)$$

EXAMPLE 9.4

The following table lists mass and coordinates of seven point masses. Find the center of mass of the system and the moments of inertia about the origin.

Point, i	Mass, m_i (kg)	x_i (m)	y_i (m)	z_i (m)
1	3	−0.5	0.2	0.3
2	7	0.2	0.75	−0.4
3	5	1	−0.8	0.9
4	6	1.2	−1.3	1.25
5	2	−1.3	1.4	−0.8
6	4	−0.3	1.35	0.75
7	1	1.5	−1.7	0.85

The total mass of this system is

$$m = \sum_{i=1}^{7} m_i = 28 \text{ kg}$$

(Example 9.4 continued)

For concentrated masses the integral in Equation 9.9 is replaced by the mass times its position vector. Therefore, in this case the three components of the position vector of the center of mass are

$$x_G = \frac{\sum_{i=1}^{7} m_i x_i}{m} = 0.35 \text{ m} \quad y_G = \frac{\sum_{i=1}^{7} m_i y_i}{m} = 0.01964 \text{ m} \quad z_G = \frac{\sum_{i=1}^{7} m_i z_i}{m} = 0.4411 \text{ m}$$

The total moment of inertia is the sum over all of the particles of Equation 9.44 evaluated at each point. Thus,

$$[I] = \overbrace{\begin{bmatrix} 0.39 & 0.3 & 0.45 \\ 0.3 & 1.02 & -0.18 \\ 0.45 & -0.18 & 0.87 \end{bmatrix}}^{(1)} + \overbrace{\begin{bmatrix} 5.0575 & -1.05 & 0.56 \\ -1.05 & 1.4 & 2.1 \\ 0.56 & 2.1 & 4.2175 \end{bmatrix}}^{(2)} + \overbrace{\begin{bmatrix} 7.25 & 4 & -4.5 \\ 4 & 9.05 & 3.6 \\ -4.5 & 3.6 & 8.2 \end{bmatrix}}^{(3)}$$

$$+ \overbrace{\begin{bmatrix} 19.515 & 9.36 & -9 \\ 9.36 & 18.015 & 9.75 \\ -9 & 9.75 & 18.78 \end{bmatrix}}^{(4)} + \overbrace{\begin{bmatrix} 5.2 & 3.64 & -2.08 \\ 3.64 & 4.66 & 2.24 \\ -2.08 & 2.24 & 7.3 \end{bmatrix}}^{(5)}$$

$$+ \overbrace{\begin{bmatrix} 9.54 & 1.62 & 0.9 \\ 1.62 & 2.61 & -4.05 \\ 0.9 & -4.05 & 7.65 \end{bmatrix}}^{(6)} + \overbrace{\begin{bmatrix} 3.6125 & 2.55 & -1.275 \\ 2.55 & 2.9725 & 1.445 \\ -1.275 & 1.445 & 5.14 \end{bmatrix}}^{(7)}$$

or

$$[I] = \begin{bmatrix} 50.56 & 20.42 & -14.94 \\ 20.42 & 39.73 & 14.90 \\ -14.94 & 14.90 & 52.16 \end{bmatrix} (\text{kg} \cdot \text{m}^2)$$

EXAMPLE 9.5

Calculate the moments of inertia of a slender, homogeneous straight rod of length l and mass m. One end of the rod is at the origin and the other has coordinates (a, b, c).

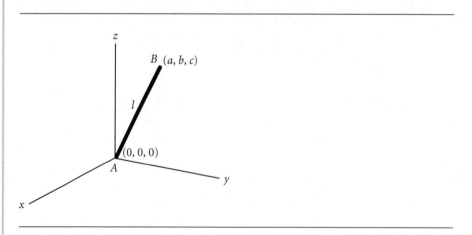

Figure 9.10 Uniform slender bar of mass m and length l.

A slender rod is one whose cross-sectional dimensions are negligible compared with its length. The mass is concentrated along its centerline. Since the rod is homogeneous, the mass per unit length ϱ is uniform and given by

$$\varrho = \frac{m}{l} \tag{a}$$

The length of the rod is

$$l = \sqrt{a^2 + b^2 + c^2}$$

Starting with I_x, we have from Equations 9.40,

$$I_x = \int_0^l (y^2 + z^2) \varrho \, ds$$

in which we replaced the element of mass dm by ϱds, where ds is the element of length along the rod. s is measured from end A of the rod, so that the x, y and z coordinates of any point along it are found in terms of s by the following relations,

$$x = \frac{s}{l} a \quad y = \frac{s}{l} b \quad z = \frac{s}{l} c$$

Thus

$$I_x = \int_0^l \left(\frac{s^2}{l^2} b^2 + \frac{s^2}{l^2} c^2 \right) \varrho \, ds = \varrho \frac{b^2 + c^2}{l^2} \int_0^l s^2 \, ds = \frac{1}{3} \varrho (b^2 + c^2) l$$

Substituting (a) yields

$$I_x = \frac{1}{3} m (b^2 + c^2)$$

In precisely the same way we find

$$I_y = \frac{1}{3} m (a^2 + c^2) \quad I_z = \frac{1}{3} m (a^2 + b^2)$$

For I_{xy} we have

$$I_{xy} = -\int_0^l xy \varrho \, ds = -\int_0^l \frac{s}{l} a \cdot \frac{s}{l} b \varrho \, ds = -\varrho \frac{ab}{l^2} \int_0^l s^2 \, ds = -\frac{1}{3} \varrho abl$$

Once again using (a),

$$I_{xy} = -\frac{1}{3} mab$$

Likewise,

$$I_{xz} = -\frac{1}{3} mac \quad I_{yz} = -\frac{1}{3} mbc$$

EXAMPLE 9.6

The gyro rotor in Example 9.3 has a mass m of 5 kg, radius r of 0.08 m, and thickness t of 0.025 m. If $N = 2.1$ rad/s, $\dot{\theta} = 4$ rad/s, $\omega = 10.5$ rad/s, and $\theta = 60°$, calculate the angular momentum of the rotor about its center of mass G in the xyz frame. What is the angle between the rotor's angular velocity vector and its angular momentum vector?

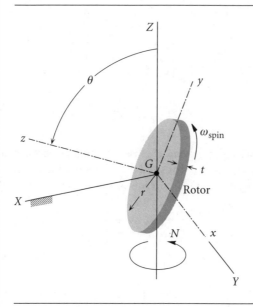

Figure 9.11 Rotor of the gyroscope in Figure 9.4.

In Example 9.3, Equation (f) gives the components of the absolute angular velocity of the rotor in the moving xyz frame,

$$\omega_x = \dot{\theta} = 4 \text{ rad/s}$$
$$\omega_y = N \sin \theta = 2.1 \cdot \sin 60° = 1.819 \text{ rad/s} \quad (a)$$
$$\omega_z = \omega_{\text{spin}} + N \cos \theta = 10.5 + 2.1 \cdot \cos 60° = 11.55 \text{ rad/s}$$

or

$$\boldsymbol{\omega} = 4\hat{\mathbf{i}} + 1.819\hat{\mathbf{j}} + 11.55\hat{\mathbf{k}} \text{ (rad/s)} \quad (b)$$

All three coordinate planes of the xyz frame contain the center of mass G and all are planes of symmetry of the circular cylindrical rotor. Therefore,

$$I_{xy} = I_{zx} = I_{yz} = 0$$

From Figure 9.9(a), we see that the non-zero diagonal entries in the moment of inertia tensor are

$$A = B = \tfrac{1}{12}mt^2 + \tfrac{1}{4}mr^2 = \tfrac{1}{12}5 \cdot 0.025^2 + \tfrac{1}{4}5 \cdot 0.08^2 = 0.008260 \text{ kg} \cdot \text{m}^2$$
$$C = \tfrac{1}{2}mr^2 = \tfrac{1}{2}5 \cdot 0.08^2 = 0.0160 \text{ kg} \cdot \text{m}^2 \quad (c)$$

9.5 Moments of inertia

We can use Equation 9.42 to calculate the angular momentum, because the origin of the *xyz* frame is the rotor's center of mass (which in this case also happens to be a fixed point of rotation, which is another reason we can use Equation 9.42). Substituting (a) and (c) into Equation 9.42 yields

$$H_x = A\omega_x = 0.008260 \cdot 4 = 0.03304 \text{ kg} \cdot \text{m}^2/\text{s}$$
$$H_y = B\omega_y = 0.008260 \cdot 1.819 = 0.0150 \text{ kg} \cdot \text{m}^2/\text{s} \quad \text{(d)}$$
$$H_z = C\omega_z = 0.0160 \cdot 11.55 = 0.1848 \text{ kg} \cdot \text{m}^2/\text{s}$$

or

$$\mathbf{H} = 0.03304\hat{\mathbf{i}} + 0.0150\hat{\mathbf{j}} + 0.1848\hat{\mathbf{k}} \text{ (kg} \cdot \text{m}^2/\text{s)} \quad \text{(e)}$$

The angle ϕ between $\mathbf{H}$ and $\boldsymbol{\omega}$ is found by using the dot product operation,

$$\phi = \cos^{-1}\left(\frac{\mathbf{H} \cdot \boldsymbol{\omega}}{H\omega}\right) = \cos^{-1}\left(\frac{2.294}{0.1883 \cdot 12.36}\right) = \underline{9.717°} \quad \text{(f)}$$

As this illustrates, the angular momentum and the angular velocity are in general not collinear.

Consider a coordinate system $x'y'z'$ with the same origin as *xyz*, but different orientation. Let $[\mathbf{Q}]$ be the orthogonal matrix ($[\mathbf{Q}]^{-1} = [\mathbf{Q}]^T$) which transforms the components of a vector *from* the *xyz* system *to* the $x'y'z'$ frame. Recall from Section 4.5 that the rows of $[\mathbf{Q}]$ are the direction cosines of the $x'y'z'$ axes relative to *xyz*. The components of the angular momentum vector transform as follows

$$\{\mathbf{H}'\} = [\mathbf{Q}]\{\mathbf{H}\}$$

From Equation 9.39 we can write this as

$$\{\mathbf{H}'\} = [\mathbf{Q}][\mathbf{I}]\{\boldsymbol{\omega}\} \quad (9.45)$$

Like the angular momentum vector, the components of the angular velocity vector in the *xyz* system are related to those in the primed system by the expression

$$\{\boldsymbol{\omega}'\} = [\mathbf{Q}]\{\boldsymbol{\omega}\}$$

The inverse relation is simply

$$\{\boldsymbol{\omega}\} = [\mathbf{Q}]^{-1}\{\boldsymbol{\omega}'\} = [\mathbf{Q}]^T\{\boldsymbol{\omega}'\} \quad (9.46)$$

Substituting this into Equation 9.45, we get

$$\{\mathbf{H}'\} = [\mathbf{Q}][\mathbf{I}][\mathbf{Q}]^T\{\boldsymbol{\omega}'\} \quad (9.47)$$

But the components of angular momentum and angular velocity in the $x'y'z'$ frame are related by an equation of the same form as Equation 9.39, so that

$$\{\mathbf{H}'\} = [\mathbf{I}']\{\boldsymbol{\omega}'\} \quad (9.48)$$

where $[\mathbf{I}']$ comprises the components of the inertia matrix in the primed system. Comparing the right-hand sides of Equations 9.47 and 9.48, we conclude that

$$[\mathbf{I}'] = [\mathbf{Q}][\mathbf{I}][\mathbf{Q}]^T \quad (9.49a)$$

that is,

$$\begin{bmatrix} I_{x'} & I_{x'y'} & I_{x'z'} \\ I_{y'x'} & I_{y'} & I_{y'z'} \\ I_{z'x'} & I_{z'y'} & I_{z'} \end{bmatrix} = \begin{bmatrix} Q_{11} & Q_{12} & Q_{13} \\ Q_{21} & Q_{22} & Q_{23} \\ Q_{31} & Q_{32} & Q_{33} \end{bmatrix} \begin{bmatrix} I_x & I_{xy} & I_{xz} \\ I_{yx} & I_y & I_{yz} \\ I_{zx} & I_{zy} & I_z \end{bmatrix} \begin{bmatrix} Q_{11} & Q_{21} & Q_{31} \\ Q_{12} & Q_{22} & Q_{32} \\ Q_{13} & Q_{32} & Q_{33} \end{bmatrix}$$

(9.49b)

This shows how to transform the components of the inertia matrix from the xyz coordinate system to any other orthogonal system with a common origin. Thus, for example,

$$I_{x'} = \overbrace{\begin{bmatrix} Q_{11} & Q_{12} & Q_{13} \end{bmatrix}}^{\lfloor \text{Row 1} \rfloor} \begin{bmatrix} I_x & I_{xy} & I_{xz} \\ I_{yx} & I_y & I_{yz} \\ I_{zx} & I_{zy} & I_z \end{bmatrix} \overbrace{\begin{Bmatrix} Q_{11} \\ Q_{12} \\ Q_{13} \end{Bmatrix}}^{\lfloor \text{Row 1} \rfloor^T}$$

$$I_{y'z'} = \overbrace{\begin{bmatrix} Q_{21} & Q_{22} & Q_{23} \end{bmatrix}}^{\lfloor \text{Row 2} \rfloor} \begin{bmatrix} I_x & I_{xy} & I_{xz} \\ I_{yx} & I_y & I_{yz} \\ I_{zx} & I_{zy} & I_{zz} \end{bmatrix} \overbrace{\begin{Bmatrix} Q_{31} \\ Q_{32} \\ Q_{33} \end{Bmatrix}}^{\lfloor \text{Row 3} \rfloor^T}$$

(9.50)

etc.

Any object represented by a square matrix whose components transform according to Equation 9.49 is called a second order tensor.

EXAMPLE 9.7

Find the mass moment of inertia of the system in Example 9.4 about an axis from the origin through the point with coordinates (2 m, −3 m, 4 m).

From Example 9.4 the moment of inertia tensor for the system of point masses is

$$[\mathbf{I}] = \begin{bmatrix} 50.56 & 20.42 & -14.94 \\ 20.42 & 39.73 & 14.90 \\ -14.94 & 14.90 & 52.16 \end{bmatrix} (\text{kg} \cdot \text{m}^2)$$

The vector connecting the origin with (2 m, −3 m, 4 m) is

$$\mathbf{V} = 2\hat{\mathbf{i}} - 3\hat{\mathbf{j}} + 4\hat{\mathbf{k}}$$

The unit vector in the direction of $\mathbf{V}$ is

$$\hat{\mathbf{u}}_V = \frac{\mathbf{V}}{\|\mathbf{V}\|} = 0.3714\hat{\mathbf{i}} - 0.5571\hat{\mathbf{j}} + 0.7428\hat{\mathbf{k}}$$

We may consider $\hat{\mathbf{u}}_V$ as the unit vector along the x' axis of a rotated cartesian coordinate system. Then, from Equation 9.50,

$$I_{V'} = \begin{bmatrix} 0.3714 & -0.5571 & 0.7428 \end{bmatrix} \begin{bmatrix} 50.56 & 20.42 & -14.94 \\ 20.42 & 39.73 & 14.90 \\ -14.94 & 14.90 & 52.16 \end{bmatrix} \begin{Bmatrix} 0.3714 \\ -0.5571 \\ 0.7428 \end{Bmatrix}$$

$$= \begin{bmatrix} 0.3714 & -0.5571 & 0.7428 \end{bmatrix} \begin{Bmatrix} -3.695 \\ -3.482 \\ 24.90 \end{Bmatrix} = \underline{19.06 \text{ kg} \cdot \text{m}^2}$$

EXAMPLE 9.8

For the satellite of Example 9.2, reproduced in Figure 9.12, the data is as follows. $N = 0.1$ rad/s and $\dot{\theta} = 0.01$ rad/s, in the directions shown. $\theta = 40°$. $d_0 = 1.5$ m. The length, width and thickness of the panel are $l = 6$ m, $w = 2$ m and $t = 0.025$ m. The uniformly distributed mass of the panel is 50 kg. Find the angular momentum of the panel relative to the center of mass O of the satellite.

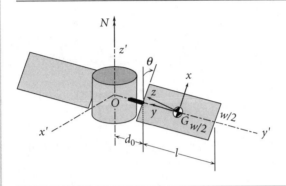

Figure 9.12 Satellite and solar panel.

We can treat the panel as a thin parallelepiped. The panel's xyz axes have their origin at the center of mass G of the panel and are parallel to its three edge directions. According to Figure 9.9(c), the moments of inertia relative to the xyz coordinate system are

$$I_{G_x} = \tfrac{1}{12} m(l^2 + t^2) = \tfrac{1}{12} \cdot 50 \cdot (6^2 + 0.025^2) = 150.0 \text{ kg} \cdot \text{m}^2$$
$$I_{G_y} = \tfrac{1}{12} m(w^2 + t^2) = \tfrac{1}{12} \cdot 50 \cdot (2^2 + 0.025^2) = 16.67 \text{ kg} \cdot \text{m}^2 \quad \text{(a)}$$
$$I_{G_z} = \tfrac{1}{12} m(w^2 + l^2) = \tfrac{1}{12} \cdot 50 \cdot (2^2 + 6^2) = 166.7 \text{ kg} \cdot \text{m}^2$$
$$I_{G_{xy}} = I_{G_{xz}} = I_{G_{yz}} = 0$$

In matrix notation,

$$[I_G] = \begin{bmatrix} 150.0 & 0 & 0 \\ 0 & 16.67 & 0 \\ 0 & 0 & 166.7 \end{bmatrix} (\text{kg} \cdot \text{m}^2) \quad \text{(b)}$$

The unit vectors of the satellite's $x'y'z'$ system are related to those of panel's xyz frame by inspection,

$$\hat{\mathbf{i}}' = -\sin\theta \hat{\mathbf{i}} + \cos\theta \hat{\mathbf{k}} = -0.6428 \hat{\mathbf{i}} + 0.7660 \hat{\mathbf{k}}$$
$$\hat{\mathbf{j}}' = -\hat{\mathbf{j}} \quad \text{(c)}$$
$$\hat{\mathbf{k}}' = \cos\theta \hat{\mathbf{i}} + \sin\theta \hat{\mathbf{k}} = 0.7660 \hat{\mathbf{i}} + 0.6428 \hat{\mathbf{k}}$$

(Example 9.8 continued)

The matrix $[\mathbf{Q}]$ of the transformation from xyz to $x'y'z'$ comprises the direction cosines of $\hat{\mathbf{i}}'$, $\hat{\mathbf{j}}'$ and $\hat{\mathbf{k}}'$:

$$[\mathbf{Q}] = \begin{bmatrix} -0.6428 & 0 & 0.7660 \\ 0 & -1 & 0 \\ 0.7660 & 0 & 0.6428 \end{bmatrix} \quad (d)$$

In Example 9.2 we found that the absolute angular velocity of the panel, in the satellite's $x'y'z'$ frame of reference, is

$$\boldsymbol{\omega} = -\dot{\theta}\hat{\mathbf{j}}' + N\hat{\mathbf{k}}' = -0.01\hat{\mathbf{j}}' + 0.1\hat{\mathbf{k}}' \text{(rad/s)}$$

That is,

$$\{\boldsymbol{\omega}'\} = \begin{Bmatrix} 0 \\ -0.01 \\ 0.1 \end{Bmatrix} \text{(rad/s)} \quad (e)$$

To find the absolute angular momentum $\{\mathbf{H}'_G\}$ in the satellite system requires using Equation 9.39,

$$\{\mathbf{H}'_G\} = [\mathbf{I}'_G]\{\boldsymbol{\omega}'\} \quad (f)$$

Before doing so, we must transform the components of the moments of inertia tensor in (b) from the unprimed system to the primed system, by means of Equation 9.49,

$$[\mathbf{I}'_G] = [\mathbf{Q}][\mathbf{I}_G][\mathbf{Q}]^T$$

$$= \begin{bmatrix} -0.6428 & 0 & 0.7660 \\ 0 & -1 & 0 \\ 0.7660 & 0 & 0.6428 \end{bmatrix} \begin{bmatrix} 150.0 & 0 & 0 \\ 0 & 16.67 & 0 \\ 0 & 0 & 166.7 \end{bmatrix} \begin{bmatrix} -0.6428 & 0 & 0.7660 \\ 0 & -1 & 0 \\ 0.7660 & 0 & 0.6428 \end{bmatrix}$$

so that

$$[\mathbf{I}'_G] = \begin{bmatrix} 159.8 & 0 & 8.205 \\ 0 & 16.67 & 0 \\ 8.205 & 0 & 156.9 \end{bmatrix} (\text{kg} \cdot \text{m}^2) \quad (g)$$

Then (f) yields

$$\{\mathbf{H}'_G\} = \begin{bmatrix} 159.8 & 0 & 8.205 \\ 0 & 16.67 & 0 \\ 8.205 & 0 & 156.9 \end{bmatrix} \begin{Bmatrix} 0 \\ -0.01 \\ 0.1 \end{Bmatrix} = \begin{Bmatrix} 0.8205 \\ -0.1667 \\ 15.69 \end{Bmatrix} (\text{kg} \cdot \text{m}^2/\text{s})$$

or, in vector notation,

$$\mathbf{H}_G = 0.8205\hat{\mathbf{i}}' - 0.1667\hat{\mathbf{j}}' + 15.69\hat{\mathbf{k}}' \, (\text{kg} \cdot \text{m}^2/\text{s}) \quad (h)$$

This is the absolute angular momentum of the panel about its own center of mass, and it is used in Equation 9.27 to calculate the angular momentum $\mathbf{H}_{O_{\text{rel}}}$ relative to the satellite's center of mass O,

$$\mathbf{H}_{O_{\text{rel}}} = \mathbf{H}_G + \mathbf{r}_{G/O} \times m\mathbf{v}_{G/O} \quad (i)$$

$\mathbf{r}_{G/O}$ is the position vector from O to G,

$$\mathbf{r}_{G/O} = \left(d_0 + \frac{l}{2}\right)\hat{\mathbf{j}}' = \left(1.5 + \frac{6}{2}\right)\hat{\mathbf{j}}' = 4.5\hat{\mathbf{j}}' \text{ (m)} \tag{j}$$

The velocity of G relative to O, $\mathbf{v}_{G/O}$, is found from Equation 9.2,

$$\mathbf{v}_{G/O} = \boldsymbol{\omega}_{\text{satellite}} \times \mathbf{r}_{G/O} = N\hat{\mathbf{k}}' \times \mathbf{r}_{G/O} = 0.1\hat{\mathbf{k}}' \times 4.5\hat{\mathbf{j}}' = -0.45\hat{\mathbf{i}}' \text{(m/s)} \tag{k}$$

Substituting (h), (j) and (k) into (i) finally yields

$$\mathbf{H}_{O_{\text{rel}}} = (0.8205\hat{\mathbf{i}}' - 0.1667\hat{\mathbf{j}}' + 15.69\hat{\mathbf{k}}') + 4.5\hat{\mathbf{j}}' \times [50(-0.45\hat{\mathbf{i}}')]$$
$$= \underline{0.8205\hat{\mathbf{i}}' - 0.1667\hat{\mathbf{j}}' + 116.9\hat{\mathbf{k}}'} \text{ (kg} \cdot \text{m}^2/\text{s)} \tag{l}$$

Note that we were unable to use Equation 9.21 to find the absolute angular momentum $\mathbf{H}_O$ because that requires knowing the absolute velocity $\mathbf{v}_G$, which in turn depends on the absolute velocity of O, which was not provided.

How can we find that transformation matrix $[\mathbf{Q}]$ such that Equation 9.49 will yield a moment of inertia matrix $[\mathbf{I}']$ which is diagonal, i.e., of the form given by Equation 9.41? In other words, how do we find the principal directions of the moment of inertia tensor? Let the angular velocity vector $\{\boldsymbol{\omega}\}$ be parallel to the principal direction defined by the vector $\{\mathbf{v}\}$, so that $\{\boldsymbol{\omega}\} = \beta\{\mathbf{v}\}$, where β is a scalar. Since $\{\boldsymbol{\omega}\}$ points in a principal direction of the inertia tensor, so must $\{\mathbf{H}\}$, which means $\{\mathbf{H}\}$ is also parallel to $\{\mathbf{v}\}$. Therefore, $\{\mathbf{H}\} = \alpha\{\mathbf{v}\}$, where α is a scalar. From Equation 9.39 it follows that

$$\alpha\{\mathbf{v}\} = [\mathbf{I}](\beta\{\mathbf{v}\})$$

or

$$[\mathbf{I}]\{\mathbf{v}\} = \lambda\{\mathbf{v}\}$$

where $\lambda = \alpha/\beta$ (a scalar). That is,

$$\begin{bmatrix} I_x & I_{xy} & I_{xz} \\ I_{xy} & I_y & I_{yz} \\ I_{xz} & I_{yz} & I_z \end{bmatrix} \begin{Bmatrix} v_x \\ v_y \\ v_z \end{Bmatrix} = \lambda \begin{Bmatrix} v_x \\ v_y \\ v_z \end{Bmatrix}$$

This can be written

$$\begin{bmatrix} I_x - \lambda & I_{xy} & I_{xz} \\ I_{xy} & I_y - \lambda & I_{yz} \\ I_{xz} & I_{yz} & I_z - \lambda \end{bmatrix} \begin{Bmatrix} v_x \\ v_y \\ v_z \end{Bmatrix} = \begin{Bmatrix} 0 \\ 0 \\ 0 \end{Bmatrix} \tag{9.51}$$

The trivial solution of Equation 9.51 is $\{\mathbf{v}\} = \{\mathbf{0}\}$, which is of no interest. The only way that Equation 9.51 will not yield the trivial solution is if the coefficient matrix on the left is singular. That will occur if its determinant vanishes, that is, if

$$\begin{vmatrix} I_x - \lambda & I_{xy} & I_{xz} \\ I_{xy} & I_y - \lambda & I_{yz} \\ I_{xz} & I_{yz} & I_z - \lambda \end{vmatrix} = 0 \tag{9.52}$$

Expanding the determinant, we find

$$\begin{vmatrix} I_x - \lambda & I_{xy} & I_{xz} \\ I_{xy} & I_y - \lambda & I_{yz} \\ I_{xz} & I_{yz} & I_z - \lambda \end{vmatrix} = -\lambda^3 + I_1\lambda^2 - I_2\lambda + I_3 \tag{9.53}$$

where

$$I_1 = I_x + I_y + I_z$$

$$I_2 = \begin{vmatrix} I_x & I_{xy} \\ I_{xy} & I_y \end{vmatrix} + \begin{vmatrix} I_x & I_{xz} \\ I_{xz} & I_z \end{vmatrix} + \begin{vmatrix} I_y & I_{yz} \\ I_{yz} & I_z \end{vmatrix} \tag{9.54}$$

$$I_3 = \begin{vmatrix} I_x & I_{xy} & I_{xz} \\ I_{xy} & I_y & I_{yz} \\ I_{xz} & I_{yz} & I_z \end{vmatrix}$$

Equations 9.52 and 9.53 yield the characteristic equation of the tensor $[\mathbf{I}]$

$$\lambda^3 - I_1\lambda^2 + I_2\lambda - I_3 = 0 \tag{9.55}$$

The three roots λ_p ($p = 1, 2, 3$) of this cubic equation are real, since $[\mathbf{I}]$ is symmetric; furthermore they are all positive, since $[\mathbf{I}]$ is a positive-definite matrix. Each root, or eigenvalue, λ_p is substituted back into Equation 9.51 to obtain

$$\begin{bmatrix} I_x - \lambda_p & I_{xy} & I_{xz} \\ I_{xy} & I_y - \lambda_p & I_{yz} \\ I_{xz} & I_{yz} & I_z - \lambda_p \end{bmatrix} \begin{Bmatrix} v_x^{(p)} \\ v_y^{(p)} \\ v_z^{(p)} \end{Bmatrix} = \begin{Bmatrix} 0 \\ 0 \\ 0 \end{Bmatrix}, \quad p = 1, 2, 3 \tag{9.56}$$

Solving this system yields the three eigenvectors $\{\mathbf{v}^{(p)}\}$ corresponding to each of the three eigenvalues λ_p. The three eigenvectors are orthogonal, also due to the symmetry of the matrix $[\mathbf{I}]$. Each eigenvalue is a principal moment of inertia, and its corresponding eigenvector is a principal direction.

EXAMPLE 9.9

Find the principal moments of inertia and the principal axes of inertia of the inertia tensor

$$[\mathbf{I}] = \begin{bmatrix} 100 & -20 & -100 \\ -20 & 300 & -50 \\ -100 & -50 & 500 \end{bmatrix} \text{kg} \cdot \text{m}^2$$

We seek the non-trivial solutions of the system

$$\begin{bmatrix} 100 - \lambda & -20 & -100 \\ -20 & 300 - \lambda & -50 \\ -100 & -50 & 500 - \lambda \end{bmatrix} \begin{Bmatrix} v_x \\ v_y \\ v_z \end{Bmatrix} = \begin{Bmatrix} 0 \\ 0 \\ 0 \end{Bmatrix} \tag{a}$$

From Equation 9.54,

$$I_1 = 100 + 300 + 500 = 900$$

$$I_2 = \begin{vmatrix} 100 & -20 \\ -20 & 300 \end{vmatrix} + \begin{vmatrix} 100 & -100 \\ -100 & 500 \end{vmatrix} + \begin{vmatrix} 300 & -50 \\ -50 & 500 \end{vmatrix} = 217\,100 \qquad (b)$$

$$I_3 = \begin{vmatrix} 100 & -20 & -100 \\ -20 & 300 & -50 \\ -100 & -50 & 500 \end{vmatrix} = 11\,350\,000$$

Thus, the characteristic equation is

$$\lambda^3 - 900\lambda^2 + 217\,100\lambda - 11\,350\,000 = 0 \qquad (c)$$

The three roots are the principal moments of inertia, which are found to be

$$\lambda_1 = 532.052 \qquad \lambda_2 = 295.840 \qquad \lambda_3 = 72.1083 \qquad (d)$$

Each of these is substituted, in turn, back into (a) to find its corresponding principal direction.

Substituting $\lambda_1 = 532.052$ kg·m^2 into (a) we obtain

$$\begin{bmatrix} -432.052 & -20.0000 & -100.0000 \\ -20.0000 & -232.052 & -50.0000 \\ -100.0000 & -50.0000 & -32.0519 \end{bmatrix} \begin{Bmatrix} v_x^{(1)} \\ v_y^{(1)} \\ v_z^{(1)} \end{Bmatrix} = \begin{Bmatrix} 0 \\ 0 \\ 0 \end{Bmatrix} \qquad (e)$$

Since the determinant of the coefficient matrix is zero, at most two of the three equations in (e) are independent. Thus, at most two of the three components of the vector $\mathbf{v}^{(1)}$ can be found in terms of the third. We can therefore arbitrarily set $v_x^{(1)} = 1$ and solve for $v_y^{(1)}$ and $v_z^{(1)}$ using any two of the independent equations in (e). With $v_x^{(1)} = 1$, the first two of Equations (e) become

$$-20.0000 v_y^{(1)} - 100.000 v_z^{(1)} = 432.052$$

$$-232.052 v_y^{(1)} - 50.000 v_z^{(1)} = 20.0000 \qquad (f)$$

Solving these two equations for $v_y^{(1)}$ and $v_z^{(1)}$ yields, together with the assumption on $v_x^{(1)}$,

$$v_x^{(1)} = 1.00000 \qquad v_y^{(1)} = 0.882793 \qquad v_z^{(1)} = -4.49708 \qquad (g)$$

To obtain the unit vector in the direction of $\mathbf{v}^{(1)}$

$$\hat{\mathbf{i}}_1 = \frac{\mathbf{v}^{(1)}}{\|\mathbf{v}^{(1)}\|} = \frac{1.00000\hat{\mathbf{i}} + 0.882793\hat{\mathbf{j}} - 4.49708\hat{\mathbf{k}}}{\sqrt{1.00000^2 + 0.882793^2 + (-4.49708)^2}}$$

$$= 0.213186\hat{\mathbf{i}} + 0.188199\hat{\mathbf{j}} - 0.958714\hat{\mathbf{k}} \qquad (h)$$

Substituting $\lambda_2 = 295.840$ into (a) and proceeding as above we find

$$\hat{\mathbf{i}}_2 = 0.176732\hat{\mathbf{i}} - 0.972512\hat{\mathbf{j}} - 0.151609\hat{\mathbf{k}} \qquad (i)$$

(Example 9.9 continued)

The two unit vectors $\hat{\mathbf{i}}_1$ and $\hat{\mathbf{i}}_2$ define two of the principal directions of the inertia tensor. Observe that $\hat{\mathbf{i}}_1 \cdot \hat{\mathbf{i}}_2 = 0$, as must be the case for symmetric matrices.

To obtain the third principal direction $\hat{\mathbf{i}}_3$, we can substitute $\lambda_3 = 72.1083$ into (a) and proceed as above. However, since the inertia tensor is symmetric, we know that the three principal directions are mutually orthogonal. That means $\hat{\mathbf{i}}_3 = \hat{\mathbf{i}}_1 \times \hat{\mathbf{i}}_2$. Substituting Equations (h) and (i) into the cross product, we find that

$$\hat{\mathbf{i}}_3 = -0.960894\hat{\mathbf{i}} - 0.137114\hat{\mathbf{j}} - 0.240587\hat{\mathbf{k}} \tag{j}$$

We can check our work by substituting λ_3 and $\hat{\mathbf{i}}_3$ into (a) and verify that it is indeed satisfied:

$$\begin{bmatrix} 100 - 72.1083 & -20 & -100 \\ -20 & 300 - 72.1083 & -50 \\ -100 & -50 & 500 - 72.1083 \end{bmatrix} \begin{Bmatrix} -0.960894 \\ -0.137114 \\ -0.240587 \end{Bmatrix} = \begin{Bmatrix} 0 \\ 0 \\ 0 \end{Bmatrix} \tag{k}$$

The components of $\hat{\mathbf{i}}_1, \hat{\mathbf{i}}_2$ and $\hat{\mathbf{i}}_3$ define the rows of the orthogonal transformation $[\mathbf{Q}]$ from the xyz system into the $x'y'z'$ system aligned along the three principal directions:

$$[\mathbf{Q}] = \begin{bmatrix} 0.213186 & 0.188199 & -0.958714 \\ 0.176732 & -0.972512 & -0.151609 \\ -0.960894 & -0.137114 & -0.240587 \end{bmatrix} \tag{l}$$

If we apply the transformation in Equation 9.49, $[\mathbf{I}'] = [\mathbf{Q}][\mathbf{I}][\mathbf{Q}]^T$, we find

$$[\mathbf{I}'] = \begin{bmatrix} 0.213186 & 0.188199 & -0.958714 \\ 0.176732 & -0.972512 & -0.151609 \\ -0.960894 & -0.137114 & -0.240587 \end{bmatrix} \begin{bmatrix} 100 & -20 & -100 \\ -20 & 300 & -50 \\ -100 & -50 & 500 \end{bmatrix}$$

$$\times \begin{bmatrix} 0.213186 & 0.176732 & -0.960894 \\ 0.188199 & -0.972512 & -0.137114 \\ -0.958714 & -0.151609 & -0.240587 \end{bmatrix}$$

$$= \begin{bmatrix} 532.052 & 0 & 0 \\ 0 & 295.840 & 0 \\ 0 & 0 & 72.1083 \end{bmatrix} (\text{kg} \cdot \text{m}^2)$$

9.5.1 PARALLEL AXIS THEOREM

Suppose the rigid body in Figure 9.13 is in pure rotation about point P. Then, according to Equation 9.39,

$$\{\mathbf{H}_{P_{\text{rel}}}\} = [\mathbf{I}_P]\{\boldsymbol{\omega}\} \tag{9.57}$$

where $[\mathbf{I}_P]$ is the moment of inertia about P, given by Equations 9.40 with

$$x = x_{G/P} + \xi \qquad y = y_{G/P} + \eta \qquad z = z_{G/P} + \zeta$$

9.5 Moments of inertia

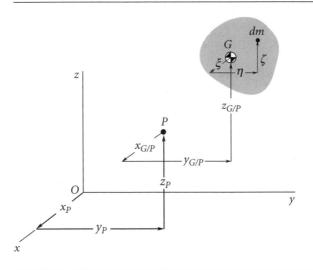

Figure 9.13 The moments of inertia are to be computed at P, given their values at G.

On the other hand, we have from Equation 9.27 that

$$\mathbf{H}_{P_{rel}} = \mathbf{H}_G + \mathbf{r}_{G/P} \times m\mathbf{v}_{G/P} \tag{9.58}$$

The vector $\mathbf{r}_{G/P} \times m\mathbf{v}_{G/P}$ is the angular momentum about P of the concentrated mass m located at G. Using matrix notation, it is computed as follows,

$$\{\mathbf{r}_{G/P} \times m\mathbf{v}_{G/P}\} \equiv \{\mathbf{H}_{m_{P_{rel}}}\} = [\mathbf{I}_{m_P}]\{\boldsymbol{\omega}\} \tag{9.59}$$

where $[\mathbf{I}_{mP}]$, the moment of inertia of m about P, is obtained from Equation 9.43, with $x = x_{G/P}, y = y_{G/P}$ and $z = z_{G/P}$. That is,

$$[\mathbf{I}_{m_P}] = \begin{bmatrix} m(y_{G/P}^2 + z_{G/P}^2) & -mx_{G/P}y_{G/P} & -mx_{G/P}z_{G/P} \\ -mx_{G/P}y_{G/P} & m(x_{G/P}^2 + z_{G/P}^2) & -my_{G/P}z_{G/P} \\ -mx_{G/P}z_{G/P} & -my_{G/P}z_{G/P} & m(x_{G/P}^2 + y_{G/P}^2) \end{bmatrix} \tag{9.60}$$

Of course, Equation 9.39 requires

$$\{\mathbf{H}_G\} = [\mathbf{I}_G]\{\boldsymbol{\omega}\}$$

Substituting this together with Equations 9.57 and 9.59 into Equation 9.58 yields

$$[\mathbf{I}_P]\{\boldsymbol{\omega}\} = [\mathbf{I}_G]\{\boldsymbol{\omega}\} + [\mathbf{I}_{m_P}]\{\boldsymbol{\omega}\} = ([\mathbf{I}_G] + [\mathbf{I}_{m_P}])\{\boldsymbol{\omega}\}$$

From this we may infer the parallel axis theorem,

$$[\mathbf{I}_P] = [\mathbf{I}_G] + [\mathbf{I}_{m_P}] \tag{9.61}$$

The moment of inertia about P is the moment of inertia about parallel axes through the center of mass plus the moment of inertia of the center of mass about P. That is,

$$I_{P_x} = I_{G_x} + m(y_{G/P}^2 + z_{G/P}^2) \quad I_{P_y} = I_{G_y} + m(x_{G/P}^2 + z_{G/P}^2) \quad I_{P_z} = I_{G_z} + m(x_{G/P}^2 + y_{G/P}^2)$$
$$I_{P_{xz}} = I_{G_{xz}} - mx_{G/P}z_{G/P} \quad I_{P_{xy}} = I_{G_{xy}} - mx_{G/P}y_{G/P} \quad I_{P_{yz}} = I_{G_{yz}} - my_{G/P}z_{G/P} \quad (9.62)$$

EXAMPLE 9.10

Find the moments of inertia of the rod in Example 9.5 (Figure 9.14) about its center of mass G.

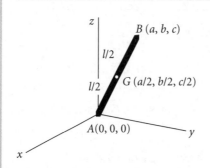

Figure 9.14 Uniform slender rod.

From Example 9.5,

$$[I_A] = \begin{bmatrix} \frac{1}{3}m(b^2 + c^2) & -\frac{1}{3}mab & -\frac{1}{3}mac \\ -\frac{1}{3}mab & \frac{1}{3}m(a^2 + c^2) & -\frac{1}{3}mbc \\ -\frac{1}{3}mac & -\frac{1}{3}mbc & \frac{1}{3}m(a^2 + b^2) \end{bmatrix}$$

Using Equation 9.62$_1$, and noting the coordinates of the center of mass in Figure 9.14,

$$I_{G_x} = I_{A_x} - m[(y_G - 0)^2 + (z_G - 0)^2]$$
$$= \frac{1}{3}m(b^2 + c^2) - m\left[\left(\frac{b}{2}\right)^2 + \left(\frac{c}{2}\right)^2\right] = \frac{1}{12}m(b^2 + c^2)$$

Equation 9.62$_4$ yields

$$I_{G_{xy}} = I_{A_{xy}} + m(x_G - 0)(y_G - 0) = -\frac{1}{3}mab + m \cdot \frac{a}{2} \cdot \frac{b}{2} = -\frac{1}{12}mab$$

The remaining four moments of inertia are found in a similar fashion, so that

$$[I_G] = \begin{bmatrix} \frac{1}{12}m(b^2 + c^2) & -\frac{1}{12}mab & -\frac{1}{12}mac \\ -\frac{1}{12}mab & \frac{1}{12}m(a^2 + c^2) & -\frac{1}{12}mbc \\ -\frac{1}{12}mac & -\frac{1}{12}mbc & \frac{1}{12}m(a^2 + b^2) \end{bmatrix} \quad (9.63)$$

EXAMPLE 9.11

Calculate the principal moments of inertia about the center of mass and the corresponding principal directions for the bent rod in Figure 9.15. Its mass is uniformly distributed at 2 kg/m.

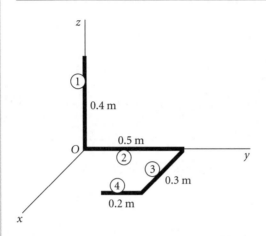

Figure 9.15 Bent rod for which the principal moments of inertia are to be determined.

The mass of each of the rod segments is

$$m^{(1)} = 2 \cdot 0.4 = 0.8 \text{ kg} \qquad m^{(2)} = 2 \cdot 0.5 = 1 \text{ kg}$$
$$m^{(3)} = 2 \cdot 0.3 = 0.6 \text{ kg} \qquad m^{(4)} = 2 \cdot 0.2 = 0.4 \text{ kg} \qquad \text{(a)}$$

The total mass of the system is

$$m = \sum_{i=1}^{4} m^{(i)} = 2.8 \text{ kg} \qquad \text{(b)}$$

The coordinates of each segment's center of mass are

$$\begin{array}{lll} x_{G_1} = 0 & y_{G_1} = 0 & z_{G_1} = 0.2 \text{ m} \\ x_{G_2} = 0 & y_{G_2} = 0.25 \text{ m} & z_{G_2} = 0.2 \text{ m} \\ x_{G_3} = 0.15 \text{ m} & y_{G_3} = 0.5 \text{ m} & z_{G_3} = 0 \\ x_{G_4} = 0.3 \text{ m} & y_{G_4} = 0.4 \text{m} & z_{G_4} = 0 \end{array} \qquad \text{(c)}$$

If the slender rod of Figure 9.14 is aligned with, say, the x axis, then $a = l$ and $b = c = 0$, so that according to Equation 9.63,

$$[\mathbf{I}_G] = \begin{bmatrix} 0 & 0 & 0 \\ 0 & \tfrac{1}{12} ml^2 & 0 \\ 0 & 0 & \tfrac{1}{12} ml^2 \end{bmatrix}$$

That is, the moment of inertia of a slender rod about axes normal to the rod at its center of mass is $\tfrac{1}{12} ml^2$, where m and l are the mass and length of the rod, respectively.

(Example 9.11 continued)

Since the mass of a slender bar is assumed to be concentrated along the axis of the bar (its cross-sectional dimensions are infinitesimal), the moment of inertia about the centerline is zero. By symmetry, the products of inertia about axes through the center of mass are all zero. Using this information and the parallel axis theorem, we find the moments and products of inertia of each rod segment about the origin O of the xyz system as follows.

Rod 1:

$$I_x^{(1)} = \left(I_{G_1}^{(1)}\right)_x + m^{(1)}(y_{G_1}^2 + z_{G_1}^2) = \frac{1}{12} \cdot 0.8 \cdot 0.4^2 + 0.8(0 + 0.2^2) = 0.04267 \text{ kg} \cdot \text{m}^2$$

$$I_y^{(1)} = \left(I_{G_1}^{(1)}\right)_y + m^{(1)}(x_{G_1}^2 + z_{G_1}^2) = \frac{1}{12} \cdot 0.8 \cdot 0.4^2 + 0.8(0 + 0.2^2) = 0.04267 \text{ kg} \cdot \text{m}^2$$

$$I_z^{(1)} = \left(I_{G_1}^{(1)}\right)_z + m^{(1)}(x_{G_1}^2 + y_{G_1}^2) = 0 + 0.8(0 + 0) = 0$$

$$I_{xy}^{(1)} = \left(I_{G_1}^{(1)}\right)_{xy} - m^{(1)} x_{G_1} y_{G_1} = 0 - 0.8(0)(0) = 0$$

$$I_{xz}^{(1)} = \left(I_{G_1}^{(1)}\right)_{xz} - m^{(1)} x_{G_1} z_{G_1} = 0 - 0.8(0)(0.2) = 0$$

$$I_{yz}^{(1)} = \left(I_{G_1}^{(1)}\right)_{yz} - m^{(1)} y_{G_1} z_{G_1} = 0 - 0.8(0)(0) = 0$$

Rod 2:

$$I_x^{(2)} = \left(I_{G_2}^{(2)}\right)_x + m^{(2)}(y_{G_2}^2 + z_{G_2}^2) = \frac{1}{12} \cdot 1.0 \cdot 0.5^2 + 1.0(0 + 0.25^2) = 0.08333 \text{ kg} \cdot \text{m}^2$$

$$I_y^{(2)} = \left(I_{G_2}^{(2)}\right)_y + m^{(2)}(x_{G_2}^2 + z_{G_2}^2) = 0 + 1.0(0 + 0) = 0$$

$$I_z^{(2)} = \left(I_{G_2}^{(2)}\right)_z + m^{(2)}(x_{G_2}^2 + y_{G_2}^2) = \frac{1}{12} \cdot 1.0 \cdot 0.5^2 + 1.0(0 + 0.5^2) = 0.08333 \text{ kg} \cdot \text{m}^2$$

$$I_{xy}^{(2)} = \left(I_{G_2}^{(2)}\right)_{xy} - m^{(2)} x_{G_2} y_{G_2} = 0 - 1.0(0)(0.5) = 0$$

$$I_{xz}^{(2)} = \left(I_{G_2}^{(2)}\right)_{xz} - m^{(2)} x_{G_2} z_{G_2} = 0 - 1.0(0)(0) = 0$$

$$I_{yz}^{(2)} = \left(I_{G_2}^{(2)}\right)_{yz} - m^{(2)} y_{G_2} z_{G_2} = 0 - 1.0(0.5)(0) = 0$$

Rod 3:

$$I_x^{(3)} = \left(I_{G_3}^{(3)}\right)_x + m^{(3)}(y_{G_3}^2 + z_{G_3}^2) = 0 + 0.6(0.5^2 + 0) = 0.15 \text{ kg} \cdot \text{m}^2$$

$$I_y^{(3)} = \left(I_{G_3}^{(3)}\right)_y + m^{(3)}(x_{G_3}^2 + z_{G_3}^2) = \frac{1}{12} \cdot 0.6 \cdot 0.3^2 + 0.6(0.15^2 + 0) = 0.018 \text{ kg} \cdot \text{m}^2$$

$$I_z^{(3)} = \left(I_{G_3}^{(3)}\right)_z + m^{(3)}(x_{G_3}^2 + y_{G_3}^2) = \frac{1}{12} \cdot 0.6 \cdot 0.3^2 + 0.6(0.15^2 + 0.5^2) = 0.1680 \text{ kg} \cdot \text{m}^2$$

$$I_{xy}^{(3)} = \left(I_{G_3}^{(3)}\right)_{xy} - m^{(3)} x_{G_3} y_{G_3} = 0 - 0.6(0.15)(0.5) = -0.045 \text{ kg} \cdot \text{m}^2$$

$$I_{xz}^{(3)} = \left(I_{G_3}^{(3)}\right)_{xz} - m^{(3)} x_{G_3} z_{G_3} = 0 - 0.6(0.15)(0) = 0$$

$$I_{yz}^{(3)} = \left(I_{G_3}^{(3)}\right)_{yz} - m^{(3)} y_{G_3} z_{G_3} = 0 - 0.6(0.5)(0) = 0$$

Rod 4:

$$I_x^{(4)} = \left(I_{G_4}^{(4)}\right)_x + m^{(4)}(y_{G_4}^2 + z_{G_4}^2) = \frac{1}{12} \cdot 0.4 \cdot 0.2^2 + 0.4(0.4^2 + 0) = 0.06533 \text{ kg} \cdot \text{m}^2$$

$$I_y^{(4)} = \left(I_{G_4}^{(4)}\right)_y + m^{(4)}(x_{G_4}^2 + z_{G_4}^2) = 0 + 0.4(0.3^2 + 0) = 0.0360 \text{ kg} \cdot \text{m}^2$$

$$I_z^{(4)} = \left(I_{G_4}^{(4)}\right)_z + m^{(4)}(x_{G_4}^2 + y_{G_4}^2) = \frac{1}{12} \cdot 0.4 \cdot 0.2^2 + 0.4(0.3^2 + 0.4^2) = 0.1013 \text{ kg} \cdot \text{m}^2$$

$$I_{xy}^{(4)} = \left(I_{G_4}^{(4)}\right)_{xy} - m^{(4)} x_{G_4} y_{G_4} = 0 - 0.4(0.3)(0.4) = -0.0480 \text{ kg} \cdot \text{m}^2$$

$$I_{xy}^{(4)} = \left(I_{G_4}^{(4)}\right)_{xy} - m^{(4)} x_{G_4} y_{G_4} = 0 - 0.4(0.3)(0.4) = -0.0480 \text{ kg} \cdot \text{m}^2$$

$$I_{xz}^{(4)} = \left(I_{G_4}^{(4)}\right)_{xz} - m^{(4)} x_{G_4} z_{G_4} = 0 - 0.4(0.3)(0) = 0$$

$$I_{yz}^{(4)} = \left(I_{G_4}^{(4)}\right)_{yz} - m^{(4)} y_{G_4} z_{G_4} = 0 - 0.4(0.4)(0) = 0$$

The total moments of inertia for all four rods are

$$I_x = \sum_{i=1}^{4} I_x^{(i)} = 0.3413 \text{ kg} \cdot \text{m}^2 \qquad I_y = \sum_{i=1}^{4} I_y^{(i)} = 0.09667 \text{ kg} \cdot \text{m}^2$$

$$I_z = \sum_{i=1}^{4} I_z^{(i)} = 0.3527 \text{ kg} \cdot \text{m}^2 \qquad I_{xy} = \sum_{i=1}^{4} I_{xy}^{(i)} = -0.0930 \text{ kg} \cdot \text{m}^2 \qquad \text{(d)}$$

$$I_{xz} = \sum_{i=1}^{4} I_{xz}^{(i)} = 0 \qquad I_{yz} = \sum_{i=1}^{4} I_{yz}^{(i)} = 0$$

The coordinates of the center of mass of the system of four rods are, from (a), (b) and (c),

$$x_G = \frac{1}{m} \sum_{i=1}^{4} m^{(i)} x_{G_i} = \frac{1}{2.8} \cdot 0.21 = 0.075 \text{ m}$$

$$y_G = \frac{1}{m} \sum_{i=1}^{4} m^{(i)} y_{G_i} = \frac{1}{2.8} \cdot 0.71 = 0.2536 \text{ m} \qquad \text{(e)}$$

$$z_G = \frac{1}{m} \sum_{i=1}^{4} m^{(i)} z_{G_i} = \frac{1}{2.8} \cdot 0.16 = 0.05714 \text{ m}$$

We use the parallel axis theorems to shift the moments of inertia in (d) to the center of mass G of the system:

$$I_{G_x} = I_x - m(y_G^2 + z_G^2) = 0.3413 - 0.1892 = 0.1522 \text{ kg} \cdot \text{m}^2$$

$$I_{G_y} = I_y - m(x_G^2 + z_G^2) = 0.09667 - 0.02489 = 0.07177 \text{ kg} \cdot \text{m}^2$$

$$I_{G_z} = I_z - m(x_G^2 + y_G^2) = 0.3527 - 0.1958 = 0.1569 \text{ kg} \cdot \text{m}^2$$

$$I_{G_{xy}} = I_{xy} + m x_G y_G = -0.093 + 0.05325 = -0.03975 \text{ kg} \cdot \text{m}^2$$

$$I_{G_{xz}} = I_{xz} + m x_G z_G = 0 + 0.012 = 0.012 \text{ kg} \cdot \text{m}^2$$

$$I_{G_{yz}} = I_{yz} + m y_G z_G = 0 + 0.04057 = 0.04057 \text{ kg} \cdot \text{m}^2$$

(Example 9.11 continued)

Therefore the inertia tensor, relative to the center of mass, is

$$[\mathbf{I}] = \begin{bmatrix} I_{G_x} & I_{G_{xy}} & I_{G_{xz}} \\ I_{G_{xy}} & I_{G_y} & I_{G_{yz}} \\ I_{G_{xz}} & I_{G_{yz}} & I_{G_z} \end{bmatrix} = \begin{bmatrix} 0.1522 & -0.03975 & 0.012 \\ -0.03975 & 0.07177 & 0.04057 \\ 0.012 & 0.04057 & 0.1569 \end{bmatrix} (\text{kg} \cdot \text{m}^2) \quad (\text{f})$$

To find the three principal moments of inertia, we set

$$\begin{bmatrix} 0.1522 - \lambda & -0.03975 & 0.012 \\ -0.03975 & 0.07177 - \lambda & 0.04057 \\ 0.012 & 0.04057 & 0.1569 - \lambda \end{bmatrix} = 0$$

from which we obtain the characteristic equation

$$-\lambda^3 + 0.3808\lambda^2 - 0.04268\lambda + 0.001166 = 0$$

The three roots are the principal moments of inertia,

$$\lambda_1 = 0.04023 \text{ kg} \cdot \text{m}^2 \qquad \lambda_2 = 0.1658 \text{ kg} \cdot \text{m}^2 \qquad \lambda_3 = 0.1747 \text{ kg} \cdot \text{m}^2 \quad (\text{g})$$

We substitute each of these principal values, in turn, into the equation,

$$\begin{bmatrix} 0.1522 - \lambda_p & -0.03975 & 0.012 \\ -0.03975 & 0.07177 - \lambda_p & 0.04057 \\ 0.012 & 0.04057 & 0.1569 - \lambda_p \end{bmatrix} \begin{Bmatrix} v_x^{(p)} \\ v_y^{(p)} \\ v_z^{(p)} \end{Bmatrix} = \begin{Bmatrix} 0 \\ 0 \\ 0 \end{Bmatrix} \quad (\text{h})$$

in order to determine the components of the three principal vectors $\mathbf{v}^{(p)}$, $p = 1, 2, 3$.

$$\lambda_1 = 0.04023:$$

Equation (h) becomes

$$\begin{bmatrix} 0.1119 & -0.03975 & 0.012 \\ -0.03975 & 0.03154 & 0.04057 \\ 0.012 & 0.04057 & 0.1166 \end{bmatrix} \begin{Bmatrix} v_x^{(1)} \\ v_y^{(1)} \\ v_z^{(1)} \end{Bmatrix} = \begin{Bmatrix} 0 \\ 0 \\ 0 \end{Bmatrix}$$

We can arbitrarily set $v_x^{(1)} = 1$, so that the first two equations become

$$-0.03975 v_y^{(1)} + 0.012 v_z^{(1)} = -0.1119$$

$$0.03154 v_y^{(1)} + 0.04057 v_z^{(1)} = 0.03975$$

Solving for $v_y^{(1)}$ and $v_z^{(1)}$ yields $v_y^{(1)} = 2.520$ and $v_z^{(1)} = -0.9794$, so that

$$\mathbf{v}^{(1)} = \hat{\mathbf{i}} + 2.520\hat{\mathbf{j}} - 0.9794\hat{\mathbf{k}} \qquad \|\mathbf{v}^{(1)}\| = 2.883$$

Normalizing this vector and calling it $\hat{\mathbf{i}}_1$, we get

$$\hat{\mathbf{i}}_1 = \frac{\mathbf{v}^{(1)}}{\|\mathbf{v}^{(1)}\|} = 0.3470\hat{\mathbf{i}} + 0.8742\hat{\mathbf{j}} - 0.3397\hat{\mathbf{k}} \quad (\text{i})$$

$$\lambda_2 = 0.1658:$$

Equation (h) becomes

$$\begin{bmatrix} -0.0137 & -0.03975 & 0.012 \\ -0.03975 & -0.09408 & 0.04057 \\ 0.012 & 0.04057 & -0.008968 \end{bmatrix} \begin{Bmatrix} v_x^{(2)} \\ v_y^{(2)} \\ v_z^{(2)} \end{Bmatrix} = \begin{Bmatrix} 0 \\ 0 \\ 0 \end{Bmatrix}$$

Repeating the above procedure, we obtain $\mathbf{v}^{(2)} = \hat{\mathbf{i}} - 0.1625\hat{\mathbf{j}} + 0.6030\hat{\mathbf{k}}$, so that

$$\hat{\mathbf{i}}_2 = \frac{\mathbf{v}^{(2)}}{\|\mathbf{v}^{(2)}\|} = 0.8482\hat{\mathbf{i}} - 0.1378\hat{\mathbf{j}} + 0.5115\hat{\mathbf{k}}$$

$$\lambda_3 = 0.1747:$$

The third principal vector $\hat{\mathbf{i}}_3$ is the cross product of the first two:

$$\hat{\mathbf{i}}_3 = \hat{\mathbf{i}}_1 \times \hat{\mathbf{i}}_2 = 0.4003\hat{\mathbf{i}} - 0.4656\hat{\mathbf{j}} - 0.7893\hat{\mathbf{k}}$$

Check this result to see that it satisfies Equation (h):

$$\begin{bmatrix} 0.1522 - 0.1747 & -0.03975 & 0.012 \\ -0.03975 & 0.07177 - 0.1747 & 0.04057 \\ 0.012 & 0.04057 & 0.1569 - 0.1747 \end{bmatrix} \begin{Bmatrix} 0.4003 \\ -0.4656 \\ -0.7893 \end{Bmatrix} \stackrel{\checkmark}{=} \begin{Bmatrix} 0 \\ 0 \\ 0 \end{Bmatrix}$$

9.6 Euler's equations

For either the center of mass G or a fixed point P about which the body is in pure rotation, we know from Equations 9.29 and 9.30 that

$$\mathbf{M}_{\text{net}} = \dot{\mathbf{H}} \tag{9.64}$$

Using a co-moving coordinate system, with angular velocity $\mathbf{\Omega}$ and its origin located at the point (G or P), the angular momentum has the analytical expression

$$\mathbf{H} = H_x\hat{\mathbf{i}} + H_y\hat{\mathbf{j}} + H_z\hat{\mathbf{k}} \tag{9.65}$$

We shall henceforth assume, for simplicity, that

(a) The moving xyz axes are the principal axes of inertia, and (9.66a)
(b) The moments of inertia relative to xyz are constant in time. (9.66b)

Equations 9.42 and 9.66a imply that

$$\mathbf{H} = A\omega_x\hat{\mathbf{i}} + B\omega_y\hat{\mathbf{j}} + C\omega_z\hat{\mathbf{k}} \tag{9.67}$$

where A, B and C are the principal moments of inertia.

According to Equation 1.28, the time derivative of **H** is $\dot{\mathbf{H}} = \dot{\mathbf{H}})_{\text{rel}} + \mathbf{\Omega} \times \mathbf{H}$, so that Equation 9.64 can be written

$$\mathbf{M}_{\text{net}} = \dot{\mathbf{H}})_{\text{rel}} + \mathbf{\Omega} \times \mathbf{H} \tag{9.68}$$

Keep in mind that, whereas $\mathbf{\Omega}$ (the angular velocity of the moving *xyz* coordinate system) and $\boldsymbol{\omega}$ (the angular velocity of the rigid body itself) are both absolute kinematic quantities, Equation 9.68 contains their components as projected onto the axes of the non-inertial *xyz* frame,

$$\boldsymbol{\omega} = \omega_x \hat{\mathbf{i}} + \omega_y \hat{\mathbf{j}} + \omega_z \hat{\mathbf{k}}$$

$$\mathbf{\Omega} = \Omega_x \hat{\mathbf{i}} + \Omega_y \hat{\mathbf{j}} + \Omega_z \hat{\mathbf{k}}$$

The absolute angular acceleration $\boldsymbol{\alpha}$ is obtained using Equation 1.28,

$$\boldsymbol{\alpha} = \dot{\boldsymbol{\omega}} = \overbrace{\frac{d\omega_x}{dt}\hat{\mathbf{i}} + \frac{d\omega_y}{dt}\hat{\mathbf{j}} + \frac{d\omega_z}{dt}\hat{\mathbf{k}}}^{\boldsymbol{\alpha}_{\text{rel}}} + \mathbf{\Omega} \times \boldsymbol{\omega}$$

that is,

$$\boldsymbol{\alpha} = (\dot{\omega}_x + \Omega_y \omega_z - \Omega_z \omega_y)\hat{\mathbf{i}} + (\dot{\omega}_y + \Omega_z \omega_x - \Omega_x \omega_z)\hat{\mathbf{j}} + (\dot{\omega}_z + \Omega_x \omega_y - \Omega_y \omega_x)\hat{\mathbf{k}} \tag{9.69}$$

Clearly, it is generally true that

$$\alpha_x \neq \dot{\omega}_x \qquad \alpha_y \neq \dot{\omega}_y \qquad \alpha_z \neq \dot{\omega}_z$$

From Equations 1.29 and 9.67

$$\dot{\mathbf{H}})_{\text{rel}} = \frac{d(A\omega_x)}{dt}\hat{\mathbf{i}} + \frac{d(B\omega_y)}{dt}\hat{\mathbf{j}} + \frac{d(C\omega_z)}{dt}\hat{\mathbf{k}}$$

Since *A*, *B* and *C* are constant, this becomes

$$\dot{\mathbf{H}})_{\text{rel}} = A\dot{\omega}_x \hat{\mathbf{i}} + B\dot{\omega}_y \hat{\mathbf{j}} + C\dot{\omega}_z \hat{\mathbf{k}} \tag{9.70}$$

Substituting Equations 9.67 and 9.70 into Equation 9.68 yields

$$\mathbf{M}_{\text{net}} = A\dot{\omega}_x \hat{\mathbf{i}} + B\dot{\omega}_y \hat{\mathbf{j}} + C\dot{\omega}_z \hat{\mathbf{k}} + \begin{vmatrix} \hat{\mathbf{i}} & \hat{\mathbf{j}} & \hat{\mathbf{k}} \\ \Omega_x & \Omega_y & \Omega_z \\ A\omega_x & B\omega_y & C\omega_z \end{vmatrix}$$

Expanding the cross product and collecting terms leads to

$$M_{x_{\text{net}}} = A\dot{\omega}_x + C\Omega_y \omega_z - B\Omega_z \omega_y$$
$$M_{y_{\text{net}}} = B\dot{\omega}_y + A\Omega_z \omega_x - C\Omega_x \omega_z \tag{9.71}$$
$$M_{z_{\text{net}}} = C\dot{\omega}_z + B\Omega_x \omega_y - A\Omega_y \omega_x$$

If the co-moving frame is a rigidly attached body frame, then its angular velocity is the same as that of the body, i.e., $\mathbf{\Omega} = \boldsymbol{\omega}$. In that case, Equations 9.68 reduce to Euler's equations of motion,

$$\mathbf{M}_{\text{net}} = \dot{\mathbf{H}})_{\text{rel}} + \boldsymbol{\omega} \times \mathbf{H} \tag{9.72a}$$

the three components of which are obtained from Equation 9.71,

$$M_{x_{net}} = A\dot{\omega}_x + (C - B)\omega_y\omega_z$$
$$M_{y_{net}} = B\dot{\omega}_y + (A - C)\omega_z\omega_x \quad (9.72b)$$
$$M_{z_{net}} = C\dot{\omega}_z + (B - A)\omega_x\omega_y$$

Equation 9.68 is sometimes referred to as the modified Euler equation.

When $\mathbf{\Omega} = \boldsymbol{\omega}$, it follows from Equation 9.69 that

$$\dot{\omega}_x = \alpha_x \quad \dot{\omega}_y = \alpha_y \quad \dot{\omega}_z = \alpha_z \quad (9.73)$$

That is, the relative angular acceleration equals the absolute angular acceleration when $\mathbf{\Omega} = \boldsymbol{\omega}$. Rather than calculating the time derivatives $\dot{\omega}_x, \dot{\omega}_y$ and $\dot{\omega}_z$ for use in Equation 9.72, we may in this case first compute $\boldsymbol{\alpha}$ in the absolute XYZ frame

$$\boldsymbol{\alpha} = \frac{d\boldsymbol{\omega}}{dt} = \frac{d\omega_X}{dt}\hat{\mathbf{I}} + \frac{d\omega_Y}{dt}\hat{\mathbf{J}} + \frac{d\omega_Z}{dt}\hat{\mathbf{K}},$$

and then project these components onto the xyz body frame, so that

$$\begin{Bmatrix} \dot{\omega}_x \\ \dot{\omega}_y \\ \dot{\omega}_z \end{Bmatrix} = [\mathbf{Q}]_{Xx} \begin{Bmatrix} d\omega_X/dt \\ d\omega_Y/dt \\ d\omega_Z/dt \end{Bmatrix} \quad (9.74)$$

where $[\mathbf{Q}]_{Xx}$ is the time-dependent orthogonal transformation from the inertial XYZ frame to the non-inertial xyz frame.

EXAMPLE 9.12

Calculate the net moment on the solar panel of Examples 9.2 and 9.8.

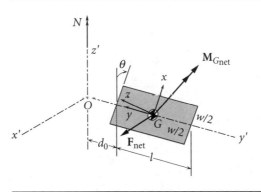

Figure 9.16 Free-body diagram of the solar panel in Examples 9.2 and 9.8.

Since the co-moving frame is rigidly attached to the panel, Euler's equation (Equation 9.72) applies to this problem,

$$\mathbf{M}_{G_{net}} = \left.\dot{\mathbf{H}}_G\right)_{rel} + \boldsymbol{\omega} \times \mathbf{H}_G \quad (a)$$

where

$$\mathbf{H}_G = A\omega_x\hat{\mathbf{i}} + B\omega_y\hat{\mathbf{j}} + C\omega_z\hat{\mathbf{k}} \quad (b)$$

(Example 9.12 continued)

and

$$\dot{\mathbf{H}}_G)_{\text{rel}} = A\dot{\omega}_x\hat{\mathbf{i}} + B\dot{\omega}_y\hat{\mathbf{j}} + C\dot{\omega}_z\hat{\mathbf{k}} \qquad (c)$$

In Example 9.2, the angular velocity of the panel in the satellite's $x'y'z'$ frame was found to be

$$\boldsymbol{\omega} = -\dot{\theta}\hat{\mathbf{j}}' + N\hat{\mathbf{k}}' \qquad (d)$$

In Example 9.8, the transformation from the panel's xyz frame to that of the satellite was shown to be represented by the matrix

$$[\mathbf{Q}] = \begin{bmatrix} -\sin\theta & 0 & \cos\theta \\ 0 & -1 & 0 \\ \cos\theta & 0 & \sin\theta \end{bmatrix} \qquad (e)$$

We use the transpose of $[\mathbf{Q}]$ to transform the components of $\boldsymbol{\omega}$ into the panel frame of reference,

$$\{\boldsymbol{\omega}\}_{xyz} = [\mathbf{Q}]^T\{\boldsymbol{\omega}\}_{x'y'z'} = \begin{bmatrix} -\sin\theta & 0 & \cos\theta \\ 0 & -1 & 0 \\ \cos\theta & 0 & \sin\theta \end{bmatrix} \begin{Bmatrix} 0 \\ -\dot{\theta} \\ N \end{Bmatrix} = \begin{Bmatrix} N\cos\theta \\ \dot{\theta} \\ N\sin\theta \end{Bmatrix}$$

or

$$\omega_x = N\cos\theta \qquad \omega_y = \dot{\theta} \qquad \omega_z = N\sin\theta \qquad (f)$$

In Example 9.2, N and $\dot{\theta}$ were said to be constant. Therefore, the time derivatives of (f) are

$$\dot{\omega}_x = \frac{d(N\cos\theta)}{dt} = -N\dot{\theta}\sin\theta$$

$$\dot{\omega}_y = \frac{d\dot{\theta}}{dt} = 0 \qquad (g)$$

$$\dot{\omega}_z = \frac{d(N\sin\theta)}{dt} = N\dot{\theta}\cos\theta$$

In Example 9.8 the moments of inertia in the panel frame of reference were listed as

$$A = \frac{1}{12}m(l^2 + t^2) \qquad B = \frac{1}{12}m(w^2 + t^2) \qquad C = \frac{1}{12}m(w^2 + l^2)$$

$$(I_{G_{xy}} = I_{G_{xz}} = I_{G_{yz}} = 0) \qquad (h)$$

Substituting (b), (c), (f), (g) and (h) into (a) yields,

$$\mathbf{M}_{G_{\text{net}}} = \tfrac{1}{12}m(l^2 + t^2)(-N\dot{\theta}\sin\theta)\hat{\mathbf{i}} + \tfrac{1}{12}m(w^2 + t^2)\cdot 0\cdot\hat{\mathbf{j}}$$
$$+ \tfrac{1}{12}m(w^2 + l^2)(N\dot{\theta}\cos\theta)\hat{\mathbf{k}}$$

$$+ \begin{vmatrix} \hat{\mathbf{i}} & \hat{\mathbf{j}} & \hat{\mathbf{k}} \\ N\cos\theta & \dot{\theta} & N\sin\theta \\ \tfrac{1}{12}m(l^2+t^2)(N\cos\theta) & \tfrac{1}{12}m(w^2+t^2)\dot{\theta} & \tfrac{1}{12}m(w^2+l^2)(N\sin\theta) \end{vmatrix}$$

Upon expanding the cross product and collecting terms, this reduces to

$$\mathbf{M}_{G_{net}} = -\tfrac{1}{6}mt^2 N\dot{\theta}\sin\theta\hat{\mathbf{i}} + \tfrac{1}{24}m(t^2-w^2)N^2\sin 2\theta\hat{\mathbf{j}} + \tfrac{1}{6}mw^2 N\dot{\theta}\cos\theta\hat{\mathbf{k}}$$

Using the numerical data of Example 9.8 ($m = 50$ kg, $N = 0.1$ rad/s, $\theta = 40°$, $\dot{\theta} = 0.01$ rad/s, $l = 6$ m, $w = 2$ m and $t = 0.025$ m), we get

$$\mathbf{M}_{G_{net}} = -3.348 \times 10^{-6}\hat{\mathbf{i}} - 0.08205\hat{\mathbf{j}} + 0.02554\hat{\mathbf{k}} \; (\text{N} \cdot \text{m})$$

EXAMPLE 9.13

Calculate the net moment on the gyro rotor of Examples 9.3 and 9.6.
Figure 9.17 is a free-body diagram of the rotor. Since in this case the co-moving frame is not rigidly attached to the rotor, we must use Equation 9.68 to find the net moment about G,

$$\mathbf{M}_{G_{net}} = \dot{\mathbf{H}}_G)_{rel} + \mathbf{\Omega} \times \mathbf{H}_G \tag{a}$$

where

$$\mathbf{H}_G = A\omega_x\hat{\mathbf{i}} + B\omega_y\hat{\mathbf{j}} + C\omega_z\hat{\mathbf{k}} \tag{b}$$

and

$$\dot{\mathbf{H}}_G)_{rel} = A\dot{\omega}_x\hat{\mathbf{i}} + B\dot{\omega}_y\hat{\mathbf{j}} + C\dot{\omega}_z\hat{\mathbf{k}} \tag{c}$$

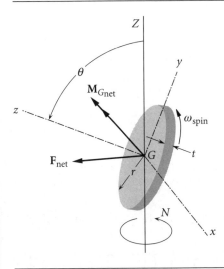

Figure 9.17 Free-body diagram of the gyro rotor of Examples 9.3 and 9.6.

From Equation (h) of Example 9.3 we know that the components of the angular velocity of the rotor in the moving reference frame are

$$\omega_x = \dot{\theta}$$
$$\omega_y = N\sin\theta \tag{d}$$
$$\omega_z = \omega_{spin} + N\cos\theta$$

(Example 9.13 continued)

Since, as specified in Example 9.3, $\dot\theta$, N and ω_{spin} are all constant, it follows that

$$\dot\omega_x = \frac{d\dot\theta}{dt} = 0$$
$$\dot\omega_y = \frac{d(N\sin\theta)}{dt} = N\dot\theta\cos\theta \qquad (e)$$
$$\dot\omega_z = \frac{d(\omega_{spin} + N\cos\theta)}{dt} = -N\dot\theta\sin\theta$$

The angular velocity $\boldsymbol{\Omega}$ of the co-moving *xyz* frame is that of the gimbal ring, which equals the angular velocity of the rotor minus its spin. Therefore,

$$\Omega_x = \dot\theta$$
$$\Omega_y = N\sin\theta \qquad (f)$$
$$\Omega_z = N\cos\theta$$

In Example 9.6 we found that

$$A = B = \frac{1}{12}mt^2 + \frac{1}{4}mr^2$$
$$C = \frac{1}{2}mr^2 \qquad (g)$$

Substituting (b) through (g) into (a), we get

$$\mathbf{M}_{G_{net}} = \left(\tfrac{1}{12}mt^2 + \tfrac{1}{4}mr^2\right)\cdot 0\hat{\mathbf{i}} + \left(\tfrac{1}{12}mt^2 + \tfrac{1}{4}mr^2\right)(N\dot\theta\cos\theta)\hat{\mathbf{j}}$$
$$+ \tfrac{1}{2}mr^2(-N\dot\theta\sin\theta)\hat{\mathbf{k}}$$

$$+ \begin{vmatrix} \hat{\mathbf{i}} & \hat{\mathbf{j}} & \hat{\mathbf{k}} \\ \dot\theta & N\sin\theta & N\cos\theta \\ \left(\tfrac{1}{12}mt^2 + \tfrac{1}{4}mr^2\right)\dot\theta & \left(\tfrac{1}{12}mt^2 + \tfrac{1}{4}mr^2\right)N\sin\theta & \tfrac{1}{2}mr^2(\omega_{spin} + N\cos\theta) \end{vmatrix}$$

Expanding the cross product, collecting terms, and simplifying leads to

$$\mathbf{M}_{G_{net}} = \left[\frac{1}{2}\omega_{spin} + \frac{1}{12}\left(3 - \frac{t^2}{r^2}\right)N\cos\theta\right]mr^2 N\sin\theta\,\hat{\mathbf{i}}$$
$$+ \left(\frac{1}{6}\frac{t^2}{r^2}N\cos\theta - \frac{1}{2}\omega_{spin}\right)mr^2\dot\theta\,\hat{\mathbf{j}} - \frac{1}{2}N\dot\theta\sin\theta\, mr^2\,\hat{\mathbf{k}} \qquad (h)$$

In Example 9.3 the following numerical data was provided: $m = 5$ kg, $r = 0.08$ m, $t = 0.025$ m, $N = 2.1$ rad/s, $\theta = 60°$, $\dot\theta = 4$ rad/s and $\omega_{spin} = 105$ rad/s. For this set of numbers, (h) becomes

$$\mathbf{M}_{G_{net}} = 0.3203\hat{\mathbf{i}} - 0.6698\hat{\mathbf{j}} - 0.1164\hat{\mathbf{k}} \text{ (N}\cdot\text{m)}$$

9.7 Kinetic energy

The kinetic energy T of a rigid body is the integral of the kinetic energy $\frac{1}{2}v^2 dm$ of its individual mass elements,

$$T = \int_m \frac{1}{2}v^2 dm = \int_m \frac{1}{2}\mathbf{v}\cdot\mathbf{v}\,dm \qquad (9.75)$$

where $\mathbf{v}$ is the absolute velocity $\dot{\mathbf{R}}$ of the element of mass dm. From Figure 9.7 we infer that $\dot{\mathbf{R}} = \dot{\mathbf{R}}_G + \dot{\boldsymbol{\varrho}}$. Furthermore, Equation 1.24 requires that $\dot{\boldsymbol{\varrho}} = \boldsymbol{\omega}\times\boldsymbol{\varrho}$. Thus, $\mathbf{v} = \mathbf{v}_G + \boldsymbol{\omega}\times\boldsymbol{\varrho}$, which means

$$\mathbf{v}\cdot\mathbf{v} = [\mathbf{v}_G + \boldsymbol{\omega}\times\boldsymbol{\varrho}]\cdot[\mathbf{v}_G + \boldsymbol{\omega}\times\boldsymbol{\varrho}] = v_G^2 + 2\mathbf{v}_G\cdot(\boldsymbol{\omega}\times\boldsymbol{\varrho}) + (\boldsymbol{\omega}\times\boldsymbol{\varrho})\cdot(\boldsymbol{\omega}\times\boldsymbol{\varrho})$$

We can apply the vector identity introduced in Equation 2.32,

$$\mathbf{A}\cdot(\mathbf{B}\times\mathbf{C}) = \mathbf{B}\cdot(\mathbf{C}\times\mathbf{A}) \qquad (9.76)$$

to the last term to get

$$\mathbf{v}\cdot\mathbf{v} = v_G^2 + 2\mathbf{v}_G\cdot(\boldsymbol{\omega}\times\boldsymbol{\varrho}) + \boldsymbol{\omega}\cdot[\boldsymbol{\varrho}\times(\boldsymbol{\omega}\times\boldsymbol{\varrho})]$$

Therefore, Equation 9.75 becomes

$$T = \int_m \frac{1}{2}v_G^2\,dm + \mathbf{v}_G\cdot\left(\boldsymbol{\omega}\times\int_m \boldsymbol{\varrho}\,dm\right) + \frac{1}{2}\boldsymbol{\omega}\cdot\int_m \boldsymbol{\varrho}\times(\boldsymbol{\omega}\times\boldsymbol{\varrho})\,dm$$

Since $\boldsymbol{\varrho}$ is measured from the center of mass, $\int_m \boldsymbol{\varrho}\,dm = 0$. Recall that, according to Equation 9.34,

$$\int_m \boldsymbol{\varrho}\times(\boldsymbol{\omega}\times\boldsymbol{\varrho})\,dm = \mathbf{H}_G$$

It follows that the kinetic energy may be written

$$T = \frac{1}{2}mv_G^2 + \frac{1}{2}\boldsymbol{\omega}\cdot\mathbf{H}_G \qquad (9.77)$$

The second term is the rotational kinetic energy T_R,

$$T_R = \frac{1}{2}\boldsymbol{\omega}\cdot\mathbf{H}_G \qquad (9.78)$$

If the body is rotating about a point P which is at rest in inertial space, we have from Equation 9.2 and Figure 9.7 that

$$\mathbf{v}_G = \mathbf{v}_P + \boldsymbol{\omega}\times\mathbf{r}_{G/P} = 0 + \boldsymbol{\omega}\times\mathbf{r}_{G/P} = \boldsymbol{\omega}\times\mathbf{r}_{G/P}$$

It follows that

$$v_G^2 = \mathbf{v}_G\cdot\mathbf{v}_G = (\boldsymbol{\omega}\times\mathbf{r}_{G/P})\cdot(\boldsymbol{\omega}\times\mathbf{r}_{G/P})$$

Making use once again of the vector identity in Equation 9.76, we find

$$v_G^2 = \boldsymbol{\omega} \cdot [\mathbf{r}_{G/P} \times (\boldsymbol{\omega} \times \mathbf{r}_{G/P})] = \boldsymbol{\omega} \cdot (\mathbf{r}_{G/P} \times \mathbf{v}_G)$$

Substituting this into Equation 9.77 yields

$$T = \frac{1}{2}\boldsymbol{\omega} \cdot [\mathbf{H}_G + \mathbf{r}_{G/P} \times m\mathbf{v}_G]$$

Equation 9.21 shows that this can be written

$$T = \frac{1}{2}\boldsymbol{\omega} \cdot \mathbf{H}_P \qquad (9.79)$$

In this case, of course, all of the kinetic energy is rotational.

In terms of the components of $\boldsymbol{\omega}$ and $\mathbf{H}$, whether it is $\mathbf{H}_P$ or $\mathbf{H}_G$, the rotational kinetic energy expression becomes, with the aid of Equation 9.39,

$$T_R = \frac{1}{2}(\omega_x H_x + \omega_y H_y + \omega_z H_z)$$

$$= \frac{1}{2}\lfloor \omega_x \quad \omega_y \quad \omega_z \rfloor \begin{bmatrix} I_x & I_{xy} & I_{xz} \\ I_{xy} & I_y & I_{yz} \\ I_{xz} & I_{yz} & I_z \end{bmatrix} \begin{Bmatrix} \omega_x \\ \omega_y \\ \omega_z \end{Bmatrix}$$

Expanding, we obtain

$$T_R = \frac{1}{2}I_x\omega_x^2 + \frac{1}{2}I_y\omega_y^2 + \frac{1}{2}I_z\omega_z^2 + I_{xy}\omega_x\omega_y + I_{xz}\omega_x\omega_z + I_{yz}\omega_y\omega_z \qquad (9.80)$$

Obviously, if the *xyz* axes are principal axes of inertia, then Equation 9.80 simplifies considerably,

$$T_R = \frac{1}{2}A\omega_x^2 + \frac{1}{2}B\omega_y^2 + \frac{1}{2}C\omega_z^2 \qquad (9.81)$$

EXAMPLE 9.14

A satellite in circular geocentric orbit of 300 km altitude has a mass of 1500 kg, and the moments of inertia relative to a body frame with origin at the center of mass *G* are

$$[\mathbf{I}] = \begin{bmatrix} 2000 & -1000 & 2500 \\ -1500 & 3000 & -1500 \\ 2500 & -1500 & 4000 \end{bmatrix} (\text{kg} \cdot \text{m}^2)$$

If at a given instant the components of angular velocity in this frame of reference are

$$\boldsymbol{\omega} = 1\hat{\mathbf{i}} - 0.9\hat{\mathbf{j}} + 1.5\hat{\mathbf{k}} \text{ (rad/s)}$$

calculate the total kinetic energy of the satellite.

The speed of the satellite in its circular orbit is

$$v = \sqrt{\frac{\mu}{r}} = \sqrt{\frac{398\,600}{6378 + 300}} = 7.7258 \text{ km/s}$$

The angular momentum of the satellite is

$$\{\mathbf{H}_G\} = [\mathbf{I}_G]\{\boldsymbol{\omega}\} = \begin{bmatrix} 2000 & -1000 & 2500 \\ -1500 & 3000 & -1500 \\ 2500 & -1500 & 4000 \end{bmatrix} \begin{Bmatrix} 1 \\ -0.9 \\ 1.5 \end{Bmatrix} = \begin{Bmatrix} 6650 \\ -5950 \\ 9850 \end{Bmatrix} \text{ (kg} \cdot \text{m}^2/\text{s)}$$

Therefore, the total kinetic energy is

$$T = \frac{1}{2}mv_G^2 + \frac{1}{2}\boldsymbol{\omega} \cdot \mathbf{H}_G = \frac{1}{2} \cdot 1500 \cdot 7725.8^2 + \frac{1}{2}[1 \ -0.9 \ 1.5] \begin{Bmatrix} 6650 \\ -5950 \\ 9850 \end{Bmatrix}$$

$$= 44.766 \times 10^6 + 13\,390$$

$$\underline{T = 44.766 \text{ MJ}}$$

Obviously, the kinetic energy is dominated by that due to the orbital motion.

9.8 THE SPINNING TOP

Let us analyze the motion of the simple axisymmetric top in Figure 9.18. It is constrained to rotate about point O.

The moving coordinate system is chosen to have its origin at O. The z axis is aligned with the spin axis of the top (the axis of rotational symmetry). The x axis is the node line, which passes through O and is perpendicular to the plane defined

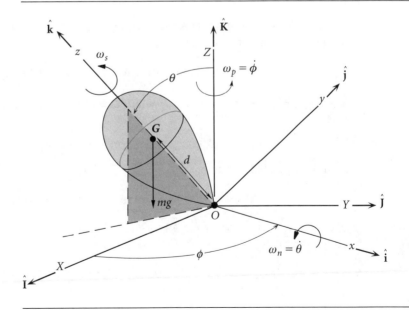

Figure 9.18 Simple top rotating about the fixed point O.

by the inertial Z axis and the spin axis of the top. The y axis is then perpendicular to x and z, such that $\hat{\mathbf{j}} = \hat{\mathbf{k}} \times \hat{\mathbf{i}}$. By symmetry, the moment of inertia matrix of the top relative to the xyz frame is diagonal, with $I_x = I_y = A$ and $I_z = C$. It is likely that $A > C$. From Equations 9.68 and 9.70, we have

$$\mathbf{M}_{O_{net}} = A\dot{\omega}_x\hat{\mathbf{i}} + A\dot{\omega}_y\hat{\mathbf{j}} + C\dot{\omega}_z\hat{\mathbf{k}} + \begin{vmatrix} \hat{\mathbf{i}} & \hat{\mathbf{j}} & \hat{\mathbf{k}} \\ \Omega_x & \Omega_y & \Omega_z \\ A\omega_x & A\omega_y & C\omega_z \end{vmatrix} \quad (9.82)$$

The angular velocity $\boldsymbol{\omega}$ of the top is the vector sum of the spin rate ω_s and the rates of precession ω_p and nutation ω_n, where

$$\omega_p = \dot{\phi} \qquad \omega_n = \dot{\theta} \quad (9.83)$$

Thus

$$\boldsymbol{\omega} = \omega_n\hat{\mathbf{i}} + \omega_p\hat{\mathbf{K}} + \omega_s\hat{\mathbf{k}}$$

From the geometry it follows that

$$\hat{\mathbf{K}} = \sin\theta\hat{\mathbf{j}} + \cos\theta\hat{\mathbf{k}} \quad (9.84)$$

Therefore, relative to the co-moving system,

$$\boldsymbol{\omega} = \omega_n\hat{\mathbf{i}} + \omega_p\sin\theta\hat{\mathbf{j}} + (\omega_s + \omega_p\cos\theta)\hat{\mathbf{k}} \quad (9.85)$$

From Equation 9.85 we see that

$$\omega_x = \omega_n \qquad \omega_y = \omega_p\sin\theta \qquad \omega_z = \omega_s + \omega_p\cos\theta \quad (9.86)$$

Computing the time rates of these three expressions yields the components of angular acceleration relative to the xyz frame,

$$\dot{\omega}_x = \dot{\omega}_n \quad \dot{\omega}_y = \dot{\omega}_p\sin\theta + \omega_p\omega_n\cos\theta \quad \dot{\omega}_z = \dot{\omega}_s + \dot{\omega}_p\cos\theta - \omega_p\omega_n\sin\theta \quad (9.87)$$

The angular velocity $\boldsymbol{\Omega}$ of the xyz system is $\boldsymbol{\Omega} = \omega_p\hat{\mathbf{K}} + \omega_n\hat{\mathbf{i}}$, so that, using Equation 9.84,

$$\boldsymbol{\Omega} = \omega_n\hat{\mathbf{i}} + \omega_p\sin\theta\hat{\mathbf{j}} + \omega_p\cos\theta\hat{\mathbf{k}} \quad (9.88)$$

From Equation 9.88 we obtain

$$\Omega_x = \omega_n \qquad \Omega_y = \omega_p\sin\theta \qquad \Omega_z = \omega_p\cos\theta \quad (9.89)$$

The moment about O in Figure 9.18 is that of the weight vector acting through the center of mass G:

$$\mathbf{M}_{O_{net}} = (d\hat{\mathbf{k}}) \times (-mg\hat{\mathbf{K}}) = -mgd\hat{\mathbf{k}} \times (\sin\theta\hat{\mathbf{j}} + \cos\theta\hat{\mathbf{k}})$$

or

$$\mathbf{M}_{O_{net}} = mgd\sin\theta\hat{\mathbf{i}} \quad (9.90)$$

Substituting Equations 9.86, 9.87, 9.89 and 9.90 into Equation 9.82, we get

$$mgd \sin\theta \hat{\mathbf{i}} = A\dot{\omega}_n \hat{\mathbf{i}} + A(\dot{\omega}_p \sin\theta + \omega_p \omega_n \cos\theta)\hat{\mathbf{j}} + C(\dot{\omega}_s + \dot{\omega}_p \cos\theta - \omega_p \omega_n \sin\theta)\hat{\mathbf{k}}$$

$$+ \begin{vmatrix} \hat{\mathbf{i}} & \hat{\mathbf{j}} & \hat{\mathbf{k}} \\ \omega_n & \omega_p \sin\theta & \omega_p \cos\theta \\ A\omega_n & A\omega_p \sin\theta & C(\omega_s + \omega_p \cos\theta) \end{vmatrix} \quad (9.91)$$

Let us consider the special case in which $\theta =$ constant, i.e., there is no nutation, so that $\omega_n = \dot{\omega}_n = 0$. Then Equation 9.91 reduces to

$$mgd \sin\theta \hat{\mathbf{i}} = A\dot{\omega}_p \sin\theta \hat{\mathbf{j}} + C(\dot{\omega}_s + \dot{\omega}_p \cos\theta)\hat{\mathbf{k}}$$

$$+ \begin{vmatrix} \hat{\mathbf{i}} & \hat{\mathbf{j}} & \hat{\mathbf{k}} \\ 0 & \omega_p \sin\theta & \omega_p \cos\theta \\ 0 & A\omega_p \sin\theta & C(\omega_s + \omega_p \cos\theta) \end{vmatrix} \quad (9.92)$$

Expanding the determinant yields

$$mgd \sin\theta \hat{\mathbf{i}} = A\dot{\omega}_p \sin\theta \hat{\mathbf{j}} + C(\dot{\omega}_s + \dot{\omega}_p \cos\theta)\hat{\mathbf{k}}$$
$$+ [C\omega_p \omega_s \sin\theta + (C - A)\omega_p^2 \cos\theta \sin\theta]\hat{\mathbf{i}}$$

Equating the coefficients of $\hat{\mathbf{i}}, \hat{\mathbf{j}}$ and $\hat{\mathbf{k}}$ on each side of the equation leads to

$$mgd \sin\theta = C\omega_p \omega_s \sin\theta + (C - A)\omega_p^2 \cos\theta \sin\theta \quad (9.93a)$$

$$0 = A\dot{\omega}_p \sin\theta \quad (9.93b)$$

$$0 = C(\dot{\omega}_s + \dot{\omega}_p \cos\theta) \quad (9.93c)$$

Equation 9.93b implies $\dot{\omega}_p = 0$, and from Equation 9.93c it follows that $\dot{\omega}_s = 0$. Therefore, the rates of spin and precession are both constant. From Equation 9.93a we find

$$(A - C)\cos\theta \omega_p^2 - C\omega_s \omega_p + mgd = 0 \quad (0 < \theta < 180°) \quad (9.94)$$

If the spin rate is zero, Equation 9.94 yields

$$\omega_p)_{\omega_s=0} = \pm\sqrt{\frac{mgd}{(C - A)\cos\theta}} \quad \text{if } (A - C)\cos\theta < 0 \quad (9.95)$$

In this case, the top rotates about O at this rate, without spinning. If $A > C$ (prolate), its symmetry axis must make an angle between 90° and 180° to the vertical; otherwise ω_p is imaginary. On the other hand, if $A < C$ (oblate), the angle lies between 0° and 90°. Thus, in steady rotation without spin, the top's axis sweeps out a cone which lies either below the horizontal plane $(A > C)$ or above the plane $(A < C)$.

In the special case $(A - C)\cos\theta = 0$, Equation 9.94 yields a steady precession rate which is inversely proportional to the spin rate,

$$\omega_p = \frac{mgd}{C\omega_s} \quad \text{if } (A - C)\cos\theta = 0 \quad (9.96)$$

If $A = C$, this precession apparently occurs irrespective of tilt angle θ. If $A \neq C$, this rate of precession occurs at $\theta = 90°$, i.e., the spin axis is perpendicular to the precession axis.

In general, Equation 9.94 is a quadratic equation in ω_p, so we can use the quadratic formula to find

$$\omega_p = \frac{C}{2(A-C)\cos\theta}\left(\omega_s \pm \sqrt{\omega_s^2 - \frac{4mgd(A-C)\cos\theta}{C^2}}\right) \quad (9.97)$$

Thus, for a given spin rate and tilt angle θ ($\theta \neq 90°$), there are two rates of precession $\dot\phi$.

Observe that if $(A-C)\cos\theta > 0$, then ω_p is imaginary when $\omega_s^2 < 4mgd(A-C)\cos\theta/C^2$. Therefore, the minimum spin rate required for steady precession at a constant inclination θ is

$$(\omega_s)_{min} = \frac{2}{C}\sqrt{mgd(A-C)\cos\theta} \quad \text{if } (A-C)\cos\theta > 0 \quad (9.98)$$

If $(A-C)\cos\theta < 0$, the radical in Equation 9.97 is real for all ω_s. In this case, as $\omega_s \to 0$, ω_p approaches the value given above in Equation 9.95.

EXAMPLE 9.15

For the top of Figure 9.18, let $m = 0.5$ kg, $A(=I_x=I_y) = 12 \times 10^{-4}$ kg·m², $C(=I_z) = 4.5 \times 10^{-4}$ kg·m² and $d = 0.05$ m. For an inclination of, say, 60°, $(A-C)\cos\theta > 0$ so that Equation 9.98 requires $(\omega_s)_{min} = 407.01$ rpm. Let us choose the spin rate to be $\omega_s = 1000$ rpm $= 104.7$ rad/s. Then, from Equation 9.97, the precession rate as a function of the inclination θ is given by either one of the following formulas

$$\omega_p = 31.42\frac{1+\sqrt{1-0.3312\cos\theta}}{\cos\theta} \quad \text{and} \quad \omega_p = 31.42\frac{1-\sqrt{1-0.3312\cos\theta}}{\cos\theta} \quad (a)$$

These are plotted in Figure 9.19.

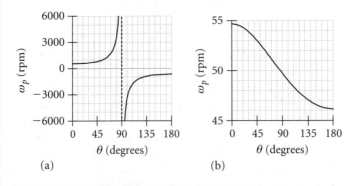

Figure 9.19 (a) High-energy precession rate (unlikely to be observed). (b) Low energy precession rate (the one most always seen).

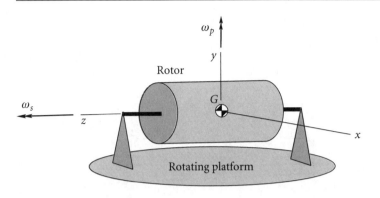

Figure 9.20 A spinning rotor on a rotating platform.

Figure 9.20 shows an axisymmetric rotor mounted so that its spin axis (z) remains perpendicular to the precession axis (y). In that case Equation 9.85 with $\theta = 90°$ yields

$$\boldsymbol{\omega} = \omega_p \hat{\mathbf{j}} + \omega_s \hat{\mathbf{k}} \tag{9.99}$$

Likewise, from Equation 9.88, the angular velocity of the co-moving xyz system is $\boldsymbol{\Omega} = \omega_p \hat{\mathbf{j}}$. If we assume that the spin rate and precession rate are constant ($d\omega_p/dt = d\omega_s/dt = 0$), then Equation 9.68, written for the center of mass G, becomes

$$\mathbf{M}_{G_{net}} = \boldsymbol{\Omega} \times \mathbf{H} = (\omega_p \hat{\mathbf{j}}) \times (A\omega_p \hat{\mathbf{j}} + C\omega_s \hat{\mathbf{k}}) \tag{9.100}$$

where A and C are the moments of inertia of the rotor about the x and z axes, respectively. Setting $C\omega_s \hat{\mathbf{k}} = \mathbf{H}_s$, the spin angular momentum, and $\omega_p \hat{\mathbf{j}} = \boldsymbol{\omega}_p$, we obtain

$$\mathbf{M}_{G_{net}} = \boldsymbol{\omega}_p \times \mathbf{H}_s \quad (\mathbf{H}_s = C\omega_s \hat{\mathbf{k}}) \tag{9.101}$$

This is the gyroscope equation, which is similar to Equation 9.96. Since the center of mass is the reference point, there is no restriction on the motion G for which Equation 9.101 is valid. Observe that the net gyroscopic moment $\mathbf{M}_{G_{net}}$ exerted on the rotor by its supports is perpendicular to the plane of the spin and precession vectors. If a spinning rotor is forced to precess, the gyroscopic moment $\mathbf{M}_{G_{net}}$ develops. Or, if a moment is applied normal to the spin axis of a rotor, it will precess so as to cause the spin axis to turn towards the moment axis.

EXAMPLE 9.16

A uniform cylinder of radius r, length L and mass m spins at a constant angular velocity ω_p. It rests on simple supports, mounted on a platform which rotates at an angular velocity of ω_p. Find the reactions at A and B. Neglect the weight (i.e., calculate the reactions due just to gyroscopic effects).

(Example 9.16 continued)

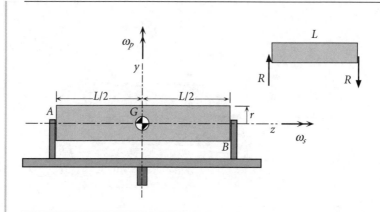

Figure 9.21 Illustration of the gyroscopic effect.

The net vertical force on the cylinder is zero, so the reactions at each end are equal and opposite in direction, as shown on the free-body diagram insert in Figure 9.21. Noting that the moment of inertia of a uniform cylinder about its axis of rotational symmetry is $\frac{1}{2}mr^2$, Equation 9.101 yields

$$RL\hat{\mathbf{i}} = (\omega_p\hat{\mathbf{j}}) \times \left(\frac{1}{2}mr^2\omega_s\hat{\mathbf{k}}\right) = \frac{1}{2}mr^2\omega_p\omega_s\hat{\mathbf{i}}$$

so that,

$$R = \frac{mr^2\omega_p\omega_s}{2L}$$

9.9 EULER ANGLES

Three angles are required to specify the orientation of a rigid body relative to an inertial frame. The choice is not unique, but there are two sets in common use: the Euler angles and the yaw, pitch and roll angles. We will discuss each of them in turn.

The three Euler angles give the orientation of a rigid, orthogonal *xyz* frame of reference relative to the *XYZ* inertial frame of reference. The orthogonal triad of unit vectors parallel to the inertial axes *XYZ* are $\hat{\mathbf{I}}, \hat{\mathbf{J}}$ and $\hat{\mathbf{K}}$, respectively. The orthogonal triad of unit vectors lying along the axes of the *xyz* frame are $\hat{\mathbf{i}}, \hat{\mathbf{j}}$ and $\hat{\mathbf{k}}$, respectively. Figure 9.22 shows the $\hat{\mathbf{I}}\hat{\mathbf{J}}\hat{\mathbf{K}}$ triad and the $\hat{\mathbf{i}}\hat{\mathbf{j}}\hat{\mathbf{k}}$ triad, along with the three successive rotations required to bring unit vectors initially aligned with $\hat{\mathbf{I}}\hat{\mathbf{J}}\hat{\mathbf{K}}$ into alignment with $\hat{\mathbf{i}}\hat{\mathbf{j}}\hat{\mathbf{k}}$. Since we are interested only in the relative orientation of the two frames, we can, for simplicity and without loss of generality, show the two frames sharing a common origin.

The *xy* plane intersects the *XY* plane along a line (the node line) defined by the unit vector $\hat{\mathbf{i}}'$ in the figure. The first rotation, ①, is around the $\hat{\mathbf{K}}$ axis, through the Euler angle ϕ. It rotates the $\hat{\mathbf{I}}, \hat{\mathbf{J}}$ directions into the $\hat{\mathbf{i}}', \hat{\mathbf{j}}'$ directions. Viewed down the

9.9 Euler angles

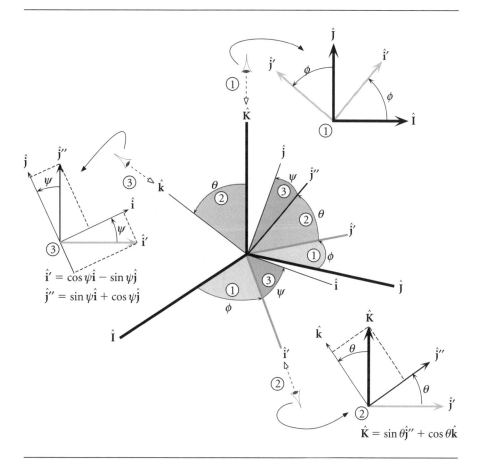

Figure 9.22 The Euler angles.

Z axis, this rotation appears as shown in the insert at the top of Figure 9.22, from which we see that

$$\hat{\mathbf{i}}' = \cos\phi \hat{\mathbf{I}} + \sin\phi \hat{\mathbf{J}}$$
$$\hat{\mathbf{j}}' = -\sin\phi \hat{\mathbf{I}} + \cos\phi \hat{\mathbf{J}}$$
$$\hat{\mathbf{k}}' = \hat{\mathbf{K}}$$

or

$$\begin{Bmatrix} \hat{\mathbf{i}}' \\ \hat{\mathbf{j}}' \\ \hat{\mathbf{k}}' \end{Bmatrix} = \begin{bmatrix} \cos\phi & \sin\phi & 0 \\ -\sin\phi & \cos\phi & 0 \\ 0 & 0 & 1 \end{bmatrix} \begin{Bmatrix} \hat{\mathbf{I}} \\ \hat{\mathbf{J}} \\ \hat{\mathbf{K}} \end{Bmatrix}$$

Therefore, the orthogonal transformation matrix associated with this rotation is

$$[\mathbf{R}_3(\phi)] = \begin{bmatrix} \cos\phi & \sin\phi & 0 \\ -\sin\phi & \cos\phi & 0 \\ 0 & 0 & 1 \end{bmatrix} \qquad (9.102)$$

Recall from Section 4.5 that the subscript on **R** denotes that the rotation is around the '3' direction, in this case the $\hat{\mathbf{K}}$ axis.

The second Euler rotation, ②, is around the node line ($\hat{\mathbf{i}}'$), through the angle θ required to bring the XY plane parallel to the xy plane. In other words, it rotates the Z axis into alignment with the z axis, and $\hat{\mathbf{j}}'$ simultaneously rotates into $\hat{\mathbf{j}}''$. The insert in the lower right of Figure 9.22 shows how this rotation appears when viewed from the $\hat{\mathbf{i}}'$ direction. From that illustration we can deduce that

$$\hat{\mathbf{i}}'' = \hat{\mathbf{i}}'$$
$$\hat{\mathbf{j}}'' = \cos\theta \hat{\mathbf{j}}' + \sin\phi \hat{\mathbf{K}}$$
$$\hat{\mathbf{k}} = -\sin\phi \hat{\mathbf{j}}' + \cos\phi \hat{\mathbf{K}}$$

or

$$\begin{Bmatrix} \hat{\mathbf{i}}'' \\ \hat{\mathbf{j}}'' \\ \hat{\mathbf{k}} \end{Bmatrix} = \begin{bmatrix} 1 & 0 & 0 \\ 0 & \cos\theta & \sin\theta \\ 0 & -\sin\theta & \cos\theta \end{bmatrix} \begin{Bmatrix} \hat{\mathbf{i}}' \\ \hat{\mathbf{j}}' \\ \hat{\mathbf{K}} \end{Bmatrix} \quad (9.103)$$

Clearly, the orthogonal transformation matrix for this rotation is

$$[\mathbf{R}_1(\theta)] = \begin{bmatrix} 1 & 0 & 0 \\ 0 & \cos\theta & \sin\theta \\ 0 & -\sin\theta & \cos\theta \end{bmatrix} \quad (9.104)$$

Since the inverse of an orthogonal matrix is just its transpose, the inverse of Equation 9.103 is

$$\begin{Bmatrix} \hat{\mathbf{i}}' \\ \hat{\mathbf{j}}' \\ \hat{\mathbf{K}} \end{Bmatrix} = \begin{bmatrix} 1 & 0 & 0 \\ 0 & \cos\theta & -\sin\theta \\ 0 & \sin\theta & \cos\theta \end{bmatrix} \begin{Bmatrix} \hat{\mathbf{i}}'' \\ \hat{\mathbf{j}}'' \\ \hat{\mathbf{k}} \end{Bmatrix}$$

from which we get the particular result needed below, namely,

$$\hat{\mathbf{K}} = \sin\theta \hat{\mathbf{j}}'' + \cos\theta \hat{\mathbf{k}} \quad (9.105)$$

The third and final Euler rotation, ③, is in the xy plane and rotates the unit vectors $\hat{\mathbf{i}}'$ and $\hat{\mathbf{j}}''$ through the angle ψ around the z axis so that they become aligned with $\hat{\mathbf{i}}$ and $\hat{\mathbf{j}}$, respectively. This rotation appears from the z direction as shown in the insert on the left of Figure 9.22. From that picture, we observe that

$$\hat{\mathbf{i}} = \cos\psi \hat{\mathbf{i}}' + \sin\psi \hat{\mathbf{j}}''$$
$$\hat{\mathbf{j}} = -\sin\psi \hat{\mathbf{i}}' + \cos\psi \hat{\mathbf{j}}''$$
$$\hat{\mathbf{k}} = \hat{\mathbf{k}}$$

or

$$\begin{Bmatrix} \hat{\mathbf{i}} \\ \hat{\mathbf{j}} \\ \hat{\mathbf{k}} \end{Bmatrix} = \begin{bmatrix} \cos\psi & \sin\psi & 0 \\ -\sin\psi & \cos\psi & 0 \\ 0 & 0 & 1 \end{bmatrix} \begin{Bmatrix} \hat{\mathbf{i}}' \\ \hat{\mathbf{j}}'' \\ \hat{\mathbf{k}} \end{Bmatrix} \quad (9.106)$$

9.9 Euler angles

From this, the orthogonal transformation matrix is seen to be

$$[\mathbf{R}_3(\psi)] = \begin{bmatrix} \cos\psi & \sin\psi & 0 \\ -\sin\psi & \cos\psi & 0 \\ 0 & 0 & 1 \end{bmatrix} \tag{9.107}$$

The inverse of Equations 9.106 is

$$\begin{Bmatrix} \hat{\mathbf{i}}' \\ \hat{\mathbf{j}}'' \\ \hat{\mathbf{k}} \end{Bmatrix} = \begin{bmatrix} \cos\psi & -\sin\psi & 0 \\ \sin\psi & \cos\psi & 0 \\ 0 & 0 & 1 \end{bmatrix} \begin{Bmatrix} \hat{\mathbf{i}} \\ \hat{\mathbf{j}} \\ \hat{\mathbf{k}} \end{Bmatrix}$$

from which we obtain

$$\hat{\mathbf{i}}' = \cos\psi\,\hat{\mathbf{i}} - \sin\psi\,\hat{\mathbf{j}} \tag{9.108a}$$

$$\hat{\mathbf{j}}'' = \sin\psi\,\hat{\mathbf{i}} + \cos\psi\,\hat{\mathbf{j}} \tag{9.108b}$$

Substituting Equation 9.108b into Equation 9.105 yields another result we will need below,

$$\hat{\mathbf{K}} = \sin\theta\sin\psi\,\hat{\mathbf{i}} + \sin\theta\cos\psi\,\hat{\mathbf{j}} + \cos\theta\,\hat{\mathbf{k}} \tag{9.109}$$

The time rates of change of the Euler angles ϕ, θ and ψ are, respectively, the precession ω_p, the nutation ω_n and the spin ω_s. That is,

$$\omega_p = \dot{\phi} \qquad \omega_n = \dot{\theta} \qquad \omega_s = \dot{\psi} \tag{9.110}$$

If the absolute angular velocity $\boldsymbol{\omega}$ of the rigid xyz frame is resolved into components ω_x, ω_y and ω_z along the xyz axes, we can express it analytically as

$$\boldsymbol{\omega} = \omega_x\hat{\mathbf{i}} + \omega_y\hat{\mathbf{j}} + \omega_z\hat{\mathbf{k}} \tag{9.111}$$

On the other hand, in terms of the precession, nutation and spin, the absolute angular velocity can be written in terms of the non-orthogonal Euler angle rates

$$\boldsymbol{\omega} = \omega_p\hat{\mathbf{K}} + \omega_n\hat{\mathbf{i}}' + \omega_s\hat{\mathbf{k}} \tag{9.112}$$

Substituting Equations 9.108a and 9.109 yields

$$\boldsymbol{\omega} = (\omega_p\sin\theta\sin\psi + \omega_n\cos\psi)\hat{\mathbf{i}} + (\omega_p\sin\theta\cos\psi - \omega_n\sin\psi)\hat{\mathbf{j}}$$
$$+ (\omega_s + \omega_p\cos\theta)\hat{\mathbf{k}} \tag{9.113}$$

Comparing Equations 9.111 and 9.113, we see that

$$\begin{aligned} \omega_x &= \omega_p\sin\theta\sin\psi + \omega_n\cos\psi \\ \omega_y &= \omega_p\sin\theta\cos\psi - \omega_n\sin\psi \\ \omega_z &= \omega_s + \omega_p\cos\theta \end{aligned} \tag{9.114}$$

We can solve these three equations to obtain the Euler rates in terms of ω_x, ω_y and ω_z:

$$\omega_p = \dot{\phi} = \frac{1}{\sin\theta}(\omega_x \sin\psi + \omega_y \cos\psi)$$

$$\omega_n = \dot{\theta} = \omega_x \cos\psi - \omega_y \sin\psi \qquad (9.115)$$

$$\omega_s = \dot{\psi} = -\frac{1}{\tan\theta}(\omega_x \sin\psi + \omega_y \cos\psi) + \omega_z$$

Observe that if ω_x, ω_y and ω_z are given functions of time, found by solving Euler's equations of motion (Equations 9.72), then Equations 9.115 are three coupled differential equations which may be solved to obtain the three time-dependent Euler angles

$$\phi = \phi(t) \qquad \theta = \theta(t) \qquad \psi = \psi(t)$$

With this solution, the orientation of the *xyz* frame, and hence the body to which it is attached, is known for any given time t. Note, however, that Equations 9.115 'blow up' when $\theta = 0$, i.e., when the *xy* plane is parallel to the *XY* plane.

Finally, let us note that the transformation matrix $[\mathbf{Q}]_{Xx}$ from the inertial *XYZ* frame into the moving *xyz* frame is just the product of the three rotation matrices given by Equations 9.102, 9.104 and 9.107, i.e.,

$$[\mathbf{Q}]_{Xx} = [\mathbf{R}_3(\psi)][\mathbf{R}_1(\theta)][\mathbf{R}_3(\phi)] \qquad (9.116)$$

Substituting the three matrices on the right and carrying out the matrix multiplications yields

$$[\mathbf{Q}]_{Xx} = \begin{bmatrix} \cos\phi\cos\psi - \sin\phi\sin\psi\cos\theta & \sin\phi\cos\psi + \cos\phi\cos\theta\sin\psi & \sin\theta\sin\psi \\ -\cos\phi\sin\psi - \sin\phi\cos\theta\cos\psi & -\sin\phi\sin\psi + \cos\phi\cos\theta\cos\psi & \sin\theta\cos\psi \\ \sin\phi\sin\theta & -\cos\phi\sin\theta & \cos\theta \end{bmatrix}$$

$$(9.117)$$

Remember, this is an orthogonal matrix, so that for the inverse transformation from *xyz* to *XYZ* we have $[\mathbf{Q}]_{xX} = ([\mathbf{Q}]_{Xx})^T$, or

$$[\mathbf{Q}]_{xX} = \begin{bmatrix} \cos\phi\cos\psi - \sin\phi\sin\psi\cos\theta & -\cos\phi\sin\psi - \sin\phi\cos\theta\cos\psi & \sin\phi\sin\theta \\ \sin\phi\cos\psi + \cos\phi\cos\theta\sin\psi & -\sin\phi\sin\psi + \cos\phi\cos\theta\cos\psi & -\cos\phi\sin\theta \\ \sin\theta\sin\psi & \sin\theta\cos\psi & \cos\theta \end{bmatrix}$$

$$(9.118)$$

EXAMPLE 9.17

At a given instant, the unit vectors of a body-fixed frame are

$$\hat{\mathbf{i}} = 0.40825\hat{\mathbf{I}} - 0.40825\hat{\mathbf{J}} + 0.8165\hat{\mathbf{K}}$$

$$\hat{\mathbf{j}} = -0.10102\hat{\mathbf{I}} - 0.90914\hat{\mathbf{J}} - 0.40406\hat{\mathbf{K}} \qquad (a)$$

$$\hat{\mathbf{k}} = 0.90726\hat{\mathbf{I}} + 0.082479\hat{\mathbf{J}} - 0.41239\hat{\mathbf{K}}$$

and the angular velocity is

$$\boldsymbol{\omega} = -3.1\hat{\mathbf{I}} + 2.5\hat{\mathbf{J}} + 1.7\hat{\mathbf{K}} \text{ (rad/s)} \qquad (b)$$

Calculate ω_p, ω_n and ω_s (the precession, nutation and spin rates) at this instant.

We will ultimately use Equations 9.115 to find ω_p, ω_n and ω_s. To do so we must first obtain the Euler angles ϕ, θ and ψ as well as the components of the angular velocity in the body frame.

The procedure for determining ϕ, θ and ψ is practically identical to that used to obtain the orbital elements i, Ω, and ω of a satellite orbit from its state vector (Algorithm 4.1). Referring to Figure 9.22, we first note that the angle between $\hat{\mathbf{k}}$ and $\hat{\mathbf{K}}$ is the inclination angle θ, so that

$$\theta = \cos^{-1} k_Z = \cos^{-1}(-0.41239) = 114.36° \tag{c}$$

θ lies between 0° and 180°.

The 'node' vector $\mathbf{N}$ points in the direction of $\hat{\mathbf{i}}'$ in Figure 9.22, and it is found by taking the cross product of $\hat{\mathbf{K}}$ into $\hat{\mathbf{k}}$,

$$\mathbf{N} = \hat{\mathbf{K}} \times \hat{\mathbf{k}} = \begin{vmatrix} \hat{\mathbf{I}} & \hat{\mathbf{J}} & \hat{\mathbf{K}} \\ 0 & 0 & 1 \\ 0.90726 & 0.082479 & -0.41239 \end{vmatrix}$$

$$= -0.082479\hat{\mathbf{I}} + 0.90726\hat{\mathbf{J}} \quad (\|\mathbf{N}\| = 0.911) \tag{d}$$

The precession angle ϕ (analogous to RA of the ascending node Ω) is measured from the X axis positive towards $\mathbf{N}$. Therefore, taking care to place ϕ in the proper quadrant,

$$\phi = \begin{cases} \cos^{-1} \dfrac{N_X}{\|\mathbf{N}\|} & \text{if } N_Y \geq 0 \\ 360° - \cos^{-1} \dfrac{N_X}{\|\mathbf{N}\|} & \text{if } N_Y < 0 \end{cases} \tag{e}$$

Substituting (d), noting that in this case $N_Y = 0.90276 > 0$, we get

$$\phi = \cos^{-1} \dfrac{-0.082479}{0.9110} = 95.194° \tag{f}$$

The spin angle ψ is measured positive from $\mathbf{N}$ to $\hat{\mathbf{i}}$ in Figure 9.22. It plays the same role here as argument of perigee does for satellites. We can find the angle between $\mathbf{N}$ and $\hat{\mathbf{i}}$ by using the dot product operation, again being careful to put ψ in the right quadrant,

$$\psi = \begin{cases} \cos^{-1} \dfrac{\mathbf{N} \cdot \hat{\mathbf{i}}}{\|\mathbf{N}\|} & \text{if } i_Z \geq 0 \\ 360° - \cos^{-1} \dfrac{\mathbf{N} \cdot \hat{\mathbf{i}}}{\|\mathbf{N}\|} & \text{if } i_Z < 0 \end{cases} \tag{g}$$

From (a) we note that $i_Z = 0.8165 > 0$, so

$$\psi = \cos^{-1} \dfrac{(-0.082479\hat{\mathbf{I}} + 0.90726\hat{\mathbf{J}}) \cdot (0.40825\hat{\mathbf{I}} - 0.40825\hat{\mathbf{J}} + 0.8165\hat{\mathbf{K}})}{0.9110}$$

$$= \cos^{-1}\left(\dfrac{-0.40406}{0.9110}\right) = 116.33° \tag{h}$$

To transform the components of the given angular velocity vector into components along the body frame, we need the matrix $[\mathbf{Q}]_{Xx}$ of the transformation from XYZ to xyz. The rows of $[\mathbf{Q}]_{Xx}$ are the direction cosines of $\hat{\mathbf{i}}, \hat{\mathbf{j}}$ and $\hat{\mathbf{k}}$, which are given in (a).

(Example 9.17 continued)

Thus

$$\{\omega\}_x = [Q]_{Xx}\{\omega\}_X$$

$$= \begin{bmatrix} 0.40825 & -0.40825 & 0.8165 \\ -0.10102 & -0.90914 & -0.40406 \\ 0.90726 & 0.082479 & -0.41239 \end{bmatrix} \begin{Bmatrix} -3.1 \\ 2.5 \\ 1.7 \end{Bmatrix} = \begin{Bmatrix} -0.89815 \\ -2.6466 \\ -3.3074 \end{Bmatrix} \text{ (rad/s)}$$

That is,

$$\omega_x = -0.89815 \text{ rad/s} \qquad \omega_y = -2.6466 \text{ rad/s} \qquad \omega_z = -3.3074 \text{ rad/s} \quad \text{(i)}$$

Finally, substituting (c), (f), (h) and (i) into Equations 9.115 yields

$$\omega_p = \frac{1}{\sin 114.36°}[-0.89815 \cdot \sin 116.33° + (-2.6466) \cdot \cos 116.33°]$$

$$= \underline{0.40492 \text{ rad/s}}$$

$$\omega_n = -0.89815 \cdot \cos 116.33° - (-2.6466) \cdot \sin 116.33° = \underline{2.7704 \text{ rad/s}}$$

$$\omega_s = -\frac{1}{\tan 114.36°}[-0.89815 \cdot \sin 116.33° + (-2.6466) \cdot \cos 116.33°]$$

$$+ (-3.3074) = \underline{-3.1404 \text{ rad/s}}$$

EXAMPLE 9.18

The mass moments of inertia of a body about the principal body frame axes with origin at the center of mass G are

$$A = 1000 \text{ kg} \cdot \text{m}^2 \qquad B = 2000 \text{ kg} \cdot \text{m}^2 \qquad C = 3000 \text{ kg} \cdot \text{m}^2 \quad \text{(a)}$$

The Euler angles in radians are given as functions of time in seconds as follows:

$$\phi = 2te^{-0.05t}$$
$$\theta = 0.02 + 0.3 \sin 0.25t \quad \text{(b)}$$
$$\psi = 0.6t$$

At $t = 0$, find (a) the net moment about G and (b) the components α_X, α_Y and α_Z of the absolute angular acceleration in the inertial frame.

(a) We must use Euler's equations (Equations 9.72) to calculate the net moment, which means we must first obtain $\omega_x, \omega_y, \omega_z, \dot{\omega}_x, \dot{\omega}_y$ and $\dot{\omega}_z$. Since we are given the Euler angles as a function of time, we can compute their time derivatives and then use Equation 9.114 to find the body frame angular velocity components and their derivatives.

Starting with (b)$_1$, we get

$$\omega_p = \frac{d\phi}{dt} = \frac{d}{dt}(2te^{-0.05t}) = 2e^{-0.05t} - 0.1te^{-0.05t}$$

$$\dot{\omega}_p = \frac{d\omega_p}{dt} = \frac{d}{dt}(2e^{-0.05t} - 0.1e^{-0.05t}) = -0.2e^{-0.05t} + 0.005te^{-0.05t}$$

9.9 Euler angles

Proceeding to the remaining two Euler angles leads to

$$\omega_n = \frac{d\theta}{dt} = \frac{d}{dt}(0.02 + 0.3\sin 0.25t) = 0.075\cos 0.25t$$

$$\dot{\omega}_n = \frac{d\omega_n}{dt} = \frac{d}{dt}(0.075\cos 0.25t) = -0.01875\sin 0.25t$$

$$\omega_s = \frac{d\psi}{dt} = \frac{d}{dt}(0.6t) = 0.6$$

$$\dot{\omega}_s = \frac{d\omega_s}{dt} = 0$$

Evaluating all of these quantities, including those in (b), at $t = 0$ yields

$$\phi = 335.03° \quad \omega_p = 0.60653 \text{ rad/s} \quad \dot{\omega}_p = -0.09098 \text{ rad/s}^2$$
$$\theta = 11.433° \quad \omega_n = -0.06009 \text{ rad/s} \quad \dot{\omega}_n = -0.011221 \text{ rad/s}^2 \quad \text{(c)}$$
$$\psi = 343.77 \quad \omega_s = 0.6 \text{ rad/s} \quad \dot{\omega}_s = 0$$

Equation 9.114 relates the Euler angle rates to the angular velocity components,

$$\omega_x = \omega_p \sin\theta \sin\psi + \omega_n \cos\psi$$
$$\omega_y = \omega_p \sin\theta \cos\psi - \omega_n \sin\psi \quad \text{(d)}$$
$$\omega_z = \omega_s + \omega_p \cos\theta$$

Taking the time derivative of each of these equations in turn leads to the following three equations,

$$\dot{\omega}_x = \omega_p \omega_n \cos\theta \sin\psi + \omega_p \omega_s \sin\theta \cos\psi - \omega_n \omega_s \sin\psi$$
$$\quad + \dot{\omega}_p \sin\theta \sin\psi + \dot{\omega}_n \cos\psi$$
$$\dot{\omega}_y = \omega_p \omega_n \cos\theta \cos\psi - \omega_p \omega_s \sin\theta \sin\psi - \omega_n \omega_s \cos\psi \quad \text{(e)}$$
$$\quad + \dot{\omega}_p \sin\theta \cos\psi - \dot{\omega}_n \sin\psi$$
$$\dot{\omega}_z = -\omega_p \omega_n \sin\theta + \dot{\omega}_p \cos\theta + \dot{\omega}_s$$

Substituting the data in (c) into (d) and (e) yields

$$\omega_x = -0.091286 \text{ rad/s} \quad \omega_y = 0.098649 \text{ rad/s} \quad \omega_z = 1.1945 \text{ rad/s}$$
$$\dot{\omega}_x = 0.063435 \text{ rad/s}^2 \quad \dot{\omega}_y = 2.2346 \times 10^{-5} \text{ rad/s}^2 \quad \dot{\omega}_z = -0.08195 \text{ rad/s}^2$$
$$\text{(f)}$$

With (a) and (f) we have everything we need for Euler's equations,

$$M_{x_{net}} = A\dot{\omega}_x + (C - B)\omega_y \omega_z$$
$$M_{y_{net}} = B\dot{\omega}_y + (A - C)\omega_z \omega_x$$
$$M_{z_{net}} = C\dot{\omega}_z + (B - A)\omega_x \omega_y$$

(Example 9.18 continued)

from which we find

$$M_{x_{\text{net}}} = 181.27 \text{ N} \cdot \text{m}$$

$$M_{y_{\text{net}}} = 218.12 \text{ N} \cdot \text{m}$$

$$M_{z_{\text{net}}} = -254.86 \text{ N} \cdot \text{m}$$

(b) Since the co-moving *xyz* frame is a body frame, rigidly attached to the solid, we know from Equation 9.73 that

$$\begin{Bmatrix} \alpha_X \\ \alpha_Y \\ \alpha_Z \end{Bmatrix} = [Q]_{xX} \begin{Bmatrix} \dot{\omega}_x \\ \dot{\omega}_y \\ \dot{\omega}_z \end{Bmatrix} \qquad (g)$$

In other words, the absolute angular acceleration and the relative angular acceleration of the body are the same. All we have to do is project the components of relative acceleration in (f) onto the axes of the inertial frame. The required orthogonal transformation matrix is given in Equation 9.118,

$$[Q]_{xX}$$
$$= \begin{bmatrix} \cos\phi\cos\psi - \sin\phi\sin\psi\cos\theta & -\cos\phi\sin\psi - \sin\phi\cos\theta\cos\psi & \sin\phi\sin\theta \\ \sin\phi\cos\psi + \cos\phi\cos\theta\sin\psi & -\sin\phi\sin\psi + \cos\phi\cos\theta\cos\psi & -\cos\phi\sin\theta \\ \sin\theta\sin\psi & \sin\theta\cos\psi & \cos\theta \end{bmatrix}$$

Upon substituting the numerical values of the Euler angles from (c), this becomes

$$[Q]_{xX} = \begin{bmatrix} -0.90855 & 0.20144 & 0.3660 \\ -0.29194 & -0.93280 & -0.21131 \\ 0.29884 & -0.29884 & 0.90631 \end{bmatrix}$$

Substituting this and the relative angular velocity rates from (c) into (g) yields

$$\begin{Bmatrix} \alpha_X \\ \alpha_Y \\ \alpha_Z \end{Bmatrix} = \begin{bmatrix} -0.90855 & 0.20144 & 0.3660 \\ -0.29194 & -0.93280 & -0.21131 \\ 0.29884 & -0.29884 & 0.90631 \end{bmatrix} \begin{Bmatrix} -0.027359 \\ -0.32619 \\ 1.4532 \end{Bmatrix}$$

$$= \begin{Bmatrix} 0.4910 \\ 0.0051972 \\ 1.4063 \end{Bmatrix} (\text{rad/s}^2)$$

EXAMPLE 9.19

Figure 9.23 shows a rotating platform on which is mounted a rectangular parallelepiped shaft (with dimensions *b*, *h* and *l*) spinning about the inclined axis *DE*. If the mass of the shaft is *m*, and the angular velocities ω_p and ω_s are constant, calculate the bearing forces at *D* and *E* as a function of ϕ and ψ. Neglect gravity, since we are interested only in the gyroscopic forces. (The small extensions shown at each end of the parallelepiped are just for clarity; the distance between the bearings at *D* and *E* is *l*.)

9.9 Euler angles

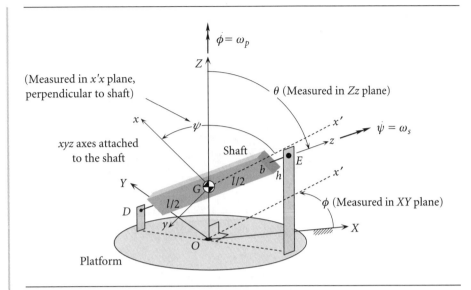

Figure 9.23 Spinning block mounted on rotating platform.

The inertial XYZ frame is centered at O on the platform, and it is right-handed ($\hat{\mathbf{I}} \times \hat{\mathbf{J}} = \hat{\mathbf{K}}$). The origin of the right-handed co-moving body frame xyz is at the shaft's center of mass G, and it is aligned with the symmetry axes of the parallelepiped. The three Euler angles ϕ, θ and ψ are shown in Figure 9.23. Since θ is constant, the nutation rate is zero ($\omega_n = 0$). Thus, Equations 9.114 reduce to

$$\omega_x = \omega_p \sin\theta \sin\psi \quad \omega_y = \omega_p \sin\theta \cos\psi \quad \omega_z = \omega_p \cos\theta + \omega_s \qquad \text{(a)}$$

Since ω_p, ω_s and θ are constant, it follows (recalling Equations 9.110) that

$$\dot{\omega}_x = \omega_p \omega_s \sin\theta \cos\psi \quad \dot{\omega}_y = -\omega_p \omega_s \sin\theta \sin\psi \quad \dot{\omega}_z = 0 \qquad \text{(b)}$$

The principal moments of inertia of the parallelepiped are [see Figure 9.9(c)]

$$\begin{aligned} A = I_x &= \frac{1}{12}m(h^2 + l^2) \\ B = I_y &= \frac{1}{12}m(b^2 + l^2) \\ C = I_z &= \frac{1}{12}m(b^2 + h^2) \end{aligned} \qquad \text{(c)}$$

Figure 9.24 is a free-body diagram of the shaft. Let us assume that the bearings at D and E are such as to exert just the six body frame components of force shown. Thus, D is a thrust bearing to which the axial torque T_D is applied from, say, a motor of some kind. At E there is a simple journal bearing.

458 Chapter 9 Rigid-body dynamics

(Example 9.19 continued)

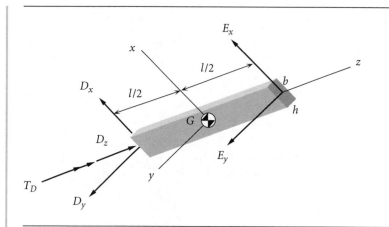

Figure 9.24 Free-body diagram of the block in Figure 9.23.

From Newton's laws of motion we have $\mathbf{F}_{net} = m\mathbf{a}_G$. But G is fixed in inertial space, so $\mathbf{a}_G = \mathbf{0}$. Thus,

$$(D_x\hat{\mathbf{i}} + D_y\hat{\mathbf{j}} + D_z\hat{\mathbf{k}}) + (E_x\hat{\mathbf{i}} + E_y\hat{\mathbf{j}}) = \mathbf{0}$$

It follows that

$$E_x = -D_x \qquad E_y = -D_y \qquad D_z = 0 \qquad (d)$$

Summing moments about G we get

$$\mathbf{M}_{G_{net}} = \frac{l}{2}\hat{\mathbf{k}} \times (E_x\hat{\mathbf{i}} + E_y\hat{\mathbf{j}}) + \left(-\frac{l}{2}\hat{\mathbf{k}}\right) \times (D_x\hat{\mathbf{i}} + D_y\hat{\mathbf{j}}) + T_D\hat{\mathbf{k}}$$

$$= \left(D_y\frac{l}{2} - E_y\frac{l}{2}\right)\hat{\mathbf{i}} + \left(-D_x\frac{l}{2} + E_x\frac{l}{2}\right)\hat{\mathbf{j}} + T_D\hat{\mathbf{k}}$$

$$= D_y l\hat{\mathbf{i}} - D_x l\hat{\mathbf{j}} + T_D\hat{\mathbf{k}}$$

where we made use of Equation (d)$_2$. Thus,

$$M_{x_{net}} = D_y l \qquad M_{y_{net}} = -D_x l \qquad M_{z_{net}} = T_D \qquad (e)$$

We substitute (a), (b), (c), and (e) into Euler's equations (Equations 9.72):

$$\begin{aligned} M_{x_{net}} &= A\dot{\omega}_x + (C - B)\omega_y\omega_z \\ M_{y_{net}} &= B\dot{\omega}_y + (A - C)\omega_x\omega_z \\ M_{z_{net}} &= C\dot{\omega}_z + (B - A)\omega_x\omega_y \end{aligned} \qquad (f)$$

After making the substitutions and simplifying, the first Euler equation, Equation (f)$_1$, becomes

$$D_x = \left\{\frac{1}{12}\frac{m}{l}[(l^2 - h^2)\omega_p\cos\theta - 2h^2\omega_s]\omega_p\sin\theta\right\}\cos\psi \qquad (g)$$

Likewise, from Equation (f)$_2$ we obtain

$$D_y = \left\{ \frac{1}{12}\frac{m}{l}[(l^2 - b^2)\omega_p \cos\theta - 2b^2\omega_s]\omega_p \sin\theta \right\} \sin\psi \tag{h}$$

Finally, Equation (f)$_3$ yields

$$T_D = \left[\frac{1}{24}m(b^2 - h^2)\omega_p^2 \sin^2\theta \right] \sin 2\psi \tag{i}$$

This completes the solution, since $E_y = -D_y$ and $E_z = -D_z$. Note that the resultant transverse bearing load V at D (and E) is

$$V = \sqrt{D_x^2 + D_y^2} \tag{j}$$

As a numerical example, let

$$l = 1\,\text{m} \qquad h = 0.1\,\text{m} \qquad b = 0.025\,\text{m} \qquad \theta = 30° \qquad m = 10\,\text{kg}$$

and

$$\omega_p = 100\,\text{rpm} = 10.47\,\text{rad/s} \qquad \omega_s = 2000\,\text{rpm} = 209.4\,\text{rad/s}$$

For these numbers, the variation of V and T_D with ψ are as shown in Figure 9.25.

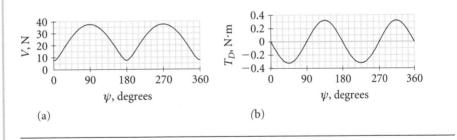

Figure 9.25 (a) Transverse bearing load. (b) Axial torque at D.

9.10 YAW, PITCH AND ROLL ANGLES

The problem of the Euler angle relations, Equations 9.114, becoming singular when the nutation angle θ is zero can be alleviated by using the yaw, pitch and roll angles illustrated in Figure 9.26. As in the Euler angles, the inertial $\hat{\mathbf{I}}\hat{\mathbf{J}}\hat{\mathbf{K}}$ triad is rotated into the body $\hat{\mathbf{i}}\hat{\mathbf{j}}\hat{\mathbf{k}}$ triad by a sequence of three rotations, detailed in Figure 9.26. The first step is to rotate the $\hat{\mathbf{I}}$ and $\hat{\mathbf{J}}$ directions through a yaw angle ϕ around the $\hat{\mathbf{K}}$ axis until they line up with the orthogonal unit vectors $\hat{\mathbf{i}}'$, $\hat{\mathbf{j}}'$. The $\hat{\mathbf{i}}'$ direction is the projection of the body x axis on the XY plane. This rotation appears as shown in insert ① at the

460 Chapter 9 Rigid-body dynamics

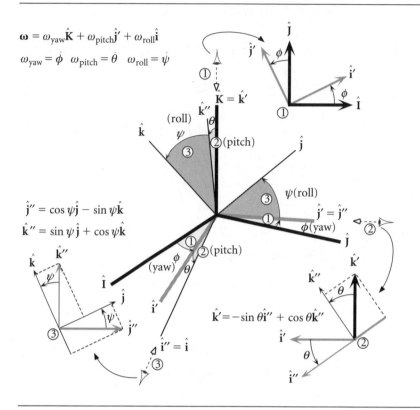

Figure 9.26 Yaw, pitch and roll angles.

top of Figure 9.26, from which it can be seen that

$$\hat{\mathbf{i}}' = \cos\phi\hat{\mathbf{I}} + \sin\phi\hat{\mathbf{J}} \qquad (9.119a)$$

$$\hat{\mathbf{j}}' = -\sin\phi\hat{\mathbf{I}} + \cos\phi\hat{\mathbf{J}} \qquad (9.119b)$$

$$\hat{\mathbf{k}}' = \hat{\mathbf{K}} \qquad (9.119c)$$

or

$$\begin{Bmatrix} \hat{\mathbf{i}}' \\ \hat{\mathbf{j}}' \\ \hat{\mathbf{k}}' \end{Bmatrix} = \begin{bmatrix} \cos\phi & \sin\phi & 0 \\ -\sin\phi & \cos\phi & 0 \\ 0 & 0 & 1 \end{bmatrix} \begin{Bmatrix} \hat{\mathbf{I}} \\ \hat{\mathbf{J}} \\ \hat{\mathbf{K}} \end{Bmatrix}$$

Clearly, the yaw rotation matrix is $[\mathbf{R}_3(\phi)]$, where

$$[\mathbf{R}_3(\phi)] = \begin{bmatrix} \cos\phi & \sin\phi & 0 \\ -\sin\phi & \cos\phi & 0 \\ 0 & 0 & 1 \end{bmatrix}$$

The second rotation is a pitch around $\hat{\mathbf{j}}'$ through the pitch angle θ. This carries $\hat{\mathbf{i}}'$ and $\hat{\mathbf{k}}'(=\hat{\mathbf{K}})$ into $\hat{\mathbf{i}}''(=\hat{\mathbf{i}})$ and $\hat{\mathbf{k}}''$, while, of course, leaving $\hat{\mathbf{j}}'$ unchanged. We see this in auxiliary view ② of Figure 9.26, from which we obtain

$$\hat{\mathbf{i}}'' = \cos\theta\hat{\mathbf{i}}' - \sin\theta\hat{\mathbf{k}}'$$
$$\hat{\mathbf{j}}'' = \hat{\mathbf{j}}'$$
$$\hat{\mathbf{k}}'' = \sin\theta\hat{\mathbf{i}}' + \cos\theta\hat{\mathbf{k}}'$$

or

$$\begin{Bmatrix} \hat{\mathbf{i}}'' \\ \hat{\mathbf{j}}'' \\ \hat{\mathbf{k}}'' \end{Bmatrix} = \begin{bmatrix} \cos\theta & 0 & -\sin\theta \\ 0 & 1 & 0 \\ \sin\theta & 0 & \cos\theta \end{bmatrix} \begin{Bmatrix} \hat{\mathbf{i}}' \\ \hat{\mathbf{j}}' \\ \hat{\mathbf{k}}' \end{Bmatrix}$$

This rotation about the intermediate y' axis is therefore represented by $[\mathbf{R}_2(\theta)]$, where

$$[\mathbf{R}_2(\theta)] = \begin{bmatrix} \cos\theta & 0 & -\sin\theta \\ 0 & 1 & 0 \\ \sin\theta & 0 & \cos\theta \end{bmatrix}$$

The inverse of this orthogonal matrix is its transpose,

$$[\mathbf{R}_2(\theta)]^{-1} = \begin{bmatrix} \cos\theta & 0 & \sin\theta \\ 0 & 1 & 0 \\ -\sin\theta & 0 & \cos\theta \end{bmatrix}$$

so that

$$\begin{Bmatrix} \hat{\mathbf{i}}' \\ \hat{\mathbf{j}}' \\ \hat{\mathbf{k}}' \end{Bmatrix} = \begin{bmatrix} \cos\theta & 0 & \sin\theta \\ 0 & 1 & 0 \\ -\sin\theta & 0 & \cos\theta \end{bmatrix} \begin{Bmatrix} \hat{\mathbf{i}}'' \\ \hat{\mathbf{j}}'' \\ \hat{\mathbf{k}}'' \end{Bmatrix}$$

From this we see that

$$\hat{\mathbf{j}}' = \hat{\mathbf{j}}''$$
$$\hat{\mathbf{k}}' = -\sin\theta\hat{\mathbf{i}}'' + \cos\theta\hat{\mathbf{k}}'' \tag{9.120}$$

Finally, we roll around the body x axis through the angle ψ, which brings $\hat{\mathbf{j}}''$ and $\hat{\mathbf{k}}''$ into alignment with body unit vectors $\hat{\mathbf{j}}$ and $\hat{\mathbf{k}}$, respectively. Auxiliary view ③ shows that

$$\hat{\mathbf{i}} = \hat{\mathbf{i}}''$$
$$\hat{\mathbf{j}} = \cos\psi\hat{\mathbf{j}}'' + \sin\psi\hat{\mathbf{k}}''$$
$$\hat{\mathbf{k}} = -\sin\psi\hat{\mathbf{j}}'' + \cos\psi\hat{\mathbf{k}}''$$

or

$$\begin{Bmatrix} \hat{\mathbf{i}} \\ \hat{\mathbf{j}} \\ \hat{\mathbf{k}} \end{Bmatrix} = \begin{bmatrix} 1 & 0 & 0 \\ 0 & \cos\psi & \sin\psi \\ 0 & -\sin\psi & \cos\psi \end{bmatrix} \begin{Bmatrix} \hat{\mathbf{i}}'' \\ \hat{\mathbf{j}}'' \\ \hat{\mathbf{k}}'' \end{Bmatrix} \tag{9.121}$$

Thus, the third and last rotation matrix is

$$[\mathbf{R}_1(\psi)] = \begin{bmatrix} 1 & 0 & 0 \\ 0 & \cos\psi & \sin\psi \\ 0 & -\sin\psi & \cos\psi \end{bmatrix}$$

Taking the transpose of this array, we find the inverse of Equations 9.121

$$\begin{Bmatrix} \hat{\mathbf{i}}'' \\ \hat{\mathbf{j}}'' \\ \hat{\mathbf{k}}'' \end{Bmatrix} = \begin{bmatrix} 1 & 0 & 0 \\ 0 & \cos\psi & -\sin\psi \\ 0 & \sin\psi & \cos\psi \end{bmatrix} \begin{Bmatrix} \hat{\mathbf{i}} \\ \hat{\mathbf{j}} \\ \hat{\mathbf{k}} \end{Bmatrix}$$

which means

$$\hat{\mathbf{i}}'' = \hat{\mathbf{i}}$$
$$\hat{\mathbf{j}}'' = \cos\psi\,\hat{\mathbf{j}} - \sin\psi\,\hat{\mathbf{k}} \qquad (9.122)$$
$$\hat{\mathbf{k}}'' = \sin\psi\,\hat{\mathbf{j}} + \cos\psi\,\hat{\mathbf{k}}$$

The matrix $[\mathbf{Q}]_{Xx}$ of the transformation from $\hat{\mathbf{I}}\hat{\mathbf{J}}\hat{\mathbf{K}}$ into $\hat{\mathbf{i}}\hat{\mathbf{j}}\hat{\mathbf{k}}$ is the product of the three rotation matrices obtained above,

$$[\mathbf{Q}]_{Xx} = [\mathbf{R}_1(\psi)][\mathbf{R}_2(\theta)][\mathbf{R}_3(\phi)]$$

Carrying out the matrix multiplications yields

$$[\mathbf{Q}]_{Xx} = \begin{bmatrix} \cos\phi\cos\theta & \sin\phi\cos\theta & -\sin\theta \\ -\sin\phi\cos\psi + \cos\phi\sin\theta\sin\psi & \cos\phi\cos\psi + \sin\phi\sin\theta\sin\psi & \cos\theta\sin\psi \\ \sin\phi\sin\psi + \cos\phi\sin\theta\cos\psi & -\cos\phi\sin\psi + \sin\phi\sin\theta\cos\psi & \cos\theta\cos\psi \end{bmatrix}$$
(9.123)

The inverse matrix which transforms xyz into XYZ is just the transpose,

$$[\mathbf{Q}]_{xX} = \begin{bmatrix} \cos\phi\cos\theta & -\sin\phi\cos\psi + \cos\phi\sin\theta\sin\psi & \sin\phi\sin\psi + \cos\phi\sin\theta\cos\psi \\ \sin\phi\cos\theta & \cos\phi\cos\psi + \sin\phi\sin\theta\sin\psi & -\cos\phi\sin\psi + \sin\phi\sin\theta\cos\psi \\ -\sin\theta & \cos\theta\sin\psi & \cos\theta\cos\psi \end{bmatrix}$$
(9.124)

The angular velocity $\boldsymbol{\omega}$, expressed in terms of the rates of yaw, pitch and roll, is

$$\boldsymbol{\omega} = \omega_{\text{yaw}}\hat{\mathbf{K}} + \omega_{\text{pitch}}\hat{\mathbf{j}}' + \omega_{\text{roll}}\hat{\mathbf{i}}$$

in which

$$\omega_{\text{yaw}} = \dot{\phi} \qquad \omega_{\text{pitch}} = \dot{\theta} \qquad \omega_{\text{roll}} = \dot{\psi}$$

Using Equation 9.119c, we can write $\boldsymbol{\omega}$ as

$$\boldsymbol{\omega} = \omega_{\text{yaw}}\hat{\mathbf{k}}' + \omega_{\text{pitch}}\hat{\mathbf{j}}' + \omega_{\text{roll}}\hat{\mathbf{i}}$$

Substituting Equation 9.120 into this expression yields

$$\boldsymbol{\omega} = \omega_{\text{yaw}}(-\sin\theta\,\hat{\mathbf{i}}'' + \cos\theta\,\hat{\mathbf{k}}'') + \omega_{\text{pitch}}\hat{\mathbf{j}}'' + \omega_{\text{roll}}\hat{\mathbf{i}}$$

Finally, with Equations 9.122, we obtain from this

$$\boldsymbol{\omega} = \omega_{\text{yaw}}[-\sin\theta\,\hat{\mathbf{i}} + \cos\theta(\sin\psi\,\hat{\mathbf{j}} + \cos\psi\,\hat{\mathbf{k}})] + \omega_{\text{pitch}}(\cos\psi\,\hat{\mathbf{j}} - \sin\psi\,\hat{\mathbf{k}}) + \omega_{\text{roll}}\hat{\mathbf{i}}$$

After collecting terms, we see that

$$\omega_x = \omega_{\text{roll}} - \omega_{\text{yaw}} \sin \theta_{\text{pitch}}$$
$$\omega_y = \omega_{\text{yaw}} \cos \theta_{\text{pitch}} \sin \psi_{\text{roll}} + \omega_{\text{pitch}} \cos \psi_{\text{roll}} \quad (9.125)$$
$$\omega_z = \omega_{\text{yaw}} \cos \theta_{\text{pitch}} \cos \psi_{\text{roll}} - \omega_{\text{pitch}} \sin \psi_{\text{roll}}$$

wherein the subscript on each symbol helps us remember the rotation it describes. The inverse of these equations is

$$\omega_{\text{yaw}} = \omega_y \frac{\sin \psi_{\text{roll}}}{\cos \theta_{\text{pitch}}} + \omega_z \frac{\cos \psi_{\text{roll}}}{\cos \theta_{\text{pitch}}}$$
$$\omega_{\text{pitch}} = \omega_y \cos \psi_{\text{roll}} - \omega_z \sin \psi_{\text{roll}} \quad (9.126)$$
$$\omega_{\text{roll}} = \omega_x + \omega_y \tan \theta_{\text{pitch}} \sin \psi_{\text{roll}} + \omega_z \tan \theta_{\text{pitch}} \cos \psi_{\text{roll}}$$

Notice that this system becomes singular ($\cos \theta_{\text{pitch}} = 0$) when the pitch angle is $\pm 90°$.

PROBLEMS

9.1 Rigid, bent shaft 1 (ABC) rotates at a constant angular velocity of $2\hat{\mathbf{K}}$ rad/s around the positive Z axis of the inertial frame. Bent shaft 2 (CDE) rotates around BC with a constant angular velocity of $3\hat{\mathbf{j}}$ rad/s, relative to BC. Spinner 3 at E rotates around DE with a constant angular velocity of $4\hat{\mathbf{i}}$ rad/s relative to DE. Calculate the magnitude of the absolute angular acceleration $\boldsymbol{\alpha}_3$ of the spinner at the instant shown.
{Ans.: $\|\boldsymbol{\alpha}_3\| = \sqrt{180 + 64 \sin^2 \theta - 144 \cos \theta}$ (rad/s²)}

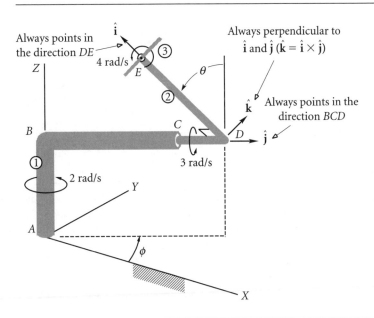

Figure P.9.1

9.2 The body-fixed *xyz* frame is attached to the cylinder as shown. The cylinder rotates around the inertial *Z* axis, which is collinear with the *z* axis, with a constant absolute angular velocity $\dot{\theta}\hat{\mathbf{k}}$. Rod *AB* is attached to the cylinder and aligned with the *y* axis. Rod *BC* is perpendicular to *AB* and rotates around *AB* with the constant angular velocity $\dot{\phi}\hat{\mathbf{j}}$ relative to the cylinder. Rod *CD* is perpendicular to *BC* and rotates around *BC* with the constant angular velocity $\dot{v}\hat{\mathbf{m}}$ relative to *BC*, where $\hat{\mathbf{m}}$ is the unit vector in the direction of *BC*. The plate *abcd* rotates around *CD* with a constant angular velocity $\dot{\psi}\hat{\mathbf{n}}$ relative to *CD*, where the unit vector $\hat{\mathbf{n}}$ points in the direction of *CD*. Thus the absolute angular velocity of the plate is $\boldsymbol{\omega}_{\text{plate}} = \dot{\theta}\hat{\mathbf{k}} + \dot{\phi}\hat{\mathbf{j}} + \dot{v}\hat{\mathbf{m}} + \dot{\psi}\hat{\mathbf{n}}$. Show that

(a) $\boldsymbol{\omega}_{\text{plate}} = (\dot{v}\sin\phi - \dot{\psi}\cos\phi\sin v)\hat{\mathbf{i}} + (\dot{\phi} + \dot{\psi}\cos v)\hat{\mathbf{j}} + (\dot{\theta} + \dot{v}\cos\phi + \dot{\psi}\sin\phi\sin v)\hat{\mathbf{k}}$

(b) $\boldsymbol{\alpha}_{\text{plate}} = \dfrac{d\boldsymbol{\omega}_{\text{plate}}}{dt} = [\dot{v}(\dot{\phi}\cos\phi - \dot{\psi}\cos\phi\cos v) + \dot{\psi}\dot{\phi}\sin\phi\sin v - \dot{\psi}\dot{\theta}\cos v - \dot{\phi}\dot{\theta}]\hat{\mathbf{i}}$
$+ [\dot{v}(\dot{\theta}\sin\phi - \dot{\psi}\sin v) - \dot{\psi}\dot{\theta}\cos\phi\sin v]\hat{\mathbf{j}}$
$+ [\dot{\psi}\dot{v}\cos v\sin\phi + \dot{\psi}\dot{\phi}\cos\phi\sin v - \dot{\phi}\dot{v}\sin\phi]\hat{\mathbf{k}}$

(c) $\mathbf{a}_C = -l(\dot{\phi}^2 + \dot{\theta}^2)\sin\phi\hat{\mathbf{i}} + (2l\dot{\phi}\dot{\theta}\cos\phi - \tfrac{5}{4}l\dot{\theta}^2)\hat{\mathbf{j}} - l\dot{\phi}^2\cos\phi\hat{\mathbf{k}}$

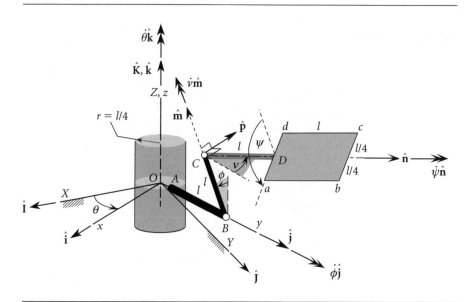

Figure P.9.2

9.3 The mass center *G* of a rigid body has a velocity $\mathbf{v} = t^3\hat{\mathbf{i}} + 4\hat{\mathbf{j}}$ m/s and an angular velocity $\boldsymbol{\omega} = 2t^2\hat{\mathbf{k}}$ rad/s, where *t* is time in seconds. The $\hat{\mathbf{i}}, \hat{\mathbf{j}}, \hat{\mathbf{k}}$ unit vectors are attached to and rotate with the rigid body. Calculate the magnitude of the acceleration $\mathbf{a}_G$ of the center of mass at $t = 2$ seconds.
{Ans.: $\mathbf{a}_G = -20\hat{\mathbf{i}} + 64\hat{\mathbf{j}}$ (m/s^2)}

9.4 The inertial angular velocity of a rigid body is $\boldsymbol{\omega} = \omega_x\hat{\mathbf{i}} + \omega_y\hat{\mathbf{j}} + \omega_z\hat{\mathbf{k}}$, where $\hat{\mathbf{i}}, \hat{\mathbf{j}}, \hat{\mathbf{k}}$ are the unit vectors of a co-moving frame whose inertial angular velocity is $\boldsymbol{\Omega} = \omega_x\hat{\mathbf{i}} + \omega_y\hat{\mathbf{j}}$. Calculate the components of angular acceleration of the rigid body in the moving frame, assuming that ω_x, ω_y and ω_z are all constant.
{Ans.: $\boldsymbol{\alpha} = \omega_y\omega_z\hat{\mathbf{i}} - \omega_x\omega_z\hat{\mathbf{j}}$}

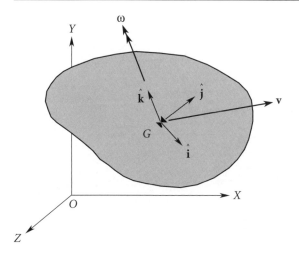

Figure P.9.3

9.5 Find the moments of inertia about the center of mass of the system of six point masses listed in the table.

Table P.9.5

Point, i	Mass m_i (kg)	x_i (m)	y_i (m)	z_i (m)
1	10	1	1	1
2	10	−1	−1	−1
3	8	4	−4	4
4	8	−2	2	−2
5	12	3	−3	−3
6	12	−3	3	3

$$\{\text{Ans.: } [\mathbf{I}_G] = \begin{bmatrix} 783.5 & 351.7 & 40.27 \\ 351.7 & 783.5 & -80.27 \\ 40.27 & -80.27 & 783.5 \end{bmatrix} (\text{kg} \cdot \text{m}^2)\}$$

9.6 Find the mass moment of inertia of the configuration of Problem 9.5 about an axis through the origin and the point with coordinates (1 m, 2 m, 2 m).
{Ans.: 621.3 kg · m^2}

9.7 A uniform slender rod of mass m and length l lies in the xy plane inclined to the x axis by the angle θ. Use the results of Example 9.10 to find the mass moments of inertia about the xyz axes passing through the center of mass G.

$$\{\text{Ans.: } [\mathbf{I}_G] = \tfrac{1}{12} ml^2 \begin{bmatrix} \sin^2 \theta & -\tfrac{1}{2} \sin 2\theta & 0 \\ -\tfrac{1}{2} \sin 2\theta & \cos^2 \theta & 0 \\ 0 & 0 & 1 \end{bmatrix}\}$$

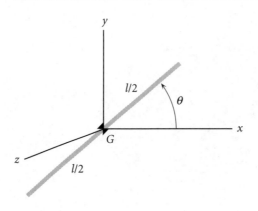

Figure P.9.7

9.8 The uniform rectangular box has a mass of 1000 kg. The dimensions of its edges are shown.
(a) Find the mass moments of inertia about the *xyz* axes.

$$\{\text{Ans.: } [\mathbf{I}_O] = \begin{bmatrix} 1666.7 & -1500 & -750 \\ -1500 & 3333.3 & -500 \\ -750 & -500 & 4333.3 \end{bmatrix} (\text{kg} \cdot \text{m}^2)\}$$

(b) Find the principal moments of inertia and the principal directions about the *xyz* axes through *O*.

{Partial ans.: $I_1 = 568.9 \text{ kg} \cdot \text{m}^2, \hat{\mathbf{v}}_1 = 0.8366\hat{\mathbf{i}} + 0.4960\hat{\mathbf{j}} + 0.2326\hat{\mathbf{k}}$}

(c) Find the moment of inertia about the line through *O* and the point with coordinates (3 m, 2 m, 1 m).

{Ans.: 583.3 kg · m²}

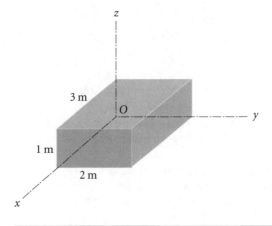

Figure P.9.8

9.9 A taxiing airplane turns about its vertical axis with an angular velocity Ω while its propeller spins at an angular velocity $\omega = \dot\theta$. Determine the components of the angular momentum of the propeller about the body-fixed xyz axes centered at P. Treat the propeller as a uniform slender rod of mass m and length l.

{Ans.: $\mathbf{H}_P = \dfrac{1}{12}m\omega l^2 \hat{\mathbf{i}} - \dfrac{1}{24}m\Omega l^2 \sin 2\theta \hat{\mathbf{j}} + \left(\dfrac{1}{12}ml^2 \cos^2\theta + md^2\right)\Omega \hat{\mathbf{k}}$}

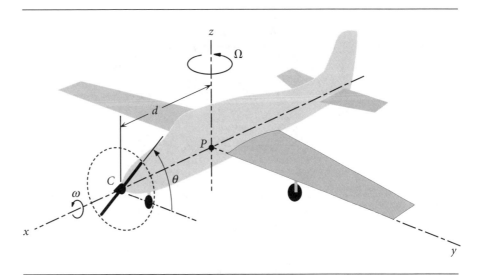

Figure P.9.9

9.10 Relative to an $\hat{\mathbf{i}}, \hat{\mathbf{j}}, \hat{\mathbf{k}}$ frame of reference the components of angular momentum $\mathbf{H}$ are given by

$$\{\mathbf{H}\} = \begin{bmatrix} 1000 & 0 & -300 \\ 0 & 1000 & 500 \\ -300 & 500 & 1000 \end{bmatrix} \begin{Bmatrix} \omega_x \\ \omega_y \\ \omega_z \end{Bmatrix} \; (\text{kg} \cdot \text{m}^2/\text{s})$$

where ω_x, ω_y and ω_z are the components of the angular velocity $\boldsymbol{\omega}$. Find the components $\boldsymbol{\omega}$ such that $\{\mathbf{H}\} = 1000\{\boldsymbol{\omega}\}$, where the magnitude of $\boldsymbol{\omega}$ is 20 rad/s.
{Ans.: $\boldsymbol{\omega} = 17.15\hat{\mathbf{i}} + 10.29\hat{\mathbf{j}}$ (rad/s)}

9.11 Relative to a body-fixed xyz frame $[I_G] = \begin{bmatrix} 10 & 0 & 0 \\ 0 & 20 & 0 \\ 0 & 0 & 30 \end{bmatrix}$ (kg · m^2) and $\boldsymbol{\omega} = 2t^2\hat{\mathbf{i}} + 4\hat{\mathbf{j}} + 3t\hat{\mathbf{k}}$ (rad/s), where t is the time in seconds. Calculate the magnitude of the net moment about the center of mass G at $t = 3$ s.
{Ans.: 3374 N · m}

9.12 In Example 9.11, the system is at rest when a 100 N force is applied to point A as shown. Calculate the inertial components of angular acceleration at that instant.
{Ans.: $\alpha_X = 143.9$ rad/s^2, $\alpha_Y = 553.1$ rad/s^2, $\alpha_Z = 7.61$ rad/s^2}

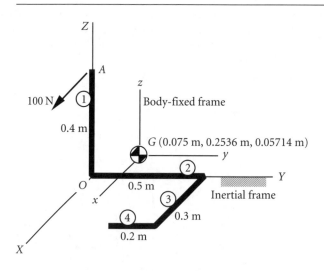

Figure P.9.12

9.13 The body-fixed *xyz* axes pass through the center of mass *G* of the airplane and are the principal axes of inertia. The moments of inertia about these axes are *A*, *B* and *C*, respectively. The airplane is in a level turn of radius *R* with a speed *v*.
   (a) Calculate the bank angle θ.
   (b) Use Euler's equations to calculate the rolling moment M_y which must be applied by the aerodynamic surfaces.
   {Ans.: (a) $\theta = \tan^{-1} v^2/Rg$; (b) $M_y = v^2 \sin 2\theta (C - A)/2R^2$}

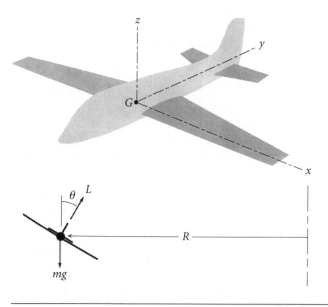

Figure P.9.13

9.14 The airplane in Problem 9.11 is spinning with an angular velocity ω_Z about the vertical Z axis. The nose is pitched down at the angle α. What external moments must accompany this maneuver?
{Ans.: $M_y = M_z = 0$, $M_x = \omega_Z^2 \sin 2\alpha (C - B)/2$}

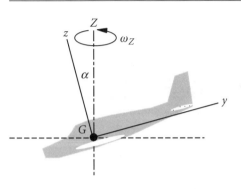

Figure P.9.14

9.15 Two identical slender rods of mass m and length l are rigidly joined together at an angle θ at point C, their 2/3 point. Determine the bearing reactions at A and B if the shaft rotates at a constant angular velocity ω. Neglect gravity and assume that the only bearing forces are normal to rod AB.
{Ans.: $||\mathbf{F}_A|| = m\omega^2 l \sin \theta (1 + 2 \cos \theta)/18$, $||\mathbf{F}_B|| = m\omega^2 l \sin \theta (1 - \cos \theta)/9$}

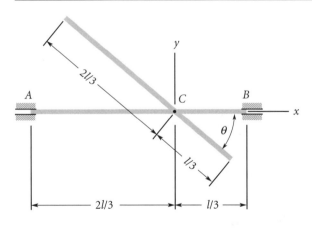

Figure P.9.15

9.16 The flywheel ($A = B = 5 \text{ kg} \cdot \text{m}^2$, $C = 10 \text{ kg} \cdot \text{m}^2$) spins at a constant angular velocity of $\boldsymbol{\omega}_s = 100\hat{\mathbf{k}}$ (rad/s). It is supported by a *massless* gimbal which is mounted on the platform as shown. The gimbal is initially stationary relative to the platform, which rotates with a constant angular velocity of $\boldsymbol{\omega}_p = 0.5\hat{\mathbf{j}}$ (rad/s). What will be the gimbal's angular acceleration when the torquer applies a torque of $600\hat{\mathbf{i}}$ (N · m) to the flywheel?
{Ans.: $70\hat{\mathbf{i}} \text{ rad/s}^2$}

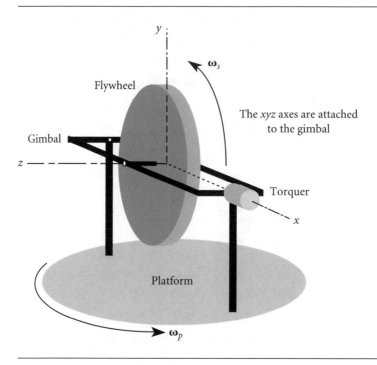

Figure P.9.16

9.17 A uniform slender rod of length L and mass m is attached by a smooth pin at O to a vertical shaft which rotates at constant angular velocity ω. Use Euler's equations and the body frame shown to calculate ω at the instant shown.
{Ans.: $\omega = \sqrt{3g/(2L\cos\theta)}$}

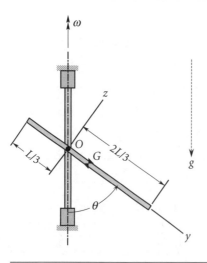

Figure P.9.17

9.18 A uniform, thin, circular disk of mass 10 kg spins at a constant angular velocity of 630 rad/s about axis OG, which is normal to the disk, and pivots about the frictionless ball joint at O. Neglecting the mass of the shaft OG, determine the rate of precession if OG remains horizontal as shown. Gravity acts down, as shown. G is the center of mass, and the y axis remains fixed in space. The moments of inertia about G are $I_{G_z} = 0.02812$ kg·m², and $I_{G_x} = I_{G_y} = 0.01406$ kg·m².
{Ans.: 1.38 rad/s}

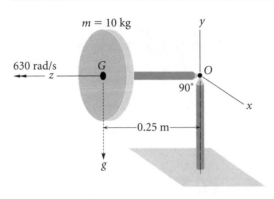

Figure P.9.18

9.19 At the end of its take-off run, an airplane with retractable landing gear leaves the runway with a speed of 130 km/hr. The gear rotates into the wing with an angular velocity of 0.8 rad/s with the wheels still spinning. Calculate the gyroscopic bending moment in the wheel bearing B. The wheels have a diameter of 0.6 m, a mass of 25 kg and a radius of gyration of 0.2 m.
{Ans.: 96.3 N·m}.

Figure P.9.19

9.20 The gyro rotor, including shaft AB, has a mass of 4 kg and a radius of gyration 7 cm around AB. The rotor spins at 10 000 revolutions per minute while also being forced to

rotate around the gimbal axis CC at 2 rad/s. What are the transverse forces exerted on the shaft at A and B? Neglect gravity.
{Ans.: 1.03 kN}

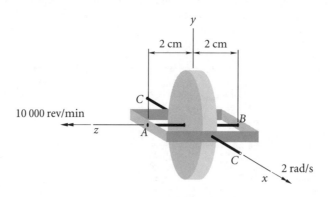

Figure P.9.20

9.21 A jet aircraft is making a level, 2.5 km radius turn to the left at a speed of 650 km/hr. The rotor of the turbojet engine has a mass of 200 kg, a radius of gyration of 0.25 m and rotates at 15 000 revolutions per minute clockwise as viewed from the front of the airplane. Calculate the gyroscopic moment that the engine exerts on the airframe and specify whether it tends to pitch the nose up or down.
{Ans.: 1.418 kN · m; pitch down}

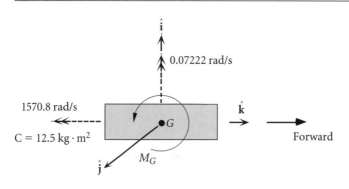

Figure P.9.21

9.22 A cylindrical rotor of mass 10 kg, radius 0.05 m and length 0.60 m is simply-supported at each end in a cradle that rotates at a constant 20 rad/s counterclockwise as viewed from above. Relative to the cradle, the rotor spins at 200 rad/s counterclockwise as viewed from the right (from B towards A). Assuming there is no gravity, calculate the bearing reactions R_A and R_B. Use the co-moving xyz frame shown, which is attached to the cradle but not to the rotor.
{Ans.: $R_A = -R_B = 83.3$ N}

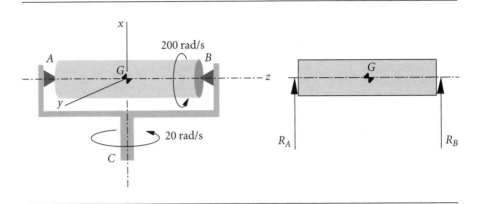

Figure P.9.22

9.23 The Euler angles of a rigid body are $\phi = 50°, \theta = 25°$ and $\psi = 70°$. Calculate the angle (a positive number) between the body-fixed x axis and the inertial X axis.
{Ans.: 115.6°}

9.24 Consider a rigid body experiencing rotational motion associated with angular velocity $\boldsymbol{\omega}$. The inertia tensor (relative to body-fixed axes through the center of mass G) is

$$\begin{bmatrix} 20 & -10 & 0 \\ -10 & 30 & 0 \\ 0 & 0 & 40 \end{bmatrix} (\text{kg} \cdot \text{m}^2)$$

and $\boldsymbol{\omega} = 10\hat{\mathbf{i}} + 20\hat{\mathbf{j}} + 30\hat{\mathbf{k}}$ (rad/s). Calculate
(a) the angular momentum $\mathbf{H}_G$, and
(b) the rotational kinetic energy (about G).
{Partial ans.: (b) $T_R = 23\,000$ J}

Chapter 10
Satellite attitude dynamics

Chapter outline

10.1	Introduction	475
10.2	Torque-free motion	476
10.3	Stability of torque-free motion	486
10.4	Dual-spin spacecraft	491
10.5	Nutation damper	495
10.6	Coning maneuver	503
10.7	Attitude control thrusters	506
10.8	Yo-yo despin mechanism	509
10.9	Gyroscopic attitude control	516
10.10	Gravity-gradient stabilization	530
Problems		543

10.1 Introduction

In this chapter we apply the equations of rigid body motion presented in Chapter 9 to the study of the attitude dynamics of satellites. We begin with spin-stabilized spacecraft. Spinning a satellite around its axis is a very simple way to keep the vehicle pointed in a desired direction. We investigate the stability of a spinning satellite to show that only oblate spinners are stable over long times. Overcoming this restriction on the shape of spin-stabilized spacecraft led to the development of dual-spin vehicles, which consist of two interconnected segments rotating at different rates about a common axis. We consider the stability of that type of configuration as well. The nutation damper and its effect on the stability of spin-stabilized spacecraft is covered next.

The rest of the chapter is devoted to some of the common means of changing the attitude or motion of a spacecraft by applying external or internal forces or torques. The coning maneuver changes the attitude of a spinning spacecraft by using thrusters to apply impulsive torque, which alters the angular momentum and hence the orientation of the spacecraft. The much-used yo-yo despin maneuver reduces or eliminates the spin rate by releasing small masses attached to cords initially wrapped around the cylindrical vehicle.

An alternative to spin stabilization is three-axis stabilization by gyroscopic attitude control. In this case, the vehicle does not continuously rotate. Instead, the desired attitude is maintained by the spin of small wheels within the spacecraft. These are called reaction wheels or momentum wheels. If allowed to pivot relative to the vehicle, they are known as control moment gyros. The attitude of the vehicle can be changed by varying the speed or orientation of these internal gyros. Small thrusters may also be used to supplement the gyroscopic attitude control and to hold the spacecraft orientation fixed when it is necessary to despin or reorient gyros that have become saturated (reached their maximum spin rate or deflection) over time.

The chapter concludes with a discussion of how the earth's gravitational field by itself can stabilize the attitude of large satellites such as the space shuttle or space station in low earth orbits.

10.2 Torque-free motion

Gravity is the only force acting on a satellite coasting in orbit (if we neglect secondary drag forces and the gravitational influence of bodies other than the planet being orbited). Unless the satellite is unusually large, the gravitational force is concentrated at the center of mass G. Since the net moment about the center of mass is zero, the satellite is 'torque-free', and according to Equation 9.30,

$$\dot{\mathbf{H}}_G = \mathbf{0} \qquad (10.1)$$

The angular momentum $\mathbf{H}_G$ about the center of mass does not depend on time. It is a vector fixed in inertial space. We will use $\mathbf{H}_G$ to define the Z axis of an inertial frame, as shown in Figure 10.1. The xyz axes in the figure comprise the principal body frame, centered at G. The angle between the z axis and $\mathbf{H}_G$ is (by definition of the Euler angles) the nutation angle θ. Let us determine the conditions for which θ is constant. From the dot product operation we know that

$$\cos\theta = \frac{\mathbf{H}_G}{\|\mathbf{H}_G\|} \cdot \hat{\mathbf{k}}$$

Differentiating this expression with respect to time, keeping in mind Equation 10.1, we get

$$\frac{d\cos\theta}{dt} = \frac{\mathbf{H}_G}{\|\mathbf{H}_G\|} \cdot \frac{d\hat{\mathbf{k}}}{dt}$$

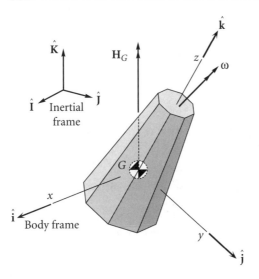

Figure 10.1 Rotationally symmetric satellite in torque-free motion.

But $d\hat{\mathbf{k}}/dt = \boldsymbol{\omega} \times \hat{\mathbf{k}}$, according to Equation 1.24, so

$$\frac{d\cos\theta}{dt} = \frac{\mathbf{H}_G \cdot (\boldsymbol{\omega} \times \hat{\mathbf{k}})}{\|\mathbf{H}_G\|} \quad (10.2)$$

Now

$$\boldsymbol{\omega} \times \hat{\mathbf{k}} = \begin{vmatrix} \hat{\mathbf{i}} & \hat{\mathbf{j}} & \hat{\mathbf{k}} \\ \omega_x & \omega_y & \omega_z \\ 0 & 0 & 1 \end{vmatrix} = \omega_y \hat{\mathbf{i}} - \omega_x \hat{\mathbf{j}}$$

Furthermore, we know from Equation 9.67 that the angular momentum is related to the angular velocity in the principal body frame by the expression

$$\mathbf{H}_G = A\omega_x \hat{\mathbf{i}} + B\omega_y \hat{\mathbf{j}} + C\omega_z \hat{\mathbf{k}}$$

Thus

$$\mathbf{H}_G \cdot (\boldsymbol{\omega} \times \hat{\mathbf{k}}) = (A\omega_x \hat{\mathbf{i}} + B\omega_y \hat{\mathbf{j}} + C\omega_z \hat{\mathbf{k}}) \cdot (\omega_y \hat{\mathbf{i}} - \omega_x \hat{\mathbf{j}}) = (A - B)\omega_x \omega_y$$

so that Equation 10.2 can be written

$$\dot{\theta} = \omega_n = -\frac{(A-B)\omega_x \omega_y}{\|\mathbf{H}_G\| \sin\theta} \quad (10.3)$$

From this we see that the nutation rate vanishes only if $A = B$. If $A \neq B$, the nutation angle θ will not in general be constant.

Relative to the body frame, Equation 10.1 is written (cf. Equation 1.28)

$$\dot{\mathbf{H}}_{G\mathrm{rel}} + \boldsymbol{\omega} \times \mathbf{H}_G = \mathbf{0}$$

This is Euler's equation with $\mathbf{M}_{G_{net}} = \mathbf{0}$, the components of which are given by Equations 9.72,

$$A\dot{\omega}_x + (C - B)\omega_z\omega_y = 0$$
$$B\dot{\omega}_y + (A - C)\omega_x\omega_z = 0 \quad (10.4)$$
$$C\dot{\omega}_z + (B - A)\omega_y\omega_x = 0$$

In the interest of simplicity, let us consider the special case illustrated in Figure 10.1, namely that in which the z axis is an axis of rotational symmetry, so that $A = B$. Then Equations 10.4 become

$$A\dot{\omega}_x + (C - A)\omega_z\omega_y = 0$$
$$A\dot{\omega}_y + (A - C)\omega_x\omega_z = 0 \quad (10.5)$$
$$C\dot{\omega}_z = 0$$

From Equation 10.5$_3$ we see that

$$\omega_z = \omega_0 \quad \text{(constant)} \quad (10.6)$$

The assumption of rotational symmetry therefore reduces the three differential equations 10.4 to just two. Substituting Equation 10.6 into Equations 10.5$_1$ and 10.5$_2$ and introducing the notation

$$\lambda = \frac{A - C}{A}\omega_0 \quad (10.7)$$

they can be written

$$\dot{\omega}_x - \lambda\omega_y = 0$$
$$\dot{\omega}_y + \lambda\omega_x = 0 \quad (10.8)$$

To reduce these two equations in ω_x and ω_y down to just one equation in ω_x, we first differentiate Equation 10.8$_1$ with respect to time to get

$$\ddot{\omega}_x - \lambda\dot{\omega}_y = 0 \quad (10.9)$$

We then solve Equation 10.8$_2$ for $\dot{\omega}_y$ and substitute the result into Equation 10.9, which leads to

$$\ddot{\omega}_x + \lambda^2\omega_x = 0 \quad (10.10)$$

The solution of this well-known differential equation is

$$\omega_x = \Omega \sin \lambda t \quad (10.11)$$

where the constant amplitude Ω ($\Omega \neq 0$) has yet to be determined. (Without loss of generality, we have set the phase angle, the other constant of integration, equal to zero.) Substituting Equation 10.11 back into Equation 10.8$_1$ yields the solution for ω_y,

$$\omega_y = \frac{1}{\lambda}\frac{d\omega_x}{dt} = \frac{1}{\lambda}\frac{d}{dt}(\Omega \sin \lambda t)$$

or

$$\omega_y = \Omega \cos \lambda t \qquad (10.12)$$

Equations 10.6, 10.11 and 10.12 give the components of the absolute angular velocity $\boldsymbol{\omega}$ along the three principal body axes,

$$\boldsymbol{\omega} = \Omega \sin \lambda t \hat{\mathbf{i}} + \Omega \cos \lambda t \hat{\mathbf{j}} + \omega_0 \hat{\mathbf{k}}$$

or

$$\boldsymbol{\omega} = \boldsymbol{\omega}_\perp + \omega_0 \hat{\mathbf{k}} \qquad (10.13)$$

where

$$\boldsymbol{\omega}_\perp = \Omega(\sin \lambda t \hat{\mathbf{i}} + \cos \lambda t \hat{\mathbf{j}}) \qquad (10.14)$$

$\boldsymbol{\omega}_\perp$ ('omega-perp') is the component of $\boldsymbol{\omega}$ normal to the z axis. It sweeps out a circle of radius Ω in the xy plane at an angular velocity λ. Thus, $\boldsymbol{\omega}$ sweeps out a cone, as illustrated in Figure 10.2.

From Equations 9.115, the three Euler orientation angles (and their rates) are related to the angular velocity components ω_x, ω_y and ω_z by

$$\omega_p = \dot{\phi} = \frac{1}{\sin \theta}(\omega_x \sin \psi + \omega_y \cos \psi)$$

$$\omega_n = \dot{\theta} = \omega_x \cos \psi - \omega_y \sin \psi$$

$$\omega_s = \dot{\psi} = -\frac{1}{\tan \theta}(\omega_x \sin \psi + \omega_y \cos \psi) + \omega_z$$

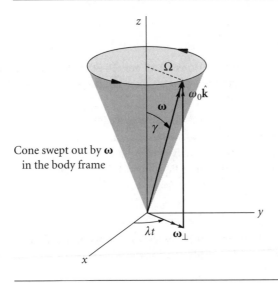

Figure 10.2 Components of the angular velocity * in the body frame.

Substituting Equations 10.6, 10.11 and 10.12 into these three equations yields

$$\omega_p = \frac{\Omega}{\sin\theta} \cos(\lambda t - \psi)$$
$$\omega_n = \Omega \sin(\lambda t - \psi) \qquad (10.15)$$
$$\omega_s = \omega_0 - \frac{\Omega}{\tan\theta} \cos(\lambda t - \psi)$$

Since $A = B$, we know from Equation 10.3 that $\omega_n = 0$. It follows from Equation 10.15_2 that

$$\psi = \lambda t \qquad (10.16)$$

(Actually, $\lambda t - \psi = n\pi, n = 0, 1, 2, \ldots$. We can set $n = 0$ without loss of generality.) Substituting Equation 10.16 into Equations 10.15_1 and 10.15_3 yields

$$\omega_p = \frac{\Omega}{\sin\theta} \qquad (10.17)$$

and

$$\omega_s = \omega_0 - \frac{\Omega}{\tan\theta} \qquad (10.18)$$

We have thus obtained the Euler angle rates ω_p and ω_s in terms of the components of the angular velocity $\boldsymbol{\omega}$.

Differentiating Equation 10.16 with respect to time shows that

$$\lambda = \dot{\psi} = \omega_s \qquad (10.19)$$

That is, the rate λ at which $\boldsymbol{\omega}$ rotates around the body z axis equals the spin rate. Substituting the spin rate for λ in Equation 10.7 shows that ω_s is related to ω_0 alone,

$$\omega_s = \frac{A - C}{A} \omega_0 \qquad (10.20)$$

Eliminating ω_s from Equations 10.18 and 10.20 yields the relationship between the magnitudes of the orthogonal components of the angular velocity in Equation 10.13,

$$\Omega = \frac{C}{A} \omega_0 \tan\theta \qquad (10.21)$$

A similar relationship exists between ω_p and ω_s, which generally are *not* orthogonal. Substitute Equation 10.21 into Equation 10.17 to obtain

$$\omega_0 = \frac{A}{C} \omega_p \cos\theta \qquad (10.22)$$

Placing this in Equation 10.20 leaves an expression involving only ω_p and ω_s, from which we get a useful formula relating the precession of a torque-free body to its spin,

$$\omega_p = \frac{C}{A - C \cos\theta} \omega_s \qquad (10.23)$$

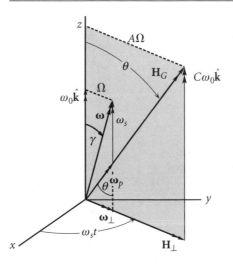

Figure 10.3 Angular velocity and angular momentum vectors in the body frame.

Observe that if $A > C$ (i.e., the body is *prolate*, like a soup can or an American football), then ω_p has the same sign as ω_s, which means the precession is *prograde*. For an *oblate* body (like a tuna fish can or a frisbee), $A < C$ and the precession is *retrograde*.

The components of angular momentum along the body frame axes are obtained from the body frame components of $\boldsymbol{\omega}$,

$$\mathbf{H}_G = A\omega_x \hat{\mathbf{i}} + A\omega_y \hat{\mathbf{j}} + C\omega_z \hat{\mathbf{k}}$$

or

$$\mathbf{H}_G = \mathbf{H}_\perp + C\omega_0 \hat{\mathbf{k}} \tag{10.24}$$

where

$$\mathbf{H}_\perp = A\Omega(\sin \omega_s t\, \hat{\mathbf{i}} + \cos \omega_s t\, \hat{\mathbf{j}}) = A\boldsymbol{\omega}_\perp \tag{10.25}$$

Since $\omega_0 \hat{\mathbf{k}}$ and $C\omega_0 \hat{\mathbf{k}}$ are colinear, as are $\boldsymbol{\omega}_\perp$ and $A\boldsymbol{\omega}_\perp$, it follows that $\hat{\mathbf{k}}$, $\boldsymbol{\omega}$ and $\mathbf{H}_G$ all lie in the same plane. $\mathbf{H}_G$ and $\boldsymbol{\omega}$ both rotate around the z axis at the same rate ω_s. These details are illustrated in Figure 10.3. See how the precession and spin angular velocities, $\boldsymbol{\omega}_p$ and $\boldsymbol{\omega}_s$, add up vectorially to give $\boldsymbol{\omega}$. Note also that from the point of view of inertial space, where $\mathbf{H}_G$ is fixed, $\boldsymbol{\omega}$ and $\hat{\mathbf{k}}$ rotate around $\mathbf{H}_G$ with angular velocity ω_p.

Let γ be the angle between $\boldsymbol{\omega}$ and the spin axis z, as shown in Figures 10.2 and 10.3. γ is sometimes referred to as the wobble angle. Then

$$\cos \gamma = \frac{\omega_z}{\|\boldsymbol{\omega}\|} = \frac{\omega_0}{\sqrt{\Omega^2 + \omega_0^2}} = \frac{\omega_0}{\sqrt{\left(\omega_0 \dfrac{C}{A} \tan \theta\right)^2 + \omega_0^2}} = \frac{A}{\sqrt{A^2 + C^2 \tan^2 \theta}}$$

γ is constant, since A, C and θ are fixed. Using trig identities, this expression can be recast as

$$\cos \gamma = \frac{\cos \theta}{\sqrt{\dfrac{C^2}{A^2} + \left(1 - \dfrac{C^2}{A^2}\right)\cos^2 \theta}} \tag{10.26}$$

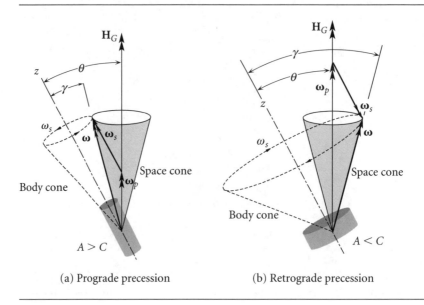

Figure 10.4 Space and body cones for a rotationally symmetric body in torque-free motion. (a) Prolate body. (b) Oblate body.

From this we conclude that if $A > C$, then $\gamma < \theta$, whereas $C > A$ means $\gamma > \theta$. That is, the angular velocity vector $\boldsymbol{\omega}$ lies between the z axis and the angular momentum vector $\mathbf{H}_G$ when $A > C$ (prolate body). On the other hand, when $C > A$ (oblate body), $\mathbf{H}_G$ lies between the z axis and $\boldsymbol{\omega}$. These two situations are illustrated in Figure 10.4, which also shows the *body cone* and *space cone*. The space cone is swept out in inertial space by the angular velocity vector as it rotates with angular velocity ω_p around $\mathbf{H}_G$, whereas the body cone is the trace of $\boldsymbol{\omega}$ in the body frame as it rotates with angular velocity ω_s about the z axis. From inertial space, the motion may be visualized as the body cone rolling on the space cone, with the line of contact being the angular velocity vector. From the body frame it appears as though the space cone rolls on the body cone. Figure 10.4 graphically confirms our deduction from Equation 10.23, namely, that precession and spin are in the same direction for prolate bodies and opposite in direction for oblate shapes.

Finally, we know from Equations 10.24 and 10.25 that the magnitude $\|\mathbf{H}_G\|$ of the angular momentum is

$$\|\mathbf{H}_G\| = \sqrt{A^2 \Omega^2 + C^2 \omega_0^2}$$

Using Equation 10.21, we can write this as

$$\|\mathbf{H}_G\| = \sqrt{A^2 \left(\omega_0 \frac{C}{A} \tan\theta\right)^2 + C^2 \omega_0^2} = C\omega_0 \sqrt{1 + \tan^2\theta} = \frac{C\omega_0}{\cos\theta}$$

Substituting Equation 10.22 into this expression yields a surprisingly simple formula for the magnitude of the angular momentum,

$$\|\mathbf{H}_G\| = A\omega_p \qquad (10.27)$$

EXAMPLE 10.1

A cylindrical shell is rotating in torque-free motion about its longitudinal axis. If the axis is wobbling slightly, determine the ratios of l/r for which the precession will be prograde or retrograde.

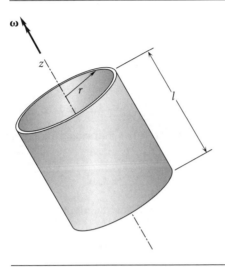

Figure 10.5 Cylindrical shell in torque-free motion.

Figure 9.9(b) shows the moments of inertia of a thin-walled circular cylinder,

$$C = mr^2 \quad A = \frac{1}{2}mr^2 + \frac{1}{12}ml^2$$

According to Equation 10.23 and Figure 10.4, direct or prograde precession exists if $A > C$, that is, if

$$\frac{1}{2}mr^2 + \frac{1}{12}ml^2 > mr^2$$

or

$$\frac{1}{12}ml^2 > \frac{1}{2}mr^2$$

Thus

$$l > 2.45r \quad \Rightarrow \quad \text{Direct precession.}$$
$$l < 2.45r \quad \Rightarrow \quad \text{Retrograde precession.}$$

EXAMPLE 10.2

In the previous example, let $r = 1$ m, $l = 3$ m, $m = 100$ kg and the nutation angle θ is 20°. How long does it take the cylinder to precess through 180° if the spin rate is 2π radians per minute?

Since $l > 2.45r$, the precession is direct. Furthermore,

$$C = mr^2 = 100 \cdot 1^2 = 100 \text{ kg} \cdot \text{m}^2$$

$$A = \frac{1}{2}mr^2 + \frac{1}{12}ml^2 = \frac{1}{2} \cdot 100 \cdot 1^2 + \frac{1}{12} 100 \cdot 3^2 = 125 \text{ kg} \cdot \text{m}^2$$

(Example 10.2 continued)

Thus, Equation 10.23 yields

$$\omega_p = \frac{C}{A - C} \frac{\omega_s}{\cos\theta} = \frac{100}{125 - 100 \cos 20°} \frac{2\pi}{} = 26.75 \text{ rad/min}$$

At this rate, the time for a precession angle of 180° is

$$t = \frac{\pi}{\omega_p} = \underline{0.1175 \text{ min}}$$

EXAMPLE 10.3

What is the torque-free motion of a satellite for which $A = B = C$?

If $A = B = C$, the satellite is spherically symmetric. Any orthogonal triad at G is a principal body frame, so $\mathbf{H}_G$ and $\boldsymbol{\omega}$ are collinear,

$$\mathbf{H}_G = C\boldsymbol{\omega}$$

Substituting this and $\mathbf{M}_{G_{net}} = \mathbf{0}$, into Euler's equations, Equation 10.72a, yields

$$C\frac{d\boldsymbol{\omega}}{dt} + \boldsymbol{\omega} \times (C\boldsymbol{\omega}) = \mathbf{0}$$

That is,

$$\boldsymbol{\omega} = \text{constant}$$

The angular velocity vector of a spherically symmetric satellite is fixed in magnitude and direction.

EXAMPLE 10.4

The inertial components of the angular momentum of a torque-free rigid body are

$$\mathbf{H}_G = 320\hat{\mathbf{I}} - 375\hat{\mathbf{J}} + 450\hat{\mathbf{K}} \text{ (kg} \cdot \text{m}^2/\text{s)} \quad \text{(a)}$$

The Euler angles are

$$\phi = 20° \quad \theta = 50° \quad \psi = 75° \quad \text{(b)}$$

If the inertia tensor in the body-fixed principal frame is

$$[\mathbf{I}_G] = \begin{bmatrix} 1000 & 0 & 0 \\ 0 & 2000 & 0 \\ 0 & 0 & 3000 \end{bmatrix} \text{ (kg} \cdot \text{m}^2) \quad \text{(c)}$$

calculate the inertial components of the (absolute) angular acceleration.

Substituting the Euler angles from (b) into Equation 9.117, we obtain the matrix of the transformation from the inertial frame to the body frame,

$$[\mathbf{Q}]_{Xx} = \begin{bmatrix} 0.03086 & 0.6720 & 0.7399 \\ -0.9646 & -0.1740 & 0.1983 \\ 0.2620 & -0.7198 & 0.6428 \end{bmatrix} \quad \text{(d)}$$

10.2 Torque-free motion

We use this to obtain the components of $\mathbf{H}_G$ in the body frame,

$$\{\mathbf{H}_G\}_x = [\mathbf{Q}]_{Xx}\{\mathbf{H}_G\}_X = \begin{bmatrix} 0.03086 & 0.6720 & 0.7399 \\ -0.9646 & -0.1740 & 0.1983 \\ 0.2620 & -0.7198 & 0.6428 \end{bmatrix} \begin{Bmatrix} 320 \\ -375 \\ 450 \end{Bmatrix}$$

$$= \begin{Bmatrix} 90.86 \\ -154.2 \\ 643.0 \end{Bmatrix} (\text{kg} \cdot \text{m}^2/\text{s}) \quad (e)$$

In the body frame $\{\mathbf{H}_G\}_x = [\mathbf{I}_G]\{\boldsymbol{\omega}\}_x$, where $\{\boldsymbol{\omega}\}_x$ are the components of angular velocity in the body frame. Thus

$$\begin{Bmatrix} 90.86 \\ -154.2 \\ 643.0 \end{Bmatrix} = \begin{bmatrix} 1000 & 0 & 0 \\ 0 & 2000 & 0 \\ 0 & 0 & 3000 \end{bmatrix} \{\boldsymbol{\omega}\}_x$$

or, solving for $\{\boldsymbol{\omega}\}_x$,

$$\{\boldsymbol{\omega}\}_x = \begin{bmatrix} 1000 & 0 & 0 \\ 0 & 2000 & 0 \\ 0 & 0 & 3000 \end{bmatrix}^{-1} \begin{Bmatrix} 90.86 \\ -154.2 \\ 643.0 \end{Bmatrix} = \begin{Bmatrix} 0.09086 \\ -0.07709 \\ 0.2144 \end{Bmatrix} (\text{rad/s}) \quad (f)$$

Euler's equations of motion (Equation 9.72a) may be written for the case at hand as

$$[\mathbf{I}_G]\{\boldsymbol{\alpha}\}_x + \{\boldsymbol{\omega}\}_x \times ([\mathbf{I}_G]\{\boldsymbol{\omega}\}_x) = \{0\} \quad (g)$$

where $\{\boldsymbol{\alpha}\}_x$ is the absolute acceleration in body frame components. Substituting (c) and (f) into this expression, we get

$$\begin{bmatrix} 1000 & 0 & 0 \\ 0 & 2000 & 0 \\ 0 & 0 & 3000 \end{bmatrix} \{\boldsymbol{\alpha}\}_x + \begin{Bmatrix} 0.09086 \\ -0.07709 \\ 0.2144 \end{Bmatrix}$$

$$\times \left(\begin{bmatrix} 1000 & 0 & 0 \\ 0 & 2000 & 0 \\ 0 & 0 & 3000 \end{bmatrix} \begin{Bmatrix} 0.09086 \\ -0.07709 \\ 0.2144 \end{Bmatrix} \right) = \begin{Bmatrix} 0 \\ 0 \\ 0 \end{Bmatrix}$$

$$\begin{bmatrix} 1000 & 0 & 0 \\ 0 & 2000 & 0 \\ 0 & 0 & 3000 \end{bmatrix} \{\boldsymbol{\alpha}\}_x + \begin{Bmatrix} -16.52 \\ -38.95 \\ -7.005 \end{Bmatrix} = \begin{Bmatrix} 0 \\ 0 \\ 0 \end{Bmatrix}$$

so that, finally,

$$\{\boldsymbol{\alpha}\}_x = -\begin{bmatrix} 1000 & 0 & 0 \\ 0 & 2000 & 0 \\ 0 & 0 & 3000 \end{bmatrix}^{-1} \begin{Bmatrix} -16.52 \\ -38.95 \\ -7.005 \end{Bmatrix} = \begin{Bmatrix} 0.01652 \\ 0.01948 \\ 0.002335 \end{Bmatrix} (\text{rad/s}^2) \quad (h)$$

(Example 10.4 continued)

These are the components of the angular acceleration in the body frame. To transform them into the inertial frame we use

$$\{\alpha\}_X = [Q]_{xX}\{\alpha\}_x = ([Q]_{Xx})^T\{\alpha\}_x$$

$$= \begin{bmatrix} 0.03086 & -0.9646 & 0.2620 \\ 0.6720 & -0.1740 & -0.7198 \\ 0.7399 & 0.1983 & 0.6428 \end{bmatrix} \begin{Bmatrix} 0.01652 \\ 0.01948 \\ 0.002335 \end{Bmatrix} = \begin{Bmatrix} -0.01766 \\ 0.006033 \\ 0.01759 \end{Bmatrix} (\text{rad/s}^2)$$

That is,

$$\underline{\boldsymbol{\alpha} = -0.01766\hat{\mathbf{I}} + 0.006033\hat{\mathbf{J}} + 0.01759\hat{\mathbf{K}}\ (\text{rad/s}^2)}$$

10.3 STABILITY OF TORQUE-FREE MOTION

Let a rigid body be in torque-free motion with its angular velocity vector directed along the principal body z axis, so that $\boldsymbol{\omega} = \omega_0 \hat{\mathbf{k}}$, where ω_0 is constant. The nutation angle is zero and there is no precession. Let us perturb the motion slightly, as illustrated in Figure 10.6, so that

$$\omega_x = \delta\omega_x \quad \omega_y = \delta\omega_y \quad \omega_z = \omega_0 + \delta\omega_z \tag{10.28}$$

As in Chapter 7, 'δ' means a very small quantity. In this case, $\delta\omega_x \ll \omega_0$ and $\delta\omega_y \ll \omega_0$. Thus, the angular velocity vector has become slightly inclined to the z axis. For torque-free motion, $M_{G_x} = M_{G_y} = M_{G_z} = 0$, so that Euler's equations (Equations 9.72b)

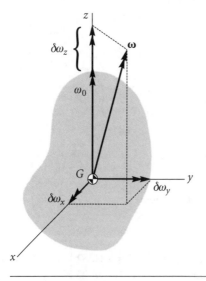

Figure 10.6 Principal body axes of a rigid body rotating primarily about the body z axis.

10.3 Stability of torque-free motion

become

$$A\dot{\omega}_x + (C - B)\omega_y\omega_z = 0$$
$$B\dot{\omega}_y + (A - C)\omega_x\omega_z = 0 \qquad (10.29)$$
$$C\dot{\omega}_z + (B - A)\omega_x\omega_y = 0$$

Observe that we have not assumed $A = B$, as we did in the previous section. Substituting Equations 10.28 into Equations 10.29 and keeping in mind our assumption that $\dot{\omega}_0 = 0$, we get

$$A\delta\dot{\omega}_x + (C - B)\omega_0\delta\omega_y + (C - B)\delta\omega_y\delta\omega_z = 0$$
$$B\delta\dot{\omega}_y + (A - C)\omega_0\delta\omega_x + (C - B)\delta\omega_x\delta\omega_z = 0 \qquad (10.30)$$
$$C\delta\dot{\omega}_z + (B - A)\delta\omega_x\delta\omega_y = 0$$

Neglecting all products of the $\delta\omega$s (because they are arbitrarily small), Equations 10.30 become

$$A\delta\dot{\omega}_x + (C - B)\omega_0\delta\omega_y = 0$$
$$B\delta\dot{\omega}_y + (A - C)\omega_0\delta\omega_x = 0 \qquad (10.31)$$
$$C\delta\dot{\omega}_z = 0$$

Equation 10.31$_3$ implies that $\delta\omega_z$ is constant. Differentiating Equation 10.31$_1$ with respect to time, we get

$$A\delta\ddot{\omega}_x + (C - B)\omega_0\delta\dot{\omega}_y = 0 \qquad (10.32)$$

Solving Equation 10.31$_2$ for $\delta\dot{\omega}_y$ yields $\delta\dot{\omega}_y = -[(A - C)/B]\omega_0\delta\omega_x$, and substituting this into Equation 10.32 gives

$$\delta\ddot{\omega}_x - \frac{(A - C)(C - B)}{AB}\omega_0^2\delta\omega_x = 0 \qquad (10.33)$$

Likewise, by differentiating Equation 10.31$_2$ and then substituting $\delta\dot{\omega}_x$ from Equation 10.31$_1$ yields

$$\delta\ddot{\omega}_y - \frac{(A - C)(C - B)}{AB}\omega_0^2\delta\omega_y = 0 \qquad (10.34)$$

If we define

$$k = \frac{(A - C)(B - C)}{AB}\omega_0^2 \qquad (10.35)$$

then both Equations 10.33 and 10.34 may be written in the form

$$\delta\ddot{\omega} + k\delta\omega = 0 \qquad (10.36)$$

If $k > 0$, then $\delta\omega \propto e^{\pm i\sqrt{k}t}$, which means $\delta\omega_x$ and $\delta\omega_y$ vary sinusoidally with small amplitude. The motion is therefore bounded and neutrally stable. That means the amplitude does not die out with time, but it does not exceed the small amplitude of the perturbation. Observe from Equation 10.35 that $k > 0$ if either $C > A$ and $C > B$

or $C < A$ and $C < B$. This means that the spin axis (z axis) is either the major axis of inertia or the minor axis of inertia. That is, if the spin axis is either the major or minor axis of inertia, the motion is stable. The stability is neutral for a rigid body, because there is no damping.

On the other hand, if $k < 0$, then $\delta\omega \propto e^{\pm\sqrt{k}t}$, which means that the initially small perturbations $\delta\omega_x$ and $\delta\omega_y$ increase without bound. The motion is unstable. From Equation 10.35 we see that $k < 0$ if either $A > C$ and $C > B$ or $A < C$ and $C < B$. This means that the spin axis is the intermediate axis of inertia ($A > C > B$ or $B > C > A$). If the spin axis is the intermediate axis of inertia, the motion is unstable.

If the angular velocity of a satellite lies in the direction of its major axis of inertia, the satellite is called a major axis spinner or oblate spinner. A minor axis spinner or prolate spinner has its minor axis of inertia aligned with the angular velocity. 'Intermediate axis spinners' are unstable and will presumably end up being major or minor axis spinners, if the satellite is a rigid body. However, the flexibility inherent in any real satellite leads to an additional instability, as we shall now see.

Consider again the rotationally symmetric satellite in torque-free motion discussed in Section 10.2. From Equations 10.24 and 10.25, we know that the angular momentum $\mathbf{H}_G$ is given by

$$\mathbf{H}_G = A\boldsymbol{\omega}_\perp + C\omega_z \hat{\mathbf{k}} \tag{10.37}$$

Hence,

$$H_G^2 = A^2 \omega_\perp^2 + C^2 \omega_z^2 \tag{10.38}$$

Differentiating this equation with respect to time yields

$$\frac{dH_G^2}{dt} = A^2 \frac{d\omega_\perp^2}{dt} + 2C^2 \omega_z \dot{\omega}_z \tag{10.39}$$

But, according to Equation 10.1, $\mathbf{H}_G$ is constant, so that $dH_G^2/dt = 0$ and Equation 10.39 can be written

$$\frac{d\omega_\perp^2}{dt} = -2\frac{C^2}{A^2} \omega_z \dot{\omega}_z \tag{10.40}$$

The rotary kinetic energy of a rotationally symmetric body ($A = B$) is found using Equation 9.81,

$$T_R = \frac{1}{2} A\omega_x^2 + \frac{1}{2} A\omega_y^2 + \frac{1}{2} C\omega_z^2 = \frac{1}{2} A(\omega_x^2 + \omega_y^2) + \frac{1}{2} C\omega_z^2$$

From Equation 10.13 we know that $\omega_x^2 + \omega_y^2 = \omega_\perp^2$, which means

$$T_R = \frac{1}{2} A\omega_\perp^2 + \frac{1}{2} C\omega_z^2 \tag{10.41}$$

The time derivative of T_R is, therefore,

$$\dot{T}_R = \frac{1}{2} A \frac{d\omega_\perp^2}{dt} + C\omega_z \dot{\omega}_z$$

Solving this for $\dot{\omega}_z$, we get

$$\dot{\omega}_z = \frac{1}{C\omega_z}\left(\dot{T}_R - \frac{1}{2}A\frac{d\omega_\perp^2}{dt}\right)$$

Substituting this expression for $\dot{\omega}_z$ into Equation 10.40 and solving for $d\omega_\perp^2/dt$ yields

$$\frac{d\omega_\perp^2}{dt} = 2\frac{C}{A}\frac{\dot{T}_R}{C-A} \qquad (10.42)$$

Real bodies are not completely rigid, and their flexibility, however slight, gives rise to small dissipative effects which cause the kinetic energy to decrease over time. That is,

$$\dot{T}_R < 0 \quad \text{For satellites with dissipation.} \qquad (10.43)$$

Substituting this inequality into Equation 10.42 leads us to conclude that

$$\frac{d\omega_\perp^2}{dt} < 0 \quad \text{if } C > A \text{ (oblate spinner)}$$

$$\frac{d\omega_\perp^2}{dt} > 0 \quad \text{if } C < A \text{ (prolate spinner)} \qquad (10.44)$$

If $d\omega_\perp^2/dt$ is negative, the spin is asymptotically stable. Should a non-zero value of $\omega_\perp$ develop for some reason, it will drift back to zero over time so that once again the angular velocity lies completely in the spin direction. On the other hand, if $d\omega_\perp^2/dt$ is positive, the spin is unstable. $\omega_\perp$ does not damp out, and the angular velocity vector drifts away from the spin axis as $\omega_\perp$ increases without bound. We pointed out above that spin about a minor axis of inertia is stable with respect to small disturbances. Now we see that only major axis spin is stable in the long run if dissipative mechanisms exist.

For some additional insight into this phenomenon, solve Equation 10.38 for $\omega_\perp^2$,

$$\omega_\perp^2 = \frac{H_G^2 - C^2\omega_z^2}{A^2}$$

and substitute this result into the expression for kinetic energy, Equation 10.41, to obtain

$$T_R = \frac{1}{2}\frac{H_G^2}{A} + \frac{1}{2}\frac{(A-C)C}{A}\omega_z^2 \qquad (10.45)$$

According to Equation 10.24,

$$\omega_z = \frac{H_{G_z}}{C} = \frac{H_G \cos\theta}{C}$$

Substituting this into Equation 10.45 yields the kinetic energy as a function of just the inclination angle θ,

$$T_R = \frac{1}{2}\frac{H_G^2}{A}\left(1 + \frac{A-C}{C}\cos^2\theta\right) \qquad (10.46)$$

The extreme values of T_R occur at $\theta = 0$ or $\theta = \pi$,

$$T_R = \frac{1}{2}\frac{H_G^2}{C} \quad \text{(major axis spinner)}$$

and $\theta = \pi/2$,

$$T_R = \frac{1}{2}\frac{H_G^2}{A} \quad \text{(minor axis spinner)}$$

Clearly, the kinetic energy of a torque-free satellite is smallest when the spin is around the major axis of inertia. We may think of a satellite with dissipation ($dT_R/dt < 0$) as seeking the state of minimum kinetic energy that occurs when it spins about its major axis.

EXAMPLE 10.5

A rigid spacecraft is modeled by the solid cylinder B which has a mass of 300 kg and the slender rod R which passes through the cylinder and has a mass of 30 kg. Which of the principal axes x, y, z can be an axis about which stable torque-free rotation can occur?

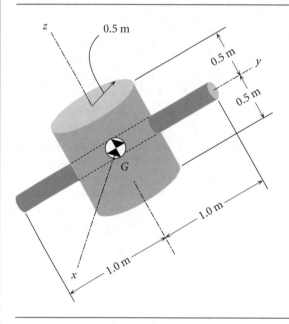

Figure 10.7 Built-up satellite structure.

For the cylindrical shell A, we have

$$r_B = 0.5\,\text{m} \quad l_B = 1.0\,\text{m} \quad m_B = 300\,\text{kg}$$

The principal moments of inertia about the center of mass are found in Figure 10.9(b),

$$I_{B_x} = \frac{1}{4}m_B r_B^2 + \frac{1}{12}m_B l_B^2 = 43.75\,\text{kg}\cdot\text{m}^2$$

$$I_{B_y} = I_{B_{xx}} = 43.75\,\text{kg}\cdot\text{m}^2$$

$$I_{B_z} = \frac{1}{2}m_B r_B^2 = 37.5\,\text{kg}\cdot\text{m}^2$$

The properties of the transverse rod are

$$l_R = 1.0 \text{ m} \quad m_R = 30 \text{ kg}$$

Figure 10.9(a), with $r = 0$, yields the moments of inertia,

$$I_{R_y} = 0$$

$$I_{R_z} = I_{R_x} = \frac{1}{12} m_A r_A^2 = 10.0 \text{ kg} \cdot \text{m}^2$$

The moments of inertia of the assembly is the sum of the moments of inertia of the cylinder and the rod,

$$I_x = I_{B_x} + I_{R_x} = 53.75 \text{ kg} \cdot \text{m}^2$$
$$I_y = I_{B_y} + I_{R_y} = 43.75 \text{ kg} \cdot \text{m}^2$$
$$I_z = I_{B_z} + I_{R_z} = 47.50 \text{ kg} \cdot \text{m}^2$$

Since I_z is the intermediate mass moment of inertia, rotation about the z axis is unstable. With energy dissipation, rotation is stable in the long term only about the major axis, which in this case is the x axis.

10.4 DUAL-SPIN SPACECRAFT

If a satellite is to be spin stabilized, it must be an oblate spinner. The diameter of the spacecraft is restricted by the cross-section of the launch vehicle's upper stage, and its length is limited by stability requirements. Therefore, oblate spinners cannot take full advantage of the payload volume available in a given launch vehicle, which after all are slender, prolate shapes for aerodynamic reasons. The dual-spin design permits spin stabilization of a prolate shape.

The axisymmetric, dual-spin configuration, or gyrostat, consists of an axisymmetric rotor and a smaller axisymmetric platform joined together along a common longitudinal spin axis at a bearing, as shown in Figure 10.8. The platform and rotor have their own components of angular velocity, ω_p and ω_r respectively, along the spin axis direction $\hat{\mathbf{k}}$. The platform spins at a much slower rate than the rotor. The assembly acts like a rigid body as far as transverse rotations are concerned; i.e., the rotor and the platform have $\omega_\perp$ in common. An electric motor integrated into the axle bearing connecting the two components acts to overcome frictional torque which would otherwise eventually cause the relative angular velocity between the rotor and platform to go to zero. If that should happen, the satellite would become a single spin unit, probably an unstable prolate spinner, since the rotor of a dual-spin spacecraft is likely to be prolate.

The first dual-spin satellite was OSO-I (Orbiting Solar Observatory), which NASA launched in 1962. It was a major-axis spinner. The first prolate dual-spin spacecraft was the two-storey tall TACSAT I (Tactical Communications Satellite). It was launched into geosynchronous orbit by the US Air Force in 1969. Typical of many of today's

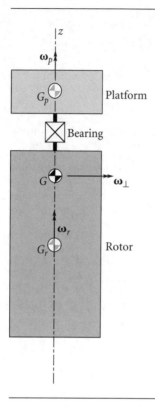

Figure 10.8 Axisymmetric, dual-spin satellite.

communications satellites, TACSAT's platform rotated at one revolution per day to keep its antennas pointing towards the earth. The rotor spun at about one revolution per second. Of course, the axis of the spacecraft was normal to the plane of its orbit. The first dual-spin interplanetary spacecraft was Galileo, which we discussed briefly in Section 8.9. Galileo's platform was completely despun to provide a fixed orientation for cameras and other instruments. The rotor spun at three revolutions per minute.

The equations of motion of a dual-spin spacecraft will be developed later in Section 10.8. Let us determine the stability of the motion by following the same 'energy sink' procedure employed in the previous section for a single-spin stabilized spacecraft. The angular momentum of the dual-spin configuration about the spacecraft's center of mass G is the sum of the angular momenta of the rotor (r) and the platform (p) about G,

$$\mathbf{H}_G = \mathbf{H}_G^{(p)} + \mathbf{H}_G^{(r)} \tag{10.47}$$

The angular momentum of the platform about the spacecraft center of mass is

$$\mathbf{H}_G^{(p)} = C_p \omega_p \hat{\mathbf{k}} + A_p \boldsymbol{\omega}_\perp \tag{10.48}$$

where C_p is the moment of inertia of the platform about the spacecraft spin axis, and A_p is its transverse moment of inertia about G (not G_p). Likewise, for the rotor,

$$\mathbf{H}_G^{(r)} = C_r \omega_r \hat{\mathbf{k}} + A_r \boldsymbol{\omega}_\perp \qquad (10.49)$$

where C_r and A_r are its longitudinal and transverse moments of inertia about axes through G. Substituting Equations 10.48 and 10.49 into 10.47 yields

$$\mathbf{H}_G = (C_r \omega_r + C_p \omega_p) \hat{\mathbf{k}} + A_\perp \boldsymbol{\omega}_\perp \qquad (10.50)$$

where $A_\perp$ is the total transverse moment of inertia,

$$A_\perp = A_p + A_r$$

From this it follows that

$$H_G^2 = (C_r \omega_r + C_p \omega_p)^2 + A_\perp^2 \omega_\perp^2$$

For torque-free motion, $\dot{\mathbf{H}}_G = \mathbf{0}$, so that $dH_G^2/dt = 0$, or

$$2(C_r \omega_r + C_p \omega_p)(C_r \dot{\omega}_r + C_p \dot{\omega}_p) + A_\perp^2 \frac{d\omega_\perp^2}{dt} = 0 \qquad (10.51)$$

Solving this for $d\omega_\perp^2/dt$ yields

$$\frac{d\omega_\perp^2}{dt} = -\frac{2}{A_\perp^2}(C_r \omega_r + C_p \omega_p)(C_r \dot{\omega}_r + C_p \dot{\omega}_p) \qquad (10.52)$$

The total rotational kinetic energy of rotation of the dual spin spacecraft is the sum of that of the rotor and the platform,

$$T = \frac{1}{2} C_r \omega_r^2 + \frac{1}{2} C_p \omega_p^2 + \frac{1}{2} A_\perp \omega_\perp^2$$

Differentiating this expression with respect to time and solving for $d\omega_\perp^2/dt$ yields

$$\frac{d\omega_\perp^2}{dt} = \frac{2}{A_\perp}(\dot{T} - C_r \omega_r \dot{\omega}_r - C_p \omega_p \dot{\omega}_p) \qquad (10.53)$$

$\dot{T}$ is the sum of the power $P^{(r)}$ dissipated in the rotor and the power $P^{(p)}$ dissipated in the platform,

$$\dot{T} = P^{(r)} + P^{(p)} \qquad (10.54)$$

Substituting Equation 10.54 into 10.53 we find

$$\frac{d\omega_\perp^2}{dt} = \frac{2}{A_\perp}(P^{(r)} - C_r \omega_r \dot{\omega}_r + P^{(p)} - C_p \omega_p \dot{\omega}_p) \qquad (10.55)$$

Equating the two expressions for $d\omega_\perp^2/dt$ in Equations 10.52 and 10.55 yields

$$\frac{2}{A_\perp}(\dot{T} - C_r \omega_r \dot{\omega}_r - C_p \omega_p \dot{\omega}_p) = -\frac{2}{A_\perp^2}(C_r \omega_r + C_p \omega_p)(C_r \dot{\omega}_r + C_p \dot{\omega}_p)$$

Solve this for $\dot{T}$ to obtain

$$\dot{T} = \frac{C_r}{A_\perp}[(A_\perp - C_r)\omega_r - C_p\omega_p]\dot{\omega}_r + \frac{C_p}{A_\perp}[(A_\perp - C_p)\omega_p - C_r\omega_r]\dot{\omega}_p \quad (10.56)$$

Following Likins (1967), we identify the terms containing $\dot{\omega}_r$ and $\dot{\omega}_p$ as the power dissipation in the rotor and platform, respectively. That is, comparing Equations 10.54 and 10.56,

$$P^{(r)} = \frac{C_r}{A_\perp}[(A_\perp - C_r)\omega_r - C_p\omega_p]\dot{\omega}_r \quad (10.57a)$$

$$P^{(p)} = \frac{C_p}{A_\perp}[(A_\perp - C_p)\omega_p - C_r\omega_r]\dot{\omega}_p \quad (10.57b)$$

Solving these two expressions for $\dot{\omega}_r$ and $\dot{\omega}_p$, respectively, yields

$$\dot{\omega}_r = \frac{A_\perp}{C_r} \frac{P^{(r)}}{(A_\perp - C_r)\omega_r - C_p\omega_p} \quad (10.58a)$$

$$\dot{\omega}_p = \frac{A_\perp}{C_p} \frac{P^{(p)}}{(A_\perp - C_p)\omega_p - C_r\omega_r} \quad (10.58b)$$

Substituting these results into Equation 10.55 leads to

$$\frac{d\omega_\perp^2}{dt} = \frac{2}{A_\perp}\left[\frac{P^{(r)}}{C_p\frac{\omega_p}{\omega_r} - (A_\perp - C_r)} + \frac{P^{(p)}}{C_r - (A_\perp - C_p)\frac{\omega_p}{\omega_r}}\right]\left(C_r + C_p\frac{\omega_p}{\omega_r}\right) \quad (10.59)$$

As pointed out above, for geosynchronous dual-spin communication satellites,

$$\frac{\omega_p}{\omega_r} \approx \frac{2\pi \text{ rad/d}}{2\pi \text{ rad/s}} \approx 10^{-5}$$

whereas for interplanetary dual-spin spacecraft, $\omega_p = 0$. Therefore, there is an important class of spin stabilized spacecraft for which $\omega_p/\omega_r \approx 0$. For a despun platform wherein ω_p is zero (or nearly so), Equation 10.59 yields

$$\frac{d\omega_\perp^2}{dt} = \frac{2}{A_\perp}\left[P^{(p)} + \frac{C_r}{C_r - A_\perp}P^{(r)}\right] \quad (10.60)$$

If the rotor is oblate ($C_r > A_\perp$), then, since $P^{(r)}$ and $P^{(p)}$ are both negative, it follows from Equation 10.60 that $d\omega_\perp^2/dt < 0$. That is, the oblate dual spin configuration with a despun platform is unconditionally stable. In practice, however, the rotor is likely to be prolate ($C_r < A_\perp$), so that

$$\frac{C_r}{C_r - A_\perp}P^{(r)} > 0$$

In that case, $d\omega_\perp^2/dt < 0$ only if the dissipation in the platform is significantly greater than that of the rotor. Specifically, for a prolate design it must be true that

$$|P^{(p)}| > \left|\frac{C_r}{C_r - A_\perp}P^{(r)}\right|$$

The platform dissipation rate $P^{(p)}$ can be augmented by adding nutation dampers, which are discussed in the next section.

For the despun prolate dual-spin configuration, Equations 10.58 imply

$$\dot{\omega}_r = \frac{P^{(r)}}{(A_\perp - C_r)} \frac{A_\perp}{C_r \omega_r}$$

$$\dot{\omega}_p = -\frac{P^{(p)}}{C_p} \frac{A_\perp}{C_r \omega_r}$$

Clearly, the signs of $\dot{\omega}_r$ and $\dot{\omega}_p$ are opposite. If $\omega_r > 0$, then dissipation causes the spin rate of the rotor to decrease and that of the platform to increase. Were it not for the action of the motor on the shaft connecting the two components of the spacecraft, eventually $\omega_p = \omega_r$. That is, the relative motion between the platform and rotor would cease and the dual-spinner would become an unstable single spin spacecraft. Setting $\omega_p = \omega_r$ in Equation 10.59 yields

$$\frac{d\omega_\perp^2}{dt} = 2 \frac{C_r + C_p}{A_\perp} \frac{P^{(r)} + P^{(p)}}{(C_r + C_p) - A_\perp}$$

which is the same as Equation 10.42, the energy sink conclusion for a single spinner.

10.5 Nutation damper

Nutation dampers are passive means of dissipating energy. A common type consists essentially of a tube filled with viscous fluid and containing a mass attached to springs, as illustrated in Figure 10.9. Dampers may contain just fluid, only partially filling the tube so it can slosh around. In either case, the purpose is to dissipate energy through fluid friction. The wobbling of the spacecraft due to non-alignment of the angular

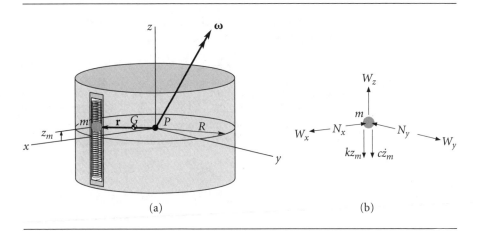

Figure 10.9 (a) Precessing oblate spacecraft with a nutation damper aligned with the z axis. (b) Free-body diagram of the moving mass in the nutation damper.

velocity with the principal spin axis induces accelerations throughout the satellite, giving rise to the sloshing of fluids, stretching and flexing of non-rigid components, etc., all of which dissipate energy to one degree or another. Nutation dampers are added to deliberately increase energy dissipation, which is desirable for stabilizing oblate single spinners and dual-spin spacecraft.

Let us focus on the motion of the mass within the nutation damper of Figure 10.9 in order to gain some insight into how relative motion and deformation are induced by the satellite's precession. Note that point P is the center of mass of the rigid satellite body itself. The center of mass G of the satellite-damper mass combination lies between P and m, as shown in Figure 10.9. We suppose that the tube is lined up with the z axis of the body-fixed xyz frame, as shown. The mass m in the tube is therefore constrained by the tube walls to move only in the z direction. When the springs are undeformed, the mass lies in the xy plane. In general, the position vector of m in the body frame is

$$\mathbf{r} = R\hat{\mathbf{i}} + z_m\hat{\mathbf{k}} \tag{10.61}$$

where z_m is the z coordinate of m and R is the distance of the damper from the centerline of the spacecraft. The velocity and acceleration of m relative to the satellite are, therefore,

$$\mathbf{v}_{rel} = \dot{z}_m\hat{\mathbf{k}} \tag{10.62}$$

$$\mathbf{a}_{rel} = \ddot{z}_m\hat{\mathbf{k}} \tag{10.63}$$

The absolute angular velocity $\boldsymbol{\omega}$ of the satellite (and, therefore, the body frame) is

$$\boldsymbol{\omega} = \omega_x\hat{\mathbf{i}} + \omega_y\hat{\mathbf{j}} + \omega_z\hat{\mathbf{k}} \tag{10.64}$$

Recall Equation 9.73, which states that when $\boldsymbol{\omega}$ is given in a body frame, we find the absolute angular acceleration by taking the time derivative of $\boldsymbol{\omega}$, holding the unit vectors fixed. Thus,

$$\dot{\boldsymbol{\omega}} = \dot{\omega}_x\hat{\mathbf{i}} + \dot{\omega}_y\hat{\mathbf{j}} + \dot{\omega}_z\hat{\mathbf{k}} \tag{10.65}$$

The absolute acceleration of m is found using Equation 1.42, which for the case at hand becomes

$$\mathbf{a} = \mathbf{a}_P + \dot{\boldsymbol{\omega}} \times \mathbf{r} + \boldsymbol{\omega} \times (\boldsymbol{\omega} \times \mathbf{r}) + 2\boldsymbol{\omega} \times \mathbf{v}_{rel} + \mathbf{a}_{rel} \tag{10.66}$$

in which $\mathbf{a}_P$ is the absolute acceleration of the reference point P. Substituting Equations 10.61 through 10.65 into Equation 10.66, carrying out the vector operations, combining terms, and simplifying leads to the following expressions for the three components of the inertial acceleration of m,

$$\begin{aligned} a_x &= a_{P_x} - R(\omega_y^2 + \omega_z^2) + z_m\dot{\omega}_y + z_m\omega_x\omega_z + 2\dot{z}_m\omega_y \\ a_y &= a_{P_y} + R\dot{\omega}_z + R\omega_x\omega_y - z_m\dot{\omega}_x + z_m\omega_y\omega_z - 2\dot{z}_m\omega_x \\ a_z &= a_{P_z} - z_m(\omega_x^2 + \omega_y^2) - R\dot{\omega}_y + R\omega_x\omega_z + \ddot{z}_m \end{aligned} \tag{10.67}$$

Figure 10.9(b) shows the free-body diagram of the damper mass m. In the x and y directions the forces on m are the components of the force of gravity (W_x and W_y)

and the components N_x and N_y of the force of contact with the smooth walls of the damper tube. The directions assumed for these components are, of course, arbitrary. In the z direction, we have the z component W_z of the weight, plus the force of the springs and the viscous drag of the fluid. The spring force $(-kz_m)$ is directly proportional and opposite in direction to the displacement z_m. k is the net spring constant. The viscous drag $(-c\dot{z}_m)$ is directly proportional and opposite in direction to the velocity $\dot{z}_m$ of m relative to the tube. c is the damping constant. Thus, the three components of the net force on the damper mass m are

$$F_{net_x} = W_x - N_x$$
$$F_{net_y} = W_y - N_y \quad (10.68)$$
$$F_{net_z} = W_z - kz_m - c\dot{z}_m$$

Substituting Equations 10.67 and 10.68 into Newton's second law, $\mathbf{F}_{net} = m\mathbf{a}$, yields

$$N_x = mR(\omega_y^2 + \omega_z^2) - mz_m\dot{\omega}_y - mz_m\omega_x\omega_y - 2m\dot{z}_m\omega_y + \overbrace{(W_x - ma_{P_x})}^{=0}$$

$$N_y = -mR\dot{\omega}_z - mR\omega_x\omega_y + mz_m\dot{\omega}_x - mz_m\omega_y\omega_z$$
$$+ 2m\dot{z}_m\omega_x + \overbrace{(W_y - ma_{P_y})}^{=0} \quad (10.69)$$

$$m\ddot{z}_m + c\dot{z}_m + [k - m(\omega_x^2 + \omega_y^2)]z_m = mR(\dot{\omega}_y - \omega_x\omega_z) + \overbrace{(W_z - ma_{P_z})}^{=0}$$

The last terms in parentheses in each of these expressions vanish if the acceleration of gravity is the same at m as at the reference point P of the spacecraft. This will be true unless the satellite is of enormous size.

If the damper mass m is vanishingly small compared to the mass M of the rigid spacecraft body, then it will have little effect on the rotary motion. If the rotational state is that of an axisymmetric satellite in torque-free motion, then we know from Equations 10.13, 10.14 and 10.19 that

$$\omega_x = \Omega \sin \omega_s t \qquad \omega_y = \Omega \cos \omega_s t \qquad \omega_z = \omega_0$$
$$\dot{\omega}_x = \Omega\omega_s \cos \omega_s t \qquad \dot{\omega}_y = -\Omega\omega_s \sin \omega_s t \qquad \dot{\omega}_z = 0$$

in which case Equations 10.69 become

$$N_x = mR(\omega_0^2 + \Omega^2 \cos^2 \omega_s t) + m(\omega_s - \omega_0)\Omega z_m \sin \omega_s t - 2m\Omega\dot{z}_m \cos \omega_s t$$
$$N_y = -mR\Omega^2 \cos \omega_s t \sin \omega_s t + m(\omega_s - \omega_0)\Omega z_m \cos \omega_s t + 2m\Omega\dot{z}_m \sin \omega_s t \quad (10.70)$$
$$m\ddot{z}_m + c\dot{z}_m + (k - m\Omega^2)z_m = -mR(\omega_s + \omega_0)\Omega \sin \omega_s t$$

Equation 10.70$_3$ is that of a single degree of freedom, damped oscillator with a sinusoidal forcing function. The precession produces a force of amplitude $m(\omega_0 + \omega_s)\Omega R$ and frequency ω_s which causes the damper mass m to oscillate back and forth in the tube, such that

$$z_m = \frac{mR\Omega(\omega_s + \omega_0)}{[k - m(\omega_s^2 + \Omega^2)]^2 + (c\omega_s)^2}\{c\omega_s \cos \omega_s t - [k - m(\omega_s^2 + \Omega^2)] \sin \omega_s t\}$$

Observe that the contact forces N_x and N_y depend exclusively on the amplitude and frequency of the precession. If the angular velocity lines up with the spin axis, so that $\Omega = 0$ (precession vanishes), then

$$N_x = m\omega_0^2 R$$
$$N_y = 0 \qquad \text{No precession.}$$
$$z_m = 0$$

If precession is eliminated, so there is pure spin around the principal axis, the time-varying motions and forces vanish throughout the spacecraft, which thereafter rotates as a rigid body with no energy dissipation.

Now, the whole purpose of a nutation damper is to interact with the rotational motion of the satellite so as to damp out any tendencies to precess. Therefore, its mass should not be ignored in the equations of motion of the satellite. We will derive the equations of motion of the rigid satellite with nutation damper to show how rigid body mechanics is brought to bear upon the problem and, simply, to discover precisely what we are up against in even this extremely simplified system. We will continue to use P as the origin of our body frame. Since a moving mass has been added to the rigid satellite and since we are not using the center of mass of the system as our reference point, we cannot use Euler's equations. Applicable to the case at hand is Equation 9.33, according to which the equation of rotational motion of the system of satellite plus damper is

$$\dot{\mathbf{H}}_{P_{\text{rel}}} + \mathbf{r}_{G/P} \times (M+m)\mathbf{a}_{P/G} = \mathbf{M}_{G_{\text{net}}} \tag{10.71}$$

The angular momentum of the satellite body plus that of the damper mass, relative to point P on the spacecraft, is

$$\mathbf{H}_{P_{\text{rel}}} = \overbrace{A\omega_x \hat{\mathbf{i}} + B\omega_y \hat{\mathbf{j}} + C\omega_z \hat{\mathbf{k}}}^{\text{body of the spacecraft}} + \overbrace{\mathbf{r} \times m\dot{\mathbf{r}}}^{\text{damper mass}} \tag{10.72}$$

where the position vector $\mathbf{r}$ is given by Equation 10.61. According to Equation 1.28,

$$\dot{\mathbf{r}} = \left.\frac{d\mathbf{r}}{dt}\right)_{\text{rel}} + \boldsymbol{\omega} \times \mathbf{r} = \dot{z}_m \hat{\mathbf{k}} + \begin{vmatrix} \hat{\mathbf{i}} & \hat{\mathbf{j}} & \hat{\mathbf{k}} \\ \omega_x & \omega_y & \omega_z \\ R & 0 & z \end{vmatrix}$$

$$= \omega_y z_m \hat{\mathbf{i}} + (\omega_z R - \omega_x z_m)\hat{\mathbf{j}} + (\dot{z}_m - \omega_y R)\hat{\mathbf{k}}$$

After substituting this into Equation 10.72 and collecting terms we obtain

$$\mathbf{H}_{P_{\text{rel}}} = [(A + mz_m^2)\omega_x - mRz_m\omega_z]\hat{\mathbf{i}}$$
$$+ [(B + mR^2 + mz_m^2)\omega_y - mR\dot{z}_m]\hat{\mathbf{j}}$$
$$+ [(C + mR^2)\omega_z - mRz_m\omega_x]\hat{\mathbf{k}} \tag{10.73}$$

To calculate $\dot{\mathbf{H}}_{P_{\text{rel}}}$, we again use Equation 1.28,

$$\dot{\mathbf{H}}_{P_{\text{rel}}} = \left.\frac{d\mathbf{H}_{P_{\text{rel}}}}{dt}\right)_{\text{rel}} + \boldsymbol{\omega} \times \mathbf{H}_{P_{\text{rel}}}$$

Carrying out the operations on the right leads eventually to

$$\begin{aligned}\dot{\mathbf{H}}_{P_{\text{rel}}} = &[(A + mz_m^2)\dot{\omega}_x - mRz_m\dot{\omega}_z + (C - B - mz_m^2)\omega_y\omega_z \\ &- mRz_m\omega_x\omega_y + 2mz_m\dot{z}_m\omega_x]\hat{\mathbf{i}} \\ &+ \{(B + mR^2 + mz_m^2)\dot{\omega}_y + mRz_m(\omega_x^2 - \omega_z^2) \\ &+ [A + mz_m^2 - (C + mR^2)]\omega_x\omega_z + 2mz_m\dot{z}_m\omega_y - mR\ddot{z}_m\}\hat{\mathbf{j}} \\ &+ [-mRz_m\dot{\omega}_x + (C + mR^2)\dot{\omega}_z + (B + mR^2 - A)\omega_x\omega_y \\ &+ mRz_m\omega_y\omega_z - 2mR\dot{z}_m\omega_x]\hat{\mathbf{k}}\end{aligned} \quad (10.74)$$

To calculate the second term on the left of Equation 10.71, we keep in mind that P is the center of mass of the body of the satellite and first determine the position vector of the center of mass G of the vehicle plus damper relative to P,

$$(M + m)\mathbf{r}_{G/P} = M(\mathbf{0}) + m\mathbf{r} \quad (10.75)$$

where $\mathbf{r}$, the position of the damper mass m relative to P, is given by Equation 10.61. Thus

$$\mathbf{r}_{G/P} = \frac{m}{m + M}\mathbf{r} = \mu\mathbf{r} = \mu(R\hat{\mathbf{i}} + z_m\hat{\mathbf{k}}) \quad (10.76)$$

in which

$$\mu = \frac{m}{m + M} \quad (10.77)$$

Thus,

$$\mathbf{r}_{G/P} \times (M + m)\mathbf{a}_{P/G} = \left(\frac{m}{M + m}\right)\mathbf{r} \times (M + m)\,\mathbf{a}_{P/G} = \mathbf{r} \times m\mathbf{a}_{P/G} \quad (10.78)$$

The acceleration of P relative to G is found with the aid of Equation 1.32,

$$\mathbf{a}_{P/G} = -\ddot{\mathbf{r}}_{G/P} = -\mu\frac{d^2\mathbf{r}}{dt^2} = -\mu\left[\left(\frac{d^2\mathbf{r}}{dt^2}\right)_{\text{rel}} + \dot{\boldsymbol{\omega}} \times \mathbf{r} + \boldsymbol{\omega} \times (\boldsymbol{\omega} \times \mathbf{r}) + 2\boldsymbol{\omega} \times \left(\frac{d\mathbf{r}}{dt}\right)_{\text{rel}}\right] \quad (10.79)$$

where

$$\left(\frac{d\mathbf{r}}{dt}\right)_{\text{rel}} = \frac{dR}{dt}\hat{\mathbf{i}} + \frac{dz_m}{dt}\hat{\mathbf{k}} = \dot{z}_m\hat{\mathbf{k}} \quad (10.80)$$

and

$$\left(\frac{d^2\mathbf{r}}{dt^2}\right)_{\text{rel}} = \frac{d^2R}{dt^2}\hat{\mathbf{i}} + \frac{d^2z_m}{dt^2}\hat{\mathbf{k}} = \ddot{z}_m\hat{\mathbf{k}} \quad (10.81)$$

Substituting Equations 10.61, 10.64, 10.65, 10.80 and 10.81 into Equation 10.79 yields

$$\begin{aligned}\mathbf{a}_{P/G} = &[-\mu z_m\dot{\omega}_y + \mu R(\omega_y^2 + \omega_z^2) - \mu z_m\omega_x\omega_z - 2\mu\dot{z}_m\omega_y]\hat{\mathbf{i}} \\ &+ (\mu z_m\dot{\omega}_x - \mu R\dot{\omega}_z - \mu R\omega_x\omega_y - \mu z_m\omega_y\omega_z + 2\mu\dot{z}_m\omega_x)\hat{\mathbf{j}} \\ &+ [\mu R\dot{\omega}_y + \mu z_m(\omega_x^2 + \omega_y^2) - \mu R\omega_x\omega_z - \mu\ddot{z}_m]\hat{\mathbf{k}}\end{aligned} \quad (10.82)$$

We move this expression into Equation 10.78 to get

$$\begin{aligned}\mathbf{r}_{G/P} &\times (M+m)\mathbf{a}_{P/G} \\ &= [-\mu m z_m^2 \dot{\omega}_x - 2\mu m \ddot{z}_m \omega_x + \mu m R(\omega_x \omega_y + \dot{\omega}_z) + \mu m z_m^2 \omega_y \omega_z]\hat{\mathbf{i}} \\ &+ [-\mu m(R^2 + z_m^2)\dot{\omega}_y - 2\mu m z_m \dot{z}_m \omega_y + \mu m R z_m (\omega_z^2 - \omega_x^2) \\ &+ \mu m(R^2 - z_m^2)\omega_x \omega_z + \mu m \ddot{R} z_m]\hat{\mathbf{j}} \\ &+ (\mu m R z_m \dot{\omega}_x - \mu m R^2 \dot{\omega}_z + 2\mu m R \dot{z}_m \omega_x \\ &- \mu m R^2 \omega_x \omega_y - \mu m R z_m \omega_y \omega_z)\hat{\mathbf{k}}\end{aligned}$$

Placing this result and Equation 10.74 in Equation 10.71, and using the fact that $\mathbf{M}_{G_\text{net}} = \mathbf{0}$, yields a vector equation whose three components are

$$A\dot{\omega}_x + (C-B)\omega_y \omega_z + (1-\mu)m z_m^2 \dot{\omega}_x - (1-\mu)m z_m^2 \omega_y \omega_z$$
$$+ 2(1-\mu)m z_m \dot{z}_m \omega_x - (1-\mu)m R z_m \omega_x \omega_y = 0$$

$$[B + (1-\mu)mR^2]\dot{\omega}_y + [A - C - (1-\mu)mR^2]\omega_x \omega_z$$
$$+ (1-\mu)m z_m^2 (\omega_x \omega_z + \dot{\omega}_y) + 2(1-\mu)m z_m \dot{z}_m \omega_y \quad (10.83)$$
$$- (1-\mu)m R \ddot{z}_m + (1-\mu)m R z_m (\omega_x^2 - \omega_z^2) = 0$$

$$[C + (1-\mu)mR^2]\dot{\omega}_z + [B - A + (1-\mu)mR^2]\omega_x \omega_y$$
$$+ (1-\mu)m R z_m \omega_y \omega_z - 2(1-\mu)m R \dot{z}_m \omega_x - (1-\mu)m R z_m \dot{\omega}_x = 0$$

These are three equations in the four unknowns $\omega_x, \omega_y, \omega_z$ and z_m. The fourth equation is that of the motion of the damper mass m in the z direction,

$$W_z - k z_m - c \dot{z}_m = m a_z \quad (10.84)$$

where a_z is given by Equation 10.67$_3$, in which $a_{P_z} = a_{P_z} - a_{G_z} + a_{G_z} = a_{P/G_z} + a_{G_z}$, so that

$$a_z = a_{P/G_z} + a_{G_z} - z_m(\omega_x^2 + \omega_y^2) - R\dot{\omega}_y + R\omega_x \omega_z + \ddot{z}_m \quad (10.85)$$

Substituting the z component of Equation 10.82 into this expression and that result into Equation 10.84 leads (with $W_z = m a_{G_z}$) to

$$(1-\mu)m\ddot{z}_m + c\dot{z}_m + [k - (1-\mu)m(\omega_x^2 + \omega_y^2)]z_m = (1-\mu)mR[\dot{\omega}_y - \omega_x \omega_z] \quad (10.86)$$

Compare Equation 10.69$_3$ with this expression, which is the fourth equation of motion we need.

Equations 10.83 and 10.86 are a rather complicated set of non-linear, second order differential equations, which must be solved (numerically) to obtain a precise description of the motion of the semirigid spacecraft. That is beyond our scope. However, to study their stability we can linearize the equations in much the same way as we did in Section 10.3. (Note that Equations 10.83 reduce to 10.29 when $m = 0$.) We assume the satellite is in pure spin with angular velocity ω_0 about the z axis and that the damper mass is at rest ($z_m = 0$). This motion is slightly perturbed, in such a way that

$$\omega_x = \delta\omega_x \quad \omega_y = \delta\omega_y \quad \omega_z = \omega_0 + \delta\omega_z \quad z_m = \delta z_m \quad (10.87)$$

10.5 Nutation damper

It will be convenient for this analysis to introduce operator notation for the time derivative, $D = d/dt$. Thus, given a function of time $f(t)$, for any integer n, $D^n f = d^n f / dt^n$, and $D^0 f(t) = f(t)$. Then the various time derivatives throughout the equations will, in accordance with Equation 10.87, be replaced as follows,

$$\dot{\omega}_x = D\delta\omega_x \quad \dot{\omega}_y = D\delta\omega_y \quad \dot{\omega}_z = D\delta\omega_z \quad \dot{z}_m = D\delta z_m \quad \ddot{z}_m = D^2 \delta z_m \quad (10.88)$$

Substituting Equations 10.87 and 10.88 into Equations 10.83 and 10.86 and retaining only those terms which are at most linear in the small perturbations leads to

$$AD\delta\omega_x + (C - B)\omega_0 \delta\omega_y = 0$$

$$[A - C - (1 - \mu)mR^2]\omega_0 \delta\omega_x + [B + (1 - \mu)mR^2]D\delta\omega_y$$
$$- (1 - \mu)mR(D^2 + \omega_0^2)\delta z_m = 0 \quad (10.89)$$

$$[C + (1 - \mu)mR^2]D\delta\omega_z = 0$$

$$(1 - \mu)mR\omega_0 \delta\omega_x - (1 - \mu)mRD\delta\omega_y + [(1 - \mu)mD^2 + cD + k]\delta z_m = 0$$

$\delta\omega_z$ appears only in the third equation, which states that $\delta\omega_z = $ constant. The first, second and fourth equations may be combined in matrix notation,

$$\begin{bmatrix} AD & (C - B)\omega_0 & 0 \\ [A - C - (1 - \mu)mR^2]\omega_0 & [B + (1 - \mu)mR^2]D & -(1 - \mu)mR(D^2 + \omega_0^2) \\ (1 - \mu)mR\omega_0 & -(1 - \mu)mRD & (1 - \mu)mD^2 + cD + k \end{bmatrix}$$

$$\times \begin{Bmatrix} \delta w_x \\ \delta \omega_y \\ \delta z_m \end{Bmatrix} = \begin{Bmatrix} 0 \\ 0 \\ 0 \end{Bmatrix} \quad (10.90)$$

This is a set of three linear differential equations in the perturbations $\delta\omega_x, \delta\omega_y$ and δz_m. We won't try to solve them, since all we are really interested in is the stability of the satellite-damper system. It can be shown that the determinant Δ of the 3 by 3 matrix in Equation 10.90 is

$$\Delta = a_4 D^4 + a_3 D^3 + a_2 D^2 + a_1 D + a_0 \quad (10.91)$$

in which the coefficients of the characteristic equation $\Delta = 0$ are

$$a_4 = (1 - \mu)mAB$$

$$a_3 = cA[B + (1 - \mu)mR^2]$$

$$a_2 = k[B + (1 - \mu)mR^2]A + (1 - \mu)m[(A - C)(B - C)$$
$$- (1 - \mu)AmR^2]\omega_0^2 \quad (10.92)$$

$$a_1 = c\{[A - C - (1 - \mu)mR^2](B - C)\}\omega_0^2$$

$$a_0 = k\{[A - C - (1 - \mu)mR^2](B - C)\}\omega_0^2 + [(B - C)(1 - \mu)^2]m^2 R^2 \omega_0^4$$

According to the Routh–Hurwitz stability criteria (see any text on control systems, e.g., Palm, 1983), the motion represented by Equations 10.90 is asymptotically stable if and only if the signs of all of the following quantities, defined in terms of the coefficients of the characteristic equation, are the same

$$r_1 = a_4 \quad r_2 = a_3 \quad r_3 = a_2 - \frac{a_4 a_1}{a_3} \quad r_4 = a_1 - \frac{a_3 a_0}{a_3 a_2 - a_4 a_1} \quad r_5 = a_0 \quad (10.93)$$

EXAMPLE 10.6

A satellite is spinning about the z axis of its principal body frame at 2π radians per second. The principal moments of inertia about its center of mass are

$$A = 300 \text{ kg} \cdot \text{m}^2 \quad B = 400 \text{ kg} \cdot \text{m}^2 \quad C = 500 \text{ kg} \cdot \text{m}^2 \quad \text{(a)}$$

For the nutation damper, the following properties are given

$$R = 1 \text{ m} \quad \mu = 0.01 \quad m = 10 \text{ kg} \quad k = 10\,000 \text{ N/m} \quad c = 150 \text{ N} \cdot \text{s/m} \quad \text{(b)}$$

Use the Routh–Hurwitz stability criteria to assess the stability of the satellite as a major-axis spinner, a minor-axis spinner, and an intermediate-axis spinner.

The data in (a) are for a major-axis spinner. Substituting into Equations 10.92 and 10.93, we find

$$\begin{aligned} r_1 &= +1.188 \times 10^6 \text{ kg}^3\text{m}^4 \\ r_2 &= +18.44 \times 10^6 \text{ kg}^3\text{m}^4/\text{s} \\ r_3 &= +1.228 \times 10^9 \text{ kg}^3\text{m}^4/\text{s}^2 \\ r_4 &= +92\,820 \text{ kg}^3\text{m}^4/\text{s}^3 \\ r_5 &= +8.271 \times 10^9 \text{ kg}^3\text{m}^4/\text{s}^4 \end{aligned} \quad \text{(c)}$$

Since the rs are all positive, spin about the major axis is asymptotically stable. As we know from Section 10.3, without the damper the motion is neutrally stable.

For spin about the minor axis,

$$A = 500 \text{ kg} \cdot \text{m}^2 \quad B = 400 \text{ kg} \cdot \text{m}^2 \quad C = 300 \text{ kg} \cdot \text{m}^2 \quad \text{(d)}$$

For these moment of inertia values, we obtain

$$\begin{aligned} r_1 &= +1.980 \times 10^6 \text{ kg}^3\text{m}^4 \\ r_2 &= +30.74 \times 10^6 \text{ kg}^3\text{m}^4/\text{s} \\ r_3 &= +2.048 \times 10^9 \text{ kg}^3\text{m}^4/\text{s}^2 \\ r_4 &= -304\,490 \text{ kg}^3\text{m}^4/\text{s}^3 \\ r_5 &= +7.520 \times 10^9 \text{ kg}^3\text{m}^4/\text{s}^4 \end{aligned} \quad \text{(e)}$$

Since the rs are not all of the same sign, spin about the minor axis is not asymptotically stable. Recall that for the rigid satellite, such a motion was neutrally stable.

Finally, for spin about the intermediate axis,

$$A = 300 \text{ kg} \cdot \text{m}^2 \quad B = 500 \text{ kg} \cdot \text{m}^2 \quad C = 400 \text{ kg} \cdot \text{m}^2 \quad \text{(f)}$$

We know this motion is unstable, even without the nutation damper, but doing the Routh–Hurwitz stability check anyway, we get

$$\begin{aligned} r_1 &= +1.485 \times 10^6 \text{ kg}^3\text{m}^4 \\ r_2 &= +22.94 \times 10^6 \text{ kg}^3\text{m}^4/\text{s} \end{aligned}$$

$$r_3 = +1.529 \times 10^9 \text{ kg}^3\text{m}^4/\text{s}^2$$
$$r_4 = -192\,800 \text{ kg}^3\text{m}^4/\text{s}^3$$
$$r_5 = -4.323 \times 10^9 \text{ kg}^3\text{m}^4/\text{s}^4$$

The motion, as we expected, is not stable.

10.6 CONING MANEUVER

Like the use of nutation dampers, the coning maneuver is an example of the attitude control of spinning spacecraft. In this case, the angular momentum is changed by the use of on-board thrusters (small rockets) to apply pure torques.

Consider a satellite in pure spin with angular momentum $\mathbf{H}_{G_0}$. Suppose we wish to maintain the magnitude of the angular momentum but change its direction by rotating the spin axis through an angle θ, as illustrated in Figure 10.10. Recall from Section 9.4 that to change the angular momentum of the spacecraft requires applying an external moment,

$$\Delta \mathbf{H}_G = \int_0^{\Delta t} \mathbf{M}_G \, dt$$

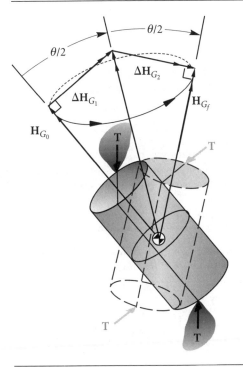

Figure 10.10 Impulsive coning maneuver.

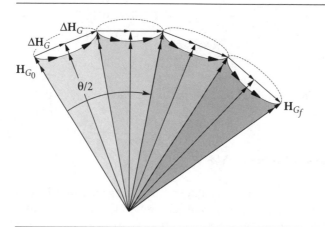

Figure 10.11 A sequence of small coning maneuvers.

Thrusters may be used to provide the external impulsive torque required to produce an angular momentum increment $\Delta\mathbf{H}_{G_1}$ normal to the spin axis. Since the spacecraft is spinning, this induces coning (precession) of the satellite about an axis at an angle $\theta/2$ to $\mathbf{H}_{G_0}$. The precession rate is given by Equation 10.23,

$$\omega_p = \frac{C}{A-C}\frac{\omega_s}{\cos\left(\frac{\theta}{2}\right)} \tag{10.94}$$

After precessing 180°, an angular momentum increment $\Delta\mathbf{H}_{G_2}$ normal to the spin axis and in the same direction relative to the spacecraft as the initial torque impulse, with $\|\Delta\mathbf{H}_{G_2}\| = \|\Delta\mathbf{H}_{G_1}\|$, stabilizes the spin vector in the desired direction. The time required for an angular reorientation θ using a single coning maneuver is found by simply dividing the precession angle, π radians, by the precession rate ω_p,

$$t_1 = \frac{\pi}{\omega_p} = \pi\frac{A-C}{C\omega_s}\cos\frac{\theta}{2} \tag{10.95}$$

Propellant expenditure is reflected in the magnitude of the individual angular momentum increments, in obvious analogy to delta-v calculations for orbital maneuvers. The total delta-H required for the single coning maneuver is therefore given by

$$\Delta H_{\text{total}} = \|\Delta\mathbf{H}_{G_1}\| + \|\Delta\mathbf{H}_{G_2}\| = 2\left(\|\mathbf{H}_{G_0}\|\tan\frac{\theta}{2}\right) \tag{10.96}$$

Figure 10.11 illustrates the fact that ΔH_{total} can be reduced by using a sequence of small coning maneuvers (small θs) rather than one big θ. The large number of small ΔHs approximates a circular arc of radius $\|\mathbf{H}_{G_0}\|$, subtended by the angle θ. Therefore, approximately,

$$\Delta H_{\text{total}} = 2\left(\|\mathbf{H}_{G_0}\|\frac{\theta}{2}\right) = \|\mathbf{H}_{G_0}\|\theta \tag{10.97}$$

This expression becomes more precise as the number of intermediate maneuvers increases. Figure 10.12 reveals the extent to which the multiple coning maneuver

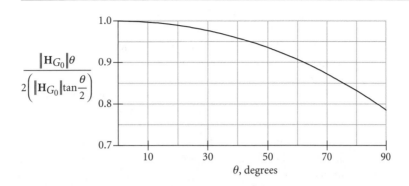

Figure 10.12 Ratio of delta-H for a sequence of small coning maneuvers to that for a single coning maneuver, as a function of the angle of swing of the spin axis.

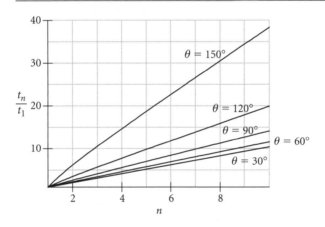

Figure 10.13 Time for a coning maneuver versus the number of intermediate steps.

strategy reduces energy requirements. The difference is quite significant for large reorientation angles.

One of the prices to be paid for the reduced energy of the multiple coning maneuver is time. (The other is the risk involved in repeating the maneuver over and over again.) From Equation 10.95, the time required for n small-angle coning maneuvers through a total angle of θ is

$$t_n = n\pi \frac{A - C}{C\omega_s} \cos \frac{\theta}{2n} \tag{10.98}$$

The ratio of this to the time t_1 required for a single coning maneuver is

$$\frac{t_n}{t_1} = n \frac{\cos \dfrac{\theta}{2n}}{\cos \dfrac{\theta}{2}} \tag{10.99}$$

The time is directly proportional to the number of intermediate coning maneuvers, as illustrated in Figure 10.13.

10.7 ATTITUDE CONTROL THRUSTERS

As mentioned above, thrusters are small jets mounted in pairs on a spacecraft to control its rotational motion about the center of mass. These thruster pairs may be mounted in principal planes (planes normal to the principal axes) passing through the center of mass. Figure 10.14 illustrates a pair of thrusters for producing a torque about the positive y axis. These would be accompanied by another pair of reaction motors pointing in the opposite directions to exert torque in the negative x direction. If the position vectors of the thrusters relative to the center of mass are $\mathbf{r}$ and $-\mathbf{r}$, and if $\mathbf{T}$ is their thrust, then the impulsive moment they exert during a brief time interval Δt is

$$\mathbf{M} = \mathbf{r} \times \mathbf{T}\Delta t + (-\mathbf{r}) \times (-\mathbf{T}\Delta t) = 2\mathbf{r} \times \mathbf{T}\Delta t \quad (10.100)$$

If the angular velocity was initially zero, then after the firing, according to Equation 10.31, the angular momentum becomes

$$\mathbf{H} = 2\mathbf{r} \times \mathbf{T}\Delta t \quad (10.101)$$

For $\mathbf{H}$ in the principal x direction, as in the figure, the corresponding angular velocity acquired by the vehicle is, from Equation 10.67,

$$\omega_y = \frac{\|\mathbf{H}\|}{B} \quad (10.102)$$

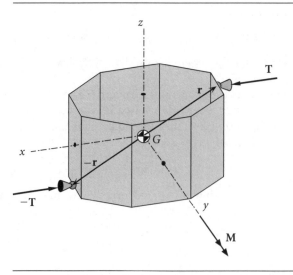

Figure 10.14 Pair of attitude control thrusters mounted in the xz plane of the principal body frame.

EXAMPLE 10.7

A spacecraft of mass m and with the dimensions shown in Figure 10.15 is spinning without precession at the rate ω_0 about the z axis of the principal body frame. At the instant shown in part (a) of the figure, the spacecraft initiates a coning maneuver to swing its spin axis through 90°, so that at the end of the maneuver the vehicle is oriented as illustrated in Figure 10.15(b). Calculate the total delta-H required, and compare it with that required for the same reorientation without coning. Motion is to be controlled exclusively by the pairs of attitude thrusters shown, all of which have identical thrust T.

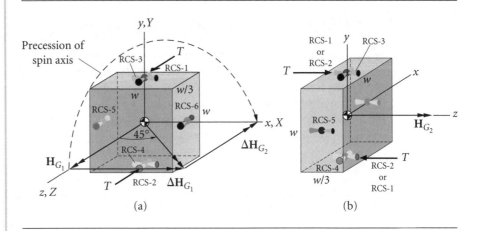

Figure 10.15 (a) Initial orientation of spinning spacecraft. (b) Final configuration, with spin axis rotated 90°.

According to Figure 9.9(c), the moments of inertia about the principal body axes are

$$A = B = \frac{1}{12}m\left[w^2 + \left(\frac{w}{3}\right)^2\right] = \frac{5}{54}mw^2 \qquad C = \frac{1}{12}m(w^2 + w^2) = \frac{1}{6}mw^2$$

The initial angular momentum $\mathbf{H}_{G_1}$ points in the spin direction, along the positive z axis of the body frame,

$$\mathbf{H}_{G_1} = C\omega_z \hat{\mathbf{k}} = \frac{1}{6}mw^2\omega_0 \hat{\mathbf{k}}$$

We can presume that in the initial orientation, the body frame happens to coincide instantaneously with inertial frame XYZ. The coning motion is initiated by briefly firing the pair of thrusters RCS-1 and RCS-2, aligned with the body z axis and lying in the yz plane. The impulsive torque will cause a change $\Delta\mathbf{H}_{G_1}$ in angular momentum directed normal to the plane of the thrusters, in the positive body x direction. The resultant angular momentum vector must lie at 45° to the x and z axes, bisecting the angle between the initial and final angular momenta. Thus,

$$\|\Delta\mathbf{H}_{G_1}\| = \|\mathbf{H}_{G_1}\| \tan 45° = \frac{1}{6}mw^2\omega_0$$

(Example 10.7 continued)

After the coning is underway, the body axes of course move away from the XYZ frame. Since the spacecraft is oblate ($C > A$), the precession of the spin axis will be opposite to the spin direction, as indicated in Figure 10.15. When the spin axis, after 180° of precession, lines up with the x axis the thrusters must fire again for the same duration as before so as to produce the angular momentum change $\Delta \mathbf{H}_{G_2}$, equal in magnitude but perpendicular to $\Delta \mathbf{H}_{G_1}$, so that

$$\mathbf{H}_{G_1} + \Delta \mathbf{H}_{G_1} + \Delta \mathbf{H}_{G_2} = \mathbf{H}_{G_2}$$

where

$$\mathbf{H}_{G_2} = \|\mathbf{H}_{G_1}\|\hat{\mathbf{I}} = \frac{1}{6} m w^2 \omega_0 \hat{\mathbf{k}}$$

For this to work, the plane of thrusters RCS-1 and RCS-2 – the yz plane – must be parallel to the XY plane when they fire, as illustrated in Figure 10.15(b). Since the thrusters can fire fore or aft, it does not matter which of them ends up on top or bottom. The vehicle must therefore spin through an integral number n of half rotations while it precesses to the desired orientation. That is, the total spin angle ψ between the initial and final configurations is

$$\psi = n\pi = \omega_s t \qquad (a)$$

where ω_s is the spin rate and t is the time for the proper final configuration to be achieved. In the meantime, the precession angle ϕ must be π or 3π or 5π, or, in general,

$$\phi = (2m-1)\pi = \omega_p t \qquad (b)$$

where m is an integer and t is, of course, the same as that in (a). Eliminating t from both (a) and (b) yields

$$n\pi = (2m-1)\pi \frac{\omega_s}{\omega_p}$$

Substituting Equation 10.94, with $\theta = \pi/2$, gives

$$n = (1-2m)\frac{4}{9}\frac{1}{\sqrt{2}} \qquad (c)$$

Obviously, this equation cannot be valid if both m and n are integers. However, by tabulating n as a function of m we find that when $m = 18, n = -10.999$. The minus sign simply reminds us that spin and precession are in opposite directions. Thus, the eighteenth time that the spin axis lines up with the x axis the thrusters may be fired to almost perfectly align the angular momentum vector with the body z axis. The slight misalignment due to the fact that $\|n\|$ is not precisely 11 would probably occur in reality anyway. Passive or active nutation damping can drive this deviation to zero.

Since $\|\mathbf{H}_{G_1}\| = \|\mathbf{H}_{G_2}\|$, we conclude that

$$\Delta H_{\text{total}} = 2\left(\frac{1}{6} m w^2 \omega_0\right) = \frac{2}{3} m w^2 \omega_0 \qquad (d)$$

An obvious alternative to the coning maneuver is to use thrusters RCS-3 and 4 to despin the craft completely, thrusters RCS-5 and 6 to initiate roll around the y axis and stop it after 90°, and then RCS-3 and 4 to respin the spacecraft to ω_0 around the z axis. The combined delta-H for the first and last steps equals that of (d). Additional fuel expenditure is required to start and stop the roll around the y axis. Hence, the coning maneuver is more fuel efficient.

10.8 YO-YO DESPIN MECHANISM

A simple, inexpensive way to despin an axisymmetric satellite is to deploy small masses attached to cords wound around the girth of the satellite near the transverse plane through the center of mass. As the masses unwrap in the direction of the satellite's angular velocity, they exert centrifugal force through the cords on the periphery of the satellite, creating a moment opposite to the spin direction, thereby slowing down the rotational motion. The cord forces are internal to the system of satellite plus weights, so as the strings unwind, the total angular momentum must remain constant. Since the total moment of inertia increases as the yo-yo masses spiral further away, the angular velocity must drop. Not only angular momentum but also rotational kinetic energy is conserved during this process. Yo-yo despin devices were introduced early in unmanned space flight (e.g., 1959 Transit 1-A) and continue to be used today (e.g., 1996 Mars Pathfinder, 1998 Mars Climate Orbiter, 1999 Mars Polar Lander, 2003 Mars Exploration Rover).

We will use the conservation of energy and momentum to determine the length of cord required to reduce the satellite's angular velocity a specified amount. To maintain the position of the center of mass, two identical yo-yo masses are wound around the spacecraft in a symmetrical fashion, as illustrated in Figure 10.16. Both masses are

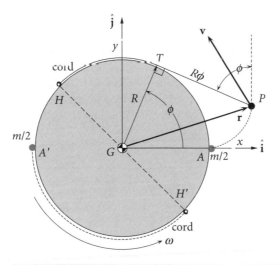

Figure 10.16 Two identical string and mass systems wrapped symmetrically around the periphery of an axisymmetric satellite. For simplicity, only one is shown being deployed.

released simultaneously by explosive bolts and unwrap in the manner shown (for only one of the weights) in the figure. In so doing, the point of tangency T moves around the circumference towards the split hinge device where the cord is attached to the spacecraft. When T and T' reach the hinges H and H', the cords automatically separate from the spacecraft.

Let each yo-yo weight have mass $m/2$. By symmetry, we need to track only one of the masses, to which we can ascribe the total mass m. Let the xyz system be a body frame rigidly attached to the satellite, as shown in Figure 10.16. As usual, the z axis lies in the spin direction, pointing out of the page. The x axis is directed from the center of mass of the system through the initial position of the yo-yo mass. The satellite and the yo-yo masses, prior to release, are rotating as a single rigid body with angular velocity $\boldsymbol{\omega}_0 = \omega_0 \hat{\mathbf{k}}$. The moment of inertia of the satellite, excluding the yo-yo mass, is C, so that the angular momentum of the satellite by itself is $C\omega_0$. The concentrated yo-yo masses are fastened a distance R from the spin axis, so their total moment of inertia is mR^2. Therefore, the initial angular momentum of the satellite plus yo-yo system is

$$H_{G_0} = C\omega_0 + mR^2\omega_0$$

It will be convenient to write this as

$$H_{G_0} = KmR^2\omega_0 \tag{10.103}$$

where the non-dimensional factor K is defined as

$$K = 1 + \frac{C}{mR^2} \tag{10.104}$$

$\sqrt{K}R$ is the initial radius of gyration of the system. The initial rotational kinetic energy of the system, before the masses are released, is

$$T_0 = \frac{1}{2}C\omega_0^2 + \frac{1}{2}mR^2\omega_0^2 = \frac{1}{2}KmR^2\omega_0^2 \tag{10.105}$$

At any state between the release of the weights and the release of the cords at the hinges, the velocity of the yo-yo mass must be found in order to compute the new angular momentum and kinetic energy. Observe that when the string has unwrapped an angle ϕ, the free length of string (between the point of tangency T and the yo-yo mass P) is $R\phi$. From the geometry shown in Figure 10.16, the position vector of the mass relative to the body frame is seen to be

$$\mathbf{r} = \overbrace{(R\cos\phi\,\hat{\mathbf{i}} + R\sin\phi\,\hat{\mathbf{j}})}^{\mathbf{r}_{T/G}} + \overbrace{(R\phi\sin\phi\,\hat{\mathbf{i}} - R\phi\cos\phi\,\hat{\mathbf{j}})}^{\mathbf{r}_{P/T}} \tag{10.106}$$
$$= (R\cos\phi + R\phi\sin\phi)\hat{\mathbf{i}} + (R\sin\phi - R\phi\cos\phi)\hat{\mathbf{j}}$$

Since $\mathbf{r}$ is measured in the moving reference, the absolute velocity $\mathbf{v}$ of the yo-yo mass is found using Equation 1.28,

$$\mathbf{v} = \left(\frac{d\mathbf{r}}{dt}\right)_{\text{rel}} + \boldsymbol{\Omega} \times \mathbf{r} \tag{10.107}$$

10.8 Yo-yo despin mechanism

where $\boldsymbol{\Omega}$ is the angular velocity of the *xyz* axes, which, of course, is the angular velocity $\boldsymbol{\omega}$ of the satellite at that instant,

$$\boldsymbol{\Omega} = \boldsymbol{\omega} \tag{10.108}$$

To calculate $d\mathbf{r}/dt)_{\text{rel}}$, we hold $\hat{\mathbf{i}}$ and $\hat{\mathbf{j}}$ constant in Equation 10.106, obtaining

$$\left.\frac{d\mathbf{r}}{dt}\right)_{\text{rel}} = (-R\dot{\phi}\sin\phi + R\dot{\phi}\sin\phi + R\phi\dot{\cos\phi})\hat{\mathbf{i}} + (R\dot{\phi}\cos\phi - R\dot{\phi}\cos\phi + R\phi\dot{\sin\phi})\hat{\mathbf{j}}$$

$$= R\phi\dot{\cos\phi}\hat{\mathbf{i}} + R\phi\dot{\sin\phi}\hat{\mathbf{j}}$$

Thus

$$\mathbf{v} = R\phi\dot{\cos\phi}\hat{\mathbf{i}} + R\phi\dot{\sin\phi}\hat{\mathbf{j}} + \begin{vmatrix} \hat{\mathbf{i}} & \hat{\mathbf{j}} & \hat{\mathbf{k}} \\ 0 & 0 & \omega \\ R\cos\phi + R\phi\sin\phi & R\sin\phi - R\phi\cos\phi & 0 \end{vmatrix}$$

or

$$\mathbf{v} = [R\phi(\omega + \dot\phi)\cos\phi - R\omega\sin\phi]\hat{\mathbf{i}} + [R\omega\cos\phi + R\phi(\omega + \dot\phi)\sin\phi]\hat{\mathbf{j}} \tag{10.109}$$

From this we find the speed of the yo-yo weights,

$$v = \sqrt{\mathbf{v}\cdot\mathbf{v}} = R\sqrt{\omega^2 + (\omega + \dot\phi)^2\phi^2} \tag{10.110}$$

The angular momentum of the satellite plus the weights at an intermediate stage of the despin process is

$$\mathbf{H}_G = C\omega\hat{\mathbf{k}} + \mathbf{r}\times m\mathbf{v}$$

$$= C\omega\hat{\mathbf{k}} + m\begin{vmatrix} \hat{\mathbf{i}} & \hat{\mathbf{j}} & \hat{\mathbf{k}} \\ R\cos\phi + R\phi\sin\phi & R\sin\phi - R\phi\cos\phi & \omega \\ R\phi(\omega+\dot\phi)\cos\phi - R\omega\sin\phi & R\omega\cos\phi + R\phi(\omega+\dot\phi)\sin\phi & 0 \end{vmatrix}$$

Carrying out the cross product, combining terms and simplifying, leads to

$$H_G = C\omega + mR^2[\omega + (\omega + \dot\phi)\phi^2]$$

which, using Equation 10.104, can be written

$$H_G = mR^2[K\omega + (\omega + \dot\phi)\phi^2] \tag{10.111}$$

The kinetic energy of the satellite plus the yo-yo mass is

$$T = \frac{1}{2}C\omega^2 + \frac{1}{2}mv^2$$

Substituting the speed from Equation 10.110 and making use again of Equation 10.104, we find

$$T = \frac{1}{2}mR^2[K\omega^2 + (\omega + \dot\phi)^2\phi^2] \tag{10.112}$$

By the conservation of angular momentum, $H_G = H_{G_0}$, we obtain from Equations 10.103 and 10.111,

$$mR^2[K\omega + (\omega + \dot\phi)\phi^2] = KmR^2\omega_0$$

which we can write as

$$K(\omega_0 - \omega) = (\omega + \dot{\phi})\phi^2 \tag{10.113}$$

Equations 10.105 and 10.112 and the conservation of kinetic energy, $T = T_0$, combine to yield

$$\frac{1}{2}mR^2[K\omega^2 + (\omega + \dot{\phi})^2\phi^2] = \frac{1}{2}KmR^2\omega_0^2$$

or

$$K(\omega_0^2 - \omega^2) = (\omega + \dot{\phi})^2\phi^2 \tag{10.114}$$

Since $\omega_0^2 - \omega^2 = (\omega_0 - \omega)(\omega_0 + \omega)$, this can be written

$$K(\omega_0 - \omega)(\omega_0 + \omega) = (\omega + \dot{\phi})^2\phi^2$$

Replacing the factor $K(\omega_0 - \omega)$ on the left using Equation 10.113 yields

$$(\omega + \dot{\phi})\phi^2(\omega_0 + \omega) = (\omega + \dot{\phi})^2\phi^2$$

After canceling terms, we find $\omega_0 + \omega = \omega + \dot{\phi}$, or, simply

$$\dot{\phi} = \omega_0 \tag{10.115}$$

In other words, the cord unwinds at a constant rate (relative to the satellite), equal to the satellite's initial angular velocity. Thus at any time t after the release of the weights,

$$\phi = \omega_0 t \tag{10.116}$$

By substituting Equation 10.115 into Equation 10.113,

$$K(\omega_0 - \omega) = (\omega + \omega_0)\phi^2$$

we find that

$$\phi = \sqrt{K\frac{\omega_0 - \omega}{\omega_0 + \omega}} \quad \text{Partial despin.} \tag{10.117}$$

Recall that the unwrapped length l of the cord is $R\phi$, which means

$$l = R\sqrt{K\frac{\omega_0 - \omega}{\omega_0 + \omega}} \quad \text{Partial despin.} \tag{10.118}$$

We use Equation 10.118 to find the length of cord required to despin the spacecraft from ω_0 to ω. To remove all of the spin ($\omega = 0$),

$$\phi = \sqrt{K} \quad \Rightarrow \quad l = R\sqrt{K} \quad \text{Complete despin.} \tag{10.119}$$

Surprisingly, the length of cord required to reduce the angular velocity to zero is independent of the initial angular velocity.

We can solve Equation 10.117 for ω in terms of ϕ,

$$\omega = \left(\frac{2K}{K + \phi^2} - 1\right)\omega_0 \tag{10.120}$$

By means of Equation 10.116, this becomes an expression for the angular velocity as a function of time

$$\omega = \left(\frac{2K}{K + \omega_0^2 t^2} - 1\right)\omega_0 \qquad (10.121)$$

Alternatively, since $\phi = l/R$, Equation 10.120 yields the angular velocity as a function of cord length,

$$\omega = \left(\frac{2KR^2}{KR^2 + l^2} - 1\right)\omega_0 \qquad (10.122)$$

Differentiating ω with respect to time in Equation 10.121 gives us an expression for the angular acceleration of the spacecraft,

$$\alpha = \frac{d\omega}{dt} = -\frac{4K\omega_0^3 t}{(K + \omega_0^2 t^2)^2} \qquad (10.123)$$

whereas integrating ω with respect to time yields the angle rotated by the satellite since release of the yo-yo mass,

$$\theta = 2\sqrt{K}\tan^{-1}\frac{\omega_0 t}{\sqrt{K}} - \omega_0 t = 2\sqrt{K}\tan^{-1}\frac{\phi}{\sqrt{K}} - \phi \qquad (10.124)$$

For complete despin, this expression, together with Equation 10.119, yields

$$\theta = \sqrt{K}\left(\frac{\pi}{2} - 1\right) \qquad (10.125)$$

From the free-body diagram of the spacecraft shown in Figure 10.17, it is clear that the torque exerted by the yo-yo weights is

$$M_{G_z} = -2RN \qquad (10.126)$$

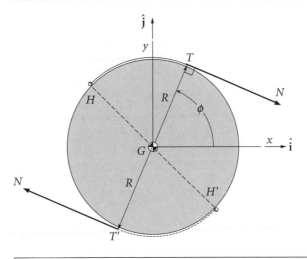

Figure 10.17 Free-body diagram of the satellite during the despin process.

where N is the tension in the cord. From Euler's equations of motion, Equation 10.72,

$$M_{G_z} = C\alpha \tag{10.127}$$

Combining Equations 10.123, 10.126 and 10.127 leads to a formula for the tension in the yo-yo cables,

$$N = \frac{C}{R}\frac{2K\omega_0^3 t}{(K+\omega_0^2 t^2)^2} = \frac{C\omega_0^2}{R}\frac{2K\phi}{(K+\phi^2)^2} \tag{10.128}$$

Radial release

Finally, we note that instead of releasing the yo-yo masses when the cables are tangent at the split hinges (H and H'), they can be forced to pivot about the hinge and released when the string is directed radially outward, as illustrated in Figure 10.18. The above analysis must be then extended to include the pivoting of the cord around the hinges. It turns out that in this case, the length of the cord as a function of the final angular velocity is

$$l = R\left(\sqrt{\frac{[(\omega_0-\omega)K+\omega]^2}{(\omega_0^2-\omega^2)K+\omega^2}} - 1\right) \quad \text{Partial despin, radial release.} \tag{10.129}$$

so that for $\omega = 0$,

$$l = R(\sqrt{K}-1) \quad \text{Complete despin, radial release.} \tag{10.130}$$

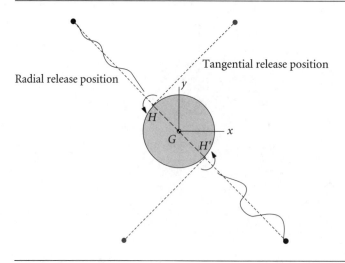

Figure 10.18 Radial versus tangential release of yo-yo masses.

EXAMPLE 10.8

A satellite is to be completely despun using a two-mass yo-yo device with tangential release. Assume the spin axis of moment of inertia of the satellite is $C = 200$ kg·m^2 and the initial spin rate is $\omega_0 = 5$ rad/s. The total yo-yo mass is 4 kg, and the radius of the spacecraft is 1 meter. Find (a) the required cord length l; (b) the time t to despin; (c) the maximum tension in the yo-yo cables; (d) the speed of the masses at release; (e) the angle rotated by the satellite during the despin; (f) the cord length required for radial release.

(a) From Equation 10.104,

$$K = 1 + \frac{C}{mR^2} = 1 + \frac{200}{4 \cdot 1^2} = 51 \qquad (a)$$

From Equation 10.118 it follows that the cord length required for complete despin is

$$l = R\sqrt{K} = 1 \cdot \sqrt{51} = \underline{7.1414 \text{ m}} \qquad (b)$$

(b) The time for complete despin is obtained from Equations 10.116 and 10.118,

$$\omega_0 t = \sqrt{K} \quad \Rightarrow \quad t = \frac{\sqrt{K}}{\omega_0} = \frac{\sqrt{51}}{5} = \underline{1.4283 \text{ s}}$$

(c) A graph of Equation 10.128 is shown in Figure 10.19. The maximum tension is $\underline{455 \text{ N}}$, which occurs at 0.825 s.

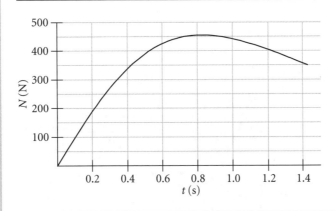

Figure 10.19 Variation of cable tension N up to point of release.

(d) From Equation 10.110, the speed of the yo-yo masses is

$$v = R\sqrt{\omega^2 + (\omega + \dot{\phi})^2 \phi^2}$$

(Example 10.8 continued)

According to Equation 10.115, $\dot{\phi}=\omega_0$ and at the time of release ($\omega=0$) Equation 10.118 states that $\dot{\phi}=\sqrt{K}$. Thus

$$v = R\sqrt{\omega^2 + (\omega+\omega_0)^2 \sqrt{K}^2} = 1 \cdot \sqrt{0^2 + (0+5)^2 \sqrt{51}^2} = \underline{35.71\text{ m/s}}$$

(e) The angle through which the satellite rotates before coming to rotational rest is given by Equation 10.124,

$$\theta = \sqrt{K}\left(\frac{\pi}{2}-1\right) = \sqrt{51}\left(\frac{\pi}{2}-1\right) = \underline{4.076\text{ rad }(233.5°)}$$

(f) Allowing the cord to detach radially reduces the cord length required for complete despin from 7.141 m to

$$l = R\left(\sqrt{K}-1\right) = 1 \cdot \left(\sqrt{51}-1\right) = \underline{6.141\text{ m}}$$

10.9 Gyroscopic attitude control

Momentum exchange systems ('gyros') are used to control the attitude of a spacecraft without throwing consumable mass overboard, as occurs with the use of thruster jets. A momentum exchange system is illustrated schematically in Figure 10.20. n flywheels, labeled 1, 2, 3, etc., are attached to the body of the spacecraft at various locations. The mass of flywheel i is m_i. The mass of the body of the spacecraft is m_0.

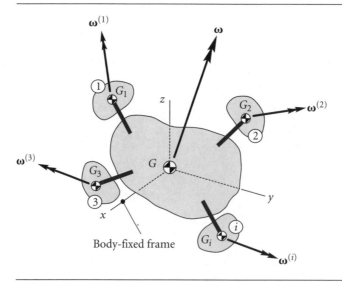

Figure 10.20 Several attitude control flywheels, each with their own angular velocity, attached to the body of a spacecraft.

10.9 Gyroscopic attitude control

The total mass of the entire system – the 'vehicle' – is m,

$$m = m_0 + \sum_{i=1}^{n} m_i$$

The vehicle's center of mass is G, through which pass the three axes xyz of the vehicle's body-fixed frame. The center of mass G_i of each flywheel is connected rigidly to the spacecraft, but the wheel, driven by electric motors, rotates more or less independently, depending on the type of gyro. The body of the spacecraft has an angular velocity $\boldsymbol{\omega}$. The angular velocity of the ith flywheel is $\boldsymbol{\omega}_i$, and differs from that of the body of the spacecraft unless the gyro is 'caged'. A caged gyro has no spin relative to the spacecraft, in which case $\boldsymbol{\omega}_i = \boldsymbol{\omega}$.

The angular momentum of the entire system about the vehicle's center of mass G is the sum of the angular momenta of the individual components of the system,

$$\mathbf{H}_G = \mathbf{H}_G^{(v)} + \mathbf{H}^{(w)} \tag{10.131}$$

$\mathbf{H}_G^{(v)}$ is the total angular momentum of the rigid body comprising the spacecraft and all of the flywheel masses concentrated at their centers of mass G_i. That system has the common vehicle angular velocity $\boldsymbol{\omega}$, which means, according to Equation 9.39, that

$$\{\mathbf{H}_G^{(v)}\} = [\mathbf{I}_G^{(v)}]\{\boldsymbol{\omega}\} \tag{10.132}$$

where $[\mathbf{I}_G^{(v)}]$ is the moment of inertia found by adding the moments of inertia of all the concentrated flywheel masses about G to that of the body of the spacecraft. On the other hand, $\mathbf{H}^{(w)}$ is the net angular momentum of the n flywheels about each of their individual centers of mass,

$$\mathbf{H}^{(w)} = \sum_{i=1}^{n} \mathbf{H}_{G_i}^{(i)} \tag{10.133}$$

$\mathbf{H}_{G_i}^{(i)}$, the angular momentum of flywheel i about its center of mass G_i, is obtained by once again using Equation 9.39,

$$\{\mathbf{H}_{G_i}^{(i)}\} = [\mathbf{I}_{G_i}^{(i)}]\{\boldsymbol{\omega}^{(i)}\} \tag{10.134}$$

$[\mathbf{I}_{G_i}^{(i)}]$ is the moment of inertia of flywheel i about G_i, relative to axes which are parallel to the body-fixed xyz axes. The mass distribution reflected in $[\mathbf{I}_G^{(v)}]$ is fixed relative to the body frame, which means this matrix does not vary with time. On the other hand, since a momentum wheel might be one that pivots on gimbals relative to the body frame, the inertia tensor $[\mathbf{I}_{G_i}^{(i)}]$ may be time dependent.

Substituting Equation 10.131 into Equation 9.30 yields the equations of rotational motion of the gyro stabilized spacecraft,

$$\mathbf{M}_{G_{\text{net}}} = \dot{\mathbf{H}}_G^{(v)} + \dot{\mathbf{H}}^{(w)} \tag{10.135}$$

Since the angular momenta are computed in the non-inertial body-fixed frame, we must use Equation 1.28 to obtain the time derivatives on the right-hand side of

Equation 10.135. Therefore,

$$\mathbf{M}_{G_{\text{net}}} = \left[\frac{d\mathbf{H}_G^{(v)}}{dt} \bigg)_{\text{rel}} + \boldsymbol{\omega} \times \mathbf{H}_G^{(v)} \right] + \left[\frac{d\mathbf{H}^{(w)}}{dt} \bigg)_{\text{rel}} + \boldsymbol{\omega} \times \mathbf{H}^{(w)} \right] \quad (10.136)$$

For torque-free motion, $\mathbf{M}_{G_{\text{net}}} = \mathbf{0}$, in which case we have the conservation of angular momentum about the vehicle center of mass,

$$\mathbf{H}_G^{(v)} + \mathbf{H}^{(w)} = \text{constant} \quad (10.137)$$

EXAMPLE 10.9

Use Equation 10.136 to obtain the equations of motion of a torque-free, axisymmetric, dual-spin satellite, such as the one shown in Figure 10.21.

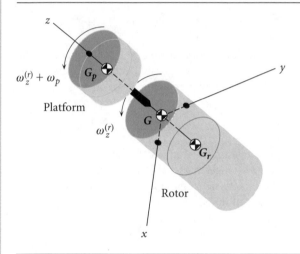

Figure 10.21 Dual-spin spacecraft.

In the dual-spin satellite, we may arbitrarily choose the rotor as the body of the vehicle, to which the body frame is attached. The coaxial platform will play the role of the single reaction wheel. The center of mass G of the satellite lies on the axis of rotational symmetry (the z axis), between the center of mass of the rotor (G_r) and that of the platform (G_p). For this torque-free system, Equation 10.136 becomes

$$\frac{d\mathbf{H}_G^{(v)}}{dt} \bigg)_{\text{rel}} + \boldsymbol{\omega}^{(r)} \times \mathbf{H}_G^{(v)} + \frac{d\mathbf{H}_{G_p}^{(p)}}{dt} \bigg)_{\text{rel}} + \boldsymbol{\omega}^{(r)} \times \mathbf{H}_{G_p}^{(p)} = 0 \quad \text{(a)}$$

in which r signifies the rotor and p the platform.

The vehicle angular momentum about G is that of the rotor plus that of the platform center of mass,

$$\{\mathbf{H}_G^{(v)}\} = [\mathbf{I}_G^{(r)}]\{\boldsymbol{\omega}^{(r)}\} + [\mathbf{I}_{m_G}^{(p)}]\{\boldsymbol{\omega}^{(r)}\} = ([\mathbf{I}_G^{(r)}] + [\mathbf{I}_{m_G}^{(p)}])\{\boldsymbol{\omega}^{(r)}\} \quad \text{(b)}$$

$\left[\mathbf{I}_{m_G}^{(p)}\right]$ is the moment of inertia tensor of the concentrated mass of the platform about the system center of mass, and it is calculated by means of Equation 9.44. The components of $\left[\mathbf{I}_G^{(r)}\right]$ and $\left[\mathbf{I}_{m_G}^{(p)}\right]$ are constants, so from (b) we obtain

$$\left.\frac{d\{\mathbf{H}_G^{(v)}\}}{dt}\right)_{\text{rel}} = \left(\left[\mathbf{I}_G^{(r)}\right] + \left[\mathbf{I}_{m_G}^{(p)}\right]\right)\{\dot{\boldsymbol{\omega}}^{(r)}\} \tag{c}$$

The angular momentum of the *platform* about its own center of mass is

$$\{\mathbf{H}_{G_p}^{(p)}\} = \left[\mathbf{I}_{G_p}^{(p)}\right]\{\boldsymbol{\omega}^{(p)}\} \tag{d}$$

For both the platform and the rotor, the z axis is an axis of rotational symmetry. Thus, even though the platform is not stationary in xyz, the moment of inertia matrix $\left[\mathbf{I}_{G_p}^{(p)}\right]$ is not time dependent. It follows that

$$\left.\frac{d\{\mathbf{H}_{G_p}^{(p)}\}}{dt}\right)_{\text{rel}} = \left[\mathbf{I}_{G_p}^{(p)}\right]\{\dot{\boldsymbol{\omega}}^{(p)}\} \tag{e}$$

Using (b) through (e), we can write the equation of motion (a) as

$$\left(\left[\mathbf{I}_G^{(r)}\right] + \left[\mathbf{I}_{m_G}^{(p)}\right]\right)\{\dot{\boldsymbol{\omega}}^{(r)}\} + \{\boldsymbol{\omega}^{(r)}\} \times \left(\left[\mathbf{I}_G^{(r)}\right] + \left[\mathbf{I}_{m_G}^{(p)}\right]\right)\{\boldsymbol{\omega}^{(r)}\}$$
$$+ \left[\mathbf{I}_{G_p}^{(p)}\right]\{\dot{\boldsymbol{\omega}}^{(p)}\} + \{\boldsymbol{\omega}^{(r)}\} \times \left[\mathbf{I}_{G_p}^{(p)}\right]\{\boldsymbol{\omega}^{(p)}\} = \{\mathbf{0}\} \tag{f}$$

The angular velocity $\boldsymbol{\omega}^{(p)}$ of the platform is that of the rotor, $\boldsymbol{\omega}^{(r)}$, plus the angular velocity of the platform relative to the rotor, $\boldsymbol{\omega}_{\text{rel}}^{(p)}$. Hence, we may replace $\{\boldsymbol{\omega}^{(p)}\}$ with $\{\boldsymbol{\omega}^{(r)}\} + \{\boldsymbol{\omega}_{\text{rel}}^{(p)}\}$, so that, after a little rearrangement, (f) becomes

$$\left(\left[\mathbf{I}_G^{(r)}\right] + \left[\mathbf{I}_G^{(p)}\right]\right)\{\dot{\boldsymbol{\omega}}^{(r)}\} + \{\boldsymbol{\omega}^{(r)}\} \times \left(\left[\mathbf{I}_G^{(r)}\right] + \left[\mathbf{I}_G^{(p)}\right]\right)\{\boldsymbol{\omega}^{(r)}\}$$
$$+ \left[\mathbf{I}_{G_p}^{(p)}\right]\{\dot{\boldsymbol{\omega}}_{\text{rel}}^{(p)}\} + \{\boldsymbol{\omega}^{(r)}\} \times \left[\mathbf{I}_{G_p}^{(p)}\right]\{\boldsymbol{\omega}_{\text{rel}}^{(p)}\} = \{\mathbf{0}\} \tag{g}$$

in which

$$\left[\mathbf{I}_G^{(p)}\right] = \left[\mathbf{I}_{m_G}^{(p)}\right] + \left[\mathbf{I}_{G_p}^{(p)}\right] \quad \text{(Parallel axis formula.)} \tag{h}$$

The components of the matrices and vectors in (g) relative to the principal xyz body frame axes are

$$\left[\mathbf{I}_G^{(r)}\right] = \begin{bmatrix} A_r & 0 & 0 \\ 0 & A_r & 0 \\ 0 & 0 & C_r \end{bmatrix} \quad \left[\mathbf{I}_G^{(p)}\right] = \begin{bmatrix} A_p & 0 & 0 \\ 0 & A_p & 0 \\ 0 & 0 & C_p \end{bmatrix} \quad \left[\mathbf{I}_{G_p}^{(p)}\right] = \begin{bmatrix} \bar{A}_p & 0 & 0 \\ 0 & \bar{A}_p & 0 \\ 0 & 0 & C_p \end{bmatrix} \tag{i}$$

and

$$\{\boldsymbol{\omega}^{(r)}\} = \begin{Bmatrix} \omega_x^{(r)} \\ \omega_y^{(r)} \\ \omega_z^{(r)} \end{Bmatrix} \quad \{\dot{\boldsymbol{\omega}}^{(r)}\} = \begin{Bmatrix} \dot{\omega}_x^{(r)} \\ \dot{\omega}_y^{(r)} \\ \dot{\omega}_z^{(r)} \end{Bmatrix} \quad \{\dot{\boldsymbol{\omega}}_{\text{rel}}^{(p)}\} = \begin{Bmatrix} 0 \\ 0 \\ \omega_p \end{Bmatrix} \quad \{\dot{\boldsymbol{\omega}}_{\text{rel}}^{(p)}\} = \begin{Bmatrix} 0 \\ 0 \\ \dot{\omega}_p \end{Bmatrix} \tag{j}$$

(Example 10.9 continued)

A_r, C_r, A_p and C_p are the rotor and platform principal moments of inertia about the vehicle center of mass G, whereas $\bar{A}_p$ is the moment of inertia of the platform about its own center of mass. We also used the fact that $\bar{C}_p = C_p$, which of course is due to the fact that G and G_p both lie on the z axis. This notation is nearly identical to that employed in our consideration of the stability of dual-spin satellites in Section 10.4 (wherein $\omega_r = \omega_z^{(r)}$ and $\boldsymbol{\omega}_\perp = \omega_x^{(r)}\hat{\mathbf{i}} + \omega_y^{(r)}\hat{\mathbf{j}}$). Substituting (i) and (j) into each of the four terms in (g), we get

$$([\mathbf{I}_G^{(r)}] + [\mathbf{I}_G^{(p)}])\{\dot{\boldsymbol{\omega}}^{(r)}\} = \begin{bmatrix} A_r + A_p & 0 & 0 \\ 0 & A_r + A_p & 0 \\ 0 & 0 & C_r + C_p \end{bmatrix} \begin{Bmatrix} \dot{\omega}_x^{(r)} \\ \dot{\omega}_y^{(r)} \\ \dot{\omega}_z^{(r)} \end{Bmatrix}$$

$$= \begin{Bmatrix} (A_r + A_p)\dot{\omega}_x^{(r)} \\ (A_r + A_p)\dot{\omega}_y^{(r)} \\ (C_r + C_p)\dot{\omega}_z^{(r)} \end{Bmatrix} \quad (k)$$

$$\{\boldsymbol{\omega}^{(r)}\} \times ([\mathbf{I}_G^{(r)}] + [\mathbf{I}_G^{(p)}])\{\boldsymbol{\omega}^{(r)}\} = \begin{Bmatrix} \omega_x^{(r)} \\ \omega_y^{(r)} \\ \omega_z^{(r)} \end{Bmatrix} \times \begin{Bmatrix} (A_r + A_p)\omega_x^{(r)} \\ (A_r + A_p)\omega_y^{(r)} \\ (C_r + C_p)\omega_z^{(r)} \end{Bmatrix}$$

$$= \begin{Bmatrix} [(C_p - A_p) + (C_r - A_r)]\omega_y^{(r)}\omega_z^{(r)} \\ [(A_p - C_p) + (A_r - C_r)]\omega_x^{(r)}\omega_z^{(r)} \\ 0 \end{Bmatrix} \quad (l)$$

$$[\mathbf{I}_{G_p}^{(p)}]\{\dot{\boldsymbol{\omega}}_{\text{rel}}^{(p)}\} = \begin{bmatrix} \bar{A}_p & 0 & 0 \\ 0 & \bar{A}_p & 0 \\ 0 & 0 & C_p \end{bmatrix} \begin{Bmatrix} 0 \\ 0 \\ \dot{\omega}_p \end{Bmatrix} = \begin{Bmatrix} 0 \\ 0 \\ C_p\dot{\omega}_p \end{Bmatrix} \quad (m)$$

$$\{\boldsymbol{\omega}^{(r)}\} \times [\mathbf{I}_{G_p}^{(p)}]\{\boldsymbol{\omega}_{\text{rel}}^{(p)}\} = \begin{Bmatrix} \omega_x^{(r)} \\ \omega_y^{(r)} \\ \omega_z^{(r)} \end{Bmatrix} \times \begin{bmatrix} \bar{A}_p & 0 & 0 \\ 0 & \bar{A}_p & 0 \\ 0 & 0 & C_p \end{bmatrix} \begin{Bmatrix} 0 \\ 0 \\ \omega_p \end{Bmatrix} = \begin{Bmatrix} C_p\omega_y^{(r)}\omega_p \\ -C_p\omega_x^{(r)}\omega_p \\ 0 \end{Bmatrix} \quad (n)$$

With these four expressions, (g) becomes

$$\begin{Bmatrix} (A_r + A_p)\dot{\omega}_x^{(r)} \\ (A_r + A_p)\dot{\omega}_y^{(r)} \\ (C_r + C_p)\dot{\omega}_z^{(r)} \end{Bmatrix} + \begin{Bmatrix} [(C_p - A_p) + (C_r - A_r)]\omega_y^{(r)}\omega_z^{(r)} \\ [(A_p - C_p) + (A_r - C_r)]\omega_x^{(r)}\omega_z^{(r)} \\ 0 \end{Bmatrix}$$

$$+ \begin{Bmatrix} 0 \\ 0 \\ C_p\dot{\omega}_p \end{Bmatrix} + \begin{Bmatrix} C_p\omega_y^{(r)}\omega_p \\ -C_p\omega_x^{(r)}\omega_p \\ 0 \end{Bmatrix} = \begin{Bmatrix} 0 \\ 0 \\ 0 \end{Bmatrix} \quad (o)$$

Combining the four vectors on the left-hand side, and then extracting the three components of the vector equation finally yields the three equations of motion of the dual-spin satellite in the body frame,

$$A\dot{\omega}_x^{(r)} + (C-A)\omega_y^{(r)}\omega_z^{(r)} + C_p\omega_y^{(r)}\omega_p = 0$$
$$A\dot{\omega}_y^{(r)} + (A-C)\omega_x^{(r)}\omega_z^{(r)} - C_p\omega_x^{(r)}\omega_p = 0 \qquad (p)$$
$$C\dot{\omega}_z^{(r)} + C_p\dot{\omega}_p = 0$$

where A and C are the combined transverse and axial moments of inertia of the dual-spin vehicle about its center of mass,

$$A = A_r + A_p \quad C = C_r + C_p \qquad (q)$$

The three equations (p) involve four unknowns, $\omega_x^{(r)}, \omega_y^{(r)}, \omega_z^{(r)}$ and ω_p. A fourth equation is required to account for the means of providing the relative velocity ω_p between the platform and the rotor. Friction in the axle bearing between the platform and the rotor would eventually cause ω_p to go to zero, as pointed out in Section 10.4. We may assume that the electric motor in the bearing acts to keep ω_p constant at a specified value, so that $\dot{\omega}_p = 0$. Then Equation (p)$_3$ implies that $\omega_z^{(r)} = $ constant as well. Thus, ω_p and $\omega_z^{(r)}$ are removed from our list of unknowns, leaving $\omega_x^{(r)}$ and $\omega_y^{(r)}$ to be governed by the first two equations in (p).

EXAMPLE 10.10

A spacecraft in torque-free motion has three identical momentum wheels with their spin axes aligned with the vehicle's principal body axes. The spin axes of momentum wheels 1, 2 and 3 are aligned with the x, y and z axes, respectively. The inertia tensors of the rotationally symmetric momentum wheels about their centers of mass are, therefore,

$$[\mathbf{I}_{G_1}^{(1)}] = \begin{bmatrix} I & 0 & 0 \\ 0 & J & 0 \\ 0 & 0 & J \end{bmatrix} \quad [\mathbf{I}_{G_2}^{(2)}] = \begin{bmatrix} J & 0 & 0 \\ 0 & I & 0 \\ 0 & 0 & J \end{bmatrix} \quad [\mathbf{I}_{G_3}^{(3)}] = \begin{bmatrix} J & 0 & 0 \\ 0 & J & 0 \\ 0 & 0 & I \end{bmatrix} \qquad (a)$$

The spacecraft moment of inertia tensor about the vehicle center of mass is

$$[\mathbf{I}_G^{(v)}] = \begin{bmatrix} A & 0 & 0 \\ 0 & B & 0 \\ 0 & 0 & C \end{bmatrix} \qquad (b)$$

Calculate the spin accelerations of the momentum wheels in the presence of external torque.

The absolute angular velocity $\boldsymbol{\omega}$ of the spacecraft and the angular velocities $\boldsymbol{\omega}_{rel}^{(1)}, \boldsymbol{\omega}_{rel}^{(2)}, \boldsymbol{\omega}_{rel}^{(3)}$ of the three flywheels *relative* to the spacecraft are

$$\{\boldsymbol{\omega}\} = \begin{Bmatrix} \omega_x \\ \omega_y \\ \omega_z \end{Bmatrix} \quad \{\boldsymbol{\omega}^{(1)}\}_{rel} = \begin{Bmatrix} \omega^{(1)} \\ 0 \\ 0 \end{Bmatrix} \quad \{\boldsymbol{\omega}^{(2)}\}_{rel} = \begin{Bmatrix} 0 \\ \omega^{(2)} \\ 0 \end{Bmatrix} \quad \{\boldsymbol{\omega}^{(3)}\}_{rel} = \begin{Bmatrix} 0 \\ 0 \\ \omega^{(3)} \end{Bmatrix}$$
$$(c)$$

(Example 10.10 continued)

Therefore, the angular momentum of the spacecraft and momentum wheels is

$$\{\mathbf{H}_G\} = [\mathbf{I}_G^{(v)}]\{\boldsymbol{\omega}\} + [\mathbf{I}_{G_1}^{(1)}](\{\boldsymbol{\omega}\} + \{\boldsymbol{\omega}^{(1)}\}_{\text{rel}}) + [\mathbf{I}_{G_2}^{(2)}](\{\boldsymbol{\omega}\} + \{\boldsymbol{\omega}^{(2)}\}_{\text{rel}})$$
$$+ [\mathbf{I}_{G_3}^{(3)}](\{\boldsymbol{\omega}\} + \{\boldsymbol{\omega}^{(3)}\}_{\text{rel}}) \tag{d}$$

Substituting Equations (a), (b) and (c) into this expression yields

$$\{\mathbf{H}_G\} = \begin{bmatrix} I & 0 & 0 \\ 0 & I & 0 \\ 0 & 0 & I \end{bmatrix} \begin{Bmatrix} \omega^{(1)} \\ \omega^{(2)} \\ \omega^{(3)} \end{Bmatrix} + \begin{bmatrix} A+I+2J & 0 & 0 \\ 0 & B+I+2J & 0 \\ 0 & 0 & C+I+2J \end{bmatrix} \begin{Bmatrix} \omega_x \\ \omega_y \\ \omega_z \end{Bmatrix} \tag{e}$$

In this case, Euler's equations are

$$\{\dot{\mathbf{H}}_G\}_{\text{rel}} + \{\boldsymbol{\omega}\} \times \{\mathbf{H}_G\} = \{\mathbf{M}_G\} \tag{f}$$

Substituting (e), we get

$$\begin{bmatrix} I & 0 & 0 \\ 0 & I & 0 \\ 0 & 0 & I \end{bmatrix} \begin{Bmatrix} \dot{\omega}^{(1)} \\ \dot{\omega}^{(2)} \\ \dot{\omega}^{(3)} \end{Bmatrix} + \begin{bmatrix} A+I+2J & 0 & 0 \\ 0 & B+I+2J & 0 \\ 0 & 0 & C+I+2J \end{bmatrix} \begin{Bmatrix} \dot{\omega}_x \\ \dot{\omega}_y \\ \dot{\omega}_z \end{Bmatrix} + \begin{Bmatrix} \omega_x \\ \omega_y \\ \omega_z \end{Bmatrix}$$

$$\times \left(\begin{bmatrix} I & 0 & 0 \\ 0 & I & 0 \\ 0 & 0 & I \end{bmatrix} \begin{Bmatrix} \omega^{(1)} \\ \omega^{(2)} \\ \omega^{(3)} \end{Bmatrix} + \begin{bmatrix} A+I+2J & 0 & 0 \\ 0 & B+I+2J & 0 \\ 0 & 0 & C+I+2J \end{bmatrix} \begin{Bmatrix} \omega_x \\ \omega_y \\ \omega_z \end{Bmatrix} \right)$$

$$= \begin{Bmatrix} M_{G_x} \\ M_{G_y} \\ M_{G_z} \end{Bmatrix} \tag{g}$$

Expanding and collecting terms yields the time rates of change of the flywheel spins (relative to the spacecraft) in terms of those of the spacecraft's absolute angular velocity components,

$$\dot{\omega}^{(1)} = \frac{M_{G_x}}{I} + \frac{B-C}{I}\omega_y\omega_z - \left(1 + \frac{A}{I} + 2\frac{J}{I}\right)\dot{\omega}_x + \omega^{(2)}\omega_z - \omega^{(3)}\omega_y$$

$$\dot{\omega}^{(2)} = \frac{M_{G_y}}{I} + \frac{C-A}{I}\omega_x\omega_z - \left(1 + \frac{B}{I} + 2\frac{J}{I}\right)\dot{\omega}_y + \omega^{(3)}\omega_x - \omega^{(1)}\omega_z \tag{h}$$

$$\dot{\omega}^{(3)} = \frac{M_{G_z}}{I} + \frac{A-B}{I}\omega_x\omega_y - \left(1 + \frac{C}{I} + 2\frac{J}{I}\right)\dot{\omega}_z + \omega^{(1)}\omega_y - \omega^{(2)}\omega_x$$

EXAMPLE 10.11

The communications satellite is in a circular earth orbit of period T. The body z axis always points towards the earth, so the angular velocity about the body y axis is $2\pi/T$. The angular velocities about the body x and z axes are zero. The attitude control system consists of three momentum wheels 1, 2 and 3 aligned with the principal x, y and z axes of the satellite. Variable torque is applied to each wheel by its own electric motor. At time $t=0$ the angular velocities of the three wheels relative to the spacecraft are all zero. A small, constant environmental torque $\mathbf{M_0}$ acts on the spacecraft.

Determine the axial torques $C^{(1)}, C^{(2)}$ and $C^{(3)}$ that the three motors must exert on their wheels so that the angular velocity $\boldsymbol{\omega}$ of the satellite will remain constant. The moment of inertia of each reaction wheel about its spin axis is I.

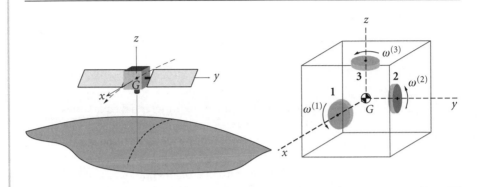

Figure 10.22 Three-axis stabilized satellite.

The absolute angular velocity of the xyz frame is given by

$$\boldsymbol{\omega} = \omega_0 \hat{\mathbf{j}} \qquad (a)$$

where $\omega_0 = 2\pi/T$, a constant. At any instant, the absolute angular velocities of the three reaction wheels are, accordingly,

$$\begin{aligned}
\boldsymbol{\omega}^{(1)} &= \omega^{(1)}\hat{\mathbf{i}} + \omega_0 \hat{\mathbf{j}} \\
\boldsymbol{\omega}^{(2)} &= [\omega^{(2)} + \omega_0]\hat{\mathbf{j}} \\
\boldsymbol{\omega}^{(3)} &= \omega_0 \hat{\mathbf{j}} + \omega^{(3)}\hat{\mathbf{k}}
\end{aligned} \qquad (b)$$

From (a) it is clear that $\omega_x = \omega_z = \dot{\omega}_x = \dot{\omega}_y = \dot{\omega}_z = 0$. Therefore, Equations (h) of Example 10.10 become, for the case at hand,

$$\dot{\omega}^{(1)} = \frac{M_{G_x}}{I} + \frac{B-C}{I}\omega_0(0) - \left(1 + \frac{A}{I} + 2\frac{J}{I}\right)(0) + \omega^{(2)}(0) - \omega^{(3)}\omega_0$$

$$\dot{\omega}^{(2)} = \frac{M_{G_y}}{I} + \frac{C-A}{I}(0)(0) - \left(1 + \frac{B}{I} + 2\frac{J}{I}\right)(0) + \omega^{(3)}(0) - \omega^{(1)}(0)$$

$$\dot{\omega}^{(3)} = \frac{M_{G_z}}{I} + \frac{A-B}{I}(0)\omega_0 - \left(1 + \frac{C}{I} + 2\frac{J}{I}\right)(0) + \omega^{(1)}\omega_0 - \omega^{(2)}(0)$$

which reduce to the following set of three first order differential equations,

$$\begin{aligned}
\dot{\omega}^{(1)} + \omega_0 \omega^{(3)} &= \frac{M_{G_x}}{I} \\
\dot{\omega}^{(2)} &= \frac{M_{G_y}}{I} \\
\dot{\omega}^{(3)} - \omega_0 \omega^{(1)} &= \frac{M_{G_z}}{I}
\end{aligned} \qquad (c)$$

(Example 10.11 continued)

Equation (c)$_2$ implies that $\omega^{(2)} = M_{G_y}t/I +$ constant, and since $\omega^{(2)} = 0$ at $t = 0$, this means that for time thereafter,

$$\omega^{(2)} = \frac{M_{G_y}}{I}t \tag{d}$$

Differentiating (c)$_3$ with respect to t and solving for $\dot{\omega}^{(1)}$ yields $\dot{\omega}^{(1)} = \ddot{\omega}^{(3)}/\omega_0$. Substituting this result into (c)$_1$ we get

$$\ddot{\omega}^{(3)} + \omega_0^2 \omega^{(3)} = \frac{\omega_0 M_{G_x}}{I}$$

The well-known solution of this differential equation is

$$\omega^{(3)} = a \cos \omega_0 t + b \sin \omega_0 t + \frac{M_{G_x}}{I\omega_0}$$

where a and b are constants of integration. According to the problem statement, $\omega^{(3)} = 0$ when $t = 0$. This initial condition requires $a = -M_{G_x}/\omega_0 I$, so that

$$\omega^{(3)} = b \sin \omega_0 t + \frac{M_{G_x}}{I\omega_0}(1 - \cos \omega_0 t) \tag{e}$$

From this we obtain $\dot{\omega}^{(3)} = b\omega_0 \cos \omega_0 t + \frac{M_{G_x}}{I} \sin \omega_0 t$, which, when substituted into (c)$_3$, yields

$$\omega^{(1)} = b \cos \omega_0 t + \frac{M_{G_x}}{I\omega_0} \sin \omega_0 t - \frac{M_{G_z}}{I\omega_0} \tag{f}$$

Since $\omega^{(1)} = 0$ at $t = 0$, this implies $b = M_{G_z}/\omega_0 I$. In summary, therefore, the angular velocities of wheels 1, 2 and 3 relative to the satellite are

$$\omega^{(1)} = \frac{M_{G_x}}{I\omega_0} \sin \omega_0 t + \frac{M_{G_z}}{I\omega_0}(\cos \omega_0 t - 1) \tag{g1}$$

$$\omega^{(2)} = \frac{M_{G_y}}{I}t \tag{g2}$$

$$\omega^{(3)} = \frac{M_{G_z}}{I\omega_0} \sin \omega_0 t + \frac{M_{G_x}}{I\omega_0}(1 - \cos \omega_0 t) \tag{g3}$$

The angular momenta of the reaction wheels are

$$\mathbf{H}_{G_1}^{(1)} = I_x^{(1)} \omega_x^{(1)} \hat{\mathbf{i}} + I_y^{(1)} \omega_y^{(1)} \hat{\mathbf{j}} + I_z^{(1)} \omega_z^{(1)} \hat{\mathbf{k}}$$
$$\mathbf{H}_{G_2}^{(2)} = I_x^{(2)} \omega_x^{(2)} \hat{\mathbf{i}} + I_y^{(2)} \omega_y^{(2)} \hat{\mathbf{j}} + I_z^{(2)} \omega_z^{(2)} \hat{\mathbf{k}} \tag{h}$$
$$\mathbf{H}_{G_3}^{(3)} = I_x^{(3)} \omega_x^{(3)} \hat{\mathbf{i}} + I_y^{(3)} \omega_y^{(3)} \hat{\mathbf{j}} + I_z^{(3)} \omega_z^{(3)} \hat{\mathbf{k}}$$

According to (b), the components of the flywheels' angular velocities are

$$\omega_x^{(1)} = \omega^{(1)} \qquad \omega_y^{(1)} = \omega_0 \qquad \omega_z^{(1)} = 0$$
$$\omega_x^{(2)} = 0 \qquad \omega_y^{(2)} = \omega^{(2)} + \omega_0 \qquad \omega_z^{(2)} = 0$$
$$\omega_x^{(3)} = 0 \qquad \omega_y^{(3)} = \omega_0 \qquad \omega_z^{(3)} = \omega^{(3)}$$

Furthermore, $I_x^{(1)} = I_y^{(2)} = I_z^{(3)} = I$, so that (h) becomes

$$\mathbf{H}_{G_1}^{(1)} = I\omega^{(1)}\hat{\mathbf{i}} + I_y^{(1)}\omega_0\hat{\mathbf{j}}$$
$$\mathbf{H}_{G_2}^{(2)} = I(\omega^{(2)} + \omega_0)\hat{\mathbf{j}} \qquad (i)$$
$$\mathbf{H}_{G_3}^{(3)} = I_y^{(3)}\omega_0\hat{\mathbf{j}} + I\omega^{(3)}\hat{\mathbf{k}}$$

Substituting (g) into these expressions yields the angular momenta of the wheels as a function of time,

$$\mathbf{H}_{G_1}^{(1)} = \left[\frac{M_{G_x}}{\omega_0}\sin\omega_0 t + \frac{M_{G_z}}{\omega_0}(\cos\omega_0 t - 1)\right]\hat{\mathbf{i}} + I_y^{(1)}\omega_0\hat{\mathbf{j}}$$
$$\mathbf{H}_{G_2}^{(2)} = (M_{G_y}t + I\omega_0)\hat{\mathbf{j}} \qquad (j)$$
$$\mathbf{H}_{G_3}^{(3)} = I_y^{(3)}\omega_0\hat{\mathbf{j}} + \left[\frac{M_{G_z}}{\omega_0}\sin\omega_0 t + \frac{M_{G_x}}{\omega_0}(1 - \cos\omega_0 t)\right]\hat{\mathbf{k}}$$

The torque on the reaction wheels is found by applying Euler's equation to each one. Thus, for wheel 1

$$\mathbf{M}_{G_{1\text{net}}} = \left.\frac{d\mathbf{H}_{G_1}^{(1)}}{dt}\right)_{\text{rel}} + \boldsymbol{\omega} \times \mathbf{H}_{G_1}^{(1)}$$
$$= (M_{G_x}\cos\omega_0 t - M_{G_z}\sin\omega_0 t)\hat{\mathbf{i}} + [M_{G_z}(1 - \cos\omega_0 t) - M_{G_x}\sin\omega_0 t]\hat{\mathbf{k}}$$

Since the axis of wheel 1 is in the x direction, the torque is the x component of this moment (the z component being a gyroscopic bending moment),

$$\underline{C^{(1)} = M_{G_x}\cos\omega_0 t - M_{G_z}\sin\omega_0 t}$$

For wheel 2,

$$\mathbf{M}_{G_{2\text{net}}} = \left.\frac{d\mathbf{H}_{G_2}^{(2)}}{dt}\right)_{\text{rel}} + \boldsymbol{\omega} \times \mathbf{H}_{G_2}^{(2)} = M_{G_y}\hat{\mathbf{j}}$$

Thus

$$\underline{C^{(2)} = M_{G_y}}$$

Finally, for wheel 3

$$\mathbf{M}_{G_{3\text{net}}} = \left.\frac{d\mathbf{H}_{G_3}^{(3)}}{dt}\right)_{\text{rel}} + \boldsymbol{\omega} \times \mathbf{H}_{G_3}^{(3)}$$
$$= [M_{G_x}(1 - \cos\omega_0 t) + M_{G_z}\sin\omega_0 t]\hat{\mathbf{i}} + (M_{G_x}\sin\omega_0 t + M_{G_z}\cos\omega_0 t)\hat{\mathbf{k}}$$

For this wheel, the torque direction is the z axis, so

$$\underline{C^{(3)} = M_{G_x}\sin\omega_0 t + M_{G_z}\cos\omega_0 t}$$

The external torques on the spacecraft of the previous example may be due to thruster misalignment or they may arise from environmental effects such as gravity gradients or solar pressure. The example assumed that these torques were constant, which is the simplest means of introducing their effects, but they actually vary with time. In any case, their magnitudes are extremely small, typically less than 10^{-3} N · m for ordinary-sized, unmanned spacecraft. Equation (g)$_2$ of the example reveals that a small torque normal to the satellite's orbital plane will cause the angular velocity of momentum wheel 2 to slowly but constantly increase. Over a long enough period of time, the angular velocity of the gyro might approach its design limits, whereupon it is said to be *saturated*. At that point, attitude jets on the satellite would have to be fired to produce a torque around the y axis while the wheel is 'caged', i.e., its angular velocity is reduced to zero or to its non-zero *bias* value. Finally, note that if all of the external torques were zero, none of the momentum wheels in the example would be required. The constant angular velocity $\boldsymbol{\omega} = (2\pi/T)\hat{\mathbf{j}}$ of the vehicle, once initiated, would continue unabated.

So far we have dealt with momentum wheels, which are characterized by the fact that their axes are rigidly aligned with the principal axes of the spacecraft, as shown in Figure 10.23. The speed of the electrically driven wheels is varied to produce the required rotation rates of the vehicle in response to external torques. Depending on the spacecraft, the nominal speed of a momentum wheel may be from zero to several thousand rpm.

Momentum wheels that are free to pivot on one or more gimbals are called control moment gyros. Figure 10.24 illustrates a double-gimbaled control moment gyro. These gyros spin at several thousand rpm. The motor-driven speed of the flywheel is constant, and moments are exerted on the vehicle when torquers (electric motors) tilt the wheel about a gimbal axis. The torque direction is normal to the gimbal axis. To simplify the analysis of high-rpm gyros, we can assume that the angular momentum is directed totally along the spin axis. That is, in calculating the angular momentum

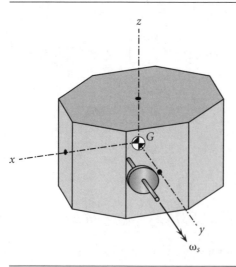

Figure 10.23 Momentum wheel aligned with a principal body axis.

$\mathbf{H}_{G_w}^{(w)}$ of a momentum wheel about its center of mass, we use the formula

$$\{\mathbf{H}_{G_w}^{(w)}\} = [\mathbf{I}_{G_w}^{(w)}]\{\boldsymbol{\omega}^{(w)}\}$$

where $\boldsymbol{\omega}^{(w)}$ is the absolute angular velocity of the spinning flywheel, which may be written

$$\boldsymbol{\omega}^{(w)} = \boldsymbol{\omega}^{(v)} + \boldsymbol{\omega}_p^{(w)} + \boldsymbol{\omega}_n^{(w)} + \boldsymbol{\omega}_s^{(w)}$$

$\boldsymbol{\omega}^{(v)}$ is the angular velocity of the vehicle to which the gyro is attached, while $\boldsymbol{\omega}_p^{(w)}$, $\boldsymbol{\omega}_n^{(w)}$ and $\boldsymbol{\omega}_s^{(w)}$ are the precession, nutation and spin rates of the gyro relative to the vehicle. The spin rate of the gyro is three or more orders of magnitude greater than any of the other rates. That is, under conditions in which a control moment gyro is designed to operate,

$$\|\boldsymbol{\omega}_s^{(w)}\| \gg \|\boldsymbol{\omega}^{(v)}\| \qquad \|\boldsymbol{\omega}_s^{(w)}\| \gg \|\boldsymbol{\omega}_p^{(w)}\| \qquad \|\boldsymbol{\omega}_s^{(w)}\| \gg \|\boldsymbol{\omega}_n^{(w)}\|$$

We may therefore accurately express the angular momentum of any high-rpm gyro as

$$\{\mathbf{H}_{G_w}^{(w)}\} = [\mathbf{I}_{G_w}^{(w)}]\{\boldsymbol{\omega}_s^{(w)}\} \tag{10.138}$$

Since the spin axis of a gyro is an axis of symmetry, about which the moment of inertia is $C^{(w)}$, this can be written

$$\mathbf{H}_{G_w}^{(w)} = C^{(w)} \omega_s^{(w)} \hat{\mathbf{n}}_s^{(w)}$$

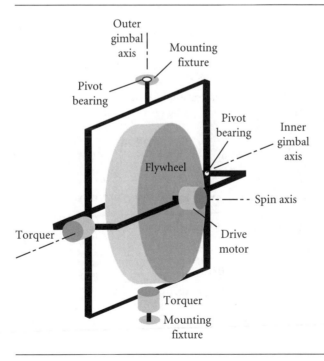

Figure 10.24 Two-gimbal control moment gyro.

Chapter 10 Satellite attitude dynamics

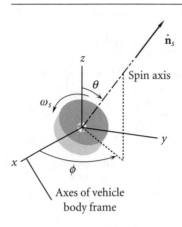

Figure 10.25 Inclination angles of the spin vector of a gyro.

where $\hat{\mathbf{n}}_s^{(w)}$ is the unit vector along the spin axis, as illustrated in Figure 10.25. Relative to the body frame axes of the spacecraft, the components of $\hat{\mathbf{n}}_s^{(w)}$ appear as follows,

$$\hat{\mathbf{n}}_s^{(w)} = \sin\theta\cos\phi\hat{\mathbf{i}} + \sin\theta\sin\phi\hat{\mathbf{j}} + \cos\theta\hat{\mathbf{k}} \tag{10.139}$$

If we let

$$H^{(w)} = C^{(w)}\omega_s^{(w)}$$

then Equation 10.138 becomes, simply,

$$\mathbf{H}_{G_w}^{(w)} = H^{(w)}\hat{\mathbf{n}}_s^{(w)} \tag{10.140}$$

Let us consider the equation of motion of a spacecraft with a single gyro. From Equation 10.136,

$$\left(\frac{d\mathbf{H}_G^{(v)}}{dt}\right)_{rel} + \boldsymbol{\omega}\times\mathbf{H}_G^{(v)} + \left(\frac{d\mathbf{H}_{G_w}^{(w)}}{dt}\right)_{rel} + \boldsymbol{\omega}\times\mathbf{H}_{G_w}^{(w)} = \mathbf{M}_{G_{net}} \tag{10.141}$$

Calculating each term on the left, we have, for the vehicle,

$$\mathbf{H}_G^{(v)} = A\omega_x\hat{\mathbf{i}} + B\omega_y\hat{\mathbf{j}} + C\omega_z\hat{\mathbf{k}}$$

$$\left(\frac{d\mathbf{H}_G^{(v)}}{dt}\right)_{rel} = A\dot{\omega}_x\hat{\mathbf{i}} + B\dot{\omega}_y\hat{\mathbf{j}} + C\dot{\omega}_z\hat{\mathbf{k}} \tag{10.142}$$

$$\boldsymbol{\omega}\times\mathbf{H}_G^{(v)} = \begin{vmatrix} \hat{\mathbf{i}} & \hat{\mathbf{j}} & \hat{\mathbf{k}} \\ \omega_x & \omega_y & \omega_z \\ A\omega_x & B\omega_y & C\omega_z \end{vmatrix} = (C-B)\omega_y\omega_z\hat{\mathbf{i}} + (A-C)\omega_x\omega_z\hat{\mathbf{j}} + (B-A)\hat{\mathbf{k}}$$

$$\tag{10.143}$$

For the gyro,

$$\mathbf{H}_{G_w}^{(w)} = H^{(w)}\hat{\mathbf{n}}_s^{(w)} = H^{(w)}\sin\theta\cos\phi\hat{\mathbf{i}} + H^{(w)}\sin\theta\sin\phi\hat{\mathbf{j}} + H^{(w)}\cos\theta\hat{\mathbf{k}}$$

$$\left(\frac{d\mathbf{H}_{G_w}^{(w)}}{dt}\right)_{rel} = (\dot{H}^{(w)}\sin\theta\cos\phi + H^{(w)}\dot\theta\cos\theta\cos\phi - H^{(w)}\dot\phi\sin\theta\sin\phi)\hat{\mathbf{i}}$$
$$+ (\dot{H}^{(w)}\sin\theta\sin\phi + H^{(w)}\dot\theta\cos\theta\sin\phi + H^{(w)}\dot\phi\sin\theta\cos\phi)\hat{\mathbf{j}}$$
$$+ (\dot{H}^{(w)}\cos\theta - H^{(w)}\dot\theta\sin\theta)\hat{\mathbf{k}} \qquad (10.144)$$

$$\boldsymbol{\omega}\times\mathbf{H}_{G_w}^{(w)} = \begin{vmatrix} \hat{\mathbf{i}} & \hat{\mathbf{j}} & \hat{\mathbf{k}} \\ \omega_x & \omega_y & \omega_z \\ H^{(w)}\sin\theta\cos\phi & H^{(w)}\sin\theta\sin\phi & H^{(w)}\cos\theta \end{vmatrix}$$
$$= (H^{(w)}\omega_y\cos\theta - H^{(w)}\omega_z\sin\phi\sin\theta)\hat{\mathbf{i}}$$
$$+ (-H^{(w)}\omega_x\cos\theta + H^{(w)}\omega_z\cos\phi\sin\theta)\hat{\mathbf{j}}$$
$$+ (-H^{(w)}\omega_y\cos\phi\sin\theta + H^{(w)}\omega_x\sin\phi\sin\theta)\hat{\mathbf{k}} \qquad (10.145)$$

Substituting Equations 10.142 through 10.145 into Equation 10.141 yields a vector equation with the following three components

$$A\dot\omega_x + H^{(w)}\dot\theta\cos\phi\cos\theta - H^{(w)}\dot\phi\sin\phi\sin\theta + \dot{H}^{(w)}\cos\phi\sin\theta$$
$$+ (H^{(w)}\cos\theta + C\omega_z)\omega_y - (H^{(w)}\sin\phi\sin\theta + B\omega_y)\omega_z = M_{G_{net_x}} \qquad (10.146a)$$

$$B\dot\omega_y + H^{(w)}\dot\theta\sin\phi\cos\theta + H^{(w)}\dot\phi\cos\phi\sin\theta + \dot{H}^{(w)}\sin\phi\sin\theta$$
$$- (H^{(w)}\cos\theta + C\omega_z)\omega_x + (H^{(w)}\cos\phi\sin\theta + A\omega_x)\omega_z = M_{G_{net_y}} \qquad (10.146b)$$

$$C\dot\omega_z - H^{(w)}\dot\theta\sin\theta + \dot{H}^{(w)}\cos\theta - (H^{(w)}\cos\phi\sin\theta + A\omega_x)\omega_y$$
$$+ (H^{(w)}\sin\phi\sin\theta + B\omega_y)\omega_x = M_{G_{net_z}} \qquad (10.146c)$$

Additional gyros are accounted for by adding the components of Equations 10.144 and 10.145 for each additional unit.

EXAMPLE 10.12

A satellite is in torque-free motion ($\mathbf{M}_{G_{net}} = \mathbf{0}$). A non-gimbaled gyro (momentum wheel) is aligned with the vehicle's x axis and is spinning at the rate ω_{s_0}. The spacecraft angular velocity is $\omega = \omega_x\hat{\mathbf{i}}$. If the spin of the gyro is increased at the rate $\dot\omega_s$, find the angular acceleration of the spacecraft.

Using Figure 10.25 as a guide, we set $\phi = 0$ and $\theta = 90°$ to align the spin axis with the x axis. Since there is no gimbaling, $\dot\theta = \dot\phi = 0$. Equations 10.146 then yield

$$A\dot\omega_x + \dot{H}^{(w)} = 0$$
$$B\dot\omega_y = 0$$
$$C\dot\omega_z = 0$$

(Example 10.12 continued)

Clearly, the angular velocities around the y and z axes remain zero, whereas,

$$\dot{\omega}_x = -\frac{\dot{H}^{(w)}}{A} = -\frac{C^{(w)}}{A}\dot{\omega}_s$$

Thus, a change in the vehicle's roll rate around the x axis can be initiated by accelerating the momentum wheel in the opposite direction.

EXAMPLE 10.13

A satellite is in torque-free motion. A control moment gyro, spinning at the constant rate ω_s, is gimbaled about the spacecraft y and z axes, with $\phi = 0$ and $\theta = 90°$ (cf. Figure 10.25). The spacecraft angular velocity is $\boldsymbol{\omega} = \omega_z \hat{\mathbf{k}}$. If the spin axis of the gyro, initially along the x direction, is rotated around the y axis at the rate $\dot{\theta}$, what is the resulting angular acceleration of the spacecraft?

Substituting $\omega_x = \omega_y = \dot{H}^{(w)} = \phi = 0$ and $\theta = 90°$ into Equations 10.146 gives

$$A\dot{\omega}_x = 0$$
$$B\dot{\omega}_y + H^{(w)}(\omega_z + \dot{\phi}) = 0$$
$$C\dot{\omega}_z - H^{(w)}\dot{\theta} = 0$$

where $H^{(w)} = C^{(w)}\omega_s$. Thus, the components of vehicle angular acceleration are

$$\dot{\omega}_x = 0 \qquad \dot{\omega}_y = -\frac{C^{(w)}}{B}\omega_s(\omega_z + \dot{\phi}) \qquad \dot{\omega}_z = \frac{C^{(w)}}{C}\omega_s\dot{\theta}$$

We see that pitching the gyro at the rate $\dot{\theta}$ around the vehicle y axis alters only ω_z, leaving ω_x unchanged. However, to keep $\omega_y = 0$ clearly requires $\dot{\phi} = -\omega_z$. In other words, for the control moment gyro to control the angular velocity about only one vehicle axis, it must therefore be able to precess around that axis (the z axis in this case). That is why the control moment gyro must have two gimbals.

10.10 GRAVITY-GRADIENT STABILIZATION

Consider a satellite in circular orbit, as shown in Figure 10.26. Let $\mathbf{r}$ be the position vector of a mass element dm relative to the center of attraction, $\mathbf{r}_0$ the position vector of the center of mass G, and $\boldsymbol{\varrho}$ the position of dm relative to G. The force of gravity on dm is

$$d\mathbf{F}_g = -G\frac{Mdm}{r^3}\mathbf{r} = -\mu\frac{\mathbf{r}}{r^3}dm \qquad (10.147)$$

where M is the mass of the central body, and $\mu = GM$. The net moment of the gravitational force around G is

$$\mathbf{M}_{G_{\text{net}}} = \int_m \boldsymbol{\varrho} \times d\mathbf{F}_g\, dm \qquad (10.148)$$

10.10 Gravity-gradient stabilization

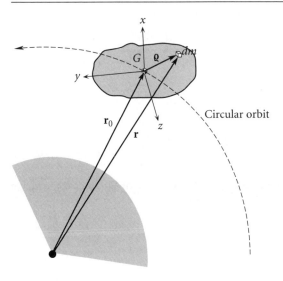

Figure 10.26 Rigid satellite in a circular orbit is the principal body frame.

Since $\mathbf{r} = \mathbf{r}_0 + \boldsymbol{\varrho}$, and

$$\mathbf{r}_0 = r_{0_x}\hat{\mathbf{i}} + r_{0_y}\hat{\mathbf{j}} + r_{0_z}\hat{\mathbf{k}}$$
$$\boldsymbol{\varrho} = x\hat{\mathbf{i}} + y\hat{\mathbf{j}} + z\hat{\mathbf{k}} \quad (10.149)$$

we have

$$\boldsymbol{\varrho} \times d\mathbf{F}_g = -\mu \frac{dm}{r^3}\boldsymbol{\varrho} \times (\mathbf{r}_0 + \boldsymbol{\varrho}) = -\mu \frac{dm}{r^3}\boldsymbol{\varrho} \times \mathbf{r}_0 = -\mu \frac{dm}{r^3}\begin{vmatrix}\hat{\mathbf{i}} & \hat{\mathbf{j}} & \hat{\mathbf{k}} \\ x & y & z \\ r_{0_x} & r_{0_y} & r_{0_z}\end{vmatrix}$$

Thus,

$$\boldsymbol{\varrho} \times d\mathbf{F}_g = -\mu \frac{dm}{r^3}(r_{0_z}y - r_{0_y}z)\hat{\mathbf{i}} - \mu \frac{dm}{r^3}(r_{0_x}z - r_{0_z}x)\hat{\mathbf{j}} - \mu \frac{dm}{r^3}(r_{0_y}x - r_{0_x}y)\hat{\mathbf{k}}$$

Substituting this back into Equation 10.148 yields

$$\mathbf{M}_{G_{\text{net}}} = \left(-\mu r_{0_z}\int_m \frac{y}{r^3}dm + \mu r_{0_y}\int_m \frac{z}{r^3}dm\right)\hat{\mathbf{i}} + \left(-\mu r_{0_x}\int_m \frac{z}{r^3}dm + \mu r_{0_z}\int_m \frac{x}{r^3}dm\right)\hat{\mathbf{j}}$$
$$+ \left(-\mu r_{0_y}\int_m \frac{x}{r^3}dm + \mu r_{0_x}\int_m \frac{y}{r^3}dm\right)\hat{\mathbf{k}}$$

or

$$M_{G_{\text{net}x}} = -\mu r_{0_z}\int_m \frac{y}{r^3}dm + \mu r_{0_y}\int_m \frac{z}{r^3}dm$$
$$M_{G_{\text{net}y}} = -\mu r_{0_x}\int_m \frac{z}{r^3}dm + \mu r_{0_z}\int_m \frac{x}{r^3}dm \quad (10.150)$$
$$M_{G_{\text{net}z}} = -\mu r_{0_y}\int_m \frac{x}{r^3}dm + \mu r_{0_x}\int_m \frac{y}{r^3}dm$$

Now, since $\|\boldsymbol{\varrho}\| \ll \|\mathbf{r}_0\|$, it follows from Equation 7.12 that
$$\frac{1}{r^3} = \frac{1}{r_0^3} - \frac{3}{r_0^5}\mathbf{r}_0 \cdot \boldsymbol{\varrho}$$
or
$$\frac{1}{r^3} = \frac{1}{r_0^3} - \frac{3}{r_0^5}(r_{0_x}x + r_{0_y}y + r_{0_z}z)$$

Therefore,
$$\int_m \frac{x}{r^3}\,dm = \frac{1}{r_0^3}\int_m x\,dm - \frac{3r_{0_x}}{r_0^5}\int_m x^2\,dm - \frac{3r_{0_y}}{r_0^5}\int_m xy\,dm - \frac{3r_{0_z}}{r_0^5}\int_m xz\,dm$$

But the center of mass lies at the origin of the xyz axes, which are principal moment of inertia directions. That means
$$\int_m x\,dm = \int_m xy\,dm = \int_m xz\,dm = 0$$

so that
$$\int_m \frac{x}{r^3}\,dm = -\frac{3r_{0_x}}{r_0^5}\int_m x^2\,dm \qquad (10.151)$$

In a similar fashion, we can show that
$$\int_m \frac{y}{r^3}\,dm = -\frac{3r_{0_y}}{r_0^5}\int_m y^2\,dm \qquad (10.152)$$

and
$$\int_m \frac{z}{r^3}\,dm = -\frac{3r_{0_y}}{r_0^5}\int_m z^2\,dm \qquad (10.153)$$

Substituting these last three expressions into Equations 10.150 leads to
$$M_{G_{net_x}} = \frac{3\mu r_{0_y} r_{0_z}}{r_0^5}\left(\int_m y^2\,dm - \int_m z^2\,dm\right)$$
$$M_{G_{net_y}} = \frac{3\mu r_{0_x} r_{0_z}}{r_0^5}\left(\int_m z^2\,dm - \int_m x^2\,dm\right) \qquad (10.154)$$
$$M_{G_{net_z}} = \frac{3\mu r_{0_x} r_{0_y}}{r_0^5}\left(\int_m x^2\,dm - \int_m y^2\,dm\right)$$

From Section 9.5 we recall that the moments of inertia are defined as
$$A = \int_m y^2\,dm + \int_m z^2\,dm \quad B = \int_m x^2\,dm + \int_m z^2\,dm \quad C = \int_m x^2\,dm + \int_m y^2\,dm$$
$$(10.155)$$

from which we may write
$$B - A = \int_m x^2\,dm - \int_m y^2\,dm \qquad A - C = \int_m z^2\,dm - \int_m x^2\,dm$$
$$C - B = \int_m y^2\,dm - \int_m z^2\,dm$$

It follows that Equations 10.154 reduce to

$$M_{G_{net_x}} = \frac{3\mu r_{0_y} r_{0_z}}{r_0^5}(C-B)$$

$$M_{G_{net_y}} = \frac{3\mu r_{0_x} r_{0_z}}{r_0^5}(A-C) \quad (10.156)$$

$$M_{G_{net_z}} = \frac{3\mu r_{0_x} r_{0_y}}{r_0^5}(B-A)$$

These are the components, in the spacecraft body frame, of the gravitational torque produced by the variation of the earth's gravitational field over the volume of the spacecraft. To get an idea of these torque magnitudes, note first of all that r_{0_x}/r_0, r_{0_y}/r_0 and r_{0_z}/r_0 are the direction cosines of the position vector of the center of mass, so that their magnitudes do not exceed 1. For a satellite in a low earth orbit of radius 6700 km, $3\mu/r_0^3 \cong 4 \times 10^{-6}$ s^{-2}, which is therefore the maximum order of magnitude of the coefficients of the inertia terms in Equation 10.156. The moments of inertia of the space shuttle are on the order of 10^6 kg·m^2, so the gravitational torques on this large vehicle are on the order of 1 N·m.

Substituting Equations 10.156 into Euler's equations of motion (Equations 9.72), we get

$$A\dot{\omega}_x + (C-B)\omega_y\omega_z = \frac{3\mu r_{0_y} r_{0_z}}{r_0^5}(C-B)$$

$$B\dot{\omega}_y + (A-C)\omega_z\omega_x = \frac{3\mu r_{0_x} r_{0_z}}{r_0^5}(A-C) \quad (10.157)$$

$$C\dot{\omega}_z + (B-A)\omega_x\omega_y = \frac{3\mu r_{0_x} r_{0_y}}{r_0^5}(B-A)$$

Now consider the orbital reference frame shown in Figure 10.27. It is actually the Clohessy–Wiltshire frame of Chapter 7, with the axes relabeled. The z' axis points radially outward from the center of the earth, the x' axis is in the direction of the local horizon, and the y' axis completes the right-handed triad by pointing in the direction of the orbit normal. This frame rotates around the y' axis with an angular velocity equal to the mean motion n of the circular orbit. Suppose we align the satellite's principal body frame axes xyz with $x'y'z'$, respectively. When the body x axis is aligned with the x' direction, it is called the *roll* axis. The body y axis, when aligned with the y' direction, is the *pitch* axis. The body z axis, pointing outward from the earth in the z' direction, is the *yaw* axis. These directions are illustrated in Figure 10.28. With the spacecraft aligned in this way, the body frame components of the inertial angular velocity $\boldsymbol{\omega}$ are $\omega_x = \omega_z = 0$ and $\omega_y = n$. The components of the position vector $\mathbf{r}_0$ are $r_{0_x} = r_{0_y} = 0$ and $r_{0_z} = r_0$. Substituting this data into Equations 10.157 yields

$$\dot{\omega}_x = \dot{\omega}_y = \dot{\omega}_z = 0$$

That is, the spacecraft will orbit the planet with its principal axes remaining aligned with the orbital frame. If this motion is stable under the influence of gravity alone, without the use of thrusters, gyros or other devices, then the spacecraft is *gravity gradient* stabilized. We need to assess the stability of this motion so we can

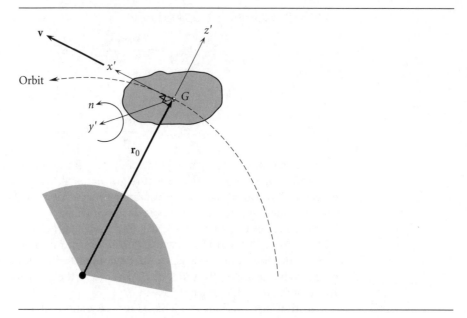

Figure 10.27 Orbital reference frame $x'y'z'$ attached to the center of mass of the satellite.

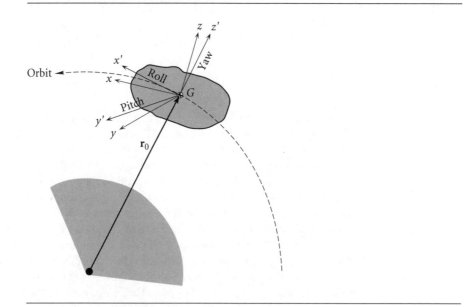

Figure 10.28 Satellite body frame slightly misaligned with the orbital frame $x'y'z'$.

determine how to orient a spacecraft to take advantage of this type of passive attitude stabilization.

Let the body frame xyz be slightly misaligned with the orbital reference frame, so that the yaw, pitch and roll angles between the xyz axes and the $x'y'z'$ axes, respectively,

10.10 Gravity-gradient stabilization

are very small, as suggested in Figure 10.28. The absolute angular velocity $\boldsymbol{\omega}$ of the spacecraft is the angular velocity $\boldsymbol{\omega}_{rel}$ relative to the orbital reference frame plus the inertial angular velocity $\boldsymbol{\Omega}$ of the $x'y'z'$ frame,

$$\boldsymbol{\omega} = \boldsymbol{\omega}_{rel} + \boldsymbol{\Omega}$$

The components of $\boldsymbol{\omega}_{rel}$ in the body frame are found using the yaw, pitch and roll relations, Equations 9.125. In so doing, it must be kept in mind that all angles and rates are assumed to be so small that their squares and products may be neglected. Recalling that $\sin\alpha = \alpha$ and $\cos\alpha = 1$ when $\alpha \ll 1$, we therefore obtain

$$\omega_{x_{rel}} = \omega_{roll} - \omega_{yaw} \overbrace{\sin\theta_{pitch}}^{=\theta_{pitch}} = \dot{\psi}_{roll} - \overbrace{\dot{\phi}_{yaw}\theta_{pitch}}^{\text{neglect product}} = \dot{\psi}_{roll} \tag{10.158}$$

$$\omega_{y_{rel}} = \omega_{yaw} \overbrace{\cos\theta_{pitch}}^{=1} \overbrace{\sin\psi_{roll}}^{=\psi_{roll}} + \omega_{pitch} \overbrace{\cos\psi_{roll}}^{=1} = \overbrace{\dot{\phi}_{yaw}\psi_{roll}}^{\text{neglect product}} + \dot{\theta}_{pitch} = \dot{\theta}_{pitch} \tag{10.159}$$

$$\omega_{z_{rel}} = \omega_{yaw} \overbrace{\cos\theta_{pitch}}^{=1} \overbrace{\cos\psi_{roll}}^{=1} - \omega_{pitch} \overbrace{\sin\psi_{roll}}^{=\psi_{roll}} = \dot{\phi}_{yaw} - \overbrace{\dot{\theta}_{pitch}\psi_{roll}}^{\text{neglect product}} = \dot{\phi}_{yaw} \tag{10.160}$$

The orbital frame's angular velocity is the mean motion n of the circular orbit, so that

$$\boldsymbol{\Omega} = n\hat{\mathbf{j}}'$$

To obtain the orbital frame's angular velocity components along the body frame, we must use the transformation rule

$$\{\boldsymbol{\Omega}\}_x = [Q]_{x'x}\{\boldsymbol{\Omega}\}_{x'} \tag{10.161}$$

where $[Q]_{x'x}$ is given by Equation 9.123. (Keep in mind that $x'y'z'$ are playing the role of XYZ in Figure 9.26.) Using the small angle approximations in Equation 9.123 leads to

$$[Q]_{x'x} = \begin{bmatrix} 1 & \phi_{yaw} & -\theta_{pitch} \\ -\phi_{yaw} & 1 & \psi_{roll} \\ \theta_{pitch} & -\psi_{roll} & 1 \end{bmatrix}$$

With this, Equation 10.161 becomes

$$\begin{Bmatrix} \Omega_x \\ \Omega_y \\ \Omega_z \end{Bmatrix} = \begin{bmatrix} 1 & \phi_{yaw} & -\theta_{pitch} \\ -\phi_{yaw} & 1 & \psi_{roll} \\ \theta_{pitch} & -\psi_{roll} & 1 \end{bmatrix} \begin{Bmatrix} 0 \\ n \\ 0 \end{Bmatrix} = \begin{Bmatrix} n\phi_{yaw} \\ n \\ -n\psi_{roll} \end{Bmatrix}$$

Now we can calculate the components of the satellite's inertial angular velocity along the body frame axes,

$$\begin{aligned} \omega_x &= \omega_{x_{rel}} + \Omega_x = \dot{\psi}_{roll} + n\phi_{yaw} \\ \omega_y &= \omega_{y_{rel}} + \Omega_y = \dot{\theta}_{pitch} + n \\ \omega_z &= \omega_{z_{rel}} + \Omega_z = \dot{\phi}_{yaw} - n\psi_{roll} \end{aligned} \tag{10.162}$$

Differentiating these with respect to time, remembering that n is constant for a circular orbit, gives the components of inertial angular acceleration in the body frame,

$$\dot{\omega}_x = \ddot{\psi}_{\text{roll}} + n\dot{\phi}_{\text{yaw}}$$
$$\dot{\omega}_y = \ddot{\theta}_{\text{pitch}} \qquad (10.163)$$
$$\dot{\omega}_z = \ddot{\phi}_{\text{yaw}} - n\dot{\psi}_{\text{roll}}$$

The position vector of the satellite's center of mass lies along the z' axis of the orbital frame,

$$\mathbf{r}_0 = r_0 \hat{\mathbf{k}}'$$

To obtain the components of $\mathbf{r}_0$ in the body frame we once again use the transformation matrix $[\mathbf{Q}]_{x'x}$

$$\begin{Bmatrix} r_{0_x} \\ r_{0_y} \\ r_{0_z} \end{Bmatrix} = \begin{bmatrix} 1 & \phi_{\text{yaw}} & -\theta_{\text{pitch}} \\ -\phi_{\text{yaw}} & 1 & \psi_{\text{roll}} \\ \theta_{\text{pitch}} & -\psi_{\text{roll}} & 1 \end{bmatrix} \begin{Bmatrix} 0 \\ 0 \\ r_0 \end{Bmatrix} = \begin{Bmatrix} -r_0 \theta_{\text{pitch}} \\ r_0 \psi_{\text{roll}} \\ r_0 \end{Bmatrix} \qquad (10.164)$$

Substituting Equations 10.162, 10.163 and 10.164, together with Equation 7.23, into Equations 10.157, and setting

$$A = I_{\text{roll}} \qquad B = I_{\text{pitch}} \qquad C = I_{\text{yaw}} \qquad (10.165)$$

yields

$$I_{\text{roll}}(\ddot{\psi}_{\text{roll}} + n\dot{\phi}_{\text{yaw}}) + (I_{\text{yaw}} - I_{\text{pitch}})(\dot{\theta}_{\text{pitch}} + n)(\dot{\phi}_{\text{yaw}} - n\psi_{\text{roll}})$$
$$= 3(I_{\text{yaw}} - I_{\text{pitch}})n^2 \psi_{\text{roll}}$$

$$I_{\text{pitch}}\ddot{\theta}_{\text{pitch}} + (I_{\text{roll}} - I_{\text{yaw}})(\dot{\psi}_{\text{roll}} + n\phi_{\text{yaw}})(\dot{\phi}_{\text{yaw}} - n\psi_{\text{roll}})$$
$$= -3(I_{\text{roll}} - I_{\text{yaw}})n^2 \theta_{\text{pitch}}$$

$$I_{\text{yaw}}(\ddot{\phi}_{\text{yaw}} - n\dot{\psi}_{\text{roll}}) + (I_{\text{pitch}} - I_{\text{roll}})(\dot{\theta}_{\text{pitch}} + n)(\dot{\psi}_{\text{roll}} + n\phi_{\text{yaw}})$$
$$= -3(I_{\text{pitch}} - I_{\text{roll}})n^2 \theta_{\text{pitch}} \psi_{\text{roll}}$$

Expanding terms and retaining terms at most linear in all angular quantities and their rates yields

$$I_{\text{yaw}}\ddot{\phi}_{\text{yaw}} + (I_{\text{pitch}} - I_{\text{roll}})n^2 \phi_{\text{yaw}} + (I_{\text{pitch}} - I_{\text{roll}} - I_{\text{yaw}})n\dot{\psi}_{\text{roll}} = 0 \qquad (10.166)$$

$$I_{\text{roll}}\ddot{\psi}_{\text{roll}} + (I_{\text{roll}} - I_{\text{pitch}} + I_{\text{yaw}})n\dot{\phi}_{\text{yaw}} + 4(I_{\text{pitch}} - I_{\text{yaw}})n^2 \psi_{\text{roll}} = 0 \qquad (10.167)$$

$$I_{\text{pitch}}\ddot{\theta}_{\text{pitch}} + 3(I_{\text{roll}} - I_{\text{yaw}})n^2 \theta_{\text{pitch}} = 0 \qquad (10.168)$$

These are the differential equations governing the influence of gravity gradient torques on the small angles and rates of misalignment of the body frame with the orbital frame.

Equation 10.168, governing the pitching motion around the y' axis, is not coupled to the other two equations. We make the classical assumption that the solution is of the form

$$\theta_{\text{pitch}} = Pe^{pt} \qquad (10.169)$$

where and P are p constants. P is the amplitude of the small disturbance that initiates the pitching motion. Substituting Equation 10.169 into Equation 10.168 yields $[I_{\text{pitch}} p^2 + 3(I_{\text{roll}} - I_{\text{yaw}}) n^2] P e^{pt} = 0$ for all t, which implies that the bracketed term must vanish, and that means p must have either of the two values

$$p_{1,2} = \pm i \sqrt{3 \frac{(I_{\text{roll}} - I_{\text{yaw}}) n^2}{I_{\text{pitch}}}} \quad (i = \sqrt{-1})$$

Thus

$$\theta_{\text{pitch}} = P_1 e^{p_1 t} + P_2 e^{p_2 t}$$

yields the stable, small-amplitude, steady-state harmonic oscillator solution only if p_1 and p_2 are imaginary, that is, if

$$I_{\text{roll}} > I_{\text{yaw}} \qquad \text{For stability in pitch.} \tag{10.170}$$

The stable pitch oscillation frequency is

$$\omega_{f_{\text{pitch}}} = n \sqrt{3 \frac{(I_{\text{roll}} - I_{\text{yaw}})}{I_{\text{pitch}}}} \tag{10.171}$$

(If $I_{\text{yaw}} > I_{\text{roll}}$, then p_1 and p_2 are both real, one positive, the other negative. The positive root causes $\theta_{\text{pitch}} \to \infty$, which is the undesirable, unstable case.)

Let us now turn our attention to Equations 10.166 and 10.167, which govern yaw and roll motion under gravity gradient torque. Again, we assume the solution is exponential in form,

$$\phi_{\text{yaw}} = Y e^{qt} \qquad \psi_{\text{roll}} = R e^{qt} \tag{10.172}$$

Substituting these into Equations 10.166 and 10.167 yields

$$[(I_{\text{pitch}} - I_{\text{roll}}) n^2 + I_{\text{yaw}} q^2] Y + (I_{\text{pitch}} - I_{\text{roll}} - I_{\text{yaw}}) nqR = 0$$

$$(I_{\text{roll}} - I_{\text{pitch}} + I_{\text{yaw}}) nqY + [4(I_{\text{pitch}} - I_{\text{yaw}}) n^2 + I_{\text{roll}} q^2] R = 0$$

In the interest of simplification, we can factor I_{yaw} out of the first equation and I_{roll} out of the second one to get

$$\left(\frac{I_{\text{pitch}} - I_{\text{roll}}}{I_{\text{yaw}}} n^2 + q^2 \right) Y + \left(\frac{I_{\text{pitch}} - I_{\text{roll}}}{I_{\text{yaw}}} - 1 \right) nqR = 0$$

$$\left(1 - \frac{I_{\text{pitch}} - I_{\text{yaw}}}{I_{\text{roll}}} \right) nqY + \left(4 \frac{I_{\text{pitch}} - I_{\text{yaw}}}{I_{\text{roll}}} n^2 + q^2 \right) R = 0 \tag{10.173}$$

Let

$$k_Y = \frac{I_{\text{pitch}} - I_{\text{roll}}}{I_{\text{yaw}}} \qquad k_R = \frac{I_{\text{pitch}} - I_{\text{yaw}}}{I_{\text{roll}}} \tag{10.174}$$

It is easy to show from Equations 10.155, 10.165 and 10.174 that

$$k_Y = \frac{\left(\int_m x^2 \, dm \big/ \int_m y^2 \, dm \right) - 1}{\left(\int_m x^2 \, dm \big/ \int_m y^2 \, dm \right) + 1} \qquad k_R = \frac{\left(\int_m z^2 \, dm \big/ \int_m y^2 \, dm \right) - 1}{\left(\int_m z^2 \, dm \big/ \int_m y^2 \, dm \right) + 1}$$

which means
$$|k_Y| < 1 \quad |k_R| < 1$$

Using the definitions in Equation 10.174, we can write Equations 10.173 more compactly as
$$(k_Y n^2 + q^2)Y + (k_Y - 1)nqR = 0$$
$$(1 - k_R)nqY + (4k_R n^2 + q^2)R = 0$$

or, using matrix notation,
$$\begin{bmatrix} k_Y n^2 + q^2 & (k_Y - 1)nq \\ (1 - k_R)nq & 4k_R n^2 + q^2 \end{bmatrix} \begin{Bmatrix} Y \\ R \end{Bmatrix} = \begin{Bmatrix} 0 \\ 0 \end{Bmatrix} \qquad (10.175)$$

In order to avoid the trivial solution ($Y = R = 0$), the determinant of the coefficient matrix must be zero. Expanding the determinant and collecting terms yields the characteristic equation for q,
$$q^4 + bn^2 q^2 + cn^4 = 0 \qquad (10.176)$$

where
$$b = 3k_R + k_Y k_R + 1 \qquad c = 4k_Y k_R \qquad (10.177)$$

This quadratic equation has four roots which, when substituted back into Equation 10.172, yield
$$\phi_{\text{yaw}} = Y_1 e^{q_1 t} + Y_2 e^{q_2 t} + Y_3 e^{q_3 t} + Y_4 e^{q_4 t}$$
$$\psi_{\text{roll}} = R_1 e^{q_1 t} + R_2 e^{q_2 t} + R_3 e^{q_3 t} + R_4 e^{q_4 t}$$

In order for these solutions to remain finite in time, the roots $q_1, \ldots, q_4$ must be negative (solution decays to zero) or imaginary (steady oscillation at initial small amplitude).

To reduce Equation 10.176 to a quadratic equation, let us introduce a new variable λ and write,
$$q = \pm n\sqrt{\lambda} \qquad (10.178)$$

Then Equation 10.176 becomes
$$\lambda^2 + b\lambda + c = 0 \qquad (10.179)$$

the familiar solution of which is
$$\lambda_1 = -\frac{1}{2}\left(b + \sqrt{b^2 - 4c}\right) \qquad \lambda_2 = -\frac{1}{2}\left(b - \sqrt{b^2 - 4c}\right) \qquad (10.180)$$

To guarantee that q in Equation 10.178 does not take a positive value, we must require that λ be real and negative (so q will be imaginary). For λ to be real requires that $b > 2\sqrt{c}$, or
$$3k_R + k_Y k_R + 1 > 4\sqrt{k_Y k_R} \qquad (10.181)$$

For λ to be negative requires $b^2 > b^2 - 4c$, which will be true if $c > 0$; i.e.,
$$k_Y k_R > 0 \qquad (10.182)$$

Equations 10.181 and 10.182 are the conditions required for yaw and roll stability under gravity gradient torques, to which we must add Equation 10.170 for pitch stability. Observe that we can solve Equations 10.174 to obtain

$$I_{\text{yaw}} = \frac{1 - k_R}{1 - k_Y k_R} I_{\text{pitch}} \qquad I_{\text{roll}} = \frac{1 - k_Y}{1 - k_Y k_R} I_{\text{pitch}}$$

By means of these relationships, the pitch stability criterion, $I_{\text{roll}}/I_{\text{yaw}} > 1$, becomes

$$\frac{1 - k_Y}{1 - k_R} > 1$$

In view of the fact that $|k_R| < 1$, this means

$$k_Y < k_R \qquad (10.183)$$

Figure 10.29 shows those regions I and II on the $k_Y - k_R$ plane in which all three stability criteria (Equations 10.181, 10.182 and 10.183) are simultaneously satisfied, along with the requirement that the three moments of inertia I_{pitch}, I_{roll} and I_{yaw} are positive.

In the small sliver of region I, and $k_Y < 0$ and $k_R < 0$; therefore, according to Equations 10.174, $I_{\text{yaw}} > I_{\text{pitch}}$ and $I_{\text{roll}} > I_{\text{pitch}}$, which together with Equation 10.170 yield $I_{\text{roll}} > I_{\text{yaw}} > I_{\text{pitch}}$. Remember that the gravity gradient spacecraft is slowly 'spinning' about the minor pitch axis (normal to the orbit plane) at an angular velocity equal to the mean motion of the orbit. So this criterion makes the spacecraft a 'minor axis spinner', the roll axis (flight direction) being the major axis of inertia. With energy dissipation, we know this orientation is not stable in the long run. On the other hand, in region II, k_Y and k_R are both positive, so that Equations 10.174 imply

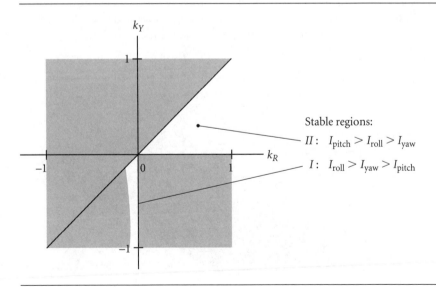

Figure 10.29 Regions in which the values of k_Y and k_R yield neutral stability in yaw, pitch and roll of a gravity gradient satellite.

$I_{pitch} > I_{yaw}$ and $I_{pitch} > I_{roll}$. Thus, along with the pitch criterion ($I_{roll} > I_{yaw}$), we have $I_{pitch} > I_{roll} > I_{yaw}$. In this, the preferred, configuration, the gravity gradient spacecraft is a 'major axis spinner' about the pitch axis, and the minor yaw axis is the minor axis of inertia. It turns out that all of the known gravity-gradient stabilized moons of the solar system, like the earth's, whose 'captured' rate of rotation equals the orbital period, are major axis spinners.

In Equation 10.171 we presented the frequency of the gravity gradient pitch oscillation. For completeness we should also point out that the coupled yaw and roll motions have two oscillation frequencies, which are obtained from Equations 10.178 and 10.180,

$$\omega_{f_{yaw/roll}})_{1,2} = n\sqrt{\frac{1}{2}(b \pm \sqrt{b^2 - 4c})} \qquad (10.184)$$

Recall that b and c are found in Equation 10.177.

We have assumed throughout this discussion that the orbit of the gravity gradient satellite is circular. Kaplan (1976) shows that the effect of a small eccentricity turns up only in the pitching motion. In particular, the natural oscillation expressed by Equation 10.170 is augmented by a forced oscillation term,

$$\theta_{pitch} = P_1 e^{p_1 t} + P_2 e^{p_2 t} + \frac{2e \sin nt}{3\left(\frac{I_{roll} - I_{yaw}}{I_{pitch}}\right) - 1} \qquad (10.185)$$

where e is the (small) eccentricity of the orbit. From this we see that there is a pitch resonance. When $(I_{roll} - I_{yaw})/I_{pitch}$ approaches 1/3, the amplitude of the last term grows without bound.

EXAMPLE 10.14

The uniform, monolithic 10 000 kg slab, having the dimensions shown in Figure 10.30, is in a circular LEO. Determine the orientation of the satellite in its orbit for gravity gradient stabilization, and compute the periods of the pitch and yaw/roll oscillations in terms of the orbital period T.

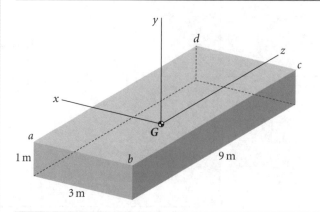

Figure 10.30 Parallelepiped satellite.

10.10 Gravity-gradient stabilization

According to Figure 9.9(c), the principal moments of inertia around the *xyz* axes through the center of mass are

$$A = \frac{10\,000}{12}(1^2 + 9^2) = 68\,333 \text{ kg} \cdot \text{m}^2$$

$$B = \frac{10\,000}{12}(3^2 + 9^2) = 75\,000 \text{ kg} \cdot \text{m}^2$$

$$C = \frac{10\,000}{12}(3^2 + 1^2) = 8333.3 \text{ kg} \cdot \text{m}^2$$

Let us first determine whether we can stabilize this object as a minor axis spinner. In that case,

$$I_{\text{pitch}} = C = 8333.3 \text{ kg} \cdot \text{m}^2 \quad I_{\text{yaw}} = A = 68\,333 \text{ kg} \cdot \text{m}^2 \quad I_{\text{roll}} = B = 75\,000 \text{ kg} \cdot \text{m}^2$$

Since $I_{\text{roll}} > I_{\text{yaw}}$, the satellite would be stable in pitch. To check yaw/roll stability, we first compute

$$k_Y = \frac{I_{\text{pitch}} - I_{\text{roll}}}{I_{\text{yaw}}} = -0.97561 \qquad k_R = \frac{I_{\text{pitch}} - I_{\text{yaw}}}{I_{\text{roll}}} = -0.8000$$

We see that $k_Y k_R > 0$, which is one of the two requirements. The other one is found in Equation 10.181, but in this case

$$1 + 3k_R + k_Y k_R - 4\sqrt{k_Y k_R} = -4.1533 < 0$$

so that condition is not met. Hence, the object cannot be gravity-gradient stabilized as a minor axis spinner.

As a major axis spinner, we must have

$$I_{\text{pitch}} = B = 75\,000 \text{ kg} \cdot \text{m}^2 \quad I_{\text{yaw}} = C = 8333.3 \text{ kg} \cdot \text{m}^2 \quad I_{\text{roll}} = A = 68\,333 \text{ kg} \cdot \text{m}^2$$

Then $I_{\text{roll}} > I_{\text{yaw}}$, so the pitch stability condition is satisfied. Furthermore, since

$$k_Y = \frac{I_{\text{pitch}} - I_{\text{roll}}}{I_{\text{yaw}}} = 0.8000 \qquad k_R = \frac{I_{\text{pitch}} - I_{\text{yaw}}}{I_{\text{roll}}} = 0.97561$$

we have

$$k_Y k_R = 0.7805 > 0$$
$$1 + 3k_R + k_Y k_R - 4\sqrt{k_Y k_R} = 1.1735 > 0$$

which means the two criteria for stability in the yaw and roll modes are met. The satellite should therefore be orbited as shown in Figure 10.31, with its minor axis aligned with the radial from the earth's center, the plane *abcd* lying in the orbital plane, and the body *x* axis aligned with the local horizon.

According to Equation 10.171, the frequency of the pitch oscillation is

$$\omega_{f_{\text{pitch}}} = n\sqrt{3\frac{I_{\text{roll}} - I_{\text{yaw}}}{I_{\text{pitch}}}} = n\sqrt{3\frac{68\,333 - 8333.3}{75\,000}} = 1.5492n$$

(Example 10.14 continued)

where n is the mean motion. Hence, the period of this oscillation, in terms of that of the orbit, is

$$T_{\text{pitch}} = \frac{2\pi}{\omega_{f_{\text{pitch}}}} = 0.6455 \frac{2\pi}{n} = \underline{0.6455T}$$

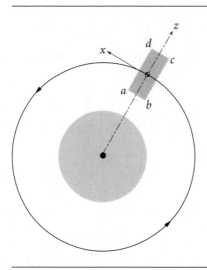

Figure 10.31 Orientation of the parallelepiped for gravity-gradient stabilization.

For the yaw/roll frequencies, we use Equation 10.184,

$$\left(\omega_{f_{\text{yaw/roll}}}\right)_1 = n\sqrt{\frac{1}{2}\left(b + \sqrt{b^2 - 4c}\right)}$$

where

$$b = 1 + 3k_R + k_Y k_R = 4.7073 \quad \text{and} \quad c = 4k_Y k_R = 3.122$$

Thus,

$$\left(\omega_{f_{\text{yaw/roll}}}\right)_1 = 2.3015n$$

Likewise,

$$\left(\omega_{f_{\text{yaw/roll}}}\right)_2 = \sqrt{\frac{1}{2}\left(b - \sqrt{b^2 - 4c}\right)} = 1.977n$$

From these we obtain

$$T_{\text{yaw/roll}_1} = \underline{0.5058T} \qquad T_{\text{yaw/roll}_2} = \underline{0.4345T}$$

Finally, observe that

$$\frac{I_{\text{roll}} - I_{\text{yaw}}}{I_{\text{pitch}}} = 0.8$$

so that we are far from the pitch resonance condition that exists if the orbit has a small eccentricity.

Problems

10.1 The axisymmetric satellite has axial and transverse mass moments of inertia about axes through the mass center G of $C = 1200$ kg·m^2 and $A = 2600$ kg·m^2, respectively. If it is spinning at $\omega_s = 6$ rad/s when it is launched, determine its angular momentum. Precession occurs about the inertial Z axis.
{Ans.: $\|\mathbf{H}_G\| = 13\,450$ kg·m^2/s}

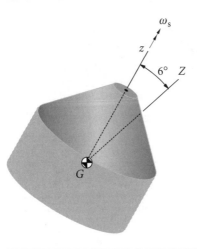

Figure P.10.1

10.2 A spacecraft is symmetrical about its body-fixed z axis. Its principal mass moments of inertia are $A = B = 300$ kg·m^2 and $C = 500$ kg·m^2. The z axis sweeps out a cone with a total vertex angle of 10° as it precesses around the angular momentum vector. If the spin velocity is 6 rad/s, compute the period of precession.
{Ans.: 0.417 s}

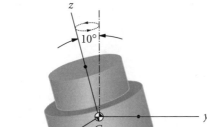

Figure P.10.2

10.3 A thin ring tossed into the air with a spin velocity of ω_s has a very small nutation angle θ (in radians). What is the precession rate ω_p?
{Ans.: $\omega_p = 2\omega_s(1 + \theta^2/2)$, retrograde}

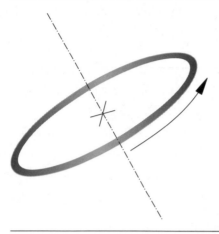

Figure P.10.3

10.4 For an axisymmetric rigid satellite,

$$[\mathbf{I}_G] = \begin{bmatrix} I_{xx} & 0 & 0 \\ 0 & I_{yy} & 0 \\ 0 & 0 & I_{zz} \end{bmatrix} = \begin{bmatrix} 1000 & 0 & 0 \\ 0 & 1000 & 0 \\ 0 & 0 & 5000 \end{bmatrix} \text{kg} \cdot \text{m}^2$$

It is spinning about the body z axis in torque-free motion, precessing around the angular momentum vector **H** at the rate of 2 rad/s. Calculate the magnitude of **H**.
{Ans.: 2000 N · m · s}

10.5 At a given instant the box-shaped 500 kg satellite (in torque-free motion) has an absolute angular velocity $\boldsymbol{\omega} = 0.01\hat{\mathbf{i}} - 0.03\hat{\mathbf{j}} + 0.02\hat{\mathbf{k}}$ (rad/s). Its moments of inertia about the principal body axes xyz are $A = 385.4$ kg · m², $B = 416.7$ kg · m² and $C = 52.08$ kg · m², respectively. Calculate the magnitude of its absolute angular acceleration.
{Ans.: 6.167×10^{-4} m/s²}

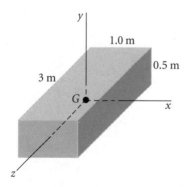

Figure P.10.5

10.6 An 8 kg thin ring in torque-free motion is spinning with an angular velocity of 30 rad/s and a constant nutation angle of 15°. Calculate the rotational kinetic energy if $A = B = 0.36$ kg·m², $C = 0.72$ kg·m².
{Ans.: 370.5 J}

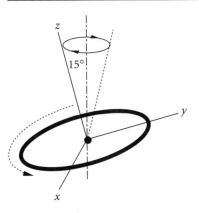

Figure P.10.6

10.7 The rectangular block has an angular velocity $\boldsymbol{\omega} = 1.5\omega_0\hat{\mathbf{i}} + 0.8\omega_0\hat{\mathbf{j}} + 0.6\omega_0\hat{\mathbf{k}}$, where ω_0 has units of rad/s.
(a) Determine the angular velocity ω of the block if it spins around the body z axis with the same rotational kinetic energy.
(b) Determine the angular velocity ω of the block if it spins around the body z axis with the same angular momentum.
{Ans.: (a) $\omega = 1.31\omega_0$, (b) $\omega = 1.04\omega_0$}

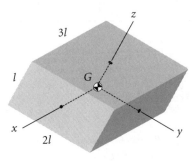

Figure P.10.7

10.8 For a rigid axisymmetric satellite, the mass moment of inertia about its long axis is 1000 kg·m², and the moment of inertia about transverse axes through the centroid is 5000 kg·m². It is spinning about the minor principal body axis in torque-free motion at 6 rad/s with the angular velocity lined up with the angular momentum vector **H**. Over time, the energy degrades due to internal effects and the satellite is eventually spinning about a major principal body axis with the angular velocity lined up with

the angular momentum vector **H**. Calculate the change in rotational kinetic energy between the two states.
{Ans.: −14.4 kJ}

10.9 Let the object in Example 9.11 be a highly dissipative torque-free satellite, whose angular velocity at the instant shown is $\boldsymbol{\omega} = 10\hat{\mathbf{i}}$ rad/s. Calculate the decrease in kinetic energy after it becomes, as eventually it must, a major axis spinner.
{Ans.: −0.487 J}

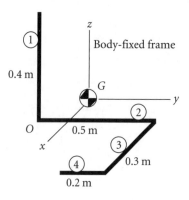

Figure P.10.9

10.10 For a non-precessing, dual-spin satellite, $C_r = 1000$ kg · m² and $C_p = 500$ kg · m². The angular velocity of the rotor is $3\hat{\mathbf{k}}$ rad/s and the angular velocity of the platform relative to the rotor is $1\hat{\mathbf{k}}$ rad/s. If the relative angular velocity of the platform is reduced to $0.5\hat{\mathbf{k}}$ rad/s, what is the new angular velocity of the rotor?
{Ans.: 3.17 rad/s}

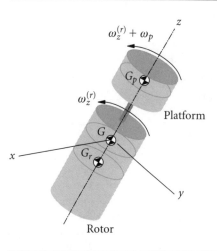

Figure P.10.10

10.11 For a rigid axisymmetric satellite, the mass moment of inertia about its long axis is 1000 kg · m^2, and the moment of inertia about transverse axes through the center of mass is 5000 kg · m^2. It is initially spinning about the *minor* principal body axis in torque-free motion at $\omega_s = 0.1$ rad/s, with the angular velocity lined up with the angular momentum vector $\mathbf{H}_0$. A pair of thrusters exert an external impulsive torque on the satellite, causing an instantaneous change $\Delta \mathbf{H}$ of angular momentum in the direction normal to $\mathbf{H}_0$ (no change in spin rate), so that the new angular momentum is $\mathbf{H}_1$, at an angle of 20° to $\mathbf{H}_0$, as shown in the figure. How long does it take the satellite to precess ('cone') through an angle of 180° around $\mathbf{H}_1$?
{Ans.: 118 s}

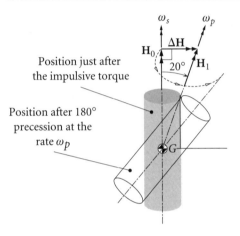

Figure P.10.11

10.12 The solid right-circular cylinder of mass 500 kg is set into torque-free motion with its symmetry axis initially aligned with the fixed spatial line a–a. Due to an injection error, the vehicle's angular velocity vector $\boldsymbol{\omega}$ is misaligned 5° (the wobble angle) from the symmetry axis. Calculate to three significant figures the maximum angle ϕ between fixed line a–a and the axis of the cylinder.
{Ans.: 31°}

10.13 A satellite is spinning at 0.01 rev/s. The moment of inertia of the satellite about the spin axis is 2000 kg · m^2. Paired thrusters are located at a distance of 1.5 m from the spin axis. They deliver their thrust in pulses, each thruster producing an impulse of 15 N · s per pulse. At what rate will the satellite be spinning after 30 pulses?
{Ans.: 0.0637 rev/s}

10.14 A satellite has moments of inertia $A = 2000$ kg · m^2, $B = 4000$ kg · m^2 and $C = 6000$ kg · m^2 about its principal body axes *xyz*. Its angular velocity is $\boldsymbol{\omega} = 0.1\hat{\mathbf{i}} + 0.3\hat{\mathbf{j}} + 0.5\hat{\mathbf{k}}$ (rad/s). If thrusters cause the angular momentum vector to undergo the change $\Delta \mathbf{H}_G = 50\hat{\mathbf{i}} - 100\hat{\mathbf{j}} + 3000\hat{\mathbf{k}}$ (kg · m^2/s), what is the magnitude of the new angular velocity?
{Ans.: 0.628 rad/s}

10.15 The body-fixed *xyz* axes are principal axes of inertia passing through the center of mass of the 300 kg cylindrical satellite, which is spinning at 1 revolution per second about the *z* axis. What impulsive torque about the *y* axis must the thrusters impart to cause the satellite to precess at 0.1 revolution per second?
{Ans.: 137 N · m · s}

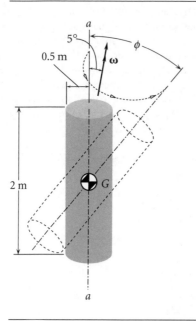

Figure P.10.12

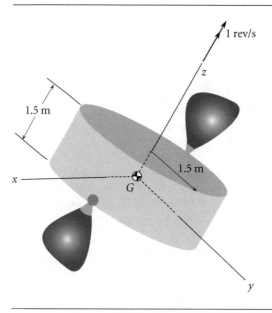

Figure P.10.15

10.16 A satellite is to be despun by means of a tangential-release yo-yo mechanism consisting of two masses, 3 kg each, wound around the mid plane of the satellite. The satellite is spinning around its axis of symmetry with an angular velocity $\omega_s = 5$ rad/s. The radius of the cylindrical satellite is 1.5 m and the moment of inertia about the spin axis is $C = 300$ kg·m².

(a) Find the cord length and the deployment time to reduce the spin rate to 1 rad/s.
(b) Find the cord length and time to reduce the spin rate to zero.
{Ans.: (a) $l = 5.902$ m, $t = 0.787$ s; (b) $l = 7.228$ m, $t = 0.964$ s}

10.17 A cylindrical satellite of radius 1 m is initially spinning about the axis of symmetry at the rate of 2 revolutions per second with a nutation angle of 15°. The principal moments of inertia are $I_x = I_y = 30$ kg · m², $I_z = 60$ kg · m². An energy dissipation device is built into the satellite, so that it eventually ends up in pure spin around the z axis.
(a) Calculate the final spin rate about the z axis.
(b) Calculate the loss of kinetic energy.
(c) A tangential release yo-yo despin device is also included in the satellite. If the two yo-yo masses are each 7 kg, what cord length is required to completely despin the satellite? Is it wrapped in the proper direction in the figure?
{Ans.: (a) 2.071 rad/s; (b) 8.62 J; (c) 2.3 m}

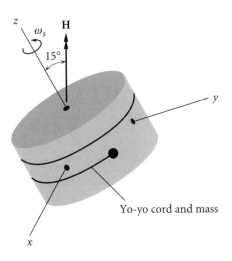

Figure P.10.17

10.18 A communications satellite is in a GEO (geostationary equatorial orbit) with a period of 24 hours. The spin rate ω_s about its axis of symmetry is 1 revolution per minute, and the moment of inertia about the spin axis is 550 kg · m². The moment of inertia about transverse axes through the mass center G is 225 kg · m². If the spin axis is initially pointed towards the earth, calculate the magnitude and direction of the applied torque $\mathbf{M}_G$ required to keep the spin axis pointed always towards the earth.
{Ans.: 0.00420 N · m, about the negative x axis}

10.19 The moments of inertia of a satellite about its principal body axes xyz are $A = 1000$ kg · m², $B = 600$ kg · m² and $C = 500$ kg · m², respectively. The moments of inertia of a momentum wheel at the center of mass of the satellite and aligned with the x axis are $I_x = 20$ kg · m² and $I_y = I_z = 6$ kg · m². The absolute angular velocity of the satellite with the momentum wheel locked is $\boldsymbol{\omega}_0 = 0.1\hat{\mathbf{i}} + 0.05\hat{\mathbf{j}}$ (rad/s). Calculate the angular velocity ω_f of the momentum wheel (relative to the satellite) required to reduce the x component of the absolute angular velocity of the satellite to 0.003 rad/s.
{Ans.: 4.95 rad/s}

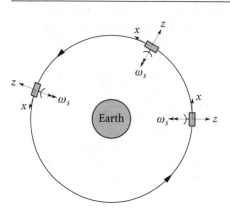

Figure P.10.18

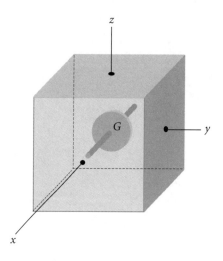

Figure P.10.19

10.20 A satellite has principal moments of inertia $I_1 = 300 \text{ kg} \cdot \text{m}^2$, $I_2 = 400 \text{ kg} \cdot \text{m}^2$, $I_3 = 500 \text{ kg} \cdot \text{m}^2$. Determine the permissible orientations in a circular orbit for gravity-gradient stabilization. Specify which axes may be aligned in the pitch, roll and yaw directions. (Recall that, relative to a Clohessy–Wiltshire frame at the center of mass of the satellite, yaw is about the x axis (outward radial from earth's center); roll is about the y axis (velocity vector); pitch is about the z axis (normal to orbital plane).

CHAPTER 11

Rocket vehicle dynamics

Chapter outline

11.1	Introduction	551
11.2	Equations of motion	552
11.3	The thrust equation	555
11.4	Rocket performance	557
11.5	Restricted staging in field-free space	560
11.6	Optimal staging	570
	11.6.1 Lagrange multiplier	570
Problems		578

11.1 Introduction

In previous chapters we have made frequent reference to delta-v maneuvers of spacecraft. These require a propulsion system of some sort whose job it is to throw vehicle mass (in the form of propellants) overboard. Newton's balance of momentum principle dictates that when mass is ejected from a system in one direction, the mass left behind must acquire a velocity in the opposite direction. The familiar and oft-quoted example is the rapid release of air from an inflated toy balloon. Another is that of a diver leaping off a small boat at rest in the water, causing the boat to acquire a motion of its own. The unfortunate astronaut who becomes separated from his ship in the vacuum of space cannot with any amount of flailing of arms and legs 'swim' back to safety. If he has tools or other expendable objects of equipment, accurately throwing them in the direction opposite to his spacecraft may do the trick. Spewing compressed gas from a tank attached to his back through to a nozzle pointed away from the spacecraft would be a better solution.

The purpose of a rocket motor is to use the chemical energy of solid or liquid propellants to steadily and rapidly produce a large quantity of hot, high pressure gas which is then expanded and accelerated through a nozzle. This large mass of combustion products flowing out of the nozzle at supersonic speed possesses a lot of momentum and, leaving the vehicle behind, causes the vehicle itself to acquire a momentum in the opposite direction. This is represented as the action of the force we know as thrust. The design and analysis of rocket propulsion systems is well beyond our scope.

This chapter contains a necessarily brief introduction to some of the fundamentals of rocket vehicle dynamics. The equations of motion of a launch vehicle in a gravity turn trajectory are presented first. This is followed by a simple development of the thrust equation, which brings in the concept of specific impulse. The thrust equation and the equations of motion are then combined to produce the rocket equation, which relates delta-v to propellant expenditure and specific impulse. The sounding rocket provides an important but relatively simple application of the concepts introduced to this point. The chapter concludes with an elementary consideration of multi-stage launch vehicles.

Those seeking a more detailed introduction to the subject of rockets and rocket performance will find the texts by Wiesel (1997) and Hale (1994), as well as references cited therein, useful.

11.2 Equations of motion

Figure 11.1 illustrates the trajectory of a satellite launch vehicle and the forces acting on it during the powered ascent. Rockets at the base of the booster produce the

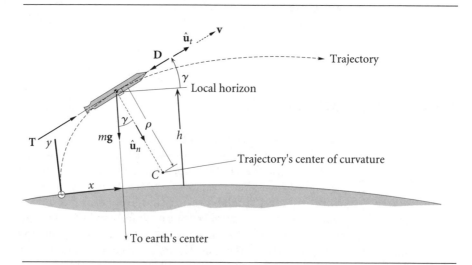

Figure 11.1 Launch vehicle boost trajectory. γ is the flight path angle.

thrust **T** which acts along the vehicle's axis in the direction of the velocity vector **v**. The aerodynamic drag force **D** is directed opposite to the velocity, as shown. Its magnitude is given by

$$D = qAC_D \qquad (11.1)$$

where $q = \frac{1}{2}\varrho v^2$ is the dynamic pressure, in which ϱ is the density of the atmosphere and v is the speed, i.e., the magnitude of **v**. A is the frontal area of the vehicle and C_D is the coefficient of drag. C_D depends on the speed and the external geometry of the rocket. The force of gravity on the booster is $m\mathbf{g}$, where m is its mass and **g** is the local gravitational acceleration, pointing towards the center of the earth. As discussed in Section 1.2, at any point of the trajectory, the velocity **v** defines the direction of the unit tangent $\hat{\mathbf{u}}_t$ to the path. The unit normal $\hat{\mathbf{u}}_n$ is perpendicular to **v** and points towards the center of curvature C. The distance of point C from the path is ϱ (not to be confused with density). ϱ is the radius of curvature.

In Figure 11.1 the vehicle and its flight path are shown relative to the earth. In the interest of simplicity we will ignore the earth's spin and write the equations of motion relative to a non-rotating earth. The small acceleration terms required to account for the earth's rotation can be added for a more refined analysis. Let us resolve Newton's second law, $\mathbf{F}_{net} = m\mathbf{a}$, into components along the path directions $\hat{\mathbf{u}}_t$ and $\hat{\mathbf{u}}_n$. Recall from Section 1.2 that the acceleration along the path is

$$a_t = \frac{dv}{dt} \qquad (11.2)$$

and the normal acceleration is $a_n = v^2/\varrho$ (where ϱ is the radius of curvature). It was shown in Example 1.4 (Equation 1.9) that for flight over a flat surface, $v/\varrho = -d\gamma/dt$, in which case the normal acceleration can be expressed in terms of the flight path angle as

$$a_n = -v\frac{d\gamma}{dt}$$

To account for the curvature of the earth, as was done in Section 1.6, one can use polar coordinates with origin at the earth's center to show that a term must be added to this expression, so that it becomes

$$a_n = -v\frac{d\gamma}{dt} + \frac{v^2}{R_E + h}\cos\gamma \qquad (11.3)$$

where R_E is the radius of the earth and h is the altitude of the rocket. Thus, in the direction of $\hat{\mathbf{u}}_t$ Newton's second law requires

$$T - D - mg\sin\gamma = ma_t \qquad (11.4)$$

whereas in the $\hat{\mathbf{u}}_n$ direction

$$mg\cos\gamma = ma_n \qquad (11.5)$$

After substituting Equations 11.2 and 11.3, these latter two expressions may be written

$$\frac{dv}{dt} = \frac{T}{m} - \frac{D}{m} - g\sin\gamma \qquad (11.6)$$

$$v\frac{d\gamma}{dt} = -\left(g - \frac{v^2}{R_E + h}\right)\cos\gamma \qquad (11.7)$$

To these we must add the equations for downrange distance x and altitude h,

$$\frac{dx}{dt} = \frac{R_E}{R_E + h} v\cos\gamma \qquad \frac{dh}{dt} = v\sin\gamma \qquad (11.8)$$

Recall that the variation of g with altitude is given by Equation 1.8. Numerical methods must be used to solve Equations 11.6, 11.7 and 11.8. To do so, one must account for the variation of the thrust, booster mass, atmospheric density, the drag coefficient, and the acceleration of gravity. Of course, the vehicle mass continuously decreases as propellants are consumed to produce the thrust, which we shall discuss in the following section.

The free-body diagram in Figure 11.1 does not include a lifting force, which, if the vehicle were an airplane, would act normal to the velocity vector. Launch vehicles are designed to be strong in lengthwise compression, like a column. To save weight they are, unlike an airplane, made relatively weak in bending, shear and torsion, which are the kinds of loads induced by lifting surfaces. Transverse lifting loads are held closely to zero during powered ascent through the atmosphere by maintaining zero angle of attack, i.e., by keeping the axis of the booster aligned with its velocity vector (the relative wind). Pitching maneuvers are done early in the launch, soon after the rocket clears the launch tower, when its speed is still low. At the high speeds acquired within a minute or so after launch, the slightest angle of attack can produce destructive transverse loads in the vehicle. The space shuttle orbiter has wings so it can act as a glider after re-entry into the atmosphere. However, the launch configuration of the orbiter is such that its wings are at the zero-lift angle of attack throughout the ascent.

Satellite launch vehicles take off vertically and, at injection into orbit, must be flying parallel to the earth's surface. During the initial phase of the ascent, the rocket builds up speed on a nearly vertical trajectory taking it above the dense lower layers of the atmosphere. While it transitions the thinner upper atmosphere, the trajectory bends over, trading vertical speed for horizontal speed so the rocket can achieve orbital perigee velocity at burnout. The gradual transition from vertical to horizontal flight, illustrated in Figure 11.1, is caused by the force of gravity, and it is called a gravity turn trajectory.

At lift off, the rocket is vertical and the flight path angle γ is 90°. After clearing the tower and gaining speed, vernier thrusters or gimbaling of the main engines produce a small, programmed pitchover, establishing an initial flight path angle γ_0, slightly less than 90°. Thereafter, γ will continue to decrease at a rate dictated by Equation 11.7. (For example, if $\gamma = 85°$, $v = 110$ m/s (250 mph), and $h = 2$ km, then $d\gamma/dt = -0.44°/$s.) As the speed v of the vehicle increases, the coefficient of $\cos\gamma$ in Equation 11.7 decreases, which means the rate of change of the flight path angle becomes increasingly smaller, tending towards zero as the booster approaches orbital speed, $v_{\text{circular orbit}} = \sqrt{g(R+h)}$. Ideally, the vehicle is flying horizontally ($\gamma = 0$) at that point.

The gravity turn trajectory is just one example of a practical trajectory, tailored for satellite boosters. On the other hand, sounding rockets fly straight up from launch through burnout. Rocket-powered guided missiles must execute high-speed pitch and yaw maneuvers as they careen towards moving targets, and require a rugged structure to withstand the accompanying side loads.

11.3 THE THRUST EQUATION

To discuss rocket performance requires an expression for the thrust T in Equation 11.6. It can be obtained by a simple one-dimensional momentum analysis. Figure 11.2(a) shows a system consisting of a rocket and its propellants. The exterior of the rocket is surrounded by the static pressure p_a of the atmosphere everywhere except at the rocket nozzle exit where the pressure is p_e. p_e acts over the nozzle exit area A_e. The value of p_e depends on the design of the nozzle. For simplicity, we assume no other forces act on the system. At time t the mass of the system is m and the absolute velocity in its axial direction is v. The propellants combine chemically in the rocket's combustion chamber, and during the small time interval Δt a small mass Δm of combustion products is forced out of the nozzle, to the left. As a result of this expulsion, the velocity of the rocket changes by the small amount Δv, to the right. The absolute velocity of Δm is v_e, assumed to be to the left. According to Newton's second law of motion,

(momentum of the system at $t + \Delta t$) − (momentum of the system at t)

= net external impulse

or

$$\left[(m - \Delta m)(v + \Delta v)\hat{\mathbf{i}} + \Delta m\left(-v_e\hat{\mathbf{i}}\right)\right] - mv\hat{\mathbf{i}} = (p_e - p_a)A_e\Delta t\hat{\mathbf{i}} \qquad (11.9)$$

Let $\dot{m}_e$ (a positive quantity) be the rate at which exhaust mass flows across the nozzle exit plane. The mass m of the rocket decreases at the rate dm/dt, and conservation of mass requires the decrease of mass to equal the mass flow rate out of the nozzle. Thus,

$$\frac{dm}{dt} = -\dot{m}_e \qquad (11.10)$$

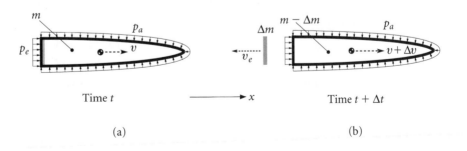

Figure 11.2 (a) System of rocket and propellant at time t. (b) The system an instant later, after ejection of a small element Δm of combustion products.

Assuming $\dot{m}_e$ is constant, the vehicle mass as a function of time (from $t=0$) may therefore be written

$$m(t) = m_0 - \dot{m}_e t \tag{11.11}$$

where m_0 is the initial mass of the vehicle. Since Δm is the mass which flows out in the time interval Δt, we have

$$\Delta m = \dot{m}_e \Delta t \tag{11.12}$$

Let us substitute this expression into Equation 11.9 to obtain

$$\left[(m - \dot{m}_e \Delta t)(v + \Delta v)\hat{\mathbf{i}} + \dot{m}_e \Delta t \left(-v_e \hat{\mathbf{i}}\right) \right] - mv\hat{\mathbf{i}} = (p_e - p_a)A_e \Delta t \hat{\mathbf{i}}$$

Collecting terms, we get

$$m\Delta v\hat{\mathbf{i}} - \dot{m}_e \Delta t(v_e + v)\hat{\mathbf{i}} - \dot{m}_e \Delta t \Delta v \hat{\mathbf{i}} = (p_e - p_a)A_e \Delta t \hat{\mathbf{i}}$$

Dividing through by Δt, taking the limit as $\Delta t \to 0$, and canceling the common unit vector leads to

$$m\frac{dv}{dt} - \dot{m}_e c_a = (p_e - p_a)A_e \tag{11.13}$$

where c_a is the speed of the exhaust relative to the rocket,

$$c_a = v_e + v \tag{11.14}$$

Rearranging terms, Equation 11.13 may be written

$$\dot{m}_e c_a + (p_e - p_a)A_e = m\frac{dv}{dt} \tag{11.15}$$

The left-hand side of this equation is the unbalanced force responsible for the acceleration dv/dt of the system in Figure 11.2. This unbalanced force is the thrust T,

$$T = \dot{m}_e c_a + (p_e - p_a)A_e \tag{11.16}$$

where $\dot{m}_e c_a$ is the jet thrust and $(p_e - p_a)A_e$ is the pressure thrust. We can write Equation 11.16 as

$$T = \dot{m}_e \left[c_a + \frac{(p_e - p_a)A_e}{\dot{m}_e} \right] \tag{11.17}$$

The term in brackets is called the effective exhaust velocity c,

$$c = c_a + \frac{(p_e - p_a)A_e}{\dot{m}_e} \tag{11.18}$$

In terms of the effective exhaust velocity, the thrust may be expressed simply as

$$T = \dot{m}_e c \tag{11.19}$$

The specific impulse I_{sp} is defined as the thrust per sea-level weight rate (per second) of propellant consumption. That is,

$$I_{sp} = \frac{T}{\dot{m}_e g_0} \tag{11.20}$$

where g_0 is the standard sea-level acceleration of gravity. The unit of specific impulse is force ÷ (force/second) or seconds. Together, Equations 11.19 and 11.20 imply that

$$c = I_{sp} g_0 \tag{11.21}$$

Obviously, one can infer the jet velocity directly from the specific impulse. Specific impulse is an important performance parameter for a given rocket engine and propellant combination. However, large specific impulse equates to large thrust only if the mass flow rate is large, which is true of chemical rocket engines. The specific impulses of chemical rockets typically lie in the range 200–300 s for solid fuels and 250–450 s for liquid fuels. Ion propulsion systems have very high specific impulse ($>10^4$ s), but their very low mass flow rates produce much smaller thrust than chemical rockets.

11.4 Rocket performance

From Equations 11.10 and 11.20 we have

$$T = -I_{sp} g_0 \frac{dm}{dt} \tag{11.22}$$

or

$$\frac{dm}{dt} = -\frac{T}{I_{sp} g_0}$$

If the thrust and specific impulse are constant, then the integral of this expression over the burn time Δt is

$$\Delta m = -\frac{T}{I_{sp} g_0} \Delta t$$

from which we obtain

$$\Delta t = \frac{I_{sp} g_0}{T}(m_0 - m_f) = \frac{I_{sp} g_0}{T} m_0 \left(1 - \frac{m_f}{m_0}\right) \tag{11.23}$$

where m_0 and m_f are the mass of the vehicle at the beginning and end of the burn, respectively. The mass ratio is defined as the ratio of the initial mass to final mass,

$$n = \frac{m_0}{m_f} \tag{11.24}$$

Clearly, the mass ratio is always greater than unity. In terms of the initial mass ratio, Equation 11.23 may be written

$$\Delta t = \frac{n-1}{n} \frac{I_{sp}}{T/m_0 g_0} \tag{11.25}$$

T/mg_0 is the thrust-to-weight ratio. The thrust-to-weight ratio for a launch vehicle at lift-off is typically in the range 1.3 to 2.

Substituting Equation 11.22 into Equation 11.6, we get

$$\frac{dv}{dt} = -I_{sp}g_0 \frac{dm/dt}{m} - \frac{D}{m} - g\sin\gamma$$

Integrating with respect to time, from t_0 to t_f, yields

$$\Delta v = I_{sp}g_0 \ln \frac{m_0}{m_f} - \Delta v_D - \Delta v_G \tag{11.26}$$

where the drag loss Δv_D and the gravity loss Δv_g are given by the integrals

$$\Delta v_D = \int_{t_0}^{t_f} \frac{D}{m} dt \qquad \Delta v_G = \int_{t_0}^{t_f} g\sin\gamma \, dt \tag{11.27}$$

Since the drag D, acceleration of gravity g, and flight path angle γ are unknown functions of time, these integrals cannot be computed. (Equations 11.6 through 11.8, together with 11.3, must be solved numerically to obtain $v(t)$ and $\gamma(t)$; but then Δv would follow from those results.) Equation 11.26 can be used for rough estimates where previous data and experience provide a basis for choosing conservative values of Δv_D and Δv_G. Obviously, if drag can be neglected, then $\Delta v_D = 0$. This would be a good approximation for the last stage of a satellite booster, for which it can also be said that $\Delta v_G = 0$, since $\gamma \cong 0°$ when the satellite is injected into orbit.

Sounding rockets are launched vertically and fly straight up to their maximum altitude before falling back to earth, usually by parachute. Their purpose is to measure remote portions of the earth's atmosphere. ('Sound' in this context means to measure or investigate.) If for a sounding rocket $\gamma = 90°$, then $\Delta v_G \approx g_0(t_f - t_0)$, since g is within 90 percent of g_0 out to 300 km altitude.

EXAMPLE 11.1

A sounding rocket of initial mass m_0 and mass m_f after all propellant is consumed is launched vertically ($\gamma = 90°$). The propellant mass flow rate $\dot{m}_e$ is constant. Neglecting drag and the variation of gravity with altitude, calculate the maximum height h attained by the rocket. For what flow rate is the greatest altitude reached?

The vehicle mass as a function of time, up to burnout, is

$$m = m_0 - \dot{m}_e t \tag{a}$$

At burnout, $m = m_f$, so the burnout time t_{bo} is

$$t_{bo} = \frac{m_0 - m_f}{\dot{m}_e} \tag{b}$$

The drag loss is assumed to be zero, and the gravity loss is

$$\Delta v_G = \int_0^{t_{bo}} g_0 \sin(90°) dt = g_0 t_{bo}$$

11.4 Rocket performance

Recalling that $I_{sp}g_0 = c$ and using (a), it follows from Equation 11.26 that, up to burnout, the velocity as a function of time is

$$v = c \ln \frac{m_0}{m_0 - \dot{m}_e t} - g_0 t \tag{c}$$

Since $dh/dt = v$, the altitude as a function of time is

$$h = \int_0^t v\, dt = \int_0^t \left(c \ln \frac{m_0}{m_0 - \dot{m}_e t} - g_0 t \right) dt$$

$$= \frac{c}{\dot{m}_e} \left[(m_0 - \dot{m}_e t) \ln \frac{m_0 - bt}{m_0} + \dot{m}_e t \right] - \frac{1}{2} g_0 t^2 \tag{d}$$

The height at burnout h_{bo} is found by substituting (b) into this expression,

$$h_{bo} = \frac{c}{\dot{m}_e} \left(m_f \ln \frac{m_f}{m_0} + m_0 - m_f \right) - \frac{1}{2} \left(\frac{m_0 - m_f}{\dot{m}_e} \right)^2 g \tag{e}$$

Likewise, the burnout velocity is obtained by substituting (b) into (c),

$$v_{bo} = c \ln \frac{m_0}{m_f} - \frac{g_0}{\dot{m}_e} (m_0 - m_f) \tag{f}$$

After burnout, the rocket coasts upward with the constant downward acceleration of gravity,

$$v = v_{bo} - g_0(t - t_{bo})$$

$$h = h_{bo} + v_{bo}(t - t_{bo}) - \frac{1}{2} g_0 (t - t_{bo})^2$$

Substituting (b), (e) and (f) into these expressions yields, for $t > t_{bo}$,

$$v = c \ln \frac{m_0}{m_f} - g_0 t$$

$$h = \frac{c}{\dot{m}_e} \left(m_0 \ln \frac{m_f}{m_0} + m_0 - m_f \right) + ct \ln \frac{m_0}{m_f} - \frac{1}{2} g_0 t^2 \tag{g}$$

The maximum height h_{max} is reached when $v = 0$,

$$c \ln \frac{m_0}{m_f} - g_0 t_{max} = 0 \quad \Rightarrow \quad t_{max} = \frac{c}{g_0} \ln \frac{m_0}{m_f}$$

Substituting t_{max} into (g) leads to our result,

$$\boxed{h_{max} = \frac{cm_0}{\dot{m}_e} (1 + \ln n - n) + \frac{1}{2} \frac{c^2}{g_0} \ln^2 n}$$

where n is the mass ratio ($n > 1$). Since $n > (1 + \ln n)$, it follows that $(1 + \ln n - n)$ is negative. Hence, h_{max} can be increased by increasing the mass flow rate $\dot{m}_e$. In fact, the greatest height is achieved when $\dot{m}_e \to \infty$, i.e., all of the propellant is expended at once, like a mortar shell.

11.5 Restricted staging in field-free space

In field-free space we neglect drag and gravitational attraction. In that case, Equation 11.26 becomes

$$\Delta v = I_{sp} g_0 \ln \frac{m_0}{m_f} \tag{11.28}$$

This is at best a poor approximation for high-thrust rockets, but it will suffice to shed some light on the rocket staging problem. Observe that we can solve this equation for the mass ratio to obtain

$$\frac{m_0}{m_f} = e^{\frac{\Delta v}{I_{sp} g_0}} \tag{11.29}$$

The amount of propellant expended to produce the velocity increment Δv is $m_0 - m_f$. If we let $\Delta m = m_0 - m_f$, then Equation 11.29 can be written as

$$\frac{\Delta m}{m_0} = 1 - e^{-\frac{\Delta v}{I_{sp} g_0}} \tag{11.30}$$

This relation is used to compute the propellant required to produce a given delta-v.

The gross mass m_0 of a launch vehicle consists of the empty mass m_E, the propellant mass m_p and the payload mass m_{PL},

$$m_0 = m_E + m_p + m_{PL} \tag{11.31}$$

The empty mass comprises the mass of the structure, the engines, fuel tanks, control systems, etc. m_E is also called the structural mass, although it embodies much more than just structure. Dividing Equation 11.31 through by m_0, we obtain

$$\pi_E + \pi_p + \pi_{PL} = 1 \tag{11.32}$$

where $\pi_E = m_E/m_0$, $\pi_p = m_p/m_0$ and $\pi_{PL} = m_{PL}/m_0$ are the structural fraction, propellant fraction and payload fraction, respectively. It is convenient to define the payload ratio

$$\lambda = \frac{m_{PL}}{m_E + m_p} = \frac{m_{PL}}{m_0 - m_{PL}} \tag{11.33}$$

and the structural ratio

$$\varepsilon = \frac{m_E}{m_E + m_p} = \frac{m_E}{m_0 - m_{PL}} \tag{11.34}$$

The mass ratio n was introduced in Equation 11.24. Assuming all of the propellant is consumed, that may now be written

$$n = \frac{m_E + m_p + m_{PL}}{m_E + m_{PL}} \tag{11.35}$$

λ, ε and n are not independent. From Equation 11.34 we have

$$m_E = \frac{\varepsilon}{1 - \varepsilon} m_p \tag{11.36}$$

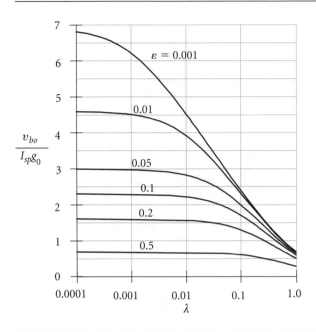

Figure 11.3 Dimensionless burnout speed versus payload ratio.

whereas Equation 11.33 gives

$$m_{PL} = \lambda(m_E + m_p) = \lambda\left(\frac{\varepsilon}{1-\varepsilon}m_p + m_p\right) = \frac{\lambda}{1-\varepsilon}m_p \tag{11.37}$$

Substituting Equations 11.36 and 11.37 into Equation 11.35 leads to

$$n = \frac{1+\lambda}{\varepsilon+\lambda} \tag{11.38}$$

Thus, given any two of the ratios λ, ε and n, we obtain the third from Equation 11.38. Using this relation in Equation 11.28 and setting Δv equal to the burnout speed v_{bo}, when the propellants have been used up, yields

$$v_{bo} = I_{sp}g_0 \ln n = I_{sp}g_0 \ln \frac{1+\lambda}{\varepsilon+\lambda} \tag{11.39}$$

This equation is plotted in Figure 11.3 for a range of structural ratios. Clearly, for a given empty mass, the greatest possible Δv occurs when the payload is zero. However, what we want to do is maximize the amount of payload while keeping the structural weight to a minimum. Of course, the mass of load-bearing structure, rocket motors, pumps, piping, etc., cannot be made arbitrarily small. Current materials technology places a lower limit on ε of about 0.1. For this value of the structural ratio and $\lambda = 0.05$, Equation 11.39 yields

$$v_{bo} = 1.94 I_{sp}g_0 = 0.019 I_{sp} \text{ (km/s)}$$

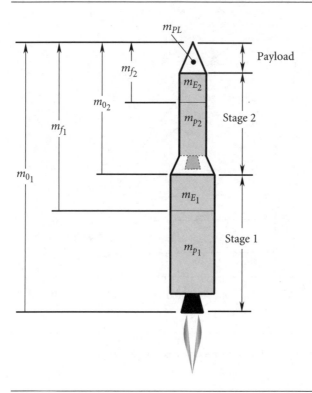

Figure 11.4 Tandem two-stage booster.

The specific impulse of a typical chemical rocket is about 300 s, which in this case would provide $\Delta v = 5.7$ km/s. However, the circular orbital velocity at the earth's surface is 7.905 km/s. So this booster by itself could not orbit the payload. The minimum specific impulse required for a single stage to orbit would be 416 s. Only today's most advanced liquid hydrogen/liquid oxygen engines, e.g., the space shuttle main engines, have this kind of performance. Practicality and economics would likely dictate going the route of a multi-stage booster.

Figure 11.4 shows a series or tandem two-stage rocket configuration, with one stage sitting on top of the other. Each stage has its own engines and propellant tanks. The dividing line between the stages is where they separate during flight. The first stage drops off first, the second stage next, etc. The payload of an N stage rocket is actually stage $N + 1$. Indeed, satellites commonly carry their own propulsion systems into orbit. The payload of a given stage is everything above it. Therefore, as illustrated in Figure 11.4, the initial mass m_{0_1} of stage 1 is that of the entire vehicle. After stage 1 expels all of its fuel, the mass m_{f_1} which remains is stage 1's empty mass m_{E_1} plus the mass of stage 2 and the payload. After separation of stage 1, the process continues likewise for stage 2, with m_{0_2} being its initial mass.

Titan II, the launch vehicle for the US Gemini program, had the two-stage, tandem configuration. So did the Saturn 1B, used to launch earth orbital flights early in the US Apollo program, as well as to send crews to Skylab and an Apollo spacecraft to dock with a Russian Soyuz spacecraft in 1975.

11.5 Restricted staging in field-free space 563

Figure 11.5 Parallel staging.

Figure 11.5 illustrates the concept of parallel staging. Two or more solid or liquid rockets are attached ('strapped on') to a core vehicle carrying the payload. Whereas in the tandem arrangement, the motors in a given stage cannot ignite until separation of the previous stage, all of the rockets ignite at once in the parallel-staged vehicle. The strap-on boosters fall away after they burn out early in the ascent. The space shuttle is the most obvious example of parallel staging. Its two solid rocket boosters are mounted on the external tank, which fuels the three 'main' engines built into the orbiter. The solid rocket boosters and the external tank are cast off after they are depleted. In more common use is the combination of parallel and tandem staging, in which boosters are strapped to the first stage of a multi-stage stack. Examples include the United States' Titan III and IV and Delta launchers, Europe's Ariane 4 and 5, Russia's Proton and Soyuz variants, Japan's H-2, and China's Long March launch vehicles.

The original Atlas, used in many variants, for among other things, to launch the orbital flights of the US Mercury program, had three main liquid-fuel engines at its base. They all fired simultaneously at launch, but several minutes into the flight, the outer two 'boosters' dropped away, leaving the central sustainer engine to burn the rest of the way to orbit. Since the booster engines shared the sustainer's propellant tanks, the Atlas exhibited partial staging, and is sometimes referred to as a one and a half stage rocket, the discarded boosters comprising the half stage.

We will for simplicity focus on tandem staging, although parallel-staged systems are handled in a similar way (Wiesel, 1997). Restricted staging involves the simple but unrealistic assumption that all stages are similar. That is, each stage has the same specific impulse I_{sp}, the same structural ratio ε, and the same payload ratio λ. From Equation 11.38 it follows that the mass ratios n are identical, too. Let us investigate the effect of restricted staging on the final burnout speed v_{bo} for a given payload mass m_{PL} and overall payload fraction

$$\pi_{PL} = \frac{m_{PL}}{m_0} \tag{11.40}$$

where m_0 is the total mass of the tandem-stacked vehicle.

For a single-stage vehicle, the payload ratio is

$$\lambda = \frac{m_{PL}}{m_0 - m_{PL}} = \frac{1}{\frac{m_0}{m_{PL}} - 1} = \frac{\pi_{PL}}{1 - \pi_{PL}} \tag{11.41}$$

so that, from Equation 11.38, the mass ratio is

$$n = \frac{1}{\pi_{PL}(1 - \varepsilon) + \varepsilon} \tag{11.42}$$

According to Equation 11.39, the burnout speed is

$$v_{bo} = I_{sp} g_0 \ln \frac{1}{\pi_{PL}(1 - \varepsilon) + \varepsilon} \tag{11.43}$$

Let m_0 be the total mass of the two-stage rocket of Figure 11.4, i.e.,

$$m_0 = m_{0_1} \tag{11.44}$$

The payload of stage 1 is the entire mass m_{0_2} of stage 2. Thus, for stage 1 the payload ratio is

$$\lambda_1 = \frac{m_{0_2}}{m_{0_1} - m_{0_2}} = \frac{m_{0_2}}{m_0 - m_{0_2}} \tag{11.45}$$

The payload ratio of stage 2 is

$$\lambda_2 = \frac{m_{PL}}{m_{0_2} - m_{PL}} \tag{11.46}$$

By virtue of the two stages' being similar, $\lambda_1 = \lambda_2$, or

$$\frac{m_{0_2}}{m_0 - m_{0_2}} = \frac{m_{PL}}{m_{0_2} - m_{PL}}$$

Solving this equation for m_{0_2} yields

$$m_{0_2} = \sqrt{m_0}\sqrt{m_{PL}}$$

But $m_0 = m_{PL}/\pi_{PL}$, so the gross mass of the second stage is

$$m_{0_2} = \sqrt{\frac{1}{\pi_{PL}} m_{PL}} \tag{11.47}$$

Putting this back into Equation 11.45 (or 11.46), we obtain the common two-stage payload ratio $\lambda = \lambda_1 = \lambda_2$,

$$\lambda_{\text{2-stage}} = \frac{\pi_{PL}^{\frac{1}{2}}}{1 - \pi_{PL}^{\frac{1}{2}}} \quad (11.48)$$

This together with Equation 11.38 and the assumption that $\varepsilon_1 = \varepsilon_2 = \varepsilon$ leads to the common mass ratio for each stage,

$$n_{\text{2-stage}} = \frac{1}{\pi_{PL}^{\frac{1}{2}}(1 - \varepsilon) + \varepsilon} \quad (11.49)$$

Assuming that stage 2 ignites immediately after burnout of stage 1, the final velocity of the two-stage vehicle is the sum of the burnout velocities of the individual stages,

$$v_{bo} = v_{bo_1} + v_{bo_2}$$

or

$$v_{bo_{\text{2-stage}}} = I_{sp}g_0 \ln n_{\text{2-stage}} + I_{sp}g_0 \ln n_{\text{2-stage}} = 2I_{sp}g_0 \ln n_{\text{2-stage}}$$

so that, with Equation 11.49, we get

$$v_{bo_{\text{2-stage}}} = I_{sp}g_0 \ln \left[\frac{1}{\pi_{PL}^{\frac{1}{2}}(1 - \varepsilon) + \varepsilon} \right]^2 \quad (11.50)$$

The empty mass of each stage can be found in terms of the payload mass using the common structural ratio ε,

$$\frac{m_{E_1}}{m_{0_1} - m_{0_2}} = \varepsilon \qquad \frac{m_{E_2}}{m_{0_2} - m_{PL}} = \varepsilon$$

Substituting Equations 11.40 and 11.44 together with 11.47 yields

$$m_{E_1} = \frac{\left(1 - \pi_{PL}^{\frac{1}{2}}\right)\varepsilon}{\pi_{PL}} m_{PL} \qquad m_{E_2} = \frac{\left(1 - \pi_{PL}^{\frac{1}{2}}\right)\varepsilon}{\pi_{PL}^{\frac{1}{2}}} m_{PL} \quad (11.51)$$

Likewise, we can find the propellant mass for each stage from the expressions

$$m_{p_1} = m_{0_1} - (m_{E_1} + m_{0_2}) \qquad m_{p_2} = m_{0_2} - (m_{E_2} + m_{PL}) \quad (11.52)$$

Substituting Equations 11.40 and 11.44, together with 11.47, 11.51 and 11.52, we get

$$m_{p_1} = \frac{\left(1 - \pi_{PL}^{\frac{1}{2}}\right)(1 - \varepsilon)}{\pi_{PL}} m_{PL} \qquad m_{p_2} = \frac{\left(1 - \pi_{PL}^{\frac{1}{2}}\right)(1 - \varepsilon)}{\pi_{PL}^{\frac{1}{2}}} m_{PL} \quad (11.53)$$

EXAMPLE 11.2

The following data is given

$$m_{PL} = 10\,000 \text{ kg}$$
$$\pi_{PL} = 0.05$$
$$\varepsilon = 0.15 \qquad \text{(a)}$$
$$I_{sp} = 350 \text{ s}$$
$$g_0 = 0.00981 \text{ km/s}^2$$

Calculate the payload velocity v_{bo} at burnout, the empty mass of the launch vehicle and the propellant mass for (a) a single stage and (b) a restricted, two-stage vehicle.

(a) From Equation 11.43 we find

$$v_{bo} = 350 \cdot 0.00981 \ln \frac{1}{0.05(1+0.15)+0.15} = \underline{5.657 \text{ km/s}}$$

Equation 11.40 yields the gross mass

$$m_0 = \frac{10\,000}{0.05} = 200\,000 \text{ kg}$$

from which we obtain the empty mass using Equation 11.34,

$$m_E = \varepsilon(m_0 - m_{PL}) = 0.15(200\,000 - 10\,000) = \underline{28\,500 \text{ kg}}$$

The mass of propellant is

$$m_p = m_0 - m_E - m_{PL} = 200\,000 - 28\,500 - 10\,000 = \underline{161\,500 \text{ kg}}$$

(b) For a restricted two-stage vehicle, the burnout speed is given by Equation 11.50,

$$v_{bo_{2\text{-stage}}} = 350 \cdot 0.00981 \ln \left[\frac{1}{0.05^{\frac{1}{2}}(1-0.15)+0.15} \right]^2 = \underline{7.407 \text{ km/s}}$$

The empty mass of each stage is found using Equations 11.51,

$$m_{E_1} = \frac{\left(1 - 0.05^{\frac{1}{2}}\right) \cdot 0.15}{0.05} \cdot 10\,000 = \underline{23\,292 \text{ kg}}$$

$$m_{E_2} = \frac{\left(1 - 0.05^{\frac{1}{2}}\right) \cdot 0.15}{0.05^{\frac{1}{2}}} \cdot 10\,000 = \underline{5208 \text{ kg}}$$

For the propellant masses, we turn to Equations 11.53

$$m_{p_1} = \frac{\left(1 - 0.05^{\frac{1}{2}}\right) \cdot (1 - 0.15)}{0.05} \cdot 10\,000 = \underline{131\,990 \text{ kg}}$$

$$m_{p_2} = \frac{\left(1 - 0.05^{\frac{1}{2}}\right) \cdot (1 - 0.15)}{0.05^{\frac{1}{2}}} \cdot 10\,000 = \underline{29\,513\,\text{kg}}$$

The total empty mass, $m_E = m_{E_1} + m_{E_2}$, and the total propellant mass, $m_p = m_{p_1} + m_{p_2}$, are the same as for the single stage rocket. The mass of the second stage, including the payload, is 22.4 percent of the total vehicle mass.

Observe in the previous example that, although the total vehicle mass was unchanged, the burnout velocity increased 31 percent for the two-stage arrangement. The reason is that the second stage is lighter and can therefore be accelerated to a higher speed. Let us determine the velocity gain associated with adding another stage, as illustrated in Figure 11.6.

The payload ratios of the three stages are

$$\lambda_1 = \frac{m_{0_2}}{m_{0_1} - m_{0_2}} \qquad \lambda_2 = \frac{m_{0_3}}{m_{0_2} - m_{0_3}} \qquad \lambda_3 = \frac{m_{PL}}{m_{0_3} - m_{PL}}$$

Since the stages are similar, these payload ratios are all the same. Setting $\lambda_1 = \lambda_2$ and recalling that $m_{01} = m_0$, we find

$$m_{0_2}^2 - m_{0_3} m_0 = 0$$

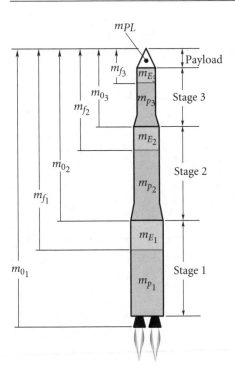

Figure 11.6 Tandem three-stage launch vehicle.

Similarly, $\lambda_1 = \lambda_3$ yields

$$m_{0_2} m_{0_3} - m_0 m_{PL} = 0$$

These two equations imply that

$$m_{0_2} = \frac{m_{PL}}{\pi_{PL}^{\frac{2}{3}}} \qquad m_{0_3} = \frac{m_{PL}}{\pi_{PL}^{\frac{1}{3}}} \qquad (11.54)$$

Substituting these results back into any one of the above expressions for λ_1, λ_2 or λ_3 yields the common payload ratio for the restricted three-stage rocket,

$$\lambda_{\text{3-stage}} = \frac{\pi_{PL}^{\frac{1}{3}}}{1 - \pi_{PL}^{\frac{1}{3}}}$$

With this result and Equation 11.38 we find the common mass ratio,

$$n_{\text{3-stage}} = \frac{1}{\pi_{PL}^{\frac{1}{3}}(1-\varepsilon) + \varepsilon} \qquad (11.55)$$

Since the payload burnout velocity is $v_{bo} = v_{bo_1} + v_{bo_2} + v_{bo_3}$, we have

$$v_{bo_{\text{3-stage}}} = 3 I_{sp} g_0 \ln n_{\text{3-stage}} = I_{sp} g_0 \ln \left(\frac{1}{\pi_{PL}^{\frac{1}{3}}(1-\varepsilon) + \varepsilon} \right)^3 \qquad (11.56)$$

Because of the common structural ratio across each stage,

$$\frac{m_{E_1}}{m_{0_1} - m_{0_2}} = \varepsilon \qquad \frac{m_{E_2}}{m_{0_2} - m_{0_3}} = \varepsilon \qquad \frac{m_{E_3}}{m_{0_3} - m_{PL}} = \varepsilon$$

Substituting Equations 11.40 and 11.54 and solving the resultant expressions for the empty stage masses yields

$$m_{E_1} = \frac{\left(1 - \pi_{PL}^{\frac{1}{3}}\right)\varepsilon}{\pi_{PL}} m_{PL} \qquad m_{E_2} = \frac{\left(1 - \pi_{PL}^{\frac{1}{3}}\right)\varepsilon}{\pi_{PL}^{\frac{2}{3}}} m_{PL} \qquad m_{E_3} = \frac{\left(1 - \pi_{PL}^{\frac{1}{3}}\right)\varepsilon}{\pi_{PL}^{\frac{1}{3}}} m_{PL}$$

$$(11.57)$$

The stage propellant masses are

$$m_{p_1} = m_{0_1} - (m_{E_1} + m_{0_2}) \qquad m_{p_2} = m_{0_2} - (m_{E_2} + m_{0_3}) \qquad m_{p_3} = m_{0_3} - (m_{E_3} + m_{PL})$$

Substituting Equations 11.40, 11.54 and 11.57 leads to

$$m_{p_1} = \frac{\left(1 - \pi_{PL}^{\frac{1}{3}}\right)(1-\varepsilon)}{\pi_{PL}} m_{PL}$$

$$m_{p_2} = \frac{\left(1 - \pi_{PL}^{\frac{1}{3}}\right)(1-\varepsilon)}{\pi_{PL}^{\frac{2}{3}}} m_{PL} \qquad (11.58)$$

$$m_{p_3} = \frac{\left(1 - \pi_{PL}^{\frac{1}{3}}\right)(1-\varepsilon)}{\pi_{PL}^{\frac{1}{3}}} m_{PL}$$

EXAMPLE 11.3

Repeat Example 11.2 for the restricted three-stage launch vehicle.

Equation 11.56 gives the burnout velocity for three stages,

$$v_{bo} = 350 \cdot 0.00981 \cdot \ln\left(\frac{1}{0.05^{\frac{1}{3}}(1-0.15)+0.15}\right)^3 = \underline{7.928 \text{ km/s}}$$

Substituting $m_{PL} = 10\,000$ kg, $\pi_{PL} = 0.05$ and $\varepsilon = 0.15$ into Equations 11.57 and 11.58 yields

$$m_{E_1} = \underline{18\,948 \text{ kg}} \qquad m_{E_2} = \underline{6980 \text{ kg}} \qquad m_{E_3} = \underline{2572 \text{ kg}}$$
$$m_{p_1} = \underline{107\,370 \text{ kg}} \qquad m_{p_2} = \underline{39\,556 \text{ kg}} \qquad m_{p_3} = \underline{14\,573 \text{ kg}}$$

Again, the total empty mass and total propellant mass are the same as for the single and two-stage vehicles. Notice that the velocity increase over the two-stage rocket is just 7 percent, which is much less than the advantage the two-stage had over the single stage vehicle.

Looking back over the velocity formulas for one, two and three stage vehicles (Equations 11.43, 11.50 and 11.56), we can induce that for an N-stage rocket,

$$v_{bo_{N\text{-stage}}} = I_{sp}g_0 \ln\left(\frac{1}{\pi_{PL}^{\frac{1}{N}}(1-\varepsilon)+\varepsilon}\right)^N$$

$$= I_{sp}g_0 N \ln\left(\frac{1}{\pi_{PL}^{\frac{1}{N}}(1-\varepsilon)+\varepsilon}\right) \quad (11.59)$$

What happens as we let N become very large? First of all, it can be shown using Taylor series expansion that, for large N,

$$\pi_{PL}^{\frac{1}{N}} \approx 1 + \frac{1}{N}\ln\pi_{PL} \quad (11.60)$$

Substituting this into Equation 11.59, we find that

$$v_{bo_{N\text{-stage}}} \approx I_{sp}g_0 N \ln\left[\frac{1}{1+\frac{1}{N}(1-\varepsilon)\ln\pi_{PL}}\right]$$

Since the term $\frac{1}{N}(1-\varepsilon)\ln\pi_{PL}$ is arbitrarily small, we can use the fact that $1/(1+x) = 1 - x + x^2 - x^3 + \cdots$ to write

$$\frac{1}{1+\frac{1}{N}(1-\varepsilon)\ln\pi_{PL}} \approx 1 - \frac{1}{N}(1-\varepsilon)\ln\pi_{PL}$$

which means

$$v_{bo_{N\text{-stage}}} \approx I_{sp}g_0 N \ln\left[1 - \frac{1}{N}(1-\varepsilon)\ln\pi_{PL}\right]$$

Finally, since $\ln(1-x) = -x - x^2/2 - x^3/3 - x^4/4 - \cdots$, we can write this as

$$v_{bo_{N\text{-stage}}} \approx I_{sp}g_0 N \left[-\frac{1}{N}(1-\varepsilon)\ln\pi_{PL}\right]$$

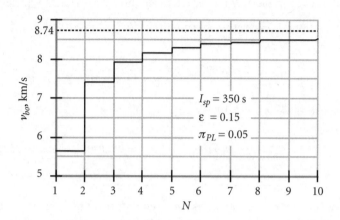

Figure 11.7 Burnout velocity versus number of stages (Equation 11.59).

Therefore, as N, the number of stages, tends towards infinity, the burnout velocity approaches

$$v_{bo_\infty} = I_{sp}g_0(1-\varepsilon)\ln\frac{1}{\pi_{PL}} \qquad (11.61)$$

Thus, no matter how many similar stages we use, for a given specific impulse, payload fraction and structural ratio, we cannot exceed this burnout speed. For example, using $I_{sp} = 350$ s, $\pi_{PL} = 0.05$ and $\varepsilon = 0.15$ from the previous two examples yields $v_{bo_\infty} = 8.743$ km/s, which is only 10 percent greater than v_{bo} of a three-stage vehicle. The trend of v_{bo} towards this limiting value is illustrated by Figure 11.7.

Our simplified analysis does not take into account the added weight and complexity accompanying additional stages. Practical reality has limited the number of stages of actual launch vehicles to rarely more than three.

11.6 Optimal staging

Let us now abandon the restrictive assumption that all stages of a tandem-stacked vehicle are similar. Instead, we will specify the specific impulse I_{sp_i} and structural ratio ε_i of each stage, and then seek the minimum-mass N-stage vehicle that will carry a given payload m_{PL} to a specified burnout velocity v_{bo}. To optimize the mass requires using the Lagrange multiplier method, which we shall briefly review.

11.6.1 Lagrange multiplier

Consider a bivariate function f on the xy plane. Then $z = f(x,y)$ is a surface lying above or below the plane, or both. $f(x,y)$ is stationary at a given point if it takes on a local maximum or a local minimum, i.e., an extremum, at that point. For f to be stationary means $df = 0$; i.e.,

$$\frac{\partial f}{\partial x}dx + \frac{\partial f}{\partial y}dy = 0 \qquad (11.62)$$

where dx and dy are independent and not necessarily zero. It follows that for an extremum to exist,

$$\frac{\partial f}{\partial x} = \frac{\partial f}{\partial y} = 0 \qquad (11.63)$$

Now let $g(x,y) = 0$ be a curve in the xy plane. Let us find the points on the curve $g = 0$ at which f is stationary. That is, rather than searching the entire xy plane for extreme values of f, we confine our attention to the curve $g = 0$, which is therefore a constraint. Since $g = 0$, it follows that $dg = 0$, or

$$\frac{\partial g}{\partial x}dx + \frac{\partial g}{\partial y}dy = 0 \qquad (11.64)$$

If Equations 11.62 and 11.64 are both valid at a given point, then

$$\frac{dy}{dx} = -\frac{\partial f/\partial x}{\partial f/\partial y} = -\frac{\partial g/\partial x}{\partial g/\partial y}$$

That is,

$$\frac{\partial f/\partial x}{\partial g/\partial x} = \frac{\partial f/\partial y}{\partial g/\partial y} = -\eta$$

From this we obtain

$$\frac{\partial f}{\partial x} + \eta\frac{\partial g}{\partial x} = 0 \qquad \frac{\partial f}{\partial y} + \eta\frac{\partial g}{\partial y} = 0$$

But these, together with the constraint $g(x,y) = 0$, are the very conditions required for the function

$$h(x, y, \eta) = f(x, y) + \eta g(x, y) \qquad (11.65)$$

to have an extremum, namely,

$$\frac{\partial h}{\partial x} = \frac{\partial f}{\partial x} + \eta\frac{\partial g}{\partial x} = 0$$

$$\frac{\partial h}{\partial y} = \frac{\partial f}{\partial y} + \eta\frac{\partial g}{\partial y} = 0 \qquad (11.66)$$

$$\frac{\partial h}{\partial \eta} = g = 0$$

η is the Lagrange multiplier. The procedure generalizes to functions of any number of variables.

One can determine mathematically whether the extremum is a maximum or a minimum by checking the sign of the second differential d^2h of the function h in Equation 11.65,

$$d^2h = \frac{\partial^2 h}{\partial x^2}dx^2 + 2\frac{\partial^2 h}{\partial x \partial y}dxdy + \frac{\partial^2 h}{\partial y^2}dy^2 \qquad (11.67)$$

If $d^2h < 0$ at the extremum for all dx and dy satisfying the constraint condition, Equation 11.64, then the extremum is a local maximum. Likewise, if $d^2h > 0$, then the extremum is a local minimum.

EXAMPLE 11.4

(a) Find the extrema of the function $z = -x^2 - y^2$. (b) Find the extrema of the same function under the constraint $y = 2x + 3$.

(a) To find the extrema we must use Equations 11.63. Since $\partial z/\partial x = -2x$ and $\partial z/\partial y = -2y$, it follows that $\partial z/\partial x = \partial z/\partial y = 0$ at $x = y = 0$, at which point $z = 0$. Since z is negative everywhere else (see Figure 11.8), it is clear that the extreme value is the maximum value.

(b) The constraint may be written $g = y - 2x - 3$. Clearly, $g = 0$. Multiply the constraint by the Lagrange multiplier η and add the result (zero!) to the function $-(x^2 + y^2)$ to obtain

$$h = -(x^2 + y^2) + \eta(y - 2x - 3)$$

This is a function of the three variables x, y and η. For it to be stationary, the partial derivatives with respect to all three of these variables must vanish. First we have

$$\frac{\partial h}{\partial x} = -2x - 2\eta$$

Setting this equal to zero yields

$$x = -\eta \qquad (a)$$

Next,

$$\frac{\partial h}{\partial y} = -2y + \eta$$

For this to be zero means

$$y = \frac{\eta}{2} \qquad (b)$$

Finally

$$\frac{\partial h}{\partial \eta} = y - 2x - 3$$

Setting this equal to zero gives us back the constraint condition,

$$y - 2x - 3 = 0 \qquad (c)$$

Substituting (a) and (b) into (c) yields $\eta = 1.2$, from which (a) and (b) imply,

$$x = -1.2 \qquad y = 0.6 \qquad (d)$$

These are the coordinates of the point on the line $y = 2x + 3$ at which $z = -x^2 - y^2$ is stationary. Using (d), we find that $z = -1.8$ at this point.

Figure 11.8 is an illustration of this problem, and it shows that the computed extremum (a maximum, in the sense that small negative numbers exceed

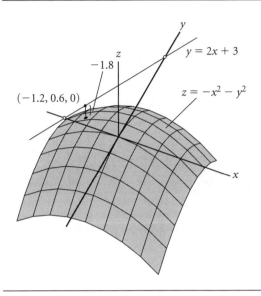

Figure 11.8 Location of the point on the line $y = 2x + 3$ at which the surface $z = -x^2 - y^2$ is closest to the xy plane.

large negative numbers) is where the surface $z = -x^2 - y^2$ is closest to the line $y = 2x + 3$, as measured in the z direction. Note that in this case, Equation 11.67 yields $d^2h = -2dx^2 - 2dy^2$, which is negative, confirming our conclusion that the extremum is a maximum.

Now let us return to the optimal staging problem. It is convenient to introduce the step mass m_i of the ith stage. The step mass is the empty mass plus the propellant mass of the stage, exclusive of all the other stages,

$$m_i = m_{E_i} + m_{p_i} \tag{11.68}$$

The empty mass of stage i can be expressed in terms of its step mass and its structural ratio ε_i as follows,

$$m_{E_i} = \varepsilon_i(m_{E_i} + m_{p_i}) = \varepsilon_i m_i \tag{11.69}$$

The total mass of the rocket excluding the payload is M, which is the sum of all of the step masses,

$$M = \sum_{i=1}^{N} m_i \tag{11.70}$$

Thus, recalling that m_0 is the total mass of the vehicle, we have

$$m_0 = M + m_{PL} \tag{11.71}$$

Our goal is to minimize m_0.

For simplicity, we will deal first with a two-stage rocket, and then generalize our results to N stages. For a two-stage vehicle, $m_0 = m_1 + m_2 + m_{PL}$, so we can write,

$$\frac{m_0}{m_{PL}} = \frac{m_1 + m_2 + m_{PL}}{m_2 + m_{PL}} \frac{m_2 + m_{PL}}{m_{PL}} \tag{11.72}$$

The mass ratio of stage 1 is

$$n_1 = \frac{m_{0_1}}{m_{E_1} + m_2 + m_{PL}} = \frac{m_1 + m_2 + m_{PL}}{\varepsilon_1 m_1 + m_2 + m_{PL}} \tag{11.73}$$

where Equation 11.69 was used. Likewise, the mass ratio of stage 2 is

$$n_2 = \frac{m_{0_2}}{\varepsilon_2 m_2 + m_{PL}} = \frac{m_2 + m_{PL}}{\varepsilon_2 m_2 + m_{PL}} \tag{11.74}$$

We can solve Equations 11.73 and 11.74 to obtain the step masses from the mass ratios,

$$m_2 = \frac{n_2 - 1}{1 - n_2 \varepsilon_2} m_{PL}$$

$$m_1 = \frac{n_1 - 1}{1 - n_1 \varepsilon_1} (m_2 + m_{PL}) \tag{11.75}$$

Now,

$$\frac{m_1 + m_2 + m_{PL}}{m_2 + m_{PL}} = \frac{1 - \varepsilon_1}{1 - \varepsilon_1} \frac{m_1 + m_2 + m_{PL}}{m_2 + m_{PL} + (\varepsilon_1 m_1 - \varepsilon_1 m_1)} \frac{\dfrac{1}{\varepsilon_1 m_1 + m_2 + m_{PL}}}{\dfrac{1}{\varepsilon_1 m_1 + m_2 + m_{PL}}}$$

These manipulations leave the right-hand side unchanged. Carrying out the multiplications proceed as follows,

$$\frac{m_1 + m_2 + m_{PL}}{m_2 + m_{PL}} = \frac{(1 - \varepsilon_1)(m_1 + m_2 + m_{PL})}{\varepsilon_1 m_1 + m_2 + m_{PL} - \varepsilon_1(m_1 + m_2 + m_{PL})} \frac{\dfrac{1}{\varepsilon_1 m_1 + m_2 + m_{PL}}}{\dfrac{1}{\varepsilon_1 m_1 + m_2 + m_{PL}}}$$

$$= \frac{(1 - \varepsilon_1) \dfrac{m_1 + m_2 + m_{PL}}{\varepsilon_1 m_1 + m_2 + m_{PL}}}{\dfrac{\varepsilon_1 m_1 + m_2 + m_{PL}}{\varepsilon_1 m_1 + m_2 + m_{PL}} - \varepsilon_1 \dfrac{m_1 + m_2 + m_{PL}}{\varepsilon_1 m_1 + m_2 + m_{PL}}}$$

Finally, with the aid of Equation 11.73, this algebraic trickery reduces to

$$\frac{m_1 + m_2 + m_{PL}}{m_2 + m_{PL}} = \frac{(1 - \varepsilon_1) n_1}{1 - \varepsilon_1 n_1} \tag{11.76}$$

Likewise,

$$\frac{m_2 + m_{PL}}{m_{PL}} = \frac{(1 - \varepsilon_2) n_2}{1 - \varepsilon_2 n_2} \tag{11.77}$$

11.6 Optimal staging

so that Equation 11.72 may be written in terms of the stage mass ratios instead of the step masses,

$$\frac{m_0}{m_{PL}} = \frac{(1-\varepsilon_1)n_1}{1-\varepsilon_1 n_1} \frac{(1-\varepsilon_2)n_2}{1-\varepsilon_2 n_2} \tag{11.78}$$

Taking the natural logarithm of both sides, we get

$$\ln \frac{m_0}{m_{PL}} = \ln \frac{(1-\varepsilon_1)n_1}{1-\varepsilon_1 n_1} + \ln \frac{(1-\varepsilon_2)n_2}{1-\varepsilon_2 n_2}$$

Expanding the logarithms on the right side leads to

$$\ln \frac{m_0}{m_{PL}} = [\ln(1-\varepsilon_1) + \ln n_1 - \ln(1-\varepsilon_1 n_1)]$$
$$+ [\ln(1-\varepsilon_2) + \ln n_2 - \ln(1-\varepsilon_2 n_2)] \tag{11.79}$$

Observe that for m_{PL} fixed, $\ln(m_0/m_{PL})$ is a monotonically increasing function of m_0,

$$\frac{d}{dm_0}\left(\ln \frac{m_0}{m_{PL}}\right) = \frac{1}{m_0} > 0$$

Therefore, $\ln(m_0/m_{PL})$ is stationary when m_0 is stationary.

From Equations 11.21 and 11.39, the burnout velocity of the two-stage rocket is

$$v_{bo} = v_{bo_1} + v_{bo_2} = c_1 \ln n_1 + c_2 \ln n_2 \tag{11.80}$$

which means that, given v_{bo}, our constraint equation is

$$v_{bo} - c_1 \ln n_1 - c_2 \ln n_2 = 0 \tag{11.81}$$

Introducing the Lagrange multiplier η, we combine Equations 11.79 and 11.81 to obtain

$$h = [\ln(1-\varepsilon_1) + \ln n_1 - \ln(1-\varepsilon_1 n_1)] + [\ln(1-\varepsilon_2) + \ln n_2 - \ln(1-\varepsilon_2 n_2)]$$
$$+ \eta(v_{bo} - c_1 \ln n_1 - c_2 \ln n_2) \tag{11.82}$$

Finding the values of n_1 and n_2 for which h is stationary will extremize $\ln(m_0/m_{PL})$ (and, hence, m_0) for the prescribed burnout velocity v_{bo}. h is stationary when $\partial h/\partial n_1 = \partial h/\partial n_2 = \partial h/\partial \eta = 0$. Thus,

$$\frac{\partial h}{\partial n_1} = \frac{1}{n_1} + \frac{\varepsilon_1}{1-\varepsilon_1 n_1} - \eta \frac{c_1}{n_1} = 0$$

$$\frac{\partial h}{\partial n_2} = \frac{1}{n_2} + \frac{\varepsilon_2}{1-\varepsilon_2 n_2} - \eta \frac{c_2}{n_2} = 0$$

$$\frac{\partial h}{\partial \eta} = v_{bo} - c_1 \ln n_1 - c_2 \ln n_2 = 0$$

These three equations yield, respectively,

$$n_1 = \frac{c_1\eta - 1}{c_1\varepsilon_1\eta} \qquad n_2 = \frac{c_2\eta - 1}{c_2\varepsilon_2\eta} \qquad v_{bo} = c_1 \ln n_1 + c_2 \ln n_2 \tag{11.83}$$

Substituting n_1 and n_2 into the expression for v_{bo}, we get

$$c_1 \ln\left(\frac{c_1\eta - 1}{c_1\varepsilon_1\eta}\right) + c_2 \ln\left(\frac{c_2\eta - 1}{c_2\varepsilon_2\eta}\right) = v_{bo} \tag{11.84}$$

This equation must be solved iteratively for η, after which η is substituted into Equations 11.83$_{1,2}$ to obtain the stage mass ratios n_1 and n_2. These mass ratios are used in Equations 11.75 together with the assumed structural ratios, exhaust velocities, and payload mass to obtain the step masses of each stage.

We can now generalize the optimization procedure to an N-stage vehicle, for which Equation 11.82 becomes

$$h = \sum_{i=1}^{N}\left[\ln(1-\varepsilon_i) + \ln n_i - \ln(1-\varepsilon_i n_i)\right] - \eta\left(v_{bo} - \sum_{i=1}^{N} c_i \ln n_i\right) \tag{11.85}$$

At the outset, we know the required burnout velocity v_{bo}, the payload mass m_{PL}, and for every stage we have the structural ratio ε_i and the exhaust velocity c_i (i.e., the specific impulse). The first step is to solve for the Lagrange parameter η using Equation 11.84, which, for N stages, is written

$$\sum_{i=1}^{N} c_i \ln \frac{c_i\eta - 1}{c_i\varepsilon_i\eta} = v_{bo}$$

Expanding the logarithm, this can be written

$$\sum_{i=1}^{N} c_i \ln(c_i\eta - 1) - \ln\eta \sum_{i=1}^{N} c_i - \sum_{i=1}^{N} c_i \ln c_i\varepsilon_i = v_{bo} \tag{11.86}$$

After solving this equation iteratively for η, we use that result to calculate the optimum mass ratio for each stage (cf. Equation 11.83),

$$n_i = \frac{c_i\eta - 1}{c_i\varepsilon_i\eta}, \quad i = 1, 2, \ldots, N \tag{11.87}$$

Of course, each n_i must be greater than 1.

Referring to Equations 11.75, we next obtain the step masses of each stage, beginning with stage N and working our way down the stack to stage 1,

$$m_N = \frac{n_N - 1}{1 - n_N\varepsilon_N} m_{PL}$$

$$m_{N-1} = \frac{n_{N-1} - 1}{1 - n_{N-1}\varepsilon_{N-1}}(m_N + m_{PL})$$

$$m_{N-2} = \frac{n_{N-2} - 1}{1 - n_{N-2}\varepsilon_{N-2}}(m_{N-1} + m_N + m_{PL}) \tag{11.88}$$

$$\vdots$$

$$m_1 = \frac{n_1 - 1}{1 - n_1\varepsilon_1}(m_2 + m_3 + \cdots m_{PL})$$

Having found each step mass, each empty stage mass is

$$m_{E_i} = \varepsilon_i m_i \tag{11.89}$$

and each stage propellant mass is

$$m_{p_i} = m_i - m_{E_i} \tag{11.90}$$

For the function h in Equation 11.85 it is easily shown that

$$\frac{\partial^2 h}{\partial n_i \partial n_j} = 0, \quad i,j = 1, \ldots, N (i \neq j)$$

It follows that the second differential of h is

$$d^2 h = \sum_{i=1}^{N} \sum_{j=1}^{N} \frac{\partial^2 h}{\partial n_i \partial n_j} dn_i dn_j = \sum_{i=1}^{N} \frac{\partial^2 h}{\partial n_i^2} (dn_i)^2 \tag{11.91}$$

where it can be shown, again using Equation 11.85, that

$$\frac{\partial^2 h}{\partial n_i^2} = \frac{\eta c_i (\varepsilon_i n_i - 1)^2 + 2\varepsilon_i n_i - 1}{(\varepsilon_i n_i - 1)^2 n_i^2} \tag{11.92}$$

For h to be minimum at the mass ratios n_i given by Equation 11.87, it must be true that $d^2 h > 0$. Equations 11.91 and 11.92 indicate that this will be the case if

$$\eta c_i (\varepsilon_i n_i - 1)^2 + 2\varepsilon_i n_i - 1 > 0, \quad i = 1, \ldots, N \tag{11.93}$$

EXAMPLE 11.5

Find the optimal mass for a three-stage launch vehicle which is required to lift a 5000 kg payload to a speed of 10 km/s. For each stage, we are given that

Stage 1	$I_{sp_1} = 400$ s ($c_1 = 3.924$ km/s)	$\varepsilon_1 = 0.10$
Stage 2	$I_{sp_2} = 350$ s ($c_2 = 3.434$ km/s)	$\varepsilon_2 = 0.15$
Stage 3	$I_{sp_3} = 300$ s ($c_3 = 2.943$ km/s)	$\varepsilon_3 = 0.20$

Substituting this data into Equation 11.86, we get

$$3.924 \ln(3.924\eta - 1) + 3.434 \ln(3.434\eta - 1) + 2.943 \ln(2.943\eta - 1)$$
$$- 10.30 \ln \eta + 7.5089 = 10$$

As can be checked by substitution, the iterative solution of this equation is

$$\eta = 0.4668$$

Substituting η into Equations 11.87 yields the optimum mass ratios,

$$n_1 = 4.541 \quad n_2 = 2.507 \quad n_3 = 1.361$$

For the step masses, we appeal to Equations 11.88 to obtain

$$m_1 = 165\,700 \text{ kg} \quad m_2 = 18\,070 \text{ kg} \quad m_3 = 2477 \text{ kg}$$

(Example 11.5 continued)

Using Equations 11.89 and 11.90, the empty masses and propellant masses are found to be

$$m_{E_1} = 16\,570\,\text{kg} \quad m_{E_2} = 2710\,\text{kg} \quad m_{E_3} = 495.4\,\text{kg}$$
$$m_{P_1} = 149\,100\,\text{kg} \quad m_{P_2} = 15\,360\,\text{kg} \quad m_{P_3} = 1982\,\text{kg}$$

The payload ratios for each stage are

$$\lambda_1 = \frac{m_2 + m_3 + m_{PL}}{m_1} = 0.1542$$

$$\lambda_2 = \frac{m_3 + m_{PL}}{m_2} = 0.4139$$

$$\lambda_3 = \frac{m_{PL}}{m_3} = 2.018$$

The total mass of the vehicle is

$$m_0 = m_1 + m_2 + m_3 + m_{PL} = 191\,200\,\text{kg}$$

and the overall payload fraction is

$$\pi_{PL} = \frac{m_{PL}}{m_0} = \frac{5000}{191\,200} = 0.0262$$

Finally, let us check Equation 11.93,

$$\eta c_1(\varepsilon_1 n_1 - 1)^2 + 2\varepsilon_1 n_1 - 1 = 0.4541$$
$$\eta c_2(\varepsilon_2 n_2 - 1)^2 + 2\varepsilon_2 n_2 - 1 = 0.3761$$
$$\eta c_3(\varepsilon_3 n_3 - 1)^2 + 2\varepsilon_3 n_3 - 1 = 0.2721$$

A positive number in every instance means we have indeed found a local minimum of the function in Equation 11.85.

Problems

11.1 Suppose a spacecraft in permanent orbit around the earth is to be used for delivering payloads from low earth orbit (LEO) to geostationary equatorial orbit (GEO). Before each flight from LEO, the spacecraft is refueled with propellant which it uses up in its round trip to GEO. The outbound leg requires four times as much propellant as the inbound return leg. The delta-v for transfer from LEO to GEO is 4.22 km/s (see Example 6.12). The specific impulse of the propulsion system is 430 s. If the payload mass is 3500 kg, calculate the empty mass of the vehicle.
{Ans.: 2733 kg}

11.2 A two stage, solid-propellant sounding rocket has the following properties:

First stage: $m_0 = 249.5\,\text{kg} \quad m_f = 170.1\,\text{kg} \quad \dot{m}_e = 10.61\,\text{kg/s} \quad I_{sp} = 235\,\text{s}$

Second stage: $m_0 = 113.4\,\text{kg} \quad m_f = 58.97\,\text{kg} \quad \dot{m}_e = 4.053\,\text{kg/s} \quad I_{sp} = 235\,\text{s}$

Delay time between burnout of first stage and ignition of second stage: 3 seconds.

As a preliminary estimate, neglect drag and the variation of earth's gravity with altitude to calculate the maximum height reached by the second stage after burnout.
{Ans.: 322 km}

11.3 A two-stage launch vehicle has the following properties:

First stage: 2 solid propellant rockets. Each one has a total mass of 525 000 kg, 450 000 kg of which is propellant. $I_{sp} = 290$ s.

Second stage: 2 liquid rockets with $I_{sp} = 450$ s. Dry mass $= 30\,000$ kg, propellant mass $= 600\,000$ kg.

Calculate the payload mass to a 300 km orbit if launched due east from KSC. Let the total gravity and drag loss be 2 km/s.
{Ans.: 114 000 kg}

11.4 Consider a rocket comprising three similar stages (i.e., each stage has the same specific impulse, structural ratio and payload ratio). The common specific impulse is 310 s. The total mass of the vehicle is 150 000 kg, the total structural mass (empty mass) is 20 000 kg and the payload mass is 10 000 kg. Calculate
(a) The mass ratio n and the total Δv for the three-stage rocket.
{Ans.: $n = 2.04$, $\Delta v = 6.50$ km/s}
(b) $m_{p_1}, m_{p_2},$ and m_{p_3}.
(c) m_{E_1}, m_{E_2} and m_{E_3}.
(d) m_{0_1}, m_{0_2} and m_{0_3}.

11.5 A small two-stage vehicle is to propel a 10 kg payload to a speed of 6.2 km/s. The properties of the stages are: for the first stage, $I_{sp} = 300$ s and $\varepsilon = 0.2$; for the second stage, $I_{sp} = 235$ s and $\varepsilon = 0.3$. Estimate the optimum mass of the vehicle.
{Ans.: 1125 kg}

11.6 Find the extrema of the function $z = x^2 + y^2 + 2xy$ subject to the constraint $x^2 - 2x + y^2 = 0$.
{Ans.: $z_{min} = 0.1716$ at $(x, y) = (0.2929, -0.7071)$ and $z_{max} = 5.828$ at $(x, y) = (1.707, 0.7071)$}

REFERENCES

Bate, R. R., Mueller, D., and White, J. E. (1971). *Fundamentals of Astrodynamics*, Dover Publications.

Battin, R. H. (1999). *An Introduction to the Mathematics and Methods of Astrodynamics*, Revised Edition, AIAA Education Series.

Beyer, W. H., ed. (1991). *Standard Mathematical Tables and Formulae*, 29th Edition, CRC Press.

Bond, V. R. and Allman, M. C. (1996). *Modern Astrodynamics: Fundamentals and Perturbation Methods*, Princeton University Press.

Boulet, D. L. (1991). *Methods of Orbit Determination for the Microcomputer*, Willmann-Bell.

Chobotov, V. A., ed. (2002). *Orbital Mechanics*, Third Edition, AIAA Education Series, AIAA.

Coriolis, G. (1835). 'On the Equations of Relative Motion of a System of Bodies', *J. École Polytechnique*, Vol. 15, No. 24, 142–54.

Hahn, B. D. (2002). *Essential MATLAB® for Scientists and Engineers*, Second Edition, Butterworth-Heinemann.

Hale, F. J. (1994). *Introduction to Space Flight*, Prentice-Hall, Englewood Cliffs, New Jersey.

Hohmann, W. (1925). *The Attainability of Celestial Bodies* (in German), R. Oldenbourg.

Kaplan, M. H. (1976). *Modern Spacecraft Dynamics and Control*, Wiley.

Kermit, S. and Davis, T. A. (2002). *MATLAB Primer*, Sixth Edition, Chapman & Hall/CRC.

Likins, P. W. (1967). 'Attitude Stability Criteria for Dual Spin Spacecraft', *Journal of Spacecraft and Rockets*, Vol. 4, No. 12, 1638–43.

Magrab, E. B., ed. (2000). *An Engineer's Guide to MATLAB®*, Prentice-Hall.

NASA Goddard Space Flight Center (2003). *National Space Science Data Center*, http://nssdc.gsfc.nasa.gov.

Nise, N. S. (2003). *Control Systems Engineering*, Fourth Edition, Wiley.

Ogata, K. (2001). *Modern Control Engineering*, Fourth Edition, Prentice-Hall.

Palm, W. J. (1983). *Modeling, Analysis and Control of Dynamic Systems*, Wiley.

Prussing, J. E. and Conway, B. A. (1993). *Orbital Mechanics*, Oxford University Press.

Seidelmann, P. K., ed. (1992). *Explanatory Supplement to the Astronomical Almanac*, University Science Books.

Standish, E. M., Newhall, X.X., Williams, J. G., and Yeomans, D. K. (1992). 'Orbital Ephemerides of the Sun, Moon and Planets'. In *Explanatory Supplement to the Astronomical Almanac* (P. K. Seidelmann, ed.), p. 316, University Science Books.

US Naval Observatory (2004). *The Astronomical Almanac*, GPO.

Wiesel, W. E. (1997). *Spacecraft Dynamics*, Second Edition, McGraw-Hill.

FURTHER READING

Brown, C. D. (1992). *Spacecraft Mission Design*, AIAA Education Series.

Chobotov, V. A. (1991). *Spacecraft Attitude Dynamics and Control*, Krieger.

Chobotov, V. A., ed. (2002). *Orbital Mechanics*, Third Edition, AIAA Education Series.

Danby, J. M. A. (1988). *Fundamentals of Celestial Mechanics*, Second Edition, Willmann-Bell.

Escobal, P. R. (1976). *Methods of Orbit Determination*, Second Edition, Krieger.

Griffin, M. D. and French, J. R. (1991). *Space Vehicle Design*, AIAA Education Series.

Hill, P. P. and Peterson, C. R. (1992). *Mechanics and Thermodynamics of Propulsion*, Addison-Wesley.

Kane, T. R., Likins, P. W., and Levinson, D. A. (1983). *Spacecraft Dynamics*, McGraw-Hill.

Larson, W. J. and Wertz, J. R., ed. (1992). *Space Mission Analysis and Design*, Second Edition, Microcosm Press and Kluwer Academic Publishers.

Logsdon, T. (1998). *Orbital Mechanics: Theory and Applications*, Wiley.

McCuskey, S. W. (1963). *Introduction to Celestial Mechanics*, Addison-Wesley.

Meeus, J. (1998). *Astronomical Algorithms*, Second Edition, Willmann-Bell.

Moulton, F. R. (1970). *An Introduction to Celestial Mechanics*, Second Edition, Dover Publications.

Schaub, S. and Junkins, J. L. (2003). *Analytical Mechanics of Space Systems*, AIAA Education Series.

Sellers, J. J. (1994). *Understanding Space: An Introduction to Astronautics*, McGraw-Hill.

Sutton, G. P. and Biblarz, O. (2001). *Rocket Propulsion Elements*, Seventh Edition, Wiley.

Thomson, W. T. (1986). *Introduction to Space Dynamics*, Dover Publications.

Vallado, D. A. (2001). *Fundamentals of Astrodynamics and Applications*, Second Edition, Microcosm Press and Kluwer Academic Publishers.

Wertz, J. R. (1978). *Spacecraft Attitude Determination and Control*, Kluwer Academic Publishers.

Appendix A

Physical data

The following tables contain information that is commonly available and may be found in the literature and on the world wide web. See, for example, the *Astronomical Almanac* (US Naval Observatory, 2004) and *National Space Science Data Center* (NASA Goddard Space Flight Center, 2003).

Table A.1 Astronomical data for the sun, the planets and the moon

Object	Radius (km)	Mass (kg)	Sidereal rotation period	Inclination of equator to orbit plane	Semimajor axis of orbit (km)	Orbit eccentricity	Inclination of orbit to the ecliptic plane	Orbit sidereal period
Sun	696 000	1.989×10^{30}	25.38d	7.25°	–	–	–	–
Mercury	2440	330.2×10^{21}	58.65d	0.01°	57.91×10^6	0.2056	7.00°	87.97d
Venus	6052	4.869×10^{24}	243d*	177.4°	108.2×10^6	0.0067	3.39°	224.7d
Earth	6378	5.974×10^{24}	23.9345h	23.45°	149.6×10^6	0.0167	0.00°	365.256d
(Moon)	1737	73.48×10^{21}	27.32d	6.68°	384.4×10^3	0.0549	5.145°	27.322d
Mars	3396	641.9×10^{21}	24.62h	25.19°	227.9×10^6	0.0935	1.850°	1.881y
Jupiter	71 490	1.899×10^{27}	9.925h	3.13°	778.6×10^6	0.0489	1.304°	11.86y
Saturn	60 270	568.5×10^{24}	10.66h	26.73°	1.433×10^9	0.0565	2.485°	29.46y
Uranus	25 560	86.83×10^{24}	17.24h*	97.77°	2.872×10^9	0.0457	0.772°	84.01y
Neptune	24 760	102.4×10^{24}	16.11h	28.32°	4.495×10^9	0.0113	1.769°	164.8y
Pluto	1195	12.5×10^{21}	6.387d*	122.5°	5.870×10^9	0.2444	17.16°	247.7y

* Retrograde

Table A.2 Gravitational parameter (μ) and sphere of influence (SOI) radius for the sun, the planets and the moon

Celestial body	μ (km³/s²)	SOI radius (km)
Sun	132 712 000 000	–
Mercury	22 030	112 000
Venus	324 900	616 000
Earth	398 600	925 000
Earth's moon	4903	66 200
Mars	42 828	577 000
Jupiter	126 686 000	48 200 000
Saturn	37 931 000	54 800 000
Uranus	5 794 000	51 800 000
Neptune	6 835 100	86 600 000
Pluto	830	3 080 000

Table A.3 Some conversion factors

1 ft = 0.3048 m
1 mile (mi) = 1.609 km
1 nautical mile (n mi) = 1.151 mi = 1.852 km
1 mi/h = 0.0004469 km/s
1 lb (mass) = 0.4536 kg
1 lb (force) = 4.448 N
1 psi = 6895 kPa

Appendix B

A Road Map

Figure B.1 is a road map through Chapters 1, 2 and 3. Those who from time to time feel they have lost their bearings may find it useful to refer to this flow chart, which shows how the various concepts and results are interrelated. The pivotal influence of Sir Isaac Newton is obvious. All of the equations of classical orbital mechanics (the two-body problem) are derived from those listed here.

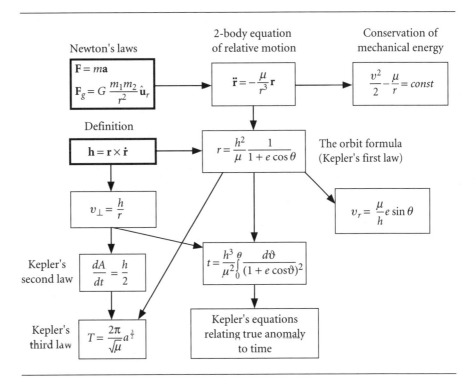

Figure B.1 Logic flow for the major outcomes of Chapters 1, 2 and 3.

585

APPENDIX C

NUMERICAL INTEGRATION OF THE n-BODY EQUATIONS OF MOTION

APPENDIX OUTLINE

C.1 FUNCTION FILE `accel_3body.m` 590
C.2 SCRIPT FILE `threebody.m` 592

Without loss of generality we shall derive the equations of motion of the three-body system illustrated in Figure C.1. The equations of motion for n bodies can easily be generalized from those of a three-body system.

Each mass of a three-body system experiences the force of gravitational attraction from the other members of the system. As shown in Figure C.1, the forces exerted on body 1 by bodies 2 and 3 are $\mathbf{F}_{12}$ and $\mathbf{F}_{13}$, respectively. Likewise, body 2 experiences the forces $\mathbf{F}_{21}$ and $\mathbf{F}_{23}$ whereas the forces $\mathbf{F}_{31}$ and $\mathbf{F}_{32}$ act on body 3. These gravitational forces can be inferred from Equation 2.6:

$$\mathbf{F}_{12} = -\mathbf{F}_{21} = \frac{Gm_1 m_2 (\mathbf{R}_2 - \mathbf{R}_1)}{\|\mathbf{R}_2 - \mathbf{R}_1\|^3} \tag{C.1a}$$

$$\mathbf{F}_{13} = -\mathbf{F}_{31} = \frac{Gm_1 m_3 (\mathbf{R}_3 - \mathbf{R}_1)}{\|\mathbf{R}_3 - \mathbf{R}_1\|^3} \tag{C.1b}$$

$$\mathbf{F}_{23} = -\mathbf{F}_{32} = \frac{Gm_2 m_3 (\mathbf{R}_3 - \mathbf{R}_2)}{\|\mathbf{R}_3 - \mathbf{R}_2\|^3} \tag{C.1c}$$

Appendix C Numerical integration of the n-body equations of motion

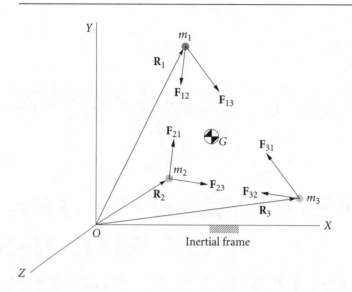

Figure C.1 Three-body problem.

Relative to an inertial frame of reference the accelerations of the bodies are

$$\mathbf{a}_i = \ddot{\mathbf{R}}_i \qquad i = 1, 2, 3$$

where $\mathbf{R}_i$ is the absolute position vector of body i. The equation of motion of body 1 is

$$\mathbf{F}_{12} + \mathbf{F}_{13} = m_1 \mathbf{a}_1$$

Substituting Equations C.1a and C.1b yields

$$\mathbf{a}_1 = \frac{G m_2 (\mathbf{R}_2 - \mathbf{R}_1)}{\|\mathbf{R}_2 - \mathbf{R}_1\|^3} + \frac{G m_3 (\mathbf{R}_3 - \mathbf{R}_1)}{\|\mathbf{R}_3 - \mathbf{R}_1\|^3} \tag{C.2a}$$

For bodies 2 and 3 we find in a similar fashion that

$$\mathbf{a}_2 = \frac{G m_1 (\mathbf{R}_1 - \mathbf{R}_2)}{\|\mathbf{R}_1 - \mathbf{R}_2\|^3} + \frac{G m_3 (\mathbf{R}_3 - \mathbf{R}_2)}{\|\mathbf{R}_3 - \mathbf{R}_2\|^3} \tag{C.2b}$$

$$\mathbf{a}_3 = \frac{G m_1 (\mathbf{R}_1 - \mathbf{R}_3)}{\|\mathbf{R}_1 - \mathbf{R}_3\|^3} + \frac{G m_2 (\mathbf{R}_2 - \mathbf{R}_3)}{\|\mathbf{R}_2 - \mathbf{R}_3\|^3} \tag{C.2c}$$

The velocities are related to the accelerations by

$$\frac{d\mathbf{v}_i}{dt} = \mathbf{a}_i \qquad i = 1, 2, 3 \tag{C.3}$$

and the position vectors are likewise related to the velocities,

$$\frac{d\mathbf{R}_i}{dt} = \mathbf{v}_i \qquad i = 1, 2, 3 \tag{C.4}$$

Equations C.2 through C.4 constitute a system of ordinary differential equations (ODEs) in the variable time.

Appendix C *Numerical integration of the n-body equations of motion*

Since there are no external forces on the system, the acceleration of the center of mass is zero

$$\mathbf{a}_G = 0 \tag{C.5a}$$

so that

$$\frac{d\mathbf{v}_G}{dt} = 0 \tag{C.5b}$$

and

$$\frac{d\mathbf{R}_G}{dt} = \mathbf{v}_G \tag{C.5c}$$

Given the initial positions $\mathbf{R}_{i0}$ and initial velocities $\mathbf{v}_{i0}$, we must integrate Equation C.3 to find $\mathbf{v}_i$ as a function of time and substitute those results into Equations C.4 to obtain $\mathbf{R}_i$ as a function of time. The integrations must be done numerically.

To do this using MATLAB, we first resolve all of the vectors into their three components along the XYZ axes of the inertial frame and write them as column vectors,

$$\{\mathbf{R}_1\} = \begin{Bmatrix} R_{1X} \\ R_{1Y} \\ R_{1Z} \end{Bmatrix} \quad \{\mathbf{R}_2\} = \begin{Bmatrix} R_{2X} \\ R_{2Y} \\ R_{2Z} \end{Bmatrix} \quad \{\mathbf{R}_3\} = \begin{Bmatrix} R_{3X} \\ R_{3Y} \\ R_{3Z} \end{Bmatrix} \quad \{\mathbf{R}_G\} = \begin{Bmatrix} R_{GX} \\ R_{GY} \\ R_{GZ} \end{Bmatrix} \tag{C.6}$$

$$\{\mathbf{v}_1\} = \begin{Bmatrix} v_{1X} \\ v_{1Y} \\ v_{1Z} \end{Bmatrix} \quad \{\mathbf{v}_2\} = \begin{Bmatrix} v_{2X} \\ v_{2Y} \\ v_{2Z} \end{Bmatrix} \quad \{\mathbf{v}_3\} = \begin{Bmatrix} v_{3X} \\ v_{3Y} \\ v_{3Z} \end{Bmatrix} \quad \{\mathbf{v}_G\} = \begin{Bmatrix} v_{GX} \\ v_{GY} \\ v_{GZ} \end{Bmatrix} \tag{C.7}$$

According to Equations C.2,

$$\{\mathbf{a}_1\} = \begin{Bmatrix} a_{1X} \\ a_{1Y} \\ a_{1Z} \end{Bmatrix} = \begin{Bmatrix} \dfrac{Gm_2(R_{2X}-R_{1X})}{R_{12}} + \dfrac{Gm_3(R_{3X}-R_{1X})}{R_{13}} \\ \dfrac{Gm_2(R_{2Y}-R_{1Y})}{R_{12}} + \dfrac{Gm_3(R_{3Y}-R_{1Y})}{R_{13}} \\ \dfrac{Gm_2(R_{2Z}-R_{1Z})}{R_{12}} + \dfrac{Gm_3(R_{3Z}-R_{1Z})}{R_{13}} \end{Bmatrix} \tag{C.8a}$$

$$\{\mathbf{a}_2\} = \begin{Bmatrix} a_{2X} \\ a_{2Y} \\ a_{2Z} \end{Bmatrix} = \begin{Bmatrix} \dfrac{Gm_1(R_{1X}-R_{2X})}{R_{12}} + \dfrac{Gm_3(R_{3X}-R_{2X})}{R_{13}} \\ \dfrac{Gm_1(R_{1Y}-R_{2Y})}{R_{12}} + \dfrac{Gm_3(R_{3Y}-R_{2Y})}{R_{13}} \\ \dfrac{Gm_1(R_{1Z}-R_{2Z})}{R_{12}} + \dfrac{Gm_3(R_{3Z}-R_{2Z})}{R_{13}} \end{Bmatrix} \tag{C.8b}$$

$$\{\mathbf{a}_3\} = \begin{Bmatrix} a_{3X} \\ a_{3Y} \\ a_{3Z} \end{Bmatrix} = \begin{Bmatrix} \dfrac{Gm_1(R_{1X}-R_{3X})}{R_{12}} + \dfrac{Gm_2(R_{2X}-R_{3X})}{R_{13}} \\ \dfrac{Gm_1(R_{1Y}-R_{3Y})}{R_{12}} + \dfrac{Gm_2(R_{2Y}-R_{3Y})}{R_{13}} \\ \dfrac{Gm_1(R_{1Z}-R_{3Z})}{R_{12}} + \dfrac{Gm_2(R_{2Z}-R_{3Z})}{R_{13}} \end{Bmatrix} \tag{C.8c}$$

where

$$R_{12} = \|\{\mathbf{R}_2\} - \{\mathbf{R}_1\}\|^3 \qquad R_{13} = \|\{\mathbf{R}_3\} - \{\mathbf{R}_1\}\|^3 \qquad R_{23} = \|\{\mathbf{R}_3\} - \{\mathbf{R}_2\}\|^3 \quad (C.9)$$

Next, we form the 24-component column vector

$$\{\mathbf{f}\} = \lfloor \{\mathbf{R}_1\} \ \{\mathbf{R}_2\} \ \{\mathbf{R}_3\} \ \{\mathbf{R}_G\} \ \{\mathbf{v}_1\} \ \{\mathbf{v}_2\} \ \{\mathbf{v}_3\} \ \{\mathbf{v}_G\} \rfloor^T \quad (C.10)$$

The first derivatives of the components of this vector comprise the column vector

$$\left\{\frac{d\mathbf{f}}{dt}\right\} = \lfloor \{\mathbf{v}_1\} \ \{\mathbf{v}_2\} \ \{\mathbf{v}_3\} \ \{\mathbf{v}_G\} \ \{\mathbf{a}_1\} \ \{\mathbf{a}_2\} \ \{\mathbf{a}_3\} \ \{0\} \rfloor^T \quad (C.11)$$

If the vector $\{\mathbf{f}\}$ is given at time t, then $\{d\mathbf{f}/dt\}$ is used to obtain an accurate estimate of $\{\mathbf{f}\}$ at time $t + \Delta t$ by means of a procedure such as that due originally to the German mathematicians Carle Runge (1856–1927) and Martin Kutta (1867–1944). Sophisticated Runge–Kutta algorithms are implemented in MATLAB in the form of the solvers `ode23` and `ode45`. `ode45` is the more accurate of the two and is recommended as a first try for solving most ODEs.

For simplicity, we will use MATLAB to solve the three-body problem in the plane. That is, we will restrict ourselves to only the *XY* components of the vectors **R**, **v** and **a**. The reader can use these scripts as a starting point for investigating more complex n-body problems.

The M-function `accel_3body.m` is used by `ode45` to calculate the accelerations of each of the masses from Equations C.8.

C.1 Function file `accel_3body.m`

```
% ~~~~~~~~~~~~~~~~~~~~~~~~~~~~~~~~~~~~~~~~~~~~~~~~~~~~~~~~~~~~~~
function dfdt = accel_3body(t,f)
% ~~~~~~~~~~~~~~~~~~~~~~~~~~~~~~~~~~~
%
% This function evaluates the acceleration of each member of a
% planar 3-body system at time t from their positions and
% velocities at that time.
%
% G                             - gravitational constant
%                                  (km^3/kg/s^2)
% m                             - vector [m1, m2, m3] containing
%                                  the masses m1, m2, m3 of the
%                                  three bodies (kg)
% r1x, r1y; r2x, r2y; r3x, r3y  - components of the position
%                                  vectors of each mass (km)
% v1x, v1y; v2x, v2y; v3x, v3y  - components of the velocity
%                                  vectors of each mass (km/s)
% a1x, a1y; a2x, a2y; a3x, a3y  - components of the acceleration
%                                  vectors of each mass (km/s^2)
% rGx, rGy; vGx, vGy; aGx, aGy  - components of the position,
%                                  velocity and acceleration of
%                                  the center of mass
```

```
%   t                           - time (s)
%   f                           - column vector containing the
%                                 position and velocity
%                                 components of the three
%                                 masses and the center of
%                                 mass at time t
%   dfdt                        - column vector containing the
%                                 velocity and acceleration
%                                 components of the three
%                                 masses and the center of
%                                 mass at time t
%
% User M-functions required: none
% --------------------------------------------------------------

global G m

%...Initialize the 16 by 1 column vector dfdt:
dfdt = zeros(16,1);

%...For ease of reading the code, assign each component of f
%...to a mnemonic variable:

r1x = f( 1);
r1y = f( 2);

r2x = f( 3);
r2y = f( 4);

r3x = f( 5);
r3y = f( 6);

rGx = f( 7);
rGy = f( 8);

v1x = f( 9);
v1y = f(10);

v2x = f(11);
v2y = f(12);

v3x = f(13);
v3y = f(14);

vGx = f(15);
vGy = f(16);

%...Equations C.9:
r12 = norm([r2x - r1x, r2y - r1y])^3;
r13 = norm([r3x - r1x, r3y - r1y])^3;
r23 = norm([r3x - r2x, r3y - r2y])^3;
```

```
%...Equations C.8:
a1x = G*m(2)*(r2x - r1x)/r12 + G*m(3)*(r3x - r1x)/r13;
a1y = G*m(2)*(r2y - r1y)/r12 + G*m(3)*(r3y - r1y)/r13;
a2x = G*m(1)*(r1x - r2x)/r12 + G*m(3)*(r3x - r2x)/r23;
a2y = G*m(1)*(r1y - r2y)/r12 + G*m(3)*(r3y - r2y)/r23;
a3x = G*m(1)*(r1x - r3x)/r13 + G*m(2)*(r2x - r3x)/r23;
a3y = G*m(1)*(r1y - r3y)/r13 + G*m(2)*(r2y - r3y)/r23;

%...Equation C.5a:
aGx = 0;
aGy = 0;

%...Place the evaluated velocity and acceleration components
%...into the vector dfdt, to be returned to the calling
%...program:

dfdt = [v1x; v1y; ...
        v2x; v2y; ...
        v3x; v3y; ...
        vGx; vGy; ...
        a1x; a1y; ...
        a2x; a2y; ...
        a3x; a3y; ...
        aGx; aGy];

% ~~~~~~~~~~~~~~~~~~~~~~~~~~~~~~~~~~~~~~~~~~~~~~~~~~~~~~~~~~~~~~~
```

The script `threebody.m` defines the initial conditions, passes that information to ode45 and finally plots the solutions. The results of this program were used to create Figures 2.5 and 2.6. Similar scripts can obviously be written for the two-body problem and may be used to produce Figures 2.3 and 2.4.

C.2 SCRIPT FILE `threebody.m`

```
% ~~~~~~~~~~~~~~~~~~~~~~~~~~~~~~~~~~~~~~~~~~~~~~~~~~~~~~~~~~~~~~~
% threebody
% ~~~~~~~~~~
%
% This program presents the graphical solution of the motion of
% three bodies in the plane for data provided in the input
% definitions below.
%
% G                      - gravitational constant (km^3/kg/s^2)
% t_initial, t_final     - initial and final times (s)
% m                      - vector [m1, m2, m3] containing the
%                          masses m1, m2, m3 of the three
%                          bodies (kg)
% r0                     - 3 by 2 matrix each row of which
%                          contains the initial x and y components
%                          of the position vector of the
%                          respective mass (km)
```

```
% v0                    - 3 by 2 matrix each row of which contains
%                         the initial x and y components of the
%                         velocity of the respective mass (km/s)
% rG0                   - vector containing the initial x and y
%                         components of the center of mass (km)
% vG0                   - vector containing the initial x and y
%                         components of the velocity of the
%                         center of mass (km/s)
% f0                    - column vector of the initial conditions
%                         passed to the Runge-Kutta solver ode45
% t                     - column vector of times at which the
%                         solution was computed
% f                     - matrix the columns of which contain the
%                         position and velocity components
%                         evaluated at the times t(:):
%                         f(:,1) , f(:,2) =   x1(:), y1(:)
%                         f(:,3) , f(:,4) =   x2(:), y2(:)
%                         f(:,5) , f(:,6) =   x3(:), y3(:)
%                         f(:,7) , f(:,8) =   xG(:), yG(:)
%
%                         f(:,9) , f(:,10) = v1x(:), v1y(:)
%                         f(:,11), f(:,12) = v2x(:), v2y(:)
%                         f(:,13), f(:,14) = v3x(:), v3y(:)
%                         f(:,15), f(:,16) = vGx(:), vGy(:)
%
% User M-function required: accel_3body
% ---------------------------------------------------------------

clear
global G m
G = 6.67259e-20;

%...Input data:
t_initial = 0; t_final = 67000;
m  = [1.e29 1.e29 1.e29];
r0 = [[     0 0]
      [300000 0]
      [600000 0]];

v0 = [[  0   0]
      [250 250]
      [  0   0]];
%...

%...Initial position and velocity of center of mass:
rG0 = m*r0/sum(m);
vG0 = m*v0/sum(m);

%...Initial conditions must be passed to ode45 in a column
%...vector:
f0 = [r0(1,:)'; r0(2,:)'; r0(3,:)'; rG0'; ...
      v0(1,:)'; v0(2,:)'; v0(3,:)'; vG0']
```

```
%...Pass the initial conditions and time interval to ode45,
%...which calculates the position and velocity at discrete
%...times t, returning the solution in the column vector f.
%...ode45 uses the m-function 'accel_3body' to evaluate the
%...acceleration at each integration time step.
[t,f] = ode45('accel_3body', [t_initial t_final], f0);

close all

%...Plot the motion relative to the inertial frame
%...(Figure 2.5):
figure
title('Figure 2.5: Motion relative to the inertial frame', ...
      'Fontweight', 'bold', 'FontSize', 12)
hold on

%...x1 vs y1:
plot(f(:,1), f(:,2),    'r', 'LineWidth', 0.5)

%...x2 vs y2:
plot(f(:,3), f(:,4),    'g', 'LineWidth', 1.0)

%...x3 vs y3:
plot(f(:,5), f(:,6),    'b', 'LineWidth', 1.5)

%...xG vs yG:
plot(f(:,7), f(:,8),   '--k', 'LineWidth', 0.25)

xlabel('X'); ylabel('Y')
grid on
axis('equal')

%...Plot the motion relative to the center of mass
%...(Figure 2.6):
figure
title('Figure 2.6: Motion relative to the center of mass', ...
      'Fontweight', 'bold', 'FontSize', 12)
hold on

%...(x1 - xG) vs (y1 - yG):
plot(f(:,1) - f(:,7), f(:,2) - f(:,8),    'r', 'LineWidth', 0.5)

%...(x2 - xG) vs (y2 - yG):
plot(f(:,3) - f(:,7), f(:,4) - f(:,8),  '--g', 'LineWidth', 1.0)

%...(x3 - xG) vs (y3 - yG):
plot(f(:,5) - f(:,7), f(:,6) - f(:,8),    'b', 'LineWidth', 1.5)

xlabel('X'); ylabel('Y')
grid on
axis('equal')
% ~~~~~~~~~~~~~~~~~~~~~~~~~~~~~~~~~~~~~~~~~~~~~~~~~~~~~~~~~~~~~~~~~~~~~
```

APPENDIX D

MATLAB ALGORITHMS

APPENDIX OUTLINE

D.1	Introduction	596
D.2	Algorithm 3.1: solution of Kepler's equation by Newton's method	596
D.3	Algorithm 3.2: solution of Kepler's equation for the hyperbola using Newton's method	598
D.4	Calculation of the Stumpff functions $S(z)$ and $C(z)$	600
D.5	Algorithm 3.3: solution of the universal Kepler's equation using Newton's method	601
D.6	Calculation of the Lagrange coefficients f and g and their time derivatives	603
D.7	Algorithm 3.4: calculation of the state vector $(\mathbf{r}, \mathbf{v})$ given the initial state vector $(\mathbf{r}_0, \mathbf{v}_0)$ and the time lapse Δt	604
D.8	Algorithm 4.1: calculation of the orbital elements from the state vector	606
D.9	Algorithm 4.2: calculation of the state vector from the orbital elements	610
D.10	Algorithm 5.1: Gibbs' method of preliminary orbit determination	613
D.11	Algorithm 5.2: solution of Lambert's problem	616
D.12	Calculation of Julian day number at 0 hr UT	621
D.13	Algorithm 5.3: calculation of local sidereal time	623
D.14	Algorithm 5.4: calculation of the state vector from measurements of range, angular position and their rates	626

D.15	ALGORITHMS 5.5 AND 5.6: GAUSS'S METHOD OF PRELIMINARY ORBIT DETERMINATION WITH ITERATIVE IMPROVEMENT	631
D.16	CONVERTING THE NUMERICAL DESIGNATION OF A MONTH OR A PLANET INTO ITS NAME	640
D.17	ALGORITHM 8.1: CALCULATION OF THE STATE VECTOR OF A PLANET AT A GIVEN EPOCH	641
D.18	ALGORITHM 8.2: CALCULATION OF THE SPACECRAFT TRAJECTORY FROM PLANET 1 TO PLANET 2	648

D.1 INTRODUCTION

This appendix lists MATLAB scripts which implement all of the numbered algorithms presented throughout the text. The programs use only the most basic features of MATLAB and are liberally commented so as to make reading the code as easy as possible. To 'drive' the various algorithms, one can use MATLAB to create graphical user interfaces (GUIs). However, in the interest of simplicity and keeping our focus on the algorithms rather than elegant programming techniques, GUIs were not developed. Furthermore, the scripts do not use files to import and export data. Data is defined in declaration statements within the scripts. All output is to the screen, i.e., to the MATLAB command window. It is hoped that interested students will embellish these simple scripts or use them as a springboard towards generating their own programs.

Each algorithm is illustrated by a MATLAB coding of a related example problem in the text. The actual output of each of these examples is also listed.

It would be helpful to have MATLAB documentation at hand. There are a number of practical references on the subject, including Hahn (2002), Kermit and Davis (2002) and Magrab (2000). MATLAB documentation may also be found at The MathWorks web site (www.mathworks.com). Should it be necessary to do so, it is a fairly simple matter to translate these programs into other software languages.

These programs are presented solely as an alternative to carrying out otherwise lengthy hand computations and are intended for academic use only. They are all based exclusively on the introductory material presented in this text and therefore do not include the effects of perturbations of any kind.

D.2 ALGORITHM 3.1: SOLUTION OF KEPLER'S EQUATION BY NEWTON'S METHOD

FUNCTION FILE kepler_E.m

```
% ~~~~~~~~~~~~~~~~~~~~~~~~~~~~~~~~~~~~~~~~~~~~~~~~~~~~~~~~~~~~~~~~
  function E = kepler_E(e, M)
% ~~~~~~~~~~~~~~~~~~~~~~~~~~~~
%
```

D.2 Algorithm 3.1: solution of Kepler's equation by Newton's method

```
% This function uses Newton's method to solve Kepler's
% equation   E - e*sin(E) = M   for the eccentric anomaly,
% given the eccentricity and the mean anomaly.
%
% E  - eccentric anomaly (radians)
% e  - eccentricity, passed from the calling program
% M  - mean anomaly (radians), passed from the calling program
% pi - 3.1415926...
%
% User M-functions required: none
% ------------------------------------------------------------

%...Set an error tolerance:
error = 1.e-8;

%...Select a starting value for E:
if M < pi
    E = M + e/2;
else
    E = M - e/2;
end

%...Iterate on Equation 3.14 until E is determined to within
%...the error tolerance:
ratio = 1;
while abs(ratio) > error
    ratio = (E - e*sin(E) - M)/(1 - e*cos(E));
    E = E - ratio;
end
% ~~~~~~~~~~~~~~~~~~~~~~~~~~~~~~~~~~~~~~~~~~~~~~~~~~~~~~~~~~~~
```

SCRIPT FILE Example_3_02.m

```
% ~~~~~~~~~~~~~~~~~~~~~~~~~~~~~~~~~~~~~~~~~~~~~~~~~~~~~~~~~~~~
% Example_3_02
%  ~~~~~~~~~~~~
%
% This program uses Algorithm 3.1 and the data of Example 3.2
% to solve Kepler's equation.
%
% e  - eccentricity
% M  - mean anomaly (rad)
% E  - eccentric anomaly (rad)
%
% User M-function required: kepler_E
% ------------------------------------------------------------

clear

%...Input data for Example 3.2:
e = 0.37255;
M = 3.6029;
%...

%...Pass the input data to the function kepler_E, which returns E:
E = kepler_E(e, M);
```

```
%...Echo the input data and output to the command window:
fprintf('-----------------------------------------------------')
fprintf('\n Example 3.2\n')
fprintf('\n Eccentricity               = %g',e)
fprintf('\n Mean anomaly (radians)     = %g\n',M)
fprintf('\n Eccentric anomaly (radians) = %g',E)
fprintf('\n-----------------------------------------------------\n')

% ~~~~~~~~~~~~~~~~~~~~~~~~~~~~~~~~~~~~~~~~~~~~~~~~~~~~~~~~~~~~~~~
```

Output from `Example_3_02`

```
-----------------------------------------------------
Example 3.2

Eccentricity                = 0.37255
Mean anomaly (radians)      = 3.6029

Eccentric anomaly (radians) = 3.47942
-----------------------------------------------------
```

D.3 Algorithm 3.2: solution of Kepler's equation for the hyperbola using Newton's method

Function file `kepler_H.m`

```
% ~~~~~~~~~~~~~~~~~~~~~~~~~~~~~~~~~~~~~~~~~~~~~~~~~~~~~~~~~~~~~~
  function F = kepler_H(e, M)
% ~~~~~~~~~~~~~~~~~~~~~~~~~~~~
%
% This function uses Newton's method to solve Kepler's
% equation for the hyperbola  e*sinh(F) - F = M  for the
% hyperbolic eccentric anomaly, given the eccentricity and
% the hyperbolic mean anomaly.
%
% F - hyperbolic eccentric anomaly (radians)
% e - eccentricity, passed from the calling program
% M - hyperbolic mean anomaly (radians), passed from the
%     calling program
%
% User M-functions required: none
% ---------------------------------------------------------------

%...Set an error tolerance:
error = 1.e-8;

%...Starting value for F:
F = M;
```

```
%...Iterate on Equation 3.42 until F is determined to within
%...the error tolerance:
ratio = 1;
while abs(ratio) > error
    ratio = (e*sinh(F) - F - M)/(e*cosh(F) - 1);
    F = F - ratio;
end
% ~~~~~~~~~~~~~~~~~~~~~~~~~~~~~~~~~~~~~~~~~~~~~~~~~~~~~~~~~~~~
```

SCRIPT FILE Example_3_05.m

```
% ~~~~~~~~~~~~~~~~~~~~~~~~~~~~~~~~~~~~~~~~~~~~~~~~~~~~~~~~~~~~
% Example_3_05
%  ~~~~~~~~~~~~
%
% This program uses Algorithm 3.2 and the data of
% Example 3.5 to solve Kepler's equation for the hyperbola.
%
% e - eccentricity
% M - hyperbolic mean anomaly (dimensionless)
% F - hyperbolic eccentric anomaly (dimensionless)
%
% User M-function required: kepler_H
% -------------------------------------------------------------

clear

%...Input data for Example 3.5:
e = 2.7696;
M = 40.69;
%...

%...Pass the input data to the function kepler_H, which returns F:
F = kepler_H(e, M);

%...Echo the input data and output to the command window:
fprintf('-----------------------------------------------------')
fprintf('\n Example 3.5\n')
fprintf('\n Eccentricity                   = %g',e)
fprintf('\n Hyperbolic mean anomaly        = %g\n',M)
fprintf('\n Hyperbolic eccentric anomaly = %g',F)
fprintf('\n-----------------------------------------------------\n')
% ~~~~~~~~~~~~~~~~~~~~~~~~~~~~~~~~~~~~~~~~~~~~~~~~~~~~~~~~~~~~
```

Output from Example_3_05

```
-----------------------------------------------
 Example 3.5

 Eccentricity                  = 2.7696
 Hyperbolic mean anomaly       = 40.69

 Hyperbolic eccentric anomaly = 3.46309
-----------------------------------------------
```

D.4 Calculation of the Stumpff functions $S(z)$ and $C(z)$

The following scripts implement Equations 3.49 and 3.50 for use in other programs.

FUNCTION FILE stumpS.m

```matlab
% ~~~~~~~~~~~~~~~~~~~~~~~~~~~~~~~~~~~~~~~~~~~~~~~~~~~~~~~~~~~~~~
  function s = stumpS(z)
% ~~~~~~~~~~~~~~~~~~~~~~
%
% This function evaluates the Stumpff function S(z) according
% to Equation 3.49.
%
% z - input argument
% s - value of S(z)
%
% User M-functions required: none
% ---------------------------------------------------------------

if z > 0
    s = (sqrt(z) - sin(sqrt(z)))/(sqrt(z))^3;
elseif z < 0
    s = (sinh(sqrt(-z)) - sqrt(-z))/(sqrt(-z))^3;
else
    s = 1/6;
end
% ~~~~~~~~~~~~~~~~~~~~~~~~~~~~~~~~~~~~~~~~~~~~~~~~~~~~~~~~~~~~~~
```

FUNCTION FILE stumpC.m

```matlab
% ~~~~~~~~~~~~~~~~~~~~~~~~~~~~~~~~~~~~~~~~~~~~~~~~~~~~~~~~~~~~~~
  function c = stumpC(z)
% ~~~~~~~~~~~~~~~~~~~~~~
%
% This function evaluates the Stumpff function C(z) according
% to Equation 3.50.
%
% z - input argument
% c - value of C(z)
%
% User M-functions required: none
% ---------------------------------------------------------------

if z > 0
    c = (1 - cos(sqrt(z)))/z;
elseif z < 0
    c = (cosh(sqrt(-z)) - 1)/(-z);
else
    c = 1/2;
end
% ~~~~~~~~~~~~~~~~~~~~~~~~~~~~~~~~~~~~~~~~~~~~~~~~~~~~~~~~~~~~~~
```

D.5 Algorithm 3.3: solution of the universal Kepler's equation using Newton's method

FUNCTION FILE kepler_U.m

```
%  ~~~~~~~~~~~~~~~~~~~~~~~~~~~~~~~~~~~~~~~~~~~~~~~~~~~~~~~~~~~~~~
   function x = kepler_U(dt, ro, vro, a)
%  ~~~~~~~~~~~~~~~~~~~~~~~~~~~~~~~~~~~~~
%
% This function uses Newton's method to solve the universal
% Kepler equation for the universal anomaly.
%
% mu   - gravitational parameter (km^3/s^2)
% x    - the universal anomaly (km^0.5)
% dt   - time since x = 0 (s)
% ro   - radial position (km) when x = 0
% vro  - radial velocity (km/s) when x = 0
% a    - reciprocal of the semimajor axis (1/km)
% z    - auxiliary variable (z = a*x^2)
% C    - value of Stumpff function C(z)
% S    - value of Stumpff function S(z)
% n    - number of iterations for convergence
% nMax - maximum allowable number of iterations
%
% User M-functions required: stumpC, stumpS
% -----------------------------------------------------------
global mu

%...Set an error tolerance and a limit on the number of
%   iterations:
error = 1.e-8;
nMax  = 1000;

%...Starting value for x:
x = sqrt(mu)*abs(a)*dt;

%...Iterate on Equation 3.62 until convergence occurs within
%...the error tolerance:
n     = 0;
ratio = 1;
while abs(ratio) > error & n <= nMax
    n    = n + 1;
    C    = stumpC(a*x^2);
    S    = stumpS(a*x^2);
    F    = ro*vro/sqrt(mu)*x^2*C + (1 - a*ro)*x^3*S + ro*x-...
           sqrt(mu)*dt;
    dFdx = ro*vro/sqrt(mu)*x*(1 - a*x^2*S)+...
           (1 - a*ro)*x^2*C+ro;

ratio = F/dFdx;
    x    = x - ratio;
end

%...Deliver a value for x, but report that nMax was reached:
```

```
            if n > nMax
                fprintf('\n **No. iterations of Kepler''s equation')
                fprintf(' = %g', n)
                fprintf('\n    F/dFdx                      = %g\n', F/dFdx)
            end
            % ~~~~~~~~~~~~~~~~~~~~~~~~~~~~~~~~~~~~~~~~~~~~~~~~~~~~~~~~~~~~~
```

SCRIPT FILE Example_3_06.m

```
            % ~~~~~~~~~~~~~~~~~~~~~~~~~~~~~~~~~~~~~~~~~~~~~~~~~~~~~~~~~~~~~
            % Example_3_06
            % ~~~~~~~~~~~~
            %
            % This program uses Algorithm 3.3 and the data of Example 3.6
            % to solve the universal Kepler's equation.
            %
            % mu  - gravitational parameter (km^3/s^2)
            % x   - the universal anomaly (km^0.5)
            % dt  - time since x = 0 (s)
            % ro  - radial position when x = 0 (km)
            % vro - radial velocity when x = 0 (km/s)
            % a   - semimajor axis (km)
            %
            % User M-function required: kepler_U
            % ---------------------------------------------------------------

            clear
            global mu
            mu = 398600;

            %...Input data for Example 3.6:
            ro  = 10000;
            vro = 3.0752;
            dt  = 3600;
            a   = -19655;
            %...

            %...Pass the input data to the function kepler_U, which returns x
            %...(Universal Kepler's requires the reciprocal of
            %    semimajor axis):
            x   = kepler_U(dt, ro, vro, 1/a);

            %...Echo the input data and output the results to the command window:
            fprintf('-----------------------------------------------------')
            fprintf('\n Example 3.6\n')
            fprintf('\n Initial radial coordinate (km) = %g',ro)
            fprintf('\n Initial radial velocity (km/s) = %g',vro)
            fprintf('\n Elapsed time (seconds)         = %g',dt)
            fprintf('\n Semimajor axis (km)            = %g\n',a)
            fprintf('\n Universal anomaly (km^0.5)     = %g',x)
            fprintf('\n-----------------------------------------------------\n')

            % ~~~~~~~~~~~~~~~~~~~~~~~~~~~~~~~~~~~~~~~~~~~~~~~~~~~~~~~~~~~~~
```

Output from `Example_3_06`

```
-----------------------------------------------------
Example 3.6

Initial radial coordinate (km) = 10000
Initial radial velocity (km/s) = 3.0752
Elapsed time (seconds)         = 3600
Semimajor axis (km)            = -19655

Universal anomaly (km^0.5)     = 128.511
-----------------------------------------------------
```

D.6 Calculation of the Lagrange coefficients f and g and their time derivatives

The following scripts implement Equations 3.66 for use in other programs.

FUNCTION FILE f_and_g.m

```
% ~~~~~~~~~~~~~~~~~~~~~~~~~~~~~~~~~~~~~~~~~~~~~~~~~~~~~~~~~~~~~~~~
  function [f, g] = f_and_g(x, t, ro, a)
% ~~~~~~~~~~~~~~~~~~~~~~~~~~~~~~~~~~~~~~
%
% This function calculates the Lagrange f and g coefficients.
%
% mu - the gravitational parameter (km^3/s^2)
% a  - reciprocal of the semimajor axis (1/km)
% ro - the radial position at time t (km)
% t  - the time elapsed since t (s)
% x  - the universal anomaly after time t (km^0.5)
% f  - the Lagrange f coefficient (dimensionless)
% g  - the Lagrange g coefficient (s)
%
% User M-functions required:   stumpC, stumpS
% ----------------------------------------------------------------

global mu

z = a*x^2;

%...Equation 3.66a:
f = 1 - x^2/ro*stumpC(z);

%...Equation 3.66b:
g = t - 1/sqrt(mu)*x^3*stumpS(z);
% ~~~~~~~~~~~~~~~~~~~~~~~~~~~~~~~~~~~~~~~~~~~~~~~~~~~~~~~~~~~~~~~~
```

FUNCTION FILE fDot_and_gDot.m

```
% ~~~~~~~~~~~~~~~~~~~~~~~~~~~~~~~~~~~~~~~~~~~~~~~~~~~~~~~~~~~~~~~~
  function [fdot, gdot] = fDot_and_gDot(x, r, ro, a)
```

```
%  ~~~~~~~~~~~~~~~~~~~~~~~~~~~~~~~~~~~~~~~~~~~~~~~~~~~~~~~
%
% This function calculates the time derivatives of the
% Lagrange f and g coefficients.
%
% mu     - the gravitational parameter (km^3/s^2)
% a      - reciprocal of the semimajor axis (1/km)
% ro     - the radial position at time t (km)
% t      - the time elapsed since initial state vector (s)
% r      - the radial position after time t (km)
% x      - the universal anomaly after time t (km^0.5)
% fDot   - time derivative of the Lagrange f coefficient (1/s)
% gDot   - time derivative of the Lagrange g coefficient
%          (dimensionless)
%
% User M-functions required:   stumpC, stumpS
% -------------------------------------------------------------

global mu

z = a*x^2;

%...Equation 3.66c:
fdot = sqrt(mu)/r/ro*(z*stumpS(z) - 1)*x;

%...Equation 3.66d:
gdot = 1 - x^2/r*stumpC(z);

%  ~~~~~~~~~~~~~~~~~~~~~~~~~~~~~~~~~~~~~~~~~~~~~~~~~~~~~~~
```

D.7 Algorithm 3.4: calculation of the state vector (r, v) given the initial state vector (r_0, v_0) and the time lapse Δt

Function File rv_from_r0v0.m

```
%  ~~~~~~~~~~~~~~~~~~~~~~~~~~~~~~~~~~~~~~~~~~~~~~~~~~~~~~~~~~~
   function [R,V] = rv_from_r0v0(R0, V0, t)
%  ~~~~~~~~~~~~~~~~~~~~~~~~~~~~~~~~~~~~~~~~~
% This function computes the state vector (R,V) from the
% initial state vector (R0,V0) and the elapsed time.
%
% mu - gravitational parameter (km^3/s^2)
% R0 - initial position vector (km)
% V0 - initial velocity vector (km/s)
% t  - elapsed time (s)
% R  - final position vector (km)
% V  - final velocity vector (km/s)
%
% User M-functions required: kepler_U, f_and_g, fDot_and_gDot
% -------------------------------------------------------------
```

```
            global  mu

            %...Magnitudes of R0 and V0:
            r0         = norm(R0);
            v0         = norm(V0);

            %...Initial radial velocity:
            vr0        = dot(R0, V0)/r0;

            %...Reciprocal of the semimajor axis (from the energy equation):
            alpha      = 2/r0 - v0^2/mu;

            %...Compute the universal anomaly:
            x          = kepler_U(t, r0, vr0, alpha);

            %...Compute the f and g functions:
            [f, g]     = f_and_g(x, t, r0, alpha);

            %...Compute the final position vector:
            R          = f*R0 + g*V0;

            %...Compute the magnitude of R:
            r          = norm(R);

            %...Compute the derivatives of f and g:
            [fdot, gdot] = fDot_and_gDot(x, r, r0, alpha);

            %...Compute the final velocity:
            V          = fdot*R0 + gdot*V0;
            % ~~~~~~~~~~~~~~~~~~~~~~~~~~~~~~~~~~~~~~~~~~~~~~~~~~~~~~~~~~~~~~~
```

SCRIPT FILE Example_3_07.m

```
            % ~~~~~~~~~~~~~~~~~~~~~~~~~~~~~~~~~~~~~~~~~~~~~~~~~~~~~~~~~~~~~~~
            % Example_3_07
            % ~~~~~~~~~~~~
            %
            % This program computes the state vector (R,V) from the
            % initial state vector (R0,V0) and the elapsed time using the
            % data in Example 3.7.
            %
            % mu - gravitational parameter (km^3/s^2)
            % R0 - the initial position vector (km)
            % V0 - the initial velocity vector (km/s)
            % R  - the final position vector (km)
            % V  - the final velocity vector (km/s)
            % t  - elapsed time (s)
            %
            % User M-functions required: rv_from_r0v0
            % -----------------------------------------------------------

            clear
            global mu
            mu = 398600;
```

```
%...Input data for Example 3.7:
R0 = [  7000 -12124 0];
V0 = [2.6679 4.6210 0];
t  = 3600;
%...

%...Algorithm 3.4:
[R V] = rv_from_r0v0(R0, V0, t);

%...Echo the input data and output the results to the command window:
fprintf('-----------------------------------------------------')
fprintf('\n Example 3.7\n')
fprintf('\n Initial position vector (km):')
fprintf('\n   r0 = (%g, %g, %g)\n', R0(1), R0(2), R0(3))
fprintf('\n Initial velocity vector (km/s):')
fprintf('\n   v0 = (%g, %g, %g)', V0(1), V0(2), V0(3))
fprintf('\n\n Elapsed time = %g s\n',t)
fprintf('\n Final position vector (km):')
fprintf('\n   r = (%g, %g, %g)\n', R(1), R(2), R(3))
fprintf('\n Final velocity vector (km/s):')
fprintf('\n   v = (%g, %g, %g)', V(1), V(2), V(3))
fprintf('\n-----------------------------------------------------\n')
% ~~~~~~~~~~~~~~~~~~~~~~~~~~~~~~~~~~~~~~~~~~~~~~~~~~~~~~~~~~~~~
```

Output from `Example_3_07`

```
-----------------------------------------------------
 Example 3.7

 Initial position vector (km):
   r0 = (7000, -12124, 0)

 Initial velocity vector (km/s):
   v0 = (2.6679, 4.621, 0)

 Elapsed time = 3600 s

 Final position vector (km):
   r = (-3297.77, 7413.4, 0)

 Final velocity vector (km/s):
   v = (-8.2976, -0.964045, -0)
-----------------------------------------------------
```

D.8 ALGORITHM 4.1: CALCULATION OF THE ORBITAL ELEMENTS FROM THE STATE VECTOR

FUNCTION FILE `coe_from_sv.m`

```
% ~~~~~~~~~~~~~~~~~~~~~~~~~~~~~~~~~~~~~~~~~~~~~~~~~~~~~~~~~~~~
  function coe = coe_from_sv(R,V)
% ~~~~~~~~~~~~~~~~~~~~~~~~~~~~~~~
%
```

D.8 Algorithm 4.1: calculation of the orbital elements from the state vector

```
% This function computes the classical orbital elements (coe)
% from the state vector (R,V) using Algorithm 4.1.
%
% mu    - gravitational parameter (km^3/s^2)
% R     - position vector in the geocentric equatorial frame
%         (km)
% V     - velocity vector in the geocentric equatorial frame
%         (km)
% r, v  - the magnitudes of R and V
% vr    - radial velocity component (km/s)
% H     - the angular momentum vector (km^2/s)
% h     - the magnitude of H (km^2/s)
% incl  - inclination of the orbit (rad)
% N     - the node line vector (km^2/s)
% n     - the magnitude of N
% cp    - cross product of N and R
% RA    - right ascension of the ascending node (rad)
% E     - eccentricity vector
% e     - eccentricity (magnitude of E)
% eps   - a small number below which the eccentricity is
%         considered to be zero
% w     - argument of perigee (rad)
% TA    - true anomaly (rad)
% a     - semimajor axis (km)
% pi    - 3.1415926...
% coe   - vector of orbital elements [h e RA incl w TA a]
%
% User M-functions required: None
% -------------------------------------------------------------

global mu;
eps = 1.e-10;

r     = norm(R);
v     = norm(V);

vr    = dot(R,V)/r;

H     = cross(R,V);
h     = norm(H);

%...Equation 4.7:
incl = acos(H(3)/h);

%...Equation 4.8:
N     = cross([0 0 1],H);
n     = norm(N);

%...Equation 4.9:
if n ~= 0
    RA = acos(N(1)/n);
    if N(2) < 0
        RA = 2*pi - RA;
    end
else
    RA = 0;
end
```

```
%...Equation 4.10:
E = 1/mu*((v^2 - mu/r)*R - r*vr*V);
e = norm(E);

%...Equation 4.12 (incorporating the case e = 0):
if n ~= 0
    if e > eps
        w = acos(dot(N,E)/n/e);
        if E(3) < 0
            w = 2*pi - w;
        end
    else
        w = 0;
    end
else
    w = 0;
end

%...Equation 4.13a (incorporating the case e = 0):
if e > eps
    TA = acos(dot(E,R)/e/r);
    if vr < 0
        TA = 2*pi - TA;
    end
else
    cp = cross(N,R);
    if cp(3) >= 0
        TA = acos(dot(N,R)/n/r);
    else
        TA = 2*pi - acos(dot(N,R)/n/r);
    end
end

%...Equation 2.61 (a < 0 for a hyperbola):
a = h^2/mu/(1 - e^2);

coe = [h e RA incl w TA a];
% ~~~~~~~~~~~~~~~~~~~~~~~~~~~~~~~~~~~~~~~~~~~~~~~~~~~~~~~~~~~~~~~~
```

SCRIPT FILE Example_4_03.m

```
% ~~~~~~~~~~~~~~~~~~~~~~~~~~~~~~~~~~~~~~~~~~~~~~~~~~~~~~~~~~~~~~~~
% Example_4_03
% ~~~~~~~~~~~~~
%
% This program uses Algorithm 4.1 to obtain the orbital
% elements from the state vector provided in Example 4.3.
%
% pi    - 3.1415926...
% deg   - factor for converting between degrees and radians
% mu    - gravitational parameter (km^3/s^2)
% r     - position vector (km) in the geocentric equatorial
%         frame
% v     - velocity vector (km/s) in the geocentric equatorial
%         frame
```

D.8 Algorithm 4.1: calculation of the orbital elements from the state vector

```
% coe   - orbital elements [h e RA incl w TA a]
%         where h    = angular momentum (km^2/s)
%               e    = eccentricity
%               RA   = right ascension of the ascending node
%                      (rad)
%               incl = orbit inclination (rad)
%               w    = argument of perigee (rad)
%               TA   = true anomaly (rad)
%               a    = semimajor axis (km)
% T     - Period of an elliptic orbit (s)
%
% User M-function required: coe_from_sv
% -----------------------------------------------------------

clear
global mu
deg = pi/180;
mu  = 398600;

%...Input data:
r = [ -6045  -3490   2500];
v = [-3.457   6.618  2.533];
%...

%...Algorithm 4.1:
coe = coe_from_sv(r,v);

%...Echo the input data and output results to the command window:
fprintf('-------------------------------------------------')
fprintf('\n Example 4.3\n')
fprintf('\n Gravitational parameter (km^3/s^2) = %g\n', mu)
fprintf('\n State vector:\n')
fprintf('\n r (km)                      = [%g  %g  %g]', ...
                                           r(1), r(2), r(3))
fprintf('\n v (km/s)                    = [%g  %g  %g]', ...
                                           v(1), v(2), v(3))
disp(' ')
fprintf('\n Angular momentum (km^2/s)   = %g', coe(1))
fprintf('\n Eccentricity                = %g', coe(2))
fprintf('\n Right ascension (deg)       = %g', coe(3)/deg)
fprintf('\n Inclination (deg)           = %g', coe(4)/deg)
fprintf('\n Argument of perigee (deg)   = %g', coe(5)/deg)
fprintf('\n True anomaly (deg)          = %g', coe(6)/deg)
fprintf('\n Semimajor axis (km):        = %g', coe(7))

%...if the orbit is an ellipse, output its period:
if coe(2)<1
    T = 2*pi/sqrt(mu)*coe(7)^1.5; % Equation 2.73
    fprintf('\n Period:')
    fprintf('\n   Seconds                 = %g', T)
    fprintf('\n   Minutes                 = %g', T/60)
    fprintf('\n   Hours                   = %g', T/3600)
    fprintf('\n   Days                    = %g', T/24/3600)
end

fprintf('\n-------------------------------------------------\n')
%  ~~~~~~~~~~~~~~~~~~~~~~~~~~~~~~~~~~~~~~~~~~~~~~~~~~~~~~~~~~
```

Output from Example_4_03

```
-----------------------------------------------------
Example 4.3

Gravitational parameter (km^3/s^2) = 398600

State vector:

r (km)                                = [-6045  -3490  2500]
v (km/s)                              = [-3.457  6.618  2.533]

Angular momentum (km^2/s)             = 58311.7
Eccentricity                          = 0.171212
Right ascension (deg)                 = 255.279
Inclination (deg)                     = 153.249
Argument of perigee (deg)             = 20.0683
True anomaly (deg)                    = 28.4456
Semimajor axis (km):                  = 8788.1
Period:
  Seconds                             = 8198.86
  Minutes                             = 136.648
  Hours                               = 2.27746
  Days                                = 0.0948942
-----------------------------------------------------
```

D.9 Algorithm 4.2: calculation of the state vector from the orbital elements

Function file sv_from_coe.m

```
% ~~~~~~~~~~~~~~~~~~~~~~~~~~~~~~~~~~~~~~~~~~~~~~~~~~~~~~~~~~~~~~~
  function [r, v] = sv_from_coe(coe)
% ~~~~~~~~~~~~~~~~~~~~~~~~~~~~~~~~~~
% This function computes the state vector (r,v) from the
% classical orbital elements (coe).
%
% mu   - gravitational parameter (km^3; s^2)
% coe  - orbital elements [h e RA incl w TA]
%        where
%          h    = angular momentum (km^2/s)
%          e    = eccentricity
%          RA   = right ascension of the ascending node (rad)
%          incl = inclination of the orbit (rad)
%          w    = argument of perigee (rad)
%          TA   = true anomaly (rad)
% R3_w - Rotation matrix about the z-axis through the angle w
% R1_i - Rotation matrix about the x-axis through the angle i
% R3_W - Rotation matrix about the z-axis through the angle RA
% Q_pX - Matrix of the transformation from perifocal to
%        geocentric equatorial frame
% rp   - position vector in the perifocal frame (km)
% vp   - velocity vector in the perifocal frame (km/s)
```

```
%   r    - position vector in the geocentric equatorial frame
%          (km)
%   v    - velocity vector in the geocentric equatorial frame
%          (km/s)
%
% User M-functions required: none
% -------------------------------------------------------------

global mu

h    = coe(1);
e    = coe(2);
RA   = coe(3);
incl = coe(4);
w    = coe(5);
TA   = coe(6);

%...Equations 4.37 and 4.38 (rp and vp are column vectors):
rp = (h^2/mu) * (1/(1 + e*cos(TA))) * (cos(TA)*[1;0;0] ...
     + sin(TA)*[0;1;0]);
vp = (mu/h) * (-sin(TA)*[1;0;0] + (e + cos(TA))*[0;1;0]);

%...Equation 4.39:
R3_W = [ cos(RA)   sin(RA)   0
        -sin(RA)   cos(RA)   0
             0         0     1];

%...Equation 4.40:
R1_i = [1      0            0
        0   cos(incl)   sin(incl)
        0  -sin(incl)   cos(incl)];

%...Equation 4.41:
R3_w = [ cos(w)   sin(w)   0
        -sin(w)   cos(w)   0
            0        0     1];

%...Equation 4.44:
Q_pX = R3_W'*R1_i'*R3_w';

%...Equations 4.46 (r and v are column vectors):
r = Q_pX*rp;
v = Q_pX*vp;

%...Convert r and v into row vectors:
r = r';
v = v';
% ~~~~~~~~~~~~~~~~~~~~~~~~~~~~~~~~~~~~~~~~~~~~~~~~~~~~~~~~~~~~~~
```

SCRIPT FILE Example_4_05.m

```
% ~~~~~~~~~~~~~~~~~~~~~~~~~~~~~~~~~~~~~~~~~~~~~~~~~~~~~~~~~~~~~~
% Example_4_05
```

```
% ~~~~~~~~~~~
%
% This program uses Algorithm 4.2 to obtain the state vector
% from the orbital elements provided in Example 4.5.
%
% pi   - 3.1415926...
% deg  - factor for converting between degrees and radians
% mu   - gravitational parameter (km^3/s^2)
% coe  - orbital elements [h e RA incl w TA a]
%        where h    = angular momentum (km^2/s)
%              e    = eccentricity
%              RA   = right ascension of the ascending node
%                     (rad)
%              incl = orbit inclination (rad)
%              w    = argument of perigee (rad)
%              TA   = true anomaly (rad)
%              a    = semimajor axis (km)
% r    - position vector (km) in geocentric equatorial frame
% v    - velocity vector (km) in geocentric equatorial frame
%
% User M-functions required: sv_from_coe
% -------------------------------------------------------------

clear
global mu
deg = pi/180;
mu  = 398600;

%...Input data (angles in degrees):
h    = 80000;
e    = 1.4;
RA   = 40;
incl = 30;
w    = 60;
TA   = 30;
%...

coe = [h, e, RA*deg, incl*deg, w*deg, TA*deg];

%...Algorithm 4.2 (requires angular elements be in radians):
[r, v] = sv_from_coe(coe);

%...Echo the input data and output the results to the command window:
fprintf('-----------------------------------------------')
fprintf('\n Example 4.5\n')
fprintf('\n Gravitational parameter (km^3/s^2)  = %g\n', mu)
fprintf('\n Angular momentum (km^2/s)           = %g', h)
fprintf('\n Eccentricity                        = %g', e)
fprintf('\n Right ascension (deg)               = %g', RA)
fprintf('\n Argument of perigee (deg)           = %g', w)
fprintf('\n True anomaly (deg)                  = %g', TA)
fprintf('\n\n State vector:')
fprintf('\n   r (km)   = [%g  %g  %g]', r(1), r(2), r(3))
fprintf('\n   v (km/s) = [%g  %g  %g]', v(1), v(2), v(3))
fprintf('\n-----------------------------------------------\n')

% ~~~~~~~~~~~~~~~~~~~~~~~~~~~~~~~~~~~~~~~~~~~~~~~~~~~~~~~~~~~~~
```

Output from `Example_4_05`

```
--------------------------------------------------------
 Example 4.5

 Gravitational parameter (km^3/s^2)   = 398600

 Angular momentum (km^2/s)            = 80000
 Eccentricity                         = 1.4
 Right ascension (deg)                = 40
 Argument of perigee (deg)            = 60
 True anomaly (deg)                   = 30

 State vector:
   r (km)   = [-4039.9   4814.56   3628.62]
   v (km/s) = [-10.386   -4.77192  1.74388]
--------------------------------------------------------
```

D.10 Algorithm 5.1: Gibbs' method of preliminary orbit determination

FUNCTION FILE `gibbs.m`

```
% ~~~~~~~~~~~~~~~~~~~~~~~~~~~~~~~~~~~~~~~~~~~~~~~~~~~~~~~~~~
  function [V2, ierr] = gibbs(R1, R2, R3)
% ~~~~~~~~~~~~~~~~~~~~~~~~~~~~~~~~~~~~~~~
%
% This function uses the Gibbs method of orbit determination
% to compute the velocity corresponding to the second of
% three supplied position vectors.
%
% mu             - gravitational parameter (km^3/s^2)
% R1, R2, R3     - three coplanar geocentric position vectors
%                  (km)
% r1, r2, r3     - the magnitudes of R1, R2 and R3 (km)
% c12, c23, c31  - three independent cross products among
%                  R1, R2 and R3
% N, D, S        - vectors formed from R1, R2 and R3 during
%                  the Gibbs' procedure
% tol            - tolerance for determining if R1, R2 and R3
%                  are coplanar
% ierr           - = 0 if R1, R2, R3 are found to be coplanar
%                  = 1 otherwise
% V2             - the velocity corresponding to R2 (km/s)
%
% User M-functions required: none
% ----------------------------------------------------------

global mu
tol  = 1e-4;
ierr = 0;
```

```
%...Magnitudes of R1, R2 and R3:
r1 = norm(R1);
r2 = norm(R2);
r3 = norm(R3);

%...Cross products among R1, R2 and R3:
c12 = cross(R1,R2);
c23 = cross(R2,R3);
c31 = cross(R3,R1);

%...Check that R1, R2 and R3 are coplanar; if not set error flag:
if abs(dot(R1,c23)/r1/norm(c23)) > tol
    ierr = 1;
end

%...Equation 5.13:
N = r1*c23 + r2*c31 + r3*c12;

%...Equation 5.14:
D = c12 + c23 + c31;

%...Equation 5.21:
S = R1*(r2 - r3) + R2*(r3 - r1) + R3*(r1 - r2);

%...Equation 5.22:
V2 = sqrt(mu/norm(N)/norm(D))*(cross(D,R2)/r2 + S);
%
```

SCRIPT FILE Example_5_01.m

```
% ~~~~~~~~~~~~~~~~~~~~~~~~~~~~~~~~~~~~~~~~~~~~~~~~~~~~~~~~~~~~~~~~~
% Example_5_01
%  ~~~~~~~~~~~~
%
% This program uses Algorithm 5.1 (Gibbs' method) and
% Algorithm 4.1 to obtain the orbital elements from the data
% provided in Example 5.1.
%
% deg         - factor for converting between degrees and
%               radians
% pi          - 3.1415926...
% mu          - gravitational parameter (km^3/s^2)
% r1, r2, r3  - three coplanar geocentric position vectors (km)
% ierr        - 0 if r1, r2, r3 are found to be coplanar
%               1 otherwise
% v2          - the velocity corresponding to r2 (km/s)
% coe         - orbital elements [h e RA incl w TA a]
%               where h    = angular momentum (km^2/s)
%                     e    = eccentricity
%                     RA   = right ascension of the ascending
%                            node (rad)
%                     incl = orbit inclination (rad)
%                     w    = argument of perigee (rad)
```

```
%                       TA    = true anomaly (rad)
%                       a     = semimajor axis (km)
% T              - period of elliptic orbit (s)
%
% User M-functions required: gibbs, coe_from_sv
% ---------------------------------------------------------

clear
deg = pi/180;
global mu

%...Input data for Example 5.1:
mu = 398600;
r1 = [-294.32 4265.1 5986.7];
r2 = [-1365.4 3637.6 6346.8];
r3 = [-2940.3 2473.7 6555.8];
%...

%...Echo the input data to the command window:
fprintf('---------------------------------------------------')
fprintf('\n Example 5.1: Gibbs Method\n')
fprintf('\n\n Input data:\n')
fprintf('\n  Gravitational parameter (km^3/s^2)  = %g\n', mu)
fprintf('\n  r1 (km) = [%g  %g  %g]', r1(1), r1(2), r1(3))
fprintf('\n  r2 (km) = [%g  %g  %g]', r2(1), r2(2), r2(3))
fprintf('\n  r3 (km) = [%g  %g  %g]', r3(1), r3(2), r3(3))
fprintf('\n\n')
%...Algorithm 5.1:
[v2, ierr] = gibbs(r1, r2, r3);

%...If the vectors r1, r2, r3, are not coplanar, abort:
if ierr == 1
    fprintf('\n  These vectors are not coplanar.\n\n')
    return
end

%...Algorithm 4.1
coe  = coe_from_sv(r2,v2);

h    = coe(1);
e    = coe(2);
RA   = coe(3);
incl = coe(4);
w    = coe(5);
TA   = coe(6);
a    = coe(7);

%...Output the results to the command window:
fprintf(' Solution:')
fprintf('\n');
fprintf('\n  v2 (km/s) = [%g  %g  %g]', v2(1), v2(2), v2(3))
fprintf('\n\n  Orbital elements:')
fprintf('\n    Angular momentum (km^2/s)     = %g', h)
fprintf('\n    Eccentricity                  = %g', e)
fprintf('\n    Inclination (deg)             = %g', incl/deg)
fprintf('\n    RA of ascending node (deg)    = %g', RA/deg)
fprintf('\n    Argument of perigee (deg)     = %g', w/deg)
fprintf('\n    True anomaly (deg)            = %g', TA/deg)
```

```
            fprintf('\n   Semimajor axis (km)        = %g', a)
%...If the orbit is an ellipse, output the period:
if e < 1
    T = 2*pi/sqrt(mu)*coe(7)^1.5;
    fprintf('\n   Period (s)                 = %g', T)
end
fprintf('\n-----------------------------------------------\n')
% ~~~~~~~~~~~~~~~~~~~~~~~~~~~~~~~~~~~~~~~~~~~~~~~~~~~~~~~~~~~~~
```

Output from Example_5_01

```
-----------------------------------------------------
 Example 5.1: Gibbs Method

 Input data:

  Gravitational parameter (km^3/s^2)  = 398600

  r1 (km) = [-294.32   4265.1   5986.7]
  r2 (km) = [-1365.4   3637.6   6346.8]
  r3 (km) = [-2940.3   2473.7   6555.8]
 Solution:

  v2 (km/s) = [-6.2176   -4.01237   1.59915]

  Orbital elements:
    Angular momentum (km^2/s)  = 56193
    Eccentricity               = 0.100159
    Inclination (deg)          = 60.001
    RA of ascending node (deg) = 40.0023
    Argument of perigee (deg)  = 30.1093
    True anomaly (deg)         = 49.8894
    Semimajor axis (km)        = 8002.14
    Period (s)                 = 7123.94
-----------------------------------------------------
```

D.11 Algorithm 5.2: solution of Lambert's problem

FUNCTION FILE lambert.m

```
% ~~~~~~~~~~~~~~~~~~~~~~~~~~~~~~~~~~~~~~~~~~~~~~~~~~~~~~~~~~~~~
  function [V1, V2] = lambert(R1, R2, t, string)
% ~~~~~~~~~~~~~~~~~~~~~~~~~~~~~~~~~~~~~~~~~~~~~~~~~~~~~~~~~~~~~
%
% This function solves Lambert's problem.
%
% mu         - gravitational parameter (km^3/s^2)
% R1, R2     - initial and final position vectors (km)
% r1, r2     - magnitudes of R1 and R2
% t          - the time of flight from R1 to R2
%              (a constant) (s)
```

```
% V1, V2     - initial and final velocity vectors (km/s)
% c12        - cross product of R1 into R2
% theta      - angle between R1 and R2
% string     - 'pro' if the orbit is prograde
%              'retro' if the orbit is retrograde
% A          - a constant given by Equation 5.35
% z          - alpha*x^2, where alpha is the reciprocal of the
%              semimajor axis and x is the universal anomaly
% y(z)       - a function of z given by Equation 5.38
% F(z,t)     - a function of the variable z and constant t,
%              given by Equation 5.40
% dFdz(z)    - the derivative of F(z,t), given by
%              Equation 5.43
% ratio      - F/dFdz
% tol        - tolerance on precision of convergence
% nmax       - maximum number of iterations of Newton's
%              procedure
% f, g       - Lagrange coefficients
% gdot       - time derivative of g
% C(z), S(z) - Stumpff functions
% dum        - a dummy variable
%
% User M-functions required: stumpC and stumpS
% --------------------------------------------------------

global mu
global r1 r2 A

%...Magnitudes of R1 and R2:
r1 = norm(R1);
r2 = norm(R2);

c12   = cross(R1, R2);
theta = acos(dot(R1,R2)/r1/r2);

%...Determine whether the orbit is prograde or retrograde:
if strcmp(string, 'pro')
    if c12(3) <= 0
        theta = 2*pi - theta;
    end
elseif strcmp(string,'retro')
    if c12(3) >= 0
        theta = 2*pi - theta;
    end
else
    string = 'pro'
    fprintf('\n ** Prograde trajectory assumed.\n')
end

%...Equation 5.35:
A = sin(theta)*sqrt(r1*r2/(1 - cos(theta)));

%...Determine approximately where F(z,t) changes sign, and
%...use that value of z as the starting value for Equation 5.45:
z = -100;
while F(z,t) < 0
    z = z + 0.1;
end
```

```
%...Set an error tolerance and a limit on the number of iterations:
tol   = 1.e-8;
nmax  = 5000;

%...Iterate on Equation 5.45 until z is determined to within
%...the error tolerance:
ratio = 1;
n     = 0;
while (abs(ratio) > tol) & (n <= nmax)
    n     = n + 1;
    ratio = F(z,t)/dFdz(z);
    z     = z - ratio;
end

%...Report if the maximum number of iterations is exceeded:
if n >= nmax
    fprintf('\n\n **Number of iterations exceeds')
    fprintf(' %g \n\n ', nmax)
end

%...Equation 5.46a:
f    = 1 - y(z)/r1;

%...Equation 5.46b:
g    = A*sqrt(y(z)/mu);

%...Equation 5.46d:
gdot = 1 - y(z)/r2;

%...Equation 5.28:
V1   = 1/g*(R2 - f*R1);

%...Equation 5.29:
V2   = 1/g*(gdot*R2 - R1);

return

% ~~~~~~~~~~~~~~~~~~~~~~~~~~~~~~~~~~~~~~~~~~~~~~~~~~~~~~~~~~~~~~~
% Subfunctions used in the main body:

%...Equation 5.38:
function dum = y(z)
    global r1 r2 A
    dum = r1 + r2 + A*(z*S(z) - 1)/sqrt(C(z));
return

%...Equation 5.40:
function dum = F(z,t)
    global mu A
    dum = (y(z)/C(z))^1.5*S(z) + A*sqrt(y(z)) - sqrt(mu)*t;
return

%...Equation 5.43:
function dum = dFdz(z)
    global A
    if z == 0
        dum = sqrt(2)/40*y(0)^1.5 + A/8*(sqrt(y(0)) ...
              + A*sqrt(1/2/y(0)));
```

```
        else
            dum = (y(z)/C(z))^1.5*(1/2/z*(C(z) - 3*S(z)/2/C(z)) ...
                  + 3*S(z)^2/4/C(z)) ...
                  + A/8*(3*S(z)/C(z)*sqrt(y(z)) ...
                  + A*sqrt(C(z)/y(z))));
        end
return

%...Stumpff functions:
function dum = C(z)
    dum = stumpC(z);
return
function dum = S(z)
    dum = stumpS(z);
return

% ~~~~~~~~~~~~~~~~~~~~~~~~~~~~~~~~~~~~~~~~~~~~~~~~~~~~~~~~~~~~~~~
```

SCRIPT FILE Example_5_02.m

```
% ~~~~~~~~~~~~~~~~~~~~~~~~~~~~~~~~~~~~~~~~~~~~~~~~~~~~~~~~~~~~~~~
% Example_5_02
%  ~~~~~~~~~~~~
%
% This program uses Algorithm 5.2 to solve Lambert's problem
% for the data provided in Example 5.2.
%
% deg    - factor for converting between degrees and radians
% pi     - 3.1415926...
% mu     - gravitational parameter (km^3/s^2)
% r1, r2 - initial and final position vectors (km)
% dt     - time between r1 and r2 (s)
% string - = 'pro' if the orbit is prograde
%          = 'retro' if the orbit is retrograde
% v1, v2 - initial and final velocity vectors (km/s)
% coe    - orbital elements [h e RA incl w TA a]
%          where h    = angular momentum (km^2/s)
%                e    = eccentricity
%                RA   = right ascension of the ascending node
%                       (rad)
%                incl = orbit inclination (rad)
%                w    = argument of perigee (rad)
%                TA   = true anomaly (rad)
%                a    = semimajor axis (km)
% TA1    - Initial true anomaly
% TA2    - Final true anomaly
% T      - period of an elliptic orbit
%
% User M-functions required: lambert, coe_from_sv
% ----------------------------------------------------------------

clear
global mu
deg = pi/180;
mu  = 398600;
```

```
%...Input data from Example 5.2:
r1     = [  5000   10000   2100];
r2     = [-14600    2500   7000];
dt     = 3600;
string = 'pro';
%...

%...Algorithm 5.2:
[v1, v2] = lambert(r1, r2, dt, string);

%...Algorithm 4.1 (using r1 and v1):
coe      = coe_from_sv(r1, v1);
%...Save the initial true anomaly:
TA1      = coe(6);

%...Algorithm 4.1 (using r2 and v2):
coe      = coe_from_sv(r2, v2);
%...Save the final true anomaly:
TA2      = coe(6);

%...Echo the input data and output the results to the command window:
fprintf('-------------------------------------------------')
fprintf('\n Example 5.2: Lambert''s Problem\n')
fprintf('\n\n Input data:\n');
fprintf('\n   Gravitational parameter (km^3/s^2) = %g\n', mu)
fprintf('\n   r1 (km)                       = [%g  %g  %g]', ...
                                               r1(1), r1(2), r1(3))
fprintf('\n   r2 (km)                       = [%g  %g  %g]', ...
                                               r2(1), r2(2), r2(3))
fprintf('\n   Elapsed time (s)              = %g', dt);
fprintf('\n\n Solution:\n')

fprintf('\n   v1 (km/s)                     = [%g  %g  %g]', ...
                                               v1(1), v1(2), v1(3))
fprintf('\n   v2 (km/s)                     = [%g  %g  %g]', ...
                                               v2(1), v2(2), v2(3))

fprintf('\n\n Orbital elements:')
fprintf('\n   Angular momentum (km^2/s)     = %g', coe(1))
fprintf('\n   Eccentricity                  = %g', coe(2))
fprintf('\n   Inclination (deg)             = %g', coe(4)/deg)
fprintf('\n   RA of ascending node (deg)    = %g', coe(3)/deg)
fprintf('\n   Argument of perigee (deg)     = %g', coe(5)/deg)
fprintf('\n   True anomaly initial (deg)    = %g', TA1/deg)
fprintf('\n   True anomaly final   (deg)    = %g', TA2/deg)
fprintf('\n   Semimajor axis (km)           = %g', coe(7))
fprintf('\n   Periapse radius (km)          = %g', ...
                   coe(1)^2/mu/(1 + coe(2)))
if coe(2)<1
    T = 2*pi/sqrt(mu)*coe(7)^1.5;
    fprintf('\n   Period:')
    fprintf('\n     Seconds                 = %g', T)
    fprintf('\n     Minutes                 = %g', T/60)
    fprintf('\n     Hours                   = %g', T/3600)
    fprintf('\n     Days                    = %g', T/24/3600)
end
fprintf('\n-------------------------------------------------\n')

% ~~~~~~~~~~~~~~~~~~~~~~~~~~~~~~~~~~~~~~~~~~~~~~~~~~~~~~~~~
```

Output from `Example_5_02`

```
-----------------------------------------------------
Example 5.2: Lambert's Problem

 Input data:

   Gravitational parameter (km^3/s^2) = 398600

   r1 (km)                 = [5000   10000   2100]
   r2 (km)                 = [-14600  2500   7000]
   Elapsed time (s)        = 3600

 Solution:

   v1 (km/s)               = [-5.99249  1.92536   3.24564]
   v2 (km/s)               = [-3.31246  -4.19662  -0.385288]

 Orbital elements:
   Angular momentum (km^2/s) = 80466.8
   Eccentricity              = 0.433488
   Inclination (deg)         = 30.191
   RA of ascending node (deg) = 44.6002
   Argument of perigee (deg) = 30.7062
   True anomaly initial (deg) = 350.83
   True anomaly final   (deg) = 91.1223
   Semimajor axis (km)       = 20002.9
   Periapse radius (km)      = 11331.9
   Period:
     Seconds                 = 28154.7
     Minutes                 = 469.245
     Hours                   = 7.82075
     Days                    = 0.325865
-----------------------------------------------------
```

D.12 CALCULATION OF JULIAN DAY NUMBER AT 0 HR UT

The following script implements Equation 5.48 for use in other programs.

FUNCTION FILE `J0.m`

```
% ~~~~~~~~~~~~~~~~~~~~~~~~~~~~~~~~~~~~~~~~~~~~~~~~~~~~~~~~~~~~~~
  function j0 = J0(year, month, day)
% ~~~~~~~~~~~~~~~~~~~~~~~~~~~~~~~~~~~
%
% This function computes the Julian day number at 0 UT for any
% year between 1900 and 2100 using Equation 5.48.
%
% j0    - Julian day at 0 hr UT (Universal Time)
% year  - range: 1901 - 2099
% month - range: 1 - 12
% day   - range: 1 - 31
```

```
%
% User M-functions required: none
% -----------------------------------------------------------

j0 = 367*year - fix(7*(year + fix((month + 9)/12))/4) ...
    + fix(275*month/9) + day + 1721013.5;
% ~~~~~~~~~~~~~~~~~~~~~~~~~~~~~~~~~~~~~~~~~~~~~~~~~~~~~~~~~~~
```

SCRIPT FILE Example_5_04.m

```
% ~~~~~~~~~~~~~~~~~~~~~~~~~~~~~~~~~~~~~~~~~~~~~~~~~~~~~~~~~~~
% Example_5_04
%   ~~~~~~~~~~~~
%
% This program computes J0 and the Julian day number using the
% data in Example 5.4.
%
% year    - range: 1901 - 2099
% month   - range: 1 - 12
% day     - range: 1 - 31
% hour    - range: 0 - 23 (Universal Time)
% minute  - range: 0 - 60
% second  - range: 0 - 60
% ut      - universal time (hr)
% j0      - Julian day number at 0 hr UT
% jd      - Julian day number at specified UT
%
% User M-function required: J0
% -----------------------------------------------------------

clear

%...Input data from Example 5.4:
year   = 2004;
month  = 5;
day    = 12;

hour   = 14;
minute = 45;
second = 30;
%...

ut = hour + minute/60 + second/3600;

%...Equation 5.48:
j0 = J0(year, month, day);

%...Equation 5.47:
jd = j0 + ut/24;

%...Echo the input data and output the results to the command window:
fprintf('-------------------------------------------------')
fprintf('\n Example 5.4: Julian day calculation\n')
fprintf('\n Input data:\n');
fprintf('\n   Year            = %g',   year)
```

```
                fprintf('\n   Month       = %g',   month)
                fprintf('\n   Day         = %g',   day)
                fprintf('\n   Hour        = %g',   hour)
                fprintf('\n   Minute      = %g',   minute)
                fprintf('\n   Second      = %g\n', second)

                fprintf('\n Julian day number = %11.3f', jd);
                fprintf('\n-----------------------------------------------\n')
              % ~~~~~~~~~~~~~~~~~~~~~~~~~~~~~~~~~~~~~~~~~~~~~~~~~~~~~~~~~~~~~
```

Output from Example_5_04

```
-----------------------------------------------------
 Example 5.4: Julian day calculation

 Input data:

   Year              = 2004
   Month             = 5
   Day               = 12
   Hour              = 14
   Minute            = 45
   Second            = 30

 Julian day number = 2453138.115
-----------------------------------------------------
```

D.13 ALGORITHM 5.3: CALCULATION OF LOCAL SIDEREAL TIME

FUNCTION FILE LST.m

```
              % ~~~~~~~~~~~~~~~~~~~~~~~~~~~~~~~~~~~~~~~~~~~~~~~~~~~~~~~~~~~~~
                function lst = LST(y, m, d, ut, EL)
              % ~~~~~~~~~~~~~~~~~~~~~~~~~~~~~~~~~~
              %
              % This function calculates the local sidereal time.
              %
              % lst - local sidereal time (degrees)
              % y   - year
              % m   - month
              % d   - day
              % ut  - Universal Time (hours)
              % EL  - east longitude (degrees)
              % j0  - Julian day number at 0 hr UT
              % j   - number of centuries since J2000
              % g0  - Greenwich sidereal time (degrees) at 0 hr UT
              % gst - Greenwich sidereal time (degrees) at the specified UT
              %
              % User M-function required: J0
              % ---------------------------------------------------------------

              %...Equation 5.48;
```

```matlab
            j0 = J0(y, m, d);

            %...Equation 5.49:
            j = (j0 - 2451545)/36525;

            %...Equation 5.50:
            g0 = 100.4606184 + 36000.77004*j + 0.000387933*j^2 ...
                - 2.583e-8*j^3;

            %...Reduce g0 so it lies in the range 0 - 360 degrees
            g0 = zeroTo360(g0);

            %...Equation 5.51:
            gst = g0 + 360.98564724*ut/24;

            %...Equation 5.52:
            lst = gst + EL;

            %...Reduce lst to the range 0 - 360 degrees:
            lst = lst - 360*fix(lst/360);

            return
            % ~~~~~~~~~~~~~~~~~~~~~~~~~~~~~~~~~~~~~~~~~~~~~~~~~~~~~~~~~~~~~


            % Subfunction used in the main body:

            % ~~~~~~~~~~~~~~~~~~~~~~~~~~~~~~~~~~~~~~~~~~~~~~~~~~~~~~~~~~~~~
               function y = zeroTo360(x)
            % ~~~~~~~~~~~~~~~~~~~~~~~~~~
            %
            % This subfunction reduces an angle to the range
            % 0 - 360 degrees.
            %
            % x - The angle (degrees) to be reduced
            % y - The reduced value
            %
            % ---------------------------------------------------------------
            if (x >= 360)
                x = x - fix(x/360)*360;
            elseif (x < 0)
                x = x - (fix(x/360) - 1)*360;
            end
            y = x;
            return
            % ~~~~~~~~~~~~~~~~~~~~~~~~~~~~~~~~~~~~~~~~~~~~~~~~~~~~~~~~~~~~~
```

SCRIPT FILE Example_5_06.m

```matlab
            % ~~~~~~~~~~~~~~~~~~~~~~~~~~~~~~~~~~~~~~~~~~~~~~~~~~~~~~~~~~~~~
            % Example_5_06
            % ~~~~~~~~~~~~~
            %
            % This program uses Algorithm 5.3 to obtain the local sidereal
```

```
% time from the data provided in Example 5.6.
%
% lst    - local sidereal time (degrees)
% EL     - east longitude of the site (west longitude is
%          negative):
%              degrees (0 - 360)
%              minutes (0 - 60)
%              seconds (0 - 60)
% WL     - west longitude
% year   - range: 1901 - 2099
% month  - range: 1 - 12
% day    - range: 1 - 31
% ut     - universal time
%              hour (0 - 23)
%              minute (0 - 60)
%              second (0 - 60)
%
% User M-function required: LST
% ---------------------------------------------------------

clear

%...Input data for Example 5.6:

% East longitude:

degrees = 139;
minutes = 47;
seconds = 0;

% Date:
year    = 2004;
month   = 3;
day     = 3;

% Universal time:
hour    = 4;
minute  = 30;
second  = 0;

%...

%...Convert negative (west) longitude to east longitude:
if degrees < 0
    degrees = degrees + 360;
end

%...Express the longitudes as decimal numbers:
EL = degrees + minutes/60 + seconds/3600;
WL = 360 - EL;

%...Express universal time as a decimal number:
ut = hour + minute/60 + second/3600;

%...Algorithm 5.3:
lst = LST(year, month, day, ut, EL);

%...Echo the input data and output the results to the command window:
fprintf('-----------------------------------------------------')
fprintf('\n Example 5.6: Local sidereal time calculation\n')
```

```
                fprintf('\n Input data:\n');
                fprintf('\n   Year                       = %g', year)
                fprintf('\n   Month                      = %g', month)
                fprintf('\n   Day                        = %g', day)
                fprintf('\n   UT (hr)                    = %g', ut)
                fprintf('\n   West Longitude (deg)       = %g', WL)
                fprintf('\n   East Longitude (deg)       = %g', EL)
                fprintf('\n\n');

                fprintf(' Solution:')

                fprintf('\n');
                fprintf('\n   Local Sidereal Time (deg) = %g', lst)
                fprintf('\n   Local Sidereal Time (hr)  = %g', lst/15)

                fprintf('\n-----------------------------------------\n')
                % ~~~~~~~~~~~~~~~~~~~~~~~~~~~~~~~~~~~~~~~~~~~~~~~~~~~~~~~~~
```

Output from Example_5_06

```
-----------------------------------------------------
 Example 5.6: Local sidereal time calculation

 Input data:

   Year                      = 2004
   Month                     = 3
   Day                       = 3
   UT (hr)                   = 4.5
   West Longitude (deg)      = 220.217
   East Longitude (deg)      = 139.783
 Solution:

   Local Sidereal Time (deg) = 8.57688
   Local Sidereal Time (hr)  = 0.571792
-----------------------------------------------------
```

D.14 ALGORITHM 5.4: CALCULATION OF THE STATE VECTOR FROM MEASUREMENTS OF RANGE, ANGULAR POSITION AND THEIR RATES

FUNCTION FILE rv_from_observe.m

```
% ~~~~~~~~~~~~~~~~~~~~~~~~~~~~~~~~~~~~~~~~~~~~~~~~~~~~~~~~~~~~~~~
  function [r,v] = rv_from_observe(rho, rhodot, A, Adot, a,...
                                   adot, theta, phi, H)
% ~~~~~~~~~~~~~~~~~~~~~~~~~~~~~~~~~~~~~~~~~~~~~~~~~~~~~~~~~~~~~~~
%
% This function calculates the geocentric equatorial position
% and velocity vectors of an object from radar observations of
% range, azimuth, elevation angle and their rates.
%
% deg    - conversion factor between degrees and radians
% pi     - 3.1415926...
%
```

D.14 Algorithm 5.4: calculation of the state vector

```
% Re     - equatorial radius of the earth (km)
% f      - earth's flattening factor
% wE     - angular velocity of the earth (rad/s)
% omega  - earth's angular velocity vector (rad/s) in the
%          geocentric equatorial frame
%
% theta  - local sidereal time (degrees) of tracking site
% phi    - geodetic latitude (degrees) of site
% H      - elevation of site (km)
% R      - geocentric equatorial position vector (km) of
%          tracking site
% Rdot   - inertial velocity (km/s) of site
% rho    - slant range of object (km)
% rhodot - range rate (km/s)
% A      - azimuth (degrees) of object relative to observation
%          site
% Adot   - time rate of change of azimuth (degrees/s)
% a      - elevation angle (degrees) of object relative to
%          observation site
% adot   - time rate of change of elevation angle (degrees/s)
% dec    - topocentric equatorial declination of object (rad)
% decdot - declination rate (rad/s)
% h      - hour angle of object (rad)
% RA     - topocentric equatorial right ascension of object
%          (rad)
% RAdot  - right ascension rate (rad/s)
%
% Rho    - unit vector from site to object
% Rhodot - time rate of change of Rho (1/s)
% r      - geocentric equatorial position vector of object (km)
% v      - geocentric equatorial velocity vector of object (km)
%
% User M-functions required: none
% -------------------------------------------------------------

global f Re wE
deg   = pi/180;
omega = [0 0 wE];

%...Convert angular quantities from degrees to radians:
A     = A    *deg;
Adot  = Adot *deg;
a     = a    *deg;
adot  = adot *deg;
theta = theta*deg;
phi   = phi  *deg;

%...Equation 5.56:
R     = [(Re/sqrt(1-(2*f - f*f)*sin(phi)^2) + H) ...
         *cos(phi)*cos(theta), ...
         (Re/sqrt(1-(2*f - f*f)*sin(phi)^2) + H) ...
         *cos(phi)*sin(theta), ...
         (Re*(1 - f)^2/sqrt(1-(2*f - f*f) ...
         *sin(phi)^2) + H)*sin(phi)];

%...Equation 5.66:
Rdot  = cross(omega, R);
```

```matlab
%...Equation 5.83a:
dec   = asin(cos(phi)*cos(A)*cos(a) + sin(phi)*sin(a));

%...Equation 5.83b:
h = acos((cos(phi)*sin(a) - sin(phi)*cos(A)*cos(a))/cos(dec));
if (A > 0) & (A < pi)
    h = 2*pi - h;
end

%...Equation 5.83c:
RA = theta - h;

%...Equations 5.57:
Rho = [cos(RA)*cos(dec)   sin(RA)*cos(dec)   sin(dec)];

%...Equation 5.63:
r   = R + rho*Rho;

%...Equation 5.84:
decdot = (-Adot*cos(phi)*sin(A)*cos(a) ...
         + adot*(sin(phi)*cos(a) ...
         - cos(phi)*cos(A)*sin(a)))/cos(dec);

%...Equation 5.85:
RAdot   = wE ...
        + (Adot*cos(A)*cos(a) - adot*sin(A)*sin(a) ...
        + decdot*sin(A)*cos(a)*tan(dec)) ...
         /(cos(phi)*sin(a) - sin(phi)*cos(A)*cos(a));

%...Equations 5.69 and 5.72:
Rhodot = [-RAdot*sin(RA)*cos(dec) - decdot*cos(RA)*sin(dec),...
           RAdot*cos(RA)*cos(dec) - decdot*sin(RA)*sin(dec),...
           decdot*cos(dec)];

%...Equation 5.64:
v = Rdot + rhodot*Rho + rho*Rhodot;
% ~~~~~~~~~~~~~~~~~~~~~~~~~~~~~~~~~~~~~~~~~~~~~~~~~~~~~~~~~~~~~
```

SCRIPT FILE Example_5_10.m

```matlab
% ~~~~~~~~~~~~~~~~~~~~~~~~~~~~~~~~~~~~~~~~~~~~~~~~~~~~~~~~~~~~~
% Example_5_10
%  ~~~~~~~~~~~~
%
% This program uses Algorithms 5.4 and 4.1 to obtain the
% orbital elements from the observational data provided in
% Example 5.10.
%
% deg     - conversion factor between degrees and radians
% pi      - 3.1415926...
% mu      - gravitational parameter (km^3/s^2)
%
% Re      - equatorial radius of the earth (km)
% f       - earth's flattening factor
% wE      - angular velocity of the earth (rad/s)
```

```
% omega   - earth's angular velocity vector (rad/s) in the
%           geocentric equatorial frame
%
% rho     - slant range of object (km)
% rhodot  - range rate (km/s)
% A       - azimuth (deg) of object relative to observation
%           site
% Adot    - time rate of change of azimuth (deg/s)
% a       - elevation angle (deg) of object relative to
%           observation site
% adot    - time rate of change of elevation angle
%           (degrees/s)
%
% theta   - local sidereal time (deg) of tracking site
% phi     - geodetic latitude (deg) of site
% H       - elevation of site (km)
%
% r       - geocentric equatorial position vector of object (km)
% v       - geocentric equatorial velocity vector of object (km)
%
% coe     - orbital elements [h e RA incl w TA a]
%           where h    = angular momentum (km^2/s)
%                 e    = eccentricity
%                 RA   = right ascension of the ascending node
%                        (rad)
%                 incl = inclination of the orbit (rad)
%                 w    = argument of perigee (rad)
%                 TA   = true anomaly (rad)
%                 a    = semimajor axis (km)
% rp      - perigee radius (km)
% T       - period of elliptical orbit (s)
%
% User M-functions required: rv_from_observe, coe_from_sv
% ---------------------------------------------------------

clear
global  f Re wE mu

deg    = pi/180;
f      = 1/298.256421867;
Re     = 6378.13655;
wE     = 7.292115e-5;
mu     = 398600.4418;

%...Input data for Example 5.10:
rho    = 2551;
rhodot = 0;
A      = 90;
Adot   = 0.1130;
a      = 30;
adot   = 0.05651;
theta  = 300;
phi    = 60;
H      = 0;
%...
%...Algorithm 5.4:
[r,v] = rv_from_observe(rho, rhodot, A, Adot, a, adot, theta, ...
        phi, H);
```

```
%...Algorithm 4.1:
coe  = coe_from_sv(r,v);

h    = coe(1);
e    = coe(2);
RA   = coe(3);
incl = coe(4);
w    = coe(5);
TA   = coe(6);
a    = coe(7);

%...Equation 2.40
rp   = h^2/mu/(1 + e);

%...Echo the input data and output the solution to
%   the command window:
fprintf('-----------------------------------------------')
fprintf('\n Example 5.10')
fprintf('\n\n Input data:\n')
fprintf('\n Slant range (km)               = %g', rho)
fprintf('\n Slant range rate (km/s)        = %g', rhodot)
fprintf('\n Azimuth (deg)                  = %g', A)
fprintf('\n Azimuth rate (deg/s)           = %g', Adot)
fprintf('\n Elevation (deg)                = %g', a)
fprintf('\n Elevation rate (deg/s)         = %g', adot)
fprintf('\n Local sidereal time (deg)      = %g', theta)
fprintf('\n Latitude (deg)                 = %g', phi)
fprintf('\n Altitude above sea level (km)  = %g', H)
fprintf('\n\n')

fprintf(' Solution:')

fprintf('\n\n State vector:\n')
fprintf('\n r (km)                   = [%g, %g, %g]', ...
                                        r(1), r(2), r(3))
fprintf('\n v (km/s)                 = [%g, %g, %g]', ...
                                        v(1), v(2), v(3))

fprintf('\n\n Orbital elements:\n')
fprintf('\n   Angular momentum (km^2/s)  = %g', h)
fprintf('\n   Eccentricity               = %g', e)
fprintf('\n   Inclination (deg)          = %g', incl/deg)
fprintf('\n   RA of ascending node (deg) = %g', RA/deg)
fprintf('\n   Argument of perigee (deg)  = %g', w/deg)
fprintf('\n   True anomaly (deg)         = %g\n', TA/deg)
fprintf('\n   Semimajor axis (km)        = %g', a)
fprintf('\n   Perigee radius (km)        = %g', rp)
%...If the orbit is an ellipse, output its period:
if e < 1
    T = 2*pi/sqrt(mu)*a^1.5;
    fprintf('\n   Period:')
    fprintf('\n     Seconds                = %g', T)
    fprintf('\n     Minutes                = %g', T/60)
    fprintf('\n     Hours                  = %g', T/3600)
    fprintf('\n     Days                   = %g', T/24/3600)
end
fprintf('\n-----------------------------------------------\n')

% ~~~~~~~~~~~~~~~~~~~~~~~~~~~~~~~~~~~~~~~~~~~~~~~~~~~~~~~~~~
```

D.15 Algorithms 5.5 and 5.6: Gauss method with iterative improvement

Output from `Example_5_10`

```
---------------------------------------------------------
Example 5.10

Input data:

Slant range (km)                 = 2551
Slant range rate (km/s)          = 0
Azimuth (deg)                    = 90
Azimuth rate (deg/s)             = 0.113
Elevation (deg)                  = 5168.62
Elevation rate (deg/s)           = 0.05651
Local sidereal time (deg)        = 300
Latitude (deg)                   = 60
Altitude above sea level (km)    = 0

Solution:

State vector:

  r (km)                         = [3830.68, -2216.47, 6605.09]
  v (km/s)                       = [1.50357, -4.56099, -0.291536]

Orbital elements:

  Angular momentum (km^2/s)      = 35621.4
  Eccentricity                   = 0.619758
  Inclination (deg)              = 113.386
  RA of ascending node (deg)     = 109.75
  Argument of perigee (deg)      = 309.81
  True anomaly (deg)             = 165.352

  Semimajor axis (km)            = 5168.62
  Perigee radius (km)            = 1965.32
  Period:
    Seconds                      = 3698.05
    Minutes                      = 61.6342
    Hours                        = 1.02724
    Days                         = 0.0428015
---------------------------------------------------------
```

D.15 ALGORITHMS 5.5 AND 5.6: GAUSS'S METHOD OF PRELIMINARY ORBIT DETERMINATION WITH ITERATIVE IMPROVEMENT

FUNCTION FILE gauss.m

```
% ~~~~~~~~~~~~~~~~~~~~~~~~~~~~~~~~~~~~~~~~~~~~~~~~~~~~~~~~~~~~~~
  function [r, v, r_old, v_old] = ...
          gauss(Rho1, Rho2, Rho3, R1, R2, R3, t1, t2, t3)
% ~~~~~~~~~~~~~~~~~~~~~~~~~~~~~~~~~~~~~~~~~~~~~~~~~~~~~~~~~~~~~~
% This function uses the Gauss method with iterative
% improvement (Algorithms 5.5 and 5.6) to calculate the state
```

```
% vector of an orbiting body from angles-only observations at
% three closely-spaced times.
%
% mu                - the gravitational parameter (km^3/s^2)
% t1, t2, t3        - the times of the observations (s)
% tau, tau1, tau3   - time intervals between observations (s)
% R1, R2, R3        - the observation site position vectors
%                     at t1, t2, t3 (km)
% Rho1, Rho2, Rho3  - the direction cosine vectors of the
%                     satellite at t1, t2, t3
% p1, p2, p3        - cross products among the three direction
%                     cosine vectors
% Do                - scalar triple product of Rho1, Rho2 and
%                     Rho3
% D                 - Matrix of the nine scalar triple products
%                     of R1, R2 and R3 with p1, p2 and p3
% E                 - dot product of R2 and Rho2
% A, B              - constants in the expression relating
%                     slant range to geocentric radius
% a,b,c             - coefficients of the 8th order polynomial
%                     in the estimated geocentric radius x
% x                 - positive root of the 8th order polynomial
% rho1, rho2, rho3  - the slant ranges at t1, t2, t3
% r1, r2, r3        - the position vectors at t1, t2, t3 (km)
% r_old, v_old      - the estimated state vector at the end of
%                     Algorithm 5.5 (km, km/s)
% rho1_old,
% rho2_old, and
% rho3_old          - the values of the slant ranges at t1, t2,
%                     t3 at the beginning of iterative
%                     improvement (Algorithm 5.6) (km)
% diff1, diff2,
% and diff3         - the magnitudes of the differences between
%                     the old and new slant ranges at the end
%                     of each iteration
% tol               - the error tolerance determining
%                     convergence
% n                 - number of passes through the
%                     iterative improvement loop
% nmax              - limit on the number of iterations
% ro, vo            - magnitude of the position and
%                     velocity vectors (km, km/s)
% vro               - radial velocity component (km)
% a                 - reciprocal of the semimajor axis (1/km)
% v2                - computed velocity at time t2 (km/s)
% r, v              - the state vector at the end of
%                     Algorithm 5.6 (km, km/s)
%
% User M-functions required:  kepler_U, f_and_g
% User subfunctions required: posroot
% -----------------------------------------------------------

global mu

%...Equations 5.98:
tau1 = t1 - t2;
tau3 = t3 - t2;
```

D.15 Algorithms 5.5 and 5.6: Gauss method with iterative improvement

```
%...Equation 5.101:
tau  = tau3 - tau1;

%...Independent cross products among the direction cosine vectors:
p1 = cross(Rho2,Rho3);
p2 = cross(Rho1,Rho3);
p3 = cross(Rho1,Rho2);

%...Equation 5.108:
Do = dot(Rho1,p1);

%...Equations 5.109b, 5.110b and 5.111b:
D  = [[dot(R1,p1) dot(R1,p2) dot(R1,p3)]
      [dot(R2,p1) dot(R2,p2) dot(R2,p3)]
      [dot(R3,p1) dot(R3,p2) dot(R3,p3)]];

%...Equation 5.115b:
E = dot(R2,Rho2);

%...Equations 5.112b and 5.112c:
A = 1/Do*(-D(1,2)*tau3/tau + D(2,2) + D(3,2)*tau1/tau);
B = 1/6/Do*(D(1,2)*(tau3^2 - tau^2)*tau3/tau ...
             + D(3,2)*(tau^2 - tau1^2)*tau1/tau);

%...Equations 5.117:
a = -(A^2 + 2*A*E + norm(R2)^2);
b = -2*mu*B*(A + E);
c = -(mu*B)^2;

%...Calculate the roots of Equation 5.116 using MATLAB's
%   polynomial 'roots' solver:
Roots = roots([1 0 a 0 0 b 0 0 c]);

%...Find the positive real root:
x = posroot(Roots);

%...Equations 5.99a and 5.99b:
f1 =    1 - 1/2*mu*tau1^2/x^3;
f3 =    1 - 1/2*mu*tau3^2/x^3;

%...Equations 5.100a and 5.100b:
g1 = tau1 - 1/6*mu*(tau1/x)^3;
g3 = tau3 - 1/6*mu*(tau3/x)^3;

%...Equation 5.112a:
rho2 = A + mu*B/x^3;

%...Equation 5.113:
rho1 = 1/Do*((6*(D(3,1)*tau1/tau3 + D(2,1)*tau/tau3)*x^3 ...
              + mu*D(3,1)*(tau^2 - tau1^2)*tau1/tau3) ...
              /(6*x^3 + mu*(tau^2 - tau3^2)) - D(1,1));

%...Equation 5.114:
rho3 = 1/Do*((6*(D(1,3)*tau3/tau1 - D(2,3)*tau/tau1)*x^3 ...
              + mu*D(1,3)*(tau^2 - tau3^2)*tau3/tau1) ...
              /(6*x^3 + mu*(tau^2 - tau3^2)) - D(3,3));
```

```
%...Equations 5.86:
r1 = R1 + rho1*Rho1;
r2 = R2 + rho2*Rho2;
r3 = R3 + rho3*Rho3;

%...Equation 5.118:
v2 = (-f3*r1 + f1*r3)/(f1*g3 - f3*g1);

%...Save the initial estimates of r2 and v2:
r_old = r2;
v_old = v2;

%...End of Algorithm 5.5

%...Use Algorithm 5.6 to improve the accuracy of the initial estimates.

%...Initialize the iterative improvement loop and set error tolerance:
rho1_old = rho1;   rho2_old = rho2;   rho3_old = rho3;
diff1    = 1;      diff2    = 1;      diff3    = 1;
n    = 0;
nmax = 1000;
tol  = 1.e-8;

%...Iterative improvement loop:
while ((diff1 > tol) & (diff2 > tol) & (diff3 > tol)) ...
        & (n < nmax)
    n = n+1;

%...Compute quantities required by universal kepler's equation:
    ro  = norm(r2);
    vo  = norm(v2);
    vro = dot(v2,r2)/ro;
    a   = 2/ro - vo^2/mu;

%...Solve universal Kepler's equation at times tau1 and tau3
%   for universal anomalies x1 and x3:
    x1 = kepler_U(tau1, ro, vro, a);
    x3 = kepler_U(tau3, ro, vro, a);

%...Calculate the Lagrange f and g coefficients at times tau1 and tau3:
    [ff1, gg1] = f_and_g(x1, tau1, ro, a);
    [ff3, gg3] = f_and_g(x3, tau3, ro, a);

%...Update the f and g functions at times tau1 and tau3 by
%   averaging old and new:
    f1   = (f1 + ff1)/2;
    f3   = (f3 + ff3)/2;
    g1   = (g1 + gg1)/2;
    g3   = (g3 + gg3)/2;

%...Equations 5.96 and 5.97:
    c1   =  g3/(f1*g3 - f3*g1);
    c3   = -g1/(f1*g3 - f3*g1);

%...Equations 5.109a, 5.110a and 5.111a:
    rho1 = 1/Do*(      -D(1,1) + 1/c1*D(2,1) - c3/c1*D(3,1));
    rho2 = 1/Do*(   -c1*D(1,2) +      D(2,2) -    c3*D(3,2));
    rho3 = 1/Do*(-c1/c3*D(1,3) + 1/c3*D(2,3) -       D(3,3));
```

D.15 Algorithms 5.5 and 5.6: Gauss method with iterative improvement

```
%...Equations 5.86:
    r1    = R1 + rho1*Rho1;
    r2    = R2 + rho2*Rho2;
    r3    = R3 + rho3*Rho3;

%...Equation 5.118:
    v2    = (-f3*r1 + f1*r3)/(f1*g3 - f3*g1);

%...Calculate differences upon which to base convergence:
    diff1 = abs(rho1 - rho1_old);
    diff2 = abs(rho2 - rho2_old);
    diff3 = abs(rho3 - rho3_old);

%...Update the slant ranges:
    rho1_old = rho1;   rho2_old = rho2;   rho3_old = rho3;
end
%...End iterative improvement loop

fprintf('\n( **Number of Gauss improvement iterations')
fprintf(' = %g)\n\n', n)

if n >= nmax

fprintf('\n\n **Number of iterations exceeds %g \n\n ', nmax);
end

%...Return the state vector for the central observation:
r = r2;
v = v2;

return

% ~~~~~~~~~~~~~~~~~~~~~~~~~~~~~~~~~~~~~~~~~~~~~~~~~~~~~~~~~~~~~~~~
% Subfunction used in the main body:

% ~~~~~~~~~~~~~~~~~~~~~~~~~~~~~~~~~~~~~~~~~~~~~~~~~~~~~~~~~~~~~~~~
   function x = posroot(Roots)

% ~~~~~~~~~~~~~~~~~~~~~~~~~~~~~
%
% This subfunction extracts the positive real roots from
% those obtained in the call to MATLAB's 'roots' function.
% If there is more than one positive root, the user is
% prompted to select the one to use.
%
% x         - the determined or selected positive root
% Roots     - the vector of roots of a polynomial
% posroots  - vector of positive roots
%
% User M-functions required: none
% ----------------------------------------------------------

%...Construct the vector of positive real roots:
posroots  = Roots(find(Roots>0 & ~imag(Roots)));
npositive = length(posroots);

%...Exit if no positive roots exist:
if npositive == 0
```

```
            fprintf('\n\n ** There are no positive roots.   \n\n')
            return
        end

        %...If there is more than one positive root, output the
        %...roots to the command window and prompt the user to
        %...select which one to use:
        if npositive == 1
            x = posroots;
        else
            fprintf('\n\n ** There are two or more positive roots.\n')
            for i = 1:npositive
                fprintf('\n root #%g = %g', i, posroots(i))
            end
            fprintf('\n\n Make a choice:\n')
            nchoice = 0;
            while nchoice < 1 | nchoice > npositive
                nchoice = input(' Use root #? ');
            end
            x = posroots(nchoice);
            fprintf('\n We will use %g .\n', x)
        end

        return
        % ~~~~~~~~~~~~~~~~~~~~~~~~~~~~~~~~~~~~~~~~~~~~~~~~~~~~~~~~~~~~~
```

SCRIPT FILE Example_5_11.m

```
        % ~~~~~~~~~~~~~~~~~~~~~~~~~~~~~~~~~~~~~~~~~~~~~~~~~~~~~~~~~~~~~
        % Example_5_11
        % ~~~~~~~~~~~~
        %
        % This program uses Algorithms 5.5 and 5.6 (Gauss's method) to
        % compute the state vector from the data provided in
        % Example 5.11.
        %
        % deg         - factor for converting between degrees and
        %               radians
        % pi          - 3.1415926...
        % mu          - gravitational parameter (km^3/s^2)
        % Re          - earth's radius (km)
        % f           - earth's flattening factor
        % H           - elevation of observation site (km)
        % phi         - latitude of site (deg)
        % t           - vector of observation times t1, t2, t3 (s)
        % ra          - vector of topocentric equatorial right
        %               ascensions at t1, t2, t3 (deg)
        % dec         - vector of topocentric equatorial right
        %               declinations at t1, t2, t3 (deg)
        % theta       - vector of local sidereal times for t1, t2, t3
        %               (deg)
        % R           - matrix of site position vectors at t1, t2, t3
        %               (km)
        % rho         - matrix of direction cosine vectors at t1,
```

```
%                     t2, t3
% fac1, fac2    - common factors
% r_old, v_old  - the state vector without iterative improvement
%                 (km, km/s)
% r, v          - the state vector with iterative improvement
%                 (km, km/s)
% coe           - vector of orbital elements for r, v:
%                 [h, e, RA, incl, w, TA, a]
%                 where h    = angular momentum (km^2/s)
%                       e    = eccentricity
%                       incl = inclination (rad)
%                       w    = argument of perigee (rad)
%                       TA   = true anomaly (rad)
%                       a    = semimajor axis (km)
% coe_old       - vector of orbital elements for r_old, v_old
%
% User M-functions required: gauss, coe_from_sv
% -------------------------------------------------------------

clear
global mu

deg = pi/180;
mu  = 398600;
Re  = 6378;
f   = 1/298.26;

%...Input data:
H     = 1;
phi   = 40*deg;
t     = [      0     118.104    237.577];
ra    = [ 43.5365     54.4196    64.3178]*deg;
dec   = [-8.78334    -12.0739   -15.1054]*deg;
theta = [ 44.5065     45.000     45.4992]*deg;
%...

%...Equations 5.56 and 5.57:
fac1 = Re/sqrt(1-(2*f - f*f)*sin(phi)^2);
fac2 = (Re*(1-f)^2/sqrt(1-(2*f - f*f)*sin(phi)^2) + H) ...
        *sin(phi);
for i = 1:3
   R(i,1) = (fac1 + H)*cos(phi)*cos(theta(i));
   R(i,2) = (fac1 + H)*cos(phi)*sin(theta(i));
   R(i,3) = fac2;
   rho(i,1) = cos(dec(i))*cos(ra(i));
   rho(i,2) = cos(dec(i))*sin(ra(i));
   rho(i,3) = sin(dec(i));
end

%...Algorithms 5.5 and 5.6:
[r, v, r_old, v_old] = gauss(rho(1,:), rho(2,:), rho(3,:), ...
                             R(1,:),   R(2,:),   R(3,:), ...
                             t(1),     t(2),     t(3));

%...Algorithm 4.1 for the initial estimate of the state vector
%   and for the iteratively improved one:
```

```matlab
coe_old = coe_from_sv(r_old,v_old);
coe     = coe_from_sv(r,v);

%...Echo the input data and output the solution to
%   the command window:
fprintf('-----------------------------------------------------')
fprintf('\n Example 5.11: Orbit determination by the Gauss
        method\n')
fprintf('\n Radius of earth (km)                = %g', Re)
fprintf('\n Flattening factor                   = %g', f)
fprintf('\n Gravitational parameter (km^3/s^2) = %g', mu)
fprintf('\n\n Input data:\n');
fprintf('\n Latitude (deg)                      = %g', phi/deg);
fprintf('\n Altitude above sea level (km)       = %g', H);
fprintf('\n\n Observations:')
fprintf('\n    Time (s)    Right ascension (deg)   Declination
        (deg)')
fprintf('   Local sidereal time (deg)')
for i = 1:3
    fprintf('\n %9.4g %17.4f %19.4f %23.4f', ...
                t(i), ra(i)/deg, dec(i)/deg, theta(i)/deg)
end

fprintf('\n\n Solution:\n')

fprintf('\n Without iterative improvement...\n')
fprintf('\n');
fprintf('\n r (km)                       = [%g, %g, %g]', ...
                                r_old(1), r_old(2), r_old(3))
fprintf('\n v (km/s)                     = [%g, %g, %g]', ...
                                v_old(1), v_old(2), v_old(3))
fprintf('\n');
fprintf('\n Angular momentum (km^2/s)   = %g', coe_old(1))
fprintf('\n Eccentricity                = %g', coe_old(2))
fprintf('\n RA of ascending node (deg)  = %g', coe_old(3)/deg)
fprintf('\n Inclination (deg)           = %g', coe_old(4)/deg)
fprintf('\n Argument of perigee (deg)   = %g', coe_old(5)/deg)
fprintf('\n True anomaly (deg)          = %g', coe_old(6)/deg)
fprintf('\n Semimajor axis (km)         = %g', coe_old(7))
fprintf('\n Periapse radius (km)        = %g', coe_old(1)^2 ...
                                            /mu/(1 + coe_old(2)))
%...If the orbit is an ellipse, output the period:
if coe_old(2)<1
    T = 2*pi/sqrt(mu)*coe_old(7)^1.5;
    fprintf('\n   Period:')
    fprintf('\n     Seconds                 = %g', T)
    fprintf('\n     Minutes                 = %g', T/60)
    fprintf('\n     Hours                   = %g', T/3600)
    fprintf('\n     Days                    = %g', T/24/3600)
end

fprintf('\n\n With iterative improvement...\n')
fprintf('\n');
fprintf('\n r (km)                       = [%g, %g, %g]', ...
                                    r(1), r(2), r(3))
fprintf('\n v (km/s)                     = [%g, %g, %g]', ...
                                    v(1), v(2), v(3))

fprintf('\n');
```

D.15 Algorithms 5.5 and 5.6: Gauss method with iterative improvement

```
    fprintf('\n    Angular momentum (km^2/s)   = %g', coe(1))
    fprintf('\n    Eccentricity                = %g', coe(2))
    fprintf('\n    RA of ascending node (deg)  = %g', coe(3)/deg)
    fprintf('\n    Inclination (deg)           = %g', coe(4)/deg)
    fprintf('\n    Argument of perigee (deg)   = %g', coe(5)/deg)
    fprintf('\n    True anomaly (deg)          = %g', coe(6)/deg)
    fprintf('\n    Semimajor axis (km)         = %g', coe(7))
    fprintf('\n    Periapse radius (km)        = %g', coe(1)^2 ...
                                                     /mu/(1 + coe(2)))
    %...If the orbit is an ellipse, output the period:
    if coe(2)<1
        T = 2*pi/sqrt(mu)*coe(7)^1.5;
        fprintf('\n    Period:')
        fprintf('\n      Seconds               = %g', T)
        fprintf('\n      Minutes               = %g', T/60)
        fprintf('\n      Hours                 = %g', T/3600)
        fprintf('\n      Days                  = %g', T/24/3600)
    end
    fprintf('\n-----------------------------------------------\n')
    % ~~~~~~~~~~~~~~~~~~~~~~~~~~~~~~~~~~~~~~~~~~~~~~~~~~~~~~~~~~~~
```

Output from Example_5_11

```
( **Number of Gauss improvement iterations = 14)

-----------------------------------------------------
Example 5.11: Orbit determination by the Gauss method

Radius of earth (km)              = 6378
Flattening factor                 = 0.00335278
Gravitational parameter (km^3/s^2) = 398600

Input data:

Latitude (deg)                = 40
Altitude above sea level (km) = 1

Observations:
              Right                                Local
   Time (s)   Ascension (deg)  Declination (deg)   Sidereal
                                                   time (deg)
        0       43.5365          -8.7833           44.5065
    118.1       54.4196         -12.0739           45.0000
    237.6       64.3178         -15.1054           45.4992

Solution:

Without iterative improvement...

  r (km)                       = [5659.03, 6533.74, 3270.15]
  v (km/s)                     = [-3.90774, 5.05735, -2.22224]
     Angular momentum (km^2/s)   = 62426.4
     Eccentricity                = 0.084887
     RA of ascending node (deg)  = 270.375
     Inclination (deg)           = 29.8362
     Argument of perigee (deg)   = 87.6835
     True anomaly (deg)          = 46.9821
     Semimajor axis (km)         = 9847.83
     Periapse radius (km)        = 9011.88
```

```
            Period:
              Seconds                      = 9725.73
              Minutes                      = 162.095
              Hours                        = 2.70159
              Days                         = 0.112566

         With iterative improvement...
         r (km)                            = [5662.04, 6537.95, 3269.05]
         v (km/s)                          = [-3.88542, 5.12141, -2.2434]


            Angular momentum (km^2/s)      = 62816.7
            Eccentricity                   = 0.0999909
            RA of ascending node (deg)     = 269.999
            Inclination (deg)              = 30.001
            Argument of perigee (deg)      = 89.9723
            True anomaly (deg)             = 45.0284
            Semimajor axis (km)            = 9999.48
            Periapse radius (km)           = 8999.62
            Period:
              Seconds                      = 9951.24
              Minutes                      = 165.854
              Hours                        = 2.76423
              Days                         = 0.115176
-----------------------------------------------------
```

D.16 Converting the numerical designation of a month or a planet into its name

The following simple script can be used in programs that input the numerical values for a month and/or a planet.

Function file month_planet_names.m

```
% ~~~~~~~~~~~~~~~~~~~~~~~~~~~~~~~~~~~~~~~~~~~~~~~~~~~~~~~~~~~~~~
  function [month, planet] = month_planet_names(month_id,
                             planet_id)
% ~~~~~~~~~~~~~~~~~~~~~~~~~~~~~~~~~~~~~~~~~~~~~~~~~~~~~~~~~~~~~~
%
% This function returns the name of the month and the planet
% corresponding, respectively, to the numbers ''month_id'' and
% ''planet_id''.
%
% month     - name of the month
% planet    - name of the planet
% months    - a vector containing the names of the 12 months
% planets   - a vector containing the names of the 9 planets
% month_id  - the month number (1 - 12)
% planet_id - the planet number (1 - 9)
%
% User M-functions required: none
% --------------------------------------------------------------
```

```
              months   = ['January  '
                         'February '
                         'March    '
                         'April    '
                         'May      '
                         'June     '
                         'July     '
                         'August   '
                         'September'
                         'October  '
                         'November '
                         'December '];
              planets  = ['Mercury'
                          'Venus  '
                          'Earth  '
                          'Mars   '
                          'Jupiter'
                          'Saturn '
                          'Uranus '
                          'Neptune'
                          'Pluto  '];
              month    = months(month_id,  1:9);
              planet   = planets(planet_id, 1:7);
              % ~~~~~~~~~~~~~~~~~~~~~~~~~~~~~~~~~~~~~~~~~~~~~~~~~~~~~~~~~~~~~
```

D.17 ALGORITHM 8.1: CALCULATION OF THE STATE VECTOR OF A PLANET AT A GIVEN EPOCH

FUNCTION FILE planet_elements_and_sv.m

```
              % ~~~~~~~~~~~~~~~~~~~~~~~~~~~~~~~~~~~~~~~~~~~~~~~~~~~~~~~~~~~~~
              function [coe, r, v, jd] = planet_elements_and_sv ...
                      (planet_id, year, month, day, hour, minute, second)
              % ~~~~~~~~~~~~~~~~~~~~~~~~~~~~~~~~~~~~~~~~~~~~~~~~~~~~~~~~~~~~~
              %
              % This function calculates the orbital elements and the state
              % vector of a planet from the date (year, month, day)
              % and universal time (hour, minute, second).
              %
              % mu       - gravitational parameter of the sun (km^3/s^2)
              % deg      - conversion factor between degrees and radians
              % pi       - 3.1415926...
              %
              % coe      - vector of heliocentric orbital elements
              %            [h  e  RA  incl  w  TA  a  w_hat  L  M  E],
              %            where
              %                h     = angular momentum           (km^2/s)
              %                e     = eccentricity
              %                RA    = right ascension            (deg)
              %                incl  = inclination                (deg)
              %                w     = argument of perihelion     (deg)
              %                TA    = true anomaly               (deg)
```

```
%             a       = semimajor axis                          (km)
%             w_hat   = longitude of perihelion
%                       ( = RA + w)                             (deg)
%             L       = mean longitude ( = w_hat + M)           (deg)
%             M       = mean anomaly                            (deg)
%             E       = eccentric anomaly                       (deg)
%
% planet_id - planet identifier:
%             1 = Mercury
%             2 = Venus
%             3 = Earth
%             4 = Mars
%             5 = Jupiter
%             6 = Saturn
%             7 = Uranus
%             8 = Neptune
%             9 = Pluto
%
% year       - range: 1901 - 2099
% month      - range: 1 - 12
% day        - range: 1 - 31
% hour       - range: 0 - 23
% minute     - range: 0 - 60
% second     - range: 0 - 60
%
% j0         - Julian day number of the date at 0 hr UT
% ut         - universal time in fractions of a day
% jd         - julian day number of the date and time
%
% J2000_coe  - row vector of J2000 orbital elements from
%              Table 8.1
% rates      - row vector of Julian centennial rates from
%              Table 8.1
% t0         - Julian centuries between J2000 and jd
% elements   - orbital elements at jd
%
% r          - heliocentric position vector
% v          - heliocentric velocity vector
%
% User M-functions required:  J0, kepler_E, sv_from_coe
% User subfunctions required: planetary_elements, zero_to_360
% ------------------------------------------------------------

global mu
deg    = pi/180;

%...Equation 5.48:
j0     = J0(year, month, day);

ut     = (hour + minute/60 + second/3600)/24;

%...Equation 5.47
jd     = j0 + ut;
%...Obtain the data for the selected planet from Table 8.1:
[J2000_coe, rates] = planetary_elements(planet_id);

%...Equation 8.104a:
t0     = (jd - 2451545)/36525;
```

D.17 Algorithm 8.1: calculation of the state vector of a planet at a given epoch

```
%...Equation 8.104b:
elements = J2000_coe + rates*t0;

a       = elements(1);
e       = elements(2);

%...Equation 2.61:
h       = sqrt(mu*a*(1 - e^2));

%...Reduce the angular elements to within the range 0 - 360 degrees:
incl    = elements(3);
RA      = zero_to_360(elements(4));
w_hat   = zero_to_360(elements(5));
L       = zero_to_360(elements(6));
w       = zero_to_360(w_hat - RA);
M       = zero_to_360((L - w_hat));

%...Algorithm 3.1 (for which M must be in radians)
E       = kepler_E(e, M*deg);

%...Equation 3.10 (converting the result to degrees):
TA      = zero_to_360...
          (2*atan(sqrt((1 + e)/(1 - e))*tan(E/2))/deg);

coe     = [h e RA incl w TA a w_hat L M E/deg];

%...Algorithm 4.2 (for which all angles must be in radians):
[r, v] = sv_from_coe([h e RA*deg incl*deg w*deg TA*deg]);

return

% ~~~~~~~~~~~~~~~~~~~~~~~~~~~~~~~~~~~~~~~~~~~~~~~~~~~~~~~~~~~~~~~~

% Subfunctions used in the main body:

% ~~~~~~~~~~~~~~~~~~~~~~~~~~~~~~~~~~~~~~~~~~~~~~~~~~~~~~~~~~~~~~~~
    function [J2000_coe, rates] = planetary_elements(planet_id)
% ~~~~~~~~~~~~~~~~~~~~~~~~~~~~~~~~~~~~~~~~~~~~~~~~~~~~~~~~~~~~~~~~
%
% This function extracts a planet's J2000 orbital elements and
% centennial rates from Table 8.1.
%
% planet_id     - 1 through 9, for Mercury through Pluto
%
% J2000_elements - 9 by 6 matrix of J2000 orbital elements for
%                  the nine planets Mercury through Pluto. The
%                  columns of each row are:
%                         a     = semimajor axis (AU)
%                         e     = eccentricity
%                         i     = inclination (degrees)
%                         RA    = right ascension of the ascending
%                                 node (degrees)
%                         w_hat = longitude of perihelion (degrees)
%                         L     = mean longitude (degrees)
%
```

```
% cent_rates     - 9 by 6 matrix of the rates of change of the
%                  J2000_elements per Julian century (Cy).
%                  Using ''dot'' for time derivative, the
%                  columns of each row are:
%                     a_dot      (AU/Cy)
%                     e_dot      (1/Cy)
%                     i_dot      (arcseconds/Cy)
%                     RA_dot     (arcseconds/Cy)
%                     w_hat_dot  (arcseconds/Cy)
%                     Ldot       (arcseconds/Cy)
%
% J2000_coe       - row vector of J2000_elements corresponding
%                   to ''planet_id'', with au converted to km
% rates           - row vector of cent_rates corresponding
%                   to ''planet_id'', with au converted to km
%                   and arcseconds converted to degrees
%
% au              - astronomical unit (km)
%
% User M-functions required: none
% --------------------------------------------------------------

J2000_elements = ...
[ 0.38709893   0.20563069    7.00487     48.33167     77.45645    252.25084
  0.72333199   0.00677323    3.39471     76.68069    131.53298    181.97973
  1.00000011   0.01671022    0.00005    -11.26064    102.94719    100.46435
  1.52366231   0.09341233    1.85061     49.57854    336.04084    355.45332
  5.20336301   0.04839266    1.30530    100.55615     14.75385     34.40438
  9.53707032   0.05415060    2.48446    113.71504     92.43194     49.94432
 19.19126393   0.04716771    0.76986     74.22988    170.96424    313.23218
 30.06896348   0.00858587    1.76917    131.72169     44.97135    304.88003
 39.48168677   0.24880766   17.14175    110.30347    224.06676    238.92881];

cent_rates = ...
[ 0.00000066   0.00002527   -23.51      -446.30       573.57    538101628.29
  0.00000092  -0.00004938    -2.86      -996.89      -108.80    210664136.06
 -0.00000005  -0.00003804   -46.94     -18228.25     1198.28    129597740.63
 -0.00007221   0.00011902   -25.47      -1020.19     1560.78     68905103.78
  0.00060737  -0.00012880    -4.15       1217.17      839.93     10925078.35
 -0.00301530  -0.00036762     6.11      -1591.05    -1948.89      4401052.95
  0.00152025  -0.00019150    -2.09      -1681.4      1312.56      1542547.79
 -0.00125196   0.00002514    -3.64       -151.25     -844.43       786449.21
 -0.00076912   0.00006465    11.07        -37.33     -132.25       522747.90];

J2000_coe     = J2000_elements(planet_id,:);
rates         = cent_rates(planet_id,:);

%...Convert from AU to km:
au            = 149597871;
J2000_coe(1)  = J2000_coe(1)*au;
rates(1)      = rates(1)*au;

%...Convert from arcseconds to fractions of a degree:
rates(3:6)    = rates(3:6)/3600;
```

```
   return
%  ~~~~~~~~~~~~~~~~~~~~~~~~~~~~~~~~~~~~~~~~~~~~~~~~~~~~~~~~~~~~~~~~



%  ~~~~~~~~~~~~~~~~~~~~~~~~~~~~~~~~~~~~~~~~~~~~~~~~~~~~~~~~~~~~~~~~
   function y = zero_to_360(x)
%  ~~~~~~~~~~~~~~~~~~~~~~~~~~~
%
% This function reduces an angle to lie in the range
% 0 - 360 degrees.
%
% x - the original angle in degrees
% y - the angle reduced to the range 0 - 360 degrees
%
% User M-functions required: none
% ------------------------------------------------------------

if x >= 360
    x = x - fix(x/360)*360;
elseif x < 0
    x = x - (fix(x/360) - 1)*360;

end

y = x;

return
%  ~~~~~~~~~~~~~~~~~~~~~~~~~~~~~~~~~~~~~~~~~~~~~~~~~~~~~~~~~~~~~~~~
```

SCRIPT FILE Example_8_07.m

```
%  ~~~~~~~~~~~~~~~~~~~~~~~~~~~~~~~~~~~~~~~~~~~~~~~~~~~~~~~~~~~~~~~~
% Example_8_07
%  ~~~~~~~~~~~~
%
% This program uses Algorithm 8.1 to compute the orbital
% elements and state vector of the earth at the date and time
% specified in Example 8.7. To obtain the same results for
% Mars, set planet_id = 4.
%
% mu         - gravitational parameter of the sun (km^3/s^2)
% deg        - conversion factor between degrees and radians
% pi         - 3.1415926...
%
% coe        - vector of heliocentric orbital elements
%              [h e RA incl w TA a w_hat L M E],
%              where
%                  h     = angular momentum            (km^2/s)
%                  e     = eccentricity
%                  RA    = right ascension             (deg)
%                  incl  = inclination                 (deg)
%                  w     = argument of perihelion      (deg)
%                  TA    = true anomaly                (deg)
%                  a     = semimajor axis              (km)
```

```
%                  w_hat  = longitude of perihelion
%                           ( = RA + w)                       (deg)
%                  L      = mean longitude ( = w_hat + M)     (deg)
%                  M      = mean anomaly                      (deg)
%                  E      = eccentric anomaly                 (deg)
%
% r         - heliocentric position vector (km)
% v         - heliocentric velocity vector (km/s)
%
% planet_id - planet identifier:
%                  1 = Mercury
%                  2 = Venus
%                  3 = Earth
%                  4 = Mars
%                  5 = Jupiter
%                  6 = Saturn
%                  7 = Uranus
%                  8 = Neptune
%                  9 = Pluto
%
% year      - range: 1901 - 2099
% month     - range: 1 - 12
% day       - range: 1 - 31
% hour      - range: 0 - 23
% minute    - range: 0 - 60
% second    - range: 0 - 60
%
% User M-functions required: planet_elements_and_sv,
%                            month_planet_names
% -----------------------------------------------------------

global mu
mu  = 1.327124e11;
deg = pi/180;

%...Input data
planet_id = 3;
year      = 2003;
month     = 8;
day       = 27;
hour      = 12;
minute    = 0;
second    = 0;
%...

%...Algorithm 8.1:
[coe, r, v, jd] = planet_elements_and_sv ...
          (planet_id, year, month, day, hour, minute, second);

%...Convert the planet_id and month numbers into names for output:
[month_name, planet_name] = month_planet_names(month, ...
                            planet_id);

%...Echo the input data and output the solution to
%   the command window:
fprintf('-----------------------------------------------------')
fprintf('\n Example 8.7')
```

D.17 Algorithm 8.1: calculation of the state vector of a planet at a given epoch

```
fprintf('\n\n Input data:\n');
fprintf('\n   Planet: %s', planet_name)
fprintf('\n   Year  : %g', year)
fprintf('\n   Month : %s', month_name)
fprintf('\n   Day   : %g', day)
fprintf('\n   Hour  : %g', hour)
fprintf('\n   Minute: %g', minute)
fprintf('\n   Second: %g', second)
fprintf('\n\n   Julian day: %11.3f', jd)

fprintf('\n\n');
fprintf(' Orbital elements:')
fprintf('\n');

fprintf('\n  Angular momentum (km^2/s)           = %g', coe(1));
fprintf('\n  Eccentricity                        = %g', coe(2));
fprintf('\n  Right ascension of the ascending node')
fprintf(' (deg) = %g', coe(3));
fprintf('\n  Inclination to the ecliptic (deg)   = %g', coe(4));
fprintf('\n  Argument of perihelion (deg)        = %g', coe(5));
fprintf('\n  True anomaly (deg)                  = %g', coe(6));
fprintf('\n  Semimajor axis (km)                 = %g', coe(7));

fprintf('\n');

fprintf('\n  Longitude of perihelion (deg)       = %g', coe(8));
fprintf('\n  Mean longitude (deg)                = %g', coe(9));
fprintf('\n  Mean anomaly (deg)                  = %g', coe(10));
fprintf('\n  Eccentric anomaly (deg)             = %g', coe(11));

fprintf('\n\n');
fprintf(' State vector:')
fprintf('\n');

fprintf('\n  Position vector (km) = [%g  %g  %g]', ...
                                    r(1), r(2), r(3))
fprintf('\n  Magnitude            = %g\n', norm(r))
fprintf('\n  Velocity (km/s)      = [%g  %g  %g]', ...
                                    v(1), v(2), v(3))
fprintf('\n  Magnitude            = %g', norm(v))

fprintf('\n---------------------------------------------\n')
% ~~~~~~~~~~~~~~~~~~~~~~~~~~~~~~~~~~~~~~~~~~~~~~~~~~~~~~~~~~~
```

Output from Example_8_07

```
-----------------------------------------------------
 Example 8.7

 Input data:

   Planet: Earth
   Year  : 2003
   Month : August
```

```
                    Day    : 27
                    Hour   : 12
                    Minute: 0
                    Second: 0

                    Julian day: 2452879.000

                Orbital elements:

                    Angular momentum (km^2/s)                    = 4.4551e+09
                    Eccentricity                                 = 0.0167088
                    Right ascension of the ascending node (deg)  = 348.554
                    Inclination to the ecliptic (deg)            = -0.000426218
                    Argument of perihelion (deg)                 = 114.405
                    True anomaly (deg)                           = 230.812
                    Semimajor axis (km)                          = 1.49598e+08

                    Longitude of perihelion (deg)                = 102.959
                    Mean longitude (deg)                         = 335.267
                    Mean anomaly (deg)                           = 232.308
                    Eccentric anomaly (deg)                      = 231.558

                State vector:

                    Position vector (km) = [1.35589e+08  -6.68029e+07  286.909]
                    Magnitude            = 1.51152e+08
                    Velocity (km/s)      = [12.6804  26.61  -0.000212731]
                    Magnitude            = 29.4769
                -----------------------------------------------------------
```

D.18 ALGORITHM 8.2: CALCULATION OF THE SPACECRAFT TRAJECTORY FROM PLANET 1 TO PLANET 2

FUNCTION FILE interplanetary.m

```
% ~~~~~~~~~~~~~~~~~~~~~~~~~~~~~~~~~~~~~~~~~~~~~~~~~~~~~~~~~~~~~~~~~
  function [planet1, planet2, trajectory] = interplanetary ...
                                                (depart, arrive)
% ~~~~~~~~~~~~~~~~~~~~~~~~~~~~~~~~~~~~~~~~~~~~~~~~~~~~~~~~~~~~~~~~~
%
% This function determines the spacecraft trajectory from the
% sphere of influence of planet 1 to that of planet 2 using
% Algorithm 8.2.
%
% mu         - gravitational parameter of the sun (km^3/s^2)
% dum        - a dummy vector not required in this procedure
%
% planet_id  - planet identifier:
%              1 = Mercury
%              2 = Venus
%              3 = Earth
%              4 = Mars
```

```
%                       5 = Jupiter
%                       6 = Saturn
%                       7 = Uranus
%                       8 = Neptune
%                       9 = Pluto
%
% year              - range: 1901 - 2099
% month             - range: 1 - 12
% day               - range: 1 - 31
% hour              - range: 0 - 23
% minute            - range: 0 - 60
% second            - range: 0 - 60
%
% jd1, jd2          - Julian day numbers at departure and arrival
% tof               - time of flight from planet 1 to planet 2 (s)
%
% Rp1, Vp1          - state vector of planet 1 at departure (km, km/s)
% Rp2, Vp2          - state vector of planet 2 at arrival (km, km/s)
% R1, V1            - heliocentric state vector of spacecraft at
%                     departure (km, km/s)
% R2, V2            - heliocentric state vector of spacecraft at
%                     arrival (km, km/s)
%
% depart            - [planet_id, year, month, day, hour, minute,
%                     second] at departure
% arrive            - [planet_id, year, month, day, hour, minute,
%                     second] at arrival

% planet1           - [Rp1, Vp1, jd1]
% planet2           - [Rp2, Vp2, jd2]
% trajectory        - [V1, V2]
%
% User M-functions required: planet_elements_and_sv, lambert
% -------------------------------------------------------------
global mu

planet_id = depart(1);
year      = depart(2);
month     = depart(3);
day       = depart(4);
hour      = depart(5);
minute    = depart(6);
second    = depart(7);

%...Use Algorithm 8.1 to obtain planet 1's state vector (don't
%...need its orbital elements [''dum'']):
[dum, Rp1, Vp1, jd1] = planet_elements_and_sv ...
          (planet_id, year, month, day, hour, minute, second);

planet_id = arrive(1);
year      = arrive(2);
month     = arrive(3);
day       = arrive(4);
hour      = arrive(5);
minute    = arrive(6);
second    = arrive(7);

%...Likewise use Algorithm 8.1 to obtain planet 2's state vector:
```

```
            [dum, Rp2, Vp2, jd2] = planet_elements_and_sv ...
                    (planet_id, year, month, day, hour, minute, second);

            tof = (jd2 - jd1)*24*3600;

            %...Patched conic assumption:
            R1 = Rp1;
            R2 = Rp2;

            %...Use Algorithm 5.2 to find the spacecraft's velocity at
            %   departure and arrival, assuming a prograde trajectory:
            [V1, V2] = lambert(R1, R2, tof, 'pro');

            planet1    = [Rp1, Vp1, jd1];
            planet2    = [Rp2, Vp2, jd2];
            trajectory = [V1, V2];

            % ~~~~~~~~~~~~~~~~~~~~~~~~~~~~~~~~~~~~~~~~~~~~~~~~~~~~~~~~~~~~
```

SCRIPT FILE Example_8_08.m

```
            % ~~~~~~~~~~~~~~~~~~~~~~~~~~~~~~~~~~~~~~~~~~~~~~~~~~~~~~~~~~~~
            % Example_8_08
            % ~~~~~~~~~~~~
            %
            % This program uses Algorithm 8.2 to solve Example 8.8.
            %
            % mu            - gravitational parameter of the sun (km^3/s^2)
            % deg           - conversion factor between degrees and radians
            % pi            - 3.1415926...
            %
            % planet_id     - planet identifier:
            %                   1 = Mercury
            %                   2 = Venus
            %                   3 = Earth
            %                   4 = Mars
            %                   5 = Jupiter
            %                   6 = Saturn
            %                   7 = Uranus
            %                   8 = Neptune
            %                   9 = Pluto
            % planet_name   - name of the planet
            %
            % year          - range: 1901 - 2099
            % month         - range: 1 - 12
            % month_name    - name of the month
            % day           - range: 1 - 31
            % hour          - range: 0 - 23
            % minute        - range: 0 - 60
            % second        - range: 0 - 60
            %
            % depart        - [planet_id, year, month, day, hour, minute,
            %                  second] at departure
            % arrive        - [planet_id, year, month, day, hour, minute,
            %                  second] at arrival
            %
            % planet1       - [Rp1, Vp1, jd1]
```

```
% planet2        - [Rp2, Vp2, jd2]
% trajectory     - [V1, V2]
%
% coe            - orbital elements [h e RA incl w TA]
%                  where
%                    h    = angular momentum (km^2/s)
%                    e    = eccentricity
%                    RA   = right ascension of the ascending
%                           node (rad)
%                    incl = inclination of the orbit (rad)
%                    w    = argument of perigee (rad)
%                    TA   = true anomaly (rad)
%                    a    = semimajor axis (km)
%
% jd1, jd2       - Julian day numbers at departure and arrival
% tof            - time of flight from planet 1 to planet 2
%                  (days)
%
% Rp1, Vp1       - state vector of planet 1 at departure
%                  (km, km/s)
% Rp2, Vp2       - state vector of planet 2 at arrival
%                  (km, km/s)
% R1, V1         - heliocentric state vector of spacecraft at
%                  departure (km, km/s)
% R2, V2         - heliocentric state vector of spacecraft at
%                  arrival (km, km/s)
%
% vinf1, vinf2   - hyperbolic excess velocities at departure
%                  and arrival (km/s)
%
% User M-functions required: interplanetary, coe_from_sv,
%                            month_planet_names
% --------------------------------------------------------------

clear
global mu
mu  = 1.327124e11;
deg = pi/180;

%...Data for planet 1:
planet_id = 3; % (earth)
year      = 1996;
month     = 11;
day       = 7;
hour      = 0;
minute    = 0;
second    = 0;
%...

depart = [planet_id year month day hour minute second];

%...Data for planet 2:
planet_id = 4; % (Mars)
year      = 1997;
month     = 9;
day       = 12;
```

```
hour     = 0;
minute   = 0;
second   = 0;
%...

arrive = [planet_id  year  month  day  hour  minute  second];

[planet1, planet2, trajectory] = interplanetary ...
                                   (depart, arrive);
R1  = planet1(1,1:3);
Vp1 = planet1(1,4:6);
jd1 = planet1(1,7);

R2  = planet2(1,1:3);
Vp2 = planet2(1,4:6);
jd2 = planet2(1,7);

V1 = trajectory(1,1:3);
V2 = trajectory(1,4:6);

tof = jd2 - jd1;

%...Use Algorithm 4.1 to find the orbital elements of the
%   spacecraft trajectory based on [Rp1, V1]...
coe  = coe_from_sv(R1, V1);
%    ... and [R2, V2]
coe2 = coe_from_sv(R2, V2);

%...Equations 8.102 and 8.103:
vinf1 = V1 - Vp1;
vinf2 = V2 - Vp2;

%...Echo the input data and output the solution to
%   the command window:
fprintf('-----------------------------------------------------')
fprintf('\n Example 8.8')
fprintf('\n\n Departure:\n');
[month_name, planet_name] = month_planet_names(depart(3), ...
                             depart(1));
fprintf('\n   Planet: %s', planet_name)
fprintf('\n   Year  : %g', depart(2))
fprintf('\n   Month : %s', month_name)
fprintf('\n   Day   : %g', depart(4))
fprintf('\n   Hour  : %g', depart(5))
fprintf('\n   Minute: %g', depart(6))
fprintf('\n   Second: %g', depart(7))
fprintf('\n\n   Julian day: %11.3f\n', jd1)
fprintf('\n   Planet position vector (km)   = [%g  %g  %g]', ...
                                     R1(1), R1(2), R1(3))

fprintf('\n     Magnitude                  = %g\n', norm(R1))

fprintf('\n   Planet velocity (km/s)        = [%g  %g  %g]', ...
                                     Vp1(1), Vp1(2), Vp1(3))

fprintf('\n     Magnitude                  = %g\n', norm(Vp1))
```

D.18 Algorithm 8.2: calculation of the spacecraft trajectory

```
fprintf('\n   Spacecraft velocity (km/s)    = [%g  %g  %g]', ...
                                V1(1), V1(2), V1(3))

fprintf('\n   Magnitude                     = %g\n', norm(V1))

fprintf('\n   v-infinity at departure (km/s) = [%g  %g  %g]', ...
                                vinf1(1), vinf1(2), vinf1(3))

fprintf('\n   Magnitude                     = %g\n', norm(vinf1))

fprintf('\n\n Time of flight = %g days\n', tof)

fprintf('\n\n Arrival:\n');
[month_name, planet_name] = month_planet_names(arrive(3), ...
                                arrive(1));
fprintf('\n   Planet: %s', planet_name)
fprintf('\n   Year  : %g', arrive(2))
fprintf('\n   Month : %s', month_name)
fprintf('\n   Day   : %g', arrive(4))
fprintf('\n   Hour  : %g', arrive(5))
fprintf('\n   Minute: %g', arrive(6))
fprintf('\n   Second: %g', arrive(7))
fprintf('\n\n   Julian day: %11.3f\n', jd2)
fprintf('\n   Planet position vector (km)   = [%g  %g  %g]', ...
                                R2(1), R2(2), R2(3))

fprintf('\n   Magnitude                     = %g\n', norm(R1))

fprintf('\n   Planet velocity (km/s)        = [%g  %g  %g]', ...
                                Vp2(1), Vp2(2), Vp2(3))

fprintf('\n   Magnitude                     = %g\n', norm(Vp2))

fprintf('\n   Spacecraft Velocity (km/s)    = [%g  %g  %g]', ...
                                V2(1), V2(2), V2(3))

fprintf('\n   Magnitude                     = %g\n', norm(V2))

fprintf('\n   v-infinity at arrival (km/s)  = [%g  %g  %g]', ...
                                vinf2(1), vinf2(2), vinf2(3))
fprintf('\n   Magnitude                     = %g', norm(vinf2))

fprintf('\n\n\n Orbital elements of flight trajectory:\n')

fprintf('\n   Angular momentum (km^2/s)     = %g', coe(1))
fprintf('\n   Eccentricity                  = %g', coe(2))
fprintf('\n   Right ascension of the ascending node')
fprintf(' (deg) = %g', coe(3)/deg)
fprintf('\n   Inclination to the ecliptic (deg)  = %g', ...
                                coe(4)/deg)
fprintf('\n   Argument of perihelion (deg)  = %g', ...
                                coe(5)/deg)
fprintf('\n   True anomaly at departure (deg) = %g', ...
                                coe(6)/deg)
fprintf('\n   True anomaly at arrival (deg) = %g\n', ...
                                coe2(6)/deg)
fprintf('\n   Semimajor axis (km)           = %g', coe(7))
```

```
    if coe(2) < 1
        fprintf('\n  Period  (days)                  = %g', ...
                                  2*pi/sqrt(mu)*coe(7)^1.5/24/3600)
end
fprintf('\n-----------------------------------------------\n')
% ~~~~~~~~~~~~~~~~~~~~~~~~~~~~~~~~~~~~~~~~~~~~~~~~~~~~~~~~~~~~~~
```

Output from `Example_8_08`

```
-----------------------------------------------------
 Example 8.8

 Departure:

   Planet: Earth
   Year   : 1996
   Month  : November
   Day    : 7
   Hour   : 0
   Minute : 0
   Second : 0

   Julian day: 2450394.500

   Planet position vector (km)    = [1.04994e+08 1.04655e+08 988.331]
   Magnitude                      = 1.48244e+08

   Planet velocity (km/s)         = [-21.515 20.9865 0.000132284]
   Magnitude                      = 30.0554

   Spacecraft velocity (km/s)     = [-24.4282 21.7819 0.948049]
   Magnitude                      = 32.7427

   v-infinity at departure (km/s) = [-2.91321 0.79542 0.947917]
   Magnitude                      = 3.16513


 Time of flight = 309 days


 Arrival:

   Planet: Mars
   Year   : 1997
   Month  : September
   Day    : 12
   Hour   : 0
   Minute : 0
   Second : 0

   Julian day: 2450703.500
   Planet position vector (km)    = [-2.08329e+07 -2.18404e+08 -4.06287e+06]
   Magnitude                      = 1.48244e+08

   Planet velocity (km/s)         = [25.0386 -0.220288 -0.620623]
   Magnitude                      = 25.0472
```

```
    Spacecraft Velocity (km/s)    = [22.1581 -0.19666 -0.457847]
    Magnitude                     = 22.1637

    v-infinity at arrival (km/s)  = [-2.88049 0.023628 0.162776]
    Magnitude                     = 2.88518


Orbital elements of flight trajectory:

    Angular momentum (km^2/s)                       = 4.84554e+09
    Eccentricity                                    = 0.205785
    Right ascension of the ascending node (deg)     = 44.8942
    Inclination to the ecliptic (deg)               = 1.6621
    Argument of perihelion (deg)                    = 19.9738
    True anomaly at departure (deg)                 = 340.039
    True anomaly at arrival (deg)                   = 199.695

    Semimajor axis (km)                             = 1.84742e+08
    Period (days)                                   = 501.254
-----------------------------------------------------------------
```

APPENDIX E

Gravitational potential energy of a sphere

Figure E.1 shows a point mass m with cartesian coordinates (x, y, z) as well as a system of N point masses $m_1, m_2, m_3, \ldots, m_N$. The ith one of these particles has mass m_i and coordinates (x_i, y_i, z_i). The total mass of the N particles is M,

$$M = \sum_{i=1}^{N} m_i \tag{E.1}$$

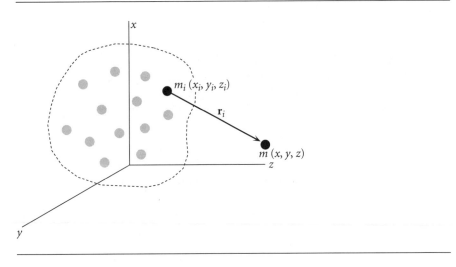

Figure E.1 A system of point masses and a neighboring test mass m.

Appendix E Gravitational potential energy of a sphere

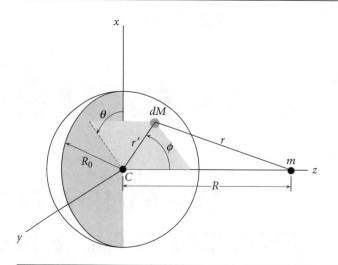

Figure E.2 Sphere with a spherically symmetric mass distribution.

The position vector drawn from m_i to m is $\mathbf{r}_i$ and the unit vector in the direction of $\mathbf{r}_i$ is

$$\hat{\mathbf{u}}_i = \frac{\mathbf{r}_i}{r_i}$$

The gravitational force exerted on m by m_i is opposite in direction to $\mathbf{r}_i$, and is given by

$$\mathbf{F}_i = -\frac{Gmm_i}{r_i^2}\hat{\mathbf{u}}_i = -\frac{Gmm_i}{r_i^3}\mathbf{r}_i$$

The potential energy of this force is

$$V_i = -G\frac{mm_i}{r_i} \tag{E.2}$$

The total gravitational potential energy of the system due to the gravitational attraction of all of the N particles is

$$V = \sum_{i=1}^{N} V_i \tag{E.3}$$

Therefore, the total force of gravity $\mathbf{F}$ on the mass m is

$$\mathbf{F} = -\nabla V = -\left(\frac{\partial V}{\partial x}\hat{\mathbf{i}} + \frac{\partial V}{\partial y}\hat{\mathbf{j}} + \frac{\partial V}{\partial z}\hat{\mathbf{k}}\right) \tag{E.4}$$

Consider the solid sphere of mass M and radius R_0 illustrated in Figure E.2. Instead of a discrete system as above, we have a continuum with mass density ϱ. Each 'particle'

Appendix E *Gravitational potential energy of a sphere*

is a differential element $dM = \varrho\,dv$ of the total mass M. Equation E.1 becomes

$$M = \iiint_v \varrho\,dv \tag{E.5}$$

where dv is the volume element and v is the total volume of the sphere. In this case, Equation E.2 becomes

$$dV = -G\frac{m\,dM}{r} = -Gm\frac{\varrho\,dv}{r}$$

where r is the distance from the differential mass dM to the finite point mass m. Equation E.3 is replaced by

$$V = -Gm \iiint_v \frac{\varrho\,dv}{r} \tag{E.6}$$

Let the mass of the sphere have a spherically symmetric distribution, which means that the mass density ϱ depends only on r', the distance from the center C of the sphere. An element of mass dM has spherical coordinates (r', θ, ϕ), where the angle θ is measured in the xy plane of a cartesian coordinate system with origin at C, as shown in Figure E.2. In spherical coordinates the volume element is

$$dv = r'^2 \sin\phi\,d\phi\,dr'\,d\theta \tag{E.7}$$

Therefore Equation E.5 becomes

$$M = \int_{\theta=0}^{2\pi} \int_{r'=0}^{R_0} \int_{\phi=0}^{\pi} \varrho r'^2 \sin\phi\,d\phi\,dr'\,d\theta = \left(\int_0^{2\pi} d\theta\right)\left(\int_0^{\pi} \sin\phi\,d\phi\right)\left(\int_0^{R_0} \varrho r'^2\,dr'\right)$$

$$= (2\pi)(2)\left(\int_0^{R_0} \varrho r'^2\,dr'\right)$$

so that the mass of the sphere is given by

$$M = 4\pi \int_{r'=0}^{R_0} \varrho r'^2\,dr' \tag{E.8}$$

Substituting Equation E.7 into Equation E.6 yields

$$V = -Gm \int_{\theta=0}^{2\pi} \int_{r'=0}^{R_0} \int_{\phi=0}^{\pi} \frac{\varrho r'^2 \sin\phi\,d\phi\,dr'\,d\theta}{r}$$

$$= -2\pi Gm \left[\int_0^{R_0} \left(\int_0^{\pi} \frac{\sin\phi\,d\phi}{r}\right) \varrho r'^2\,dr'\right] \tag{E.9}$$

The distance r is found by using the law of cosines,

$$r = (R^2 + r'^2 - 2r'R\cos\phi)^{\frac{1}{2}}$$

where R is the distance from the center of the sphere to the mass m. Differentiating this equation with respect to ϕ, holding r' constant, yields

$$\frac{dr}{d\phi} = \frac{1}{2}(R^2 + r'^2 - 2r'R\cos\phi)^{-\frac{1}{2}}(2r'R\sin\phi\,d\phi) = \frac{r'R\sin\phi}{r}$$

so that
$$\sin\phi\, d\phi = \frac{r\, dr}{r'R}$$

It follows that
$$\int_{\phi=0}^{\pi} \frac{\sin\phi\, d\phi}{r} = \frac{1}{r'R} \int_{R-r'}^{R+r'} dr = \frac{2}{R}$$

Substituting this result along with Equation E.8 into Equation E.9 yields
$$V = -\frac{GMm}{R}$$

We conclude that the gravitational potential energy, and hence (from Equation E.4) the gravitational force, of a sphere with a spherically symmetric mass distribution M is the same as that of a point mass M located at the center of the sphere.

Index

Absolute acceleration
  angular 402–8, 436–40, 484–6
  nutation dampers 496
  point masses 20–9
  rigid-body kinematics 401–8, 410, 411
  two-body motion 35
Absolute angular momentum 411–14
Absolute angular velocity
  gyroscopic attitude control 521
  nutation dampers 496
  rigid-body dynamics 402–7, 451
  torque-free motion 479–80
Absolute position vectors 20–9
Absolute velocity
  close-proximity circular orbits 340
  rigid-body kinematics 401–8, 451
  two-body motion 35
  two-impulse maneuvers 331
  vectors 20–9
Acceleration
  see also absolute...; angular...; relative
  Coriolis 21
  five-term 21, 23
  gravitational 7–10, 177
  gyroscopic attitude control 521–5
  oblateness 177–8
  point masses 2–7, 16–18, 20–9
  preliminary orbit determination 228

relative motion and rendezvous 317–20
restricted three-body motion 91
rocket vehicle dynamics 553
three-body systems 590–4
Advance of perigee 178–80, 184
Aiming radius
  hyperbolic trajectories 71–2, 75, 382
  planetary rendezvous 370–1, 373
Altitude
  equation 554
  gravity-gradient stabilization 534
  perigee 64–5, 208–10, 211–12
  preliminary orbit determination 208–10, 211–12, 231–5
  rocket performance 554, 558–9
  Sun-synchronous three dimensional orbits 181
  two-body motion 53–4
Amplitude 478
Angles
  see also flight path...
  auxiliary 111–15
  azimuth 227–8, 231–6, 626–31
  dihedral 290, 293
  elevation 227–8, 232, 235–6, 626–31
  Euler's 158–9, 448–59, 480
  to periapse 373, 374–5
  phase 350–3
  preliminary orbit determination 228–50
  of rotation 282–5

spin 508
tilt 446
turn 70, 75, 369–70, 378, 382
wobble 481–2
Angular acceleration
  absolute 402–8, 436–40, 484–6
  gyroscopic attitude control 529–30
  point masses 16–18, 20–9
  relative 317, 319, 437
  rendezvous 317, 319
  rigid-body kinematics 401–8
  satellite attitude dynamics 513, 529–30, 484–6
  torque-free motion 484–6
  yo-yo despin 513
Angular momentum
  chase maneuvers 286–9
  conservation of 79–80
  double-gimbaled control moment gyros 528–9
  Hohmann transfers 261
  hyperbolic trajectories 74–5, 130
  Lagrange coefficients 78–89
  moments of inertia 414–15, 420–1, 423–5
  orbit formulas 42–50
  plane change maneuvers 302–3
  planetary departure 361–2
  planetary flyby 379
  point masses 13–15
  preliminary orbit determination 195–201, 204–5, 236–8
  rigid-body dynamics 414–15, 420–1, 423–5, 435–40

661

Angular momentum (continued)
    satellite attitude dynamics
        coning maneuvers 503–5
        dual-spin spacecraft 491–2
        gyroscopic control 517,
            518–22, 524–5
        nutation dampers 498–9
        thrusters 506–9
        torque-free motion 476–7,
            481–2, 484–6, 488
        yo-yo despin 509–10, 511
    spinning tops 447
    three dimensional orbits
        158–60, 161–2
    torque-free motion 476–7,
        481–2, 484–6, 488
    two-body motion 42–50,
        74–5, 78–89
Angular position 232, 629–34
Angular velocity
    close-proximity circular orbits
        338
    Euler angles 453–6
    Euler's equations 436–40
    moments of inertia 420–1
    pitch 462–3
    point masses 16–18, 21, 25–7
    relative motion and
        rendezvous 316–17, 319
    rigid-body dynamics 420–1,
        436–40, 453–6
    roll 462–3
    satellite attitude dynamics
        dual-spin spacecraft 493–5
        gravity-gradient
            stabilization 535–6
        gyroscopic control 516–17,
            521, 523–5
        thrusters 506
        torque-free motion
            477–80, 484, 486–9
        yo-yo despin 511–13
    spinning tops 444
    two-body motion 47–8
    yaw 462–3
Angular-impulse 13–15,
    413–14
Apoapse 56, 290–303, 373

Apogee
    kick 258–60
    radius 62, 70–1, 183–7
    towards the sun 117–19
    velocity 62
Applied torque 413–14
Approach trajectories 368–75,
    379–86, 397
    see also two-body motion
Apse lines 260, 273–85
Arcseconds 387, 388
Areal velocity 44
Argument of perigee
    oblateness 178–81, 183–7
    orbital elements 159, 161, 163
Arrival phase 391, 393–4
Astronomical units 387, 388
Attitude dynamics see satellite...
Auxiliary angles 111–15
Axial bearing loads 459
Axial torques 523–5
Axis of rotation 150
Axisymmetric dual-spin 491–5,
    518–21, 529–30
Axisymmetric tops 443–8
Azimuth
    angles 227–8, 231–6, 626–31
    averaged radius 62
    plane change maneuvers
        294–5

Bearing forces 456–9
Bent rods 431–5
Bessel functions 121–2
Bi-elliptic Hohmann transfers
    264–8
Bias values 526
Bivariate functions 570
Body cones 482
Body frames 18
Burnout
    Jacobi constant 100–1
    rocket vehicle dynamics
        558–9, 561, 564–78
    sensitivity analysis 366–8

Capture orbits 372–3, 375,
    396–7
Capture radius 372
Cartesian coordinates
    elliptical orbits 58
    equation of a parabola 68
    hyperbolic trajectories 72–3
    rotation 169–72
    three dimensional orbits
        164–5
Cassini gravity assist maneuvers
    386
Celestial bodies 149–54
Center of mass
    inertial frames 34–7
    moving reference frames
        38–42
    rigid-body dynamics
        Euler angles 454–6
        Euler's equations 435–40
        moments of inertia 417–18,
            420–1, 423–5, 430–5
        parallel axis theorem 430–5
        rotational motion 412–14
        translational motion
            408–10
    two-body motion 34–7,
        38–42
Chase maneuvers 285–9,
    322–40
Chasles' theorem 399–400
Circular orbits
    close-proximity relative
        motion 338–40
    Hohmann transfers 257, 264
    parking 362, 394–6
    position as a function of time
        108–9
    rigid-body kinetic energy
        442–3
    two-body motion 51–5
Classical orbital elements
    159–61, 175, 199–201,
    607–14
Clohessy–Wiltshire (CW)
    frames
        equations 324–30, 336–7,
            338, 340

gravity-gradient stabilization 533–4
matrices 329, 333, 336
Close-proximity circular orbits 338–40
Co-moving reference frames 316–22, 324–30
Coaxial elliptic orbits 260, 273–4
Common apse lines 273–9
Common focus 290
Conics 359–60, 391–7
Coning maneuvers 503–5, 507
Conservation of…
angular momentum 79–80
energy 65, 509–10
momentum 79–80, 509–10
Constant amplitude 478
Continuous three dimensional bodies 408–10
Control moment gyros 526–30
Coordinate systems 218–28
see also Cartesian…; topocentric…
polar 48
Coordinate transformations
geocentric equatorial 172–6, 186–7, 224–8
perifocal frames 172–6, 186–7
rotation 169–72
three dimensional orbits 164–76, 186–7
topocentric 224–8
Coplanar orbits 257, 297–8, 340
Cord lengths 509–16
Cord unwind rates 512–13
Coriolis acceleration 21
Cosine vectors 242–3, 244
Cruise phase 391–2, 393–6
Curvature of the earth 553–4
Curvilinear motion 1–7
CW *see* Clohessy–Wiltshire frames

Dark side approaches 379–84, 385

Declination
preliminary orbit determination 222–3, 225–6, 230–2
state vectors 155–8
three dimensional orbits 149–54
Delta-H requirements 504–5, 507–9
Delta-v requirements
bi-elliptic Hohmann transfers 264–8
chase maneuvers 285–9
Hohmann transfers 257–73, 348–50
impulsive orbital maneuvers 256–7, 330–7
interplanetary trajectories 348–50, 362, 364–6
non-Hohmann transfers 273–82, 394–7
phasing maneuvers 268–73
plane change maneuvers 290–303
planetary rendezvous 362, 364–6, 371–5
rocket vehicle dynamics 551–2
two-impulse maneuvers 330–7
Departure trajectories 360–6, 391, 393–6
Despin mechanisms 509–16
Diagonal moment of inertia matrices 425–8
Dihedral angles 290, 293
Direct ascent trajectories 363
Direction cosine vectors 242–3, 244
Distances between planets 389–91
Double-gimbaled control moment gyros 526–30
Downrange equations 554
Drag force 553, 558
Dual-spin satellites 518–21, 529–30
Dual-spin spacecraft 491–5

Earth
centered inertial frames 23–9
earth orbits 52–3, 149, 258–60
earth satellites 149, 183–7
earth-moon systems 98–101
earth's curvature 553–4
earth's gravitational parameter 52
earth's oblateness 177–87
earth's shadow 117–19
earth's sphere of influence 358–9
low earth orbits 52–3, 297–8, 300
East longitude 218–21, 222–3
East-North-Zenith (ENZ) frame 223
Easterly launches 294–5
Eccentric anomaly
hyperbolic trajectories 126–33
Kepler's equation 113–17, 130, 596–600
MATLAB algorithms 115, 130, 596–600
oblateness 184
orbit equation 135
position as a function of time 111–15, 126–33, 135
Eccentricity
chase maneuvers 286–9
elliptical orbits 55–65
hyperbolic trajectories 125–6
interplanetary trajectories 361–2, 387, 388
limiting values 120–1
non-Hohmann transfers 282–5
orbit formulas 46–7
orbital elements 158–9, 160–1, 163
plane change maneuvers 302–3
planetary departure 361–2
planetary ephemeris 387, 388
planetary flyby 379

Eccentricity (continued)
    planetary rendezvous 369
    position as a function of time
        113, 120–1, 125–6
    preliminary orbit
        determination 219
Ecliptic plane 150
Effective exhaust velocity 556–7
Eigenvalues 426–8
Eigenvectors 426–8
Elevation angles 227–8, 232,
    235–6, 626–31
Elliptical orbits
    Hohmann transfers 257–68
    non-Hohmann trajectories
        396–7
    position as a function of time
        109–23, 134–5
    two-body motion 55–65
Empty masses 560, 561, 566–9
Energy
    circular orbits 52
    conservation of 65, 509–10
    dissipation 495–503
    elliptical orbits 59
    Hohmann transfers 257
    hyperbolic trajectories 73
    kinetic 441–3, 488–91, 493–4,
        509–12
    law 50–1, 52, 59, 73
    non-Hohmann transfers
        275–6
    orbital elements 158–9
    plane change maneuvers 293
    position as a function of time
        135
    potential 36–7, 658–61
    sinks 492–5
    three dimensional orbits
        158–9
ENZ *see* East-North-Zenith
Ephemeris 152–3, 387–91
Epochs 388, 641–8
Equations of motion
    double-gimbaled control
        moment gyros 528–9
    dual-spin spacecraft 492
    inertial frames 34–7

integration 587–94
interplanetary trajectories
    356–7
linearization of relative
    motion 322–4
numerical integration 587–94
relative 37–42, 322–4
rocket vehicle dynamics
    552–5
rotational 410–14, 517–21
satellite attitude dynamics
    496–503
translational 408–10
Equations of parabolas 68
Equatorial frames
    see also geocentric...
    plane change maneuvers
        293–4, 301–2
    state vectors 154–8
    three dimensional orbits 150
    topocentric coordinates
        221–3, 225–7
Equilibrium points 92–6
Escape velocity 66, 73
Euler, Leonhard 16
Euler rotations 448–59
Euler's angles 158–9, 448–59,
    480
Euler's equation
    rigid-body dynamics 435–40
    satellite attitude dynamics
        478, 485–7, 525, 533
Excess speed 73–4
Excess velocity 360–6, 368–75,
    392–4
Exhaust 555–7
Extremum 571–2

Field-free space restricted
    staging 560–70
Five-term acceleration 21, 23
Flattening *see* oblateness
Flight path angles
    elliptical orbits 63–4
    hyperbolic flyby 381
    Newton's law of gravitation
        9–10

non-Hohmann transfers
    274–5
parabolic trajectories 66
rocket vehicle dynamics
    552–5
Flight time 265–8
Floor 120
Flow rates 558–9
Fluids 410
Flyby 375–86
Flywheels 516–30
Forces
    see also gravitational...
    bearing 456–9
    drag 553, 558
    gyroscopic 456–9
    lifting 554
    net 27–9
    nutation dampers 497
    point masses 7–15
    sphere of influence 355–9
    units of 10–15
Free-fall 9–10

Gases 410
Gauss's method of preliminary
    orbit determination
    235–50, 631–41
GEO *see* geostationary
    equatorial orbits
Geocentric...
    latitude 219–21
    orbits 115–17, 130–3, 442–3
    position vectors 194–201
    right ascension-declination
        149–54, 155–8
    satellites 276–9
Geocentric equatorial frames
    coordinate transformations
        172–6, 186–7, 224–8
    MATLAB algorithms 232,
        626–31
    orbital elements 158–64,
        175–6
    perifocal frame 172–6, 186–7
    state vectors 154–8, 175–6

topocentric transformations 224–8
transformations 172–6, 186–7, 224–8
Geodetic latitude 220–1
Geostationary equatorial orbits (GEO) 53–6
  phasing maneuvers 271–3
  plane change maneuvers 293–4, 297–8, 300
Geosynchronous dual-spin communication satellites 494
Gibb's method 194–201, 614–18
Gimbals 406–8, 526–30
Gradient operator 36–7
Gravitation
  acceleration 7–10, 177
  attraction 33–105
  geocentric right ascension-declination 151–2
  point masses 7–10
  potential energy 658–61
  restricted three-body motion 91–2
  satellite attitude dynamics 530–1
  sphere of influence 355–9
Gravity assist maneuvers 385–6
Gravity gradient stabilization 530–42
Gravity turn trajectories 552–5
Greenwich sidereal time 214, 216, 218
Ground track 296–7
Guided missiles 554
Gyros
  gyroscope equation 447
  gyroscopic attitude control 516–30
  gyroscopic forces 456–9
  gyroscopic moment 447
  motors 439–40
  rotors 406–8, 420–1
  satellite attitude dynamics 491–5

Heliocentric trajectories 359
  approach velocity 368
  post-flyby 375–86
  speed 360, 363
  velocity 368, 375–86
High-energy precession rates 446–7
Hohmann transfers
  bi-elliptic transfers 264–8
  common apse line 274
  interplanetary trajectories 348–50, 391–7
  non-Hohmann trajectories 391–7
  orbital maneuvers 257–73, 274
  phasing maneuvers 268–73
  plane change maneuvers 297–9, 300–1
  planetary rendezvous 368–9, 373
Horizon coordinate system 223–8
Hyperbolas 130, 598–600
Hyperbolic trajectories
  approach 368–75, 397
  departure 360–6
  excess velocity 360–6, 368–86, 392–4
  flyby 375–86
  position as a function of time 125–35
  rotations 370–1
  two-body motion 69–76

Identity matrices 167
Impulse
  angular 13–15, 413–14
  coning maneuvers 503–5
  rendezvous maneuvers 257–73, 330–7
  rocket vehicle dynamics 552, 557–9, 562–4, 570–8
Impulsive orbital maneuvers 255–73
Inclination

double-gimbaled control moment gyros 528
plane change maneuvers 294–301
planetary ephemeris 387–8
Sun-synchronous orbits 181
three dimensional orbits 159, 160, 162
Inertia
  see also moments of inertia
  angular velocity 89, 402–3, 535–6
  equations of two-body motion 34–7
  gravity-gradient stabilization 531–2, 535–8
  matrices 416, 421–8, 519–25
  rigid-body dynamics 414–35, 457
  tensors 421–8, 434
  torque-free motion stability 491
  velocity 89, 91, 402–3, 535–6
Insertion points 293–4
Integration, equations of motion 587–94
Intercept trajectories 285–9
Intermediate-axis spinners 502–3
Interplanetary dual-spin spacecraft 494
Interplanetary trajectories 347–98
  ephemeris 387–91
  flyby 375–86
  Hohmann transfers 348–50
  method of patched conics 359–60
  non-Hohmann 391–7
  patched conics 359–60
  planetary departure 360–6
  planetary ephemeris 387–91
  planetary flyby 375–86
  planetary rendezvous 368–75
  rendezvous 349–54, 368–75
  sensitivity analysis 366–8
  sphere of influence 354–9
  three dimensional orbits 149

Iterations 242–3, 245–50, 631–40

Jacobi constant 96–101
Julian centuries 388–91
Julian days (JD) 214–18
   numbers 214–18, 388–91, 621–3, 641–8
Jupiter's right ascension 225–6

Kepler, Johannes 44
Kepler's equation
   Bessel functions 121–2
   eccentric anomaly 113–17, 130, 596–600
   hyperbola eccentric anomaly 115, 130, 598–600
   hyperbolic trajectories 128–30
   MATLAB algorithms 115, 130, 596–600, 601–3
   Newton's method 596–600, 601–3
   position as a function of time 1, 34–5, 113–17, 121–2, 128–30, 134–44
   universal variables 134–5, 136–44
Kepler's second law 44
Kilograms 10–15
Kinematics 2–7, 400–8
Kinetic energy 441–3, 488–91, 493–4, 509–12

Lagrange coefficients
   MATLAB algorithms 603–5
   position as a function of time 141–4
   preliminary orbit determination 204, 207–10, 237–9, 249
   two-body motion 78–89
Lagrange multiplier method 570–8
Lagrange points 92–6
Lambert's problem

chase maneuvers 285, 288–9
   MATLAB algorithms 208, 616–22
   patched conics 391–7
   preliminary orbit determination 202–13, 616–22
Laplace limit 120–1
Latitude 54–5, 218–23, 231, 294–7
Latus rectum 49, 302–3
Launch azimuth 294–7
Launch vehicle boost trajectories 552–5
Leading-side flyby 375–6, 378–9
LEO see low-earth orbits
Libration points 92–6
Lifting forces 554
Limiting values 120–1
Linear momentum 412
Linearized equations of relative motion 322–4
Local horizon 49
Local sidereal time 214, 216–18, 623–6
Longitude of perihelion 388
Low earth orbits (LEO) 52–3, 297–8, 300
Low-energy precession rates 446–7
Lunar trajectories 359

Major-axis spinners 502–3, 541
Mars missions 354–5
Mass
   gravitational potential energy 657–60
   moments of inertia 422
   nutation dampers 496–503
   point masses 7–15
   ratios 557–9, 560–1, 564
   rocket vehicle dynamics 573–8
MATLAB algorithms 595–656
   acceleration 590–4
   angular position 232, 626–31
   chase maneuvers 287–8

classical orbital elements 159–61, 175, 606–13
eccentric anomaly 115, 130, 138–9, 596–600
epochs 388, 641–8
Gauss's method of preliminary orbit determination 242–3, 245–50, 631–41
geocentric equatorial position 232, 626–31
Gibbs method of preliminary orbit determination 613–16
hyperbola eccentric anomaly 130, 598–600
Julian day number 388, 621–3, 641–8
Kepler's equation 115, 130, 138–9, 596–603
Lagrange coefficients 603–5
Lambert's problem 208, 616–22
local sidereal time 217, 623–7
month identity conversions 640–1
Newton's method 115, 130, 138–9, 596–603
non-Hohmann trajectories 393
numerical designation conversions 640–1
orbital elements from the state vector 159–61, 175, 606–13
planet identity conversions 640–1
planet state vector calculation 388, 641–8
planetary ephemeris 388–9
position as a function of time 142, 604–6
preliminary orbit determination 198, 208, 217, 232, 242–50, 613–40
range 232, 626–31
sidereal time 217, 623–6

spacecraft trajectories 393, 648–55
sphere of influence 393, 648–55
state vectors 159–61, 175, 232, 604–13, 626–31
Stumpff functions 600–1
three-body systems 589–94
time lapse 604–6
transformation matrices 175
universal anomaly 138–9, 601–3
universal Kepler's equation 138–9, 601–3
Universal Time 388, 641–8
Matrices
see also transformation...
Clohessy–Wiltshire frames 329, 333, 336
diagonal 425–8
direction cosines 166–72, 174–6, 186–7
identity matrices 167
inertia 416, 421–8, 519–25
moments of inertia 421–8, 519–25
orthogonal 320, 421–8, 449–50
rotation 460–3
unit 167
Mean...
anomaly 110–15, 124–6, 134–5, 159
distance 61
longitude 388
motion 110, 184, 326, 338
Mercator projections 296–7
Method of patched conics 359–60, 391–7
Minor-axis spinners 502–3, 541
Missiles 554
Molniya orbit 182–3
Moments 410–14, 435–40, 454–6
Moments of inertia
gravity-gradient stabilization 531–2
matrices 421–8, 519–25

parallel axis theorem 428–35
principal 419, 426–8, 431–6, 457
rigid-body dynamics 414–40, 457
torque-free motion stability 491
Momentum
see also angular...
absolute angular 411–14
conservation of 509–10
exchange systems 406–8, 420–1, 439–40, 491–5
linear 412
rigid-body rotational motion 412
rocket vehicle dynamics 555–7
yo-yo despin 509–10
Month identity conversions 640–1
Moon ephemeris 152–3
Moving reference frames 37–42, 316–22, 324–30
Moving vectors 15–20
Multi-stage vehicles 552, 562, 563–78
Mutual gravitational attraction 33–105
see also two-body motion

n body equations of motion 587–94
Net forces 27–9
Net moments 437–40, 454–6
Newton's law of gravitation 7–10, 355–9
Newton's laws of motion 10–15, 409
Newton's method
Kepler's equation 596–600, 601–3
MATLAB algorithms 138–9, 596–600, 601–3
preliminary orbit determination 206, 207, 209

roots 114–15
universal Kepler's equation 138–9, 601–3
Newton's second law of motion 10–15, 409
Node regression 178–80
Non-coplanar orbits 290–303
Non-Hohmann transfers 273–85, 391–7
Non-rotating inertial frames 23–9
Numerical designation conversions 640–1
Numerical integration, equations of motion 587–94
Nutation
dampers 495–503, 509
double-gimbaled control moment gyros 527
rigid-body dynamics 451–4
spinning tops 445
torque-free motion 476–7

Oblateness
preliminary orbit determination 219
satellite attitude dynamics 481–2, 494, 495–6
spinner stability 475, 491
three dimensional orbits 177–87
One-dimensional momentum analysis 555–7
Optimal staging 570–8
Orbit formulas 42–50, 135
Orbit rotation 302–3
Orbital elements
geocentric equatorial frame 158–64
interplanetary trajectories 387, 388, 392
non-Hohmann trajectories 392
oblateness 184–7
planet state vectors 388, 641–8

Orbital elements *(continued)*
  planetary flyby 379
  preliminary orbit
      determination 199–201,
      208–11, 232–5, 250
  state vectors 158–64, 175,
      607–14
  three dimensional orbits
      158–64
Orbital maneuvers 255–314
  apse line rotation 279–85
  bi-elliptic Hohmann transfers
      264–8
  chase maneuvers 285–9
  common apse line 273–9
  Hohmann transfers 257–73,
      274
  impulsive 255–314
  non-Hohmann transfers
      273–85
  phasing maneuvers 268–73
  plane change 290–303
  two-impulse rendezvous
      330–7
Orbital parameters 286–9
Orbiting Solar Observatory
    (OSO-1) 491–2
Orientation
  delta-v maneuver 276–9,
      280–2
  gravity-gradient stabilization
      540–2
  rigid-body dynamics 448
Orthogonal transformation
    matrices 320, 421–8,
    449–50
Orthogonal unit vectors 5–7
Orthonormal basis vectors 165
Overall payload fractions 564

Parabolic trajectories 65–9,
    124–5
Parallel axis theorem 428–35
Parallel staging 563
Parallelepipeds 456–9, 540–2
Parameter of the orbit 49
Parking orbits 360–6, 394–6

Particles 1–7
Passive altitude stabilization 534
Passive energy dissipation
    495–503
Patched conics 359–60, 391–7
Payloads
  masses 560, 564–70
  ratios 560–1, 564, 567–8
  velocity 566–7
Periapse
  angle to 373, 374–5
  orbit formulas 49
  plane change maneuvers
      290–303
  radius 360–2, 369, 370, 372–3
  speed 362
  time since 108–9
  two-body motion 49, 55–6
Perifocal frame 76–8, 172–6,
    186–7
Perigee
  advance 178–80, 184
  altitude 64–5, 208–10, 211–12
  argument of 159, 161, 163,
      178–81, 183–7
  location 364–6
  orbit equation 68–9
  passage 115–17
  radius 71, 75, 183–7
  time since 131, 158–9, 184–5,
      208–11, 287–8
  time to 211, 212–13
  towards the sun 117–19
  velocity 61–2
Perihelion radius 384, 385
Period of orbit
  circular orbits 51, 53
  elliptical orbits 59, 65
  orbital elements 158–9
  rendezvous opportunities
      351–2, 354
  restricted three-body motion
      89
Perturbations
  gravitation 151–2
  oblateness 177–8
  sphere of influence 357–8

torque-free motion stability
    488
Phase angles 350–3
Phasing maneuvers 268–73, 350
Physical data 583–4
Pitch 459–63, 533, 534–42
Pitchover 554
Pivots 514, 526–30
Plane change maneuvers
    290–303
Planetary...
  see also interplanetary
      trajectories
  departure 360–6
  ephemeris 387–91
  flyby 375–86
  rendezvous 368–75
Planets
  geocentric right
      ascension-declination
      152–4
  identity conversions 644–5
  state vectors 645–53
Planning Hohmann transfers
    262–4
Point masses 1–32
  absolute vectors 20–9
  force 7–15
  gravitational potential energy
      661–4
  kinematics 2–7
  mass 7–15
  moments of inertia 417–18
  moving vector time
      derivatives 15–20
  Newton's law of gravitation
      7–10
  Newton's law of motion
      10–15
  relative motion 20–9
  relative vectors 20–9
Polar coordinates 48
Position errors 366–8
Position as a function of time
    107–47
  circular orbits 108–9
  elliptical orbits 109–23, 134–5

hyperbolic trajectories 125–35
MATLAB algorithms 142, 601–6
parabolic trajectories 124–5
universal variables 134–44
Position vectors
absolute 20–9
equatorial frames 175
geocentric 175, 194–201
Gibb's method 194–201
gravitational potential energy 657
gravity-gradient stabilization 530–1, 536
inertial frames 34–7
Lagrange coefficients 78–89, 141–4
MATLAB algorithms 159–61, 175, 232, 604–13, 626–31, 641–8
nutation dampers 496
orbit formulas 47–9
perifocal frame 76–7
point masses 2–7, 20–9
preliminary orbit determination 218–19, 223–4, 228, 236–8, 242–3, 247–9
restricted three-body motion 90–1
rigid-body dynamics 400–8, 410–14
satellite attitude dynamics 496, 510, 530–1, 536
three dimensional geocentric orbits 156–8
two-body motion 34–7, 47–9, 78–89
two-impulse maneuvers 330, 336
yo-yo despin 510
Post-flyby orbits 379–86
Potential energy 36–7, 657–60
Pound 10
Powered ascent phase 293
Pre-flyby ellipse 380–1
Precession

double-gimbaled control moment gyros 527
nutation dampers 497–8
rigid-body dynamics 451–4
satellite attitude dynamics 480–4, 497–8, 508
spinning tops 444–8
thrusters 508
torque-free motion 480–4
Preliminary orbit determination 193–254
angle measurements 228–50
Gauss's method 235–50, 631–41
Gibbs method 194–201, 613–16
Lagrange coefficients 204, 207–10, 237–9, 249
Lambert's problem 202–13, 616–21
MATLAB algorithms 198, 208, 217, 232, 242–50, 613–41
range measurements 228–35
sidereal time 213–18
topocentric coordinate systems 218–28
Primed systems 165, 168, 424–5
Principal directions 425–8, 431–5
Principal moments of inertia 419, 426–8, 431–6, 457
Prograde...
coasting flights 393–4
precession 481–4
trajectories 203
Prolate bodies 481–2, 494
Propellant
field-free space restricted staging 560–70
Lagrange multiplier method 573–8
mass 256–7, 364–6
rocket vehicle dynamics 555–9, 560–70, 573–8
thrust equation 555–7
Propellers 403–4
Propulsion 551–79

r-bars 316
Radar observations 232, 626–31
Radial distances 85–8
Radial release 514, 515–16
Radius
aiming 71–2, 75, 370–1, 373, 382
apoapse 373
azimuth 62
capture 372
earth's sphere of influence 358–9
gravitational potential energy 658–60
periapse 360–2, 369, 370, 372–3
perigee 71, 75, 183–7
perihelion 384, 385
true-anomaly-averaged 61, 62–3
Range measurements 228–35, 626–31
Rates of precession 444–8, 451–4
Rates of spin 444–8, 451–4
Regulus 153–4
Relative acceleration
angular 437
point masses 23, 25–6
relative motion and rendezvous 317–20
rigid-body kinematics 401–8
two-body motion 38
Relative angular...
acceleration 437
momentum 42–4
velocity 350–1
Relative linear momentum 412
Relative motion 315–40
Clohessy–Wiltshire equations 324–30, 336–7
close-proximity circular orbits 338–40
co-moving reference frames 316–22, 324–30

Relative motion (continued)
   linearization of equations of relative motion 322–4
   point masses 20–9
   restricted three-body motion 37, 38, 91
   two-impulse maneuvers 330–7
Relative position
   point masses 22, 24–5
   preliminary orbit determination 230–1
   sphere of influence 356
   two-body motion 37
Relative vectors 20–9, 37, 230–1, 356
Relative velocity
   Clohessy–Wiltshire equations 328–9
   close-proximity circular orbits 338–40
   point masses 22, 25
   relative motion and rendezvous 317–20
   rigid-body kinematics 401–8
   two-body motion 38
   two-impulse maneuvers 330, 332–3
Rendezvous 315–40
   Clohessy–Wiltshire equations 324–30, 336–7
   close-proximity circular orbits 338–40
   co-moving reference frames 316–22, 324–30
   equations of relative motion 322–4
   Hohmann transfers 262–4
   interplanetary trajectories 349–54, 368–75
   relative motion equations 322–4
   two-impulse maneuvers 330–7
Restricted staging 560–70
Restricted three-body motion 89–101
Retrofire 262–4

Retrograde orbits 203, 295–6, 481–4
Right ascension
   oblateness 178–81, 185–6
   planetary ephemeris 388
   preliminary orbit determination 222–3, 225–6, 230–2
   state vectors 155–8
   three dimensional orbits 149–54, 159, 160, 162
Rigid-body dynamics 399–463
   Chasles' theorem 399–400
   equations of rotational motion 410–14
   equations of translational motion 408–10
   Euler angles 448–59
   Euler's equations 435–40
   inertia 414–35
   kinematics 400–8
   kinetic energy 441–3
   moments of inertia 414–35
   moving vector time derivatives 15–16
   parallel axis theorem 428–35
   pitch 459–63
   plane change maneuvers 302–3
   roll 459–63
   rotation of the ellipse 302–3
   rotational motion 410–14
   satellite attitude dynamics 498–503
   spinning tops 443–8
   translational motion 408–10
   yaw 459–63
Rocket equation 552
Rocket vehicle dynamics 551–79
   equations of motion 552–5
   field-free space restricted staging 560–70
   impulsive orbital maneuvers 256–7
   Lagrange multiplier method 570–8
   motors 256–7

   optimal staging 570–8
   restricted staging 560–70
   rocket performance 555–60
   staging 560–78
   thrust equation 555–7
Rods 418–19, 430–5
Roll 459–63, 533, 534–42
Roots 426, 427, 434
Rotating platforms 447–8, 456–9
Rotation
   axis of 150
   Cartesian coordinate systems 169–72
   coordinate transformations 169–72
   geocentric equatorial frames 173–6
   matrices 460–3
   perifocal frames 173–6
   three dimensional orbits 150, 169–72
   true anomaly 284–5
Rotational...
   equations of motion 410–14, 517–21
   kinetic energy 488–91, 493–4, 509–12
   motion equations 410–14, 517–21
Rotationally symmetric satellites 477
Round-trip missions 353–5
Routh–Hurwitz stability criteria 501–3

Satellite attitude dynamics 475–550
   axisymmetric dual-spin satellites 518–21, 529–30
   coning maneuvers 503–5, 507
   control thrusters 504–9
   despin mechanisms 509–16
   dual-spin spacecraft 491–5
   gravity-gradient stabilization 530–42

gyroscopic attitude control 516–30
gyrostats 491–5
nutation dampers 495–503, 509
passive energy dissipaters 495–503
rigid-body dynamics 399
thrusters 504–9
torque-free motion 476–86, 487–91, 518–21, 529–30
yo-yo despin 509–16
Satellites
   dual-spin 518–21, 529–30
   earth 149, 183–7
   geocentric 276–9
   orientation 540–2
Saturation 526
Second order differential equations 326–8
Second zonal harmonics 177
Semi-latus rectum 49
Semimajor axis
   elliptical orbits 62, 65
   equation 134–5
   hyperbolic trajectories 75
   phasing maneuvers 269
   planetary ephemeris 387, 388
   three dimensional orbits 158–9, 184
Semiminor axis equation 135
Sensitivity analysis 366–8
Series two-stage rockets 562, 563–70
SEZ *see* South-East-Zenith
Shafts on rotating platforms 456–9
Sidereal time 213–18, 231, 623–6
Single stage rockets 566–7
Single-spin stabilized spacecraft 492–5
Slant ranges 239–41, 246, 249
Slug 11–12
Sounding rockets 552, 554–5, 558–9
South-East-Zenith (SEZ) frame 223

Space cones 482
Spacecraft trajectories 393, 648–55
Specific energy
   circular orbits 52
   elliptical orbits 59
   Hohmann transfers 257
   hyperbolic trajectories 73
   non-Hohmann transfers 275–6
   three dimensional orbits 158–9
Specific impulse
   impulsive orbital maneuvers 256–7
   rocket vehicle dynamics 552, 557, 559, 562, 564, 570–8
Speed
   circular orbits 53–4
   elliptical orbits 63
   excess 73–4
   hyperbolic trajectories 133
   parabolic trajectories 65
   planetary departure 362
   yo-yo despin 511, 515–16
Sphere of influence 354–9, 366–75, 392–4, 648–55
Spheres 657–60
Spherically symmetric distribution 657–60
Spin
   accelerations 521–5
   angles 508
   rates 451–4, 480, 527
   stabilized spacecraft 475
Spinning rotors 447–8
Spinning tops 443–8
Stability
   dual-spin spacecraft 492–5
   gravity-gradient stabilization 533–4
   nutation dampers 500–3
   spinning satellites 475
   torque-free motion 487–91
Stable pitch oscillation frequency 537
Staging 552, 560–78

Stars 152–3
State vectors
   geocentric equatorial frame 154–8, 175–6
   MATLAB algorithms 159–61, 175, 232, 604–13, 626–31, 641–8
   non-Hohmann trajectories 393–4
   orbital elements 159–61, 175, 606–13
   planetary ephemeris 387–9
   preliminary orbit determination 228–9, 232, 237–9, 244–50
   three dimensional orbits 154–64, 175–6, 184
   two-impulse maneuvers 330–2
Step mass 573–5, 576–8
Strap-on boosters 563
Structural ratios 560–1, 564, 568, 570–8
Stumpff functions 135–6, 142, 204–7, 600–1
Sun-synchronous orbits 180–7
Sunlit side approaches 379, 382, 384–6
Synodic period 351–2, 354

Tandem two-stage rockets 562, 563–70
Tangential release 514–16
Target vehicles 322–40
Tension 514, 515–16
Three dimensional curvilinear motion 1–7
Three dimensional orbits 149–91
   celestial sphere 149–54
   coordinate transformations 164–76
   declination 149–54
   earth's oblateness 177–87
   geocentric equatorial frame 154–8, 172–6

Three dimensional orbits
    (continued)
  geocentric right
      ascension-declination
      149–54
  oblateness 177–87
  orbital elements 158–64
  patched conics 391–7
  perifocal frame
      transformations 172–6
  right ascension 149–54
  state vectors 154–64
Three-body systems 41–2,
    355–9, 587–94
Three-stage launch vehicles
    577–8
Thrust equation 555–7
Thrust-to-weight ratio 558
Thrusters 504–9
Tilt angles 446
Time
  see also position as a function
      of time
  dependent vectors 18–20
  derivatives
    Lagrange coefficients 80–3,
        85–7, 603–5
    moving vectors 15–20
    relative motion 25–9
  Hohmann transfers 265–8
  lapse 601–6
  manned Mars missions 354–5
  to perigee 211, 212–13
  satellite attitude dynamics
      505, 515–16
  since periapse 108–9
  since perigee 131, 158–9,
      184–5, 208–11, 287–8
Titan II 562
Topocentric coordinates
    218–28, 230–5
Torque
  axial 523–5
  free motion 476–86, 487–91,
      518–21, 529–30
  rigid-body dynamics 413–14
  satellite attitude dynamics
      513–14, 521–5, 533

Trailing-side flyby 375, 376
Transfer ellipses 297–8, 348–50
Transfer times 203, 204, 352,
    354
Transformation matrices
  MATLAB algorithms 175
  moments of inertia 421–8
  orthogonal 320, 421–8,
      449–50
  pitch 460–3
  relative motion and
      rendezvous 320
  rigid-body dynamics 421–8,
      449–50, 452, 456
  roll 460–3
  satellite attitude dynamics
      536
  three dimensional orbits
      166–72, 174–6, 186–7
  topocentric horizon system
      225–8
  torque-free motion 484–6
  two-impulse maneuvers
      330–2
  yaw 460–3
Translational motion equations
    408–10
Transverse bearing loads 459
True anomalies
  averaged orbital radius 61,
      62–3
  elliptical orbits 110, 135
  hyperbolic flyby 382
  hyperbolic trajectories 69, 75,
      125–6, 132–3
  Lagrange coefficients 80–2,
      83–5
  non-Hohmann transfers
      279–85
  parabolic trajectories 68–9,
      124–5
  plane change maneuvers 301
  position as a function of time
      108–9, 110, 124–6,
      132–3, 135, 139–42
  preliminary orbit
      determination 202–13
  rendezvous opportunities 350

three dimensional orbits
    158–9, 161, 163–4, 184
  time since periapse 108–9
  universal variables 139–42
Turn angles 70, 75, 369–70, 378,
    382
Two-body motion
  angular momentum 42–50
  energy law 50–1
  equations of motion 34–42
  equations of relative motion
      37–42
  hyperbolic trajectories 69–76
  inertial frame equations of
      motion 34–7
  Lagrange coefficients 78–89
  mutual gravitational
      attraction 33–105
  orbit formulas 42–50
  parabolic trajectories 65–9
  perifocal frame 76–8
  restricted three-body motion
      89–101
  three dimensional orbits 149
Two-impulse maneuvers
    257–73, 330–7
Two-stage rockets 562, 563–70

Unit matrices 167
Unit triads 168–9
Unit vectors
  gravitational potential energy
      657
  Lagrange coefficients 78–9
  moments of inertia 422
  point masses 5–7, 21
  three dimensional orbits
      164–72
Units of force 10–15
Universal anomaly 134–44,
    601–3
Universal Kepler's equation
    138–9, 601–3
Universal Time (UT) 213–18,
    388, 621–8, 641–8
Universal variables 134–44
Unprimed systems 168, 424–5

UT *see* Universal time

Vectors 33–105
see also position...; state...; two-body motion; unit...; velocity...
 direction cosine 242–3, 244
 eigenvectors 426–8
 moving 15–20
 orthogonal unit 5–7
 orthonormal basis 165
 preliminary orbit determination 194–201
 relative 20–9, 37, 230–1, 356
 time dependent 18–20
 time derivatives 15–20
 weight 444–5
Velocity
 see also delta-v...
 errors 366–8
 escape 66, 73
 excess 360–6, 368–75, 392–4
 geocentric orbits 156–8

Hohmann transfers 261
 non-Hohmann transfers 274–5
 plane change maneuvers 290–1, 301–2
 relative motion and rendezvous 316–20
 rocket vehicle dynamics 560–70
Vectors
 absolute 20–9
 geocentric equatorial frame 175
 Lagrange coefficients 78–89, 141–4
 MATLAB algorithms 159–61, 175, 604–13, 626–31, 641–8
 perifocal frame 76–7
 point masses 2–7, 16–18, 20–9
 preliminary orbit determination 194–201, 203–4, 228, 241–2, 250

restricted three-body motion 91
 rotations 292–3, 299
 satellite attitude dynamics 496, 510–11
 two-impulse maneuvers 330
Venus ephemeris 152–3
Venus flyby 379–80
Vernal equinox 150–4
Visible surface areas 54–5

Wait time 353–4
Weight 7–10, 36
Weight vectors 444–5
Wobble angles 481–2

Yaw 403, 459–63, 533–42
Yo-yo despin 509–16

Zonal variation 177–87